Cellular Signaling and Innate Immune Responses to RNA Virus Infections

Cellular Signaling and Innate Immune Responses to RNA Virus Infections

Edited by

Allan R. Brasier
University of Texas Medical Branch, Galveston, Texas

Adolfo García-Sastre
Mount Sinai School of Medicine, New York, New York

Stanley M. Lemon
University of Texas Medical Branch, Galveston, Texas

ASM PRESS *Washington, D.C.*

Library of Congress Cataloging-in-Publication Data

Cellular signaling and innate immune responses to RNA virus infections / edited by
Allan R. Brasier, Adolfo García-Sastre, Stanley M. Lemon.
  p. ; cm.
 Includes bibliographical references and index.
 ISBN 978-1-55581-436-6
 1. RNA viruses.   2. Natural immunity.   3. Cellular signal transduction.   4. Virus
diseases—Immunological aspects.   I. Brasier, Allan R.   II. García-Sastre, Adolfo.
III. Lemon, Stanley M.
 [DNLM: 1. RNA Viruses—immunology.   2. Immunity, Natural—immunology.
3. Signal Transduction—immunology. QW 168 C393 2009]
 QR395.C45 2009
 579.2′5—dc22
                                                                    2008020320

Address editorial correspondence to: ASM Press, 1752 N St., N.W., Washington, DC
20036-2904, U.S.A.

Send orders to: ASM Press, P.O. Box 605, Herndon, VA 20172, U.S.A.
Phone: 800-546-2416; 703-661-1593
Fax: 703-661-1501
Email: Books@asmusa.org
Online: estore.asm.org

Contents

Contributors

SHIZUO AKIRA
Laboratory of Host Defense, WPI Immunology Frontier Research Center, Osaka University, 3-1 Yamada-oka, Suita, Osaka 565-0871, Japan

RANDY A. ALBRECHT
Dept. of Microbiology, Mount Sinai School of Medicine, New York, NY 10029

MEZTLI ARGUELLO
Lady Davis Institute for Medical Research, Jewish General Hospital, 3999 Chemin de la Cote-Ste-Catherine, Montreal, Canada H3T 1E2

XIAOYONG BAO
Dept. of Pediatrics, University of Texas Medical Branch, Galveston, TX 77555

CHRISTOPHER F. BASLER
Dept. of Microbiology, Box 1124, Mount Sinai School of Medicine, 1 Gustave L. Levy Place, New York, NY 10029

AMER A. BEG
Dept. of Interdisciplinary Oncology, H. Lee Moffitt Cancer Center & Research Institute at the University of South Florida, 12902 Magnolia Drive, Mail Stop SRB-2, Tampa, FL 33612

VIJAY G. BHOJ
Dept. of Molecular Biology, University of Texas Southwestern Medical Center, Dallas, TX 75390-9148

CHRISTINE A. BIRON
Dept. of Molecular Microbiology and Immunology, Division of Biology and Medicine, Brown University, Providence, RI 02912

ANDREW G. BOWIE
School of Biochemistry and Immunology, Trinity College Dublin, Dublin 2,
Ireland

ALLAN R. BRASIER
Dept. of Medicine and the Sealy Center for Molecular Medicine, The University
of Texas Medical Branch, Galveston, TX 77555-1060

KRZYSZTOF BRZÓZKA
Max von Pettenkofer Institute & Gene Center, Ludwig-Maximilians-University
Munich, Feodor-Lynen-Str. 25, 81377 Munich, Germany

ANTONELLA CASOLA
Dept. of Pediatrics and Dept. of Microbiology & Immunology, Sealy Center for
Vaccine Development, and Sealy Center for Molecular Medicine, University of
Texas Medical Branch, Galveston, TX 77555

ZHIJIAN J. CHEN
Howard Highes Medical Institute, Dept. of Molecular Biology, University of
Texas Southwestern Medical Center, Dallas, TX 75390-9148

GENHONG CHENG
Dept. of Microbiology, Immunology and Molecular Genetics, University of
California Los Angeles, Los Angeles, CA 90095

KARL-KLAUS CONZELMANN
Max von Pettenkofer Institute & Gene Center, Ludwig-Maximilians-University
Munich, Feodor-Lynen-Str. 25, 81377 Munich, Germany

JUAN C. DE LA TORRE
Dept. of Immunology and Microbial Science, IMM-6, The Scripps Research
Institute, 10550 North Torrey Pines Rd., La Jolla, CA 92037

TERENCE S. DERMODY
Dept. of Pediatrics, Dept. of Microbiology and Immunology, and Elizabeth B.
Lamb Center for Pediatric Research, Vanderbilt University School of Medicine,
Nashville, TN 37232

JOAN E. DURBIN
Center for Vaccines and Immunity, The Research Institute at Nationwide
Children's Hospital, Dept. of Pediatrics, The Ohio State University College of
Medicine, Columbus, OH 43205

RICHARD M. ELLIOTT
Centre for Biomolecular Sciences, School of Biology, University of St. Andrews,
St. Andrews, KY16 9ST, Scotland

MARY K. ESTES
Dept. of Molecular Virology and Microbiology, Baylor College of Medicine,
Houston, TX 77030

BRENDA L. FREDERICKSEN
Dept. of Cell Biology & Molecular Genetics and Maryland Pathogen Research
Institute, University of Maryland, College Park, MD 20742

MICHAEL GALE, JR.
Dept. of Immunology, University of Washington School of Medicine, Seattle,
WA 98195

ADOLFO GARCÍA-SASTRE
Dept. of Microbiology, Dept. of Medicine/Div. Of Infectious Diseases, Emerging Pathogens Institute, Mount Sinai School of Medicine, New York, NY 10029

ROBERTO P. GAROFALO
Dept. of Pediatrics, Dept. of Microbiology & Immunology, Sealy Center for Vaccine Development, and Sealy Center for Molecular Medicine, University of Texas Medical Branch, Galveston, TX 77555

SUSANA GUIX
Enteric Virus Laboratory, Dept. of Microbiology, University of Barcelona, Barcelona 08028, Spain

BEICHU GUO
Dept. of Microbiology, Immunology and Molecular Genetics, University of California Los Angeles, Los Angeles, CA 90095

JOHN HISCOTT
Lady Davis Institute for Medical Research, Jewish General Hospital, and Dept. of Microbiology and Immunology and Dept. of Experimental Medicine, McGill University, Montreal, Canada

SINÉAD E. KEATING
School of Biochemistry and Immunology, Trinity College Dublin, Dublin 2, Ireland

ELIN KINDBERG
Division of Molecular Virology, Dept. of Clinical and Experimental Medicine, Medical Faculty, University of Linköping, 58185 Linköping, Sweden

WILLIAM B. KLIMSTRA
Dept. of Microbiology & Immunology, Center for Molecular & Tumor Virology, Louisiana State University Health Sciences Center, Shreveport, LA 71130

STANLEY M. LEMON
Dept. of Microbiology and Institute for Human Infections and Immunity, The University of Texas Medical Branch, Galveston, TX 77555-1073

SHINJI MAKINO
Dept. of Microbiology and Immunology, The University of Texas Medical Branch at Galveston, Galveston, TX 77555-1019

PETER W. MASON
Dept. of Pathology and Dept. of Microbiology & Immunology, University of Texas Medical Branch, 301 University Blvd., Galveston, TX 77555-0436

KRISHNA NARAYANAN
Dept. of Microbiology and Immunology, The University of Texas Medical Branch at Galveston, Galveston, TX 77555-1019

JOHN T. PATTON
Laboratory of Infectious Diseases, National Institute of Allergy and Infectious Diseases, National Institutes of Health, Bethesda, MD 20892

SUZANNE PAZ
Lady Davis Institute for Medical Research, Jewish General Hospital, and Dept. of Microbiology and Immunology, McGill University, Montreal, Canada

BAHRAM RAZANI
Dept. of Microbiology, Immunology and Molecular Genetics, University of
California Los Angeles, Los Angeles, CA 90095

KATE D. RYMAN
Dept. of Microbiology & Immunology, Center for Molecular & Tumor
Virology, Louisiana State University Health Sciences Center, Shreveport,
LA 71130

FRANK SCHOLLE
Dept. of Microbiology, North Carolina State University, 4512 Gardner Hall,
Box 7615, Raleigh, NC 27695

GANES C. SEN
Dept. of Molecular Genetics, NE20, Cleveland Clinic, 9500 Euclid Ave.,
Cleveland, OH 44195

ARASH SHAHANGIAN
Dept. of Microbiology, Immunology and Molecular Genetics, University of
California Los Angeles, Los Angeles, CA 90095

BARBARA SHERRY
Dept. of Molecular Biomedical Sciences, North Carolina State University,
Raleigh, NC 27606

LENNART SVENSSON
Division of Molecular Virology, Dept. of Clinical and Experimental Medicine,
Medical Faculty, University of Linköping, 58185 Linköping, Sweden

OSAMU TAKEUCHI
Laboratory of Host Defense, WPI Immunology Frontier Research Center,
Osaka University, 3-1 Yamada-oka, Suita, Osaka 565-0871, Japan

XINGYU WANG
Dept. of Interdisciplinary Oncology, H. Lee Moffitt Cancer Center & Research
Institute at the University of South Florida, 12902 Magnolia Drive, Mail Stop
SRB-2, Tampa, FL 33612

FRIEDEMANN WEBER
Department of Virology, University of Freiburg, D-79008 Freiburg, Germany

CHRISTINE L. WHITE
Dept. of Molecular Genetics, NE20, Cleveland Clinic, 9500 Euclid Ave.,
Cleveland, OH 44195

Preface

Infections with RNA viruses represent a significant cause of morbidity and mortality in vertebrates. This volume is predicated on the belief that a careful examination of the early host responses to these RNA virus infections, as well as of the mechanisms adopted by these viruses to evade early host antiviral responses, will provide a platform of knowledge on which we will be able to develop new and more effective methods for controlling the spread of RNA viruses or modifying their disease course.

The initial response to infection with an RNA virus typically involves activation of specific arms of the innate, or "nonspecific," immune response. Although certain aspects of these innate responses have been known for years, interest in innate immunity as a field has exploded since the discovery in 2001 of specific pattern recognition receptors (PRRs) as pathogen sensors and mediators of innate signaling. Since this discovery, the field has expanded enormously as workers have identified and elucidated high-resolution structures of PRRs and begun to learn the details of how these cellular alarms send signals to activate the innate immune response.

The motivation for assembling this volume, therefore, is to summarize these exciting discoveries in a timely fashion and to serve as a yardstick for significant future advances in the field, as well as to articulate areas that need further investigation. To this end, we have enlisted contributions from leading, active investigators, all experts in their individual fields.

We wish to acknowledge the contributions of those who have authored the chapters, without which this work would not be possible. We also grateful for the support of Gregory Payne of ASM Press, who helped to champion the concept of this work and who provided timely feedback and advice.

Allan R. Brasier
Adolfo García-Sastre
Stanley M. Lemon

Introduction

I

Cellular Signaling and Innate Immune
Responses to RNA Virus Infections
Edited by A. R. Brasier et al.
© 2009 ASM Press, Washington, D.C.

Allan R. Brasier
Adolfo García-Sastre
Stanley M. Lemon

Introduction and Overview

1

Infections caused by RNA viruses represent some of the most significant diseases of vertebrates in terms of both morbidity and economic impact. With regard to human infections, RNA virus infections are responsible for a spectrum of important acute, chronic, and emerging infections. The hepatitis C virus is a flavivirus responsible for ~170 million chronic infections worldwide, many resulting in cirrhosis, liver failure, and hepatocellular carcinoma. In Central and South America, dengue virus has reemerged due to reinfestation by its mosquito vector and global travel. As a result, dengue is now epidemic (or pandemic) in 100 countries, producing dengue fever in 50 million humans annually, which in some instances develops into severe hemorrhagic fever or shock syndrome and death. The paramyxovirus respiratory syncytial virus is responsible for epidemic pediatric upper respiratory tract disease, perhaps the most common significant infection in children in developed countries. In the United States, 100,000 infant hospitalizations annually are due to this agent. Despite vaccines being available for a number of RNA viruses, including polioviruses, measles virus, mumps virus, rubella virus, influenza viruses, rabies virus, yellow fever virus, Japanese

encephalitis virus, and more recently rotaviruses, none of these viruses have been eradicated, and many of them still cause high levels of mortality in unvaccinated populations. Moreover, there are many others, like hepatitis C virus and respiratory syncytial virus, for which no vaccine is available.

In addition to those with stable human reservoirs, RNA viruses are frequently responsible for zoonotic infections. Such infections account for some of the most feared and least understood infectious diseases. Zoonoses represent viral infections of animal hosts that are spread to humans as a result of inadvertent contact. RNA viruses are predominant among the causes of zoonotic diseases. This has been suggested to be due to the high error rate of their RNA-dependent RNA polymerases, which coupled with the plasticity of their small RNA genomes, contributes to a remarkable ability of these viruses to jump species. Alternatively, as in the case of influenza virus, an orthomyxovirus of birds and mammals with a multisegment genome, new viruses may arise from reassortment of genome segments between strains from different hosts. Generation of novel human influenza strains with antigenic determinants derived from avian

Allan R. Brasier, Dept. of Medicine and the Sealy Center for Molecular Medicine, University of Texas Medical Branch, Galveston, TX 77555–1060. **Adolfo García-Sastre,** Dept. of Microbiology, Mount Sinai School of Medicine, New York, NY 10029. **Stanley M. Lemon,** Institute for Human Infections and Immunity, University of Texas Medical Branch, Galveston, TX 77555–0428.

strains may result in pandemics, such as the Spanish flu epidemic of 1918, which infected one-fifth of the world's population and was responsible for at least 20 to 40 million deaths.

Other examples of zoonotic RNA viruses abound. Endemic in the southwestern United States, hantaviruses are bunyaviruses that are nonpathogenic in rodents but produce hemorrhagic pulmonary syndrome in humans who come into contact with rodent excreta. West Nile virus is an arthropod-transmitted virus whose natural reservoir is birds; it emerged in around 2000 as an important new cause of significant human neurological disease, including death, in North America. Lassa virus infects an estimated 100,000 to 300,000 humans in West Africa, causing approximately 5,000 deaths. The infection arises from contact with an animal reservoir known as the multimammate rat. The paramyxoviruses Nipah and Hendra produce encephalitis and are acquired by contact with infected pigs or bats and horses, respectively. The highly lethal Ebola and Marburg viruses have been postulated to be maintained in a bat reservoir, causing recurrent outbreaks of severe hemorrhagic fever in humans in Central Africa. The relatively recent emergence of severe acute respiratory syndrome (SARS) coronavirus, also from a bat reservoir, should serve as a reminder of the potential of zoonotic infections to result in the generation of new human viruses and novel human diseases. Many more examples of emerging or reemerging viral infections can be found. Even previously unrecognized RNA viruses are being discovered today that have been infecting humans for many years, such as human metapneumovirus and new coronaviruses and rhinoviruses.

RNA viruses are clearly worthy of careful examination to understand mechanisms of disease pathogenesis, early and late immune responses, and immune evasion, as such knowledge will help us develop new and more effective methods for controlling their spread or modifying their disease course. To this end, we know that RNA viruses are diverse in their modes of transmission, disease pathogenesis, and mechanisms of genomic replication. Upon initial exposure to RNA viruses, critical early immune responses are elicited within infected cells whose functions serve to limit the spread of infection. Although this innate immune response is frequently referred to as a "nonspecific" immunity, research into the molecular mechanisms of how such RNA viral infections are detected is revealing that this response is anything but nonspecific. A critical emerging concept is that eukaryotic cells express a series of proteins, collectively termed pattern recognition receptors (PRRs), that sense extracellular, cytoplasmic, and endosomal compartments for molecular signatures that indicate the presence of an infectious agent. In many ways, these PRRs act as cellular burglar alarms, alerting the cell to the presence of an invader. The molecular signatures they recognize, pathogen-associated molecular patterns (PAMPs), allow the distinction between self and nonself. Sensing the presence of PAMPs, PRRs initiate signaling cascades that limit viral replication via the production of type I interferons (IFNs) and many dozens if not hundreds of other downstream effector proteins. Cell surface, endosomal, and cytoplasmic PRRs have begun to be identified and their molecular ligands defined. This has resulted in definition of two major classes of transcriptional activators that play key roles in these early responses, interferon regulatory factors (IRFs) and nuclear factor-κB (NF-κB), and dissection of the upstream pathways that lead to their activation. Emerging concepts of the consequences of the activation of innate immune responses are expanding to include critically important roles in T-cell activation and the acquisition of adaptive immunity, lymphocyte apoptosis, inflammation, and even changes in metabolic pathways.

ORGANIZATION

This book is divided into two major parts. The first part focuses on the structure, interconnectivity, and mechanisms of activation of the PRR-initiated signaling pathways. These PRRs include the Toll-like receptors (TLRs) displayed on the cell surface and within the lumen of endosomes, and intracellular cytoplasmic RNA helicases (increasingly referred to as the retinoic acid-inducible gene I [RIG-I]-like receptors, or RLRs). Each of these PRRs, upon binding their cognate PAMP-containing ligand, recruits specific signaling adaptor molecules, leading to the activation of the kinases that are in turn responsible for activation of IRFs and NF-κB. These transcription factors induce the synthesis of type I IFNs (IFN-α/β), whose action is to activate separate signals in a paracrine and autocrine manner that converge on the signal transducer and activator of transcription (STAT) pathway. Type I IFNs reduce viral replication by the induction of proteins (interferon-stimulated genes, or ISGs) that antagonize many steps of the viral life cycle, shape the acquisition of adaptive immunity, and are responsible for organismal metabolic changes. These signaling pathways are highly regulated in ways that are only partially understood today, with multiple complex and seemingly redundant mechanisms in place not only for their activation but also for their suppression. Not all individuals are susceptible to viral infections; exciting work on the effects of individual gene polymorphisms has begun to identify some sources of inter-individual differences in the response to RNA virus infections.

The second part of this book examines the ways in which the major RNA virus families interact with these signaling pathways. Here, the diversity of RNA virus life cycles, their impact on disease, and the mechanisms by which they induce (or block the induction of) innate immune responses are discussed. This exploding body of knowledge includes the identification of specific viral proteins and types of nucleic acid patterns that are recognized by PRRs and activate signaling cascades. The replication of most RNA viruses is restricted in the presence of IFN, and for this reason, most if not all of these viruses have evolved strategies to antagonize or evade this response. For each viral family, mechanism(s) that

are known to modulate the innate immune response and, where known, their consequences on viral replication or pathogenesis are discussed. The understanding of the specific pathways that will be gained from a review of the the first part of this volume will help to place these diverse mechanisms of immune activation and evasion in context.

Finally, we summarize overarching mechanisms and attempt to tie the work together with unifying perspectives. This is a difficult task in any field that is moving as fast as this, however, and we do so with the full realization that this summary might look quite different if written 18 or 36 months from now!

Antiviral Signaling Pathways and Their Contributions to Virus Control and Pathogenesis

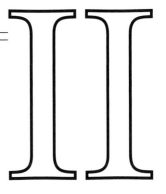

Cellular Signaling and Innate Immune
Responses to RNA Virus Infections
Edited by A. R. Brasier et al.
© 2009 ASM Press, Washington, D.C.

Andrew G. Bowie
Sinéad E. Keating

Role of Toll-Like Receptors in the Innate Immune Response to RNA Viruses

2

Innate immunity is based on an intricate system of host pattern recognition receptors (PRRs) that specifically recognize pathogen-associated molecular patterns (PAMPs) (3, 78, 117). The effective sensing of these PAMPs, generally associated with essential structural components of infectious agents, is key to the induction of an appropriate immune response upon infection of an invading pathogen. Toll-like receptors (TLRs), 13 of which have been identified thus far, represent an essential family of these PRRs (114). These receptors are part of the interleukin-1 receptor (IL-1R)/TLR superfamily, a family of proteins defined by the presence of a common cytoplasmic signaling domain (Toll/IL-1R, or TIR domain) that also includes the IL-1, IL-18, and IL-33 receptors. The importance of these receptors in host innate immunity is underscored by the fact that they are highly conserved throughout evolution from plants and insects to mammals. While historically the role of TLRs in the host immune response to bacteria and fungi was initially more apparent, the importance of the complex interplay between TLRs and viruses is now emerging. A number of viral structural entities have been identified as targets for TLR recognition. Examples include viral single-stranded (ss) RNA, which is sensed by both

TLR7 and TLR8; double-stranded (ds) RNA (TLR3); viral DNA (TLR9); and a number of specific viral structural proteins which have been shown to activate signal transduction through TLR2 and TLR4. Thus, the panel of TLR signaling pathways triggered is determined by the repertoire of PAMPs presented by the invading virus. The specific PAMPs presented to TLRs by RNA viruses are illustrated in Table 1.

TLRs are ubiquitously and constitutively expressed on many different cell types including both immune and nonimmune cells (3). Thus, they are well placed to trigger a rapid and immediate immune response to viral infection regardless of the site of initial infection. The cascade of events that occurs following engagement of TLRs by viral components may be considered twofold. First, the ensuing signaling events lead to the activation of key antiviral transcription factors such as nuclear factor-κB (NF-κB) and interferon regulatory factor (IRF) 3. These then switch on expression of a specific panel of proinflammatory cytokines and chemokines such as tumor necrosis factor (TNF)-α, IL-6, IL-8, and RANTES, and, crucially, the activation of TLRs by viruses results in the production and release of type I interferons (IFN-α/β). These antiviral cytokines function specifically

Sinéad E. Keating and Andrew G. Bowie, School of Biochemistry and Immunology, Trinity College Dublin, Dublin 2, Ireland.

Table 1 PAMPs presented by RNA viruses and the TLRs that specifically detect them

TLR	Virus	PAMP	Reference(s)
TLR4	RSV	Fusion protein	62
	VSV	Glycoprotein G	30
	MMTV	Env protein	87
TLR2	HCV	Core/NS3 proteins	17
	Measles	HA	6
TLR3	Various	dsRNA	4
TLR7	HIV	ssRNA	16, 38, 50
	IAV		
	VSV		
TLR8	HIV	ssRNA	38

to inhibit viral replication and de novo synthesis of viral proteins, for example, through the activation of dsRNA-dependent kinase (PKR) and $2',5'$-oligoadenylate synthetase (54, 93). Thus, TLRs instigate an initial rapid response to viral infection in a bid to limit viral spread within the host. A second critical function of TLR activation by viruses is the priming of the adaptive arm of the host immune response. This is not entirely separate from the IFN-inducing function of TLRs, as the potent activation of cytotoxic T lymphocytes (CTLs) seen on triggering TLR signaling has been shown to be dependent on IFN-α/β at least for TLR3, -4, -7, and -9 (2). CTLs are potentially the most important protective component of host immunity to an array of viruses. Thus, the significance of TLRs in preventing virus-induced disease may lie in this cross talk between the innate immune response and antigen-specific adaptive immunity.

TLRs AS SENSORS OF RNA VIRUS INFECTION

RNA Virus-Encoded Proteins as PAMPs for TLRs

The first event that must occur within the host upon virus infection is that the virus must gain entry to host cells at the site of infection. For most RNA viruses, this generally occurs either by direct fusion of the viral membrane to the host cell membrane, releasing the virus genome directly into the cytoplasm, or by receptor-mediated endocytosis of whole virions into endosomal compartments (92). Either way, virus envelopes and the virions of non-enveloped RNA viruses comprise numerous structural proteins, so these are potentially the first structural entity

presented to the host as potential PAMPs upon viral infection. It is known that virus particles can trigger an IFN response in the absence of virus replication and, hence, de novo protein synthesis (12). Thus, this may very well be down to the immediate sensing of structural proteins already present within the infecting virus particle.

TLR2 and TLR4

Given their localization at the plasma membrane, TLR2 and TLR4 are very well placed to act as immediate sensors of invading viruses as early as the viral entry stage of a virus infection. In support of this, the first description of a TLR recognizing a viral PAMP was that of the respiratory syncytial virus (RSV)-encoded fusion (F) protein being specifically recognized by TLR4 (63). Exposure of wild-type monocytes to F protein induced high levels of IL-6 production. Macrophages from CD14-null, C3H/HeJ mice (carry a mutation in TLR4) or C57BL10/ScCr mice (TLR4 gene deleted entirely) were insensitive to F protein treatment, thus implicating TLR4 in the response to RSV infection and pointing to RSV F protein as a direct ligand for TLR4. C57BL10/ScCr mice also have defects in the IL-12Rβ2 gene, which complicates the interpretation of phenotypes observed in these mice (87), and one study has suggested that mice lacking TLR4 can still mount an efficient immune response to RSV infection (19). This casts some doubt over the importance of TLR4 in the induction of an antiviral immune response to RSV infection in mice. However, RSV infection has been associated with the upregulation of cell surface TLR4 expression on peripheral blood lymphocytes and monocytes in infants with RSV bronchiolitis (28). Furthermore, a clear correlation between a number of common TLR4 mutations and RSV disease severity strongly suggests that TLR4 expression does indeed influence the outcome of the immune response to RSV infection in humans (110). This may be indicative of a key difference between the host immune response to RSV infection in mice and humans.

Vesicular stomatitis virus (VSV) stimulates the production of type I IFNs via a pathway shown to be dependent on both CD14 and TLR4 (46). The VSV-encoded glycoprotein G (gpG) has now been identified as the viral PAMP responsible for triggering this signaling pathway, which requires IRF7 induction and occurs in myeloid dendritic cells (mDCs) and macrophages rather than plasmacytoid DCs (pDCs) (29). The detection of viral glycoproteins by TLR2 and TLR4 is typically associated with the induction of NF-κB-dependent proinflammatory cytokine expression. Thus, this study represents the first description of TLR-mediated sensing of a viral protein leading to IRF activation and subsequent type I IFN upregulation. The core and nonstructural 3 (NS3) proteins of hepatitis C virus (HCV) both trigger potent inflammatory responses in monocytes via TLR2 (16). A direct interaction between TLR2 and these HCV proteins has not yet been verified. Nevertheless, it seems likely that TLR2 plays a key role in sensing this viral pathogen through binding of one or both of these viral PAMPs. Crucially, internalization of core and NS3 proteins did not require intact TLR2 expression, indicating that HCV does not use TLR2 to infect its target host cell. The RNA viruses described here are all from distinct virus families, highlighting the overall importance of TLR2 and TLR4 in sensing RNA viral proteins as immediate danger signals of virus infection. Thus, while TLRs may be considered a first line of defense against viruses in the overall immune response to viruses, TLR2 and TLR4 may represent the first battalion in the TLR platoon.

Coxsackievirus, a member of the *Picornavirus* genus of enteroviruses, is recognised by TLR4 (115). Coxsackievirus B4 (CVB4) is associated with insulin-dependent diabetes mellitus, and CVB4 infection triggers the secretion of proinflammatory cytokines such as IL-1β and TNF-α. It is thought that this inflammation may be responsible for the destruction of insulin-producing cells and, hence, the pathogenesis of insulin-dependent diabetes mellitus seen following CVB4 infection. This cytokine production has been specifically attributed to TLR4 expression. As TLR4-reactive antibodies specifically blocked the induction of IL-6 production by both untreated and UV-inactivated CVB4, a direct interaction between TLR4 and a component of the CVB4 virion is likely to mediate this response. Critically, anti-TLR4 antibodies did not inhibit overall binding of virus to the surface of target cells and had no effect on virus infectivity as determined by plaque assays. Thus, TLR4 appears to play a central role in sensing CVB4 and mounting an inflammatory response to this virus but is not implicated in any stage of viral entry or the infection cycle of the virus. The role of TLR4 in responding to enterovirus infection is emphasized by the finding that enteroviral replication is associated with TLR4 expression in the myocardium of patients suffering from dilated cardiomyopathy (96).

Likewise, use of TLR2- and myeloid differentiation factor 88 (MyD88)-deficient mice demonstrated a role for TLR2 signaling in mounting a proinflammatory response to lymphocytic choriomeningitis virus (LCMV), but the specific viral PAMP sensed by TLR2 has yet to be identified (123). Reduced expression of IL-6 and monocyte chemoattractant protein-1 was observed in both TLR2-null and MyD88-null mice following infection with LCMV. Conversely, while induction of type I

IFN in response to LCMV was dependent on TLR2 expression, it was found to be MyD88 independent. Despite this, MyD88-deficient mice failed to induce an efficient CD8$^+$ CTL response to the virus and defective MyD88 expression was associated with reduced viral clearance. Surprisingly, TLR2-null mice developed a CD8$^+$ CTL response comparable to the wild-type response and cleared viral infection normally. Together, these data suggest that while TLR2 plays a critical role in the proinflammatory response to LCMV, it is the CD8$^+$ CTL response that is crucial for effective clearance of the virus, and that MyD88 plays an essential role in priming this CTL response.

TLRs Can Detect Viral Nucleic Acid

If viral entry occurs by fusion of the viral membrane to the host cell plasma membrane, this results in the immediate release of naked viral genetic material directly into the cytosol. Conversely, if viral entry requires endocytosis of the viral particle, uncoating typically takes place within endosomal compartments, exposing viral nucleic acid first within the endosome and then releasing it into the cytosol. This exposure of viral nucleic acid is critical for viral gene expression and replication and, hence, propagation of progeny virus. However, this poses something of a risk to the virus as viral nucleic acid provides further material to act as PAMPs that can be detected by PRRs of the antiviral innate immune response. TLR3, TLR7, TLR8, and TLR9 constitute a subfamily of TLRs that can specifically recognize viral nucleic acid (36). These TLRs are primarily expressed intracellularly in endosomal compartments. Hence, they are likely extremely important in the detection of viruses that gain entry to host cells via endocytosis. The expression of this subgroup of TLRs in endosomal compartments allows for the rapid detection of viral nucleic acid immediately upon virus uncoating and demonstrates how these PRRs have evolved to cope very specifically with viral infection.

While TLR9 senses the presence of viral DNA, TLR7 and TLR8 respond to ssRNA and TLR3 detects dsRNA. TLR9 was originally shown to be activated by bacterial DNA sequences containing unmethylated CpG dinucleotides (39). The discovery that these motifs are also found in abundance in some viral genomes, such as herpes simplex virus (HSV), has revealed that TLR9 is a critical PRR in the host innate immune response to viruses. Most viruses, DNA and RNA viruses alike, will produce dsRNA at some stage of their life cycle, and so dsRNA may be considered a major marker of viral infection. Thus, TLR3 is a key sensor of viral infection. Guanine- or uridine-rich ssRNAs produced by replicating RNA viruses such as human immunodeficiency virus (HIV) and influenza A virus (IAV) are natural ligands for TLR7 and TLR8 (37). Positive- and negative-strand ssRNA viruses represent an extensive class of infectious agents, so TLR7 and TLR8 are similarly central PRRs in host innate immunity to viruses.

TLR3

The dsRNA-dependent protein kinase, PKR, has long been recognized as an intracellular sensor of dsRNA and as such represents the first antiviral PRR. By the same token, dsRNA represents the first identified viral PAMP. Critically, it was the observation that mice lacking the PKR gene could still respond to poly(I:C), a synthetic analogue of dsRNA often used to mimic viral infection, that led to the discovery of TLR3 as a second PRR for dsRNA (4). Alexopoulou et al. showed that TLR3 could activate NF-κB and stimulate the production of type I IFNs in response to poly(I:C). Knocking out TLR3 expression resulted in a much attenuated inflammatory response to poly(I:C) treatment and significantly protected mice from poly(I:C)-induced shock. Crucially, cellular activation induced by native viral genomic dsRNA purified from a mammalian reovirus was impaired in TLR3-deficient cells.

Despite its being the first TLR to be implicated in sensing a viral nucleic acid PAMP, demonstrating a role for TLR3 in the induction of an antiviral response to natural viral infection has proved difficult. Nevertheless, combined use of cell lines stably expressing TLR3 and small interfering RNA blockade of TLR3 protein expression have defined a role for TLR3 in the production of chemokines in response to RSV infection (90). Interestingly, in this study introducing TLR3 into cells or downregulating TLR3 expression did not affect RSV replication in any way. However, whether the same holds true in vivo remains to be determined. RSV appears to trigger both TLR3- and TLR4-dependent responses, so it would be of major interest to assess the influence of both of these inflammatory pathways on RSV infection in parallel in vivo studies. Targeting either or both of these TLR signaling pathways may be a useful therapeutic approach in the treatment of RSV-associated pulmonary disease.

The three-dimensional crystal structure of the extracellular domain of human TLR3, containing 23 leucine-rich repeat motifs, has now been elucidated (11). It forms a large horseshoe-shaped solenoid and is heavily glycosylated, but interestingly, the outer convex surface revealed a face that is free from carbohydrate moieties and contains two clusters of positively charged residues. It had been predicted that TLR ligands would bind to the inner concave surface of the TLR ectodomain. However, the identification of numerous glycosylation sites within the concave surface of TLR3 contradicted this model, and the two positively charged patches on the outer surface seemed likely ligand-binding domains for negatively charged dsRNA. Consistent with this, another study by Bell et al., which showed direct binding of poly(I:C) to the TLR3 ectodomain, used mutational analysis to confirm experimentally that the predicted sugar-free surface was the target site of ligand binding (5). While this represents the first molecular definition of ligand binding to TLR3 extracellular domains, the specific recognition of native viral dsRNA by TLR3 remains to be shown. The demonstration of TLR3 binding dsRNA during a live virus infection would clearly define the importance of TLR3 in antiviral innate immunity.

TLR7 and TLR8

The observation that imidazoquinoline compounds such as imiquimod and resiquimod (or R-848) could induce an antiviral state by triggering the production of IFN-α through an unknown mechanism led to the identification of TLR7 and TLR8 as antiviral TLRs. The stimulation of IFN-α release from murine macrophages by these agents was dependent on both TLR7 and MyD88 (38). The genes encoding TLR7 and TLR8 are highly homologous and both TLRs are expressed in mice and humans (17). However, while independent expression of either human TLR7 or TLR8 resulted in the activation of cellular signaling pathways by R-848, expression of murine TLR8 did not (49). It was then shown that synthetic ssRNA oligonucleotides based on those produced during HIV-1 infection could induce the stimulation of IFN-α and proinflammatory cytokines from DCs and macrophages via human TLR8 and murine TLR7 (37). The demonstration that the potent induction of IFN-α production by pDCs in response to IAV infection was dependent on the endosomal recognition of influenza RNA by TLR7 represented one of the first direct lines of evidence that TLR7 and TLR8 did indeed respond to true viral PAMPs (37).

Furthermore, the induction of IFN-α by pDCs in response to IAV or VSV infection was also dependent on the recognition of viral ssRNA by TLR7 in endosomes (14, 70). Diebold et al. also found that nonviral ssRNA molecules (such as polyU) could induce TLR7-dependent secretion of proinflammatory cytokines. This is very interesting in the context of a follow-on study that showed that the recognition of ssRNAs by TLR7 occurred independently of sequence and that the sole requirement for any ssRNA molecule to act as a ligand for TLR7 was the presence of multiple uridines, a nucleotide exclusive to

RNA, within a short distance of each other (15). The origin of these RNA molecules had no bearing on the efficiency of the ensuing TLR7-mediated signaling cascade induced.

TLR9

The importance of TLR9 as a PRR pivotal in the sensing of invading microorganisms was originally defined on the basis of its response to bacterial DNA sequences containing unmethylated CpG dinucleotides (39). Unlike small eukaryotic viruses, the genomes of large eukaryotic DNA viruses contain a high degree of CpG motifs. The dsDNA genomes of HSV-1 and -2, two α-herpesvirus members of the *Herpesviridae* family of large dsDNA eukaryotic viruses, show a high frequency of these motifs, and infection of mice with these viruses resulted in the induction of type I IFN in pDCs through TLR9 (62, 69). Mice lacking TLR9 produced no IFN-α in response to HSV-2 infection, while normal pDCs treated with purified HSV-2 DNA showed a massive release of IFN-α (69). Recognition of HSV-2 by pDCs occurred independently of virus replication and was inhibited by chloroquine and bafilomycin A, indicating a dependence on endocytic compartment formation. TLR9 is located solely in intracellular endosomal compartments (1, 64) and so, like TLR3, encounters its partner PAMP largely in endosomes. The importance of TLR9 as a PRR in the antiviral response specifically to RNA viruses is of course limited, but this TLR cannot be ignored in the consideration of TLRs as antiviral PRRs in general. It is part of a subgroup of TLRs that specifically recognize viral nucleic acids from within endosomal compartments and, as such, anything learned from studying the molecular mechanism of nucleic acid detection by TLR9 is potentially applicable to the detection of RNA by TLR3, TLR7, and TLR8.

COMPARISON OF TLRs TO ANOTHER FAMILY OF PRRs, THE RIG-I-LIKE RNA HELICASES

A recent explosion of research into the mechanisms of cellular detection of invading pathogens has revealed that TLRs are not the only family of PRRs involved in the sensing of PAMPs presented by infectious foreign agents. There are several families of these PRRs, and one that is now emerging as extremely important in alerting the host to viral infection at a cellular level is the retinoic acid-inducible gene I (RIG-I)-like RNA helicases (RLHs) (122). The RLH family of PRRs, thus far comprising RIG-I and melanoma differentiation-associated gene 5 (MDA-5), are undoubtedly an essential component of

host innate immunity to RNA viruses (56). The fact that they are expressed solely in the cytosol, where no TLR is expressed, arms the host with an alternative defense mechanism for this subcellular fraction. Hence, it is not by accident that different classes of PRRs have evolved to survey separate subcellular compartments for foreign material. The list of RNA viruses sensed by RIG-I and MDA-5 is extensive, and dsRNA, a common replication intermediate produced by many RNA viruses, was originally believed to be the viral PAMP detected by these PRRs (114). RIG-I has been implicated in the molecular sensing of flaviviruses, orthomyxoviruses, paramyxoviruses, and rhabdoviruses (53). The uncapped 5′-triphosphate end of the ssRNA produced by these viruses has now been identified as the exact structural moiety recognized by RIG-I (44). Conversely, MDA-5 could specifically detect the picornaviruses. The exact motif of *Picornavirus* RNA identified by MDA-5 has not yet been identified. The role of RLH molecules in fighting viral infection will be dealt with extensively in a later chapter. Hence, they are discussed here briefly for completeness to compare their relative role to that of TLRs in the response to RNA virus infection.

Knockout mouse studies revealed that in the absence of either MyD88 (the downstream signaling molecule used by all TLRs except TLR3) or TLR3, thus potentially covering most TLR signaling pathways, mice still displayed normal protection from a number of viral infections (43, 99). Thus, in most studies using mice with impaired TLR signaling, the mice did not succumb dramatically to virus infection. These studies now appear to make more sense in light of the discovery of another cellular detection mechanism for sensing viral PAMPs. While TLRs play an important role in mounting an immune response to viral infection, RIG-I and MDA-5 function to support the TLRs in this role. In the same way that viruses encode numerous mechanisms to evade recognition by PRRs, host innate immunity has evolved different systems to counteract virus infection. In fact, RIG-I and MDA-5 may have acquired immune surveillance functions as a result of evolutionary pressure from constant attack from viral infection. DExD/H box RNA helicases are highly conserved throughout evolution (111). They possess an intrinsic ATPase activity that allows them to unwind dsRNA, and this function sees them participating in a wide range of important cellular processes including transcription, translation, mRNA splicing, and RNA interference. RLHs, however, couple this RNA unwinding activity to tandem caspase activation and recruitment domains (CARDs), allowing them to interact with at least one other CARD-containing protein, CARD adaptor inducing IFN-β (Cardif; also known as MAVS, IPS-1, or VISA), to stimulate activation of the antiviral transcription factors NF-κB and IRF3 (114). Thus, whether a similar innate immune recognition system is evolutionarily conserved remains to be observed. It may be that RNA interference represents the equivalent immune function of RNA helicases in lower organisms (111). It is interesting to note in this context that the DExD/H box RNA helicase LGP2, which shares 31% sequence identity with RIG-I in its helicase region but does not possess a CARD motif, was found to be inhibitory (122).

One key difference identified between RIG-I and the TLRs is that RIG-I function appears to be most important in cells of nonmyeloid origin, for example, fibroblasts, macrophages, and conventional DCs (cDCs) (52). In stark contrast, RIG-I antiviral defense plays no role in pDCs. Crucially, the induction of type I IFNs in response to viral infection is solely dependent on TLR7/8 and TLR9 in these cells. This observation highlights a key difference between RIG-I and TLRs and therefore suggests that these PRRs are not functionally redundant.

ANTIVIRAL SIGNALING PATHWAYS TRIGGERED BY TLRs

The antiviral signaling pathways triggered by an individual viral PAMP upon engagement of its target TLR is determined by the repertoire of TIR domain-containing adaptor molecules recruited to that TLR following ligand binding. Only five of these adaptors have been identified, and four of these proteins alone mediate all downstream signaling functions of all TLRs described to date (83). Interestingly, all TLRs except TLR3 signal through the adaptor molecule MyD88 (75). For TLR7, 8, and 9, adaptor usage appears to be limited to MyD88, but in the case of TLR2 and TLR4 signaling another adaptor, MyD88 adaptor-like (MAL), is required to recruit MyD88 to the TLR cytoplasmic domain (22, 83). TLR3 signals solely via the adaptor TIR-domain-containing adaptor inducing IFN-β (TRIF) (41). TLR4 is unique in the fact that it can recruit either MyD88 or TRIF to trigger two independent signaling cascades. In the same way that MAL is required as a bridging molecule from TLR4 to MyD88, the fourth known signal-inducing adaptor, TRIF-related adaptor molecule (TRAM), links TRIF to this receptor (85, 120). Very simply, TLR4 signaling through MyD88 acts as a switch to NF-κB activation and subsequent upregulation of proinflammatory cytokine expression. Likewise, TLR4 recruitment of TRIF stimulates NF-κB activity. However, supplementary to this, TLR4 signal

transduction via TRIF is a potent trigger of the IRF family of transcription factors and, hence, type I IFN induction (23). TRIF activation by TLR3 similarly activates both NF-κB and IRFs. Somewhat paradoxically, engagement of TLR2 signaling, which proceeds only through a MAL-MyD88 interaction, does not lead to IFN-α and IFN-β production, yet TLR7, TLR8, and TLR9 can activate both NF-κB and IRFs via MyD88 alone. In any case, TLR-induced antiviral immunity can be considered in terms of the antiviral effects of these two classes of transcription factors.

TLRs Trigger NF-κB Activation

Following ligand engagement and concomitant receptor dimerization, MyD88 recruitment to the IL-1R/TLR cytoplasmic tail is mediated via a receptor:adaptor TIR:TIR protein interaction. MyD88 is the only TLR adaptor protein that contains both a TIR domain and a death domain, and it is through this death domain that MyD88 recruits the IL-1R-associated kinase (IRAK) family of serine/threonine kinases, a key group of kinases required for TLR signaling (71). Four IRAKs have been identified to date, namely, IRAK-1, IRAK-2, IRAK-M, and IRAK-4. IRAK-M has been identified as an inhibitor of TLR signaling, preventing the dissociation of the IRAK-1/IRAK-4 complex from MyD88 and hence disrupting downstream functioning of IRAK-1 (60). Both IRAK-1 and IRAK-4 are active serine/threonine kinases, and phosphorylation of IRAK-1 by IRAK-4 is crucial for IRAK-1 function during TLR signaling (9, 68).

A great deal of our information regarding the downstream signaling pathways triggered upon receptor recruitment of MyD88 has been obtained in the context of IL-1R signaling. Here, IRAK-4 is recruited to the activated receptor complex (71), and it, in turn, recruits IRAK-1, found associated with another protein called Toll-interacting protein (Tollip) (8). IRAK-1 has been shown to interact with both MyD88 and IRAK-4, and when bound it is activated via phosphorylation by IRAK-4, which in turn triggers its autophosphorylation (9, 68). Hyperphosphorylation of IRAK-1 leads to its dissociation from the receptor signalosome, allowing it to bind its downstream target, TNF receptor-associated factor 6 (TRAF6) (8). Although mice lacking IRAK-4 are completely resistant to lipopolysaccharide (LPS) treatment and, furthermore, display attenuated IL-1R-, TLR2-, TLR3-, and TLR9-induced cytokine responses (107), the role of IRAK-1 in NF-κB activation is now more controversial. IRAK-1 was originally thought to be critical for TLR-induced NF-κB activation; however, IRAK-1-deficient mice show only partial defects in

IL-1-, IL-18-, and LPS-induced signaling (50, 108, 113), while the induction of NF-κB-dependent proinflammatory cytokine production by TLR7 and TLR9 proceeds unaffected (116). It has now been shown that IRAK-2 may be more important than IRAK-1 for NF-κB activation by TLRs. Similar to IRAK-M, IRAK-2 lacks any kinase activity due to the absence of a critical aspartate residue within the catalytic site of its so-called "kinase" domain. Nevertheless, upon its discovery IRAK-2 was assigned a positive role in IL-1R signaling and shown to interact with IL-1R and MyD88 (80). The requirement for IRAK-2 in TLR signaling pathways, beyond the observation that it could also interact with MAL (22), has remained uncertain. However, a recent study carried out in this laboratory has shown that IRAK-2 is indeed important in the induction of NF-κB by all TLRs, a function that is dependent on the triggering of TRAF6 ubiquitination by IRAK-2 (58). Lysine-63 (K63)-linked ubiquitination of TRAF6 is a critical event in NF-κB activation (see below). In our hands IRAK-1 was not sufficient to induce this ubiquitination of TRAF6, so this potentially points to a divergence in the functions of these two proteins, indicating that these proteins are not functionally redundant, as perhaps previously assumed. Further, in an independent study, the Epstein-Barr virus-encoded latent membrane protein 1 (LMP1) triggered TRAF6 ubiquitination in a human cell line lacking IRAK-1, but p65/RelA phosphorylation was impaired in this setting (105). This implies that IRAK-1 is indeed dispensable for the classical pathway to NF-κB activation but perhaps plays a role in the noncanonical pathway. Also, it is now believed that IRAK-1 may be more important in TLR signaling pathways to IRF activation (see "TLR7" below). The importance of IRAK-2 in TLR signaling was revealed in a study examining the mechanism whereby A52, a potent vaccinia virus antagonist of TLR signaling to NF-κB, functioned (35). A52 was found to block NF-κB activation by multiple human and murine TLRs by binding to IRAK-2. Further, small interfering RNA targeting IRAK-2 suppressed TLR-induced NF-κB activation in a human cell line while inhibiting NF-κB-dependent chemokine induction by TLR4 in primary human cells (58). The specific targeting of IRAK-2 by a virus-encoded immunoregulatory molecule such as A52 highlights the importance of the TLR/IRAK-2/NF-κB signaling axis in antiviral immunity. This study also underscores the validity of investigating viral immune evasion strategies to decipher the molecular mechanisms mediating host immune responses.

TRAF6 has been identified as an E3 ubiquitin ligase (13, 118) and it is now apparent that this ubiquitinating activity is critical for TLR signal transduction (10). TRAF6 works in combination with an E2 enzyme complex known as TRAF6-regulated IKK activator-1, or TRIKA1, consisting of the E2 ubiquitin ligase (or ubiquitin conjugating enzyme) UBC13 and the UBC-like protein UEV1A (13). Together these proteins catalyze the synthesis of K63-linked polyubiquitin chains on target components of TLR signaling pathways and on TRAF6 itself. This polyubiquitination of TRAF6 leads to activation of a downstream trimeric complex containing transforming growth factor-β-activated kinase 1 (TAK1) and the TAK1-binding proteins, TAB1 and either TAB2 or its related protein TAB3 (118). The highly conserved novel zinc finger domains of TAB2 and TAB3 act as ubiquitin receptors and specifically associate with the K63-linked polyubiquitin chains conjugated to TRAF6 (51). This binding of TRAF6 K63-linked polyubiquitin chains by TAB2/3, as well as a direct interaction with TRAF6, are essential for the induction of TAK1 kinase activity and, hence, the subsequent TAK1-mediated phosphorylation of the IκB kinase (IKK) IKK-β and MAP kinase kinase 6 (MKK6). Within the IKK complex, containing the IKK-α, IKK-β, and IKK-γ subunits, IKK-γ (or NF-κB essential modulator [NEMO]) also undergoes K63-linked polyubiquitination, which is crucial for signaling (13). Thus, the E3 ligase activity of the TRAF6 RING domain mediates the transfer of K63-linked polyubiquitin chains to IKK-γ to activate the IKK complex. Activated IKK-α and IKK-β then phosphorylate the IκB inhibitors of NF-κB, which bind the active NF-κB dimer in the cytosol, preventing its translocation to the nucleus. Since serine phosphorylation of IκB is recognized by a distinct ubiquitinating complex, this phosphorylation acts as a degradation signal leading to the addition of K48-linked polyubiquitin chains onto IκB proteins and their rapid removal by the 26S proteasome (10). This frees NF-κB dimers to enter the nucleus and upregulate the expression of a wide array of proinflammatory cytokines and other proteins required to mount an antiviral immune response. The activation of NF-κB by key antiviral TLRs is illustrated in Fig. 1.

TLR3 can trigger NF-κB activation in a MyD88-independent fashion via the TIR adaptor TRIF (41, 84). Similarly, the recruitment of TRIF, via TRAM, to TLR4 stimulates MyD88-independent induction of NF-κB, a pathway showing delayed kinetics of NF-κB activation compared to that of the MyD88-dependent TLR4 pathway to NF-κB (85). The molecular intricacies of TRIF-mediated NF-κB activation are perhaps less clear than those of MyD88 signaling to NF-κB. Like MyD88, TRIF is thought to interact with TRAF6. A number of

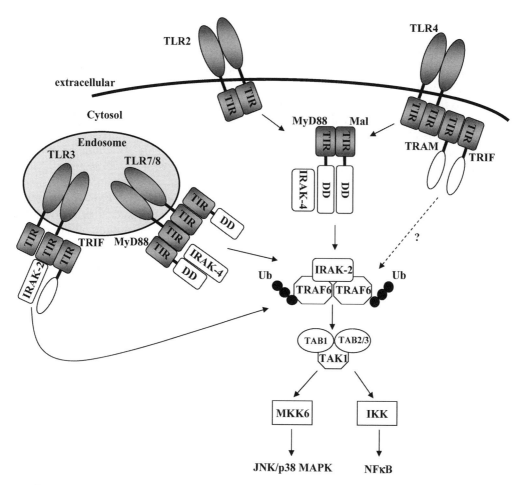

Figure 1   Activation of NF-κB by TLRs. Upon engagement of viral PAMPs at the surface of the cell (for TLR2 and TLR4) or within endosomal compartments (for TLR3 and TLR7/8), TLR dimerization occurs. This induces conformational changes within the receptor TIR domains, allowing them to recruit the appropriate downstream TIR adaptor via TIR:TIR domain associations. This is followed by activation of the IRAKs and, critically, triggering of TRAF6 ubiquitination by IRAK-2. Lysine-63-linked polyubiquitin chains conjugated to TRAF6 are specifically recognized by TAB2/3, which results in activation of the TAK1 complex and subsequent phosphorylation of the IKK complex by TAK1. TAK1-mediated phosphorylation of MKK6 leads to JNK and p38 MAP kinase activation. Phosphorylation of IκB-α by the activated IKK complex is coupled to lysine-48-linked ubiquitination of IκB-α and its subsequent proteasomal degradation. This allows NF-κB dimers to translocate into the nucleus and induce transcription of proinflammatory cytokines.

putative TRAF6 binding motifs have been defined within the N-terminal region of TRIF, and mutation of these sites completely abolished TRAF6 binding (95). In contrast, NF-κB activation by TRIF was only partially reduced and the C-terminus of TRIF was found to contribute to TRIF/NF-κB signaling through an unidentified mechanism. RIP-1 has since been identified as a component of the TRIF signaling pathway to NF-κB via its binding to the C-terminal region of TRIF (76). The relative requirement for TRAF6 in TRIF/RIP-1 signal transduction to NF-κB remains to be absolutely deci-

phered. It is possible that TRIF controls two separate pathways to NF-κB, one being TRAF6 independent via RIP-1. TLR3 signaling to NF-κB in the absence of TRAF6 has been described for murine macrophages (30). Conversely, murine fibroblasts deficient in TRAF6 were unable to stimulate NF-κB via TLR3 (47). This apparent anomaly may be reflective of two such independent pathways to NF-κB activation triggered by TRIF, with a TRAF6-dependent pathway, likely involving IRAK-2 (58), being important in fibroblasts but TRAF6-independent induction of NF-κB mediated through RIP-1 harnessed in macrophages. Studies using *TRAF6⁻ rip1⁻* double knockout mice could potentially address this issue.

IRF Induction by TLRs

IRF3

The discovery of TRIF filled a major void in our understanding of how TLRs could facilitate the induction of type I IFNs. TLR3 and TLR4 could still stimulate the expression of type I IFN and IFN-inducible genes in MyD88-deficient cells (4, 55), and while ectopic expression of either MyD88 or MAL had no effect, TRIF strongly enhanced transcriptional activity from an IFN-β promoter (84, 121). Critically, activation of IRF3 and subsequent IFN-β production following TLR3 and TLR4 engagement were both severely impaired in cells isolated from TRIF-null mice. Thus, TRIF was recognized as the major switch regulating TLR-mediated IRF3 induction and, hence, type I IFN production. However, this was only part of the story as to how TRIF triggered IRF3 phosphorylation to activate this transcription factor and so stimulate type I IFN production. In 2003, two kinases were identified as the key kinases mediating TRIF-induced phosphorylation of IRF3 (21, 101), thus shedding light on this mechanism. These kinases, inducible IKK (IKK-I or IKK-ε) and TRAF family member-associated NF-κB activator (TANK)-binding kinase 1 (TBK1; also called NF-κB-activating kinase [NAK] or TRAF2-associated kinase [T2K]), are IKK kinase family members and both can directly phosphorylate IRF3, triggering its nuclear translocation and DNA binding (74, 101). Attempts to assess their relative contribution to IRF3 activation and subsequent downstream type I IFN induction in vivo using knockout mice initially suggested that TBK1 was likely the critical upstream kinase in most signaling pathways to IRF3 (40, 74). When compared to wild-type mice, IKK-ε⁻/⁻ mice challenged with increasing doses of IAV showed no obvious defects in type I IFN and proinflammatory cytokine production (112). Significantly, microarray analysis revealed a subset of IFN-stimulated genes that were poorly induced in the absence of IKK-ε. This correlated with an increased susceptibility to IAV infection at lower titers and was attributed to a critical role for IKK-ε in type I IFN signaling via the direct phosphorylation of Stat1 (signal transducer and activator of transcription 1). Thus, the activation of IKK-ε in a positive feedback loop mechanism is in fact critical for the IFN-inducible antiviral state (112). The observation that TBK1 kinase activity was rapidly induced after only 15 min of LPS stimulation in primary human macrophages while IKK-ε activity was

not seen until 8 h post-LPS treatment potentially supports this, as these late kinetics may reflect a secondary IKK-ε activation pathway by IFN-α/β (104). Initially it was thought that TRIF associated with TBK1 via its N terminus to facilitate IRF3 activation. Conversely, it is now believed that TRIF engages TBK1 through associations with NAK-associated protein 1 (NAP1) (94) and TRAF3 (33, 82). Important antiviral TLR signal transduction pathways to IRF activation are described in Fig. 2.

IRF7

pDCs are key effector cells of the innate immune response to viruses due to their extraordinary capacity to produce type I IFNs upon viral infection and as such are specifically referred to as IFN-producing cells. These cells do not express TLR3 or TLR4 but express high levels of TLR7 and TLR9, and while wild-type pDCs demonstrated potent IFN-α production on infection with IAV and VSV, this cytokine release was grossly impaired in pDCs isolated from TLR7-deficient mice (14, 70). Similarly, pDCs from TLR9-null mice were unable to stimulate IFN-α release when infected with HSV-1 and HSV-2 (57, 62). How these TLRs could facilitate the induction of IFN in a MyD88-dependent manner posed a major question. Unlike IRF3, IRF7 expression is restricted. pDCs, however, express high levels and, in keeping with this, a pivotal role for IRF7 in TLR7- and TLR9-mediated type I IFN production has emerged. MyD88 was shown to interact with IRF7, and cotransfection of IRF7 with MyD88 led to a huge increase in IFN-α transcription (57). Furthermore, compared to wild-type cells, IFN-α promoter activity was vastly reduced when IRF7 and MyD88 were coexpressed in TRAF6-deficient murine embryonic fibroblasts, thus implicating TRAF6 in this signaling complex. This work has led to the elucidation of an essential role for IRAK-1 in the activation of IRF7 by both TLR7 and TLR9 (116). Like MyD88 and TRAF6, IRAK-1 physically associated with IRF7 and, critically, could catalyze its phosphorylation in vitro. The TLR7- and TLR9-mediated IFN-α response was almost completely ablated in IRAK-1-deficient mice, while IFN-α induction in response to poly(I:C) was normal (116). Thus, while TRIF-dependent IFN induction does not rely on IRAK-1, the activation of an IFN response by MyD88 requires this kinase. This represents one of a number of alternative functions emerging for IRAK-1, emphasizing that it is likely less important for the canonical NF-κB activation pathway than initially believed (see "TLRs Trigger NF-κB Activation" above). Although IRAK-4 could not directly phosphorylate IRF7 in vitro (116), stimulation of IFN-α production by TLR7 and TLR9 was grossly impaired in

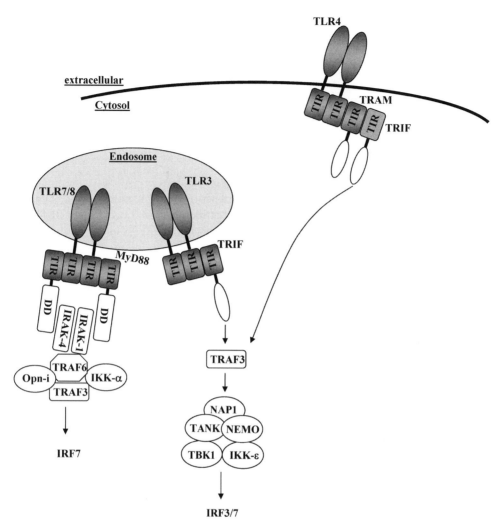

Figure 2   Activation of IRFs by TLRs. Upon binding of their associated viral PAMPs, TLR3 and TLR4 trigger activation of IRF3 and IRF7 via the TIR adaptor TRIF. TLR3 recruits TRIF directly while TLR4 engages TRIF through the bridging adaptor TRAM. TRIF stimulates activation of the noncanonical IKK kinases, TBK1 and IKK-ε, through associations with TRAF3 and NAP1. These kinases mediate phosphorylation of IRF3 and IRF7, facilitating their dimerization and translocation to the nucleus, where they upregulate the transcription of type I interferons (IFN-α and IFN-β). TLR7/8 recruits the TIR adaptor MyD88, which then activates the serine/threonine kinases IRAK-4 and IRAK-1. IRAK-1 can then go on to directly phosphorylate IRF7 in a process that requires TRAF6 and TRAF3. Opsonin and IKK-α have also been implicated in this signaling process, but the exact details of their involvement have yet to be deciphered.

IRAK-4-deficient mice (59, 61). This was attributed to defective downstream IRF7 activation, and the kinase activity of IRAK-4 was required to restore this activation in "knockin" mice studies (59). Thus, although IRF7 is not a direct substrate for IRAK-4 phosphorylation, activation of IRAK-1 by this kinase is essential in this pathway.

Furthermore, IKK-α has also been found in the MyD88/IRF7 signaling complex (45). Like IRAK-1, it also phosphorylated IRF7 directly and is now believed

to be a crucial player in the triggering of a type I IFN response by TLR7 and TLR9. Stimulation of IFN-α release on infection with VSV was severely reduced in pDCs lacking IKK-α. Finally, the phosphoprotein osteopontin has been linked to the TLR9-mediated IRF7 activation pathway (103). Triggering TLR9 signal transduction led to the upregulation of osteopontin gene expression in a T-bet transcription factor-dependent process that occurred exclusively in pDCs. Osteopontin colocalized with MyD88 on TLR9 ligand engagement, and this was found to be essential for efficient nuclear translocation of IRF7 and subsequent IFN-α gene induction. The requirement for osteopontin in TLR9 signaling was restricted to the IFN pathway, as the induction of NF-κB-dependent proinflammatory cytokines, such as IL-6 and TNF-α, was normal in osteopontin-deficient pDCs treated with TLR9 ligands.

IRF7 appears to be essential for both the MyD88-dependent and the TRIF-dependent pathways to IFN production, so a global role for IRF7 as a major regulator of type I IFN induction has been unveiled (43). The IFN response to ssRNA (encephalomyocarditis virus [EMCV], VSV) and dsDNA (HSV-1) viruses was almost completely abolished in IRF7-deficient murine embryonic fibroblasts, while MyD88 was not required. VSV- and HSV-1-induced IFN in pDCs displayed a dual dependency on IRF7 and MyD88. Importantly, the in vivo IFN-α response to HSV-1 and EMCV infection was completely abrogated in IRF7-deficient mice. Interestingly, knocking out MyD88 expression only diminished IFN-α production slightly, suggesting that although MyD88 plays a role in the IFN response to these viruses, the MyD88-independent pathway can compensate in its absence. IRF7, on the other hand, is absolutely essential for all type I IFN production in response to these viruses. A study by Honda et al. has shed some light on the question of how pDCs can induce type I IFN production through the MyD88/IRF7-dependent pathway so efficiently. In pDCs, the CpG ligand for TLR9 was retained in endosomal compartments for longer periods than in other cells, so the rapid shuttling of CpG to lysosomes in cDCs may explain the lack of a robust IFN response in these cells (42).

Thus, it would appear that while both IRF3 and IRF7 are essential for the activation of type I IFNs in response to viral infection, IRF7 plays a more global role in this function (43). IRF3 is potentially more critical for TLR3- and TLR4-mediated IFN-β induction through TRIF. Notably, although IRF5 is induced in response to TLR7 and TLR9 ligands along with IRF7, it is redundant for the activation of antiviral IFNs. In contrast, IRF5 would appear to be a central regulator of proinflammatory cytokine expression on stimulation of a range of TLRs (98, 109).

TARGETED EVASION OF TLR-MEDIATED ANTIVIRAL IMMUNITY BY RNA VIRUSES

RNA viruses have evolved specific mechanisms to inhibit TLR signaling pathways, and this provides compelling evidence for the importance of TLR-induced antiviral immunity in controlling RNA virus replication. HCV provides an excellent example of this. The NS3/4A protein encoded by this ssRNA flavivirus exhibits serine protease activity and is essential for virus survival. Cleavage of a virally encoded precursor polypetide by NS3/4A is required to generate mature structural proteins during replication. Intriguingly, NS3/4A has the ability to block host-mediated IRF3 activation upon infection of an unrelated virus such as Senda virus, a function that is solely dependent on its protease activity (25). In a bid to identify the host cell target of NS3/4A, potential NS3/4A cleavage sites were identified within TLR3, TRIF, IKK-ε, and TBK1, indicating that HCV could mediate targeted inhibition of TLR3 signaling pathways (67). However, of all these potential targets only TRIF was specifically degraded by HCV. Individually, the two polypeptides generated by proteolysis of TRIF were no longer able to activate transcription of the IFN-β promoter. NS3/4A expression potently reduced poly(I:C)-induced IFN-β promoter activity as well as poly(I:C)-mediated stimulation of the NF-κB-dependent PRDII promoter. Similarly, upregulation of two IRF3-dependent genes, *ISG56* and *ISG15*, in response to poly(I:C) stimulation was delayed in the presence of NS3/4A. Thus, by targeting TRIF, a receptor-proximal component of TLR3 signaling, HCV broadly inhibits both NF-κB- and IRF3-mediated TLR3 antiviral signaling. The importance of TLR3 in the host immune response to HCV infection is currently unclear (78), but the fact that this virus can specifically inhibit this component of antiviral innate immunity suggests that this PRR does indeed play a central role in the host response to HCV infection.

Like so many other virally encoded immunomodulatory proteins, NS3/4A employs a multifaceted approach to inhibit host innate immunity. While TLR3 delivers a transmembrane signal alerting the cell to the presence of viral dsRNA in endosomal compartments, RIG-I is now deemed a major cytoplasmic radar for this viral PAMP. Compellingly, HCV NS3/4A targets both of these arms of host innate immune response (24, 67). NS3/4A completely abolished the nuclear translocation of IRF3 induced by overexpression of the RIG-I CARD motifs along with RIG-I-mediated IRF3 dimerization and phosphorylation at serine-386 (24). Given that RIG-I can detect HCV genomes directly (106), it makes sense that this virus would develop an evasion strategy to inhibit this arm of the host innate immune response. In support of this, Foy et al. demonstrated an absolute requirement for RIG-I in the induction of an interferon response to HCV dsRNA and that RIG-I could in fact repress replication of an HCV replicon (24). Major recent advances in our understanding of RIG-I signaling have exposed the mechanism of inhibition of this innate immune signaling PRR by NS3/4A. The discovery of the primary adaptor molecule utilized by RIG-I, Cardif, simultaneously revealed it as the downstream component of RIG-I signaling pathways targeted by NS3/4A (77).

The importance of RIG-I and MDA-5 in triggering an immune response to RNA viruses is emphasized by the number of immunomodulatory strategies evolved by these viruses to target these PRRs. IAV virus-encoded NS1 protein can sequester dsRNA in the cytoplasm to block IFN induction. More recently, this viral inhibitor has been shown to directly bind RIG-I, resulting in the inhibition of RIG-I-mediated IRF3 activation (79). Similar to HCV, the Nipah paramyxovirus-encoded W protein, a viral protein previously known to antagonize IFN function, specifically inhibits TRIF-induced IRF3 phosphorylation (102).

SUBVERSION OF THE TLR RESPONSE BY RNA VIRUSES

Large DNA viruses, such as the herpesviruses and the poxviruses, with as many as 200 open reading frames at their disposal, appear to devote large fractions of their genomes to the expression of a vast array of immunomodulatory proteins that function to specifically inhibit particular branches of the host immune response. RNA viruses, however, do not have this luxury and, perhaps as a result of this, have become masters of subversion, harnessing TLR-induced signaling pathways to taper them to their own advantage, thus using host responses to facilitate efficient viral replication. This adds further complexity to the relationship between TLRs and viruses and the importance of the balance of the ensuing response, which can be tipped in favor of the host or the virus depending on the interaction in question. A fascinating example of this is the role of TLR3 in the recognition of the ssRNA virus West Nile virus (WNV) (119). Here, TLR3 was shown to initiate an inflammatory response to WNV, triggering IL-6 and TNF-α production upon virus infection. Critically, this led to disruption of the blood-brain barrier and in fact facilitated passage of the virus into the brain. TLR3-deficient mice demonstrated enhanced survival compared to wild-type mice when exposed to a lethal dose of WNV, and this correlated to a reduced viral load and dampened inflammatory responses in the brains of these animals. While this study suggests that TLR3 can detect WNV infection and go on to induce a proinflammatory response, this appears to favor the virus rather than the host.

In the case of IAV infection, knocking out TLR3 led to a reduced inflammatory response and enhanced host survival (66). Thus, IAV also triggers TLR3 signaling to induce massive inflammation that is detrimental to the host and contributes to the pathology of the virus-induced disease. Likewise, EMCV infection appears to elicit potent proinflammatory cytokine and chemokine induction through TLR3, as TLR3-deficient mice were more susceptible to EMCV infection and the hearts of these mice displayed an increased viral load compared to wild-type mice (34). It would seem, therefore, that a pattern is emerging that implies that the detection of certain RNA viruses by TLR3 has a negative impact on disease outcome for the host. However, it is not necessarily clear why it would be beneficial for a virus to trigger such a potent immune response when, in fact, it is this virus-induced inflammation that leads to death of the host. Surely, this would then limit the lateral spread of the virus from host to host. Thus, the suggestion of Gowen et al. that the overall negative effect of triggering TLR3 signaling in experimental models of virus infection may be due to the vastly higher doses of virus used leading to an overpronounced inflammatory response warrants consideration (31). In naturally occurring virus infections, TLR3 may in fact play a key role in controlling virus spread through the induction of a less damaging proinflammatory response to much lower viral doses. Therefore, it may be necessary to readdress the question of the role of TLR3 in the antiviral response to the RNA viruses described above using much lower doses of virus to see if this is indeed the case.

It is not only TLR3-mediated antiviral responses that RNA viruses appear to manipulate for their own means. TLR2 has been implicated in the response to wild-type measles virus. It detects the presence of the hemagglutinin (HA) protein of this virus and induces the expression of proinflammatory cytokines such as IL-6 in human monocytic cells (6). However, cellular activation by measles virus via TLR2 also led to the upregulation of cell surface expression of CD150. This molecule acts as a receptor for measles virus entry, hence, triggering TLR2 signaling likely makes cells more permissive to measles virus infection and thus enhances virus spread. Furthermore, mouse mammary tumor virus (MMTV) infection triggers IL-10 production via TLR4 activation (48). This is potentially through the engagement of TLR4 by the MMTV-encoded envelope (Env) protein (88). IL-10 is of course an immunosuppressive cytokine, and the induction of TLR4 signaling by MMTV infection was shown to help sustain virus infection rather than limit it. Furthermore, in support of TLR4 actually promoting MMTV persistence, C3H/HeJ mice, which lack functional TLR4, eliminated the virus completely via a CD8$^+$ cytotoxic-T-lymphocyte response (48). The advantage of manipulating host induction of IL-10 as an immunomodulatory mechanism employed by viruses is highlighted by two separate studies showing that increased IL-10 production is essential for LCMV persistence in vivo (7, 20). As in the case of measles virus infection and TLR2, engagement of TLR4 by MMTV led to the upregulation of the host-encoded MMTV receptor molecule CD71 (48). Thus, MMTV likely uses the induction of TLR4 signaling to promote viral spread.

EVIDENCE FOR A PHYSIOLOGICAL ROLE FOR TLRs IN INNATE IMMUNITY TO RNA VIRUSES

The exact physiological role of TLR3 in host innate immunity to RNA viruses has remained somewhat elusive. In an extensive in vivo study, mice deficient in TLR3 expression displayed a similar degree of susceptibility to a number of RNA viruses tested, including VSV and LCMV, compared to wild-type mice (18). This would suggest that TLR3 does not offer any protection against infection with these viruses. This is surprising because when TLR3 was first identified as a sensor of dsRNA, it was implicated in the cellular activation induced by exposure to purified reoviral genomic dsRNA (4, 73). Moreover, a role for TLR3 in the response to IAV infection in humans has been suggested (32). Infection of bronchial epithelial cells with IAV led to the simultaneous upregulation of TLR3 protein expression, a modulation that was similar to that seen with poly(I:C) stimulation. A role for TRIF in NF-κB activation by IAV was established, while MyD88 was found to be redundant, confirming that TLR3 is likely involved in mediating an immune response to this virus in humans. Also, a study by Rudd et al. elegantly demonstrated that although TLR3 deficiency in mice had no effect on viral load following RSV infection, these mice displayed dramatic increases in airway mucus production and this was coupled to a Th2-like immune response with the accompanying induction of more aggressive cytokines such as IL-13 and IL-5 (91). Conversely, RSV infection in TLR3-expressing mice triggered a less pathogenic Th1-like response, indicating that while TLR3 may not be required for effective viral clearance during RSV infection, it does contribute to the regulation of pathogenic responses in the lung environment.

Thus, while TLR3 does appear to function as an antiviral innate immune receptor and can mount a strong inflammatory response on detection of viral dsRNA, it may be that other cytoplasmic PRRs, such as RIG-I, are more suitably located to act as major sensors of dsRNA released into the cytoplasm during viral infection. The cell type-specific subcellular localization of TLR3 might reflect a divergence in TLR3 function in different tissue types. Human fibroblasts and epithelial cells display both a cell surface and an intracellular pattern of TLR3 expression, while monocyte-derived immature dendritic cells (iDCs) and CD11c$^+$ blood DCs express TLR3 solely in intracellular vesicles (72). Thus, it may be that where TLR3 is exposed at the cell surface, it can serve to detect the presence of dsRNA of both cellular and viral origin released upon lysis of virally infected cells. This is supported by the fact that anti-TLR3 antibodies could block activation of fibroblasts in response to poly(I:C) treatment (73), indicating that this is a cell surface event. The observation that TLR3-mediated IFN induction by DCs was inhibited by chloroquine or bafilomycin strongly implies that this signaling cascade is initiated in endosomal compartments (27). While endosome-localized TLR3 may serve to alert the host innate immunity to invading endocytic viruses, a more clear-cut critical role for TLR3 expressed within the endosomal compartments of DCs is in the activation of the adaptive cellular immune response to viral infection (100).

TLRs present an initial barrier to invading pathogens, alerting the host to their presence and mounting an initial rapid inflammatory response in a bid to curtail any further spread of the infectious agent. Of course, a major role of innate immunity is to prime the adaptive immune response to activate specific cell-mediated immunity to foreign antigens. DCs are the front line of the innate immune system, providing a link between the innate and adaptive responses. DCs acquire viral antigens indirectly via the phagocytosis of virally infected cells. They can then present this foreign antigenic material in a major histocompatibility complex class I-restricted fashion to CD8$^+$ CTLs, thus priming naïve CTLs to mount a specific cytotoxic response. Nevertheless, the cross talk between DCs and this cellular arm of the adaptive immune response does not always result in CTL activation. The stimulatory signals required to allow DCs to prime CD8$^+$ CTLs to mount a cytotoxic immune response, as opposed to simply inducing cross tolerance (inactivation of CTLs), have remained ambiguous. However, it would now appear that signals induced via the engagement of TLR3 by dsRNA in DCs strongly enhance cross priming of naïve CD8$^+$ CTLs, while the absence of this TLR3-mediated signal transduction results in a much weaker CTL response (100). This constitutes a completely novel role for TLR3 in host antiviral immunity. Poly(I:C)-facilitated DC activation occurred independently of PKR and MyD88, but an absolute requirement for TLR3 was identified for both poly(I:C)- and virus-induced effects. Poly(I:C) has been implicated in the induction of CD8$^+$ CTL responses to specific HIV antigens (26). Whether or not this enhanced cross priming facilitated by poly(I:C) is TLR3 dependent has not been demonstrated.

As implied for TLR3 above, cell-type specificity is a critical consideration in assessing the physiological contribution of TLRs to the innate immune response to RNA viruses. TLR7 and TLR9 are constitutively expressed in DCs and, in humans, primarily in the plasmacytoid subset of DCs (89). They are unique in that they can trigger IRF activation solely through MyD88. The induction of IRF7 by TLR7 and TLR9, via MyD88, IRAK-1, and TRAF6, occurs only in pDCs (116). Thus, in other cell types TLR7 and TLR9 engagement leads only to the upregulation of proinflammatory cytokine production and not IFN-α, as NF-κB and MAP kinases are activated in the absence of IRF7 activation. This may be attributable to the high level of constitutive IRF7 expression particular to pDCs. IRF7 expression is generally very low in most cell types, and this may explain why, in murine cDCs, triggering TLR9 signaling leads to IRF1 activation and not IRF7 activation, and thus subsequent IFN-β production but not IFN-α production (81, 97). The novel discovery that autophagy in pDCs allows these cells to mop up VSV RNA in the cytosol during replication and present it to endosomes may explain why these cells appear to be so effective at mounting TLR7-mediated antiviral responses (65, 86).

CONCLUSIONS

The fact that viruses have evolved specific strategies to inhibit host IFN induction has long been recognized. The discovery of TLRs and other PRRs has allowed us to understand the underlying mechanisms behind these immune evasion processes. By the same token, it has been known for a long time that viruses trigger cellular activation upon infection, stimulating cell signaling pathways and ultimately activating host transcription factors such as NF-κB and IRF3. Again, huge advances made in understanding how the host scans both the extracellular and intracellular environment, detecting the presence of invading pathogens via TLRs and other PRRs to trigger an immune response even as early in the virus infectious cycle as viral entry, have bridged a large gap in our knowledge of the workings of innate immunity.

A number of TLR-virus interactions have been described where it is known that a particular RNA virus triggers cellular activation via a specific TLR but the viral PAMP in question has not been identified. The examples of coxsackievirus B4 interaction with TLR4 and LCMV detection by TLR2 described earlier illustrate this. Thus, while our understanding of the TLR-mediated antiviral response has expanded immensely, there is still much to be discovered. Further, in cases of virus infection where the signal-inducing PAMP has been identified, there is still a great deal lacking in our knowledge of how specifically these PAMPs are detected by TLRs, and indeed other PRRs. Recent structural studies of the TLR3 ectodomain have formidably enhanced our understanding of how this TLR recognizes its cognate PAMP, dsRNA. Comparable studies on other TLRs will be essential to further unfold the complex interplay between viruses and TLRs.

The potential benefits of understanding TLR-virus interactions cannot be overstated. It is becoming increasingly apparent that many viruses have evolved elegant strategies to evade detection by different arms of the host immune response, and TLR-induced antiviral innate immunity is no different in this regard. These viral immune evasion strategies are perfectly fine-tuned to potently suppress or taper host innate immunity to serve the virus in question. Studying the interactions between TLRs and viruses and the mechanisms of action of virus-encoded TLR antagonists offers us remarkable tools to probe the molecular intricacies of innate immunity. Further unfolding the molecular detail of TLR function will open up new and exciting ways to target TLR signaling pathways therapeutically not only in the treatment of viral infection but also in the treatment of inflammatory and autoimmune diseases. Investigation of the mode of action of A52, a vaccinia virus-encoded antagonist of TLR signaling, revealed that it specifically targeted IRAK-2 to potently block NF-κB activation by a broad range of TLRs (58). This revealed an important role for IRAK-2 in TLR signaling to NF-κB but implied that IRAK-2 is not required for TLR-mediated IRF activation. Therefore, targeting IRAK-2 could prove a very effective way to dampen an overactive proinflammatory response in a variety of disease situations while leaving IRF-mediated immunity intact. Similarly, the identification of the HCV-encoded NS3/4A protein as an inhibitor of both TRIF- and RIG-I-mediated antiviral signaling offers significant therapeutic potential in the treatment of HCV-associated disease. The use of protease inhibitors to target this molecule, thus relieving its inhibitory effect, may restore both TLR3- and RIG-I-induced innate immunity to this virus (78, 114). Priming an efficient immune response would potentially allow the host to limit replication of the virus and hence limit disease progression and even potentially lead to viral clearance. While huge advances have been made in our understanding of the complex interplay between TLRs and viruses, clearly there is a great deal left to be discovered that will likely lead to novel therapeutic strategies.

References

1. Ahmad-Nejad, P., H. Hacker, M. Rutz, S. Bauer, R. M. Vabulas, and H. Wagner. 2002. Bacterial CpG-DNA and lipopolysaccharides activate Toll-like receptors at distinct cellular compartments. *Eur. J. Immunol.* 32:1958–1968.

2. Ahonen, C. L., C. L. Doxsee, S. M. McGurran, T. R. Riter, W. F. Wade, R. J. Barth, J. P. Vasilakos, R. J. Noelle, and R. M. Kedl. 2004. Combined TLR and CD40 triggering induces potent CD8+ T cell expansion with variable dependence on type I IFN. *J. Exp. Med.* 199:775–784.

3. Akira, S., S. Uematsu, and O. Takeuchi. 2006. Pathogen recognition and innate immunity. *Cell* 124:783–801.

4. Alexopoulou, L., A. C. Holt, R. Medzhitov, and R. A. Flavell. 2001. Recognition of double-stranded RNA and activation of NF-kappaB by Toll-like receptor 3. *Nature* 413:732–738.

5. Bell, J. K., I. Botos, P. R. Hall, J. Askins, J. Shiloach, D. M. Segal, and D. R. Davies. 2005. The molecular structure of the Toll-like receptor 3 ligand-binding domain. *Proc. Natl. Acad. Sci. USA* 102:10976–10980.

6. Bieback, K., E. Lien, I. M. Klagge, E. Avota, J. Schneider-Schaulies, W. P. Duprex, H. Wagner, C. J. Kirschning, V. Ter Meulen, and S. Schneider-Schaulies. 2002. Hemagglutinin protein of wild-type measles virus activates Toll-like receptor 2 signaling. *J. Virol.* 76:8729–8736.

7. Brooks, D. G., M. J. Trifilo, K. H. Edelmann, L. Teyton, D. B. McGavern, and M. B. Oldstone. 2006. Interleukin-10 determines viral clearance or persistence in vivo. *Nat. Med.* 12:1301–1309.

8. Burns, K., J. Clatworthy, L. Martin, F. Martinon, C. Plumpton, B. Maschera, A. Lewis, K. Ray, J. Tschopp, and F. Volpe. 2000. Tollip, a new component of the IL-1RI pathway, links IRAK to the IL-1 receptor. *Nat. Cell. Biol.* 2:346–351.

9. Cao, Z., W. J. Henzel, and X. Gao. 1996. IRAK: a kinase associated with the interleukin-1 receptor. *Science* 271:1128–1131.

10. Chen, Z. J. 2005. Ubiquitin signalling in the NF-kappaB pathway. *Nat. Cell Biol.* 7:758–765.

11. Choe, J., M. S. Kelker, and I. A. Wilson. 2005. Crystal structure of human Toll-like receptor 3 (TLR3) ectodomain. *Science* 309:581–585.

12. Collins, S. E., R. S. Noyce, and K. L. Mossman. 2004. Innate cellular response to virus particle entry requires IRF3 but not virus replication. *J. Virol.* 78:1706–1717.

13. Deng, L., C. Wang, E. Spencer, L. Yang, A. Braun, J. You, C. Slaughter, C. Pickart, and Z. J. Chen. 2000. Activation of the IkappaB kinase complex by TRAF6 requires a dimeric ubiquitin-conjugating enzyme complex and a unique polyubiquitin chain. *Cell* 103:351–361.

14. Diebold, S. S., T. Kaisho, H. Hemmi, S. Akira, and C. Reis e Sousa. 2004. Innate antiviral responses by means of TLR7-mediated recognition of single-stranded RNA. *Science* 303:1529–1531.

15. Diebold, S. S., C. Massacrier, S. Akira, C. Paturel, Y. Morel, and C. Reis e Sousa. 2006. Nucleic acid agonists for Toll-like receptor 7 are defined by the presence of uridine ribonucleotides. *Eur. J. Immunol.* 36: 3256–3267.

16. Dolganiuc, A., S. Oak, K. Kodys, D. T. Golenbock, R. W. Finberg, E. Kurt-Jones, and G. Szabo. 2004. Hepatitis C core and nonstructural 3 proteins trigger Toll-like receptor 2-mediated pathways and inflammatory activation. *Gastroenterology* 127:1513–1524.

17. Du, X., A. Poltorak, Y. Wei, and B. Beutler. 2000. Three novel mammalian Toll-like receptors: gene structure, expression, and evolution. *Eur. Cytokine Netw.* 11: 362–371.

18. Edelmann, K. H., S. Richardson-Burns, L. Alexopoulou, K. L. Tyler, R. A. Flavell, and M. B. Oldstone. 2004. Does Toll-like receptor 3 play a biological role in virus infections? *Virology* 322:231–238.

19. Ehl, S., R. Bischoff, T. Ostler, S. Vallbracht, J. Schulte-Monting, A. Poltorak, and M. Freudenberg. 2004. The role of Toll-like receptor 4 versus interleukin-12 in immunity to respiratory syncytial virus. *Eur. J. Immunol.* 34:1146–1153.

20. Ejrnaes, M., C. M. Filippi, M. M. Martinic, E. M. Ling, L. M. Togher, S. Crotty, and M. G. von Herrath. 2006. Resolution of a chronic viral infection after interleukin-10 receptor blockade. *J. Exp. Med.* 203:2461–2472.

21. Fitzgerald, K. A., S. M. McWhirter, K. L. Faia, D. C. Rowe, E. Latz, D. T. Golenbock, A. J. Coyle, S. M. Liao, and T. Maniatis. 2003. IKKepsilon and TBK1 are essential components of the IRF3 signaling pathway. *Nat. Immunol.* 4:491–496.

22. Fitzgerald, K. A., E. M. Palsson-McDermott, A. G. Bowie, C. A. Jefferies, A. S. Mansell, G. Brady, E. Brint, A. Dunne, P. Gray, M. T. Harte, D. McMurray, D. E. Smith, J. E. Sims, T. A. Bird, and L. A. O'Neill. 2001. Mal (MyD88-adapter-like) is required for Toll-like receptor-4 signal transduction. *Nature* 413:78–83.

23. Fitzgerald, K. A., D. C. Rowe, B. J. Barnes, D. R. Caffrey, A. Visintin, E. Latz, B. Monks, P. M. Pitha, and D. T. Golenbock. 2003. LPS-TLR4 signaling to IRF-3/7 and NF-kappaB involves the Toll adapters TRAM and TRIF. *J. Exp. Med.* 198:1043–1055.

24. Foy, E., K. Li, R. Sumpter, Jr., Y. M. Loo, C. L. Johnson, C. Wang, P. M. Fish, M. Yoneyama, T. Fujita, S. M. Lemon, and M. Gale, Jr. 2005. Control of antiviral defenses through hepatitis C virus disruption of retinoic acid-inducible gene-I signaling. *Proc. Natl. Acad. Sci. USA* 102:2986–2991.

25. Foy, E., K. Li, C. Wang, R. Sumpter, Jr., M. Ikeda, S. M. Lemon, and M. Gale, Jr. 2003. Regulation of interferon regulatory factor-3 by the hepatitis C virus serine protease. *Science* 300:1145–1148.

26. Fujimoto, C., Y. Nakagawa, K. Ohara, and H. Takahashi. 2004. Polyriboinosinic polyribocytidylic acid [poly(I:C)]/TLR3 signaling allows class I processing of exogenous protein and induction of HIV-specific CD8+ cytotoxic T lymphocytes. *Int. Immunol.* 16:55–63.

27. Funami, K., M. Matsumoto, H. Oshiumi, T. Akazawa, A. Yamamoto, and T. Seya. 2004. The cytoplasmic 'linker region' in Toll-like receptor 3 controls receptor localization and signaling. *Int. Immunol.* 16:1143–1154.

28. Gagro, A., M. Tominac, V. Krsulovic-Hresic, A. Bace, M. Matic, V. Drazenovic, G. Mlinaric-Galinovic, E. Kosor, K. Gotovac, I. Bolanca, S. Batinica, and S. Rabatic. 2004. Increased Toll-like receptor 4 expression in infants with respiratory syncytial virus bronchiolitis. *Clin. Exp. Immunol.* **135**:267–272.

29. Georgel, P., Z. Jiang, S. Kunz, E. Janssen, J. Mols, K. Hoebe, S. Bahram, M. B. Oldstone, and B. Beutler. 2007. Vesicular stomatitis virus glycoprotein G activates a specific antiviral Toll-like receptor 4-dependent pathway. *Virology* **362**:304–313.

30. Gohda, J., T. Matsumura, and J. Inoue. 2004. Cutting edge: TNFR-associated factor (TRAF) 6 is essential for MyD88-dependent pathway but not Toll/IL-1 receptor domain-containing adaptor-inducing IFN-beta (TRIF)-dependent pathway in TLR signaling. *J. Immunol.* **173**:2913–2917.

31. Gowen, B. B., J. D. Hoopes, M. H. Wong, K. H. Jung, K. C. Isakson, L. Alexopoulou, R. A. Flavell, and R. W. Sidwell. 2006. TLR3 deletion limits mortality and disease severity due to phlebovirus infection. *J. Immunol.* **177**:6301–6307.

32. Guillot, L., R. Le Goffic, S. Bloch, N. Escriou, S. Akira, M. Chignard, and M. Si-Tahar. 2005. Involvement of Toll-like receptor 3 in the immune response of lung epithelial cells to double-stranded RNA and influenza A virus. *J. Biol. Chem.* **280**:5571–5580.

33. Hacker, H., V. Redecke, B. Blagoev, I. Kratchmarova, L. C. Hsu, G. G. Wang, M. P. Kamps, E. Raz, H. Wagner, G. Hacker, M. Mann, and M. Karin. 2006. Specificity in Toll-like receptor signalling through distinct effector functions of TRAF3 and TRAF6. *Nature* **439**:204–207.

34. Hardarson, H. S., J. S. Baker, Z. Yang, E. Purevjav, C. H. Huang, L. Alexopoulou, N. Li, R. A. Flavell, N. E. Bowles, and J. G. Vallejo. 2007. Toll-like receptor 3 is an essential component of the innate stress response in virus-induced cardiac injury. *Am. J. Physiol. Heart Circ. Physiol.* **292**:H251–H258.

35. Harte, M. T., I. R. Haga, G. Maloney, P. Gray, P. C. Reading, N. W. Bartlett, G. L. Smith, A. Bowie, and L. A. O'Neill. 2003. The poxvirus protein A52R targets Toll-like receptor signaling complexes to suppress host defense. *J. Exp. Med.* **197**:343–351.

36. Heil, F., P. Ahmad-Nejad, H. Hemmi, H. Hochrein, F. Ampenberger, T. Gellert, H. Dietrich, G. Lipford, K. Takeda, S. Akira, H. Wagner, and S. Bauer. 2003. The Toll-like receptor 7 (TLR7)-specific stimulus loxoribine uncovers a strong relationship within the TLR7, 8 and 9 subfamily. *Eur. J. Immunol.* **33**:2987–2997.

37. Heil, F., H. Hemmi, H. Hochrein, F. Ampenberger, C. Kirschning, S. Akira, G. Lipford, H. Wagner, and S. Bauer. 2004. Species-specific recognition of single-stranded RNA via Toll-like receptor 7 and 8. *Science* **303**:1526–1529.

38. Hemmi, H., T. Kaisho, O. Takeuchi, S. Sato, H. Sanjo, K. Hoshino, T. Horiuchi, H. Tomizawa, K. Takeda, and S. Akira. 2002. Small anti-viral compounds activate immune cells via the TLR7 MyD88-dependent signaling pathway. *Nat. Immunol.* **3**:196–200.

39. Hemmi, H., O. Takeuchi, T. Kawai, T. Kaisho, S. Sato, H. Sanjo, M. Matsumoto, K. Hoshino, H. Wagner, K. Takeda, and S. Akira. 2000. A Toll-like receptor recognizes bacterial DNA. *Nature* **408**:740–745.

40. Hemmi, H., O. Takeuchi, S. Sato, M. Yamamoto, T. Kaisho, H. Sanjo, T. Kawai, K. Hoshino, K. Takeda, and S. Akira. 2004. The roles of two IkappaB kinase-related kinases in lipopolysaccharide and double stranded RNA signaling and viral infection. *J. Exp. Med.* **199**:1641–1650.

41. Hoebe, K., X. Du, P. Georgel, E. Janssen, K. Tabeta, S. O. Kim, J. Goode, P. Lin, N. Mann, S. Mudd, K. Crozat, S. Sovath, J. Han, and B. Beutler. 2003. Identification of Lps2 as a key transducer of MyD88-independent TIR signalling. *Nature* **424**:743–748.

42. Honda, K., Y. Ohba, H. Yanai, H. Negishi, T. Mizutani, A. Takaoka, C. Taya, and T. Taniguchi. 2005. Spatiotemporal regulation of MyD88-IRF-7 signalling for robust type-I interferon induction. *Nature* **434**:1035–1040.

43. Honda, K., H. Yanai, H. Negishi, M. Asagiri, M. Sato, T. Mizutani, N. Shimada, Y. Ohba, A. Takaoka, N. Yoshida, and T. Taniguchi. 2005. IRF-7 is the master regulator of type-I interferon-dependent immune responses. *Nature* **434**:772–777.

44. Hornung, V., J. Ellegast, S. Kim, K. Brzozka, A. Jung, H. Kato, H. Poeck, S. Akira, K. K. Conzelmann, M. Schlee, S. Endres, and G. Hartmann. 2006. 5'-Triphosphate RNA is the ligand for RIG-I. *Science* **314**:994–997.

45. Hoshino, K., T. Sugiyama, M. Matsumoto, T. Tanaka, M. Saito, H. Hemmi, O. Ohara, S. Akira, and T. Kaisho. 2006. IkappaB kinase-alpha is critical for interferon-alpha production induced by Toll-like receptors 7 and 9. *Nature* **440**:949–953.

46. Jiang, Z., P. Georgel, X. Du, L. Shamel, S. Sovath, S. Mudd, M. Huber, C. Kalis, S. Keck, C. Galanos, M. Freudenberg, and B. Beutler. 2005. CD14 is required for MyD88-independent LPS signaling. *Nat. Immunol.* **6**:565–570.

47. Jiang, Z., T. W. Mak, G. Sen, and X. Li. 2004. Toll-like receptor 3-mediated activation of NF-kappaB and IRF3 diverges at Toll-IL-1 receptor domain-containing adapter inducing IFN-beta. *Proc. Natl. Acad. Sci. USA* **101**:3533–3538.

48. Jude, B. A., Y. Pobezinskaya, J. Bishop, S. Parke, R. M. Medzhitov, A. V. Chervonsky, and T. V. Golovkina. 2003. Subversion of the innate immune system by a retrovirus. *Nat. Immunol.* **4**:573–578.

49. Jurk, M., F. Heil, J. Vollmer, C. Schetter, A. M. Krieg, H. Wagner, G. Lipford, and S. Bauer. 2002. Human TLR7 or TLR8 independently confer responsiveness to the antiviral compound R-848. *Nat. Immunol.* **3**:499.

50. Kanakaraj, P., P. H. Schafer, D. E. Cavender, Y. Wu, K. Ngo, P. F. Grealish, S. A. Wadsworth, P. A. Peterson, J. J. Siekierka, C. A. Harris, and W. P. Fung-Leung. 1998. Interleukin (IL)-1 receptor-associated kinase (IRAK) requirement for optimal induction of multiple IL-1 signaling pathways and IL-6 production. *J. Exp. Med.* **187**:2073–2079.

51. Kanayama, A., R. B. Seth, L. Sun, C. K. Ea, M. Hong, A. Shaito, Y. H. Chiu, L. Deng, and Z. J. Chen. 2004. TAB2 and TAB3 activate the NF-kappaB pathway through binding to polyubiquitin chains. *Mol. Cell* **15**:535–548.

52. Kato, H., O. Takeuchi, and S. Akira. 2006. [Cell type specific involvement of RIG-I in antiviral responses.] *Nippon Rinsho* **64**:1244–1247.

53. Kato, H., O. Takeuchi, S. Sato, M. Yoneyama, M. Yamamoto, K. Matsui, S. Uematsu, A. Jung, T. Kawai, K. J. Ishii, O. Yamaguchi, K. Otsu, T. Tsujimura, C. S. Koh,

C. Reis e Sousa, Y. Matsuura, T. Fujita, and S. Akira. 2006. Differential roles of MDA5 and RIG-I helicases in the recognition of RNA viruses. *Nature* 441:101–105.

54. Katze, M. G., Y. He, and M. Gale, Jr. 2002. Viruses and interferon: a fight for supremacy. *Nat. Rev. Immunol.* 2: 675–687.

55. Kawai, T., O. Adachi, T. Ogawa, K. Takeda, and S. Akira. 1999. Unresponsiveness of MyD88-deficient mice to endotoxin. *Immunity* 11:115–122.

56. Kawai, T., and S. Akira. 2007. Antiviral signaling through pattern recognition receptors. *J. Biochem. (Tokyo)* 141: 137–145.

57. Kawai, T., S. Sato, K. J. Ishii, C. Coban, H. Hemmi, M. Yamamoto, K. Terai, M. Matsuda, J. Inoue, S. Uematsu, O. Takeuchi, and S. Akira. 2004. Interferon-alpha induction through Toll-like receptors involves a direct interaction of IRF7 with MyD88 and TRAF6. *Nat. Immunol.* 5: 1061–1068.

58. Keating, S. E., G. M. Maloney, E. M. Moran, and A. G. Bowie. 2007. IRAK-2 participates in multiple Toll-like receptor signaling pathways to NFκB via activation of TRAF6 ubiquitination. *J. Biol. Chem.* 282:33435–33443.

59. Kim, T. W., K. Staschke, K. Bulek, J. Yao, K. Peters, K. H. Oh, Y. Vandenburg, H. Xiao, W. Qian, T. Hamilton, B. Min, G. Sen, R. Gilmour, and X. Li. 2007. A critical role for IRAK4 kinase activity in Toll-like receptor-mediated innate immunity. *J. Exp. Med.* 204:1025–1036.

60. Kobayashi, K., L. D. Hernandez, J. E. Galan, C. A. Janeway, Jr., R. Medzhitov, and R. A. Flavell. 2002. IRAK-M is a negative regulator of Toll-like receptor signaling. *Cell* 110:191–202.

61. Koziczak-Holbro, M., C. Joyce, A. Gluck, B. Kinzel, M. Muller, C. Tschopp, J. C. Mathison, C. N. Davis, and H. Gram. 2007. IRAK-4 kinase activity is required for interleukin-1 (IL-1) receptor- and Toll-like receptor 7-mediated signaling and gene expression. *J. Biol. Chem.* 282: 13552–13560.

62. Krug, A., G. D. Luker, W. Barchet, D. A. Leib, S. Akira, and M. Colonna. 2004. Herpes simplex virus type 1 activates murine natural interferon-producing cells through Toll-like receptor 9. *Blood* 103:1433–1437.

63. Kurt-Jones, E. A., L. Popova, L. Kwinn, L. M. Haynes, L. P. Jones, R. A. Tripp, E. E. Walsh, M. W. Freeman, D. T. Golenbock, L. J. Anderson, and R. W. Finberg. 2000. Pattern recognition receptors TLR4 and CD14 mediate response to respiratory syncytial virus. *Nat. Immunol.* 1: 398–401.

64. Latz, E., A. Schoenemeyer, A. Visintin, K. A. Fitzgerald, B. G. Monks, C. F. Knetter, E. Lien, N. J. Nilsen, T. Espevik, and D. T. Golenbock. 2004. TLR9 signals after translocating from the ER to CpG DNA in the lysosome. *Nat. Immunol.* 5:190–198.

65. Lee, H. K., J. M. Lund, B. Ramanathan, N. Mizushima, and A. Iwasaki. 2007. Autophagy-dependent viral recognition by plasmacytoid dendritic cells. *Science* 315:1398–1401.

66. Le Goffic, R., V. Balloy, M. Lagranderie, L. Alexopoulou, N. Escriou, R. Flavell, M. Chignard, and M. Si-Tahar. 2006. Detrimental contribution of the Toll-like receptor (TLR)3 to influenza A virus-induced acute pneumonia. *PLoS Pathog.* 2:e53.

67. Li, K., Z. Chen, N. Kato, M. Gale, Jr., and S. M. Lemon. 2005. Distinct poly(I-C) and virus-activated signaling pathways leading to interferon-β production in hepatocytes. *J. Biol. Chem.* 280:16739–16747.

68. Li, S., A. Strelow, E. J. Fontana, and H. Wesche. 2002. IRAK-4: a novel member of the IRAK family with the properties of an IRAK-kinase. *Proc. Natl. Acad. Sci. USA* 99:5567–5572.

69. Lund, J., A. Sato, S. Akira, R. Medzhitov, and A. Iwasaki. 2003. Toll-like receptor 9-mediated recognition of Herpes simplex virus-2 by plasmacytoid dendritic cells. *J. Exp. Med.* 198:513–520.

70. Lund, J. M., L. Alexopoulou, A. Sato, M. Karow, N. C. Adams, N. W. Gale, A. Iwasaki, and R. A. Flavell. 2004. Recognition of single-stranded RNA viruses by Toll-like receptor 7. *Proc. Natl. Acad. Sci. USA* 101:5598–5603.

71. Martin, M. U., and H. Wesche. 2002. Summary and comparison of the signaling mechanisms of the Toll/interleukin-1 receptor family. *Biochim. Biophys. Acta* 1592:265–280.

72. Matsumoto, M., K. Funami, M. Tanabe, H. Oshiumi, M. Shingai, Y. Seto, A. Yamamoto, and T. Seya. 2003. Subcellular localization of Toll-like receptor 3 in human dendritic cells. *J. Immunol.* 171:3154–3162.

73. Matsumoto, M., S. Kikkawa, M. Kohase, K. Miyake, and T. Seya. 2002. Establishment of a monoclonal antibody against human Toll-like receptor 3 that blocks double-stranded RNA-mediated signaling. *Biochem. Biophys. Res. Commun.* 293:1364–1369.

74. McWhirter, S. M., K. A. Fitzgerald, J. Rosains, D. C. Rowe, D. T. Golenbock, and T. Maniatis. 2004. IFN-regulatory factor 3-dependent gene expression is defective in Tbk1-deficient mouse embryonic fibroblasts. *Proc. Natl. Acad. Sci. USA* 101:233–238.

75. Medzhitov, R., P. Preston-Hurlburt, E. Kopp, A. Stadlen, C. Chen, S. Ghosh, and C. A. Janeway, Jr. 1998. MyD88 is an adaptor protein in the hToll/IL-1 receptor family signaling pathways. *Mol. Cell* 2:253–258.

76. Meylan, E., K. Burns, K. Hofmann, V. Blancheteau, F. Martinon, M. Kelliher, and J. Tschopp. 2004. RIP1 is an essential mediator of Toll-like receptor 3-induced NF-kappa B activation. *Nat. Immunol.* 5:503–507.

77. Meylan, E., J. Curran, K. Hofmann, D. Moradpour, M. Binder, R. Bartenschlager, and J. Tschopp. 2005. Cardif is an adaptor protein in the RIG-I antiviral pathway and is targeted by hepatitis C virus. *Nature* 437:1167–1172.

78. Meylan, E., and J. Tschopp. 2006. Toll-like receptors and RNA helicases: two parallel ways to trigger antiviral responses. *Mol. Cell* 22:561–569.

79. Mibayashi, M., L. Martinez-Sobrido, Y. M. Loo, W. B. Cardenas, M. Gale, Jr., and A. Garcia-Sastre. 2007. Inhibition of retinoic acid-inducible gene I-mediated induction of beta interferon by the NS1 protein of influenza A virus. *J. Virol.* 81:514–524.

80. Muzio, M., J. Ni, P. Feng, and V. M. Dixit. 1997. IRAK (Pelle) family member IRAK-2 and MyD88 as proximal mediators of IL-1 signaling. *Science* 278:1612–1615.

81. Negishi, H., Y. Fujita, H. Yanai, S. Sakaguchi, X. Ouyang, M. Shinohara, H. Takayanagi, Y. Ohba, T. Taniguchi, and K. Honda. 2006. Evidence for licensing of IFN-gamma-induced IFN regulatory factor 1 transcription factor by MyD88 in Toll-like receptor-dependent gene induction program. *Proc. Natl. Acad. Sci. USA* 103:15136–15141.

82. Oganesyan, G., S. K. Saha, B. Guo, J. Q. He, A. Shahangian, B. Zarnegar, A. Perry, and G. Cheng. 2006.

Critical role of TRAF3 in the Toll-like receptor-dependent and -independent antiviral response. *Nature* **439:** 208–211.

83. O'Neill, L. A., and A. G. Bowie. 2007. The family of five: TIR-domain-containing adaptors in Toll-like receptor signalling. *Nat. Rev. Immunol.* **7:**353–364.

84. Oshiumi, H., M. Matsumoto, K. Funami, T. Akazawa, and T. Seya. 2003. TICAM-1, an adaptor molecule that participates in Toll-like receptor 3-mediated interferon-beta induction. *Nat. Immunol.* **4:**161–167.

85. Oshiumi, H., M. Sasai, K. Shida, T. Fujita, M. Matsumoto, and T. Seya. 2003. TIR-containing adapter molecule (TICAM)-2, a bridging adapter recruiting to Toll-like receptor 4 TICAM-1 that induces interferon-beta. *J. Biol. Chem.* **278:**49751–49762.

86. Pichlmair, A., and C. Reis e Sousa. 2007. Innate recognition of viruses. *Immunity* **27:**370–383.

87. Poltorak, A., T. Merlin, P. J. Nielsen, O. Sandra, I. Smirnova, I. Schupp, T. Boehm, C. Galanos, and M. A. Freudenberg. 2001. A point mutation in the IL-12R beta 2 gene underlies the IL-12 unresponsiveness of Lps-defective C57BL/10ScCr mice. *J. Immunol.* **167:**2106–2111.

88. Rassa, J. C., J. L. Meyers, Y. Zhang, R. Kudaravalli, and S. R. Ross. 2002. Murine retroviruses activate B cells via interaction with Toll-like receptor 4. *Proc. Natl. Acad. Sci. USA* **99:**2281–2286.

89. Reis e Sousa, C. 2004. Toll-like receptors and dendritic cells: for whom the bug tolls. *Semin. Immunol.* **16:**27–34.

90. Rudd, B. D., E. Burstein, C. S. Duckett, X. Li, and N. W. Lukacs. 2005. Differential role for TLR3 in respiratory syncytial virus-induced chemokine expression. *J. Virol.* **79:**3350–3357.

91. Rudd, B. D., J. J. Smit, R. A. Flavell, L. Alexopoulou, M. A. Schaller, A. Gruber, A. A. Berlin, and N. W. Lukacs. 2006. Deletion of TLR3 alters the pulmonary immune environment and mucus production during respiratory syncytial virus infection. *J. Immunol.* **176:**1937–1942.

92. Saito, T., and M. Gale, Jr. 2007. Principles of intracellular viral recognition. *Curr. Opin. Immunol.* **19:**17–23.

93. Samuel, C. E. 2001. Antiviral actions of interferons. *Clin. Microbiol. Rev.* **14:**778–809.

94. Sasai, M., H. Oshiumi, M. Matsumoto, N. Inoue, F. Fujita, M. Nakanishi, and T. Seya. 2005. Cutting edge: NF-kappaB-activating kinase-associated protein 1 participates in TLR3/Toll-IL-1 homology domain-containing adapter molecule-1-mediated IFN regulatory factor 3 activation. *J. Immunol.* **174:**27–30.

95. Sato, S., M. Sugiyama, M. Yamamoto, Y. Watanabe, T. Kawai, K. Takeda, and S. Akira. 2003. Toll/IL-1 receptor domain-containing adaptor inducing IFN-beta (TRIF) associates with TNF receptor-associated factor 6 and TANK-binding kinase 1, and activates two distinct transcription factors, NF-kappa B and IFN-regulatory factor-3, in the Toll-like receptor signaling. *J. Immunol.* **171:** 4304–4310.

96. Satoh, M., M. Nakamura, T. Akatsu, Y. Shimoda, I. Segawa, and K. Hiramori. 2004. Toll-like receptor 4 is expressed with enteroviral replication in myocardium from patients with dilated cardiomyopathy. *Lab. Invest.* **84:**173–181.

97. Schmitz, F., A. Heit, S. Guggemoos, A. Krug, J. Mages, M. Schiemann, H. Adler, I. Drexler, T. Haas, R. Lang, and H. Wagner. 2007. Interferon-regulatory-factor 1 controls Toll-like receptor 9-mediated IFN-beta production in myeloid dendritic cells. *Eur. J. Immunol.* **37:** 315–327.

98. Schoenemeyer, A., B. J. Barnes, M. E. Mancl, E. Latz, N. Goutagny, P. M. Pitha, K. A. Fitzgerald, and D. T. Golenbock. 2005. The interferon regulatory factor, IRF5, is a central mediator of TLR7 signaling. *J. Biol. Chem.* **280:**17005–17012.

99. Schroder, M., and A. G. Bowie. 2005. TLR3 in antiviral immunity: key player or bystander? *Trends Immunol.* **26:**462–468.

100. Schulz, O., S. S. Diebold, M. Chen, T. I. Naslund, M. A. Nolte, L. Alexopoulou, Y. T. Azuma, R. A. Flavell, P. Liljestrom, and C. Reis e Sousa. 2005. Toll-like receptor 3 promotes cross-priming to virus-infected cells. *Nature* **433:**887–892.

101. Sharma, S., B. R. tenOever, N. Grandvaux, G. P. Zhou, R. Lin, and J. Hiscott. 2003. Triggering the interferon antiviral response through an IKK-related pathway. *Science* **300:**1148–1151.

102. Shaw, M. L., W. B. Cardenas, D. Zamarin, P. Palese, and C. F. Basler. 2005. Nuclear localization of the Nipah virus W protein allows for inhibition of both virus- and Toll-like receptor 3-triggered signaling pathways. *J. Virol.* **79:**6078–6088.

103. Shinohara, M. L., L. Lu, J. Bu, M. B. Werneck, K. S. Kobayashi, L. H. Glimcher, and H. Cantor. 2006. Osteopontin expression is essential for interferon-alpha production by plasmacytoid dendritic cells. *Nat. Immunol.* **7:**498–506.

104. Solis, M., R. Romieu-Mourez, D. Goubau, N. Grandvaux, T. Mesplede, I. Julkunen, A. Nardin, M. Salcedo, and J. Hiscott. 2007. Involvement of TBK1 and IKKepsilon in lipopolysaccharide-induced activation of the interferon response in primary human macrophages. *Eur. J. Immunol.* **37:**528–539.

105. Song, Y. J., K. Y. Jen, V. Soni, E. Kieff, and E. Cahir-McFarland. 2006. IL-1 receptor-associated kinase 1 is critical for latent membrane protein 1-induced p65/RelA serine 536 phosphorylation and NF-kappaB activation. *Proc. Natl. Acad. Sci. USA* **103:**2689–2694.

106. Sumpter, R., Jr., Y. M. Loo, E. Foy, K. Li, M. Yoneyama, T. Fujita, S. M. Lemon, and M. Gale, Jr. 2005. Regulating intracellular antiviral defense and permissiveness to hepatitis C virus RNA replication through a cellular RNA helicase, RIG-I. *J. Virol.* **79:**2689–2699.

107. Suzuki, N., S. Suzuki, G. S. Duncan, D. G. Millar, T. Wada, C. Mirtsos, H. Takada, A. Wakeham, A. Itie, S. Li, J. M. Penninger, H. Wesche, P. S. Ohashi, T. W. Mak, and W. C. Yeh. 2002. Severe impairment of interleukin-1 and Toll-like receptor signalling in mice lacking IRAK-4. *Nature* **416:**750–756.

108. Swantek, J. L., M. F. Tsen, M. H. Cobb, and J. A. Thomas. 2000. IL-1 receptor-associated kinase modulates host responsiveness to endotoxin. *J. Immunol.* **164:** 4301–4306.

109. Takaoka, A., H. Yanai, S. Kondo, G. Duncan, H. Negishi, T. Mizutani, S. Kano, K. Honda, Y. Ohba, T. W. Mak, and T. Taniguchi. 2005. Integral role of IRF-5 in the gene induction programme activated by Toll-like receptors. *Nature* **434:**243–249.

110. Tal, G., A. Mandelberg, I. Dalal, K. Cesar, E. Somekh, A. Tal, A. Oron, S. Itskovich, A. Ballin, S. Houri, A. Beigelman, O. Lider, G. Rechavi, and N. Amariglio. 2004. Association between common Toll-like receptor 4 mutations and severe respiratory syncytial virus disease. *J. Infect. Dis.* **189:**2057–2063.

111. Tanner, N. K., and P. Linder. 2001. DExD/H box RNA helicases: from generic motors to specific dissociation functions. *Mol. Cell* **8:**251–262.

112. tenOever, B. R., S. L. Ng, M. A. Chua, S. M. McWhirter, A. Garcia-Sastre, and T. Maniatis. 2007. Multiple functions of the IKK-related kinase IKKepsilon in interferon-mediated antiviral immunity. *Science* **315:**1274–1278.

113. Thomas, J. A., J. L. Allen, M. Tsen, T. Dubnicoff, J. Danao, X. C. Liao, Z. Cao, and S. A. Wasserman. 1999. Impaired cytokine signaling in mice lacking the IL-1 receptor-associated kinase. *J. Immunol.* **163:**978–984.

114. Thompson, A. J., and S. A. Locarnini. 2007. Toll-like receptors, RIG-I-like RNA helicases and the antiviral innate immune response. *Immunol. Cell. Biol.* **85:** 435–445.

115. Triantafilou, K., and M. Triantafilou. 2004. Coxsackievirus B4-induced cytokine production in pancreatic cells is mediated through Toll-like receptor 4. *J. Virol.* **78:**11313–11320.

116. Uematsu, S., S. Sato, M. Yamamoto, T. Hirotani, H. Kato, F. Takeshita, M. Matsuda, C. Coban, K. J. Ishii, T. Kawai, O. Takeuchi, and S. Akira. 2005. Interleukin-1 receptor-associated kinase-1 plays an essential role for Toll-like receptor (TLR)7- and TLR9-mediated interferon-α induction. *J. Exp. Med.* **201:**915–923.

117. Unterholzner, L., and A. G. Bowie. 2007. The interplay between viruses and innate immune signaling: recent insights and therapeutic opportunities. *Biochem. Pharmacol.* **75:**589–602.

118. Wang, C., L. Deng, M. Hong, G. R. Akkaraju, J. Inoue, and Z. J. Chen. 2001. TAK1 is a ubiquitin-dependent kinase of MKK and IKK. *Nature* **412:**346–351.

119. Wang, T., T. Town, L. Alexopoulou, J. F. Anderson, E. Fikrig, and R. A. Flavell. 2004. Toll-like receptor 3 mediates West Nile virus entry into the brain causing lethal encephalitis. *Nat. Med.* **10:**1366–1373.

120. Yamamoto, M., S. Sato, H. Hemmi, S. Uematsu, K. Hoshino, T. Kaisho, O. Takeuchi, K. Takeda, and S. Akira. 2003. TRAM is specifically involved in the Toll-like receptor 4-mediated MyD88-independent signaling pathway. *Nat. Immunol.* **4:**1144–1150.

121. Yamamoto, M., S. Sato, K. Mori, K. Hoshino, O. Takeuchi, K. Takeda, and S. Akira. 2002. Cutting edge: a novel Toll/IL-1 receptor domain-containing adapter that preferentially activates the IFN-beta promoter in the Toll-like receptor signaling. *J. Immunol.* **169:**6668–6672.

122. Yoneyama, M., M. Kikuchi, T. Natsukawa, N. Shinobu, T. Imaizumi, M. Miyagishi, K. Taira, S. Akira, and T. Fujita. 2004. The RNA helicase RIG-I has an essential function in double-stranded RNA-induced innate antiviral responses. *Nat. Immunol.* **5:**730–737.

123. Zhou, S., E. A. Kurt-Jones, L. Mandell, A. Cerny, M. Chan, D. T. Golenbock, and R. W. Finberg. 2005. MyD88 is critical for the development of innate and adaptive immunity during acute lymphocytic choriomeningitis virus infection. *Eur. J. Immunol.* **35:**822–830.

Cellular Signaling and Innate Immune
Responses to RNA Virus Infections
Edited by A. R. Brasier et al.
© 2009 ASM Press, Washington, D.C.

Osamu Takeuchi
Shizuo Akira

3

Cytoplasmic Pattern Receptors (RIG-I and MDA-5) and Signaling in Viral Infections

Virus infections are recognized by the innate immune system, and this recognition elicits initial antiviral responses. Production of type I interferons (IFNs), proinflammatory cytokines, and chemokines by innate immune cells is essential for mounting rapid innate immune responses as well as activating adaptive immunity (1, 4). The innate immune system senses viral invasion via germline-encoded pattern recognition receptors (PRRs). Currently, two classes of PRRs, namely Toll-like receptors (TLRs) and retinoic acid-inducible gene I (RIG-I)-like RNA helicases (RLHs), have been shown to be involved in the recognition of virus-specific components by innate immune cells. Recognition of viral components by these PRRs triggers intracellular signaling cascades that lead to the expression of genes encoding type I IFNs and proinflammatory cytokines. TLRs are transmembrane proteins suitable for detecting microbial components on the outside of cells as well as in the cytoplasmic vacuole after phagocytosis or endocytosis. In contrast, RLHs, which lack a transmembrane domain, are localized in the cytosol and recognize viral RNAs that are translocated into the cytosol or generated during the course of virus replication. Type I IFNs are produced not only by professional innate immune cells, such as dendritic cells (DCs) and macrophages, but also by nonprofessional cells, such as fibroblasts.

Type I IFNs—which comprise multiple IFN-α isoforms, a single IFN-β, and other members, such as IFN-ω, -ε, and -κ—are pleiotropic cytokines that are essential for antiviral immune responses (18). Type I IFNs activate intracellular signaling pathways and regulate the expression of a set of genes by inducing nuclear translocation of transcription factors known as signal transducers and activators of transcription (Stat) proteins. The IFN-inducible genes include protein kinase R (PKR) and 2′,5′-oligoadenylate synthase, among others (62). These IFN-inducible proteins are believed to suppress virus replication by both cleaving viral nucleotides and suppressing the proliferation of virus-infected cells. Importantly, type I IFNs as well as proinflammatory cytokines and chemokines are essential for antibody production in B cells and the induction of cytotoxic T cells and natural killer (NK) cells.

In this review, we will focus on the roles of RLHs in RNA virus recognition and RLH signaling pathways. We also describe the roles of the TLR system with respect to the relationship between these two virus recognition

Osamu Takeuchi and Shizuo Akira, Laboratory of Host Defense, WPI Immunology Frontier Research Center, Osaka University, 3–1 Yamada-oka, Suita, Osaka 565-0871, Japan.

mechanisms during the course of RNA virus infections in vivo.

RLHs

RIG-I (also known as DDX58) was identified as a candidate for a cytoplasmic viral RNA detector responsible for the production of type I IFNs (74). RIG-I is composed of two N-terminal caspase recruitment domains (CARDs) followed by a DExD/H box RNA helicase domain. RIG-I forms a family, designated the RLH family, with melanoma differentiation-associated gene 5 (MDA-5; also known as helicard or IFIH1) and LGP2 based on the high similarities among their helicase domains (25, 32, 51, 75). The helicase domains of the RLH family members are highly similar to that of mammalian Dicer. The expression of RLH genes is strongly induced by IFNs. RLHs interact with double-stranded (ds) RNAs through their helicase domains, and this dsRNA stimulation induces their ATP catalytic activity (40). A C-terminal portion of RIG-I, designated the repressor domain (RD), was found to inhibit the triggering of RIG-I signaling in the steady state (54). It has been suggested that RIG-I and LGP2, but not MDA-5, possess a C-terminal RD. The N-terminal CARDs are responsible for activating downstream signaling pathways that mediate dsRNA-induced type I IFN production.

RECOGNITION OF RNA VIRUSES BY RLHs

dsRNA is present in cells infected with dsRNA viruses and also generated during the course of single-stranded (ss) RNA virus replication. Since host cells do not produce dsRNA, the host innate immune system is able to discriminate between host and viral RNAs by the presence of dsRNA. Initially, both RIG-I and MDA-5 were implicated in the recognition of poly(I:C), a synthetic analogue of viral dsRNA. However, analyses of RIG-I$^{-/-}$ and MDA-5$^{-/-}$ mice revealed that MDA-5, but not RIG-I, is responsible for the IFN response to poly(I:C) stimulation (10, 27). Reciprocally, RIG-I, but not MDA-5, recognizes 5′-triphosphate ssRNA (19, 50). RNAs from some viruses are 5′-triphosphorylated and uncapped, whereas the 5′ ends of host mRNAs are capped. Thus, RIG-I discriminates between viral and host RNAs based on differences in the 5′ ends of RNAs. However, it remains unclear whether RIG-I also recognizes dsRNAs in addition to 5′-triphosphate ssRNA. PKR, a serine/threonine kinase containing dsRNA-binding domains, is activated by 5′-triphosphate RNA in a RIG-I-independent manner. PKR phosphorylates eukaryotic initiation factor 2 (eIF2), which inhibits the initiation of translation (46).

RNA viruses are also differentially recognized by RIG-I and MDA-5. RIG-I$^{-/-}$ cells do not produce type I IFNs in response to various RNA viruses, including paramyxoviruses, vesicular stomatitis virus (VSV), and influenza virus (26, 27). In contrast, MDA-5$^{-/-}$ cells do not respond to infections with picornaviruses, such as encephalomyocarditis virus (EMCV) and Theiler's virus. Cells infected with EMCV, but not influenza virus, generate dsRNA (50). Dephosphorylation of the 5′-triphosphate RNA of the influenza genome results in loss of its ability to induce IFNs, suggesting that recognition by RIG-I is mediated through 5′-triphosphate ssRNA. Consistent with the defect in type I IFN production, RIG-I$^{-/-}$ and MDA-5$^{-/-}$ mice are highly susceptible to inoculation with VSV and EMCV, respectively (27). Japanese encephalitis virus and hepatitis C virus (HCV), both of which belong to the *Flaviviridae* family, are recognized by RIG-I (27). Huh7 cells harboring a mutant RIG-I are permissive to HCV infection (64). GB virus, a small primate model of human HCV, is also recognized by RIG-I (6). However, dengue virus and West Nile virus, which also belong to the *Flaviviridae* family, still induce type I IFN production in the absence of RIG-I or MDA-5 (8, 27, 38, 64). Small interfering RNA experiments suggested that dengue virus is recognized by a combination of RIG-I and MDA-5. Vaccine strains of measles virus activate cells in a RIG-I/MDA-5-dependent manner, whereas wild-type measles virus fails to induce type I IFN production (60). It has been suggested that the presence of defective interfering RNAs, which potentially form dsRNA, in the vaccine strains of measles virus is critical for activating innate immune cells. The receptor responsible for detecting poliovirus has not been identified. However, poliovirus is expected to be recognized by MDA-5 because this virus is a member of the *Picornaviridae* family. Interestingly, MDA-5 was reported to be cleaved by a poliovirus-induced protease in infected cells (3). Thus, poliovirus may subvert the MDA-5-mediated recognition system to establish its infection.

RIG-I-mediated signaling is positively and negatively controlled by ubiquitination of RIG-I. First, the CARDs of RIG-I undergo Lys-63-linked ubiquitination by tripartite motif (TRIM) 25, a ubiquitin E3 ligase composed of a RING finger domain, B box/coiled-coil domain, and SPRY domain (9). This ubiquitination is necessary for efficient activation of the RIG-I signaling pathway, and TRIM25$^{-/-}$ cells display impaired production of type I IFNs in response to viral infection. RIG-I also undergoes ubiquitination by the ubiquitin ligase RNF125, which leads to its proteasomal degradation (2). Thus, RIG-I ubiquitination by RNF125 is considered to inhibit aberrant activation of RIG-I signaling.

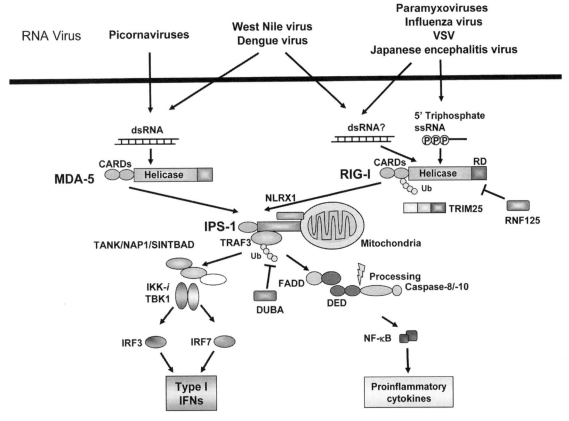

Figure 1 RLH-mediated recognition of RNA viruses. RIG-I and MDA-5 recognize 5'-triphosphate RNA and dsRNA from RNA viruses and interact with IPS-1. TRAF3 is recruited to IPS-1, and Lys-63-type polyubiquitination, which is controlled by the presence of DUBA, is induced. Subsequently, TRAF3 recruits TANK/NAP1/SINTBAD and TBK1/IKK-*i*, which phosphorylate IRF3/IRF7. The phosphorylated IRFs translocate into the nucleus and induce the expression of type I IFN genes. NF-κB is also activated by IPS-1 via the FADD and caspase-8/10-dependent pathway.

RNase L, an endonuclease originally thought to cleave viral ssRNA, was reported to be involved in the production of IFN-β in response to RNA virus infection or dsRNA stimulation (39). Furthermore, 2',5'-linked oligoadenylate generated by virus infection was found to activate RNase L for cleavage of self-RNA, resulting in the generation of small RNA products that are responsible for RIG-I/MDA-5-mediated recognition and subsequent production of type I IFNs. However, the precise structures of these small RNAs generated by RNase L require further investigation.

Since LGP2 lacks a CARD, it is suggested to function as a negative regulator of RIG-I/MDA-5 signaling. Overexpression of LGP2 inhibits Sendai virus and Newcastle disease virus (NDV) signaling (51, 54, 75). LGP2 also contains an RD, which was found to interact with the RD of RIG-I and suppress RIG-1 self-association. Recently, *Lgp2*⁻/⁻ mice were generated and analyzed by

Barber and colleagues (71). *Lgp2*⁻/⁻ mice show highly elevated induction of type I IFNs in response to poly(I:C) stimulation, as well as modestly increased IFN production in response to VSV infection. On the other hand, these mice exhibit partially impaired type I IFN production in response to EMCV infection. The authors proposed that LGP2 is a negative regulator of RIG-I, but not MDA-5. However, given that both poly(I:C) and EMCV are recognized by MDA-5, the difference cannot simply be explained by differential usage of LGP2 for RIG-I and MDA-5 signaling.

RLH SIGNALING PATHWAYS

The CARDs of RIG-I and MDA-5 are responsible for initiating signaling cascades (Fig. 1). RIG-I and MDA-5 associate with an adaptor protein, IFN-β promoter stimulator-1 (IPS-1; also known as MAVS, VISA, or

Cardif), which also contains a CARD (29, 44, 58, 72). Overexpression of IPS-1 induces the activation of IFN promoters as well as nuclear factor-κB (NF-κB). IPS-1$^{-/-}$ mice are defective in the production of type I IFNs and proinflammatory cytokines in response to all RNA viruses recognized by either RIG-I or MDA-5 (35, 65). These findings indicate that IPS-1 plays an essential role in RIG-I/MDA-5 signaling. Interestingly, this protein is present in the outer mitochondrial membrane, suggesting that mitochondria may be important for IFN responses, in addition to their roles in metabolism and cell death (58). Recently, IPS-1 and RIG-I were shown to associate with a conjugate of Atg5 and Atg12, which are essential components for the autophagic process (23). Atg5$^{-/-}$ mouse embryonic fibroblasts (MEFs) show increased type I IFN production in response to RNA virus infection, suggesting that the autophagic machinery directly affects the RIG-I signaling pathway in addition to autophagosome formation. IPS-1 is known to be cleaved by an HCV protease, NS3/4A, in HCV-infected cells.

IPS-1 associates with tumor necrosis factor receptor-associated factor (TRAF) 3, an E3 ubiquitin ligase assembling the Lys-63-linked polyubiquitin chain, through its C-terminal TRAF domain (12, 48, 53). Both the N-terminal RING finger domain and C-terminal TRAF domain are required for the function of TRAF3 to activate the type I IFN promoter (53). TRAF3$^{-/-}$ cells exhibit severely impaired production of type I IFNs in response to virus infection. In addition to the regulation of IFN responses, TRAF3 negatively regulates the noncanonical NF-κB pathway (14). TRAF3$^{-/-}$ mice show early postnatal lethality, and this phenotype is rescued by codisruption of the noncanonical NF-κB p100 gene (13). Recently, a deubiquitinase designated DUBA was found to deubiquitinate TRAF3 and suppress RLH signaling (30).

TRAF3 recruits and activates two IκB kinase (IKK)-related kinases, namely TRAF family member-associated NF-κB activator (TANK)-binding kinase 1 (TBK1) and inducible IκB kinase (IKK-i; also known as IKK-ε), which phosphorylate IFN regulatory factor (IRF) 3 and IRF7 (7, 15, 59). Although TBK1, but not IKK-i, is important for controlling IFN responses in MEFs, these two kinases function redundantly for inducing type I IFN production in macrophages and conventional DCs (cDCs) (41, 49). However, IKK-i/IKK-ε was reported to be additionally activated by IFN-β to directly phosphorylate Stat1, thereby controlling a set of IFN-inducible genes, such as dsRNA-activated adenosine deaminase gene (Adar1) (69).

Recently, an IPS-1/MAVS-interacting protein, designated NLRX1 (also known as NOD-9), was identified (45). NLRX1 is composed of a nucleotide-binding domain and leucine-rich repeats, and localizes on the mitochondrial outer membrane. Overexpression of NLRX1 inhibits virus-induced IFN-β promoter activation by disrupting the interaction of RIG-I or MDA-5 with IPS-1. Reciprocally, knockdown of NLRX1 leads to augmentation of virus-induced type I IFN production. Although the precise role of NLRX1 in RLH signaling remains to be clarified, NLRX1 is suggested to function as a modifier of IPS-1/MAVS, rather than an innate immune receptor, in contrast to other NOD-like receptors reported to directly recognize bacterial components in the cytoplasm.

TBK1 and IKK-i interact with TANK, NAK-associated protein 1 (NAP1), and similar to NAP1 TBK1 adaptor (SINTBAD) (11, 52, 55). These molecules contain a TBK1-binding motif and show similarities among their coiled-coil domains. Although knockdown of either TANK, NAP1, or SINTBAD impairs RLH signaling, the relationship between these molecules in RLH signaling is not yet fully understood.

Phosphorylation of IRF3 or IRF7 by these kinases induces the formation of homodimers and/or heterodimers (17), which translocate into the nucleus and bind to IFN-stimulated response elements (ISREs), resulting in the expression of type I IFNs and a set of IFN-inducible genes.

Fas-associated death domain-containing protein (FADD) interacts with caspase-8, caspase-10, and IPS-1, and the FADD-dependent pathway is responsible for the activation of NF-κB downstream of IPS-1 (67). Although FADD has also been implicated in RLH-mediated IFN responses, FADD$^{-/-}$ MEFs are still capable of producing IFN-β in response to NDV infection (75).

RECOGNITION OF VIRAL COMPONENTS BY THE TLR SYSTEM

In addition to RLHs, TLRs are also important in recognizing virus infections. TLRs comprise leucine-rich repeats, a transmembrane domain, and a cytoplasmic domain designated the Toll/interleukin-1 receptor (IL-1R) homology (TIR) domain (1, 4, 42). To date, 12 and 13 TLRs have been reported in humans and mice, respectively, and the microbial components recognized by each TLR have mostly been identified. Among the TLRs, TLR3, TLR7, and TLR9 are localized on cytoplasmic vesicles, such as endosomes and the endoplasmic reticulum, and recognize microbial nucleotides. Specifically, TLR3 detects dsRNA, while TLR7 and TLR9 recognize ssRNA and DNA with a CpG motif, respectively. In addition, TLR2 and TLR4, located on the cell surface, recognize RNA virus envelope proteins. Since the leucine-rich repeats, which are responsible for

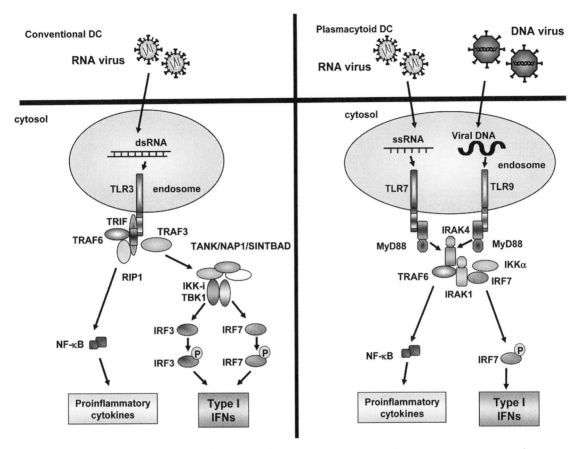

Figure 2 TLR signaling pathways leading to the production of type I IFNs. TLR3 and TLR7/9 activate distinct intracellular signaling pathways. TLR3 recruits TRIF as an adaptor molecule. TRIF associates with TRAF3, TRAF6, and RIP-1, which are responsible for the activation of IRF3, IRF7, and NF-κB. Downstream of TRAF3, TLR3 and RLHs share a common signaling cascade. On the other hand, TLR7 and TLR9 activate specialized signaling cascades in pDCs. In response to RNA and DNA viruses, TLR7 and TLR9 recruit MyD88 as an adaptor and activate IRAK-4, IRAK-1, and TRAF6, resulting in activation of the IKK complex and nuclear translocation of NF-κB, which in turn initiates the expression of proinflammatory cytokine genes. IRAK-1 and IKK-α activate IRF7 together with TRAF3 and IKK-α, and induce nuclear translocation of IRF7. Finally, transcription of type I IFN genes occurs.

microbial component recognition, face the extracellular space or endosomal lumen, TLRs are not suitable for recognizing viruses that have infected cells and are localized in the cytoplasm or nucleus.

The TIR domains of TLRs are responsible for triggering intracellular signaling pathways that lead to the expression of genes encoding proinflammatory cytokines and type I IFNs (Fig. 2). Upon stimulation of TLRs, TIR domain-containing adaptor molecules are recruited to TLRs. Currently, five adaptors (MyD88, TRIF, TIRAP/MAL, TRAM, and SARM) have been identified, and all of these except SARM are involved in TLR signaling in mice (1, 31).

In response to stimulation with poly(I:C), TLR3 recruits TIR-domain-containing adaptor inducing IFN-β

(TRIF; also known as TICAM-1) (73). Subsequently, TRIF associates with TRAF3, TRAF6, and receptor-interacting protein 1 (RIP-1) (43, 48, 56). The RLH and TLR3 signaling pathways share TRAF3 for inducing the expression of type I IFNs. Simultaneously, TRAF6 and RIP-1 are responsible for activating NF-κB, thereby leading to the expression of proinflammatory cytokines. However, the role of TLR3 in type I IFN production in vivo in response to dsRNA stimulation is questionable, since mice lacking TLR3 or TRIF show normal IFN-α production in response to systemic poly(I:C) administration (27). On the other hand, the production of IL-12p40 in serum is dependent on TLR3 signaling. Reciprocally, mice lacking MDA-5 show abrogated IFN production together with normal IL-12p40 production, suggesting that TLR3 and MDA-5 function to control proinflammatory cytokines and type I IFNs, respectively.

Plasmacytoid DCs (pDCs) produce extremely large amounts of type I IFNs in response to virus infections (61). TLR7 and TLR9 are highly expressed on pDCs, and stimulation with viral RNA or DNA efficiently induces the production of type I IFNs (22). TLR7 and TLR9 recruit MyD88, an adaptor protein comprising a death domain and TIR domain. MyD88 then interacts with IL-1R-associated kinase-1 (IRAK-1), IRAK-4, and IRF7. IRAK-1 and IκB kinase a (IKK-α) have been identified as potential IRF7 kinases (16, 20, 28, 70). Phosphorylated IRF7 dissociates from the MyD88-containing complex and translocates into the nucleus to initiate the expression of IFN-inducible genes. TRAF3 is also required for this signaling cascade. TLR9 stimulation also activates cDCs for IFN-β production, although the amounts of IFN-β are not comparable to those produced by pDCs. In cDCs, IRF1, but not IRF7, is responsible for the IFN responses (47, 57).

An endoplasmic reticulum membrane protein, UNC-93B, was identified as an essential molecule for signaling by TLR3, TLR7, and TLR9 by forward genetic screening of mice (66). Recently, an autosomal recessive mutation in UNC-93B in humans was found to result in impaired immune responses against herpes simplex virus (HSV)-1 encephalitis (5).

Autophagy is also responsible for the induction of type I IFNs in response to VSV or HSV infections in pDCs (37). Some ssRNA viruses, such as VSV, appeared to require live virus infection to induce the production of type I IFNs. Autophagosomes are constitutively formed in pDCs, and pDCs lacking *Atg5*, a gene critical for autophagosome formation, show severely impaired production of type I IFNs in response to VSV and HSV infections, indicating that autophagosome formation is critical for the recognition of some viruses in pDCs (37). Since TLR7 is responsible for the production of type I IFNs in response to VSV infection in pDCs, autophagosomes containing VSV components may fuse to organelles containing TLR7. Interestingly, CpG-DNA-induced IFN-α production, but not IL-12p40 production, is also impaired in Atg5$^{-/-}$ pDCs, suggesting that Atg5 modifies the TLR9 signaling pathway leading to the production of type I IFNs.

TYPE I IFN-PRODUCING CELLS IN RESPONSE TO VIRAL INFECTION

Although RLHs play essential roles in the production of type I IFNs and cytokines in various cell types, such as fibroblasts and cDCs, pDCs produce these cytokines in the absence of RLH signaling (26). pDCs produce huge amounts of type I IFNs in response to virus infections, and TLR signaling is essential for this IFN production. Although the importance of pDCs as a source of type I IFNs in vivo has been emphasized, direct identification of IFN-producing cells in vivo has not been carried out. Analysis of reporter mice expressing green fluorescent protein (GFP) under the control of the IFN-α6 gene (*Ifna6*$^{GFP/+}$) revealed that distinct cell types produce IFN-α in response to systemic versus local RNA virus infection (34). Although pDCs were highly potent in expressing GFP upon systemic NDV infection, lung infection of *Ifna6*$^{GFP/+}$ mice with NDV resulted in increased numbers of GFP$^+$ alveolar macrophages and cDCs, but not pDCs (34). Thus, cells other than pDCs can be a source of type I IFNs depending on the route of infection. pDCs started to produce IFN-α when alveolar macrophages were depleted, suggesting that pDCs function when the first line of defense is broken.

ROLES OF RLHs AND TLRs IN THE ACTIVATION OF ADAPTIVE IMMUNE RESPONSES TO VIRUSES

Innate immediate immune responses are important for mounting acquired immune responses to viral infections. However, it is not clear how the innate PRRs are involved in the activation of acquired immunity. Recently, two different virus infection models have been analyzed to examine the roles of RLHs and TLRs in the activation of acquired immune responses. The first model virus is lymphocytic choriomeningitis virus (LCMV), an ambisense ssRNA virus belonging to the *Arenaviridae* family, which is known to induce a cytotoxic-T-lymphocyte (CTL) response in a type I IFN-dependent manner (24). Analyses of MyD88$^{-/-}$ and IPS-1$^{-/-}$ mice revealed that the serum levels of type I IFNs and proinflammatory cytokines are mainly dependent on the presence of MyD88, but not IPS-1. Furthermore, the generation of virus-specific CTLs is critically dependent on MyD88, but not IPS-1. Analysis of *Ifna6*$^{+/GFP}$ reporter mice revealed that pDCs are the major source of IFN-α in LCMV infection. These results suggest that recognition of LCMV by pDCs via TLRs is responsible for the production of type I IFNs in vivo. Furthermore, TLRs, but not RLHs, appear to be important for mounting CTL responses to LCMV infection.

Influenza virus has also been used to study the activation of adaptive immune responses (33). Induction of type I IFNs in response to intranasal influenza A virus infection was found to be abrogated in the absence of both MyD88 and IPS-1, although mice lacking either of these molecules were capable of producing IFNs. Induction of B cells or CD4 T cells specific to viral proteins was dependent on the presence of MyD88, but not IPS-1, whereas induction of nuclear protein antigen-specific CD8 T cells was not impaired in the absence of either MyD88 or IPS-1. These results suggest that the adaptive immune responses to influenza A virus are governed by TLRs.

As described above, poly(I:C) is recognized by MDA-5 and TLR3. The contributions of these two systems to the activation of T-cell responses have been examined using mice deficient in IPS-1 or TRIF, adaptor molecules responsible for the signaling of MDA-5 and TRIF, respectively (36). Enhancement of antigen-specific antibody responses as well as CD8 T-cell expansion in response to poly(I:C) stimulation is impaired in IPS-1-deficient mice. Although the responses of TRIF-deficient mice are modestly impaired, IPS-1/TRIF doubly deficient mice are almost unresponsive to poly(I:C) treatment, suggesting that both MDA-5 and IPS-1 contribute to mounting acquired immune responses to poly(I:C) stimulation.

The virus infection models tested to date support roles for TLRs, rather than RLHs, in instructing the adaptive immune system. However, further studies are required since these two PRR systems contribute differently depending on the viruses involved, and their contributions may also depend on the route of infection.

CONCLUSIONS

Recent progress in studies of RLHs and their signaling pathways has revealed that the RLH system is essential for inducing innate immune responses in response to RNA viruses infecting cells. On the other hand, TLRs play critical roles in producing type I IFNs in pDCs—professional type I IFN-producing cells—in response to virus infection. Although recent studies have clarified the functions and signaling pathways of RLHs, the molecular structures of the RNAs recognized by MDA-5 are not fully understood. Although recognition of 5′-triphosphate RNA by RIG-I is established, it is unclear whether RIG-I also detects dsRNA without a 5′-triphosphate end. Thus, further investigations are required to fully clarify the structures of the RIG-I and MDA-5 ligands.

Furthermore, the mechanisms of DNA virus recognition are not well understood. Although TLR9 is essential for detecting HSV and mouse cytomegalovirus, the presence of a TLR-independent DNA virus recognition mechanism has been predicted. The receptors involved are supposed to recognize viral genomic DNA (21, 63),

and a protein named DAI has been proposed as a candidate for such a sensor (68). However, MEFs and DCs from DAI/ZBP1-deficient mice show normal IFN-β production as well as IFN-inducible gene expression in response to dsDNA stimulation. Thus, another unknown receptor system may be responsible for the detection of viral DNA. Although we focused on the mechanisms of innate immune responses and T-cell activation in response to RNA viruses in this review, many other cell types, such as NK cells and NK T cells, are involved in antiviral responses in vivo. Furthermore, various immune cells cooperate in order to establish optimized antiviral immune responses. Thus, studies monitoring immune responses in vivo could be vital to fully clarify the mechanisms of antiviral immune responses.

We thank M. Hashimoto for excellent secretarial assistance. This work was supported in part by grants from the Special Coordination Funds of the Japanese Ministry of Education, Culture, Sports, Science and Technology; the Ministry of Health, Labour and Welfare in Japan; the 21st Century Center of Excellence Program of Japan; and the National Institute of Health (P01 AI070167).

References

1. Akira, S., S. Uematsu, and O. Takeuchi. 2006. Pathogen recognition and innate immunity. *Cell* **124:**783–801.
2. Arimoto, K., H. Takahashi, T. Hishiki, H. Konishi, T. Fujita, and K. Shimotohno. 2007. Negative regulation of the RIG-I signaling by the ubiquitin ligase RNF125. *Proc. Natl. Acad. Sci. USA* **104:**7500–7505.
3. Barral, P. M., J. M. Morrison, J. Drahos, P. Gupta, D. Sarkar, P. B. Fisher, and V. R. Racaniello. 2007. MDA-5 is cleaved in poliovirus-infected cells. *J. Virol.* **81:**3677–3684.
4. Beutler, B., C. Eidenschenk, K. Crozat, J. L. Imler, O. Takeuchi, J. A. Hoffmann, and S. Akira. 2007. Genetic analysis of resistance to viral infection. *Nat. Rev. Immunol.* **7:**753–766.
5. Casrouge, A., S. Y. Zhang, C. Eidenschenk, E. Jouanguy, A. Puel, K. Yang, A. Alcais, C. Picard, N. Mahfoufi, N. Nicolas, L. Lorenzo, S. Plancoulaine, B. Senechal, F. Geissmann, K. Tabeta, K. Hoebe, X. Du, R. L. Miller, B. Heron, C. Mignot, T. B. de Villemeur, P. Lebon, O. Dulac, F. Rozenberg, B. Beutler, M. Tardieu, L. Abel, and J. L. Casanova. 2006. Herpes simplex virus encephalitis in human UNC-93B deficiency. *Science* **314:**308–312.
6. Chen, Z., Y. Benureau, R. Rijnbrand, J. Yi, T. Wang, L. Warter, R. E. Lanford, S. A. Weinman, S. M. Lemon, A. Martin, and K Li. 2007. GB virus B disrupts RIG-I signaling by NS3/4A-mediated cleavage of the adaptor protein MAVS. *J. Virol.* **81:**964–976.
7. Fitzgerald, K. A., S. M. McWhirter, K. L. Faia, D. C. Rowe, E. Latz, D. T. Golenbock, A. J. Coyle, S. M. Liao, and T. Maniatis. 2003. IKKepsilon and TBK1 are essential components of the IRF3 signaling pathway. *Nat. Immunol.* **4:**491–496.
8. Fredericksen, B. L., B. C. Keller, J. Fornek, M. G. Katze, and M. Gale, Jr. 2008. Establishment and maintenance of

the innate antiviral response to West Nile virus involves both RIG-I and MDA-5 signaling through IPS-1. *J. Virol.* **82:**609–616.

9. **Gack, M. U., Y. C. Shin, C. H. Joo, T. Urano, C. Liang, L. Sun, O. Takeuchi, S. Akira, Z. Chen, S. Inoue, and J. U. Jung.** 2007. TRIM25 RING-finger E3 ubiquitin ligase is essential for RIG-I-mediated antiviral activity. *Nature* **446:**916–920.

10. **Gitlin, L., W. Barchet, S. Gilfillan, M. Cella, B. Beutler, R. A. Flavell, M. S. Diamond, and M. Colonna.** 2006. Essential role of mda-5 in type I IFN responses to polyriboinosinic:polyribocytidylic acid and encephalomyocarditis picornavirus. *Proc. Natl. Acad. Sci. USA* **103:** 8459–8464.

11. **Guo, B., and G. Cheng.** 2007. Modulation of the interferon antiviral response by the TBK1/IKKi adaptor protein TANK. *J. Biol. Chem.* **282:**11817–11826.

12. **Hacker, H., V. Redecke, B. Blagoev, I. Kratchmarova, L. C. Hsu, G. G. Wang, M. P. Kamps, E. Raz, H. Wagner, G. Hacker, M. Mann, and M. Karin.** 2006. Specificity in Toll-like receptor signalling through distinct effector functions of TRAF3 and TRAF6. *Nature* **439:**204–207.

13. **He, J. Q., B. Zarnegar, G. Oganesyan, S. K. Saha, S. Yamazaki, S. E. Doyle, P. W. Dempsey, and G. Cheng.** 2006. Rescue of TRAF3-null mice by p100 NF-kappa B deficiency. *J. Exp. Med.* **203:**2413–2418.

14. **He, J. Q., S. K. Saha, J. R. Kang, B. Zarnegar, and G. Cheng.** 2007. Specificity of TRAF3 in its negative regulation of the noncanonical NF-kappa B pathway. *J. Biol. Chem.* **282:**3688–3694.

15. **Hemmi, H., O. Takeuchi, S. Sato, M. Yamamoto, T. Kaisho, H. Sanjo, T. Kawai, K. Hoshino, K. Takeda, and S. Akira.** 2004. The roles of two IkappaB kinase-related kinases in lipopolysaccharide and double stranded RNA signaling and viral infection. *J. Exp. Med.* **199:**1641–1650.

16. **Honda, K., H. Yanai, T. Mizutani, H. Negishi, N. Shimada, N. Suzuki, Y. Ohba, A. Takaoka, W. C. Yeh, and T. Taniguchi.** 2004. Role of a transductional-transcriptional processor complex involving MyD88 and IRF-7 in Toll-like receptor signaling. *Proc. Natl. Acad. Sci. USA* **101:**15416–15421.

17. **Honda, K., H. Yanai, H. Negishi, M. Asagiri, M. Sato, T. Mizutani, N. Shimada, Y. Ohba, A. Takaoka, N. Yoshida, and T. Taniguchi.** 2005. IRF-7 is the master regulator of type-I interferon-dependent immune responses. *Nature* **434:**772–777.

18. **Honda, K., A. Takaoka, and T. Taniguchi.** 2006. Type I interferon gene induction by the interferon regulatory factor family of transcription factors. *Immunity* **25:**349–360.

19. **Hornung, V., J. Ellegast, S. Kim, K. Brzozka, A. Jung, H. Kato, H. Poeck, S. Akira, K. K. Conzelmann, M. Schlee, S. Endres, and G. Hartmann.** 2006. 5′-Triphosphate RNA is the ligand for RIG-I. *Science* **314:**994–997.

20. **Hoshino, K., T. Sugiyama, M. Matsumoto, T. Tanaka, M. Saito, H. Hemmi, O. Ohara, S. Akira, and T. Kaisho.** 2006. IkappaB kinase-alpha is critical for interferon-alpha production induced by Toll-like receptors 7 and 9. *Nature* **440:**949–953.

21. **Ishii, K. J., C. Coban, H. Kato, K. Takahashi, Y. Torii, F. Takeshita, H. Ludwig, G. Sutter, K. Suzuki, H. Hemmi, S. Sato, M. Yamamoto, S. Uematsu, T. Kawai, O. Takeuchi, and S. Akira.** 2006. A Toll-like receptor-independent antiviral response induced by double-stranded B-form DNA. *Nat. Immunol.* **7:**40–48.

22. **Iwasaki, A., and R. Medzhitov.** 2004. Toll-like receptor control of the adaptive immune responses. *Nat. Immunol* **5:**987–995.

23. **Jounai, N., F. Takeshita, K. Kobiyama, A. Sawano, A. Miyawaki, K. Q. Xin, K. J. Ishii, T. Kawai, S. Akira, K. Suzuki, and K. Okuda.** 2007. The Atg5-Atg12 conjugate associates with innate antiviral immune responses. *Proc. Natl. Acad. Sci. USA* **104:**14050–14055.

24. **Jung, A., H. Kato, Y. Kumagai, H. Kumar, T. Kawai, O. Takeuchi, and S. Akira.** 2008. Lymphocytoid choriomeningitis virus activates plasmacytoid dendritic cells and induces a cytotoxic T-cell response via MyD88. *J. Virol.* **82:**196–206.

25. **Kang, D. C., R. V. Gopalkrishnan, Q. Wu, E. Jankowsky, A. M. Pyle, and P. B. Fisher.** 2002. *mda-5*: an interferon-inducible putative RNA helicase with double-stranded RNA-dependent ATPase activity and melanoma growth-suppressive properties. *Proc. Natl. Acad. Sci. USA* **99:** 637–642.

26. **Kato, H., S. Sato, M. Yoneyama, M. Yamamoto, S. Uematsu, K. Matsui, T. Tsujimura, K. Takeda, T. Fujita, O. Takeuchi, and S. Akira.** 2005. Cell type-specific involvement of RIG-I in antiviral response. *Immunity* **23:** 19–28.

27. **Kato, H., O. Takeuchi, S. Sato, M. Yoneyama, M. Yamamoto, K. Matsui, S. Uematsu, A. Jung, T. Kawai, K. J. Ishii, O. Yamaguchi, K. Otsu, T. Tsujimura, C. S. Koh, C. Reis e Sousa, Y. Matsuura, T. Fujita, and S. Akira.** 2006. Differential roles of MDA-5 and RIG-I helicases in the recognition of RNA viruses. *Nature* **441:**101–105.

28. **Kawai, T., S. Sato, K. J. Ishii, C. Coban, H. Hemmi, M. Yamamoto, K. Terai, M. Matsuda, J. Inoue, S. Uematsu, O. Takeuchi, and S. Akira.** 2004. Interferon-alpha induction through Toll-like receptors involves a direct interaction of IRF7 with MyD88 and TRAF6. *Nat. Immunol.* **5:**1061–1068.

29. **Kawai, T., K. Takahashi, S. Sato, C. Coban, H. Kumar, H. Kato, K. J. Ishii, O. Takeuchi, and S. Akira.** 2005. IPS-1, an adaptor triggering RIG-I- and MDA-5-mediated type I interferon induction. *Nat. Immunol.* **6:**981–988.

30. **Kayagaki, N., Q. Phung, S. Chan, R. Chaudhari, C. Quan, K. M. O'Rourke, M. Eby, E. Pietras, G. Cheng, J. F. Bazan, Z. Zhang, D. Arnott, and V. M. Dixit.** 2007. DUBA: a deubiquitinase that regulates type I interferon production. *Science* **318:**1628–1632.

31. **Kim, Y., P. Zhou, L. Qian, J. Z. Chuang, J. Lee, C. Li, C. Iadecola, C. Nathan, and A. Ding.** 2007. MyD88–5 links mitochondria, microtubules, and JNK3 in neurons and regulates neuronal survival. *J. Exp. Med.* **204:**2063–2074.

32. **Kovacsovics, M., F. Martinon, O. Micheau, J. L. Bodmer, K. Hofmann, and J. Tschopp.** 2002. Overexpression of Helicard, a CARD-containing helicase cleaved during apoptosis, accelerates DNA degradation. *Curr. Biol.* **12:** 838–843.

33. **Koyama, S., K. J. Ishii, H. Kumar, T. Tanimoto, C. Coban, S. Uematsu, T. Kawai, and S. Akira.** 2007. Differential role of TLR- and RLR-signaling in the immune responses to influenza A virus infection and vaccination. *J. Immunol.* **179:**4711–4720.

34. Kumagai, Y., O. Takeuchi, H. Kato, H. Kumar, K. Matsui, E. Morii, K. Aozasa, T. Kawai, and S. Akira. 2007. Alveolar macrophages are the primary interferon-alpha producer in pulmonary infection with RNA viruses. *Immunity* 27:240–252.

35. Kumar, H., T. Kawai, H. Kato, S. Sato, K. Takahashi, C. Coban, M. Yamamoto, S. Uematsu, K. J. Ishii, O. Takeuchi, and S. Akira. 2006. Essential role of IPS-1 in innate immune responses against RNA viruses. *J. Exp. Med.* 203:1795–1803.

36. Kumar, H., S. Koyama, K. J. Ishii, T. Kawai, and S. Akira. 2008. Cutting edge: cooperation of IPS-1- and TRIF-dependent pathways in poly IC-enhanced antibody production and cytotoxic T cell responses. *J. Immunol.* 180:683–687.

37. Lee, H. K., J. M. Lund, B. Ramanathan, N. Mizushima, and A. Iwasaki. 2007. Autophagy-dependent viral recognition by plasmacytoid dendritic cells. *Science* 315:1398–1401.

38. Loo, Y. M., J. Fornek, N. Crochet, G. Bajwa, O. Perwitasari, L. Martinez-Sobrido, S. Akira, M. A. Gill, A. Garcia-Sastre, M. G. Katze, and M. Gale, Jr. 2008. Distinct RIG-I and MDA5 signaling by RNA viruses in innate immunity. *J. Virol.* 82:335–345.

39. Malathi, K., B. Dong, M. Gale, Jr., and R. H. Silverman. 2007. Small self-RNA generated by RNase L amplifies antiviral innate immunity. *Nature* 448:816–819.

40. Marques, J. T., T. Devosse, D. Wang, M. Zamanian-Daryoush, P. Serbinowski, R. Hartmann, T. Fujita, M. A. Behlke, and B. R. Williams. 2006. A structural basis for discriminating between self and nonself double-stranded RNAs in mammalian cells. *Nat. Biotechnol.* 24:559–565.

41. Matsui, K., Y. Kumagai, H. Kato, S. Sato, T. Kawagoe, S. Uematsu, O. Takeuchi, and S. Akira. 2006. Cutting edge: role of TANK-binding kinase 1 and inducible IκB kinase in IFN responses against viruses in innate immune cells. *J. Immunol.* 177:5785–5789.

42. Medzhitov, R. 2007. Recognition of microorganisms and activation of the immune response. *Nature* 449:819–826.

43. Meylan, E., K. Burns, K. Hofmann, V. Blancheteau, F. Martinon, M. Kelliher, and J. Tschopp. 2004. RIP1 is an essential mediator of Toll-like receptor 3-induced NF-kappa B activation. *Nat. Immunol.* 5:503–507.

44. Meylan, E., J. Curran, K. Hofmann, D. Moradpour, M. Binder, R. Bartenschlager, and J. Tschopp. 2005. Cardif is an adaptor protein in the RIG-I antiviral pathway and is targeted by hepatitis C virus. *Nature* 437:1167–1172.

45. Moore, C. B., D. T. Bergstralh, J. A. Duncan, Y. Lei, T. E. Morrison, A. G. Zimmermann, M. A. Accavitti-Loper, V. J. Madden, L. Sun, Z. Ye, J. D. Lich, M. T. Heise, Z. Chen, and J. P. Ting. 2008. NLRX1 is a regulator of mitochondrial antiviral immunity. *Nature* 451:573–577.

46. Nallagatla, S. R., J. Hwang, R. Toroney, X. Zheng, C. E. Cameron, and P. C. Bevilacqua. 2007. 5′-triphosphate-dependent activation of PKR by RNAs with short stem-loops. *Science* 318:1455–1458.

47. Negishi, H., Y. Fujita, H. Yanai, S. Sakaguchi, X. Ouyang, M. Shinohara, H. Takayanagi, Y. Ohba, T. Taniguchi, and K. Honda. 2006. Evidence for licensing of IFN-gamma-induced IFN regulatory factor 1 transcription factor by MyD88 in Toll-like receptor-dependent gene induction program. *Proc. Natl. Acad. Sci. USA* 103:15136–15141.

48. Oganesyan, G., S. K. Saha, B. Guo, J. Q. He, A. Shahangian, B. Zarnegar, A. Perry, and G. Cheng. 2006. Critical role of TRAF3 in the Toll-like receptor-dependent and -independent antiviral response. *Nature* 439:208–211.

49. Perry, A. K., E. K. Chow, J. B. Goodnough, W. C. Yeh, and G. Cheng. 2004. Differential requirement for TANK-binding kinase-1 in type I interferon responses to Toll-like receptor activation and viral infection. *J. Exp. Med.* 199:1651–1658.

50. Pichlmair, A., O. Schulz, C. P. Tan, T. I. Naslund, P. Liljestrom, F. Weber, and C. Reis e Sousa. 2006. RIG-I-mediated antiviral responses to single-stranded RNA bearing 5′-phosphates. *Science* 314:997–1001.

51. Rothenfusser, S., N. Goutagny, G. DiPerna, M. Gong, B. G. Monks, A. Schoenemeyer, M. Yamamoto, S. Akira, and K. A. Fitzgerald. 2005. The RNA helicase Lgp2 inhibits TLR-independent sensing of viral replication by retinoic acid-inducible gene-I. *J. Immunol.* 175:5260–5268.

52. Ryzhakov, G., and F. Randow. 2007. SINTBAD, a novel component of innate antiviral immunity, shares a TBK1-binding domain with NAP1 and TANK. *EMBO J.* 26:3180–3190.

53. Saha, S. K., E. M. Pietras, J. Q. He, J. R. Kang, S. Y. Liu, G. Oganesyan, A. Shahangian, B. Zarnegar, T. L. Shiba, Y. Wang, and G. Cheng. 2006. Regulation of antiviral responses by a direct and specific interaction between TRAF3 and Cardif. *EMBO J.* 25:3257–3263.

54. Saito, T., R. Hirai, Y. M. Loo, D. Owen, C. L. Johnson, S. C. Sinha, S. Akira, T. Fujita, and M. Gale. Jr. 2007. Regulation of innate antiviral defenses through a shared repressor domain in RIG-I and LGP2. *Proc. Natl. Acad. Sci. USA* 104:582–587.

55. Sasai, M., M. Shingai, K. Funami, M. Yoneyama, T. Fujita, M. Matsumoto, and T. Seya. 2006. NAK-associated protein 1 participates in both the TLR3 and the cytoplasmic pathways in type I IFN induction. *J. Immunol.* 177:8676–8683.

56. Sato, S., M. Sugiyama, M. Yamamoto, Y. Watanabe, T. Kawai, K. Takeda, and S. Akira. 2003. Toll/IL-1 receptor domain-containing adaptor inducing IFN-beta (TRIF) associates with TNF receptor-associated factor 6 and TANK-binding kinase 1, and activates two distinct transcription factors, NF-kappa B and IFN-regulatory factor-3, in the Toll-like receptor signaling. *J. Immunol.* 171:4304–4310.

57. Schmitz, F., A. Heit, S. Guggemoos, A. Krug, J. Mages, M. Schiemann, H. Adler, I. Drexler, T. Haas, R. Lang, and H. Wagner. 2007. Interferon-regulatory-factor 1 controls Toll-like receptor 9-mediated IFN-beta production in myeloid dendritic cells. *Eur. J. Immunol.* 37:315–327.

58. Seth, R. B., L. Sun, C. K. Ea, and Z. J. Chen. 2005. Identification and characterization of MAVS, a mitochondrial antiviral signaling protein that activates NF-kappaB and IRF 3. *Cell* 122:669–682.

59. Sharma, S., B. R. tenOever, N. Grandvaux, G. P. Zhou, R. Lin, and J. Hiscott. 2003. Triggering the interferon antiviral response through an IKK-related pathway. *Science* 300:1148–1151.

60. Shingai, M., T. Ebihara, N. A. Begum, A. Kato, T. Honma, K. Matsumoto, H. Saito, H. Ogura, M. Matsumoto, and T. Seya. 2007. Differential type I IFN-inducing abilities of wild-type versus vaccine strains of measles virus. *J. Immunol.* 179:6123–6133.

61. Shortman, K., and Y. J. Liu. 2002. Mouse and human dendritic cell subtypes. *Nat. Rev. Immunol.* **2:**151–161.

62. Stark, G. R., I. M. Kerr, B. R. Williams, R. H. Silverman, and R. D. Schreiber. 1998. How cells respond to interferons. *Annu. Rev. Biochem.* **67:**227–264.

63. Stetson, D. B., and R. Medzhitov. 2006. Recognition of cytosolic DNA activates an IRF3-dependent innate immune response. *Immunity* **24:**93–103.

64. Sumpter, R., Jr., Y. M. Loo, E. Foy, K. Li, M. Yoneyama, T. Fujita, S. M. Lemon, and M. Gale, Jr. 2005. Regulating intracellular antiviral defense and permissiveness to hepatitis C virus RNA replication through a cellular RNA helicase, RIG-I. *J. Virol.* **79:**2689–2699.

65. Sun, Q., L. Sun, H. H. Liu, X. Chen, R. B. Seth, J. Forman, and Z. J. Chen. 2006. The specific and essential role of MAVS in antiviral innate immune responses. *Immunity* **24:**633–642.

66. Tabeta, K., K. Hoebe, E. M. Janssen, X. Du, P. Georgel, K. Crozat, S. Mudd, N. Mann, S. Sovath, J. Goode, L. Shamel, A. A. Herskovits, D. A. Portnoy, M. Cooke, L. M. Tarantino, T. Wiltshire, B. E. Steinberg, S. Grinstein, and B. Beutler. 2006. The Unc93b1 mutation 3d disrupts exogenous antigen presentation and signaling via Toll-like receptors 3, 7 and 9. *Nat. Immunol.* **7:**156–164.

67. Takahashi, K., T. H. Kawai, S. Sato, S. Yonehara, and S. Akira. 2006. Roles of caspase-8 and caspase-10 in innate immune responses to double-stranded RNA. *J. Immunol.* **176:**4520–4524.

68. Takaoka, A., Z. Wang, M. K. Choi, H. Yanai, H. Negishi, T. Ban, Y. Lu, M. Miyagishi, T. Kodama, K. Honda, Y. Ohba, and T. Taniguchi. 2007. DAI (DLM-1/ZBP1) is a cytosolic DNA sensor and an activator of innate immune response. *Nature* **448:**501–505.

69. tenOever, B. R., S. L. Ng, M. A. Chua, S. M. McWhirter, A. Garcia-Sastre, and T. Maniatis. 2007. Multiple functions of the IKK-related kinase IKKepsilon in interferon-mediated antiviral immunity. *Science* **315:**1274–1278.

70. Uematsu, S., S. Sato, M. Yamamoto, T. Hirotani, H. Kato, F. Takeshita, M. Matsuda, C. Coban, K. J. Ishii, T. Kawai, O. Takeuchi, and S. Akira. 2005. Interleukin-1 receptor-associated kinase-1 plays an essential role for Toll-like receptor (TLR)7- and TLR9-mediated interferon-α induction. *J. Exp. Med.* **201:**915–923.

71. Venkataraman, T., M. Valdes, R. Elsby, S. Kakuta, G. Caceres, S. Saijo, Y. Iwakura, and G. N. Barber. 2007. Loss of DExD/H box RNA helicase LGP2 manifests disparate antiviral responses. *J. Immunol.* **178:**6444–6455.

72. Xu, L. G., Y. Y. Wang, K. J. Han, L. Y. Li, Z. Zhai, and H. B. Shu. 2005. VISA is an adapter protein required for virus-triggered IFN-beta signaling. *Mol. Cell* **19:**727–740.

73. Yamamoto, M., S. Sato, H. Hemmi, K. Hoshino, T. Kaisho, H. Sanjo, O. Takeuchi, M. Sugiyama, M. Okabe, K. Takeda, and S. Akira. 2003. Role of adaptor TRIF in the MyD88-independent Toll-like receptor signaling pathway. *Science* **301:**640–643.

74. Yoneyama, M., M. Kikuchi, T. Natsukawa, N. Shinobu, T. Imaizumi, M. Miyagishi, K. Taira, S. Akira, and T. Fujita. 2004. The RNA helicase RIG-I has an essential function in double-stranded RNA-induced innate antiviral responses. *Nat. Immunol.* **5:**730–737.

75. Yoneyama, M., M. Kikuchi, K. Matsumoto, T. Imaizumi, M. Miyagishi, K. Taira, E. Foy, Y. M. Loo, M. Gale, Jr., S. Akira, S. Yonehara, A. Kato, and T. Fujita. 2005. Shared and unique functions of the DExD/H-box helicases RIG-I, MDA-5, and LGP2 in antiviral innate immunity. *J. Immunol.* **175:**2851–2858.

Cellular Signaling and Innate Immune
Responses to RNA Virus Infections
Edited by A. R. Brasier et al.
© 2009 ASM Press, Washington, D.C.

Vijay G. Bhoj
Zhijian J. Chen

Mitochondrial Antiviral Signaling

4

In the race to stay ahead of the microbes that infect us, our immune system has evolved multiple methods of detecting infectious organisms. In the case of RNA viruses, for instance, our body employs receptors that detect nucleic acid ligands as signatures of viral invasion or "danger." Once danger is detected, the signals from these receptors activate the transcription factors nuclear factor-κB (NF-κB) and interferon regulatory factor 3 (IRF3) (49). These transcription factors, along with activating transcription factor-2 (ATF-2)/c-Jun, cooperate in the nucleus to induce hundreds of genes, which serve to inhibit viral replication and spread. Type I interferons (IFNs) such as IFN-α and IFN-β constitute an important family of antiviral cytokines that are induced. Type I IFNs, along with other induced factors, set in motion mechanisms to interfere with various steps of the viral life cycle including viral transcription, translation, and assembly (52).

Since our cells are constantly transcribing our own genes and thus also produce RNA molecules, it is essential that we have the capacity to distinguish between host- and virus-derived RNA. To this end, we are equipped with Toll-like receptors (TLRs 3, 7, and 8) that detect RNA in endosomal compartments, where host RNA is normally absent (24). In addition, a set of cytosolic retinoic acid-inducible gene I (RIG-I)-like receptors (RLRs), RIG-I and melanoma differentiation-associated gene 5 (MDA-5), detect double stranded (ds) RNA and 5′-triphosphate portions of RNA, both of which are unique motifs produced during a viral life cycle (Fig. 1) (3).

TLR3 and TLR7/8, which bind dsRNA and single-stranded (ss) RNA, respectively, reside in intracellular compartments such as the endoplasmic reticulum (ER) and endosomes, with their ligand-binding regions facing into the luminal side and their signaling portions facing the cytosol. When viruses, such as influenza virus, enter a cell through the endosomal route, viral RNA exposed during vesicle acidification is detected by the cognate TLR. Once bound to their ligands, TLR3 and TLR7/8 activate the cytoplasmic adaptor proteins, TRIF (TIR-domain-containing adaptor inducing interferon) and MyD88 (myeloid differentiation primary response gene 88), respectively (30). Signaling through each adaptor results in the activation of two groups of kinases (17). One is the IκB kinase (IKK) complex, which comprises two catalytic subunits, IKK-α and IKK-β, and a regulatory subunit, NEMO (NF-κB essential modulator; also

Vijay G. Bhoj, Department of Molecular Biology, University of Texas Southwestern Medical Center, Dallas, TX 75390-9148. **Zhijian J. Chen,** Department of Molecular Biology, Howard Hughes Medical Institute, University of Texas Southwestern Medical Center, Dallas, TX 75390-9148.

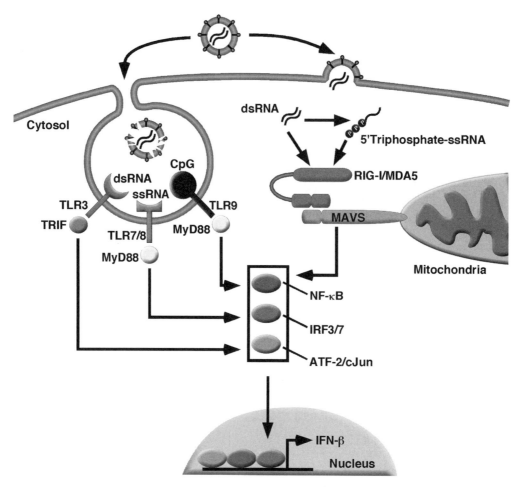

Figure 1 Mammalian immune systems employ TLRs and RLRs in order to detect viral nucleic acids. TLRs 3, 7/8, and 9 bind to viral dsRNA, ssRNA, and CpG DNA within endosomes. Upon engagement, these receptors signal through TRIF (TLR3) or MyD88 (TLR7/8 and TLR9) to activate the transcription factors NF-κB, IRF3/7, and ATF-2/c-Jun. These activated factors enter the nucleus and induce the transcription of antiviral genes including IFN-β. Alternatively, viral dsRNA and uncapped 5'-triphosphate RNA in the cytosol are detected by the RNA helicases (RLRs) RIG-I and MDA-5, respectively. Once bound to their ligands, these receptors transmit an activation signal to their common adaptor, MAVS, located on the mitochondrial surface. MAVS relays the signal to ultimately activate NF-κB, IRF3/7, and ATF-2/c-Jun, also resulting in antiviral gene induction.

known as IKK-γ/IKKAP). Upon activation, the IKK complex phosphorylates the inhibitory protein IκB, which sequesters NF-κB in the cytoplasm. Phosphorylated IκB is then ubiquitinated and degraded by the proteasome. This relieves inhibition on NF-κB so that it enters the nucleus to participate in gene activation. The other group that is activated consists of the IKK-related kinases TBK1 (TRAF family member-associated NF-κB activator [TANK]-binding kinase-1) and IKK-ε. These are thought to function redundantly to directly phosphorylate IRF3. Phosphorylated IRF3 homodimerizes and then enters the nucleus to induce gene transcription (52).

In the cytosol, MDA-5 and RIG-I bind dsRNA and uncapped 5'-triphosphate regions of RNA formed during viral replication and transcription (3, 18, 41). Upon binding their ligands through C-terminal RNA helicase domains, they relay an activation signal to an adaptor named mitochondrial antiviral signaling protein (MAVS; also known as IPS-1, VISA, or Cardif) (25, 35, 50, 55). MAVS-mediated signaling then leads to the activation of NF-κB and IRF3 through IKK and TBK1/IKK-ε, culminating in the induction of antiviral genes including IFN-I. This chapter discusses our current knowledge of the properties of MAVS, its role in signaling, as well as its role in the in vivo host response to infection with RNA viruses.

FUNCTIONAL DOMAINS AND MITOCHONDRIAL LOCALIZATION OF MAVS

MAVS was discovered independently by four different groups, and so it is also called IPS-1, VISA, and Cardif. A defining feature of MAVS is the presence of an N-terminal caspase activation and recruitment domain (CARD), similar to those present in RIG-I and MDA-5. Closer examination reveals a stretch of hydrophobic sequence at the C terminus of MAVS that resembles the mitochondrial targeting domains of a family of proteins including Bcl-2 and Bcl-xL, which are known to inhibit apoptosis (8, 50). Indeed, biochemical and imaging experiments have shown that MAVS is localized to the mitochondrial outer membrane. Importantly, the mitochondrial localization of MAVS is critical for it to induce IFN-I (50). As the name MAVS reflects both its location and function, we will use this name throughout this chapter.

The *mavs* gene, located on human chromosome 20, encodes a 540-amino-acid protein, which is constitutively expressed in a broad array of immune and non-immune tissues and cell types. The N-terminal CARD domain comprises approximately 100 amino acids and is highly conserved from fish to human (Fig. 2A and B). CARD domains belong to a larger domain family called the death domain (DD) superfamily (40). In addition to the CARD subfamily, the DD superfamily includes the DD, death effector domain (DED), and pyrin domain (PYD) subfamilies. Members of the superfamily share an overall structural fold characterized by a six-α-helical bundle. These proteins interact with members within their own subfamily through homotypic interactions. Proteins that contain DD folds participate in various signaling pathways involved in innate and adaptive immunity, inflammation, and apoptosis. As will be discussed later, the N-terminal CARD domain is essential for MAVS signaling. Deletion or even specific single-amino-acid mutations of the CARD abolish the protein's activity.

The next approximately 100 amino acids (100 to 200) of MAVS has been termed the poly-proline domain (PP) or proline-rich region (PRR). Although containing no predictable secondary structure, this region contains a high frequency of the amino acid proline (21 out of 100 amino acids). Unlike the CARD domain, no specific function has been attributed to this region, although proline-rich motifs are known to interact with the Src homology 3 (SH3) domain or WW domain (26). The PRR region also contains putative binding sites for members of the tumor necrosis factor (TNF) receptor-associated factor (TRAF) family of signaling adaptors (46, 50, 55). Current evidence suggests a role for TRAF2/6 and TRAF3 in the activation of NF-κB and

IRF3, respectively, by MAVS. A more detailed description of TRAFs and their role in the RIG-I/MDA-5 pathway will be presented later in this chapter.

Between the PRR and the mitochondrial transmembrane (TM) domain of MAVS lies a region of approximately 300 amino acids that contains no recognizable secondary folds such as helices or β-sheets. Additionally, no putative protein-binding motifs have been identified in this region until amino acids 455 to 460, which contains a binding site for TRAF6. The sequences in this intermediate region are not conserved in MAVS from different species, and over-expression experiments showed that deletion of this region together with PRR did not have much effect on type I IFN induction by MAVS (50). However, we have found that MAVS lacking PRR and the intermediate region cannot rescue type I IFN signaling in MAVS-deficient fibroblasts (V. G. Bhoj and Z. J. Chen, unpublished data), suggesting that the sequence between the CARD domain and the C-terminal transmembrane domain is important for signaling. The discrepant results between the over-expression and complementation experiments raise the interesting possibility that the truncated MAVS can associate with and activate endogenous MAVS, which then activates the downstream kinases.

The C-terminal mitochondrial targeting domain of MAVS is particularly interesting (Fig. 2A and C). MAVS is the first known example of a mitochondrial protein involved in innate immune responses against pathogens (34). The mitochondrion is best known for its crucial roles in energy (ATP) production, calcium homeostasis, and apoptosis (14, 32, 33). The similarity of the mitochondrial localization of MAVS to that of Bcl-2 family members immediately suggests a potential connection between immunity and apoptosis. However, so far there is no strong evidence indicating a role of MAVS in apoptosis. In fact, MAVS-deficient mice are phenotypically normal and cells derived from these mice do not exhibit abnormality in their response to several apoptotic stimuli such as UV and TNF-α stimulation. However, this does not exclude the possibility that MAVS may be involved in other cell death programs.

ROLES OF THE MITOCHONDRIAL LOCALIZATION OF MAVS IN HOST-PATHOGEN INTERACTION

The mitochondrial localization of MAVS is crucial for antiviral signaling because removal of the C-terminal mitochondrial targeting domain (TM) of MAVS abolishes its ability to induce IFNs (50). Furthermore, replacement of the TM domain of MAVS with the mitochondrial targeting sequences of Bcl-2 or Bcl-xL completely restores IFN induction. In contrast, mislocalization of

A.

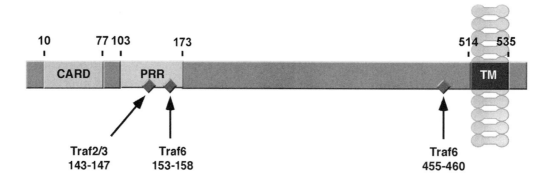

B.

C.

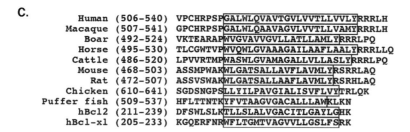

Figure 2   The functional domains of MAVS. (A) The MAVS protein contains an N-terminal CARD domain, which is thought to interact with RLRs. Amino acids 103 to 173 contain an abundance of prolines termed a PRR. The C terminus contains a single-pass TM domain, which localizes the protein to the outer mitochondrial membrane with the N terminus facing the cytosol. (B) Evolutionary conservation of the MAVS CARD domain. An alignment of CARD domains of MAVS from several species reveals a high degree of evolutionary conservation of this region. (C) Evolutionary conservation of the mitochondrial TM domain of MAVS and its similarity to the TM domain of the Bcl-2 family members Bcl-2 and Bcl-xL.

MAVS to plasma membrane or ER drastically diminished its signaling function.

Remarkably, our reliance on the mitochondrial localization of MAVS is exploited by several RNA viruses in their attempt to evade the host immune system. A pivotal example is provided by hepatitis C virus (HCV). HCV, a member of the *Flaviviridae* family, primarily infects the liver and is responsible for approximately 120 to 170 million infections worldwide (2, 51). In 55 to 85% of infected individuals, the virus establishes a state of chronic infection. In some instances, chronically infected individuals go on to develop liver cancer.

The HCV genome is transcribed and then translated into a large (>3,000 amino acids) polyprotein. This polyprotein is then cleaved into 10 mature proteins by cellular proteases as well as the virus's own serine protease complex NS3/4A. The NS4A subunit localizes NS3 to mitochondria-associated membranes, allowing the protease to also cleave MAVS at a conserved cysteine (C508) residue just N-terminal to the start of its TM domain (31, 35). Cleavage at this site releases MAVS from the mitochondrial membrane, causing it to diffuse into the cytoplasm. The resulting cytoplasmic form of MAVS loses its ability to signal and initiate the antiviral response. The ability to suppress host immune responses is thought to be essential for HCV to establish a chronic infection. Like HCV, the closely related flavivirus GB virus (also known as hepatitis G virus), which can also cause a chronic hepatic infection, encodes an NS3/4A protease complex that cleaves MAVS at the C508 site preceding the TM domain (5).

Hepatitis A virus (HAV), a ssRNA picornavirus, also infects the liver, but unlike HCV usually results in an acute, self-limiting disease. In parallel to HCV, HAV encodes a polyprotein that is cleaved by a mature viral protease, 3C^{pro}, which, in turn, is derived from cleavage of the intermediate protein, 3ABC. Interestingly, 3ABC, but not the mature 3C^{pro}, can cleave MAVS to disrupt its function (57). Like the HCV protease, the A subunit localizes 3ABC to mitochondrial membranes where it cleaves MAVS at a site preceding the TM domain (Q428). Although HAV does not establish itself chronically in the host, evasion of the innate immune response may be necessary to permit replication of the virus early in the infection cycle, as it is sensitive to the effects of type I IFNs.

Given the unique location of MAVS with respect to viral detection pathways, it is tempting to make teleological speculations regarding its role there in relation to virus replication. One such argument for MAVS's location on the mitochondrial surface is that, at this location, the RIG-I/MDA-5/MAVS detection and signaling complex resides close to sites of viral transcription and replication where viral RNA levels are highest. In support of this hypothesis are examples of viruses that organize concentrated replication and assembly centers termed "virus factories" on various intracellular membranes, including those of the ER and mitochondria (38). Members of various RNA virus families including *Flaviviridae*, *Bunyaviridae*, and *Coronaviridae* have been shown to cause structural alterations of membranes during their replication. HCV, for example, alters ER-derived membranes into structures termed "membrane webs" (37). These webs contain all of the proteins required for its genomic replication and transcription. The replication complex of another RNA virus, poliovirus, has been shown to associate with ER-derived membranes and with membranes of the autophagosome (9, 20). A study of flock house virus showed localization of the viral RNA polymerase protein A to the outer mitochondrial membrane (36). Replicase complexes of other positive-sense RNA viruses have also been localized to lysosomes, endosomes, and Golgi. It is thought that these "replication/virus factories" help bring the viral polymerases and associated factors in close proximity to the genomic material for rapid and efficient replication. Additionally, it has been shown that ER membranes may come in close proximity to mitochondria (10, 43). Therefore, it can be envisioned that we have evolved to effectively detect these RNA intermediates by placing the detection machinery (RLRs-MAVS) on nearby membranes (mitochondria) where their viral ligands may be most abundant.

SIGNALING TO AND FROM MAVS

With the discovery of MAVS in 2005, it was clear that the protein played a critical role in the signaling pathway of RLRs. Through a combination of overexpression and small interfering RNA-mediated "knockdown" of members of the RLR signaling cascade, including MAVS, those initial studies also concluded that MAVS acted downstream of RLR-RNA binding and upstream of TBK1 and IKK, as illustrated in Fig. 3. In the short time since those publications, significant progress has been made toward understanding the signal transduction cascade in the RIG-I/MAVS pathway. Here we will summarize the current knowledge of MAVS signaling and point out some outstanding questions that demand further dissection.

Regulation of MAVS by RLRs

MAVS functions downstream of RIG-I and MDA-5 and is required for signaling by both viral sensors. Upon binding of 5′-triphosphate regions of RNA or dsRNA to the helicase regions of RIG-I and MDA-5, respectively, both RLRs presumably undergo conformational changes and become released from autoinhibition (47). It is currently thought that this event is followed by the direct interaction of their tandem CARD domains with the CARD domain of MAVS. In support of this hypothesis is the fact that MAVS has been shown to coimmunoprecipitate with both RIG-I and MDA-5. Specifically, when over-expressed in mammalian cells, just the N-terminal CARD region of MAVS can be coimmunoprecipitated with the minimal tandem CARD region of RIG-I (56). Signal relay through a CARD-CARD interaction is reminiscent of the cytosolic bacterial sensors NOD-1 (nucleotide-binding oligomerization domain-containing 1)

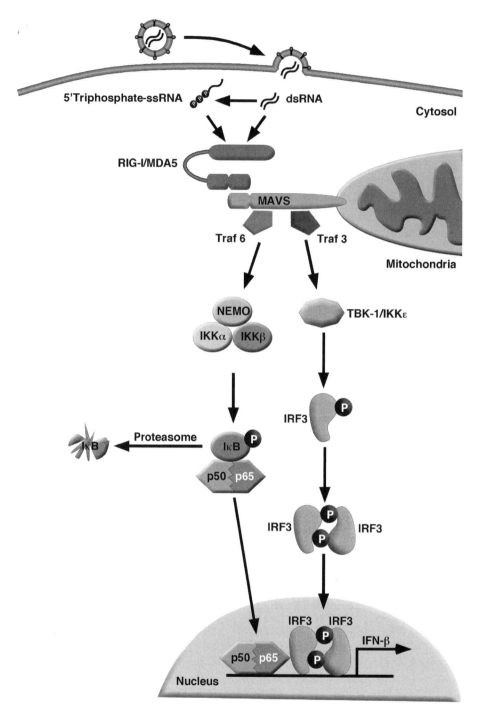

Figure 3   MAVS transmits activation signals from RLRs to activate NF-κB and IRF3, resulting in transcription of IFN-β. RIG-I and MDA-5 bind viral RNA containing 5'-triphosphate and dsRNA, respectively, in the cytosol. The ligand-bound receptors then interact with MAVS through CARD-CARD interactions. MAVS subsequently binds to the adaptors TRAF6 and TRAF3. TRAF6 then activates the IKK complex, leading to the phosphorylation of IκB. IκB is then ubiquitinated and degraded by the proteasome, releasing NF-κB from inhibition. NF-κB translocates to the nucleus to induce genes including IFN-β. TRAF3 induces the activation of the IKK-related kinases TBK1 and IKK-ε, which, in turn, phosphorylate IRF3. Phosphorylated IRF3 forms a homodimer, which enters the nucleus to form an enhanceosome complex together with NF-κB to induce IFN-β transcription.

and NOD-2 (19). Upon binding to bacterial components, these receptors oligomerize and subsequently recruit their signaling adaptor RIP-2 (receptor-interacting protein 2) (RICK) through CARD-CARD interactions (11). In the case of MAVS, the molecular result of the CARD-CARD interaction with RLRs is unknown. It is possible that the CARD-mediated interaction with the RLR causes binding sites for downstream signaling adaptors to become accessible in the MAVS protein. It should be noted, however, that it has been difficult to detect interaction between endogenous RIG-I and MAVS proteins either before or after viral infection. There is also no clear evidence that RIG-I is recruited to the mitochondria following viral infection. Therefore, the possibility that MAVS is activated by RIG-I through an indirect mechanism cannot be excluded at this time.

A recent study suggests an interesting mechanism of MAVS regulation by polyubiquitination of RIG-I (13). Ubiquitin is a 76-amino-acid polypeptide that can be covalently conjugated to other proteins through an enzymatic cascade consisting of E1, E2, and E3 (42). The best-known function of ubiquitin is to target protein destruction by the proteasome. However, nonproteolytic functions of ubiquitin have been uncovered at a rapid pace. These include roles of ubiquitin in DNA repair, vesicle trafficking, chromatin regulation, and protein kinase regulation. The role of ubiquitin in protein kinase regulation was discovered in the course of studying the NF-κB signaling pathway (6). It was shown that TRAF2 (TNF receptor associated factor 2) and TRAF6, which are crucial for IKK activation in the TNF-α and interleukin-1 (IL-1)/TLR pathways, function as RING (Really Interesting New Gene) domain ubiquitin ligases (E3). These E3s function together with a specific ubiquitin-conjugating enzyme (E2) complex consisting of Ubc13 and Uev1A to catalyze the synthesis of Lys-63 (K63)-linked polyubiquitin chains. K63 polyubiquitination leads to IKK activation through a proteasome-independent mechanism (1). The role of ubiquitin in MAVS signaling will be described in more detail below.

In the RLR pathway, RIG-I was shown to undergo K63 polyubiquitination at Lys-172 by TRIM25 (tripartite motif-containing 25), a RING-containing E3 ligase (13). Ubiquitination of RIG-I enhanced its ability to associate with MAVS. Mutation of Lys-172 of RIG-I impaired its ability to induce IFN-β, supporting a role of ubiquitination in RIG-I activation. Cells lacking TRIM25 were more susceptible to viral infection, and their ability to produce IFN-β was compromised. In contrast to RIG-I, there is no evidence of MDA-5 polyubiquitination. As both receptors signal to MAVS, it is

not clear why they appear to use distinct mechanisms in signaling to MAVS.

Regulation of IKK and TBK1 by MAVS
MAVS is essential for the activation of IKK and TBK1 in response to RNA virus infection, but the mechanism by which MAVS activates these kinases is still not clear at present. MAVS contains no predicted catalytic domains, and it likely serves as a platform to recruit downstream signaling molecules. Indeed, MAVS contains binding sites for TRAF2, 3, and 6. The binding site for TRAF2 and TRAF3 is located within the proline-rich domain (PVQET; amino acids 143 to 147), whereas the binding sites for TRAF6 are located in the PRR domain (PGENSE; amino acids 153 to 158) and in a region close to the C-terminal transmembrane domain (PEENEY; amino acids 455 to 460). In addition to TRAF proteins, MAVS also binds to RIP-1 and FADD (FAS-associated via death domain), which have been implicated in NF-κB activation (25).

Role of TRAF Proteins in Linking MAVS to IKK and TBK1 Activation
TRAFs are a family of seven proteins that have been implicated in signaling in diverse biological pathways leading to the activation of NF-κB. Members of the family contain a C-terminal TRAF domain made up of a coiled-coil and TRAF-C region, which participate in interactions with receptors or other signaling proteins. All TRAFs except TRAF1 also contain a RING domain in the N terminus followed by several zinc fingers (7). TRAF2 and TRAF6, which have been most extensively studied, signal in the TNF and IL-1/TLR pathways, respectively. Both TRAFs have been shown to function as ubiquitin ligases (E3s), a function that depends on their RING and zinc-finger regions (1).

In the IL-1/TLR pathway, the binding of a ligand to a cognate receptor containing the intracellular TIR (Toll/interleukin-1 receptor) domain leads to the recruitment of MyD88, which also contains a TIR domain. MyD88 binds to IRAK-4 (interleukin-1 receptor-associated kinase 4) and IRAK-1, which in turn recruit TRAF6. TRAF6 catalyzes K63 polyubiquitination of target proteins including TRAF6 itself (1). The K63 polyubiquitin chains serve as a scaffold to recruit and activate the TAK1 complex, which consists of the kinase subunit TAK1 as well as adaptor proteins TAB1 (TAK1-binding protein 1), TAB2, and TAB3. Both TAB2 and TAB3 contain highly conserved zinc-finger domains that bind preferentially to K63 polyubiquitin chains. This binding recruits the TAK1 complex to polyubiquitinated TRAF6, resulting in the activation of TAK1, presumably by facilitating the autophosphorylation of TAK1 at Thr-187 within its activation loop. The K63 polyubiquitin chains also recruit the IKK complex through binding to NEMO, which contains a NEMO ubiquitin-binding (NUB) domain. The recruitment of the IKK complex to the TAK1 complex through the K63 polyubiquitin scaffold facilitates the phosphorylation of IKK-β by TAK1, leading to IKK activation. With respect to the RLR/MAVS pathway, coimmunoprecipitation has shown that TRAF6 can bind to MAVS in a manner that depends on at least one intact TRAF6 binding site (56). Further, cells deficient in TRAF6 lose their ability to activate an NF-κB-responsive promoter but retain the ability to activate IRF3 and produce IFN-β (50, 56). These results support the idea that TRAF6 may be required to activate NF-κB downstream of MAVS but may be dispensable for IRF3 activation.

TRAF2 also functions as a ubiquitin E3 that catalyzes polyubiquitination of RIP-1 in the TNF-α pathway (54). The role of TRAF2 in MAVS signaling is less clear, although it has been reported that TRAF2 and MAVS interact in transfection experiments in mammalian cells and in yeast two-hybrid assays (56). The TRAF2 binding site on MAVS appears to overlap with that of the TRAF3 binding sites. Unlike TRAF2 and TRAF6, which are known to activate NF-κB, TRAF3 is required for the activation of IRF3 and IRF7 by both TLR and RLR pathways. TRAF3 binds not only to MAVS, but also to MyD88 and TRIF, explaining why it plays an important role in IRF3 activation in TLR and RLR pathways (39, 45). Interestingly, TRAF3 molecules with deletions of the RING–zinc-finger regions or of the RING domain alone could not rescue signaling when used to reconstitute TRAF3-deficient cells, suggesting that the ubiquitin E3 activity of TRAF3 may be important for signaling (46).

Role of Adaptor Proteins in IKK and TBK1 Activation

The IKK complex consists of the kinase subunits IKK-α and IKK-β and the essential regulatory subunit NEMO. Similarly, the activation of TBK1 appears to require several adaptor proteins, including TANK, NAP1 (NAK-associated protein 1), and SINTBAD (similar to NAP1 TBK1 adaptor) (12, 16, 44). These proteins bind to TBK1 through a common TBK1-binding domain (TBD), which is predicted to form an α-helix. These adaptor proteins, as well as NEMO, contain multiple coiled-coil domains, which are known for mediating oligomerization. Among these proteins, TANK has been shown to interact with TRAF3 and NEMO as well as TBK1 (4, 16, 58). Therefore, TANK not only serves as

a link from TRAF3 to TBK1 activation, but it may also couple TBK1 and IKK activation. Significantly, NEMO is not only essential for IKK activation, but it is also indispensable for the activation of TBK1 and IRF3 (58). Exactly how NEMO mediates TBK1 activation in the viral pathway is still not clear, but mutations within the ubiquitin-binding domain of NEMO abrogated IFN promoter activation by MAVS (58). RNA interference of TANK, NAP1, or SINTBAD partially inhibits type I IFN-1 induction by RNA viruses or TLR stimulation, consistent with the role of these adaptors in positively regulating TBK1 activity (16, 44, 48). The mechanism by which these adaptors mediate TBK1 activation by MAVS is not understood. A recent report showed that TANK underwent K63 polyubiquitination in response to stimulation by lipopolysaccharide, a TLR4 ligand. However, there is no evidence that K63 polyubiquitination of TANK is required for TBK1 activation or type I IFN induction by RNA viruses. It is also not clear why TBK1 requires three different adaptors to regulate its activity. One possibility is that these adaptors impart specificity to TBK1 activation by different pathways.

In summary, an outline that emerges from the recent studies is that MAVS activates IKK and TBK1 through TRAF proteins as well as several kinase adaptors. However, much more work is needed to clearly delineate the pathways and mechanisms by which MAVS signals from the mitochondrial membrane to activate the kinases in the cytosol.

ROLE OF MAVS IN VIVO

The earliest innate immune response to RNA viruses is primarily mediated by two viral detection systems, the TLRs and the RLRs. Recent studies that have explored the role of MAVS signaling in vivo have shed some light on the relative contribution of RLR-mediated responses to the overall host response to infection by RNA viruses. This has been done by genetic disruption of the *mavs* gene in mice followed by in vivo challenges with RNA viruses. Prior to these efforts, both RIG-I and MDA-5 had also been disrupted in mice with very distinct results, which will be briefly mentioned here.

Murine embryos lacking RIG-I showed a severe developmental defect, with only few mice surviving until a few weeks after birth (22). In vitro tests using cells from these mice showed a cell-type-specific involvement of RIG-I in the detection and response to RNA virus infection. In fibroblasts and conventional dendritic cells (cDCs), the response to RNA virus infection relied completely on RIG-I-mediated signaling. Wild-type cells exposed to Newcastle disease virus (NDV) activated

NF-κB and IRF3 and induced cytokine production, whereas these responses were completely abolished in the absence of RIG-I. In another subset of dendritic cells, plasmacytoid dendritic cells (pDCs), however, loss of RIG-I had no effect on cytokine production in response to virus. pDCs instead seem to depend completely on TLR signaling via MyD88, as cells lacking this adaptor were completely devoid of a cytokine response when challenged with NDV or Sendai virus (22).

Genetic deletion of MDA-5 showed a nonredundant role for this protein in mice. In contrast to RIG-I-deficient mice, those lacking MDA-5 developed normally (15, 23). In vitro infection of MDA-5-deficient DCs and macrophages and in vivo challenge with encephalomyocarditis virus (EMCV) identified MDA-5 as the receptor for this virus. MDA-5, however, played little or no role in the response to other RNA viruses tested. Using a large panel of RLR stimulators, it was found that MDA-5 serves as the receptor for poly(I:C) and members of the picornavirus family while RIG-I senses several other RNA viruses (influenza, West Nile virus, herpes simplex virus type 1, Sindbis virus, murine cytomegalovirus, and reovirus) (15, 23).

Unlike the RIG-I-deficient mice, the MAVS-deficient mice were developmentally normal and fertile (29, 53). Most cell types derived from the MAVS-deficient mice, including fibroblasts, macrophages, and CDCs, were found to be completely deficient in cytokine response to viruses and dsRNA stimuli detected by either RIG-I or MDA-5. pDCs lacking MAVS signaled normally upon virus infection, consistent with their dependence on TLRs to detect RNA viruses. In vivo, the role of MAVS signaling in the response to RNA-virus infection is a more complex picture.

Intravenous (IV) infection with EMCV, recognized by MDA-5, resulted in the production of antiviral and inflammatory cytokines (IFN-α, IFN-β, IL-6, IP-10, and monocyte chemoattractant protein-1) in wild-type mice (29). This response was completely lost in mice lacking MAVS. In addition, lack of MAVS signaling permitted the virus to reach higher titers and also resulted in earlier death of EMCV-infected mice. Similarly, IV injection of poly (I:C) resulted in cytokine production that was almost entirely dependent on MAVS (53). In response to IV infection with vesicular stomatits virus (VSV), MAVS-deficient mice produced normal levels of cytokines, although like EMCV-infected mice, they harbored greater viral loads and succumbed more readily to infection (29, 53). This important observation suggested that, for VSV, RLR-dependent signaling was not required for type I IFN production. Redundancy with the TLRs is one explanation for this effect. However, the poor survival of MAVS-

deficient mice underscores the importance of the RLR system in the host's defense against RNA virus infection.

An unexpected result of these in vivo challenges was that full immune defense against VSV required both functional copies of the *mavs* gene (53). Mice containing only one functional copy of *mavs* harbored viral loads between that of wild-type and *mavs*-deficient mice. The survival of *mavs*-heterozygous mice also showed an intermediate sensitivity to VSV infection. An implication of this result is that humans with mutations of one of their *mavs* genes may be more sensitive to viral infection.

Three recent papers have evaluated the relative roles of the TLR pathway and RLR pathway in the response to influenza virus, NDV, and lymphocytoid choriomeningitis virus (LCMV) (21, 27, 28). Using mice deficient in MAVS, MyD88, or both, it was shown that, upon influenza virus infection, type I IFN induction and viral titers were not affected when the individual adaptors were deleted. When both were deleted, the type I IFN response was absent and virus levels were nearly 1 log higher. Thus, in contrast to the anti-EMCV response, either TLR or RLR signaling can activate innate immune responses against influenza virus and either one alone is sufficient to limit the viral load. Compensation in the IFN response was also observed after lung infection with NDV. Wild-type mice infected with NDV mounted an early IFN response mediated by MAVS signaling within alveolar macrophages. However, when this component was eliminated by disruption of *mavs*, the IFN response still occurred but was delayed compared to wild-type mice. Further, the delayed IFN response came from pDCs presumably through TLR-mediated recognition. Adding to this complex picture of responses, IV infection with NDV resulted in partially reduced IFN-α responses in both MAVS$^{-/-}$ and MyD88$^{-/-}$ mice, suggesting that both pathways contributed but did not compensate for full type I IFN production. In stark contrast, IV challenge with LCMV indicated that type I IFN and proinflammatory cytokine (IL-6, IL12p40, RANTES) responses were primarily dependent on MyD88 signaling, presumably in pDCs. MAVS-deficient mice resembled wild-type mice in response to this virus. Viral clearance and survival were also significantly diminished in MyD88$^{-/-}$ mice only. Interestingly, mice lacking both TLR7 and TLR9 were able to control viral titer normally and showed only partial reduction in type I IFN production, suggesting partial compensation by another TLR.

Together, in vivo experiments examining the role of MAVS in the innate antiviral response to RNA viruses show a complicated picture. Depending on the virus, innate cytokine responses may be completely dependent on MAVS-mediated signaling alone (EMCV), dependent on MAVS but compensated by MyD88 signaling in its absence (VSV), partially dependent on both MAVS and MyD88 signaling (NDV), or primarily dependent on MyD88 signaling (LCMV).

To date, two studies have examined the activation of adaptive immune parameters in MAVS$^{-/-}$ mice (21, 27). In response to influenza virus infection, MAVS$^{-/-}$ mice were able to activate CD4$^+$ and CD8$^+$ T cells normally. Antibody production was also normal in these mice. MyD88$^{-/-}$ mice, on the other hand, showed deficient CD4$^+$ T-cell and immunoglobulin (IgG2a) responses to the virus. Similarly, CD8$^+$ T cells were activated normally in MAVS$^{-/-}$ mice infected with LCMV, whereas activation was significantly attenuated in MyD88$^{-/-}$ mice. These results show that RLR-mediated signaling does not contribute to the development of adaptive responses to influenza virus and LCMV. A more complete picture of adaptive immune activation requires examination in mice lacking both MAVS and MyD88 following infection with different RNA viruses.

CONCLUSIONS AND PERSPECTIVES

It is now well established that MAVS serves as an essential signaling adaptor in the cytosolic antiviral signaling pathway. Genetic studies have shown that MAVS is required for innate immune responses to RNA virus infection in the majority of cell types, whereas MyD88 is involved in type I IFN induction by some RNA viruses in PDCs. The relative contributions of these two different pathways in antiviral responses in vivo depend on the viruses as well as the route of infection. It is likely that we have evolved to rely on two partially redundant antiviral pathways to successfully defeat many RNA virus infections; without such redundancy, inactivation of one antiviral pathway by genetic mutations or viral evasion would have had fatal consequences.

A novel aspect of MAVS is its localization in the mitochondrial outer membrane. Although it is clear that this mitochondrial localization is essential for MAVS signaling, it is still a mystery why MAVS has to reside on the mitochondrial membrane. One possibility is that activation of MAVS relies on its association with unique components of the mitochondrial outer membrane. The definition of "activated" MAVS is not clear, although it is known that MAVS becomes more resistant to detergent extraction after viral infection (50). The mechanisms by which MAVS is activated by RIG-I and how MAVS activates downstream kinases are still not well resolved, although significant progress has been made in establishing the role of several proteins in this pathway, such as TRIM25, TRAF3, and NEMO. Interestingly, both

TRIM25 and TRAF3 have ubiquitin E3 activities that rely on their RING domains, which are important for type I IFN induction. Furthermore, NEMO contains a ubiquitin-binding domain that appears to be important for both IKK and TBK1 activation. Therefore, ubiquitin may be key to understanding the signaling cascades emanating from both RLRs and TLRs.

Accumulating evidence indicates that MAVS is a target of viral evasion, as illustrated in its cleavage by the HCV and HAV proteases. It is likely that more examples of viral inactivation of MAVS and other components of the RLR pathway will be uncovered in the near future. Further studies of the dynamic interactions between the host and virus that center around the RLR and TLR pathways should translate into more-effective therapies in a variety of viral diseases.

References

1. Adhikari, A., M. Xu, and Z. J. Chen. 2007. Ubiquitin-mediated activation of TAK1 and IKK. *Oncogene* **26:**3214–3226.
2. Afdhal, N. H. 2004. The natural history of hepatitis C. *Semin, Liver Dis.* **24** (Suppl. 2):3–8.
3. Bowie, A. G., and K. A. Fitzgerald. 2007. RIG-I: tri-ing to discriminate between self and non-self RNA. *Trends Immunol.* **28:**147–150.
4. Chariot, A., A. Leonardi, J. Muller, M. Bonif, K. Brown, and U. Siebenlist. 2002. Association of the adaptor TANK with the I kappa B kinase (IKK) regulator NEMO connects IKK complexes with IKK epsilon and TBK1 kinases. *J. Biol. Chem.* **277:**37029–37036.
5. Chen, Z., Y. Benureau, R. Rijnbrand, J. Yi, T. Wang, L. Warter, R. E. Lanford, S. A. Weinman, S. M. Lemon, A. Martin, and K. Li. 2007. GB virus B disrupts RIG-I signaling by NS3/4A-mediated cleavage of the adaptor protein MAVS. *J. Virol.* **81:**964–976.
6. Chen, Z. J. Ubiquitin signalling in the NF-kappaB pathway. 2005. *Nat. Cell. Biol.* **7:**758–765.
7. Chung, J. Y., Y. C. Park, H. Ye, and H. Wu. 2002. All TRAFs are not created equal: common and distinct molecular mechanisms of TRAF-mediated signal transduction. *J. Cell. Sci.* **115:**679–688.
8. Cory, S., and J. M. Adams. 2002. The Bcl2 family: regulators of the cellular life-or-death switch. *Nat. Rev. Cancer* **2:**647–656.
9. Egger, D., and K. Bienz. 2005. Intracellular location and translocation of silent and active poliovirus replication complexes. *J. Gen. Virol.* **86:**707–718.
10. Franzini-Armstrong, C. 2007. ER-mitochondria communication. How privileged? *Physiology (Bethesda)* **22:** 261–268.
11. Fritz, J. H., R. L. Ferrero, D. J. Philpott, and S. E. Girardin. 2006. Nod-like proteins in immunity, inflammation and disease. *Nat. Immunol.* **7:**1250–1257.
12. Fujita, F., Y. Taniguchi, T. Kato, Y. Narita, A. Furuya, T. Ogawa, H. Sakurai, T. Joh, M. Itoh, M. Delhase, M. Karin, and M. Nakanishi. 2003. Identification of NAP1,

a regulatory subunit of IkappaB kinase-related kinases that potentiates NF-kappaB signaling. *Mol. Cell. Biol.* **23**:7780–7793.

13. Gack, M. U., Y. C. Shin, C. H. Joo, T. Urano, C. Liang, L. Sun, O. Takeuchi, S. Akira, Z. Chen, S. Inoue, and J. U. Jung. 2007. TRIM25 RING-finger E3 ubiquitin ligase is essential for RIG-I-mediated antiviral activity. *Nature* **446**:916–920.

14. Giacomello, M., I. Drago, P. Pizzo, and T. Pozzan. 2007. Mitochondrial Ca^{2+} as a key regulator of cell life and death. *Cell Death Differ.* **14**:1267–1274.

15. Gitlin, L., W. Barchet, S. Gilfillan, M. Cella, B. Beutler, R. A. Flavell, M. S. Diamond, and M. Colonna. 2006. Essential role of mda-5 in type I IFN responses to polyriboinosinic:polyribocytidylic acid and encephalomyocarditis picornavirus. *Proc. Natl. Acad. Sci. USA* **103**: 8459–8464.

16. Guo, B., and G. Cheng. 2007. Modulation of the interferon antiviral response by the TBK1/IKKi adaptor protein TANK. *J. Biol. Chem.* **282**:11817–11826.

17. Hacker, H., and M. Karin. 2006. Regulation and function of IKK and IKK-related kinases. *Sci. STKE* **2006**:re13.

18. Hornung, V., J. Ellegast, S. Kim, K. Brzozka, A. Jung, H. Kato, H. Poeck, S. Akira, K. K. Conzelmann, M. Schlee, S. Endres, and G. Hartmann. 2006. 5′-Triphosphate RNA is the ligand for RIG-I. *Science* **314**:994–997.

19. Inohara, N., M. Chamaillard, C. McDonald, and G. Nunez. 2005. NOD-LRR proteins: role in host-microbial interactions and inflammatory disease. *Annu. Rev. Biochem.* **74**:355–383.

20. Jackson, W. T., T. H. Giddings, Jr., M. P. Taylor, S. Mulinyawe, M. Rabinovitch, R. R. Kopito, and K. Kirkegaard. 2005. Subversion of cellular autophagosomal machinery by RNA viruses. *PLoS Biol.* **3**:e156.

21. Jung, A., H. Kato, Y. Kumagai, H. Kumar, T. Kawai, O. Takeuchi, and S. Akira. 2008. Lymphocytoid choriomeningitis virus activates plasmacytoid dendritic cells and induces cytotoxic T cell response via MyD88. *J. Virol.* **82**:196–206.

22. Kato, H., S. Sato, M. Yoneyama, M. Yamamoto, S. Uematsu, K. Matsui, T. Tsujimura, K. Takeda, T. Fujita, O. Takeuchi, and S. Akira. 2005. Cell type-specific involvement of RIG-I in antiviral response. *Immunity* **23**:19–28.

23. Kato, H., O. Takeuchi, S. Sato, M. Yoneyama, M. Yamamoto, K. Matsui, S. Uematsu, A. Jung, T. Kawai, K. J. Ishii, O. Yamaguchi, K. Otsu, T. Tsujimura, C. S. Koh, C. Reis e Sousa, Y. Matsuura, T. Fujita, and S. Akira. 2006. Differential roles of MDA5 and RIG-I helicases in the recognition of RNA viruses. *Nature* **441**:101–105.

24. Kawai, T., and S. Akira. 2006. TLR signaling. *Cell Death Differ.* **13**:816–825.

25. Kawai, T., K. Takahashi, S. Sato, C. Coban, H. Kumar, H. Kato, K. J. Ishii, O. Takeuchi, and S. Akira. 2005. IPS-1, an adaptor triggering RIG-I- and Mda5-mediated type I interferon induction. *Nat. Immunol.* **6**:981–988.

26. Kay, B. K., M. P. Williamson, and M. Sudol. 2000. The importance of being proline: the interaction of proline-rich motifs in signaling proteins with their cognate domains. *FASEB J.* **14**:231–241.

27. Koyama, S., K. J. Ishii, H. Kumar, T. Tanimoto, C. Coban, S. Uematsu, T. Kawai, and S. Akira. 2007. Differential role of TLR- and RLR-signaling in the immune responses to influenza A virus infection and vaccination. *J. Immunol.* **179**:4711–4720.

28. Kumagai, Y., O. Takeuchi, H. Kato, H. Kumar, K. Matsui, E. Morii, K. Aozasa, T. Kawai, and S. Akira. 2007. Alveolar macrophages are the primary interferon-alpha producer in pulmonary infection with RNA viruses. *Immunity* **27**:240–252.

29. Kumar, H., T. Kawai, H. Kato, S. Sato, K. Takahashi, C. Coban, M. Yamamoto, S. Uematsu, K. J. Ishii, O. Takeuchi, and S. Akira. 2006. Essential role of IPS-1 in innate immune responses against RNA viruses. *J. Exp. Med.* **203**:1795–1803.

30. Lee, M. S., and Y. J. Kim. 2007. Signaling pathways downstream of pattern-recognition receptors and their cross talk. *Annu. Rev. Biochem.* **76**:447–480.

31. Li, X. D., L. Sun, R. B. Seth, G. Pineda, and Z. J. Chen. 2005. Hepatitis C virus protease NS3/4A cleaves mitochondrial antiviral signaling protein off the mitochondria to evade innate immunity. *Proc. Natl. Acad. Sci. USA* **102**:17717–17722.

32. Marsden, V. S., and A. Strasser. 2003. Control of apoptosis in the immune system: Bcl-2, BH3-only proteins and more. *Annu. Rev. Immunol.* **21**:71–105.

33. McBride, H. M., M. Neuspiel, and S. Wasiak. 2006. Mitochondria: more than just a powerhouse. *Curr. Biol.* **16**:R551–R560.

34. McWhirter, S. M., B. R. tenOever, and T. Maniatis. 2005. Connecting mitochondria and innate immunity. *Cell* **122**: 645–647.

35. Meylan, E., J. Curran, K. Hofmann, D. Moradpour, M. Binder, R. Bartenschlager, and J. Tschopp. 2005. Cardif is an adaptor protein in the RIG-I antiviral pathway and is targeted by hepatitis C virus. *Nature* **437**: 1167–1172.

36. Miller, D. J., M. D. Schwartz, and P. Ahlquist. 2001. Flock house virus RNA replicates on outer mitochondrial membranes in *Drosophila* cells. *J. Virol.* **75**:11664–11676.

37. Moradpour, D., R. Gosert, D. Egger, F. Penin, H. E. Blum, and K. Bienz. Membrane association of hepatitis C virus nonstructural proteins and identification of the membrane alteration that harbors the viral replication complex. *Antiviral Res.* **60**:103–109.

38. Novoa, R. R., G. Calderita, R. Arranz, J. Fontana, H. Granzow, and C. Risco. 2005. Virus factories: associations of cell organelles for viral replication and morphogenesis. *Biol. Cell.* **97**:147–172.

39. Oganesyan, G., S. K. Saha, B. Guo, J. Q. He, A. Shahangian, B. Zarnegar, A. Perry, and G. Cheng. 2006. Critical role of TRAF3 in the Toll-like receptor-dependent and -independent antiviral response. *Nature* **439**:208–211.

40. Park, H. H., Y. C. Lo, S. C. Lin, L. Wang, J. K. Yang, and H. Wu. 2007. The death domain superfamily in intracellular signaling of apoptosis and inflammation. *Annu. Rev. Immunol.* **25**:561–586.

41. Pichlmair, A., O. Schulz, C. P. Tan, T. I. Naslund, P. Liljestrom, F. Weber, and C. Reis e Sousa. 2006. RIG-I-mediated antiviral responses to single-stranded RNA bearing 5′-phosphates. *Science* **314**:997–1001.

42. Pickart, C. M. 2001. Mechanisms underlying ubiquitination. *Annu. Rev. Biochem.* **70:**503–533.

43. Pizzo, P., and T. Pozzan. 2007. Mitochondria-endoplasmic reticulum choreography: structure and signaling dynamics. *Trends Cell Biol.* **17:**511–517.

44. Ryzhakov, G., and F. Randow. 2007. SINTBAD, a novel component of innate antiviral immunity, shares a TBK1-binding domain with NAP1 and TANK. *EMBO J.* **26:**3180–3190.

45. Saha, S. K., and G. Cheng. 2006.TRAF3: a new regulator of type I interferons. *Cell Cycle* **5:**804–807.

46. Saha, S. K., E. M. Pietras, J. Q. He, J. R. Kang, S. Y. Liu, G. Oganesyan, A. Shahangian, B. Zarnegar, T. L. Shiba, Y. Wang, and G. Cheng. 2006. Regulation of antiviral responses by a direct and specific interaction between TRAF3 and Cardif. *EMBO J.* **25:**3257–3263.

47. Saito, T., R. Hirai, Y. M. Loo, D. Owen, C. L. Johnson, S. C. Sinha, S. Akira, T. Fujita, and M. Gale, Jr. 2007. Regulation of innate antiviral defenses through a shared repressor domain in RIG-I and LGP2. *Proc. Natl. Acad. Sci. USA* **104:**582–587.

48. Sasai, M., M. Shingai, K. Funami, M. Yoneyama, T. Fujita, M. Matsumoto, and T. Seya. 2006. NAK-associated protein 1 participates in both the TLR3 and the cytoplasmic pathways in type I IFN induction. *J. Immunol.* **177:**8676–8683.

49. Seth, R. B., L. Sun, and Z. J. Chen. 2006. Antiviral innate immunity pathways. *Cell Res.* **16:**141–147.

50. Seth, R. B., L. Sun, C. K. Ea, and Z. J. Chen. 2005. Identification and characterization of MAVS, a mitochondrial antiviral signaling protein that activates NF-kappaB and IRF 3. *Cell* **122:**669–682.

51. Shepard, C. W., L. Finelli, and M. J. Alter. 2005. Global epidemiology of hepatitis C virus infection. *Lancet Infect. Dis.* **5:**558–567.

52. Stark, G. R., I. M. Kerr, B. R. Williams, R. H. Silverman, and R. D. Schreiber. 1998. How cells respond to interferons. *Annu. Rev. Biochem.* **67:**227–264.

53. Sun, Q., L. Sun, H. H. Liu, X. Chen, R. B. Seth, J. Forman, and Z. J. Chen. 2006. The specific and essential role of MAVS in antiviral innate immune responses. *Immunity* **24:**633–642.

55. Xia, Z. P., and Z. J. Chen. 2005. TRAF2: a double-edged sword? *Sci. STKE* **2005:**pe7.

56. Xu, L. G., Y. Y. Wang, K. J. Han, L. Y. Li, Z. Zhai, and H. B. Shu. 2005. VISA is an adapter protein required for virus-triggered IFN-beta signaling. *Mol. Cell* **19:**727–740.

57. Yang, Y., Y. Liang, L. Qu, Z. Chen, M. Yi, K. Li, and S. M. Lemon. 2007. Disruption of innate immunity due to mitochondrial targeting of a picornaviral protease precursor. *Proc. Natl. Acad. Sci. USA* **104:**7253–7258.

58. Zhao, T., L. Yang, Q. Sun, M. Arguello, D. W. Ballard, J. Hiscott, and R. Lin. 2007. The NEMO adaptor bridges the nuclear factor-kappaB and interferon regulatory factor signaling pathways. *Nat. Immunol.* **8:**592–600.

Cellular Signaling and Innate Immune
Responses to RNA Virus Infections
Edited by A. R. Brasier et al.
© 2009 ASM Press, Washington, D.C.

Meztli Arguello
Suzanne Paz
John Hiscott

5

Interferon Regulatory Factors and the Atypical IKK-Related Kinases TBK1 and IKK-ε: Essential Players in the Innate Immune Response to RNA Virus Infection

Rapid induction of type I interferon (IFN) expression is a central event in the establishment of the innate immune response against viral infection that requires recognition of pathogen-associated molecular patterns (PAMPs) through Toll-like receptor (TLR)-dependent and -independent pathways. Following engagement of viral sensors by their cognate PAMPs, the host cell activates multiple signaling cascades that stimulate an innate antiviral response, resulting in the disruption of viral replication at multiple levels and the mobilization of the adaptive arm of the immune system (72, 163). Multiple TLRs (TLR3, 4, 7, and 9) and cytosolic sensors (retinoic acid-inducible gene I [RIG-I] and melanoma differentiation-associated gene 5 [MDA-5]) are involved in the cell-specific recognition of viral components and regulation of type I IFNs, with evidence accumulating that cooperation between different pathways is required to ensure a robust and controlled activation of antiviral response (72). Signal-induced activation of latent transcription factors such as nuclear factor-κB (NF-κB) and interferon regulatory factors (IRFs) via posttranslational

modifications—primarily phosphorylation events—leads to the recruitment of these factors to the type I IFN promoters in a temporally and spatially coordinated manner (105). This chapter opens with a brief overview of the TLR-dependent and -independent pathways of activation (covered in greater detail in other chapters) and then focuses on the activation of transcription factors IRF3 and IRF7 by the atypical IκB kinases (IKKs) TBK1 (TANK-binding kinase 1) and IKK-ε (IκB kinase ε) and their role in the induction of type I IFNs.

TLR AND RIG-I SIGNALING

The Endosomal Pathway—Sensing Viruses through TLRs
TLRs belong to the Toll/interleukin-1 (IL-1) receptor (TIR) superfamily of receptors, which share a common TIR domain in their cytoplasmic region. The original Toll receptor was identified as a single-pass transmembrane protein essential for the establishment of a dorsoventral pattern in *Drosophila* (50) and was later recognized to

Meztli Arguello, Lady Davis Institute for Medical Research, Jewish General Hospital, Montreal, Quebec, Canada H3T 1E2. **Suzanne Paz,** Lady Davis Institute for Medical Research, Jewish General Hospital, Montreal, Quebec, Canada H3T 1E2, and Department of Microbiology and Immunology, McGill University, Montreal, Quebec, Canada. **John Hiscott,** Lady Davis Institute for Medical Research, Jewish General Hospital, Montreal, Quebec, Canada H3T 1E2, and Department of Microbiology and Immunology and Department of Experimental Medicine, McGill University, Montreal, Quebec, Canada.

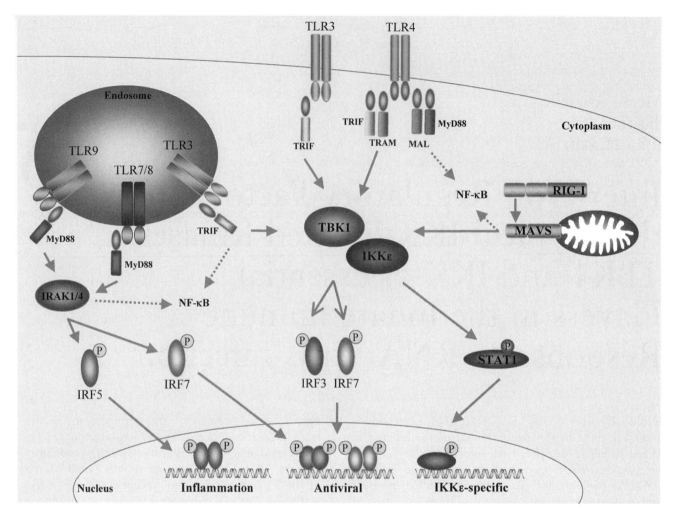

Figure 1   TLR-dependent and -independent sensing of viral infection leads to TBK1/IKK-ε activation. Simplified schematic representation of TLR3, 4, 7/8, and 9 and RIG-I/MDA-5 signaling pathways, leading to TBK1 and IKK-ε activation.

be involved in host defense against fungal infections (86, 87). Subsequently, multiple homologues of Toll were identified in mammalian cells and designated Toll-like receptors or TLRs (reviewed in references 20 and 112). The TLR family now consists of more than 13 members, each recognizing distinct PAMPs derived from various microbial pathogens (132). The recognition of viral components is mediated by TLR3, TLR4, hTLR7/mTLR8, and TLR9, which recognize double-stranded (ds)RNA, viral proteins, single-stranded RNA, and unmethylated CpG DNA, respectively (160). A schematic representation of nucleic acid-sensing TLRs is shown in Fig. 1.

The extracellular domain of TLRs contains multiple leucine-rich repeats that are involved in pathogen structure recognition. Within the intracellular Toll-interacting region (TIR) domain, there are three well-conserved regions termed boxes 1, 2, and 3 that are essential for TLR signaling (129, 130, 170). Upon stimulation with their cognate ligand, TLRs form active homodimers or heterodimers in order to induce an effective signaling cascade (35), although in the case of TLR9, homodimerization appears to be constitutive and ligand binding induces a global conformational change that brings the cytoplasmic tails into close contact, initiating signaling (85). Whether this mechanism is generalized or a particularity of TLR9 remains unknown at present. Expression of TLRs is compartmentalized: TLR 1, 2, 4, 5, and 6 are expressed at the cell surface, whereas TLR 3, 7, 8, and 9 localize to acidic endosomes (73, 103). The subcellular localization of nucleic acid-sensing TLRs is of great importance for the discrimination of virus- from self-derived components (12). Furthermore, expression of TLRs is restricted to certain cell types, mainly of hematopoietic origin.

TLR signaling is mediated by homotypic interaction with the TIR-domain-containing adaptors myeloid differentiation primary response gene 88 (MyD88), MyD88 adaptor-like molecule (MAL), TIR-domain-containing adaptor inducing IFN-β/TIR-domain-containing adaptor molecule 1 (TRIF/TICAM-1), and TIR-domain-containing adaptor molecule 2/TRIF-related adaptor molecule (TICAM-2/TRAM) (73, 112). In general, TLR signaling can be divided in MyD88-dependent and TRIF-dependent pathways. TLR5, TLR7, and TLR9 exclusively use MyD88 as the adaptor to relay downstream signaling. In contrast, TLR3 does not require MyD88 but rather uses TRIF for signaling. TLR4 uses all four adaptors: MyD88, MAL, TRIF, and TRAM. MyD88 and TIRAP act in concert, whereas TRIF requires TRAM for interaction with TLR4 (73, 112).

The Cytosolic Pathway—Sensing Viruses through RIG-I/MDA-5

Although viral and microbial pathogens are detected by the TLR family via the recognition of PAMPs, viral infection is also detected through TLR-independent mechanisms (2). Early viral RNA replicative intermediates are detected by RIG-I and MDA-5, two recently characterized cytosolic viral RNA receptors belonging to the DExD/H box RNA helicase family (3, 175) (Fig. 1). The functions of RIG-I and MDA-5 are nonredundant, as RIG-I specifically recognizes 5′-triphosphate-containing viral RNA, while MDA-5 is stimulated by dsRNA structures such as synthetic poly(I:C). As a result, RIG-I is essential for the production of IFN in response to RNA viruses including paramyxoviruses, influenza virus, and Japanese encephalitis virus, whereas MDA-5 is critical for picornavirus detection; RIG-I$^{-/-}$ and MDA-5$^{-/-}$ mice are highly susceptible to infection with these respective RNA viruses compared to control mice (71). The RIG-I knockout model demonstrated the pivotal importance of the RIG-I pathway in antiviral immunity in most cell types—as fibroblastic, epithelial, and conventional dendritic cells (DG) utilized RIG-I and not the TLR system to detect RNA virus infection (70). In contrast, plasmacytoid DCs (pDCs) utilize TLR-mediated signaling in preference to RIG-I (70).

RIG-I and MDA-5 are highly homologous in their structure and transmit signals through a common signaling pathway that induces type I IFN to generate a host-protective innate immune response against viral infection (61, 123). RIG-I contains two caspase activation and recruitment domains (CARDs) at its N terminus and RNA helicase activity in the C-terminal portion (175). The helicase domain mediates recognition of viral RNA structures, triggering a conformational change that exposes the N-terminal CARD domains. This conformational alteration allows dimerization of RIG-I through homotypic CARD interaction and binding to downstream signaling molecules (139).

The mitochondrial adaptor providing a link between RIG-I sensing of incoming viral RNA and downstream activation events was identified as mitochondrial antiviral signaling adaptor (MAVS), also known as IPS-1/VISA/Cardif (74, 106, 150, 168). Like RIG-I, MAVS contains an amino-terminal CARD domain, while a C-terminal transmembrane domain localizes MAVS to the outer mitochondrial membrane. MAVS functions as a central adaptor in RIG-I and MDA-5 signaling, as demonstrated by studies in which ectopic MAVS expression is sufficient to activate the IFN-α, IFN-β, and NF-κB promoters in a TBK1-dependent manner (150). Conversely, MAVS knockdown by small interfering RNA (siRNA) inhibits RIG-I-dependent antiviral responses (90, 106). The localization of MAVS to the mitochondria is essential for signaling, as MAVS deletions in the transmembrane domain are nonfunctional. Remarkably, the NS3/4A protease of hepatitis C virus (HCV) cleaves MAVS at its C-terminal end—adjacent to the mitochondrial targeting domain (106)—arguing that disruption of the mitochondrial association of MAVS is part of the innate immune evasion strategy used by HCV (90, 95).

MAVS-deficient mice were viable and fertile, but failed to induce IFN in response to poly(I:C) stimulation and were severely compromised in immune defense against viral infection (80, 153). Loss of MAVS abolished viral induction of IFN and prevented the activation of NF-κB and IRF3 in multiple cell types, except pDCs. MAVS was critically required for the host response to RNA viruses but was not required for IFN production by cytosolic DNA (64) or by *Listeria monocytogenes*. These results provide the in vivo evidence that the viral signaling pathway through mitochondrial-bound MAVS is specifically required for innate immune responses against viral infection (80, 153).

Following recognition of viral RNA, interaction of RIG-I/MDA-5 with MAVS via their respective CARD domains leads to the recruitment of multiple downstream signaling components and activation of the classical IKK complex and the atypical IKK-related kinases TBK1 and IKK-ε, resulting in induction of the antiviral response (113). The exact mechanism by which TBK1 and IKK-ε become activated is not clearly understood, but involves (at a minimum) TRAF3 (TNF receptor-associated factor 3), NEMO (NF-κB essential modulator), TANK (TRAF family member-associated NF-κB activator), and possibly FADD (FAS-associated death domain protein) (7, 46, 138). K63-linked ubiquitination of TRAF3, NEMO, and TANK are important (39, 75, 178); NEMO is ubiquitinated by TRAF6, whereas TRAF3 probably undergoes autoubiquitination (75, 83, 146). The identity of the E3 ligase for TANK is not yet known. Nonetheless, activation of TBK1 and/or IKK-ε results in direct phosphorylation of IRF3/7 and induction of antiviral genes. Activation of the NF-κB pathway occurs through a better-characterized pathway involving recruitment of TRAF2/6 to MAVS, followed by K63-linked ubiquitination of RIP-1 and interaction with NEMO—the regulatory subunit of the IKK complex. IKK-β then phosphorylates the inhibitor of NF-κB (IκB-α), causing its proteasomal degradation and release of active NF-κB dimers. NF-κB contributes to the induction of antiviral genes but also triggers the inflammatory response.

THE IRF FAMILY

Induction of the broad spectrum of type I IFNs and IFN-stimulated genes is achieved by members of the IRF transcription factor family (58, 59). In all, nine human IRFs have been identified (IRF1 to IRF9); each member shares extensive homology in the N-terminal DNA-binding domain (DBD), characterized by five tryptophan repeat elements located within the first 150 amino acids of the protein (98). The majority of IRFs are involved in distinct aspects of the antiviral response (59), while two members—IRF4 and IRF8—function mainly as regulators of hematopoiesis in concert with the Ets transcription factor PU.1 (115, 119). The functions of all members of the IRF family are summarized in Table 1. Typically, IRFs form homo- or heterodimers upon activation, mediated by interaction between their C-terminal IRF-association domains (IADs). The dimers translocate into the nucleus and bind to consensus GAAAN-NGAAANN and G(A)AAG/CT/CGAAAG/CT/C sequences through their DBD. These highly similar sequences have been termed the IFN-stimulated regulatory element (ISRE) and IRF-binding element, respectively, and can be found in all IFN-stimulated genes (ISGs) (54, 98). In addition to their role in immune regulation, IRFs are also involved in regulation of cell cycle, apoptosis, and tumor suppression (58). The unique function of a particular IRF is attributed to the ability of the IAD to interact with other members of the IRF family and other factors, its intrinsic transactivation potential, and cell-type-specific expression of the IRFs.

IRF3

IRF3, a critical player in the induction of type I IFNs following virus infection, is a constitutively expressed phosphoprotein of 427 amino acids (reviewed in references 5, 54, and 67). A schematic representation of IRF3 is shown in Color Plate 1A. IRF3 activation is achieved through extensive C-terminal phosphorylation at a serine/threonine cluster found between amino acids 382 and 405; virus-induced phosphorylation forces a conformational change in IRF3 that allows homo- and heterodimerization, nuclear localization, and association with the histone acetyltransferase coactivator CBP (CREB binding protein)/p300 (57, 92). Inactive IRF3 constitutively shuttles in and out of the nucleus, whereas phosphorylation-dependent association with CBP/p300 retains IRF3 in the nucleus and induces transcription of IFN-β and other target genes such as *RANTES*, *IP-10*, and *NOXA* (81, 82, 109; D. Goubau, unpublished observation).

The importance of IRF3 in regulating the early and late phases of IFN expression (91, 100) was demonstrated through the generation of IRF3 knockout mice (144). IRF3$^{-/-}$ mice were more susceptible to virus infection, and serum IFN levels from encephalomyocarditis virus-infected mice were significantly lower in the IRF3$^{-/-}$ mice than in wild-type mice. IRF3$^{-/-}$ mice also exhibited an abnormal profile of IFN-α subtype expression. The impaired IFN production observed in IRF3$^{-/-}$ murine embryonic fibroblasts (MEFs) could be reestablished by ectopic expression of IRF7, demonstrating that IRF7 induction during the late phase of the IFN response is key for enhanced IFN expression. However, IRF3$^{-/-}$ MEFs overexpressing IRF7 also had an abnormal pattern of IFN induction, and only cells reconstituted with both proteins exhibited a normal IFN profile, clearly demonstrating that IRF3 and IRF7 have essential and distinct roles in ensuring transcriptional efficiency of the IFN-α/β genes (144).

The three-dimensional crystal structures of the IRF3 DBD and C-terminal domain have been independently reported (Color Plate 2B) (116, 126, 154). These studies describe an important mechanism for the regulation of IRF3 activity, whereby N- and C-terminal autoinhibitory domains within the C-terminal IAD interact to form a highly condensed hydrophobic core. This interaction buries several key residues within the IAD required for IRF3 dimerization, effectively preventing IRF3 nuclear accumulation, DNA binding, and transactivation. It is proposed that virus-inducible, C-terminal phosphorylation events abolish autoinhibitory interactions by introducing charge repulsions within this region that unmask the IAD active site and realign the DBD to form the transcriptionally active IRF3 protein with the capacity to recruit the CBP histone acetyltransferase (126, 154).

Subsequently, the crystal structure of a complex between the IAD of IRF3 and a portion of histone

Table 1 Biological properties of IRF proteins[a]

Name	Expression	Activators	Knockout phenotype	Function

IRF1 (ISG2)	Most cell types
	Constitutive
	Inducible by
	  IFN-γ
		IFNs IL-1, IL-12, TNF
		Virus infection and
		  LPS
		Prolactin
		Concanavalin A
			Decreased number of CD8+ cells and
			  increased susceptibility to
			  pathogen due to abnormal type I
			  IFN gene induction
			Il-12 dysregulated in macrophages
				Antibacterial and
				  antiviral response

IRF2	Most cell types
	Constitutive
		Type I IFNs
		Virus infection
			Increased susceptibility to viruses as
			  well as skin disorders
			Il-12 dysregulated in macrophages
				Repressor of IRF1 and IRF9
				Oncogene and antiviral
				  defense/immune regulator
				Promotes Th1 responses and
				  natural killer cell
				  development
				May regulate nitric oxides

IRF3	Most cell types
	Constitutive
		Virus infection
		dsRNA
		TLR3, TLR4
			Increased susceptibility to virus
			  infection
			Impaired IFN response
				Antibacterial and antiviral
				  immunity
				Regulation of *IFNB*,
				  NOXA, and *ISG56* genes

IRF4
(Pip/LSIRF/
ICSAT)
	Activated T cells,
	  B cells
	Constitutive
	  and inducible
	  by TLRs
		PMA
		HTLV-1 Tax
		CD3-specific antibody
		  (T cells)
		IL-4, antibody specific for
		  IgM or DC40
		  (splenic B cells)
			Defect in B-cell maturation
			  (lack of germinal center formation)
			Profound defects in function of B
			  and T cells
			Impaired Th2 response
			Impaired CD4+ DCs
				Promotes B-cell proliferation
				Promotes differentiation of
				  mitogen-activated T cells
				Controls DC development

IRF5	B cells and DCs:
	  constitutive
	Inducible by IFNs
	  and TLR
		Viruses (NDV, VSV, HSV-1)
		Type I IFNs
		TLR7, TLR9
			Impaired secretion levels of IL-6,
			  IL-12, and TNF-α by TLR activation
			Resistance to lethal shock induced
			  by unmethylated DNA or LPS
				Induction of
				  proinflammatory cytokines

IRF6	Mesodermal in
	  early development
	Epithelial, muscular
	  skeletal cells
		ND	Abnormal skin, limb, and
			  craniofacial development
			Defect in keratinocyte differentiation
			  and proliferation
				Development of lip, palate,
				  and skin (formation of
				  cleft lip and/or palate—Van
				  der Woude's syndrome)

IRF7	B cells, monocyte,
	  and pDCs
	  constitutive
	Other cell types:
	  inducible by IFNs
		IFN
		Virus infection
		CpG-containing
		  oligodeoxynucleotides
		TLR7, TLR8, and TLR9
			Increased susceptibility to virus
			  infection
			Complete lack of IFN-α production
			Reduced efficiency of
			  CD8+ T-cell cross-priming
				Antiviral response
				Regulator of *IFNA* genes

IRF8
(ICSBP)
	Hematopoietic cells
	  and immune cells
	Constitutive and
	  inducible by
	  IFN-γ
		Type II IFNs and LPS
		  (macrophages)
		TCR engagement
		  (T cells)
			Impaired Thl response
			Defect in DC generation and
			  maturation
			Block in the pre-B to
			  B cell transition
			Susceptible to pathogens that
			  depend on IFN for clearance
				Regulating expression
				  of IFN-α and IFN-β
				Negative regulator role

IRF9
(ISG3γ/P48)
	Constitutive	Type II IFNs	Susceptible to virus infection	Act as DNA recognition
				  subunit
				Activating transcripton
				  in response to IFN-α

[a] PMA, phorbol myristate acetate; HTLV, human T-cell leukemia virus; Ig, immunoglobulin; NDV, Newcastle disease virus; ND, not determined; TCR, T-cell receptor.

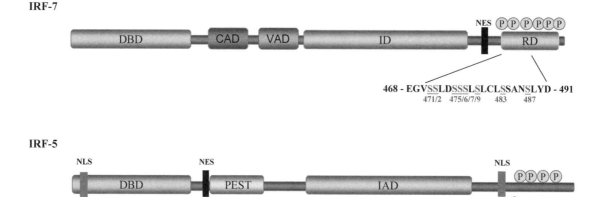

Figure 2 Schematic representation of IRF7 and IRF5. Schematic diagram of the principal domains of IRF7 (A) and IRF5 (B). The signal response domain is amplified and important amino acids are shown as gray letters with their respective position number. Pro, proline-rich region; PEST, proline-glutamic acid-serine-threonine domain; CAD, constitutive activation domain; VAD, virus-activated domain; P, phosphorylation site.

acetyltransferase CBP/p300 was solved (Color Plate 2C) (127). The IRF3-binding domain (IBiD) of CBP/p300 maps to a 46-residue segment in the C-terminal glutamine-rich region of CBP and binds to a hydrophobic surface on IAD, which is buried by intramolecular interactions in latent IRF3 (126). The IBiD covers the same surface as the autoinhibitory elements in latent IRF3, indicating that the condensed conformation of latent IRF3 and the interaction with IBiD are mutually exclusive (127); consequently, these autoinhibitory elements must be displaced during activation to allow CBP interactions, consistent with phosphorylation-induced unfolding of the C-terminal segment. Interestingly, the C-terminal domain of IRF3 exhibits sequence similarity to the Mad homology 2 (MH2) domain of the Smad family of transcriptional proteins, suggesting common molecular mechanisms of action and a common evolutionary relationship between the IRF and Smad proteins (126, 128, 154). Recent work (117) argues that IRF3 is activated by a dual phosphorylation-dependent switch in which TBK1 initially phosphorylates Ser residues 396, 398, and 405. This activation event opens the C-terminal structure to permit interaction with CBP and facilitates phosphorylation at Ser-385 and -386, required for IRF3 dimerization. This model is in agreement with previous biochemical data demonstrating that Ser-396, -398 and -405 are important targets following viral infection (92, 149).

IRF7

IRF7 was first described to bind and repress the Epstein-Barr virus (EBV) Qp promoter regulating EBV nuclear antigen 1 (EBVA1) (176, 177), but its importance in virus-induced IFN-α gene regulation was quickly recognized (4, 100, 143). IRF7 is a potent activator of IFN gene transcription that, unlike IRF3, can induce the expression of all IFN-α and IFN-λ gene family members as well as IFN-β, leading to the amplification of the IFN response (59, 60, 114). The structure of IRF7 (see schematic representation in Fig. 2A) reveals the presence of a DBD in the N terminus and multiple regulatory domains in the C-terminal region that control IRF7 activity, including a constitutive activation domain, a virus-activated domain, an inhibitory domain, and a signal response domain (89, 91, 101). Biochemical data reveal that the C-terminal virus-activated domain between amino acids 471 and 487 is the target of extensive virus-induced phosphorylation. Ser-471/472 and -477/479 are important phosphorylation targets required for activation, since their substitution to Ala results in an IRF7 that is nonresponsive to virus infection, as demonstrated by the lack of nuclear translocation, DNA binding, or transactivation of the IFN-α4 promoter (158). Conversely, substitution of the Ser-477/479 with the phosphomimetic amino acid Asp leads to the generation of a constitutively active IRF7 (158). Although no crystallography data are yet available, it has been proposed that IRF7 is activated in a manner homologous to IRF3: the inactive protein is folded on itself, so that the C-terminus inhibitory domain masks the DBD and/or constitutive activation domain. Upon activation, phosphorylation of key residues leads to electrostatic changes that force opening up of the structure, allowing dimerization and DNA binding.

Unlike IRF3, constitutive IRF7 expression is restricted to B cells and DCs; in other cells, IRF7 expression is rapidly induced by the first wave of IFN following viral infection. Indeed, IFN, lipopolysaccharide (LPS), 12-O-tetradecanoylphorbol-13-acetate (TPA), and tumor necrosis factor (TNF)-α can all modulate transcription of the *IRF7* gene (96, 97). Investigation of the *IRF7* promoter revealed the presence of an ISRE motif in the first intron capable of binding interferon-stimulated gene factor 3 complex (ISGF3)—a complex composed of IRF9, signal transducer and activator of transcription 1 (Stat1), and Stat2—that confers on the *IRF7* gene its responsiveness to IFN. Immediately upstream of the TATA box, the presence of a κB site makes it responsive to NF-κB activators such as TNF-α and TPA (97). Also distinct from IRF3, IRF7 has a rapid turnover, with a half-life of approximately 30 minutes, which may represent a mechanism to ensure a quick shutdown of IFN induction (4, 100, 143).

Using IRF7$^{-/-}$ mice, Honda et al. demonstrated that IRF7 is essential for the induction of type I IFNs via virus-mediated, TLR-dependent and -independent signaling pathways (59, 60). In IRF7$^{-/-}$ MEFs, IFN-α production was completely ablated and IFN-β production severely impaired following viral infection. This phenotype was due to a defect in MyD88-independent signaling, as MyD88$^{-/-}$ MEFs had normal IFN levels (60). In IRF7$^{-/-}$ mice, serum levels of IFN following virus infection were significantly lower than in control littermates owing to defective IFN production by pDCs (59, 60). pDCs are professional IFN-producing cells and stand out amongst other cells in their ability to produce large amounts of IFN-α/β after TLR7 or TLR9 engagement and signaling (26, 28, 51, 79). pDCs—unlike fibroblastic, epithelial, and most hematopoietic cells that employ a RIG-I- or MDA-5-dependent pathway of IFN induction after RNA virus infection (44, 70)—utilize a TLR- and MyD88-dependent pathway of IFN-α/β induction, which is controlled exclusively by IRF7 (26, 28, 51, 79, 161). Comparison between IRF3$^{-/-}$ pDCs and IRF7$^{-/-}$ pDCs revealed that IFN induction by TLR7, 8, or 9 was normal in IRF3$^{-/-}$ cells, but completely ablated in IRF7$^{-/-}$ cells, indicating that IRF7 is essential and IRF3 dispensable for MyD88-dependent induction of IFN-α/β genes in pDCs (60). Furthermore, Honda et al. demonstrated that mediation of CD8$^+$ T-cell responses by pDCs was

completely dependent on IRF7 (58). Thus, IRF7 is pivotal in the establishment of the antiviral response, whether locally at the site of infection or systemically through the recruitment of the adaptive immunity by pDCs.

Target Gene Activation by IRF3 and IRF7

Both IRF3 and IRF7 play distinct and essential roles in the IFN-α/β response to virus infection (58, 59). Biochemical and gene profiling studies have delineated dsRNA and virus-induced genes (29, 40) and have attempted to dissect IRF3 and IRF7 target genes (11, 45). Human IFN-β, IFN-α1, and RANTES promoters are stimulated by IRF3 coexpression, whereas IFN-α4, -α7, and -α14 promoters are preferentially induced by IRF7 (41, 89). A tetracycline-inducible expression system expressing a constitutively active form of IRF3, IRF3 (5D), was used in an mRNA profiling study of 8,556 genes to identify IRF3 target genes (45). Among the genes upregulated by IRF3 were transcripts for several known ISGs, indicating that in addition to its role in IFN-β regulation, IRF3 discriminated among ISRE-containing genes involved in the establishment of the antiviral state. Similarly, a microarray analysis of infected BJAB cells overexpressing IRF7 revealed that IRF7 activates a subset of antiviral, inflammatory, and proapoptotic proteins, as well as a group of mitochondrial genes and genes affecting DNA structure (11).

Evidence gathered from numerous transactivation assays revealed distinct DNA- and coactivator-binding properties for IRF3 and IRF7 (89, 107, 172). DNA binding site selection studies demonstrated that IRF3 and IRF7 bound to the 5'-GAAANNGAAANN-3' consensus motif found in many virus-inducible genes; however, single nucleotide substitutions in either of the hexameric GAAANN half-site motifs eliminate IRF3 binding and transactivation activity, but do not affect IRF7 interaction or transactivation (25, 89). The NN residues of the hexameric sequences found in the *IFNA4* gene define the preferential binding sites for IRF3 or IRF7 (107); thus, the consensus motif for IRF7 has been identified as 5'-GAAA/TNC/TGAAANT/C-3' (89). Association with CBP/p300 is also distinct between the two IRFs: IRF3 strongly associates with the C terminus but not with the N terminus of CBP/p300 (89), whereas IRF7 interacts weakly with the C terminus and robustly with the N terminus of CBP/p300 (172). Furthermore, association of IRF3 and IRF7 is required for maximal activation of IFN-β, but not other type I IFN promoters, suggesting that specificity in activation is dictated by promoter context (173). Together, the studies demonstrated that the ability of IRF3 and IRF7 to stimulate common as well as specific genes owes to a broader DNA-binding specificity for IRF7 as well as to a different ability to interact with other transactivators.

To examine differential activation of target genes by IRF3 and IRF7 in primary cells, adenoviral vectors encoding constitutively active forms of IRF3 or IRF7—IRF3(5D) and IRF7(ΔID)—were transduced into human macrophages. Active IRF3(5D) induced a limited expression of IFNα/β but resulted in rapid cell death of IRF3-transduced macrophages (135), possibly through IRF3-mediated induction of the proapoptotic genes such as the BH3-only protein NOXA (82). Interestingly, IRF7(ΔID)-transduced macrophages produced type I IFNs and displayed increased expression of genes encoding TRAIL (TNF-related apoptosis-inducing ligand), IL-12, IL-15, and CD80. Furthermore, IRF7-transduced macrophages exerted a strong cytostatic activity on different cancer cells, thus functionally demonstrating that transduction of active IRF7 in human macrophages leads to the acquisition of novel antitumor effector functions (135).

IRF5

IRF5 is a recently characterized participant in the IFN activation pathway; IRF5 is constitutively expressed in lymphocytes and DCs, but is induced in other cell types in response to IFN (10, 65). A schematic representation of IRF5 is shown in Fig. 2B. The 57-kDa IRF5 protein is localized to the cytoplasm of unstimulated cells and accumulates in the nucleus following virus infection. Two nuclear localization signal (NLS) elements, necessary for virus translocation to the nucleus, are found in the IRF5 protein (8, 9). Lin et al. also demonstrated that IRF5 possesses a nuclear export signal (NES) element that controls the dynamic shuttling between the cytoplasm and the nucleus, much like IRF3 (93). Although both NLS and NES are active in unstimulated cells, the NES element is dominant, resulting in the cytoplasmic accumulation of IRF5. The IRF5 gene can be transcribed into nine alternatively spliced mRNAs, depending on the cell type (99). In human pDCs four different isoforms (V1 through V4) are transcribed, whereas in peripheral blood mononuclear cells two additional isoforms (V5 and V6) are generated. Other isoforms can only be detected in human cancers (V7, V8, and V9). Expression of IRF5 is controlled by at least two discrete promoters: the V1 promoter is constitutively active and IFN inducible, whereas the V3 promoter is active only when cells are stimulated by IFN (99).

Activation of IRF5 is mediated by phosphorylation by TBK1/IKK-ε, which results in dimerization, nuclear accumulation, and transactivation of target genes (99,

145). However, aspects of IRF5 activation may be distinct from the virus-induced mechanisms used by IRF3 and/or IRF7, as IRF5 is not activated following Sendai virus (SeV) infection and, in some instances, phosphorylation by TBK1/IKK-ε does not lead to nuclear accumulation and activation. It has been postulated that isoforms V3/4 and V5, and not V1 and V2 used by Lin et al., are the targets for functional phosphorylation by TBK1/IKK-ε (99, 145). Whether these kinases truly phosphorylate in vivo IRF5 remains to be clarified.

In contrast to IRF3 and IRF7, which are robustly activated following viral infection or dsRNA or LPS stimulation, the triggers of IRF5 activation had been elusive. IRF5 activation was observed following infection with viruses such as Newcastle disease virus, vesicular stomatitis virus (VSV), and herpes simplex virus (HSV) type 1 (8), but it was not consistently reported (23). Similarly, the target genes of IRF5 are ill defined, as IRF5 activation induced a limited number of IFN-α subspecies (10); in IRF5-expressing cells, only IFN-α8 was preferentially produced following viral infection. A subsequent global analysis of IRF5-inducible genes revealed over 500 genes regulated by IRF5 (11).

IRF5 is a central mediator of TLR7/8 signaling (145). Indeed, stimulation of murine macrophages with the TLR7/8 ligand R848 led to nuclear translocation of both IRF5 and IRF7. Furthermore, siRNA against IRF5 dramatically reduced the induction of IFNA following R848 stimulation, although levels of induction were not compared to IRF7. Activation of IRF5 was MyD88 dependent and required IRAK-1 (interleukin-1 receptor-associated kinase 1) and TRAF6 as well as TBK1 and IKK-ε (145, 155). It has also been suggested that TLR4 and TLR9 engagement results in IRF5 translocation and cytokine gene activation (155). In a surprising twist, assessment of IRF5$^{-/-}$ hematopoietic cells in the context of TLR7 and MyD88 signaling revealed that induction of IL-6, IL-12p40, and TNF-α were severely impaired, while the induction of IFN-α remained unaffected in the knockout background (155). This result indicated that IRF5 was primarily involved in the induction of proinflammatory cytokines following TLR engagement and not in type I IFN production. However, Yanai et al. demonstrated that IRF5 was also necessary in antiviral immunity by challenging IRF5$^{-/-}$ mice with VSV or HSV-1 (171). IRF5 knockout animals were highly vulnerable to virus infection, with rates of mortality that were 50% higher than for control littermates. Although MEFs showed normal type I IFN production, mice challenged with virus exhibited a significant decrease in serum levels of type I IFN (IFN-α, IFN-β), as well as IL-6 measured by enzyme-linked immunosorbent assay. This reduction correlated with in vitro studies in macrophages, suggesting that the role of IRF5 in antiviral immunity might be cell specific. IRF5-deficient cells were more resistant to cell death induced by VSV or HSV-1 infection, ascribing a role for IRF5 in virus-induced apoptosis (171).

Further studies revealed that IRF5 had intrinsic proapoptotic properties, stemming from the modulation of proapoptotic genes such as *p21*, *Caspase-8*, *Bak*, and *Bax* (9). IRF5 expression is strongly induced by p53 when cells undergo genotoxic stress, but the proapoptotic properties of IRF5 are independent of p53 (9, 171). In keeping with the role of IRF5 as a tumor suppressor, IRF5$^{-/-}$ MEFs formed colonies in soft agar with similar efficiency as p53$^{-/-}$ MEFs and exhibited decreased levels of apoptosis compared to wild-type MEFs when activated Ras was overexpressed (171), thus demonstrating that lack of IRF5 predisposes to tumorigenic transformation.

THE VIRUS-ACTIVATED KINASES TBK1 AND IKK-ε

The requirement for NF-κB activation as a component of IFN-β regulation was recognized (55, 88, 164) soon after the initial discovery of NF-κB (108, 147, 148) and antedated many other important discoveries concerning the myriad of roles of NF-κB in immunomodulation, lymphocyte development, cell growth, inflammation, and cancer (43, 68, 69). Inducible activation of canonical NF-κB signaling requires phosphorylation of the IκB inhibitor by the 700- to 900-kDa multiprotein canonical IKK complex (47), composed of two catalytic kinase subunits, IKK-α and IKK-β, and a nonenzymatic regulatory subunit, NEMO or IKK-γ. Based on sequence homology and the potential to stimulate NF-κB signaling, the family of IKK kinases expanded with the discovery of two additional IKKs members: TBK1/NAK/T2K and IKK-ε/IKK-*i* (reviewed in reference 122). Inducible IKK (IKK-ε/IKK-*i*) was isolated using a subtractive hybridization technique from LPS-stimulated RAW264.7 murine macrophages (152), as well as from an expressed sequence tag database search for IKK homologues (121). The TBK1/NAK/T2K kinase was characterized as an IKK homologue involved in TNF-α-mediated NF-κB activation (15, 124), acting upstream of IKK-β (159). Despite the homology and functional similarities between TBK1 and IKK-ε, the expression patterns of these two IKK-related kinases are distinct. TBK1 expression is ubiquitous and constitutive in a wide variety of cells, while IKK-ε expression is relegated to cells of the immune compartment (122, 152), but is inducible in nonhematopoietic cells by stimulation with activating agents

such as TNF, phorbolmyristate acetate, LPS, and virus infection (6, 52, 152) through an NF-κB-dependent mechanism involving upregulation of IKK-ε expression by C/EBPβ (CCAT/enhancer protein β) and C/EBPδ (78).

Analysis of the response to virus infection in TBK1$^{-/-}$ and IKK-ε$^{-/-}$ mice demonstrated that TBK1 is principally involved in downstream signaling to IRF3 and IRF7 phosphorylation and development of the antiviral response (52, 120), with an accessory role associated to date with IKK-ε (52, 120). Ongoing studies, however, suggest that TBK1 and IKK-ε are not redundant but that IKK-ε selectively regulates a subset of interferon-responsive antiviral genes during influenza virus infection (see below) (102, 157).

Structure of TBK1/IKK-ε

A schematic representation of the IKK kinase family is detailed in Fig. 3. Like the classical IKK kinases IKK-α and IKK-β, the IKK-related kinases IKK-ε and TBK1 contain a catalytic kinase domain, a leucine zipper domain, and a helix-loop-helix domain involved in protein-protein interactions (121, 124, 152, 159). IKK-ε and TBK1 are 64% homologous to each other, but exhibit limited homology to the classical IKK-α and -β kinases, with about 33% sequence identity within the kinase domain (121, 124, 152, 159). IKK-ε and TBK1 also possess a kinase activation loop located between kinase subdomains VII and VIII; serine phosphorylation within the activation loop is required for kinase autophosphorylation and induction of enzymatic activity (76, 121, 152). In contrast to IKK-α and IKK-β, the primary amino acid sequence of the IKK-ε and TBK1 activation loop does not fit a canonical MEK consensus motif (76, 121, 152); that is, the Ser-176 residue is replaced with glycine—^{172}SXXXG^{176} sequence—and an S172A mutation abolishes IKK-ε and TBK1 catalytic activation. Differences within the activation loop sequence are consistent with the failure of MEKK1 (mitogen-activated protein kinase kinase kinase 1) or NIK expression to augment IKK-related kinase activity, as occurs with canonical IKKs (152). In contrast to IKK-α and IKK-β, an S172E phosphomimetic mutation does not generate a constitutively active IKK-ε and TBK1 kinase, but decreases catalytic activity (121, 152).

Bioinformatic analysis of the TBK1 and IKK-ε sequences (63) revealed that both kinases contain a ubiquitin-like domain (ULD), distantly related to typical ULDs such as ubiquitin, Parkin, and Rad23 (Fig. 4). The same domain was identified in the classical IKKs IKK-α and -β. The ULD in TBK1 consists of 79 amino acids, starting at position 305, and contains a characteristic hydrophobic patch centered on a conserved hydrophobic leucine residue. The ULD in TBK1 and

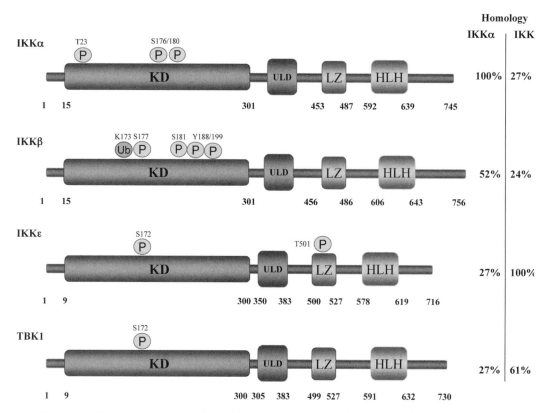

Figure 3   Schematic representation of the classical and nonclassical IKK kinases. Shown are schematic diagrams of the principal known domains of IKK-α, IKK-β, TBK1, and IKK-ε. The phosphorylated and ubiquitinated residues are indicated. Sequence homologies of each kinase compared with either IKK-α or IKK-ε are shown as percentages to the right of the schematic. KD, kinase domain; LZ, leucine zipper motif; HLH, helix-loop-helix domain; P, phosphorylation site; Ub, ubiquitination site.

IKK-ε did not interact with ubiquitin-binding proteins but rather with the kinase domain and substrates of TBK1/IKK-ε. Mutation of the ULD, which is located adjacent to the kinase domain, impaired kinase activation and resulted in the loss of IRF3 phosphorylation, dimerization, nuclear translocation, and *IFNB* or *RANTES* gene activation. Binding of IRF3 to the TBK1 ULD was abolished once IRF3 became phosphorylated by the kinase, supporting the model that phosphorylation releases IRF3 from the TBK1 complex, leading to dimerization and nuclear translocation of phosphorylated IRF3. Thus, the ULD appears to be necessary for full activation by the kinases, substrate presentation, and downstream signaling.

Phosphorylation of IRF3 and IRF7 by TBK1 and IKK-ε

An important breakthrough in the understanding of the physiological function of TBK1 and IKK-ε was the demonstration that the IRF3 and IRF7 transcription factors are the primary in vivo targets of the IKK-related kinases (34, 151). Both TBK1 and IKK-ε directly phosphorylate IRF3 and IRF7 at key resides within their C-terminal signal-responsive domain in vitro (104, 158), and both kinases target identical serine residues (118, 158). Alignment of the primary sequence of the C-terminal domains of IRF3 and IRF7 revealed an extended SxSxxxS consensus motif that appears to be the target for TBK1 and IKK-ε (118). Importantly, expression of the IKK-related kinases is essential to initiate IRF signaling in response to de novo SeV, VSV, or measles virus infection, and treatment with RNA interference directed against either IKK-ε or TBK1 reduces VSV-inducible IRF3 phosphorylation and IRF-dependent gene expression in human cells. Furthermore, expression of the IKK-related kinases generates an IRF3-dependent antiviral state in vivo that inhibited de novo VSV replication (151, 158).

Analysis of TBK1⁻/⁻ and IKK-ε⁻/⁻ mice suggested that TBK1 was the principal mediator of IRF3 and IRF7

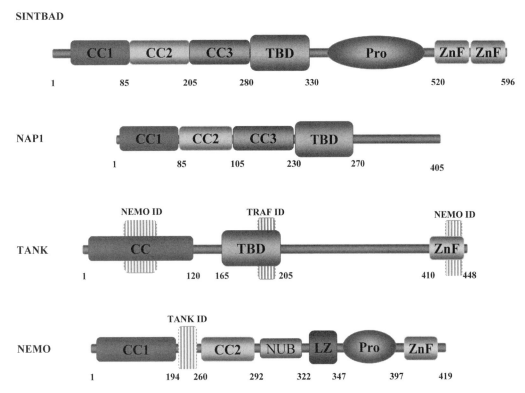

Figure 4 Adaptor molecules of the IFN signaling pathway. Schematic representation of SINT-BAD, NAP1, TANK, and NEMO adaptor molecules. CC, coiled-coil domain; Pro, proline-rich domain; ZnF, zinc-finger domain; NEMO ID, NEMO interaction domain; TRAF ID, TRAF interaction domain; TANK ID, TANK interaction domain; LZ, leucine zipper motif.

phosphorylation (52, 120). The role of IKK-ε in the development of the antiviral response was thought to be secondary to and redundant with TBK1; the inducibility of IKK-ε by IFN further suggested that it was principally involved in the amplification of the IFN response. Further studies, however, demonstrated that IKK-ε selectively regulates a subset of IFN-responsive antiviral genes during virus infection (102, 157). Mice lacking IKK-ε produced normal amounts of IFN-β, but were hypersusceptible to viral infection because of a defect in the IFN signaling pathway; a subset of type I IFN-stimulated genes were not activated in the absence of IKK-ε because ISGF3 failed to bind to the promoters of affected genes. About 30% of the IFN-inducible ISGs were poorly induced in *Ikbke*$^{-/-}$ MEFs, including *Adar1*, *Ifit3*, and *Ifi203*, whereas other ISGs such as *Irf7*, *Prkra* (RNA-activated protein kinase), and *Stat1* were unaffected. ISGF3 binding to IKK-ε-dependent promoters required phosphorylation of Stat1 by IKK-ε at Ser-708, Ser-744, and Ser-747 (157). Although the structural and functional consequences of this phosphorylation remain to be determined, IKK-ε-dependent phosphorylation of Stat1 appears to guide the transcriptional machinery to a subset of ISGs required for a direct antiviral response. The subcellular localization of TBK1 and IKK-ε following viral infection also differs. Confocal microscopy studies revealed that VSV infection resulted in recruitment of IKK-ε to the mitochondria but not TBK1, which remained dispersed throughout the cytosol (56, 90). Recruitment of IKK-ε was mediated by MAVS, as overexpression of a truncated form of MAVS lacking the transmembrane domain prevented IKK-ε from localizing to the mitochondria. Similarly, the NS3/4A protease of HCV—known to cleave MAVS—disrupted the mitochondrial localization of IKK-ε (90). Altogether, these observations indicate that the specific role for IKK-ε at the mitochondria has yet to be elucidated.

Activation of NF-κB by TBK1/IKK-ε

The identification of IKK-ε and TBK1 suggested novel roles for these kinases in the activation of the NF-κB pathway (121, 152, 159). Sequence homology with the classical IKKs certainly promoted this view and the knockout model for TBK1 further supported it, as the TBK1$^{-/-}$ mouse shared major physiological defects—such as massive liver degeneration and apoptosis—with the IKK-β$^{-/-}$ and RelA$^{-/-}$ mice (15). At first, a direct role for IKK-ε/TBK1 in phosphorylating IκB-α to activate NF-κB was suggested based on overexpression or in vitro assays (121, 152), but later experiments under more physiological conditions did not confirm this theory. Bonnard et al., however, suggested that TBK1 regulated NF-κB activity in a direct manner yet distinct from IκB-α phosphorylation (15).

TBK1 was originally characterized as a TANK-associated kinase participating in TRAF2-mediated NF-κB and JNK (c-Jun N-terminal kinase) activation (124). TBK1 functioned in complex with TANK (I-TRAF) and TRAF2, which feeds into the NIK-IKK cascade but is not required for TNF-α, IL-1, or CD40 signaling (124). However, Nomura and colleagues proposed a different model: IKK-ε/TBK1 phosphorylation of TANK led to the release of TRAF2 and NF-κB activation (111). This observation is in agreement with other studies that showed an inhibitory role for TANK in NF-κB activation following CD40 or TNF-α stimulation (136). As an additional twist, TBK1 and IKK-ε were shown to directly phosphorylate cRel and RelA(p65) at key serine residues, leading to dissociation from IκB-α and enhanced transcriptional activity (37, 49). Although far from complete, a picture is emerging in which TBK1 and IKK-ε are central regulators of all arms of the antiviral response, modulating the activation of not only the IRFs but also the NF-κB and AP-1 pathways.

ADAPTORS TO IKK-ε/TBK1 SIGNALING

The interaction between TBK1/IKK-ε and the receptor-adaptors TLR3-TRIF and RIG-I-MAVS does not appear to be direct but rather is mediated by TRAF3 and a second set of adaptors that constitutively bind the kinases. Three adaptors have been described to date: TANK, NAP1 (*NAK-associated protein 1*), and SINTBAD (similar to Nap1 TBK1 adaptor) (see the schematic representation in Fig. 4), all of which share a TBK1/IKK-ε-binding domain (TBD) predicted to form an α-helical structure with residues essential for kinase binding exposed on one side of the helix. Mutation of Tyr-236 to Ala in the NAP1 TBD disrupts the interaction of NAP1 with IKK-ε but not TBK1 (137), suggesting that the TBD recruits the two kinases via different surface contacts. Furthermore, the affinity of the different TBDs for TBK1 seems to differ, as all TBDs could compete with TANK and SINTBAD for binding of TBK1—thereby acting as dominant negatives—yet the TBD of TANK

was not efficient at competing with NAP1 and SINTBAD for binding (137). NAP1 and SINTBAD share three sets of coiled-coil (CC) domains in their N terminus, whereas SINTBAD and TANK, respectively, have two and one zinc-finger domains at their C terminus. As a unique feature, SINTBAD has a proline-rich domain, but its function remains unknown.

TANK

TANK was initially identified as a TRAF-interacting protein that regulated NF-κB activation (22). A schematic depiction of the structure of TANK is shown in Fig. 4. Initial studies demonstrated that TANK bound to the C-terminal region (TRAF-C domain) of TRAF1, TRAF2, and TRAF3 through its N-terminal portion (amino acids 1 to 212). The interaction between TANK and the TRAFs following engagement of TNF receptors was described to have both activating and inhibitory effects on the NF-κB pathway (22, 136). Elucidation of the crystal structures of TANK, CD40, lymphotoxin-β receptor, and BAFF (B-cell activating factor) receptor bound to TRAF3 revealed that interaction of TANK or the receptors with TRAF3 was mutually exclusive, as the TANK consensus sequence PxQxT bound to the same crevice in TRAF3 as the TNF family receptors (30). Following the discovery of the atypical IKK kinases, a ternary complex between TANK, TBK1/IKK-ε, and TRAF2 was described as an inducer of NF-κB activation (124), although this result was controversial (111).

More recent work revealed TANK as an essential mediator of IRF activation in TLR- or RIG-I-dependent signaling (46). Indeed, forced expression of TANK in target cells induced IFN-β promoter activity over 40-fold, but it had only a minor effect on NF-κB activation (5-fold) (46). Knockdown of TANK using siRNA technology resulted in decreased IFN production, increased viral titers, and enhanced cell sensitivity to VSV infection. Furthermore, TANK was shown to be required for IRF3/7 phosphorylation and IFN-β induction following TLR stimulation by LPS, CpG, poly(I:C), and R848 (39, 46). TANK therefore emerges as a ubiquitous mediator of TBK1/IKK-ε activity in both TLR-dependent and -independent signaling.

In MEFs, endogenous TANK was part of a multimeric complex with TRAF3, MAVS, and TBK1 following poly(I:C) or SeV stimulation. Interaction between TANK and IKK-ε was mediated by the region of TANK comprising amino acids 111 to 169 (39, 111) and the C-terminal region of TBK1/IKK-ε, downstream of the CC domain (39). TANK also interacted with the C-terminal regulatory domain of IRF3/7 through its C-terminal region upstream of the zinc-finger motif. Alain Chariot's group (14) had already mapped the region of interaction between TANK and NEMO—the regulatory subunit of the IKK complex—to amino acids 200 to 250 in NEMO, and a NEMO construct mutated in this region was unable to mediate NF-κB activation. A large multimeric complex composed of TANK, IKK-ε/TBK1, NEMO, and IKK-α could be detected in coimmunoprecipitation experiments. TANK therefore acts as a multidomain scaffolding protein that brings together elements of the IRF and NF-κB pathway, promoting activation of both the IKK complex and IKK-related kinases.

The activity of TANK appears to be regulated by multiple posttranslational modifications. Stimulation of cells with LPS but not TNF-α led to extensive TANK phosphorylation by IKK-ε/TBK1 between amino acids 192 and 247 (39). The interaction between TANK and TBK1/IKK-ε also triggered K63-linked polyubiquitination of TANK, independently of TBK1/IKK-ε kinase activity (39). TANK polyubiquitination required TRAF3 as depletion of TRAF3 by siRNA prevented this process, yet a direct TRAF3-TANK association was not necessary. TANK was also phosphorylated by IKK-β and this phosphorylation disrupted the interaction between TANK and NEMO, possibly as a negative regulatory mechanism (14).

NAP1

NAP1 (Fig. 4) was initially identified in a yeast two-hybrid screen using full-length TBK1 as bait (37). NAP1 is a ubiquitously expressed cytoplasmic protein that interacts with both IKK-ε and TBK1, as demonstrated in endogenous and overexpression assays. The region of NAP1 comprising amino acids 158 to 270 interacts with TBK1/IKK-ε and can act as a dominant negative of MAVS-mediated signaling. This region of NAP1 interacts with RIG-I and MDA-5, placing this adaptor molecule downstream of the viral RNA receptors. siRNA against NAP1 prevents IRF3 activation and IFN-β induction following VSV infection (142). Similarly, in the TLR3 pathway, NAP1 acts downstream of TRIF and upstream of TBK1 to mediate IRF3 activation (142). Recruitment of NAP1 is spatiotemporally regulated by TRIF following TLR3 engagement (38); upon dsRNA stimulation, there is a transient recruitment of TRIF to the TLR receptor followed by relocalization of TRIF to the cytoplasm and recruitment of NAP1, RIP-1, and TBK1. Formation of the cytoplasmic complex leads to the activation of IRF3. NAP1 also participates in NF-κB activation following TNF receptor, TLR3, or RIG-I engagement (37, 141). Activation of the NF-κB pathway by NAP1 requires TBK1/IKK-ε, although the mechanistic details have not been revealed yet.

SINTBAD

The new TBK1 adaptor SINTBAD was discovered in 2007 by performing BLAST searches with components of the IRF pathway. Because of homology to NAP1 in the first CC region, it was named SINTBAD (similar to NAP1 *TBK1 ad*aptor) (Fig. 4) (137). As with its two predecessors, SINTBAD constitutively interacts with TBK1 and IKK-ε but not with the canonical IKKs. Inhibition of SINTBAD expression by siRNA results in over 50% inhibition of IFN induction following SeV infection. Mechanistically, SINTBAD seems to oligomerize and interact with NAP1 but not with TANK, placing this adaptor upstream of NAP1 and the TBK1/IKK-ε kinases. However, information on SINTBAD is scarce and only further studies can determine its exact role in IFN signaling.

NEMO

NEMO or IKK-γ (Fig. 4) is better known for its involvement in the activation of the NF-κB pathway than for its role in IFN signaling. NEMO is a 419-amino-acid protein composed of two N-terminal CC domains (CC1 and CC2), a NEMO ubiquitin-binding domain (NUB), a leucine zipper, a proline-rich domain, and a C-terminal zinc-finger domain. NEMO oligomers constitutively bind the two classical kinases IKK-α and IKK-β to form what is known as the IKK complex, the central regulator of NF-κB activation (36, 156). A small fraction of NEMO can also be found as a free monomer, and it is thought to interact with additional signaling complexes (36, 167). The interaction between NEMO and TANK was first described in the context of NF-κB activation (21). As described above, TANK and TBK1 were part of a multimeric complex containing NEMO and the canonical IKKs and were required for full NF-κB activation. Termination of signaling by TANK involved phosphorylation by IKK-β and dissociation from NEMO (14). Further studies demonstrated that TANK and TBK1/IKK-ε were essential for IRF3/7 activation and only marginally involved in the NF-κB pathway (46, 151). The NEMO-TANK link became secondary and remained largely unexplored. However, a recent study demonstrated that the NEMO adaptor also plays an essential role in virus- and RIG-I-induced activation of IRF3 and IRF7 (178). Indeed, virus-mediated production of endogenous IFN and IFN-inducible gene expression including Stat1 phosphorylation was severely impaired in *Ikbkg*$^{-/-}$MEFs. Expression of MAVS or ΔRIG-I also failed to stimulate phosphorylation of IRF3 and IRF7 and ISRE reporter gene activity. In contrast, exogenous expression of TBK1 was able to stimulate the IFN response in NEMO knockout cells as well as in wild-type MEFs. Studies using truncated or single-point mutants of NEMO demonstrated that the leucine zipper and the ubiquitin-binding domains of NEMO were essential for virus-induced ISRE activity. The importance of the NEMO-TANK interaction was highlighted by the fact that NEMO mutants lacking the TANK-binding domain failed to interact with IKK-ε and TBK1 or transduce virus- and MAVS-mediated signals. Therefore, NEMO appears to act downstream of MAVS and RIG-I by physically interacting with the TANK adapter to mediate activation of TBK1/IKK-ε following RIG-I engagement.

NEMO mutations have been directly linked to rare human genetic disorders, including incontinentia pigmenti and ectodermal dysplasia with immunodeficiency (ED-ID) (reviewed in references 110, 125, and 162). Most incontinentia pigmenti patients have an identical genomic deletion from exon 4 to 10 (loss of function) in NEMO, whereas ED-ID patients carry NEMO hypomorphic mutations that impair but do not abolish NF-κB activation. NEMO mutations result in ED-ID patients with a combined, variable but profound immunodeficiency characterized by increased susceptibility to bacterial and viral infections (18, 27, 33). Strikingly, most NEMO mutations associated with ED-ID were also severely impaired in their ability to activate the RIG-I signaling pathway. These observations raise the intriguing possibility that NEMO mutations with impaired production of type I IFNs via the RIG-I pathway contribute to the immunodeficiency observed in ED-ID patients.

Hsp90

Heat shock proteins (HSPs) are a conserved protein family involved in the regulation of protein folding and assembly (1, 17, 32, 66). Amongst the several mammalian HSPs, Hsp90 emerges as selective in terms of client proteins, specializing in proteins involved in transcriptional regulation and signal transduction (19, 42, 169, 179). A yeast two-hybrid screen identified a specific interaction between Hsp90 and IRF3 (174). In vivo, Hsp90 was shown to interact with both IRF3 and TBK1, resulting in protein stabilization and increased antiviral response. Treatment of cells with the Hsp90 inhibitor geldanamycin resulted in dose-dependent inhibition of an IRF3/IRF7-dependent reporter gene. Similarly, inhibition of Hsp90 expression by siRNA resulted in an impaired activation of IRF3 following SeV infection. This study proposes that Hsp90 participates in the formation of a complex containing TBK1 and IRF3. The presence of Hsp90 results in stabilization of the kinase and its substrate, leading to efficient IRF3 activation.

NEGATIVE REGULATORS OF TBK1/IKK-ε

Activation and shutdown of the antiviral response are carefully orchestrated by posttranslational modification events. Although initial focus was directed to phosphorylation, it is now clear that addition and removal of ubiquitin chains is also a major regulatory mechanism that controls IRF and NF-κB activation (47). Addition of K63-linked ubiquitin chains—as in the case of TRAF3, TRAF6, NEMO, and RIG-I—generally favors protein-protein interactions and activation of downstream signaling. Deubiquitinating enzymes that remove K63-linked ubiquitin are emerging as key negative regulators of the IFN and NF-κB pathways.

A20

A20 is a ubiquitous cytoplasmic 90-kDa protein with dual catalytic activity: the N terminal of A20 possesses deubiquitinase activity mediated by its OTU (ovarian tumor) domain (16, 31, 166), whereas the seven zinc-finger domains found in the C terminus can function as a ubiquitin E3 ligase (166). A schematic representation of A20 is shown in Fig. 5. Both the NF-κB and IRF pathways are targeted by A20—by mechanisms that appear to be both overlapping and distinct—resulting in the downregulation of IFN-α/β production. Initial studies in a murine knockout model demonstrated that A20 deficiency results in overproduction of proinflammatory genes following TNF-α or LPS challenge, because of prolonged NF-κB activation (16). The mechanism of A20-mediated shutdown of NF-κB signaling involves the removal of K63-linked ubiquitin chains—which function as docking sites for protein-protein interaction—and the addition of K48-linked ubiquitin, a signal for proteasomal degradation. The action of A20 on components of the "inflammasome" such as TRAF6, RIP-1, and NEMO effectively terminates NF-κB activation signals (16, 53, 166). Although knockout studies did not address whether the role of A20 extended to the IFN pathway, three different groups have demonstrated that A20 is also a negative regulator of IRF3/7 activation via the TLR3 and RIG-I pathways (94, 140, 165). A20 is a potent inhibitor of RIG-I activation of the IFN pathway, with A20 completely inhibiting IRF3 phosphorylation and DNA binding induced by an active form of RIG-I (94). A20 appears to act upstream of the kinases TBK1/IKK-ε, and the C-terminal region of A20 was essential for the inhibition of RIG-I signaling (94). The intricacies of A20 negative regulation of the IFN pathway are far from understood and many exciting developments in the field are yet to come.

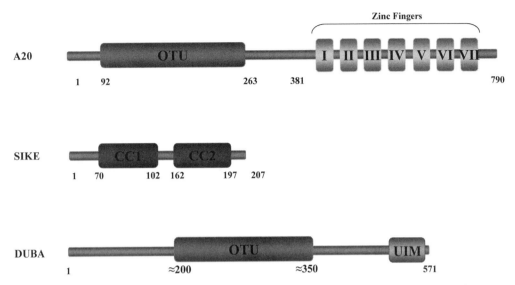

Figure 5 Negative regulators of the IFN signaling pathway. Schematic representation of A20, SIKE, and DUBA, negative regulators of the IFN pathway. UIM, ubiquitin interacting motif.

DUBA

Very recently, a genome-wide siRNA screen to search for novel regulators of IFN signaling identified DUBA (deubiquitinating enzyme A) as a novel OTU-domain DUB that negatively regulates IFN signaling following RIG-I, MDA-5, or TLR3 stimulation (75) (Fig. 6). DUBA interacted with TRAF3, and this interaction was increased after poly(I:C) stimulation of TLR3. TRAF3 undergoes K63-linked polyubiquitination, a process possibly mediated by its own RING (really interesting new gene) finger domain (138). DUBA specifically removed the K63-linked ubiquitin chains from TRAF3, resulting in the disruption of interaction between TRAF3 and the downstream kinases IKK-ε and TBK1 and blockade of IRF3 and IRF7 phosphorylation. Surprisingly, DUBA had no effect on processing of the NF-κB precursor NF-κB2/p100 into the active subunit p52, although TRAF3 is intimately involved in the noncanonical NF-κB pathway.

SIKE

Using a yeast two-hybrid screen with full-length IKK-ε as bait, SIKE (suppressor of IKK-ε) was identified as a new inhibitor of the IFN pathway (62). SIKE is an evolutionarily conserved, ubiquitously expressed protein whose orthologs can be found from *Xenopus* to humans. SIKE is a 207-amino-acid protein that contains two CC domains (Fig. 5). In the study by Huang et al., the CC domains of SIKE mediated homologous interaction with the C terminus of TBK1 and IKK-ε, which contain two potential CC domains. SIKE did not interact with RIP-1 nor with the classical IKKs IKK-α and -β. In nonstimulated cells, SIKE was constitutively associated with TBK1 but poly(I:C) or VSV treatment led to dissociation of this complex and induction of the IFN response. Forced expression of SIKE inhibited the IFN response following TLR3 or RIG-I engagement without having any effect on NF-κB activity. The mechanism of inhibition by SIKE of TBK1 or IKK-ε is based on sequestering the kinases and preventing their interaction with upstream signaling components such as TRIF and RIG-I.

THE VIRUS-ACTIVATED KINASES AND CANCER

IKK-ε

An emerging role for the atypical IKK-related kinases in cancer initiation and progression has been revealed. Analysis of breast cancer specimens with elevated activity of IKK or the serine/threonine kinase CK2 (133, 134) revealed that elevated NF-κB activity correlated with enhanced CK2 expression. IKK, an inducible protein capable of activating NF-κB, was also elevated. Furthermore, ectopic expression of CK2 was capable of inducing mRNA levels of IKK, and inhibition of CK2 via chemical reagents or ectopic expression of the dead-kinase form of CK2 reduced the levels of IKK. For the first time, these authors showed a link between enhanced CK2 and IKK-ε expression, aberrant NF-κB

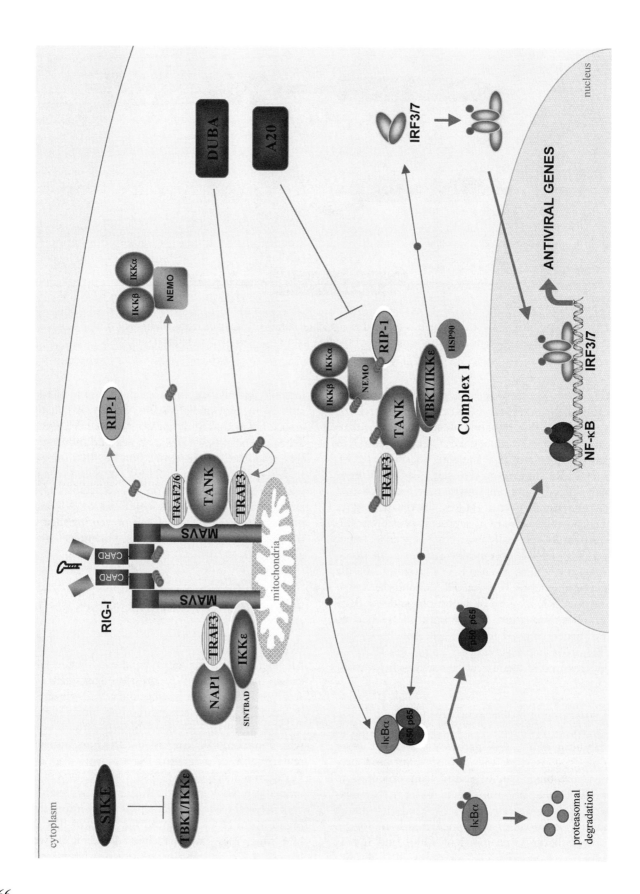

activation, and breast cancer. Subsequently, Boehm et al. demonstrated that the oncogene Ikbke, which encodes for IKK-ε, was involved in breast cancer formation using genome-wide expression assays, RNA interference, and overexpression assays (13). When IKK was overexpressed to a level comparable to breast cancer cell lines, transformation of immortalized human cells occurred, strongly supporting the idea that IKK-ε is an important player in the pathogenesis of breast cancer (13). Interestingly, the ability of IKK-ε to promote breast cancer progression was attributed to cRel phosphorylation—an NF-κB subunit previously shown to be directly phosphorylated by IKK-ε (49)—with active cRel accumulated in the nucleus of primary breast cancer specimens. This novel molecular mechanism has begun to shed light into why most breast cancers have elevated NF-κB activity and has made IKK-ε an attractive target for therapeutic intervention.

TBK1

While breast cancer cells depend on IKK-ε for their proliferation, other cancers such as prostate cancer appear to rely more on the expression of TBK1, indicating that constitutive activation of NF-κB in cancer cells may occur by distinct mechanisms depending on the tumor type. Another interesting link to cancer involves TBK1 and the exocyst pathway. Ral GTPase activation is an essential component of the Ras-induced transformation in a variety of human epithelial cells and is sufficient to drive tumorigenic phenotypes in some backgrounds (48, 131). In 2006, Chien et al. identified a novel signaling cascade connecting Ral activation to TBK1 (24). In this study, the exocyst complex subunit and Ral GTPase effector protein Sec5 bound directly to TBK1 and triggered TBK1 kinase activity in bronchial epithelial cells (24), indicating that TBK1 is an essential kinase for Ras-induced transformation. In addition to linking TBK1 to the exocyst pathway and Ras-induced transformation, this study demonstrated that the host de-

fense response requires the RalB/Sec5/TBK1 complex following dsRNA treatment or SeV infection.

TBK1 was also shown to be important for vascularization and tumor growth, based on a functional genome-wide phenotype screen to identify novel proteins that promoted proliferation of human umbilical vein endothelial cells as a functional readout for neovascularization (77). In addition to vascular endothelial growth factor and fibroblast growth factors-1 and -2—known to promote angiogenesis—the TRIF/TBK1/IRF3 pathway was also identified, thus linking this pathway to endothelial cell proliferation, seemingly through enhanced production of proangiogenic factors. Furthermore, TBK1 levels were markedly induced under hypoxic conditions and high levels were detected in many solid tumors, thus strengthening the link between TBK1 and tumor vascularization.

CONCLUDING REMARKS

The IKK-related kinases TBK1 and IKK-ε and their upstream adaptors have emerged as key players in coordinating innate antiviral immunity. The identification of TBK1 and IKK-ε as kinases directly responsible for the phosphorylation of IRF3 and IRF7 ended a long search for the virus-activated kinase but has sparked ongoing investigations into the molecular mechanisms regulating kinase activation and signaling mechanisms. The integration of available information suggests the formation of a large multisubunit complex following RIG-I engagement (Fig. 6)—a fusion between the proposed "inflammosome" and "IRFosome"—where the classical IKK-α and IKK-β kinases interact with TBK1/IKK-ε through the adaptors NEMO and TANK. The complex also contains TRAF3, MAVS, and RIG-I—and possibly TRAF6 and RIP-1. NAP1 and SINTBAD participate as additional or alternative scaffolding proteins. In addition to their well-characterized antiviral functions, TBK1 and IKK-ε also contribute to NF-κB activation

Figure 6 Global overview of the regulation of the RIG-I pathway. Schematic representation of signaling complexes formed after activation of the RIG-I pathway following viral infection. Binding of dsRNA by RIG-I leads to dimerization and exposure of CARD domains, allowing interaction with the CARD domain of MAVS. (Left) The mitochondria adaptor MAVS activates the kinases TBK1/IKK-ε through recruitment of TANK and TRAF3. (Right) The adaptors NAP1 and SINTBAD may form similar complexes with MAVS, the kinases, and TRAF3. Recruitment of TRAF6/2 to MAVS leads to NF-κB activation via RIP-1 and NEMO. Only IKK-ε is recruited directly to the mitochondria. TANK (or NAP1) may dissociate from MAVS and form a large cytoplasmic multimeric complex composed of the classical IKK kinases, NEMO, TBK1/IKK-ε, TRAF3, and RIP-1 (complex I), capable of activating both NF-κB and IRF3/7. K63 linked ubiquitination is shown as arrows with oval dots. Phosphorylation is shown as arrows with round dots. Negative regulators (A20, DUBA, and SIKE) are also shown.

via direct phosphorylation of the NF-κB subunits p65 and cRel. Future research will reveal how and when each component is recruited to the "RIGosome," a process that is likely regulated by a complex interplay between multiple posttranslational signals, including phosphorylation and ubiquitination.

We thank members of the Molecular Oncology Group, Lady Davis Institute, for helpful discussions. This research program is supported by grants from the Canadian Institutes of Health Research (CIHR), the National Cancer Institute of Canada, with funds from the Canadian Cancer Society, and the CAN-FAR (Canadian Foundation for AIDS Research). J.H. is supported by a CIHR Senior Investigator award.

References

1. Agashe, V. R., and F. U. Hartl. 2000. Roles of molecular chaperones in cytoplasmic protein folding. *Semin. Cell Dev. Biol.* **11:**15–25.
2. Akira, S. 2006. TLR signaling. *Curr. Top. Microbiol. Immunol.* **311:**1–16.
3. Andrejeva, J., K. S. Childs, D. F. Young, T. S. Carlos, N. Stock, S. Goodbourn, and R. E. Randall. 2004. The V proteins of paramyxoviruses bind the IFN-inducible RNA helicase, mda-5, and inhibit its activation of the IFN-beta promoter. *Proc. Natl. Acad. Sci. USA* **101:**17264–17269.
4. Au, W. C., P. A. Moore, D. W. LaFleur, B. Tombal, and P. M. Pitha. 1998. Characterization of the interferon regulatory factor-7 and its potential role in the transcription activation of interferon A genes. *J. Biol. Chem.* **273:**29210–29217.
5. Au, W. C., P. A. Moore, W. Lowther, Y. T. Juang, and P. M. Pitha. 1995. Identification of a member of the interferon regulatory factor family that binds to the interferon-stimulated response element and activates expression of interferon-induced genes. *Proc. Natl. Acad. Sci. USA* **92:**11657–11661.
6. Aupperle, K. R., Y. Yamanishi, B. L. Bennett, F. Mercurio, D. L. Boyle, and G. S. Firestein. 2001. Expression and regulation of inducible IkappaB kinase (IKK-i) in human fibroblast-like synoviocytes. *Cell. Immunol.* **214:**54–59.
7. Balachandran, S., E. Thomas, and G. N. Barber. 2004. A FADD-dependent innate immune mechanism in mammalian cells. *Nature* **432:**401–405.
8. Barnes, B. J., M. J. Kellum, A. E. Field, and P. M. Pitha. 2002. Multiple regulatory domains of IRF5 control activation, cellular localization, and induction of chemokines that mediate recruitment of T lymphocytes. *Mol. Cell. Biol.* **22:**5721–5740.
9. Barnes, B. J., M. J. Kellum, K. E. Pinder, J. A. Frisancho, and P. M. Pitha. 2003. Interferon regulatory factor 5, a novel mediator of cell cycle arrest and cell death. *Cancer Res.* **63:**6424–6431.
10. Barnes, B. J., P. A. Moore, and P. M. Pitha. 2001. Virus-specific activation of a novel interferon regulatory factor, IRF5, results in the induction of distinct interferon alpha genes. *J. Biol. Chem.* **276:**23382–23390.
11. Barnes, B. J., J. Richards, M. Mancl, S. Hanash, L. Beretta, and P. M. Pitha. 2004. Global and distinct targets of IRF5 and IRF7 during innate response to viral infection. *J. Biol. Chem.* **279:**45194–45207.
12. Barton, G. M., J. C. Kagan, and R. Medzhitov. 2006. Intracellular localization of Toll-like receptor 9 prevents recognition of self DNA but facilitates access to viral DNA. *Nat. Immunol.* **7:**49–56.
13. Boehm, J. S., J. J. Zhao, J. Yao, S. Y. Kim, R. Firestein, I. F. Dunn, S. K. Sjostrom, L. A. Garraway, S. Weremowicz, A. L. Richardson, H. Greulich, C. J. Stewart, L. A. Mulvey, R. R. Shen, L. Ambrogio, T. Hirozane-Kishikawa, D. E. Hill, M. Vidal, M. Meyerson, J. K. Grenier, G. Hinkle, D. E. Root, T. M. Roberts, E. S. Lander, K. Polyak, and W. C. Hahn. 2007. Integrative genomic approaches identify IKBKE as a breast cancer oncogene. *Cell* **129:**1065–1079.
14. Bonif, M., M. A. Meuwis, P. Close, V. Benoit, K. Heyninck, J. P. Chapelle, V. Bours, M. P. Merville, J. Piette, R. Beyaert, and A. Chariot. 2006. TNFalpha- and IKKbeta-mediated TANK/I-TRAF phosphorylation: implications for interaction with NEMO/IKKgamma and NF-kappaB activation. *Biochem. J.* **394:**593–603.
15. Bonnard, M., C. Mirtsos, S. Suzuki, K. Graham, J. Huang, M. Ng, A. Itie, A. Wakeham, A. Shahinian, W. J. Henzel, A. J. Elia, W. Shillinglaw, T. W. Mak, Z. Cao, and W. C. Yeh. 2000. Deficiency of T2K leads to apoptotic liver degeneration and impaired NF-kappaB-dependent gene transcription. *EMBO J.* **19:**4976–4985.
16. Boone, D. L., E. E. Turer, E. G. Lee, R. C. Ahmad, M. T. Wheeler, C. Tsui, P. Hurley, M. Chien, S. Chai, O. Hitotsumatsu, E. McNally, C. Pickart, and A. Ma. 2004. The ubiquitin-modifying enzyme A20 is required for termination of Toll-like receptor responses. *Nat. Immunol.* **5:**1052–1060.
17. Bukau, B., E. Deuerling, C. Pfund, and E. A. Craig. 2000. Getting newly synthesized proteins into shape. *Cell* **101:**119–122.
18. Bustamante, J., S. Boisson-Dupuis, E. Jouanguy, C. Picard, A. Puel, L. Abel, and J. L. Casanova. 2008. Novel primary immunodeficiencies revealed by the investigation of paediatric infectious diseases. *Curr. Opin. Immunol.* **20:**39–48.
19. Cadepond, F., G. Schweizer-Groyer, I. Segard-Maurel, N. Jibard, S. M. Hollenberg, V. Giguere, R. M. Evans, and E. E. Baulieu. 1991. Heat shock protein 90 as a critical factor in maintaining glucocorticosteroid receptor in a nonfunctional state. *J. Biol. Chem.* **266:**5834–5841.
20. Carpenter, S., and L. A. O'Neill. 2007. How important are Toll-like receptors for antimicrobial responses? *Cell. Microbiol.* **9:**1891–1901.
21. Chariot, A., A. Leonardi, J. Muller, M. Bonif, K. Brown, and U. Siebenlist. 2002. Association of the adaptor TANK with the IkappaB kinase (IKK) regulator NEMO connects IKK complexes with IKK epsilon and TBK1 kinases. *J. Biol. Chem.* **277:**37029–37036.
22. Cheng, G., and D. Baltimore. 1996. TANK, a co-inducer with TRAF2 of TNF- and CD 40L-mediated NF-kappaB activation. *Genes Dev.* **10:**963–973.
23. Cheng, T. F., S. Brzostek, O. Ando, S. Van Scoy, K. P. Kumar, and N. C. Reich. 2006. Differential activation of IFN regulatory factor (IRF)-3 and IRF5 transcription factors during viral infection. *J. Immunol.* **176:**7462–7470.

24. Chien, Y., S. Kim, R. Bumeister, Y. M. Loo, S. W. Kwon, C. L. Johnson, M. G. Balakireva, Y. Romeo, L. Kopelovich, M. Gale, Jr., C. Yeaman, J. H. Camonis, Y. Zhao, and M. A. White. 2006. RalB GTPase-mediated activation of the IkappaB family kinase TBK1 couples innate immune signaling to tumor cell survival. *Cell* 127:157–170.

25. Civas, A., P. Genin, P. Morin, R. Lin, and J. Hiscott. 2006. Promoter organization of the interferon-A genes differentially affects virus-induced expression and responsiveness to TBK1 and IKKepsilon. *J. Biol. Chem.* 281:4856–4866.

26. Colonna, M., G. Trinchieri, and Y. J. Liu. 2004. Plasmacytoid dendritic cells in immunity. *Nat. Immunol.* 5:1219–1226.

27. Courtois, G., and A. Smahi. 2006. NF-kappaB-related genetic diseases. *Cell Death Differ.* 13:843–851.

28. Diebold, S. S., T. Kaisho, H. Hemmi, S. Akira, and C. Reis e Sousa. 2004. Innate antiviral responses by means of TLR7-mediated recognition of single-stranded RNA. *Science* 303:1529–1531.

29. Elco, C. P., J. M. Guenther, B. R. Williams, and G. C. Sen. 2005. Analysis of genes induced by Sendai virus infection of mutant cell lines reveals essential roles of interferon regulatory factor 3, NF-kappaB, and interferon but not Toll-like receptor 3. *J. Virol.* 79:3920–3929.

30. Ely, K. R., R. Kodandapani, and S. Wu. 2007. Protein-protein interactions in TRAF3. *Adv. Exp. Med. Biol.* 597:114–121.

31. Evans, P. C., H. Ovaa, M. Hamon, P. J. Kilshaw, S. Hamm, S. Bauer, H. L. Ploegh, and T. S. Smith. 2004. Zinc-finger protein A20, a regulator of inflammation and cell survival, has de-ubiquitinating activity. *Biochem. J.* 378:727–734.

32. Feldman, D. E., and J. Frydman. 2000. Protein folding in vivo: the importance of molecular chaperones. *Curr. Opin. Struct. Biol.* 10:26–33.

33. Filipe-Santos, O., J. Bustamante, M. H. Haverkamp, E. Vinolo, C. L. Ku, A. Puel, D. M. Frucht, K. Christel, H. von Bernuth, E. Jouanguy, J. Feinberg, A. Durandy, B. Senechal, A. Chapgier, G. Vogt, L. de Beaucoudrey, C. Fieschi, C. Picard, M. Garfa, J. Chemli, M. Bejaoui, M. N. Tsolia, N. Kutukculer, A. Plebani, L. Notarangelo, C. Bodemer, F. Geissmann, A. Israel, M. Veron, M. Knackstedt, R. Barbouche, L. Abel, K. Magdorf, D. Gendrel, F. Agou, S. M. Holland, and J. L. Casanova. 2006. X-linked susceptibility to mycobacteria is caused by mutations in NEMO impairing CD40-dependent IL-12 production. *J. Exp. Med.* 203:1745–1759.

34. Fitzgerald, K. A., S. M. McWhirter, K. L. Faia, D. C. Rowe, E. Latz, D. T. Golenbock, A. J. Coyle, S. M. Liao, and T. Maniatis. 2003. IKKepsilon and TBK1 are essential components of the IRF3 signaling pathway. *Nat. Immunol.* 4:491–496.

35. Fitzgerald, K. A., D. C. Rowe, B. J. Barnes, D. R. Caffrey, A. Visintin, E. Latz, B. Monks, P. M. Pitha, and D. T. Golenbock. 2003. LPS-TLR4 signaling to IRF3/7 and NF-kappaB involves the Toll adapters TRAM and TRIF. *J. Exp. Med.* 198:1043–1055.

36. Fontan, E., F. Traincard, S. G. Levy, S. Yamaoka, M. Veron, and F. Agou. 2007. NEMO oligomerization in the dynamic assembly of the IkappaB kinase core complex. *FEBS J.* 274:2540–2451.

37. Fujita, F., Y. Taniguchi, T. Kato, Y. Narita, A. Furuya, T. Ogawa, H. Sakurai, T. Joh, M. Itoh, M. Delhase, M. Karin, and M. Nakanishi. 2003. Identification of NAP1, a regulatory subunit of IkappaB kinase-related kinases that potentiates NF-kappaB signaling. *Mol. Cell. Biol.* 23:7780–7793.

38. Funami, K., M. Sasai, Y. Ohba, H. Oshiumi, T. Seya, and M. Matsumoto. 2007. Spatiotemporal mobilization of Toll/IL-1 receptor domain-containing adaptor molecule-1 in response to dsRNA. *J. Immunol.* 179:6867–6872.

39. Gatot, J. S., R. Gioia, T. L. Chau, F. Patrascu, M. Warnier, P. Close, J. P. Chapelle, E. Muraille, K. Brown, U. Siebenlist, J. Piette, E. Dejardin, and A. Chariot. 2007. Lipopolysaccharide-mediated interferon regulatory factor activation involves TBK1-IKKepsilon-dependent Lys(63)-linked polyubiquitination and phosphorylation of TANK/I-TRAF. *J. Biol. Chem.* 282:31131–31146.

40. Geiss, G., G. Jin, J. Guo, R. Bumgarner, M. G. Katze, and G. C. Sen. 2001. A comprehensive view of regulation of gene expression by double-stranded RNA-mediated cell signaling. *J. Biol. Chem.* 276:30178–30182.

41. Genin, P., M. Algarte, P. Roof, R. Lin, and J. Hiscott. 2000. Regulation of RANTES chemokine gene expression requires cooperativity between NF-kappaB and IFN-regulatory factor transcription factors. *J. Immunol.* 164:5352–5361.

42. Giguere, V., S. M. Hollenberg, M. G. Rosenfeld, and R. M. Evans. 1986. Functional domains of the human glucocorticoid receptor. *Cell* 46:645–652.

43. Gilmore, T. D. 2006. Introduction to NF-kappaB: players, pathways, perspectives. *Oncogene* 25:6680–6684.

44. Gitlin, L., W. Barchet, S. Gilfillan, M. Cella, B. Beutler, R. A. Flavell, M. S. Diamond, and M. Colonna. 2006. Essential role of mda-5 in type I IFN responses to polyriboinosinic:polyribocytidylic acid and encephalomyocarditis picornavirus. *Proc. Natl. Acad. Sci. USA* 103:8459–8464.

45. Grandvaux, N., M. J. Servant, B. tenOever, G. C. Sen, S. Balachandran, G. N. Barber, R. Lin, and J. Hiscott. 2002. Transcriptional profiling of interferon regulatory factor 3 target genes: direct involvement in the regulation of interferon-stimulated genes. *J. Virol.* 76:5532–5539.

46. Guo, B., and G. Cheng. 2007. Modulation of the interferon antiviral response by the TBK1/IKKi adaptor protein TANK. *J. Biol. Chem.* 282:11817–11826.

47. Hacker, H., V. Redecke, B. Blagoev, I. Kratchmarova, L. C. Hsu, G. G. Wang, M. P. Kamps, E. Raz, H. Wagner, G. Hacker, M. Mann, and M. Karin. 2006. Specificity in Toll-like receptor signalling through distinct effector functions of TRAF3 and TRAF6. *Nature* 439:204–207.

48. Hamad, N. M., J. H. Elconin, A. E. Karnoub, W. Bai, J. N. Rich, R. T. Abraham, C. J. Der, and C. M. Counter. 2002. Distinct requirements for Ras oncogenesis in human versus mouse cells. *Genes. Dev.* 16:2045–2057.

49. Harris, J., S. Oliere, S. Sharma, Q. Sun, R. Lin, J. Hiscott, and N. Grandvaux. 2006. Nuclear accumulation of cRel following C-terminal phosphorylation by TBK1/IKK epsilon. *J. Immunol.* 177:2527–2535.

50. Hashimoto, C., K. L. Hudson, and K. V. Anderson. 1988. The *Toll* gene of Drosophila, required for dorsal-ventral embryonic polarity, appears to encode a transmembrane protein. *Cell* 52:269–279.

51. Heil, F., H. Hemmi, H. Hochrein, F. Ampenberger, C. Kirschning, S. Akira, G. Lipford, H. Wagner, and S. Bauer. 2004. Species-specific recognition of single-stranded RNA via Toll-like receptor 7 and 8. *Science* 303:1526–1529.

52. Hemmi, H., O. Takeuchi, S. Sato, M. Yamamoto, T. Kaisho, H. Sanjo, T. Kawai, K. Hoshino, K. Takeda, and S. Akira. 2004. The roles of two IkappaB kinase-related kinases in lipopolysaccharide and double stranded RNA signaling and viral infection. *J. Exp. Med.* 199:1641–1650.

53. Heyninck, K., and R. Beyaert. 1999. The cytokine-inducible zinc finger protein A20 inhibits IL-1-induced NF-kappaB activation at the level of TRAF6. *FEBS Lett.* 442:147–150.

54. Hiscott, J. 2007. Convergence of the NF-kappaB and IRF pathways in the regulation of the innate antiviral response. *Cytokine Growth Factor Rev.* 18:483–490.

55. Hiscott, J., D. Alper, L. Cohen, J. F. Leblanc, L. Sportza, A. Wong, and S. Xanthoudakis. 1989. Induction of human interferon gene expression is associated with a nuclear factor that interacts with the NF-kappaB site of the human immunodeficiency virus enhancer. *J. Virol.* 63:2557–2566.

56. Hiscott, J., J. Lacoste, and R. Lin. 2006. Recruitment of an interferon molecular signaling complex to the mitochondrial membrane: disruption by hepatitis C virus NS3–4A protease. *Biochem. Pharmacol.* 72:1477–1484.

57. Hiscott, J., P. Pitha, P. Genin, H. Nguyen, C. Heylbroeck, Y. Mamane, M. Algarte, and R. Lin. 1999. Triggering the interferon response: the role of IRF3 transcription factor. *J. Interferon Cytokine Res.* 19:1–13.

58. Honda, K., A. Takaoka, and T. Taniguchi. 2006. Type I interferon gene induction by the interferon regulatory factor family of transcription factors. *Immunity* 25:349–360.

59. Honda, K., and T. Taniguchi. 2006. IRFs: master regulators of signalling by Toll-like receptors and cytosolic pattern-recognition receptors. *Nat. Rev. Immunol.* 6:644–658.

60. Honda, K., H. Yanai, H. Negishi, M. Asagiri, M. Sato, T. Mizutani, N. Shimada, Y. Ohba, A. Takaoka, N. Yoshida, and T. Taniguchi. 2005. IRF7 is the master regulator of type-I interferon-dependent immune responses. *Nature* 434:772–777.

61. Hornung, V., J. Ellegast, S. Kim, K. Brzozka, A. Jung, H. Kato, H. Poeck, S. Akira, K. K. Conzelmann, M. Schlee, S. Endres, and G. Hartmann. 2006. 5′-Triphosphate RNA is the ligand for RIG-I. *Science* 314:994–997.

62. Huang, J., T. Liu, L. G. Xu, D. Chen, Z. Zhai, and H. B. Shu. 2005. SIKE is an IKK epsilon/TBK1-associated suppressor of TLR3- and virus-triggered IRF3 activation pathways. *EMBO J.* 24:4018–4028.

63. Ikeda, F., C. M. Hecker, A. Rozenknop, R. D. Nordmeier, V. Rogov, K. Hofmann, S. Akira, V. Dotsch, and I. Dikic. 2007. Involvement of the ubiquitin-like domain of TBK1/IKK-i kinases in regulation of IFN-inducible genes. *EMBO J.* 26:3451–3462.

64. Ishii, K. J., and S. Akira. 2006. Innate immune recognition of, and regulation by, DNA. *Trends Immunol.* 27:525–532.

65. Izaguirre, A., B. J. Barnes, S. Amrute, W. S. Yeow, N. Megjugorac, J. Dai, D. Feng, E. Chung, P. M. Pitha, and P. Fitzgerald-Bocarsly. 2003. Comparative analysis of IRF and IFN-alpha expression in human plasmacytoid and monocyte-derived dendritic cells. *J. Leukoc. Biol.* 74:1125–1138.

66. Jakob, U., H. Lilie, I. Meyer, and J. Buchner. 1995. Transient interaction of Hsp90 with early unfolding intermediates of citrate synthase. Implications for heat shock in vivo. *J. Biol. Chem.* 270:7288–7294.

67. Juang, Y. T., W. Lowther, M. Kellum, W. C. Au, R. Lin, J. Hiscott, and P. M. Pitha. 1998. Primary activation of interferon A and interferon B gene transcription by interferon regulatory factor 3. *Proc. Natl. Acad. Sci. USA* 95:9837–9842.

68. Karin, M. 2006. Nuclear factor-kappaB in cancer development and progression. *Nature* 441:431–436.

69. Karin, M., T. Lawrence, and V. Nizet. 2006. Innate immunity gone awry: linking microbial infections to chronic inflammation and cancer. *Cell* 124:823–835.

70. Kato, H., S. Sato, M. Yoneyama, M. Yamamoto, S. Uematsu, K. Matsui, T. Tsujimura, K. Takeda, T. Fujita, O. Takeuchi, and S. Akira. 2005. Cell type-specific involvement of RIG-I in antiviral response. *Immunity* 23:19–28.

71. Kato, H., O. Takeuchi, S. Sato, M. Yoneyama, M. Yamamoto, K. Matsui, S. Uematsu, A. Jung, T. Kawai, K. J. Ishii, O. Yamaguchi, K. Otsu, T. Tsujimura, C. S. Koh, C. Reis e Sousa, Y. Matsuura, T. Fujita, and S. Akira. 2006. Differential roles of MDA-5 and RIG-I helicases in the recognition of RNA viruses. *Nature* 441:101–105.

72. Kawai, T., and S. Akira. 2006. Innate immune recognition of viral infection. *Nat. Immunol.* 7:131–137.

73. Kawai, T., and S. Akira. 2007. Signaling to NF-kappaB by Toll-like receptors. *Trends Mol. Med.* 13:460–469.

74. Kawai, T., K. Takahashi, S. Sato, C. Coban, H. Kumar, H. Kato, K. J. Ishii, O. Takeuchi, and S. Akira. 2005. IPS-1, an adaptor triggering RIG-I- and MDA-5-mediated type I interferon induction. *Nat. Immunol.* 6:981–988.

75. Kayagaki, N., Q. Phung, S. Chan, R. Chaudhari, C. Quan, K. M. O'Rourke, M. Eby, E. Pietras, G. Cheng, J. F. Bazan, Z. Zhang, D. Arnott, and V. M. Dixit. 2007. DUBA: a deubiquitinase that regulates type I interferon production. *Science* 318:1628–1632.

76. Kishore, N., Q. K. Huynh, S. Mathialagan, T. Hall, S. Rouw, D. Creely, G. Lange, J. Caroll, B. Reitz, A. Donnelly, H. Boddupalli, R. G. Combs, K. Kretzmer, and C. S. Tripp. 2002. IKK-i and TBK1 are enzymatically distinct from the homologous enzyme IKK-2: comparative analysis of recombinant human IKK-i, TBK1, and IKK-2. *J. Biol. Chem.* 277:13840–13847.

77. Korherr, C., H. Gille, R. Schafer, K. Koenig-Hoffmann, J. Dixelius, K. A. Egland, I. Pastan, and U. Brinkmann. 2006. Identification of proangiogenic genes and pathways by high-throughput functional genomics: TBK1 and the IRF3 pathway. *Proc. Natl. Acad. Sci. USA* 103:4240–4245.

78. Kravchenko, V. V., J. C. Mathison, K. Schwamborn, F. Mercurio, and R. J. Ulevitch. 2003. IKKi/IKKepsilon plays a key role in integrating signals induced by proinflammatory stimuli. *J. Biol. Chem.* 278:26612–26619.

79. Krug, A., A. R. French, W. Barchet, J. A. Fischer, A. Dzionek, J. T. Pingel, M. M. Orihuela, S. Akira, W. M. Yokoyama, and M. Colonna. 2004. TLR9-dependent

recognition of MCMV by IPC and DC generates coordinated cytokine responses that activate antiviral NK cell function. *Immunity* **21**:107–119.

80. Kumar, H., T. Kawai, H. Kato, S. Sato, K. Takahashi, C. Coban, M. Yamamoto, S. Uematsu, K. J. Ishii, O. Takeuchi, and S. Akira. 2006. Essential role of IPS-1 in innate immune responses against RNA viruses. *J. Exp. Med.* **203**:1795–1803.

81. Kumar, K. P., K. M. McBride, B. K. Weaver, C. Dingwall, and N. C. Reich. 2000. Regulated nuclear-cytoplasmic localization of interferon regulatory factor 3, a subunit of double-stranded RNA-activated factor 1. *Mol. Cell. Biol.* **20**:4159–4168.

82. Lallemand, C., B. Blanchard, M. Palmieri, P. Lebon, E. May, and M. G. Tovey. 2007. Single-stranded RNA viruses inactivate the transcriptional activity of p53 but induce NOXA-dependent apoptosis via post-translational modifications of IRF-1, IRF3 and CREB. *Oncogene* **26**:328–338.

83. Lamothe, B., A. Besse, A. D. Campos, W. K. Webster, H. Wu, and B. G. Darnay. 2007. Site-specific Lys-63-linked tumor necrosis factor receptor-associated factor 6 auto-ubiquitination is a critical determinant of IkappaB kinase activation. *J. Biol. Chem.* **282**:4102–4112.

84. [Reference deleted.]

85. Latz, E., A. Verma, A. Visintin, M. Gong, C. M. Sirois, D. C. Klein, B. G. Monks, C. J. McKnight, M. S. Lamphier, W. P. Duprex, T. Espevik, and D. T. Golenbock. 2007. Ligand-induced conformational changes allosterically activate Toll-like receptor 9. *Nat. Immunol.* **8**:772–779. (Corrigendum, **8**:1266.)

86. Lemaitre, B. 2004. The road to Toll. *Nat. Rev. Immunol.* **4**:521–527.

87. Lemaitre, B., E. Nicolas, L. Michaut, J. M. Reichhart, and J. A. Hoffmann. 1996. The dorsoventral regulatory gene cassette spatzle/Toll/cactus controls the potent antifungal response in Drosophila adults. *Cell* **86**:973–983.

88. Lenardo, M. J., C. M. Fan, T. Maniatis, and D. Baltimore. 1989. The involvement of NF-kappaB in beta-interferon gene regulation reveals its role as widely inducible mediator of signal transduction. *Cell* **57**:287–294.

89. Lin, R., P. Genin, Y. Mamane, and J. Hiscott. 2000. Selective DNA binding and association with the CREB binding protein coactivator contribute to differential activation of alpha/beta interferon genes by interferon regulatory factors 3 and 7. *Mol. Cell. Biol.* **20**:6342–6353.

90. Lin, R., J. Lacoste, P. Nakhaei, Q. Sun, L. Yang, S. Paz, P. Wilkinson, I. Julkunen, D. Vitour, E. Meurs, and J. Hiscott. 2006. Dissociation of a MAVS/IPS-1/VISA/Cardif-IKKepsilon molecular complex from the mitochondrial outer membrane by hepatitis C virus NS3–4A proteolytic cleavage. *J. Virol.* **80**:6072–6083.

91. Lin, R., Y. Mamane, and J. Hiscott. 2000. Multiple regulatory domains control IRF7 activity in response to virus infection. *J. Biol. Chem.* **275**:34320–34327.

92. Lin, R., Y. Mamane, and J. Hiscott. 1999. Structural and functional analysis of interferon regulatory factor 3: localization of the transactivation and autoinhibitory domains. *Mol. Cell. Biol.* **19**:2465–2474.

93. Lin, R., L. Yang, M. Arguello, C. Penafuerte, and J. Hiscott. 2005. A CRM1-dependent nuclear export pathway is involved in the regulation of IRF5 subcellular localization. *J. Biol. Chem.* **280**:3088–3095.

94. Lin, R., L. Yang, P. Nakhaei, Q. Sun, E. Sharif-Askari, I. Julkunen, and J. Hiscott. 2006. Negative regulation of the retinoic acid-inducible gene I-induced antiviral state by the ubiquitin-editing protein A20. *J. Biol. Chem.* **281**:2095–2103.

95. Loo, Y. M., D. M. Owen, K. Li, A. K. Erickson, C. L. Johnson, P. M. Fish, D. S. Carney, T. Wang, H. Ishida, M. Yoneyama, T. Fujita, T. Saito, W. M. Lee, C. H. Hagedorn, D. T. Lau, S. A. Weinman, S. M. Lemon, and M. Gale, Jr. 2006. Viral and therapeutic control of IFN-beta promoter stimulator 1 during hepatitis C virus infection. *Proc. Natl. Acad. Sci. USA* **103**:6001–6006.

96. Lu, R., W. C. Au, W. S. Yeow, N. Hageman, and P. M. Pitha. 2000. Regulation of the promoter activity of interferon regulatory factor-7 gene. Activation by interferon and silencing by hypermethylation. *J. Biol. Chem.* **275**:31805–31812.

97. Lu, R., P. A. Moore, and P. M. Pitha. 2002. Stimulation of IRF7 gene expression by tumor necrosis factor alpha: requirement for NFkappaB transcription factor and gene accessibility. *J. Biol. Chem.* **277**:16592–16598.

98. Mamane, Y., C. Heylbroeck, P. Genin, M. Algarte, M. J. Servant, C. LePage, C. DeLuca, H. Kwon, R. Lin, and J. Hiscott. 1999. Interferon regulatory factors: the next generation. *Gene* **237**:1–14.

99. Mancl, M. E., G. Hu, N. Sangster-Guity, S. L. Olshalsky, K. Hoops, P. Fitzgerald-Bocarsly, P. M. Pitha, K. Pinder, and B. J. Barnes. 2005. Two discrete promoters regulate the alternatively spliced human interferon regulatory factor-5 isoforms. Multiple isoforms with distinct cell type-specific expression, localization, regulation, and function. *J. Biol. Chem.* **280**:21078–21090.

100. Marie, I., J. E. Durbin, and D. E. Levy. 1998. Differential viral induction of distinct interferon-alpha genes by positive feedback through interferon regulatory factor-7. *EMBO J.* **17**:6660–6669.

101. Marie, I., E. Smith, A. Prakash, and D. E. Levy. 2000. Phosphorylation-induced dimerization of interferon regulatory factor 7 unmasks DNA binding and a bipartite transactivation domain. *Mol. Cell. Biol.* **20**:8803–8814.

102. Matsui, K., Y. Kumagai, H. Kato, S. Sato, T. Kawagoe, S. Uematsu, O. Takeuchi, and S. Akira. 2006. Cutting edge: role of TANK-binding kinase 1 and inducible IkappaB kinase in IFN responses against viruses in innate immune cells. *J. Immunol.* **177**:5785–5789.

103. McCoy, C. E., and L. A. O'Neill. 2008. The role of Toll-like receptors in macrophages. *Front. Biosci.* **13**:62–70.

104. McWhirter, S. M., K. A. Fitzgerald, J. Rosains, D. C. Rowe, D. T. Golenbock, and T. Maniatis. 2004. IFN-regulatory factor 3-dependent gene expression is defective in Tbk1-deficient mouse embryonic fibroblasts. *Proc. Natl. Acad. Sci. USA* **101**:233–238.

105. Merika, M., and D. Thanos. 2001. Enhanceosomes. *Curr. Opin. Genet. Dev.* **11**:205–208.

106. Meylan, E., J. Curran, K. Hofmann, D. Moradpour, M. Binder, R. Bartenschlager, and J. Tschopp. 2005. Cardif is an adaptor protein in the RIG-I antiviral pathway and is targeted by hepatitis C virus. *Nature* **437**:1167–1172.

107. Morin, P., J. Braganca, M. T. Bandu, R. Lin, J. Hiscott, J. Doly, and A. Civas. 2002. Preferential binding sites

for interferon regulatory factors 3 and 7 involved in interferon-A gene transcription. *J. Mol. Biol.* **316:** 1009–1022.

108. **Nabel, G., and D. Baltimore.** 1987. An inducible transcription factor activates expression of human immunodeficiency virus in T cells. *Nature* **326:**711–713.

109. **Nakaya, T., M. Sato, N. Hata, M. Asagiri, H. Suemori, S. Noguchi, N. Tanaka, and T. Taniguchi.** 2001. Gene induction pathways mediated by distinct IRFs during viral infection. *Biochem. Biophys. Res. Commun.* **283:** 1150–1156.

110. **Nelson, D. L.** 2006. NEMO, NFκB signaling and incontinentia pigmenti. *Curr. Opin. Genet. Dev.* **16:**282–288.

111. **Nomura, F., T. Kawai, K. Nakanishi, and S. Akira.** 2000. NF-kappaB activation through IKK-i-dependent I-TRAF/TANK phosphorylation. *Genes Cells* **5:**191–202.

112. **O'Neill, L. A., and A. G. Bowie.** 2007. The family of five: TIR-domain-containing adaptors in Toll-like receptor signalling. *Nat. Rev. Immunol.* **7:**353–364.

113. **Onomoto, K., M. Yoneyama, and T. Fujita.** 2007. Regulation of antiviral innate immune responses by RIG-I family of RNA helicases. *Curr. Top. Microbiol. Immunol.* **316:**193–205.

114. **Osterlund, P. I., T. E. Pietila, V. Veckman, S. V. Kotenko, and I. Julkunen.** 2007. IFN regulatory factor family members differentially regulate the expression of type III IFN (IFN-lambda) genes. *J. Immunol.* **179:**3434–3442.

115. **Ozato, K., H. Tsujimura, and T. Tamura.** 2002. Toll-like receptor signaling and regulation of cytokine gene expression in the immune system. *Biotechniques* **2002**(Suppl.): 66–68, 70, 72 passim.

116. **Panne, D., T. Maniatis, and S. C. Harrison.** 2004. Crystal structure of ATF-2/c-Jun and IRF3 bound to the interferon-beta enhancer. *EMBO J.* **23:**4384–4393.

117. **Panne, D., S. M. McWhirter, T. Maniatis, and S. C. Harrison.** 2007. Interferon regulatory factor 3 is regulated by a dual phosphorylation-dependent switch. *J. Biol. Chem.* **282:**22816–22822.

118. **Paz, S., Q. Sun, P. Nakhaei, R. Romieu-Mourez, D. Goubau, I. Julkunen, R. Lin, and J. Hiscott.** 2006. Induction of IRF3 and IRF7 phosphorylation following activation of the RIG-I pathway. *Cell. Mol. Biol. (Noisy-le-grand)* **52:**17–28.

119. **Pernis, A. B.** 2002. The role of IRF-4 in B and T cell activation and differentiation. *J. Interferon Cytokine Res.* **22:**111–120.

120. **Perry, A. K., E. K. Chow, J. B. Goodnough, W. C. Yeh, and G. Cheng.** 2004. Differential requirement for TANK-binding kinase-1 in type I interferon responses to Toll-like receptor activation and viral infection. *J. Exp. Med.* **199:**1651–1658.

121. **Peters, R. T., S. M. Liao, and T. Maniatis.** 2000. IKKepsilon is part of a novel PMA-inducible IkappaB kinase complex. *Mol. Cell* **5:**513–522.

122. **Peters, R. T., and T. Maniatis.** 2001. A new family of IKK-related kinases may function as IkappaB kinase kinases. *Biochim. Biophys. Acta* **1471:**M57–M62.

123. **Pichlmair, A., O. Schulz, C. P. Tan, T. I. Naslund, P. Liljestrom, F. Weber, and C. Reis e Sousa.** 2006. RIG-I-mediated antiviral responses to single-stranded RNA bearing 5'-phosphates. *Science* **314:**997–1001.

124. **Pomerantz, J. L., and D. Baltimore.** 1999. NF-kappaB activation by a signaling complex containing TRAF2, TANK and TBK1, a novel IKK-related kinase. *EMBO J.* **18:**6694–6704.

125. **Puel, A., C. Picard, C. L. Ku, A. Smahi, and J. L. Casanova.** 2004. Inherited disorders of NF-kappaB-mediated immunity in man. *Curr. Opin. Immunol.* **16:**34–41.

126. **Qin, B. Y., C. Liu, S. S. Lam, H. Srinath, R. Delston, J. J. Correia, R. Derynck, and K. Lin.** 2003. Crystal structure of IRF3 reveals mechanism of autoinhibition and virus-induced phosphoactivation. *Nat. Struct. Biol.* **10:**913–921.

127. **Qin, B. Y., C. Liu, H. Srinath, S. S. Lam, J. J. Correia, R. Derynck, and K. Lin.** 2005. Crystal structure of IRF3 in complex with CBP. *Structure* (Camb.) **13:**1269–1277.

128. **Qing, J., C. Liu, L. Choy, R. Y. Wu, J. S. Pagano, and R. Derynck.** 2004. Transforming growth factor beta/Smad3 signaling regulates IRF7 function and transcriptional activation of the beta interferon promoter. *Mol. Cell. Biol.* **24:**1411–1425.

129. **Radons, J., S. Dove, D. Neumann, R. Altmann, A. Botzki, M. U. Martin, and W. Falk.** 2003. The interleukin 1 (IL-1) receptor accessory protein Toll/IL-1 receptor domain: analysis of putative interaction sites by in vitro mutagenesis and molecular modeling. *J. Biol. Chem.* **278:**49145–49153.

130. **Radons, J., S. Gabler, H. Wesche, C. Korherr, R. Hofmeister, and W. Falk.** 2002. Identification of essential regions in the cytoplasmic tail of interleukin-1 receptor accessory protein critical for interleukin-1 signaling. *J. Biol. Chem.* **277:**16456–16463.

131. **Rangarajan, A., S. J. Hong, A. Gifford, and R. A. Weinberg.** 2004. Species- and cell type-specific requirements for cellular transformation. *Cancer Cell* **6:** 171–183.

132. **Roach, J. C., G. Glusman, L. Rowen, A. Kaur, M. K. Purcell, K. D. Smith, L. E. Hood, and A. Aderem.** 2005. The evolution of vertebrate Toll-like receptors. *Proc. Natl. Acad. Sci. USA* **102:**9577–9582.

133. **Romieu-Mourez, R., E. Landesman-Bollag, D. C. Seldin, and G. E. Sonenshein.** 2002. Protein kinase CK2 promotes aberrant activation of nuclear factor-kappaB, transformed phenotype, and survival of breast cancer cells. *Cancer Res.* **62:**6770–6778.

134. **Romieu-Mourez, R., E. Landesman-Bollag, D. C. Seldin, A. M. Traish, F. Mercurio, and G. E. Sonenshein.** 2001. Roles of IKK kinases and protein kinase CK2 in activation of nuclear factor-kappaB in breast cancer. *Cancer Res.* **61:**3810–3818.

135. **Romieu-Mourez, R., M. Solis, A. Nardin, D. Goubau, V. Baron-Bodo, R. Lin, B. Massie, M. Salcedo, and J. Hiscott.** 2006. Distinct roles for IFN regulatory factor (IRF)-3 and IRF7 in the activation of antitumor properties of human macrophages. *Cancer Res.* **66:** 10576–10585.

136. **Rothe, M., J. Xiong, H. B. Shu, K. Williamson, A. Goddard, and D. V. Goeddel.** 1996. I-TRAF is a novel TRAF-interacting protein that regulates TRAF-mediated signal transduction. *Proc. Natl. Acad. Sci. USA* **93:** 8241–8246.

137. Ryzhakov, G., and F. Randow. 2007. SINTBAD, a novel component of innate antiviral immunity, shares a TBK1-binding domain with NAP1 and TANK. *EMBO J.* 26:3180–3190.

138. Saha, S. K., E. M. Pietras, J. Q. He, J. R. Kang, S. Y. Liu, G. Oganesyan, A. Shahangian, B. Zarnegar, T. L. Shiba, Y. Wang, and G. Cheng. 2006. Regulation of antiviral responses by a direct and specific interaction between TRAF3 and Cardif. *EMBO J.* 25:3257–3263.

139. Saito, T., R. Hirai, Y. M. Loo, D. Owen, C. L. Johnson, S. C. Sinha, S. Akira, T. Fujita, and M. Gale, Jr. 2007. Regulation of innate antiviral defenses through a shared repressor domain in RIG-I and LGP2. *Proc. Natl. Acad. Sci. USA* 104:582–587.

140. Saitoh, T., M. Yamamoto, M. Miyagishi, K. Taira, M. Nakanishi, T. Fujita, S. Akira, N. Yamamoto, and S. Yamaoka. 2005. A20 is a negative regulator of IFN regulatory factor 3 signaling. *J. Immunol.* 174:1507–1512.

141. Sasai, M., H. Oshiumi, M. Matsumoto, N. Inoue, F. Fujita, M. Nakanishi, and T. Seya. 2005. Cutting edge: NF-kappaB-activating kinase-associated protein 1 participates in TLR3/Toll-IL-1 homology domain-containing adapter molecule-1-mediated IFN regulatory factor 3 activation. *J. Immunol.* 174:27–30.

142. Sasai, M., M. Shingai, K. Funami, M. Yoneyama, T. Fujita, M. Matsumoto, and T. Seya. 2006. NAK-associated protein 1 participates in both the TLR3 and the cytoplasmic pathways in type I IFN induction. *J. Immunol.* 177: 8676–8683.

143. Sato, M., N. Hata, M. Asagiri, T. Nakaya, T. Taniguchi, and N. Tanaka. 1998. Positive feedback regulation of type I IFN genes by the IFN-inducible transcription factor IRF7. *FEBS Lett.* 441:106–110.

144. Sato, M., H. Suemori, N. Hata, M. Asagiri, K. Ogasawara, K. Nakao, T. Nakaya, M. Katsuki, S. Noguchi, N. Tanaka, and T. Taniguchi. 2000. Distinct and essential roles of transcription factors IRF-3 and IRF-7 in response to viruses for IFN-alpha/beta gene induction. *Immunity* 13:539–548.

145. Schoenemeyer, A., B. J. Barnes, M. E. Mancl, E. Latz, N. Goutagny, P. M. Pitha, K. A. Fitzgerald, and D. T. Golenbock. 2005. The interferon regulatory factor, IRF5, is a central mediator of Toll-like receptor 7 signaling. *J. Biol. Chem.* 280:17005–17012.

146. Sebban-Benin, H., A. Pescatore, F. Fusco, V. Pascuale, J. Gautheron, S. Yamaoka, A. Moncla, M. V. Ursini, and G. Courtois. 2007. Identification of TRAF6-dependent NEMO polyubiquitination sites through analysis of a new NEMO mutation causing incontinentia pigmenti. *Hum. Mol. Genet.* 16:2805–2815.

147. Sen, R., and D. Baltimore. 1986. Inducibility of kappa immunoglobulin enhancer-binding protein NF-kappaB by a posttranslational mechanism. *Cell* 47:921–928.

148. Sen, R., and D. Baltimore. 1986. Multiple nuclear factors interact with the immunoglobulin enhancer sequences. *Cell* 46:705–716.

149. Servant, M. J., N. Grandvaux, B. R. tenOever, D. Duguay, R. Lin, and J. Hiscott. 2003. Identification of the minimal phosphoacceptor site required for in vivo activation of interferon regulatory factor 3 in response to virus and double-stranded RNA. *J. Biol. Chem.* 278:9441–9447.

150. Seth, R. B., L. Sun, C. K. Ea, and Z. J. Chen. 2005. Identification and characterization of MAVS, a mitochondrial antiviral signaling protein that activates NF-kappaB and IRF 3. *Cell* 122:669–682.

151. Sharma, S., B. R. tenOever, N. Grandvaux, G. P. Zhou, R. Lin, and J. Hiscott. 2003. Triggering the interferon antiviral response through an IKK-related pathway. *Science* 300:1148–1151.

152. Shimada, T., T. Kawai, K. Takeda, M. Matsumoto, J. Inoue, Y. Tatsumi, A. Kanamaru, and S. Akira. 1999. IKK-i, a novel lipopolysaccharide-inducible kinase that is related to IkappaB kinases. *Int. Immunol.* 11:1357–1362.

153. Sun, Q., L. Sun, H. H. Liu, X. Chen, R. B. Seth, J. Forman, and Z. J. Chen. 2006. The specific and essential role of MAVS in antiviral innate immune responses. *Immunity* 24:633–642.

154. Takahasi, K., N. N. Suzuki, M. Horiuchi, M. Mori, W. Suhara, Y. Okabe, Y. Fukuhara, H. Terasawa, S. Akira, T. Fujita, and F. Inagaki. 2003. X-ray crystal structure of IRF3 and its functional implications. *Nat. Struct. Biol.* 10:922–927.

155. Takaoka, A., H. Yanai, S. Kondo, G. Duncan, H. Negishi, T. Mizutani, S. Kano, K. Honda, Y. Ohba, T. W. Mak, and T. Taniguchi. 2005. Integral role of IRF5 in the gene induction programme activated by Toll-like receptors. *Nature* 434:243–249.

156. Tegethoff, S., J. Behlke, and C. Scheidereit. 2003. Tetrameric oligomerization of IkappaB kinase gamma (IKKgamma) is obligatory for IKK complex activity and NF-kappaB activation. *Mol. Cell. Biol.* 23:2029–2041.

157. tenOever, B. R., S. L. Ng, M. A. Chua, S. M. McWhirter, A. Garcia-Sastre, and T. Maniatis. 2007. Multiple functions of the IKK-related kinase IKKepsilon in interferon-mediated antiviral immunity. *Science* 315:1274–1278.

158. tenOever, B. R., S. Sharma, W. Zou, Q. Sun, N. Grandvaux, I. Julkunen, H. Hemmi, M. Yamamoto, S. Akira, W. C. Yeh, R. Lin, and J. Hiscott. 2004. Activation of TBK1 and IKKepsilon kinases by vesicular stomatitis virus infection and the role of viral ribonucleoprotein in the development of interferon antiviral immunity. *J. Virol.* 78:10636–10649.

159. Tojima, Y., A. Fujimoto, M. Delhase, Y. Chen, S. Hatakeyama, K. Nakayama, Y. Kaneko, Y. Nimura, N. Motoyama, K. Ikeda, M. Karin, and M. Nakanishi. 2000. NAK is an IkappaB kinase-activating kinase. *Nature* 404:778–782.

160. Uematsu, S., and S. Akira. 2007. Toll-like receptors and type I interferons. *J. Biol. Chem.* 282:15319–15323.

161. Uematsu, S., S. Sato, M. Yamamoto, T. Hirotani, H. Kato, F. Takeshita, M. Matsuda, C. Coban, K. J. Ishii, T. Kawai, O. Takeuchi, and S. Akira. 2005. Interleukin-1 receptor-associated kinase-1 plays an essential role for Toll-like receptor (TLR)7- and TLR9-mediated interferon-α induction. *J. Exp. Med.* 201:915–923.

162. Uzel, G. 2005. The range of defects associated with nuclear factor kappaB essential modulator. *Curr. Opin. Allergy Clin. Immunol.* 5:513–518.

163. van Boxel-Dezaire, A. H., M. R. Rani, and G. R. Stark. 2006. Complex modulation of cell type-specific signaling in response to type I interferons. *Immunity* 25: 361–372.

164. Visvanathan, K. V., and S. Goodbourn. 1989. Double-stranded RNA activates binding of NF-kappa B to an inducible element in the human beta-interferon promoter. *EMBO J.* **8:**1129–1138.

165. Wang, Y. Y., L. Li, K. J. Han, Z. Zhai, and H. B. Shu. 2004. A20 is a potent inhibitor of TLR3- and Sendai virus-induced activation of NF-kappaB and ISRE and IFN-beta promoter. *FEBS Lett.* **576:**86–90.

166. Wertz, I. E., K. M. O'Rourke, H. Zhou, M. Eby, L. Aravind, S. Seshagiri, P. Wu, C. Wiesmann, R. Baker, D. L. Boone, A. Ma, E. V. Koonin, and V. M. Dixit. 2004. De-ubiquitination and ubiquitin ligase domains of A20 downregulate NF-kappaB signalling. *Nature* **430:**694–699.

167. Wu, Z. H., Y. Shi, R. S. Tibbetts, and S. Miyamoto. 2006. Molecular linkage between the kinase ATM and NF-kappaB signaling in response to genotoxic stimuli. *Science* **311:**1141–1146.

168. Xu, L. G., Y. Y. Wang, K. J. Han, L. Y. Li, Z. Zhai, and H. B. Shu. 2005. VISA is an adapter protein required for virus-triggered IFN-beta signaling. *Mol. Cell* **19:**727–740.

169. Xu, W., F. Yu, M. Yan, L. Lu, W. Zou, L. Sun, Z. Zheng, and X. Liu. 2004. Geldanamycin, a heat shock protein 90-binding agent, disrupts Stat5 activation in IL-2-stimulated cells. *J. Cell. Physiol.* **198:**188–196.

170. Xu, Y., X. Tao, B. Shen, T. Horng, R. Medzhitov, J. L. Manley, and L. Tong. 2000. Structural basis for signal transduction by the Toll/interleukin-1 receptor domains. *Nature* **408:**111–115.

171. Yanai, H., H. M. Chen, T. Inuzuka, S. Kondo, T. W. Mak, A. Takaoka, K. Honda, and T. Taniguchi. 2007. Role of IFN regulatory factor 5 transcription factor in antiviral immunity and tumor suppression. *Proc. Natl. Acad. Sci. USA* **104:**3402–3407.

172. Yang, H., C. H. Lin, G. Ma, M. O. Baffi, and M. G. Wathelet. 2003. Interferon regulatory factor-7 synergizes with other transcription factors through multiple interactions with p300/CBP coactivators. *J. Biol. Chem.* **278:**15495–15504.

173. Yang, H., G. Ma, C. H. Lin, M. Orr, and M. G. Wathelet. 2004. Mechanism for transcriptional synergy between interferon regulatory factor (IRF)-3 and IRF-7 in activation of the interferon-beta gene promoter. *Eur. J. Biochem.* **271:**3693–3703.

174. Yang, K., H. Shi, R. Qi, S. Sun, Y. Tang, B. Zhang, and C. Wang. 2006. Hsp90 regulates activation of interferon regulatory factor 3 and TBK1 stabilization in Sendai virus-infected cells. *Mol. Biol. Cell.* **17:**1461–1471.

175. Yoneyama, M., M. Kikuchi, T. Natsukawa, N. Shinobu, T. Imaizumi, M. Miyagishi, K. Taira, S. Akira, and T. Fujita. 2004. The RNA helicase RIG-I has an essential function in double-stranded RNA-induced innate antiviral responses. *Nat. Immunol.* **5:**730–737.

176. Zhang, L., and J. S. Pagano. 2000. Interferon regulatory factor 7 is induced by Epstein-Barr virus latent membrane protein 1. *J. Virol.* **74:**1061–1068.

177. Zhang, L., and J. S. Pagano. 1997. IRF7, a new interferon regulatory factor associated with Epstein-Barr virus latency. *Mol. Cell. Biol.* **17:**5748–5757.

178. Zhao, T., L. Yang, Q. Sun, M. Arguello, D. W. Ballard, J. Hiscott, and R. Lin. 2007. The NEMO adaptor bridges the nuclear factor-kappaB and interferon regulatory factor signaling pathways. *Nat. Immunol.* **8:**592–600.

179. Zou, J., Y. Guo, T. Guettouche, D. F. Smith, and R. Voellmy. 1998. Repression of heat shock transcription factor HSF1 activation by HSP90 (HSP90 complex) that forms a stress-sensitive complex with HSF1. *Cell* **94:**471–480.

Cellular Signaling and Innate Immune
Responses to RNA Virus Infections
Edited by A. R. Brasier et al.
© 2009 ASM Press, Washington, D.C.

Joan E. Durbin

Jak-Stat Pathway in Response to Virus Infection

6

INTRODUCTION TO THE JAK-STAT PATHWAY

Infection of susceptible cells by any viral pathogen leads to the synthesis and secretion of type I interferons (IFNs), or IFN-α/β, within hours of virus entry. Once IFN-α/β is secreted by an infected cell, these cytokines will bind to the ubiquitously expressed IFN-αR receptor (IFNAR). Signaling through this receptor leads to activation of the Jak-Stat pathway and transcriptional upregulation of the interferon-stimulated genes (ISGs) that mediate an antiviral state. Thus, IFN-α/β production by an infected cell will confer protection to its neighbors in a paracrine fashion as well as inhibiting viral replication in cells that are already infected. The Jak-Stat pathway was first described by investigators studying the coordinate regulation of ISGs in IFN-α-treated cells. These genes were found to share a promoter element termed the interferon-stimulated response element, or ISRE (AGTTTN$_3$TTTCC), which confers ligand specificity. The transcription factors binding this element, Stat1 and Stat2, were the first members of the Stat family to be identified and cloned (36, 68, 117). It was found that the Stat proteins, in addition to their role as transcription factors, were also signal transducers activated by receptor-associated tyrosine kinases called Janus kinases (Jak kinases) (21, 130). It is for this dual function that this gene family was named: signal transducers and activators of transcription.

The sequence of events, from ligand binding to transcriptional upregulation, is diagrammed in Fig. 1. Within seconds of ligand binding to the IFNAR2 subunit, the IFNAR1 subunit is recruited (71). Once this receptor-ligand complex is formed, the preassociated tyrosine kinases, Jak1 and Tyk2 (tyrosine kinase 2), respectively, transphosphorylate one another. A tyrosine phosphorylation cascade then ensues, resulting in phosphorylation of the receptor cytoplasmic tails and recruitment and tyrosine phosphorylation of Stat proteins (116, 119). In response to type I IFN activation of the IFN-αR, Stat1 and Stat2 become tyrosine phosphorylated and thereby activated to form a multimer which also includes IFN regulatory factor 9 (IRF9). IRF9 is the DNA-binding subunit of the heterotrimeric complex called interferon-stimulated gene factor 3 (ISGF3), and it is ISGF3 that translocates to the nucleus, where it binds to ISRE sequences in the promoters of IFN-α/β-regulated genes and upregulates their transcription.

An analogous series of events occurs when type II IFN, or IFN-γ, binds to its receptor. Like the type I IFN

Joan E. Durbin, Center for Vaccines and Immunity, The Research Institute at Nationwide Children's Hospital, Department of Pediatrics, The Ohio State University College of Medicine, Columbus, OH 43205.

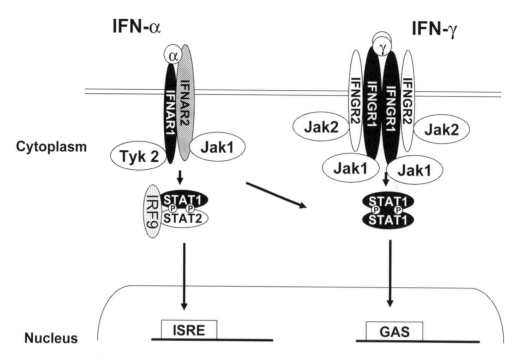

Figure 1   Stat activation and ISG induction by IFN-α/β and IFN-γ.

receptor, the IFN-γR consists of two subunits, IFNGR1 and IFNGR2. When a dimeric form of IFN-γ binds to two IFNGR1 chains, these subunits then form a complex with two IFNGR2 molecules (7). Receptor subunit association leads to transphosphorylation of the receptor-associated tyrosine kinases, Jak1 and Jak2, which then phosphorylate Y440 of the IFNGR1. Once tyrosine phosphorylation of the receptor occurs, Stat1 binds and is phosphorylated in turn on Y701. The activated Stat1 molecules dimerize to form the IFN-γ-activated site (GAS)-binding factor called GAF, which moves into the nucleus and activates transcription from GAS elements [TTC(N$_{2-4}$)GAA] present in the promoters of IFN-γ-stimulated genes (1).

Activation of Stat1 and Stat2 is mediated by IFNs, as shown in Fig. 1, but these cytokines can also activate other Stat family members. Seven Stat proteins have been identified, all with a similar protein structure that includes seven conserved domains. These are the amino-terminal domain (NH$_2$), a predominantly hydrophilic coiled-coil domain, a linker region, the SH2 domain, the tyrosine activation motif, and the transcriptional activation domain (8, 16, 35, 55, 127) (Fig. 2). These proteins vary in size from 750 to 900 amino acids and are present in an inactive state in the cytoplasm until tyrosine phosphorylation by a Jak kinase causes them to multimerize and move into the nucleus. The Stat proteins can form both homo- and heterodimers, and Stat-Stat

interactions take place via their SH2 domains that bind to the phosphorylated tyrosine at approximately residue 700 in the case of all Stat proteins (22). Stat multimers can directly bind DNA via their DNA-binding domain, although these complexes can also contain other transcription factors such as IRF9.

Signals through many different cytokine and growth factor receptors can activate Stat proteins, and a summary of Stat activation is shown in Table 1. These data suggest that there is significant redundancy in the action of these ligands, but in vivo regulation of Jak-Stat signaling is complex and cannot be reduced to the presence or absence of Stat tyrosine phosphorylation. In the case of Stat1, knockout mice lacking the *Stat1* gene are exquisitely sensitive to viral and bacterial infection and mount Th2 responses to infectious agents. This is consistent with an inability of these animals to respond to

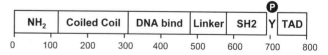

Figure 2   Conserved elements of Stat proteins. This consists of the conserved amino-terminal domain, the coiled-coil domain, the DNA-binding domain, a linker domain, the SH2 domain, and the transcriptional activation domain (TAD). The tyrosine residue necessary for Stat activation is present at approximately residue 700 in each of the Stat proteins.

Table 1 Stat activation by cytokines and growth factors[a,b]

Stat	Receptor
1	IFN-α/β, IFN-γ, IFN-λ, IL-10, IL-6, OSM, LIF, CNTF, G-CSF, IL-12, IL-13,IL-21, IL-23, GH, PRL, EGF, PDGF, CSF-1
2	IFN-α/β, IFN-λ
3	IFN-α/β, IFN-γ, IL-10, IL-19, IL-20, IL-22, IL-24, IL-26, IL-6, IL-11, IL-31, OSM, CNTF, CLC/CLF, CT1,G-CSF, IL-12, GH, EGF, PDGF, CSF-1, leptin, IL-21, IL-23, IL-27
4	IFN-α/β, IL-12, IL-21, IL-23
5a/b	IFN-α/β, IL-2, IL-7, IL-9, IL-15, IL-21, IL-5, IL-3, GM-CSF, TPO, EPO, PRL(a), GH(b)
6	IFN-α/β, IL-4, IL-13

[a] OSM, onconstatin M; LIF, leukemia inhibitory factor; CNTF, ciliary neurotrophic factor; G-CSF, granulolyte colony-stimulating factor; EGF, epidermal growth factor; PDGF, platelet-derived growth factor; CSF, colony-stimulating factor; CLC, cardiotrophin-like cytokine; CLF, cytokine-like factor; CT1, cardiotropin 1; GM-CSF, granulocyte-macrophage colony-stimulating factor; TPO, thrombopoietin; EPO, erythropoietin.
[b] Data from references 41, 45, 69, 98, 100, 102, 116, 118, 142, and 145.

Table 2 Susceptibility of Stat knockout mice to viral pathogens[a]

Virus	Stat knockout
Dengue	Stat1$^{-/-}$ mice are marginally more sensitive to infection (125).
Influenza A	The effect is virus strain dependent. WSN infection in Stat1$^{-/-}$ mice is systemic and rapidly fatal. The PR8 strain of virus is cleared similarly in both WT and Stat1$^{-/-}$ animals (37).
MNV-1	Infection is lethal and systemic in Stat1$^{-/-}$ mice (65).
RSV	Stat1$^{-/-}$ mice have elevated virus titers and exacerbated disease (32, 60).
SARS-CoV	Mortality in Stat1$^{-/-}$ but not WT mice (133).
PIV5	Infection is lethal in Stat1$^{-/-}$ but not WT mice (49).
Sindbis	Stat1$^{-/-}$ mice are marginally more sensitive to infection (40).
VSV	LD$_{50}$ for the Stat1$^{-/-}$ mouse is 6 logs lower than for WT animals, and the viral burden is significantly increased (31, 84). Stat2$^{-/-}$ animals have a similar phenotype (101).

[a] WSN, A/WSN/33 virus; WT, wild-type; MNV, murine norovirus; SARS-CoV, severe acute respiratory syndrome coronavirus; PIV5, parainfluenza virus 5; LD$_{50}$, 50% lethal dose.

IFN-α/β and IFN-γ, but signaling by other ligands found to activate Stat1 in vitro (GH, IL-6, EGF, IL-10) is intact (31, 84).

Deletion of other Stat genes in mice has shown similarly restricted phenotypes, with the exception of *Stat3*. Embryos lacking the *Stat3* gene die early in gestation (134), and only Stat3$^{-/-}$ animals with tissue-specific deletions are available for study. In contrast to other Stat genes, *Stat4* expression is limited to natural killer (NK) cells, dendritic cells (DCs), and T cells, and in its absence the biological response to IL-12 and IL-23 is deficient. NK-cell killing and NK-cell IFN-γ production are impaired in these animals, as well as Th1 differentiation of CD4$^+$ T cells (63, 88, 139). There are two highly homologous *Stat5* genes, *Stat5a* and *Stat5b*, which have significant functional overlap (89). Mice homozygous for a deletion of the *Stat5a* gene are deficient in mammary gland development, consistent with the importance of this molecule in prolactin (PRL) signaling. Mice lacking *Stat5b* are similar to animals lacking the GH signaling pathway (77, 141). In doubly deleted mice, Stat5a$^{-/-}$ and Stat5b$^{-/-}$, the phenotype is more severe, with a failure of both ovarian and mammary gland development in females, as well as an impairment of T-cell proliferation following T-cell receptor stimulation and the addition of IL-2 (137). Deficits in *Stat6* knockout mice are also quite specific and reflect the importance of this transcription factor in the IL-4, IL-13, and IL-23 signaling pathways. Th2 cell differentiation and trafficking are impaired in Stat6$^{-/-}$ animals, as is B-cell function and survival (28, 62, 124, 135). As

these studies demonstrate, only Stat1- and Stat2-deficient mice have shown a particular sensitivity to virus infection, although the degree of susceptibility appears to be pathogen dependent (Table 2). This is not to say that altered immune responses to infection do not occur in the context of other Stat deficiencies. In fact, Stat4 is required for IFN-γ production in the context of lymphocytic choriomeningitis virus (LCMV) infection (94). Nonetheless, despite obvious immune dysfunction in the absence of each Stat, the ability to contain and clear acute viral infections appears to be largely Stat1 and Stat2 dependent. This is surprising given the fact that IFN-α/β leads to the phosphorylation of every Stat, but the impact of Stat3, 4, 5, and 6 activation on early antiviral responses has yet to be demonstrated (107).

The downstream antiviral effects that result from Stat1 and Stat2 activation through the IFN-α/β or IFN-γ receptors are due to the binding of ISGF3 or GAF to the ISRE and/or GAS elements present in the promoters of the ISGs that mediate the antiviral state. Gene expression studies of transcripts upregulated by IFN-α, IFN-β, or IFN-γ treatment of cultured cells have provided a comprehensive look at the ISGs induced by these cytokines, as well as information regarding the relative responsiveness of these genes to each cytokine (25). The best studied of the ISG products preferentially induced by type I IFNs are Mx, a GTPase that sequesters viral ribonucleoproteins; protein kinase R (PKR), a serine/threonine kinase activated by double-stranded (ds) RNA binding; and the 2'-5'-oligoadenylate synthetases (OAS), also activated by dsRNA. The Mx proteins restrict movement of viral components within the cell (48), and PKR inhibits viral protein translation by phosphorylation of the elongation initiation factor eIF2α. The 2'-5'-oligos, synthesized by the 2'-5'-OAS proteins, activate RNase L, which then cleaves viral RNAs (9). A detailed discussion of the mechanisms by which these ISGs establish and maintain an antiviral state is reviewed in detail in chapter 7.

STAT ACTIVATION BY VIRUS INFECTION

The limited number of cytokines that are able to activate Stat2 confers specificity to the type I IFN pathway, but the activation of Stat1 by many ligands underscores the complexity of Stat regulation in vivo. This question has been studied intensively, and many factors have been found to impact the biological activity of tyrosine phosphorylated Stat1. It is clear that the many inflammatory pathways induced in the course of a virus infection will have a crucial role in determining the extent and timing of Stat activation. These are difficult studies, and an extensive picture of viral pathogenesis in vivo exists for only a limited number of infections. However, a common thread for essentially all virus infections is the induction of type I IFNs. Type I IFNs, all of which bind to the single type I IFN receptor, or IFN-αR, constitute a gene family, with 1 IFN-β gene and 13 IFN-α genes. Other type I IFN genes that have been found in humans include IFN-ε, IFN-τ, and IFN-ω, but little is yet known about their function (103, 104). IFN-α/β's are produced by all cell types following virus infection and/or engagement of pattern recognition receptors such as the Toll-like receptors (TLRs), and their production is among the earliest responses to viral invaders.

The source, amount, and kinetics of type I IFN production are dependent on the nature of the viral pathogen and will be determined by the susceptible cell type, the target organ(s), and the range of host sensors exposed to viral products. There are both cytoplasmic and membrane-bound sensors of viral pathogens: the helicases retinoic acid-inducible gene I (RIG-I) and melanoma differentiation-associated gene 5 (MDA-5), and the TLRs, respectively. Both RIG-I and MDA-5 recognize and bind to dsRNAs, which are produced in the process of DNA virus and positive-strand RNA virus replication (149). While it is not certain that dsRNA intermediates are produced during the course of negative-strand RNA virus infection, it has recently been demonstrated that

single-stranded RNAs with free 5'-triphosphates generated in the course of measles virus infection will bind and activate RIG-I (52, 105, 108, 149). This is consistent with new data showing that a number of negative-strand RNA viruses such as influenza A virus, vesicular stomatitis virus (VSV), and respiratory syncytial virus (RSV) require RIG-I for type I IFN induction in fibroblast and epithelial cell lines (79). Once activated, RIG-I and MDA-5 interact with the mitochondrial antiviral signaling protein, termed MAVS (or IPS-1, VISA, or Cardif) (67, 85, 120, 151), and this interaction leads to the recruitment and activation of the IKK-ε and TANK-binding kinase 1 (TBK1), which phosphorylate the transcription factors IRF3 and IRF7 (121). Simultaneously, MAVS activates the canonical IκB kinase (IKK) complex, leading to the activation of the transcription factor nuclear factor-κB (NF-κB). Although not essential for IFN-β synthesis, NF-κB binding to the IFN-β promoter is also associated with IFN-β induction (146). When cytoplasmic IRF3 is activated in the cytoplasm by serine/threonine phosphorylation, it will dimerize and move into the nucleus, where, in association with CREB-binding protein (CBP)/p300, it induces transcription from the IFN-β promoter. IRF3/IRF7 heterodimers can also upregulate IFN-α transcription, but in most cell types IRF7 must itself be induced by signaling through the IFN-αR. When IFN-β (and IFN-α1 in the human host) is secreted by the infected cell, feedback through its receptor promotes the synthesis of ISGs including IRF7 (81, 114). Several type I IFN promoters preferentially bind to IRF7 only, such that IRF7 induction serves to amplify type I IFN production (73). This pathway is diagrammed in Fig. 3.

IRF3 and IRF7 can also be activated by TLR ligands binding to their receptors. The TLRs are a family of membrane proteins that serve as receptors for pathogen-associated molecular patterns, or PAMPs. TLRs implicated in recognition of RNA virus infection are TLR3, 7, and 8, with TLR3 binding to dsRNA molecules (3) and TLR7 and 8 to single-stranded RNAs (26). TLR7/8 expression is limited to the endosomal compartment of DCs, while TLR3 expression is more widespread and can be induced by type I IFNs (23, 59, 66, 140). These molecules are thought to encounter viral RNA during uptake of viral particles via the endosomal pathway, during the fusion of autophagosomes to endosomes (72), and during uptake of apoptotic bodies from infected cells by DCs. TLR triggering by ligand binding also results in IFN-β production, but by different routes. TLR3 activation in response to dsRNA recruits the adaptor protein called Toll/IL-1 receptor domain-containing adaptor inducing IFN-β (TRIF). TRIF recruitment by TLR3 leads to NF-κB activation as well as IRF3/IRF7 activation by

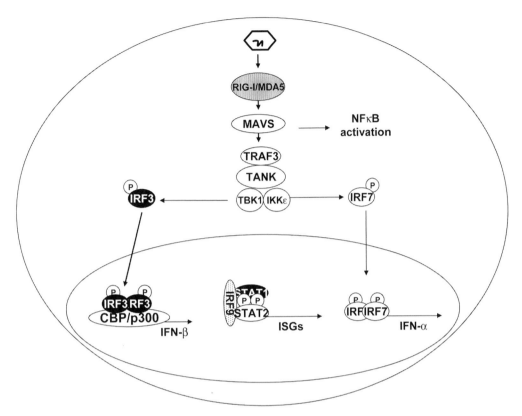

Figure 3 Type I IFN induction by virus infection mediated by RIG-I/MDA-5. MAVS acts as an adaptor protein which binds to both RIG-1/MDA-5 and TRAF3 (tumor necrosis factor receptor-associated factor 3). TRAF3 binds TBK1 and IKK-ε directly in a complex which also involves TANK (TRAF family member-associated NF-κB activator).

IKK-ε/TBK1 (115). TLR7 signaling proceeds down a different pathway involving recruitment of myeloid differentiation factor 88 (MyD88), which forms a complex including IL-1 receptor-associated kinase 4 (IRAK-4), IRAK-1, and tumor necrosis factor receptor-associated factor 6 (TRAF6). This complex directly binds to IRF7, which is then phosphorylated by IRAK-1 (51, 66). Phosphorylation of IRF7 by this mechanism is limited to the plasmacytoid DCs (pDCs), which express TLR7 as well as constitutively high levels of IRF7. In addition to TLR recognition of viral nucleic acids, there is evidence that, for a few viruses, viral proteins can activate additional TLRs expressed in macrophages and myeloid DCs (39, 70). Thus, the TLR-mediated pathways allow IFN-α/β synthesis by uninfected as well as infected cells.

Despite the identification of these many sensors of infection, the source of type I IFNs in particular viral infections is not well understood. pDCs are a major source of IFN-α/β following many viral infections, but a number of studies carried out in mouse models have demonstrated that the source of these cytokines varies with the pathogen. Dalod et al. (20a) described the murine pDC

as the IFN-producing cell in murine cytomegalovirus (MCMV) infection, but depletion of this subset did not alter type I IFN production in response to LCMV. Type I IFN induction by LCMV was found to depend upon a population of macrophages located within the marginal zone of the spleen (80). Intrigued by papers in the pediatric literature describing robust type I IFN production in influenza A virus- but not RSV-infected patients (47, 83), we looked to see whether this difference could be seen in a mouse model, and whether the source of type I IFNs would be similar in these two infections. These viruses were administered by the same intranasal route, and both infections are limited to the airway, with no viremic phase. Maximal levels of lung IFN-α/β induced by each virus were similar, but secretion was very limited following RSV infection and prolonged in influenza virus-infected animals. In situ hybridization studies carried out on tissue sections demonstrated widespread IFN-α/β synthesis by infected epithelium, and pDC depletion studies led to a large decrease only in influenza virus-infected animals (58). In vitro studies using cultured human and mouse DCs have shown that RSV can induce type I IFN production by pDCs, albeit less efficiently than influenza virus. However, and unlike influenza virus, RSV-induced pDC IFN-α/β requires active virus replication, and the pathway(s) mediating this induction appears to be independent of TLR7 (53, 58). Thus, the host recognition pathways triggered by an RNA virus will depend upon many factors, including the route of infection, susceptible cell types, the mode of virus entry, and the nature of viral RNAs produced in the course of infection.

In addition to type I IFNs, other Stat-activating cytokines have also been detected in RNA virus infection. In influenza A virus and RSV infection of mice, IFN-γ is found in lung homogenates or bronchoalveolar lavage specimens from 3 to 5 days following infection (30, 32). IL-6 and tumor necrosis factor-α have also been reported in influenza virus infection (19, 30). Like IFN-α/β, both IFN-γ and IL-6 can activate Stat1, although activation through the IFN-γ and IL-6 receptors promotes formation of Stat1:Stat1, Stat1:Stat3, and Stat3:Stat3 complexes rather than ISGF3.

MODULATION OF STAT ACTIVITY

Tyrosine phosphorylation of Stat proteins following ligand binding is essential for signal transduction to occur, but additional modifications are required for optimal function. All Stats, except Stat2, have a conserved serine residue in their C-terminal transactivation domain, and this must be phosphorylated for optimal transcription activation (24). Serine phosphorylation sites of S727 in Stats 1 and 3 and S721 in Stat4, share a PMS*P motif, and there is an additional serine phosphorylation site at S708 in Stat1 (118). The importance of serine phosphorylation for IFN signaling was noted because there are two naturally occurring splice variants of Stat1, referred to as Stat1α and Stat1β. Stat1α is a 91-kDa protein with a full-length transcriptional activation domain, including the conserved serine residue at position 727. Stat1β lacks the last 38 amino acid residues including S727, but can still be phosphorylated on Y701 and partner with Stat2 and IRF9 to form ISGF3. To determine whether Stat1α and Stat1β were functionally equivalent, cDNAs encoding each splice form were introduced into a human fibrosarcoma cell line lacking Stat1 (U3A cells). The U3A cell line is unresponsive to type I or type II IFN treatment, but antiviral activity in response to either cytokine was reconstituted by expression of Stat1α. By contrast, Stat1β expression restored the antiviral effects only of IFN-α/β. Transfection of U3A cells with the Stat1 SA substitution mutant Stat1-S727A, with alanine substituted for the serine at position 727, had a similar effect. Reconstitution of the Stat1-deficient cells with

Stat3 did not restore the antiviral activity of IFN-α/β or IFN-γ (54). When Stat1$^{-/-}$ murine fibroblasts expressing Stat1α, Stat1β, or Stat1-S727A were treated with IFN-β, all showed similar levels of tyrosine phosphorylated Stat1 and ISGF3 formation, but inducibility of ISGs was decreased by 50 to 90% in the absence of Stat1α. Consistent with this result, there was a sharp decline in type I IFN-induced antiviral activity when IFN levels dropped below 500 U/ml (106). Mice expressing Stat1-S727A are deficient in IFN-γ mediated immune responses and extremely susceptible to *Listeria monocytogenes* infection (144). A number of Stat serine kinases have been identified, and p38 MAPK (mitogen-activated kinase), PKC-δ (protein kinase C-δ), CaMKII (Ca^{2+}/calmodulin-dependent protein kinase II), and IKK-ε have all been associated with the serine phosphorylation of Stat1 (42, 92, 138). IKK-ε phosphorylates S708 of Stat1, a modification that favors the formation of ISGF3 over Stat1 homodimers and is required for efficient ISGF3 binding to a subset of ISREs (138). Serine phosphorylation therefore provides an additional mechanism for modulating Stat1 activity in response to the nature and intensity of the inflammatory stimulus.

Acetylation of the DNA-binding domains of Stat2 and IRF9 by CBP have also been shown to be essential, not for Stat1:Stat2 association but for formation of active ISGF3. Reconstitution experiments using the IFNAR2, IRF9, and Stat2 null cell lines developed by George Stark showed that transcriptional control by IFN-α could not be restored by expression of K339R-substituted IFNAR2, K81R-substituted IRF9, or K390R-substituted Stat2. Acetylation at these lysine residues and the presence of CBP are both required for antiviral signal transduction (136).

The result of Stat1 activation is also determined by exposure to other cytokines. The phenomenon of "priming" relates to the fact that the components of the IFN signaling apparatus are themselves ISGs. Concentrations of Stat1, Stat2, and IRF9 are increased by pretreatment of cells with IFN-γ, thus enhancing their response to IFN-α/β (150). As a second example of this phenomenon, IL-6 and IFN-γ activate both Stat3 and Stat1, but the ratio of the unphosphorylated Stat proteins present in cells at the time of cytokine treatment will influence which Stat complexes are formed as a result of receptor activation. In cells pretreated with IL-6, there is a large excess of unphosphorylated Stat3 protein (57), and this accumulation will alter signaling through the IFN-α/β and IFN-γ receptors (129). Both Stat1 and Stat3 bind tyrosine 419 of the IFNGR1 chain, and neither can be activated by IFN-γ when this residue is changed to phenylalanine (109), demonstrating that these proteins compete directly for binding to the IFN-γR. Consistent with these findings, there is prolonged activation of Stat3 dimers by IFN-γ in the absence of Stat1 (109), and activation of Stat1 homodimers following IL-6 treatment of Stat3 null cells (20). The importance of this mechanism in adjusting the host response to changing conditions in vivo has recently been demonstrated by Gil et al. (41) and Miyagi et al. (88), who have found that changes in ratio of Stat1 to Stat4 in T and NK cells alters their response to type I IFNs as infection progresses.

A number of pathways have been discovered that function to limit the duration of inflammatory responses triggered by cytokines. In a classic feedback loop, cytokine-inducible Src homology 2 (CIS) and suppressors of cytokine signaling (SOCS) proteins are negative regulators of Stat activation that are induced by activated Stats. This family of cytokine-inducible proteins has eight members, SOCS-1 to -7 and CIS, each with a central SH2 domain. SOCS-1 to -3 and CIS are all upregulated by type I and type II IFNs in addition to many other cytokines, and SOCS-1 and SOCS-3 have both been shown to inhibit Stat1-mediated signaling by blocking Jak kinases (128). SOCS-1 acts by binding tyrosine phosphorylated Jaks through its SH2 domain, while SOCS-3 inhibition of Jaks involves binding to the activated receptor (126). The importance of the SOCS proteins for limiting the potentially damaging effects of unchecked IFN signaling in vivo is demonstrated by the phenotype of SOCS-1$^{-/-}$ mice, which die before weaning from uncontrolled inflammation (2, 82, 93, 132). SOCS-3$^{-/-}$ deficiency is embryonic lethal secondary to abnormal placentation, making the in vivo role of this protein in viral infection more difficult to study (112). As both type I and type II IFNs induce SOCS-1 and SOCS-3, these proteins are important in limiting the duration of Stat activation, but other non-IFN cytokines also promote their expression and play a part in regulating the strength and duration of Jak-Stat signaling (131).

In addition to the feedback inhibition by SOCS family members, dephosphorylation is required for nuclear export and is therefore an important mechanism in the decay of the IFN signal. A number of phosphatases have been implicated in this process, but only SH2 domain-containing protein tyrosine phosphatase-2 (SHP-2) and T-cell protein tyrosine phosphatase (TC-PTP) are essential for Stat dephosphorylation in the nucleus (91).

IFN- AND STAT1-DEPENDENT ANTIVIRAL RESPONSES

Gene expression studies of transcripts upregulated by IFN-α, IFN-β, or IFN-γ treatment of cultured cells have provided a comprehensive look at the ISGs induced by these cytokines, as well as information regarding the relative responsiveness of these genes to each cytokine (25). A number of ISGs with known antiviral effects, such as MxA, 2′-5′OAS, and ISG56, are strongly induced by IFN-α/β but not IFN-γ treatment. This is in contrast to IRF1, which is preferentially induced by IFN-γ, or GBP1 (guanylate-binding protein 1) and Stat1, which respond to both types of IFN. This induction pattern is consistent with the finding that, although both type I and type II IFNs can mediate protection against virus infection in vitro, IFN-α/β-stimulated genes appear to be essential for maximal antiviral activity (20, 54). The availability of knockout mice lacking the type I IFN receptor (90), the type II IFN receptor (56), both receptors (143), Stat1 (31, 84), and Stat2 (101) has allowed investigators to assay the relative importance of each of these molecules in combating specific pathogens in vivo. IFNAR1-deficient mice are very susceptible to a number of virus infections, including VSV, Semliki Forest virus, vaccinia virus, LCMV (90), and Sindbis virus (40). IFNGR1$^{-/-}$ animals show increased sensitivity to vaccinia virus and LCMV.

Another perspective on the relative impact of these cytokines on host defense can now be found in the expanding body of literature describing the phenotype of patients with defects in type I or type II IFN signaling. Currently there are 13 known mutations that affect IFN-γ function, and all confer an increased susceptibility to mycobacterial infection. These patients are now classified as having a syndrome called Mendelian susceptibility to mycobacterial disease (MSMD) and are unable to secrete or respond to IFN-γ normally. Those homozygous for null mutations in the *IL12B* gene, which encodes the p40 subunit of IL-12, or the *IL12RB1* gene, which encodes the β subunit of the IL-12 and IL-23 receptors, are unable to produce IFN-γ following mycobacterial infection (33). These patients account for approximately half of those presenting with MSMD, which is generally identified when children develop systemic infection after vaccination with the bacillus Calmette-Guérin (BCG). In addition, both dominant and recessive mutations in the *IFNGR1* and *IFNGR2* genes have been identified that directly affect IFN-γ signaling. The most common of these (*818del4*) is a deletion mutation in the *IFNGR1* gene that results in the synthesis of a truncated protein with a cytoplasmic tail of only 5 amino acids (normal is 187) (27). The mutant receptor can bind IFN-γ, but is defective in both signaling and recycling such that the abnormal chains accumulate on the cell surface, where they exert a dominant negative effect. The dominant mutations result in a partial loss of function, and infected patients are treated with IFN-γ. Those with a recessive genotype have inherited two null mutations, either homozygous or compound, and therefore have a complete absence of function. While the overall prevalence of mycobacterial disease in these patients is approximately 95%, severity is tightly correlated with genotype. Infection will resolve in those with a dominant partial deficiency, whereas those with a recessive complete deficiency are at risk for chronic infection and have a 50% rate of mortality by the age of 10. The incidence of other opportunistic infections in IFN-γ-deficient patients is very low, although 3 of 22 patients with the recessive complete genotype suffered from disseminated cytomegalovirus infection and 5 of 22 reported symptomatic varicella virus infection (13, 27). Thus, the phenotype of IFN-γ-deficient patients is a surprisingly narrow one, with susceptibility to mycobacterial infection being the major phenotype.

Patients with deletions in Stat1 are much less common, but a number of these have also been identified as belonging to the MSMD cohort. A heterozygous nucleotide substitution mutation leading to a Glu→Gln substitution at position 320 (E320Q) and a heterozygous Q463H mutation are inherited in an autosomal dominant fashion and impair the ability of Stat1 to bind GAS elements. Cell lines obtained from patients had approximately 25% of normal GAS-binding activity following IFN-γ treatment. A third partial defect, corresponding to the heterozygous L706S mutation, results in the production of Stat1 molecules that cannot be tyrosine phosphorylated, and therefore impairs formation of activated Stat1 homodimers. Patients with each of the three mutant Stat1 alleles showed a partial defect in IFN-γ-stimulated GAF formation, but normal IFN-α signaling and ISGF3 formation (14).

Mutations affecting IFN-α/β signaling are much less common, and much more severe, generally resulting in death from overwhelming virus infection in infancy (61). In the few patients that have been identified, this defect is due to mutations in the *Stat1* or *Tyk2* genes, which interfere with Jak-Stat signaling. Mutations in the *UNC93B1* gene, which affects type I and type III IFN production in response to TLR ligands, result in a more limited immunodeficiency. In 2003, two unrelated children were identified with homozygous mutations in *Stat1* that resulted in a complete loss of Stat1 production (29). The children were susceptible to both mycobacterial and overwhelming virus infection, with one of the patients succumbing to herpes simplex virus type 1 encephalitis. More recently, a patient with the hyper-IgE syndrome who was susceptible to atypical mycobacterial infections, as well as repeated cutaneous viral infections, was found be homozygous for a mutation in the *Tyk2* gene. A premature termination codon resulted in the

complete absence of Tyk2 protein (87), and type I IFN treatment of this patient's T cells had no effect on tyrosine phosphorylation of Stat1, Stat2, Stat3, Stat4, or Jak1. This differs from studies in Tyk2$^{-/-}$ mice, where signaling through the IFN-αR is not abrogated and substantial residual Stat1 and Stat2 phosphorylation is found in response to IFN-α (64, 123). The hyper-IgE syndrome, which manifests as dermatitis, boils, cyst-forming pneumonias, and elevated immunoglobulin E (IgE) levels, is associated with Stat3 deficiency (50), which is also evident in this individual. However, it is not clear why this patient is not susceptible to more serious viral infections; possibly there is residual type I IFN activity in nonhematopoietic tissues and this offers some protection.

In addition to the antiviral response mediated by type I (α/β) IFN signaling, a new family of type III IFN genes was recently discovered. In the human, there are three IFN-λ genes, designated IFN-λ1 (IL-29), -λ2, and -λ3 (IL-28A/B). IL-29 is not found in mice. This cytokine family was discovered by Kotenko et al., who identified a new member of the cytokine class II family of receptors, and IFN-λ as the ligand (69). Sheppard et al. (122) originally identified the ligands, followed by the receptor. Like the type I IFNs, type III IFNs are also induced by virus infection. IFN-λ1 induction occurs by a mechanism that requires IRF3, RIG-I, MAVS, and TBK1, similar to the induction of IFN-β (96). The genes encoding IFN-λ2 and -λ3 are regulated in a manner similar to that of the IFN-α genes, which require the presence of IRF7 (97). A number of cell types have been shown to produce IFN-λ's, and one report suggests that pDCs may be the primary source of this cytokine in influenza A virus infection (17). The IFN-λ receptor consists of the IL-28Rα chain and the IL-10Rβ chain, which together signal through the associated Tyk2 and Jak1 tyrosine kinases and ISGF3. There is as yet no firm consensus regarding which cell types express the type III IFN receptor, although its distribution is thought to be more limited than that of the ubiquitously expressed type I IFN receptor (153). Comparison of IFN-α- and IFN-λ-upregulated genes demonstrated that IFN-λ targets a subset of IFN-α-stimulated genes (153), but the significance of IFN-λ production in vivo is not yet known. IL-28Rα-deficient mice have recently been generated (6) and do not show increased susceptibility to influenza A virus, LCMV, encephalomyocarditis virus, or VSV infection. However, while vaginal instillation of CpG or IFN-λ protects normal mice against herpes simplex virus type 2 challenge, either intervention requires the presence of functional IL-28Rα. It would be premature to conclude from this preliminary characterization that IFN-λ's do not play an important role in host defense against virus infection despite the fact that animals deficient in IFN-λ signaling did not show the extreme sensitivity of IFNAR1$^{-/-}$ mice to systemic VSV and encephalomyocarditis virus challenge. It remains a formal possibility that ISGF3 induction by any mechanism can offer significant protection against viral pathogens and that the presence of the IFN-λ system can explain the somewhat limited effect of type I IFN deficiency against some of the viruses studied. A better understanding of these redundant signaling pathways awaits the development of mice lacking both type I and type III IFN receptors.

STAT INACTIVATION BY VIRUS INFECTION

The mechanisms used by specific viral pathogens to defeat innate immunity are outlined in detail in later chapters of this book and will not be listed here. Nonetheless, by way of introduction to what follows, I will mention here that most viruses studied have evolved one or more mechanisms for targeting the IFN response, and many of these target Stats and Stat activation (111). Many viruses block the induction of type I IFNs. Influenza A virus does this by inhibiting RIG-I (86), and paramyxoviruses by inhibiting MDA-5 (4). Many viruses promote the inhibition or degradation of IRF3, and there are many examples of viruses directly targeting the Jak-Stat pathway. Examples include: (i) inhibition of Jak kinases by measles virus (12) and Japanese encephalitis virus (75), (ii) Stat1 and Stat2 degradation by RSV (110) and parainfluenza virus (5), (iii) Stat1 and Stat2 sequestration in the cytoplasm by Hendra virus (113), (iv) inhibition of Stat activation by Sendai virus (43), and (v) inhibition of nuclear trafficking by severe acute respiratory syndrome coronavirus (34). The existence of these mechanisms underscores the importance of the Jak-Stat pathway for host defense and the need to further our understanding of the interactions between viral proteins and the machinery of IFN signaling.

STAT1-INDEPENDENT ANTIVIRAL RESPONSES

In addition to IFN-induced Stat activation, there appear to be additional pathways leading to ISG upregulation that are likely to play a significant role in host defense. This is a relatively unexplored area, but undoubtedly an important one. As previously discussed, children homozygous for a null Stat1 mutation have a severe primary immunodeficiency, but as pointed out by Chapgier et al. (15), one of these patients was able to

clear a number of less virulent viral infections (poliovirus III, parainfluenza virus type 2, rhinovirus) without difficulty before his death following bone marrow transplantation. This is a surprising result, and yet there are data from studies using IFN-deficient mouse strains that are consistent with that observation. We have observed that Stat1$^{-/-}$ mice infected with the PR8 strain of influenza A virus can clear infection at a rate nearly comparable to that found in wild-type animals (37). That is not to say that type I IFNs have no role in controlling influenza virus infection, but only to point out that the problem is a complex one. The lack of a dramatic phenotype in IFNAR$^{-/-}$ animals must also be considered in light of a second observation: that a recombinant PR8 strain lacking the type I IFN-antagonizing NS1 gene is lethal for Stat1$^{-/-}$, but not wild-type, animals (38). Taken together, these results suggest that many viruses encountered in nature have evolved very effective anti-IFN and anti-Stat strategies and that redundant pathways are also important for resisting virus infection. Although influenza A virus is well suited to studies in the mouse model, many human pathogens are not, and analysis of in vivo data obtained in mice may be complicated by the species specificity of many steps in the IFN signaling pathway. As an example of this issue, the NS1 and NS2 proteins of RSV have been implicated in the degradation of Stat2 in RSV-infected human, but not murine, cells (78). It is therefore important to understand the limitations of each experimental model and interpret results of in vivo studies of human pathogens in light of patient data when they are available.

In searching for redundancy in the antiviral response, there are multiple examples of IFN- and/or Stat1-independent mechanisms that can upregulate the expression of ISGs. It has been demonstrated in many systems that IRFs other than IRF9 can bind to ISREs. When IRF3 is activated by virus infection, it forms homo- or heterodimers that move into the nucleus and, in association with CBP and p300, can bind and activate the promoters of a subset of ISGs (44, 74, 147, 148, 152). Karen Mossman's group has recently discovered that these events can also be triggered by fusion of cells with UV-inactivated, enveloped viruses, and that this is sufficient for direct induction of ISG54, ISG56, ISG15, and IP-10 gene expression (18). The mechanism by which virus entry alone can lead to induction of an antiviral state is not well understood, but it occurs in the absence of type I IFNs, and is independent of both RIG-I and TLRs, but requires IRF3 (18, 95, 99).

One useful approach to teasing out virus-specific mechanisms for inhibiting host defenses is the comparison of viral disease severity in strain-matched knockout mice lacking IFNAR1, IFNGR1, both IFNAR1 and IFNGR1, or Stat1. In a thought-provoking study carried out by Gil et al. (40), mice of each genotype were infected with either MCMV or Sindbis virus, and their susceptibility was compared with that of wild-type controls. In the case of MCMV, doubly deficient mice were exquisitely sensitive to infection, wild-type and IFNGR1$^{-/-}$ animals were relatively resistant, and IFNAR1- and Stat1-deficient animals had intermediate sensitivity. These data suggest that IFN-α/β-mediated protection against MCMV is Stat1 dependent, but that in the absence of Stat1, IFN-γ dependent antiviral mechanisms are revealed. This is an extremely interesting concept and is supported by additional studies by those authors showing different gene expression profiles in Stat1-sufficient or Stat1$^{-/-}$ macrophages following IFN-γ treatment. Conversely, when the same experiment was carried out using Sindbis virus, wild-type mice were completely resistant and Stat1$^{-/-}$ mice nearly so. Doubly receptor-deficient mice (lacking IFNAR1 and IFNGR1) were again exquisitely sensitive, and IFNAR1$^{-/-}$ mice had intermediate sensitivity. Thus, resistance to Sindbis virus requires IFN-αR activation but is Stat1 independent, and IFN-γ again appears to have a protective effect in the absence of Stat1. A similar study using RSV gave yet a different result. It was found that peak lung viral titers did not differ in wild-type, IFNAR1$^{-/-}$, IFNGR1$^{-/-}$, or IFNAR$^{-/-}$IFNGR1$^{-/-}$ 129SvEv animals, but increased 100-fold in the absence of Stat1 (60). Therefore, in the case of RSV infection, antiviral defenses are type I and type II IFN independent, but require the presence of Stat1. Yet another pattern of susceptibility was seen with murine norovirus 1, where IFNAR1$^{-/-}$ and IFNGR1$^{-/-}$ mice had minimal disease, Stat1$^{-/-}$ animals were very sensitive, and mice lacking both the IFN-αR and the IFN-γR showed an intermediate sensitivity (65), suggesting that for this virus Stat1 activation by either type I or type II IFNs confers some protection.

A number of reports have described the formation of noncanonical ISRE-binding transcription factors, although the biological import of these complexes is not yet known. Bluyssen et al. have described activation of ISRE-containing promoters by a heterotrimer consisting of IRF9 and two molecules of Stat1 (11), as well as a complex made up of IRF9 and Stat2 (10). The observation that type I IFN treatment of bone marrow-derived DCs upregulates OAS transcription in a Stat2-dependent, Stat1-independent process also supports the idea that some aspects of type I IFN signaling may be mediated by alternative ISRE-binding factors (46).

SUMMARY

There is a large body of literature supporting the primacy of ISGF3-mediated IFN-α/β signaling in early, innate host defense against viral infection. Nonetheless, it is also clear that other pathways are also activated by viral pathogens. Our understanding of these events in vivo is complicated by the simultaneous production of additional inflammatory mediators by both parenchymal and immune cells and by the many ways that these additional cytokines can modulate Jak-Stat signaling. The chapters that appear later in this volume review what is known about the interactions of specific viral pathogens with the innate immune response, and it is our hope that the details of these interactions will broaden our understanding of antiviral defenses.

References

1. Aaronson, D., and C. Horvath. 2002. A road map for those who don't know JAK-STAT. *Science* **296:** 1653–1655.

2. Alexander, W., R. Starr, J. Fenner, C. Scott, E. Handman, N. Sprigg, J. Corbin, A. Cornish, R. Darwiche, C. Owczarek, T. Kay, N. Nicola, P. Hertzog, D. Metcalf, and D. Hilton. 1999. SOCS1 is a critical inhibitor of interferon gamma signaling and prevents the potentially fatal neonatal actions of this cytokine. *Cell* **98:**597–608.

3. Alexopoulou, L., A. Holt, R. Medzhitov, and R. Flavell. 2001. Recognition of dsRNA and activation of NF-κB by Toll-like receptor 3. *Nature* **413:**732–738.

4. Andrejeva, J., K. S. Childs, D. F. Young, T. S. Carlos, N. Stock, S. Goodbourn, and R. E. Randall. 2004. The V proteins of paramyxoviruses bind the IFN-inducible RNA helicase, mda-5, and inhibit its activation of the IFN-beta promoter. *Proc. Natl. Acad. Sci. USA* **101:** 17264–17269.

5. Andrejeva, J., D. F. Young, S. Goodbourn, and R. E. Randall. 2002. Degradation of STAT1 and STAT2 by the V proteins of simian virus 5 and human parainfluenza virus type 2, respectively: consequences for virus replication in the presence of alpha/beta and gamma interferons. *J. Virol.* **76:**2159–2167.

6. Ank, N., M. B. Iversen, C. Bartholdy, P. Staeheli, R. Hartmann, U. B. Jensen, F. Dagnaes-Hansen, A. R. Thomsen, Z. Chen, H. Haugen, K. Klucher, and S. R. Paludan. 2008. An important role for type III interferon (IFN-λ/IL-28) in TLR-induced antiviral activity. *J. Immunol.* **180:**2474–2485.

7. Bach, E., M. Aguet, and R. Schreiber. 1997. The IFN gamma receptor: a paradigm for cytokine receptor signaling. *Annu. Rev. Immunol.* **15:**563–591.

8. Begitt, A., T. Meyer, M. van Rossum, and U. Vinkemeier. 2000. Nucleocytoplasmic translocation of Stat1 is regulated by a leucine-rich export signal in the coiled-coil domain. *Proc. Natl. Acad. Sci. USA* **97:**10418–10423.

9. Biron, C. A., and G. C. Sen. 2001. Interferons and other cytokines, p. 321–352. *In* D. M. Knipe and P. M. Howley

(ed.), *Fundamental Virology*, 4th ed., vol. 4. Lippincott-Raven Publishers, New York, NY.

10. **Bluyssen, H., and D. Levy.** 1997. Stat2 is a transcriptional activator that requires sequence-specific contacts provided by Stat1 and p48 for stable interaction with DNA. *J. Biol. Chem.* **272:**4600–4605.

11. **Bluyssen, H., R. Muzaffar, R. Vlieststra, A. van der Made, S. Leung, G. Stark, I. Kerr, J. Trapman, and D. Levy.** 1995. Combinatorial association and abundance of components of interferon-stimulated gene factor 3 dictate the selectivity of interferon responses. *Proc. Natl. Acad. Sci. USA* **92:**5645–5649.

12. **Caignard, G., M. Guerbois, J. Labernardière, Y. Jacob, L. Jones, Infections Mapping Project I-MAP, F. Wild, F. Tangy, and P. Vidalain.** 2007. Measles virus V protein blocks Jak1-mediated phosphorylation of STAT1 to escape IFN-alpha/beta signaling. *Virology* **368:**351–362.

13. **Casanova, J. L., and L. Abel.** 2002. Genetic dissection of immunity to mycobacteria: the human model. *Annu. Rev. Immunol.* **20:**581–620.

14. **Chapgier, A., S. Boisson-Dupuis, E. Jouanguy, G. Vogt, J. Feinberg, A. Prochnicka-Chalufour, A. Casrouge, K. Yang, C. Soudais, C. Fieschi, O. F. Santos, J. Bustamante, C. Picard, L. de Beaucoudrey, J. F. Emile, P. D. Arkwright, R. D. Schreiber, C. Rolinck-Werninghaus, A. Rosen-Wolff, K. Magdorf, J. Roesler, and J. L. Casanova.** 2006. Novel STAT1 alleles in otherwise healthy patients with mycobacterial disease. *PLoS Genet.* **2:**e131.

15. **Chapgier, A., R. F. Wynn, E. Jouanguy, O. Filipe-Santos, S. Zhang, J. Feinberg, K. Hawkins, J. L. Casanova, and P. D. Arkwright.** 2006. Human complete Stat-1 deficiency is associated with defective type I and II IFN responses in vitro but immunity to some low virulence viruses in vivo. *J. Immunol.* **176:**5078–5083.

16. **Chen, X., U. Vinkemeier, Y. Zhao, D. Jeruzalmi, J. J. Darnell, and J. Kuriyan.** 1998. Crystal structure of a tyrosine phosphorylated STAT-1 dimer bound to DNA. *Cell* **93:**827–839.

17. **Coccia, E. M., M. Severa, E. Giacomini, D. Monneron, M. E. Remoli, I. Julkunen, M. Cella, R. Lande, and G. Uze.** 2004. Viral infection and Toll-like receptor agonists induce a differential expression of type I and lambda interferons in human plasmacytoid and monocyte-derived dendritic cells. *Eur. J. Immunol.* **34:**796–805.

18. **Collins, S., R. Noyce, and K. Mossman.** 2004. Innate cellular response to virus particle entry requries IRF3 but not virus replication. *J. Virol.* **78:**1706–1717.

19. **Conn, C., J. McClellan, H. Massab, C. Smitka, J. Majde, and M. Kluger.** 1995. Cytokines and the acute phase response to influenza virus in mice. *Am. J. Physiol.* **268:**R78–R84.

20. **Costa-Pereira, A. P., T. M. Williams, B. Strobl, D. Watling, J. Briscoe, and I. M. Kerr.** 2002. The antiviral response to gamma interferon. *J. Virol.* **76:**9060–9068.

20a. **Dalod, M. L. Malmgaard, C. Lewis, C. Asselin-Paturel, F. Brière, G. Trinchieri, and C. A. Biron** 2002. Interferon alpha/beta and interleukin 12 responses to viral infections: pathways regulating dendritic cell cytokine expression in vivo. *J. Exp. Med.* **195:**517–528.

21. **Darnell, J., I. Kerr, and G. Stark.** 1998. Jak-STAT pathways and transcriptional activation in response to IFN

and other extracellular signaling proteins. *Science* **264:**1415–1421.

22. **Darnell, J. J.** 1997. STATs and gene regulation. *Science* **277:**1630–1635.

23. **de Bouteiller, O., E. Merck, U. A. Hasan, S. Hubac, B. Benguigui, G. Trinchieri, E. E. Bates, and C. Caux.** 2005. Recognition of double-stranded RNA by human Toll-like receptor 3 and downstream receptor signaling requires multimerization and an acidic pH. *J. Biol. Chem.* **280:**38133–38145.

24. **Decker, T., and P. Kovarik.** 2000. Serine phosphorylation of Stats. *Oncogene* **19:**2628–2637.

25. **Der, S., A. Zhou, B. Williams, and R. Silverman.** 1998. Identification of genes differentially regulated by interferon alpha, beta, or gamma using oligonucleotide arrays. *Proc. Natl. Acad. Sci. USA* **95:**15623–15628.

26. **Diebold, S. S., T. Kaisho, H. Hemmi, S. Akira, and C. Reis e Sousa.** 2004. Innate antiviral responses by means of TLR7-mediated recognition of single-stranded RNA. *Science* **303:**1529–1531.

27. **Dorman, S. E., C. Picard, D. Lammas, K. Heyne, J. T. van Dissel, R. Baretto, S. D. Rosenzweig, M. Newport, M. Levin, J. Roesler, D. Kumararatne, J. L. Casanova, and S. M. Holland.** 2004. Clinical features of dominant and recessive interferon gamma receptor 1 deficiencies. *Lancet* **364:**2113–2121.

28. **Dufort, F., B. Bleiman, M. Gumina, D. Blair, D. Wagner, M. Roberts, Y. Abu-Amer, and T. Chiles.** 2007. Cutting edge: IL-4-mediated protection of primary B lymphocytes from apoptosis via Stat6-dependent regulation of glycolytic metabolism. *J. Immunol.* **179:**4953–4957.

29. **Dupuis, S., E. Jouanguy, S. Al-Hajjar, C. Fieschi, I. Z. Al-Mohsen, S. Al-Jumaah, K. Yang, A. Chapgier, C. Eidenschenk, P. Eid, A. Al Ghonaium, H. Tufenkeji, H. Frayha, S. Al-Gazlan, H. Al-Rayes, R. D. Schreiber, I. Gresser, and J. L. Casanova.** 2003. Impaired response to interferon-alpha/beta and lethal viral disease in human STAT1 deficiency. *Nat. Genet.* **33:**388–391.

30. **Durbin, J., A. Fernandez-Sesma, C. Lee, T. Rao, A. Frey, T. Moran, S. Vukmanovic, A. García-Sastre, and D. Levy.** 2000. Type I IFN modulates innate and specific antiviral immunity. *J. Immunol.* **164:**4220–4228.

31. **Durbin, J. E., R. Hackenmiller, M. C. Simon, and D. E. Levy.** 1996. Targeted disruption of the mouse Stat1 gene results in compromised innate immunity to viral disease. *Cell* **84:**443–450.

32. **Durbin, J. E., T. R. Johnson, R. K. Durbin, S. E. Mertz, R. A. Morotti, R. S. Peebles, and B. S. Graham.** 2002. The role of IFN in respiratory syncytial virus pathogenesis. *J. Immunol.* **168:**2944–2952.

33. **Fieschi, C., M. Bosticardo, L. de Beaucoudrey, S. Boisson-Dupuis, J. Feinberg, O. Santos, J. Bustamante, J. Levy, F. Candotti, and J. Casanova.** 2004. A novel form of complete IL-12/IL-23 receptor beta1 deficiency with cell surface-expressed nonfunctional receptors. *Blood* **104:**2095–2101.

34. **Frieman, M., B. Yount, M. Heise, S. Kopecky-Bromberg, P. Palese, and R. Baric.** 2007. Severe acute respiratory syndrome coronavirus ORF6 antagonizes STAT1 function by sequestering nuclear import factors on the rough endoplasmic reticulum/Golgi membrane. *J. Virol.* **81:**9812–9824.

35. Fu, X. 1992. A transcription factor with SH2 and SH3 domains is directly activated by an interferon alpha-induced cytoplasmic protein tyrosine kinase(s). *Cell* 70:323–335.

36. Fu, X., C. Schindler, T. Improta, R. Abersold, and J. Darnell. 1992. The proteins of ISGF3, the interferon-alpha induced transcriptional activator, define a gene family involved in signal transduction. *Proc. Natl. Acad. Sci. USA* 89:7840–7843.

37. Garcia-Sastre, A., R. K. Durbin, H. Zheng, P. Palese, R. Gertner, D. E. Levy, and J. E. Durbin. 1998. The role of interferon in influenza virus tissue tropism. *J. Virol.* 72:8550–8558.

38. Garcia-Sastre, A., A. Egorov, D. Matassov, S. Brandt, D. E. Levy, J. E. Durbin, P. Palese, and T. Muster. 1998. Influenza A virus lacking the NS1 gene replicates in interferon-deficient systems. *Virology* 252:324–330.

39. Georgel, P., Z. Jiang, S. Kunz, E. Janssen, J. Mols, K. Hoebe, S. Bahram, M. Oldstone, and B. Beutler. 2007. Vesicular stomatitis virus glycoprotein G activates a specific antiviral Toll-like receptor 4-dependent pathway. *Virology* 362:304–313.

40. Gil, M., E. Bohn, A. O'Guin, C. Ramana, B. Levine, G. Stark, H. Virgin, and R. Schreiber. 2001. Biologic consequences of Stat1-independent IFN signaling. *Proc. Natl. Acad. Sci. USA* 98:6680–6685.

41. Gil, M., R. Salomon, J. Louten, and C. Biron. 2006. Modulation of STAT1 protein levels: a mechanism shaping CD8 T-cell responses in vivo. *Blood* 107:987–993.

42. Goh, K., S. Haque, and B. Williams. 1999. p38 MAP kinase is required for STAT1 serine phosphorylation and transcriptional activation induced by interferons. *EMBO J.* 18:5601–5608.

43. Gotoh, B., K. Takeuchi, T. Komatsu, and J. Yokoo. 2003. The STAT2 activation process is a crucial target of Sendai virus C protein for the blockade of alpha interferon signaling. *J. Virol.* 77:3360–3370.

44. Guo, J., K. Peters, and G. Sen. 2000. Induction of the human protein P56 by interferon, double-stranded RNA, or virus infection. *Virology* 267:209–219.

45. Gupta, S., M. Jiang, and A. Pernis. 1999. IFN-alpha activates STAT6 and leads to the formation of Stat2:Stat6 complexes in B cells. *J. Immunol.* 163:3834–3841.

46. Hahm, B., M. Trifilo, E. Zuniga, and M. Oldstone. 2005. Viruses evade the immune system through type I interferon-mediated STAT2-dependent, but STAT1-independent, signaling. *Immunity* 22:247–257.

47. Hall, C. B., R. G. J. Douglas, R. L. Simons, and J. M. Geiman. 1978. Interferon production in children with RSV, influenza and parainfluenza virus infections. *J. Pediatr.* 93:28–32.

48. Haller, O., P. Staeheli, and G. Kochs. 2007. Interferon-induced Mx proteins in antiviral host defense. *Biochimie* 89:812–818.

49. He, B., G. Y. Lin, J. E. Durbin, R. K. Durbin, and R. A. Lamb. 2001. The SH integral membrane protein of the paramyxovirus simian virus 5 is required to block apoptosis in MDBK cells. *J. Virol.* 75:4068–4079.

50. Holland, S. M., F. R. DeLeo, H. Z. Elloumi, A. P. Hsu, G. Uzel, N. Brodsky, A. F. Freeman, A. Demidowich, J. Davis, M. L. Turner, V. L. Anderson, D. N. Darnell, P. A. Welch, D. B. Kuhns, D. M. Frucht, H. L. Malech, J. I. Gallin, S. D. Kobayashi, A. R. Whitney, J. M. Voyich, J. M. Musser, C. Woellner, A. A. Schaffer, J. M. Puck, and B. Grimbacher. 2007. STAT3 mutations in the hyper-IgE syndrome. *N. Engl. J. Med.* 357:1608–1619.

51. Honda, K., H. Yanai, T. Mizutani, H. Negishi, N. Shimada, N. Suzuki, Y. Ohba, A. Takaoka, W. Yeh, and T. Taniguchi. 2004. Role of a transductional-transcriptional processor complex involving MyD88 and IRF-7 in Toll-like receptor signaling. *Proc. Natl. Acad. Sci. USA* 101:15416–15421.

52. Hornung, V., J. Ellegast, S. Kim, K. Brzozka, A. Jung, H. Kato, H. Poeck, S. Akira, K. K. Conzelmann, M. Schlee, S. Endres, and G. Hartmann. 2006. 5′-Triphosphate RNA is the ligand for RIG-I. *Science* 314:994–997.

53. Hornung, V., J. Schlender, M. Guenthner-Biller, S. Rothenfusser, S. Endres, K. Conzelmann, and G. Hartmann. 2004. Replication-dependent potent IFN-alpha induction in human plasmacytoid dendritic cells by a single-stranded RNA virus. *J. Immunol.* 173:5935–5943.

54. Horvath, C., and J. J. Darnell. 1996. The antiviral state induced by alpha interferon and gamma interferon requires transcriptionally active Stat1 protein. *J. Virol.* 70:647–650.

55. Horvath, C., Z. Wen, and J. J. Darnell. 1995. A STAT protein domain that determines DNA sequence recognition suggests a novel DNA-binding domain. *Genes. Dev.* 9:984–994.

56. Huang, S., W. Hendriks, A. Althage, S. Hemmi, H. Bluethmann, R. Kamijo, J. Vilcek, R. M. Zinkernagel, and M. Aguet. 1993. Immune response in mice that lack the interferon-gamma receptor. *Science* 259:1742–1745.

57. Ichiba, M., K. Nakajima, Y. Yamanaka, N. Kiuchi, and T. Hirano. 2001. Autoregulation of the Stat3 gene through cooperation with a cAMP-responsive element-binding protein. *J. Biol. Chem.* 273:6132–6138.

58. Jewell, N. A., N. Vaghefi, S. E. Mertz, P. Akter, R. S. Peebles, Jr., L. O. Bakaletz, R. K. Durbin, E. Flano, and J. E. Durbin. 2007. Differential type I interferon induction by respiratory syncytial virus and influenza A virus in vivo. *J. Virol.* 81:9790–9800.

59. Johnsen, I. B., T. T. Nguyen, M. Ringdal, A. M. Tryggestad, O. Bakke, E. Lien, T. Espevik, and M. W. Anthonsen. 2006. Toll-like receptor 3 associates with c-Src tyrosine kinase on endosomes to initiate antiviral signaling. *EMBO J.* 25:3335–3346.

60. Johnson, T. R., S. E. Mertz, N. Gitiban, S. Hammond, R. Legallo, R. K. Durbin, and J. E. Durbin. 2005. Role for innate IFNs in determining respiratory syncytial virus immunopathology. *J. Immunol.* 174:7234–7241.

61. Jouanguy, E., S. Y. Zhang, A. Chapgier, V. Sancho-Shimizu, A. Puel, C. Picard, S. Boisson-Dupuis, L. Abel, and J. L. Casanova. 2007. Human primary immunodeficiencies of type I interferons. *Biochimie* 89:878–883.

62. Kaplan, M., U. Schindler, S. Smiley, and M. Grusby. 1996. Stat6 is required for mediating responses to IL-4 and for the development of Th2 cells. *Immunity* 4:313–319.

63. Kaplan, M., Y. Sun, T. Hoey, and M. Grusby. 1996. Impaired IL-12 responses and enhanced development of Th2 cells in Stat4-deficient mice. *Nature* 382:174–177.

64. Karaghiosoff, M., H. Neubauer, C. Lassnig, P. Kovarik, H. Schindler, H. Pircher, B. McCoy, C. Bogdan, T. Decker, G. Brem, K. Pfeffer, and M. Muller. 2000. Partial

impairment of cytokine responses in Tyk2-deficient mice. *Immunity* 13:549–560.

65. Karst, S., C. Wobus, M. Lay, J. Davidson, and H. Virgin IV. 2003. STAT1-dependent innate immunity to a Norwalk-like virus. *Science* 299:1575–1578.

66. Kawai, T., and S. Akira. 2007. Antiviral signaling through pattern recognition receptors. *J. Biochem.* 141: 137–145.

67. Kawai, T., K. Takahashi, S. Sato, C. Coban, H. Kumar, H. Kato, K. J. Ishii, O. Takeuchi, and S. Akira. 2005. IPS-1, an adaptor triggering RIG-I- and Mda5-mediated type I interferon induction. *Nat. Immunol.* 6:981–988.

68. Kessler, D., D. Levy, and J. Darnell. 1988. Two interferon-induced nuclear factors bind a single promoter element in interferon-stimulated genes. *Proc. Natl. Acad. Sci. USA* 85:8521–8525.

69. Kotenko, S., G. Gallagher, V. Baurin, A. Lewis-Antes, M. Shen, N. Shah, J. Langer, F. Sheikh, H. Dickensheets, and R. Donnelly. 2003. IFN-lambdas mediate antiviral protection through a distinct class II cytokine receptor complex. *Nat. Immunol.* 4:69–77.

70. Kurt-Jones, E., L. Popova, L. Kwinn, L. Haynes, L. Jones, R. Tripp, E. Walsh, M. Freeman, D. Golenbock, L. Anderson, and R. Finberg. 2000. Pattern recognition receptors TLR4 and CD14 mediate response to respiratory syncytial virus. *Nat. Immunol.* 1:398–401.

71. Lamken, P., S. Lata, M. Gavutis, and J. Piehler. 2004. Ligand-induced assembling of the type I interferon receptor on supported lipid bilayers. *J. Mol. Biol.* 341:303–318.

72. Lee, H., J. Lund, B. Ramanathan, N. Mizushima, and A. Iwasaki. 2007. Autophagy-dependent viral recognition by plasmacytoid dendritic cells. *Science* 315:1398–1401.

73. Lin, R., P. Génin, Y. Mamane, and J. Hiscott. 2000. Selective DNA binding and association with the CREB binding protein coactivator contribute to differential activation of alpha/beta interferon genes by interferon regulatory factors 3 and 7. *Mol. Cell. Biol.* 20:6342–6353.

74. Lin, R., C. Heylbroeck, P. Pitha, and J. Hiscott. 1998. Virus-dependent phosphorylation of the IRF-3 transcription factor regulates nuclear translocation, transactivation potential, and proteasome-mediated degradation. *Mol. Cell. Biol.* 18:2986–2996.

75. Lin, R., C. Liao, E. Lin, and Y. Lin. 2004. Blocking of the alpha interferon-induced Jak-Stat signaling pathway by Japanese encephalitis virus infection. *J. Virol.* 78: 9285–9294.

76. [Reference deleted.]

77. Liu, X., G. W. Robinson, K. U. Wagner, L. Garrett, A. Wynshaw-Boris, and L. Hennighausen. 1997. Stat5a is mandatory for adult mammary gland development and lactogenesis. *Genes. Dev.* 11:179–186.

78. Lo, M., R. Brazas, and M. Holtzman. 2005. Respiratory syncytial virus nonstructural proteins NS1 and NS2 mediate inhibition of Stat2 expression and alpha/beta interferon responsiveness. *J. Virol.* 79:9315–9319.

79. Loo, Y. M., J. Fornek, N. Crochet, G. Bajwa, O. Perwitasari, L. Martinez-Sobrido, S. Akira, M. A. Gill, A. Garcia-Sastre, M. G. Katze, and M. Gale, Jr. 2008. Distinct RIG-I and MDA5 signaling by RNA viruses in innate immunity. *J. Virol.* 82:335–345.

80. Louten, J., N. van Rooijen, and C. Biron. 2006. Type 1 IFN deficiency in the absence of normal splenic architecture

during lymphocytic choriomeningitis virus infection. *J. Immunol.* 177:3266–3272.

81. Marie, I., J. E. Durbin, and D. E. Levy. 1998. Differential viral induction of distinct interferon-alpha genes by positive feedback through interferon regulatory factor-7. *EMBO J.* 17:6660–6669.

82. Marine, J., D. Topham, C. McKay, D. Wang, E. Parganas, D. Stravopodis, A. Yoshimura, and J. Ihle. 1999. SOCS1 deficiency causes a lymphocyte-dependent perinatal lethality. *Cell* 98:609–616.

83. McIntosh, K. 1978. Interferon in nasal secretions from infants with viral respiratory tract infections. *J. Pediatr.* 93:33–36.

84. Meraz, M. A., J. M. White, K. C. Sheehan, E. A. Bach, S. J. Rodig, A. S. Dighe, D. H. Kaplan, J. K. Riley, A. C. Greenlund, D. Campbell, K. Carver-Moore, R. N. DuBois, R. Clark, M. Aguet, and R. D. Schreiber. 1996. Targeted disruption of the Stat1 gene in mice reveals unexpected physiologic specificity in the JAK-STAT signaling pathway. *Cell* 84:431–442.

85. Meylan, E., J. Curran, K. Hofmann, D. Moradpour, M. Binder, R. Bartenschlager, and J. Tschopp. 2005. Cardif is an adaptor protein in the RIG-I antiviral pathway and is targeted by hepatitis C virus. *Nature* 437:1167–1172.

86. Mibayashi, M., L. Martinez-Sobrido, Y. Loo, W. Cardenas, M. Gale, Jr., and A. Garcia-Sastre. 2007. Inhibition of retinoic acid-inducible gene I-mediated induction of beta interferon by the NS1 protein of influenza A virus. *J. Virol.* 81:514–524.

87. Minegishi, Y., M. Saito, T. Morio, K. Watanabe, K. Agematsu, S. Tsuchiya, H. Takada, T. Hara, N. Kawamura, T. Ariga, H. Kaneko, N. Kondo, I. Tsuge, A. Yachie, Y. Sakiyama, T. Iwata, F. Bessho, T. Ohishi, K. Joh, K. Imai, K. Kogawa, M. Shinohara, M. Fujieda, H. Wakiguchi, S. Pasic, M. Abinun, H. D. Ochs, E. D. Renner, A. Jansson, B. H. Belohradsky, A. Metin, N. Shimizu, S. Mizutani, T. Miyawaki, S. Nonoyama, and H. Karasuyama. 2006. Human tyrosine kinase 2 deficiency reveals its requisite roles in multiple cytokine signals involved in innate and acquired immunity. *Immunity* 25:745–755.

88. Miyagi, T., M. P. Gil, X. Wang, J. Louten, W. M. Chu, and C. A. Biron. 2007. High basal STAT4 balanced by STAT1 induction to control type 1 interferon effects in natural killer cells. *J. Exp. Med.* 204:2383–2396.

89. Moriggl, R., V. Gouilleux-Gruart, R. Jahne, S. Berchtold, C. Gartmann, X. Liu, L. Hennighausen, A. Sotiropoulos, B. Groner, and F. Gouilleux. 1996. Deletion of the carboxyl-terminal transactivation domain of MGF-Stat5 results in sustained DNA binding and a dominant negative phenotype. *Mol. Cell. Biol.* 16:5691–5700.

90. Muller, U., U. Steinhoff, L. Reis, Hemmi, S. J. Pavlovic, R. Zinkernagel, and M. Aguet. 1994. Functional role of type I and type II interferons in antiviral defense. *Science* 264:1918–1921.

91. Mustelin, T., T. Vang, and N. Bottini. 2005. Protein tyrosine phosphatases and the immune response. *Nat. Rev. Immunol.* 5:43–57.

92. Nair, J., C. DaFonseca, A. Tjernberg, W. Sun, J. J. Darnell, B. Chait, and J. Zhang. 2002. Requirement of Ca2+ and CaMKII for Stat1 Ser-727 phosphorylation in response to IFN-gamma. *Proc. Natl. Acad. Sci. USA* 99:5971–5976.

93. Naka, T., T. Matsumoto, M. Narazaki, M. Fujimoto, Y. Morita, Y. Ohsawa, H. Saito, T. Nagasawa, Y. Uchiyama, and T. Kishimoto. 1998. Accelerated apoptosis of lymphocytes by augmented induction of Bax in SSI-1 (STAT-induced STAT inhibitor-1) deficient mice. *Proc. Natl. Acad. Sci. USA* **95:**15577–15582.

94. Nguyen, K. B., W. T. Watford, R. Salomon, S. R. Hofmann, G. C. Pien, A. Morinobu, M. Gadina, J. J. O'Shea, and C. A. Biron. 2002. Critical role for STAT4 activation by type 1 interferons in the interferon-gamma response to viral infection. *Science* **297:**2063–2066.

95. Noyce, R. S., S. E. Collins, and K. L. Mossman. 2006. Identification of a novel pathway essential for the immediate-early, interferon-independent antiviral response to enveloped virions. *J. Virol.* **80:**226–235.

96. Onoguchi, K., M. Yoneyama, A. Takemura, S. Akira, T. Taniguchi, H. Namiki, and T. Fujita. 2007. Viral infections activate types I and III interferon genes through a common mechanism. *J. Biol. Chem.* **282:**7576–7581.

97. Osterlund, P., T. Pietilä, V. Veckman, S. Kotenko, and I. Julkunen. 2007. IFN regulatory factor family members differentially regulate the expression of type III IFN (IFN-lambda) genes. *J. Immunol.* **179:**3434–3442.

98. Ozaki, K., K. Kikly, D. Michalovich, P. R. Young, and W. J. Leonard. 2000. Cloning of a type I cytokine receptor most related to the IL-2 receptor beta chain. *Proc. Natl. Acad. Sci. USA* **97:**11439–11444.

99. Paladino, P., D. T. Cummings, R. S. Noyce, and K. L. Mossman. 2006. The IFN-independent response to virus particle entry provides a first line of antiviral defense that is independent of TLRs and retinoic acid-inducible gene I. *J. Immunol.* **177:**8008–8016.

100. Parham, C., M. Chirica, J. Timans, E. Vaisberg, M. Travis, J. Cheung, S. Pflanz, R. Zhang, K. Singh, F. Vega, W. To, J. Wagner, A. O'Farrell, T. McClanahan, S. Zurawski, C. Hannum, D. Gorman, D. Rennick, R. Kastelein, R. de Waal Malefyt, and K. Moore. 2002. A receptor for the heterodimeric cytokine IL-23 is composed of IL-12Rbeta1 and a novel cytokine receptor subunit, IL-23R. *J. Immunol.* **168:**5699–5708.

101. Park, C., S. Li, E. Cha, and C. Schindler. 2000. Immune response in Stat2 knockout mice. *Immunity* **13:**795–804.

102. Parrish-Novak, J., S. R. Dillon, A. Nelson, A. Hammond, C. Sprecher, J. A. Gross, J. Johnston, K. Madden, W. Xu, J. West, S. Schrader, S. Burkhead, M. Heipel, C. Brandt, J. L. Kuijper, J. Kramer, D. Conklin, S. R. Presnell, J. Berry, F. Shiota, S. Bort, K. Hambly, S. Mudri, C. Clegg, M. Moore, F. J. Grant, C. Lofton-Day, T. Gilbert, F. Rayond, A. Ching, L. Yao, D. Smith, P. Webster, T. Whitmore, M. Maurer, K. Kaushansky, R. D. Holly, and D. Foster. 2000. Interleukin 21 and its receptor are involved in NK cell expansion and regulation of lymphocyte function. *Nature* **408:**57–63.

103. Pestka, S., C. Krause, and M. Walther. 2004. Interferons, interferon-like cytokines, and their receptors. *Immunol. Rev.* **202:**8–32.

104. Pestka, S., J. Langer, K. Zoon, and C. Samuel. 1987. Interferons and their actions. *Ann. Rev. Biochem.* **56:**727–777.

105. Pichlmair, A., O. Schulz, C. P. Tan, T. I. Naslund, P. Liljestrom, F. Weber, and C. Reis e Sousa. 2006. RIG-I-mediated antiviral responses to single-stranded RNA bearing 5′-phosphates. *Science* **314:**997–1001.

106. Pilz, A., K. Ramsauer, H. Heidari, M. Leitges, P. Kovarik, and T. Decker. 2003. Phosphorylation of the Stat1 transactivating domain is required for the response to type I IFNs. *EMBO Rep.* **4:**368–373.

107. Plantanias, L. 2005. Mechanisms of type I and type II interferon-mediated signaling. *Nat. Rev. Immunol.* **5:**375–386.

108. Plumet, S., F. Herschke, J. M. Bourhis, H. Valentin, S. Longhi, and D. Gerlier. 2007. Cytosolic 5′-triphosphate ended viral leader transcript of measles virus as activator of the RIG I-mediated interferon response. *PLoS ONE* **2:**e279.

109. Qing, Y., and G. Stark. 2004. Alternative activation of Stat1 and Stat3 in response to IFN-γ. *J. Biochem.* **279:**41679–41685.

110. Ramaswamy, M., L. Shi, M. Monick, G. Hunninghake, and D. Look. 2004. Specific inhibition of type I interferon signal transduction by respiratory syncytial virus. *Am. J. Respir. Cell Mol. Biol.* **30:**893–900.

111. Randall, R., and S. Goodbourn. 2008. Interferons and viruses: an interplay between induction, signalling, antiviral responses and virus countermeasures. *J. Gen. Virol.* **89:**1–47.

112. Roberts, A., L. Robb, S. Rakar, L. Hartley, L. Cluse, N. Nicola, D. Metcalf, D. Hilton, and W. Alexander. 2001. Placental defects and embryonic lethality in mice lacking suppressor of cytokine signaling 3. *Proc. Natl. Acad. Sci. USA* **98:**9324–9329.

113. Rodriguez, J., L. Wang, and C. Horvath. 2003. Hendra virus V protein inhibits interferon signaling by preventing Stat1 and Stat2 nuclear accumulation. *J. Virol.* **77:**11842–11845.

114. Sato, M., H. Suemori, N. Hata, M. Asagiri, K. Ogasawara, K. Nakao, T. Nakaya, M. Katsuki, S. Noguchi, N. Tanaka, and T. Taniguchi. 2000. Distinct and essential roles of transcription factors IRF-3 and IRF-7 in response to viruses for IFN-alpha/beta gene induction. *Immunity* **13:**539–548.

115. Sato, S., M. Suglyama, M. Yamamoto, Y. Watanabe, T. Kawai, K. Takeda, and S. Akira. 2003. Toll/IL-1 receptor domain-containing adaptor inducing IFN-β (TRIF) associates with TNF receptor associated factor 6 and TANK-binding kinase 1, and activates two distinct transcription factors, NF-κB and IFN-regulator factor-3, in the Toll-like receptor signaling. *J. Immunol.* **171:**4304–4310.

116. Schindler, C., and J. Darnell. 1995. Transcriptional responses to polypeptide ligands: the JAK-STAT pathway. *Annu. Rev. Biochem.* **64:**621–651.

117. Schindler, C., X. Fu, T. Improta, R. Abersold, and J. Darnell. 1992. Proteins of transcription factor ISGF-3: one gene encodes the 91 and 84 kDa ISGF-3 proteins that are activated by interferon-α. *Proc. Natl. Acad. Sci. USA* **89:**7836–7839.

118. Schindler, C., D. Levy, and T. Decker. 2007. JAK-STAT signaling: from interferons to cytokines. *J. Biol. Chem.* **282:**20059–20063.

119. Schindler, C., K. Shuai, V. Prezioso, and J. Darnell, Jr. 1992. Interferon-dependent tyrosine phosphorylation of a latent cytoplasmic transcription factor. *Science* **257:**808–813.

120. Seth, R. B., L. Sun, C. K. Ea, and Z. J. Chen. 2005. Identification and characterization of MAVS, a mitochondrial antiviral signaling protein that activates NF-kappaB and IRF 3. *Cell* 122:669–682.

121. Sharma, S., B. R. tenOever, N. Grandvaux, G. P. Zhou, R. Lin, and J. Hiscott. 2003. Triggering the interferon antiviral response through an IKK-related pathway. *Science* 300:1148–1151.

122. Sheppard, P., W. Kindsvogel, W. Xu, K. Henderson, S. Schlutsmeyer, T. E. Whitmore, R. Kuestner, U. Garrigues, C. Birks, J. Roraback, C. Ostrander, D. Dong, J. Shin, S. Presnell, B. Fox, B. Haldeman, E. Cooper, D. Taft, T. Gilbert, F. J. Grant, M. Tackett, W. Krivan, G. McKnight, C. Clegg, D. Foster, and K. M. Klucher. 2003. IL-28, IL-29 and their class II cytokine receptor IL-28R. *Nat. Immunol.* 4:63–68.

123. Shimoda, K., K. Kato, K. Aoki, T. Matsuda, A. Miyamoto, M. Shibamori, M. Yamashita, A. Numata, K. Takase, S. Kobayashi, S. Shibata, Y. Asano, H. Gondo, K. Sekiguchi, K. Nakayama, T. Nakayama, T. Okamura, S. Okamura, Y. Niho, and K. Nakayama. 2000. Tyk2 plays a restricted role in IFN alpha signaling, although it is required for IL-12-mediated T cell function. *Immunity* 13:561–571.

124. Shimoda, K., J. van Deursen, M. Y. Sangster, S. R. Sarawar, R. T. Carson, R. A. Tripp, C. Chu, F. W. Quelle, T. Nosaka, D. A. Vignali, P. C. Doherty, G. Grosveld, W. E. Paul, and J. N. Ihle. 1996. Lack of IL-4-induced Th2 response and IgE class switching in mice with disrupted Stat6 gene. *Nature* 380:630–633.

125. Shresta, S., K. Sharar, D. Prigozhin, H. Snider, P. Beatty, and E. Harris. 2005. Critical roles for both STAT1-dependent and STAT1-independent pathways in the control of primary dengue virus infection in mice. *J. Immunol.* 175:3946–3954.

126. Shuai, K., and B. Liu. 2003. Regulation of JAK-STAT signalling in the immune system. *Nat. Rev. Immunol.* 3:900–911.

127. Shuai, K., G. Stark, I. Kerr, and J. Darnell. 1993. A single phosphotyrosine residue of Stat91 required for gene activation by interferon-γ. *Science* 261:1744–1746.

128. Song, M., and K. Shuai. 1998. The suppressor of cytokine signaling (SOCS) 1 and SOCS3 but not SOCS2 proteins inhibit interferon-mediated antiviral and antiproliferative activities. *J. Biol. Chem.* 273:35056–35062.

129. Stark, G. 2007. How cells respond to interferons revisited: from early history to current complexity. *Cytokine Growth Factor Rev.* 18:419–423.

130. Stark, G., I. Kerr, B. Williams, R. Silverman, and R. Schreiber. 1998. How cells respond to interferons. *Annu. Rev. Biochem.* 87:227–264.

131. Starr, R., and D. Hilton. 2003. SOCS proteins. Negative regulators of the JAK/STAT pathway, p. 55–73. *In* P. Sehgal, D. Levy, and T. Hirano (ed.), *Signal Transducers and Activators of Transcription (STATs)*. Kluwer Academic Publishers, Dordrecht, The Netherlands.

132. Starr, R., D. Metcalf, A. Elefanty, M. Brysha, T. Willson, N. Nicola, D. Hilton, and W. Alexander. 1998. Liver degeneration and lymphoid deficiencies in mice lacking suppressor of cytokine signaling-1. *Proc. Natl. Acad. Sci. USA* 95:14395–14399.

133. Subbarao, K., and A. Roberts. 2006. Is there an ideal animal model for SARS? *Trends Microbiol.* 14:299–303.

134. Takeda, K., K. Noguchi, W. Shi, T. Tanaka, M. Matsumoto, N. Yoshida, T. Kishimoto, and S. Akira. 1997. Targeted disruption of the mouse Stat3 gene leads to early embryonic lethality. *Proc. Natl. Acad. Sci. USA.* 94:3801–3804.

135. Takeda, K., T. Tanaka, W. Shi, M. Matsumoto, M. Minami, S. Kashiwamura, K. Nakanishi, N. Yoshida, T. Kishimoto, and S. Akira. 1996. Essential role of Stat6 in IL-4 signalling. *Nature* 380:627–630.

136. Tang, X., J. S. Gao, Y. J. Guan, K. E. McLane, Z. L. Yuan, B. Ramratnam, and Y. E. Chin. 2007. Acetylation-dependent signal transduction for type I interferon receptor. *Cell* 131:93–105.

137. Teglund, S., C. McKay, E. Schuetz, J. M. van Deursen, D. Stravopodis, D. Wang, M. Brown, S. Bodner, G. Grosveld, and J. N. Ihle. 1998. Stat5a and Stat5b proteins have essential and nonessential, or redundant, roles in cytokine responses. *Cell* 93:841–850.

138. tenOever, B. R., S. L. Ng, M. A. Chua, S. M. McWhirter, A. Garcia-Sastre, and T. Maniatis. 2007. Multiple functions of the IKK-related kinase IKKepsilon in interferon-mediated antiviral immunity. *Science* 315:1274–1278.

139. Thierfelder, W., J. Van Deursen, K. Yamamoto, R. Tripp, S. Sarawar, R. Carson, M. Sangster, D. Vignali, P. Doherty, G. Grosveld, and J. Ihle. 1996. Requirement for Stat4 in IL-12 mediated response of NK and T cells. *Nature* 382:171–174.

140. Tissari, J., J. Sirén, S. Meri, I. Julkunen, and S. Matikainen. 2005. IFN-alpha enhances TLR3-mediated antiviral cytokine expression in human endothelial and epithelial cells by up-regulating TLR3 expression. *J. Immunol.* 174:4289–4294.

141. Udy, G. B., R. P. Towers, R. G. Snell, R. J. Wilkins, S. H. Park, P. A. Ram, D. J. Waxman, and H. W. Davey. 1997. Requirement of STAT5b for sexual dimorphism of body growth rates and liver gene expression. *Proc. Natl. Acad. Sci. USA* 94:7239–7244.

142. van Boxel-Dazaire, A., M. Rani, and G. Stark. 2006. Complex modulation of cell type-specific response to type I interferons. *Immunity* 25:361–372.

143. van den Broek, M. F., U. Muller, S. Huang, M. Aguet, and R. M. Zinkernagel. 1995. Antiviral defense in mice lacking both alpha/beta and gamma interferon receptors. *J. Virol.* 69:4792–4796.

144. Varinou, L., K. Ramsauer, M. Karaghiosoff, T. Kolbe, K. Pfeffer, M. Müller, and T. Decker. 2003. Phosphorylation of the Stat1 transactivation domain is required for full-fledged IFN-gamma-dependent innate immunity. *Immunity* 19:793–802.

145. Wang, I., H. Lin, S. Goldman, and M. Kobayashi. 2004. STAT-1 is activated by IL-4 and IL-13 in multiple cell types. *Mol. Immunol.* 41:873–884.

146. Wang, X., S. Hussain, E. Wang, X. Wang, M. Li, A. García-Sastre, and A. Beg. 2007. Lack of essential role of NF-kappa B p50, RelA, and cRel subunits in virus-induced type 1 IFN expression. *J. Immunol.* 178:6770–6776.

147. Wathelet, M., C. Lin, B. Parekh, L. Ronco, P. Howley, and T. Maniatis. 1998. Virus infection induces the assembly of coordinately activated transcription factors on the IFN-beta enhancer in vivo. *Mol. Cell* 1:507–518.

148. Weaver, B., K. Kumar, and N. Reich. 1998. Interferon regulatory factor 3 and CREB-binding protein/p300 are subunits of double-stranded RNA-activated transcription factor DRAF1. *Mol. Cell. Biol.* **18:**1359–1368.

149. Weber, F., V. Wagner, S. B. Rasmussen, R. Hartmann, and S. R. Paludan. 2006. Double-stranded RNA is produced by positive-strand RNA viruses and DNA viruses but not in detectable amounts by negative-strand RNA viruses. *J. Virol.* **80:**5059–5064.

150. Wong, L., I. Hatzinisiriou, R. Devenish, and S. Ralph. 1998. IFN-gamma priming up-regulates IFN-stimulated gene factor 3 (ISGF3) components, augmenting responsiveness of IFN-resistant melanoma cells to type I IFNs. *J. Immunol.* **160:**5475–5484.

151. Xu, L. G., Y. Y. Wang, K. J. Han, L. Y. Li, Z. Zhai, and H. B. Shu. 2005. VISA is an adapter protein required for virus-triggered IFN-beta signaling. *Mol. Cell.* **19:**727–740.

152. Yoneyama, M., W. Suhara, Y. Fukuhara, M. Fukuda, E. Nishida, and T. Fujita. 1998. Direct triggering of the type I interferon system by virus infection: activation of a transcription factor complex containing IRF-3 and CBP/p300. *EMBO J.* **17:**1087–1095.

153. Zhou, Z., O. J. Hamming, N. Ank, S. R. Paludan, A. L. Nielsen, and R. Hartmann. 2007. Type III interferon (IFN) induces a type I IFN-like response in a restricted subset of cells through signaling pathways involving both the Jak-STAT pathway and the mitogen-activated protein kinases. *J. Virol.* **81:**7749–7758.

Cellular Signaling and Innate Immune
Responses to RNA Virus Infections
Edited by A. R. Brasier et al.
© 2009 ASM Press, Washington, D.C.

Christine L. White
Ganes C. Sen

Interferons and Antiviral Action

7

THE INTERFERON SYSTEM

Viruses and their hosts have evolved to coexist by maintaining viral homeostasis. At the organism level, the immune system of the host plays a major role in clearing the infection or driving the viruses to enter a latent phase. In addition to the direct action of the cells of the immune system, various cytokines, most importantly the interferons (IFNs), produced by them are critically important in this process. The infected cells themselves, almost of every cell lineage, also respond to virus infection by activating many cellular regulatory processes including immediate stress response, apoptosis, autophagy, and the induction of transcription of many cellular genes, collectively called the viral stress-inducible genes, many of which directly or indirectly block virus replication. Some of them encode cytoplasmic or nuclear proteins that directly interfere with specific steps of viral gene expression, while others encode secreted proteins, such as IFNs. IFNs are not themselves antiviral, but they induce expression of antiviral genes in as yet uninfected cells to prepare them for oncoming virus infection. Many of the viral stress-inducible genes are induced by IFNs as well, and they have been originally identified as interferon-stimulated genes (ISGs).

As indicated above, many ISGs can be induced by not only IFNs, but also many viral gene products including DNA, double-stranded (ds) and single-stranded (ss) RNA, and glycoproteins. Moreover, other microbial products, such as bacterial lipopolysaccharides (LPSs), can also induce these genes, suggesting their broad antimicrobial effects. In evolution, some of these genes might have existed before the IFN system developed in the vertebrates as a complex defense mechanism against infectious agents. Indeed, they can be induced via Toll-like receptors (TLRs), which are present even in insects. Thus, the ISGs should be viewed as genes that encode antimicrobial proteins in the broadest sense.

The IFN system, viruses, and dsRNA are strongly connected to one another. Many viruses produce dsRNA during replication, and both virus infection and exogenous dsRNA induce IFN synthesis. IFN induces the synthesis of several enzymes that require dsRNA as a cofactor or a substrate. The enzyme adenosine deaminase acting on RNA (ADAR) recognizes dsRNA regions of an mRNA and edits specific nucleotide residues in it. Protein kinase RNA regulated (PKR) is a latent protein kinase that is activated by a conformational change induced by dsRNA binding. Similarly, the 2′-5′-oligoadenylate

Christine L. White and Ganes C. Sen, Department of Molecular Genetics, Cleveland Clinic, 9500 Euclid Ave., Cleveland, OH 44195.

synthetases (OAS) are enzymatically inactive without dsRNA. All of these enzymes are deleterious to cells; hence keeping them latent until required is a good survival strategy. In this context, the IFN system functions in two steps; in the first, IFN induces synthesis of these enzymes, and in the second, dsRNA, presumably produced during virus replication, activates them. This strategy allows all cells to be ready for oncoming virus infection, but only in the infected ones do the inhibitory effects of the enzymes block both viral and cellular metabolic processes, often causing premature death of the infected cells, an effective antiviral strategy for multicellular organisms.

Because most ISGs encode proteins that can inhibit cell growth and survival, their expression is tightly controlled at multiple levels. Their induction by all signaling pathways is transient, a process regulated by cessation of the initiating signal, downregulation of the receptor, inactivation or degradation of the signaling kinases, inactivation of the newly activated transcription factors, and induction of signaling inhibitors and phosphatases. The induced ISG mRNAs and proteins are also often short-lived, ensuring their transient action. However, in contrast to these elaborate mechanisms for self-limiting the duration of ISG induction and function, many signaling proteins, including receptors and transcription factors, are themselves induced by the signals they propagate, which sets up positive feedback regulatory loops.

SIGNALING PATHWAYS

Because this topic is extensively covered in other chapters, here we only outline the general features of the signaling pathways that lead to ISG induction. The majority, if not all, of the ISGs that are induced by IFN, dsRNA, and viruses contain IFN-stimulated response elements (ISREs) in their promoters. These elements are recognized by members of the IFN regulatory factor (IRF) transcription factor family, and consequently all signaling pathways that induce these genes activate one or more IRFs or transcription factor complexes that contain an IRF. The general activation patterns are similar: upon binding of the ligand to membrane-bound or cytoplasmic receptors, protein kinases are activated by their recruitment to the receptor directly or via adaptor proteins. The activated protein kinases phosphorylate the latent transcription factors present in the cytoplasm and activate them. Active transcription factors translocate to the nucleus, bind to the ISREs of ISGs, and induce their transcription. Type I IFNs use the Jak/Stat pathway to activate the transcription factor ISGF3, which is a trimeric complex of IRF9 and Tyr-phosphorylated Stat1

and Stat2. Viruses and dsRNA, on the other hand, activate IRF3, IRF5, or IRF7 by Ser phosphorylation. These activated dimeric IRFs can induce transcription. The TLRs are the proteins that initially recognize the viral gene products. TLR3, TLR7/8, and TLR9 reside primarily in the endosomal membrane and recognize viral nucleic acids. Some viral glycoproteins are recognized by TLR4 and TLR2 present in the plasma membrane. dsRNA is recognized not only by TLR3, but also by the members of the RIG-I-like RNA helicase (RLH) family of RNA helicases present in the cytoplasm. These proteins, retinoic acid-inducible gene I (RIG-I) and melanoma differentiation-associated gene 5 (MDA-5), can recognize not only dsRNA but also ssRNAs that have uncapped phosphorylated 5′ ends. The components of a similar cytoplasmic DNA recognition system have just started to be identified. TLR2, TLR4, TLR7/8, and TLR9 all use the adaptor protein MyD88, whereas TLR3 uses TRIF, which is also recruited by TLR4. RIG-I and MDA-5 use the mitochondria-bound adaptor IPS-1. Most of these adaptors, most notably TRIF and IPS-1, recruit the serine kinases TBK1 and IKK-ε to the receptors. These kinases phosphorylate and activate IRF3 and IRF7. In the case of TLR3, additional phosphorylation of IRF3 by a PI3 kinase-mediated pathway is needed for its full activation. The specific pathway used for IRF3 activation varies among viruses and the cell types. For example, influenza A virus uses primarily the RIG-I pathway in myeloid dendritic cells but the TLR7/8 pathway in plasmacytoid dendritic cells. Although it is correct to think that the primary function of the TLRs is to elicit host responses that impair virus replication, there are specific examples of TLRs enhancing viral pathogenesis in human and mouse.

ANTIVIRAL ACTIONS OF ISG PRODUCTS

No single ISG product serves as the antiviral magic bullet for any virus studied to date. The usual mechanism calls for inhibition of several steps of viral gene expression through the actions of several ISG products. Because viral DNA, RNA, and protein syntheses are interdependent, even a relatively modest inhibition of any of them gets amplified thereby causing major inhibition of virus yield. Viral entry and uncoating at the beginning of infection, and virus assembly and egress, at the end, are also affected by some ISG products. Existence of this network of antiviral mechanisms probably ensures that the emergence of resistant viral mutants remains extremely unlikely. On the other hand, functional inactivation of one ISG product by the action of a viral

A

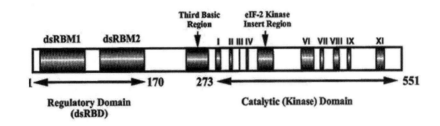

B

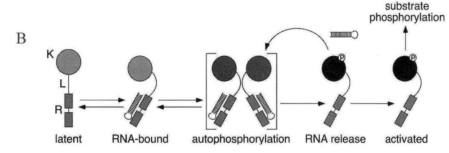

Figure 1   PKR structure and activation. (A) PKR contains two dsRNA-binding domains at its amino terminus and a kinase domain at its carboxyl terminus (126). (B) Binding of dsRNA facilitates dimerization and autophosphorylation of PKR on Thr-446 and Thr-451. Autophosphorylation activates PKR, allowing it to phosphorylate eIF2α, thereby preventing protein translation (82).

evasion mechanism does not allow the virus to escape the antiviral effects completely. The multistep antiviral process also provides the opportunity for quantitative regulation of the overall effects and positive or negative cross talk among different signaling pathways that induce the relevant antiviral genes.

Among the hundreds of ISG-encoded proteins, only a few have been functionally studied in detail. Their biochemical activities are diverse, affecting many cellular processes that are vital to virus replication. Below we discuss what is known about selected members of this family. They have been chosen because of the diversity of their functions and their perceived importance in mediating antiviral actions.

PKR and PACT
PKR was one of the earliest antiviral ISGs identified and is one of the most thoroughly investigated to date. Structurally, PKR contains two distinct domains. A dsRNA-binding domain at the amino terminus contains two dsRNA-binding motifs (reviewed in reference 35). A linker region joins the dsRNA-binding domain to the kinase domain present in the carboxyl terminus (Fig. 1). The inactive conformation of PKR is maintained by an

intramolecular interaction between the second dsRNA-binding motif and the kinase domain, which prevents activation (85). Binding of dsRNA breaks this association and causes a major conformational change resulting in exposure of a catalytic site(s). This conformational change results in dimerization of PKR monomers, which allows for autophosphorylation at Thr-446 and Thr-451. Although this dimerization and activation process can occur in the absence of dsRNA, the efficiency of the reaction is greatly enhanced by the presence of dsRNA, heparin, or other protein activators such as PACT (protein activator of PKR) (71). Once phosphorylated, PKR is able to function as a kinase. The only in vivo phosphorylation substrate identified for PKR to date is eukaryotic initiation factor 2α (eIF2α). PKR phosphorylates eIF2α at Ser-51 and prevents its protein translation initiation activity (35). As such, PKR activation ultimately inhibits protein translation.

PKR is constitutively expressed in most cell types and induced by IFN (83). Studies in $pkr^{-/-}$ mice have shown that these mice are more susceptible than wild-type mice to a select range of viruses including vesicular stomatitis virus (VSV), reovirus, and the A/WSN/33 strain of influenza virus (27). However, the mice did not show greater sensitivity to the W29 strain of influenza virus or to encephalomyocarditis virus (EMCV) unless infection was accompanied by intraperitoneal injection of dsRNA (128). Recent in vitro studies have implicated PKR in control of hepatitis C virus (HCV) replication and VSV replication (4, 11).

Finally, PKR has been implicated in regulation of apoptosis both in the presence and absence of viral infection. A variety of viruses, including vaccinia virus, influenza virus, and EMCV, induce PKR-dependent apoptosis in host cells (reviewed in reference 36). In the absence of virus, PKR-induced apoptosis has been shown in response to oncogenes (7), dsRNA or inflammatory stimuli, and stress stimuli such as tunicamycin or growth factor withdrawal (59, 89). In the case of inflammatory or stress stimuli, a number of these effects appear to be mediated by PACT (91). The ability of PKR to regulate signaling intermediates involved in pro- or antiapoptotic pathways such as nuclear factor-κB (NF-κB), activating transcription factor-3 (ATF-3), and p53 also provides another mechanism by which PKR can regulate apoptosis or cell growth. While roles for PKR in regulating apoptosis and cell growth have been shown clearly in vitro, $pkr^{-/-}$ mice develop normally and have not been shown to have a higher incidence of spontaneous tumors (1, 128). However, aberrant PKR expression has been detected in human tumors and tumor-derived cell lines (reviewed in reference 36).

PACT, or its murine homolog, RAX, is one of a number of protein activators of PKR identified to date. Like PKR, it contains two dsRNA-binding domains at its amino terminus. These domains are required for PKR binding but are not required for its activation (58). A third domain of PACT at the carboxyl terminus (domain 3) is sufficient for PKR activation; however, alone it only weakly interacts with PKR (96). PKR activation by PACT requires binding of domain 3 to a region in the kinase domain of PKR that is highly conserved but unrelated to any sequences found in other eIF2α kinases such as hemin-regulated inhibitor (HRI) or general control nondepressible 2 (Gcn2) (74) (Fig. 2).

PACT is thought to be constitutively expressed. Expression of PACT is regulated by the transcription factor Sp1, and posttranslational modification appears to be the predominant mechanism for controlling PACT activity in response to stimuli, rather than changes in mRNA or protein abundance (32). In the absence of stress stimuli, PACT is able to bind dsRNA and PKR. Phosphorylation at serine residue 287 in response to stress stimuli, along with constitutive phosphorylation of Ser-246, has also been shown to be essential for stress-induced apoptosis mediated by PACT (97).

Overexpression and knockdown studies have shown an important role for PACT in control of apoptosis in response to stress stimuli. Cells overexpressing PACT are more sensitive to apoptosis in response to stress stimuli such as arsenite, peroxide, and serum starvation (91). Cells where PACT expression was reduced using small interfering RNA showed reduced apoptosis in response to serum starvation or inflammatory stimuli (6). In vitro studies have shown a role for PACT in defense against VSV in murine embryonic fibroblasts (MEFs) (6); however, $pact^{-/-}$ MEFs did not show reduced eIF2α phosphorylation following VSV infection, suggesting that PACT may not be required for PKR activation in this system (105). In vivo studies have shown enhanced PKR phosphorylation and eIF2α phosphorylation in the cerebellum of rat pups in response to ethanol, with this effect attributed to increased PACT expression following ethanol exposure (13).

Mice lacking functional PACT expression showed unexpected developmental defects. These mice are smaller than wild-type littermates, weighing approximately 40% less than control mice at 4 to 6 weeks of age. PACT knockout mice also show craniofacial defects including a rounded rostrum, shortened nose, and prominent forehead. Ear development is also defective, with a severe reduction in the size of the outer ear and accompanying defects in the middle ear, resulting in hearing loss (105). However, viral infection experiments in these mice will provide further information regarding the role of PACT in antiviral defense.

The importance of PKR in mediating antiviral actions of IFN is manifested by the variety of strategies used by different viruses to evade PKR's activation or action (reviewed in references 36 and 113). Activation of PKR by dsRNA is often blocked by competing viral RNAs, such as adenoviral virus-associated I RNA, Epstein-Barr virus-encoded small RNA, and human immunodeficiency virus type 1 (HIV-1) transactivation responsive RNA, which bind to PKR but do not activate it. Another way of blocking PKR activation is by sequestration of the activator, dsRNA, by viral proteins such as reovirus σ3 and vaccinia virus E3L. Influenza A virus NS1 protein and herpes simplex virus type 1 (HSV-1) US11 protein, being RNA-binding proteins, were originally thought to function the same way, but in reality they block PKR activation by directly binding to the enzyme. Another HSV-1 protein, γ34.5, blocks PKR action by promoting dephosphorylation of eIF2α, the substrate of PKR. Human cytomegalovirus TRS1 binds to PKR and translocates it to the nucleus, thus preventing access to both its activator and substrate. Vaccinia virus K3L protein can function as an alternative substrate of PKR.

A

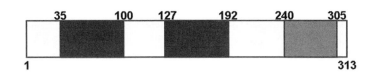

B

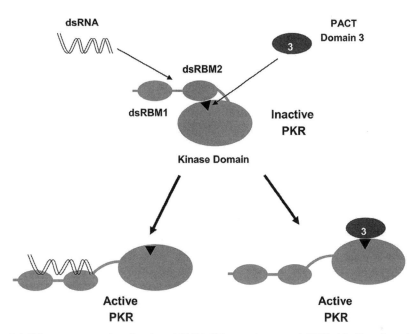

Figure 2 PACT structure and activation. (A) PACT contains two dsRNA-binding domains that also bind PKR and a third domain that is only involved in interaction with PKR. Numbers indicate amino acid sequence number (96). (B) Interaction of PACT-domain 3 with the PACT-binding motif present in the kinase domain of PKR disrupts the intramolecular interaction maintaining PKR in a latent state and leads to its activation (74).

Poliovirus and HIV-1 promote degradation of PKR, and HCV NS5A and the influenza virus-activated cellular protein P58IPK probably block PKR activation by blocking its dimerization. These examples indicate that evading PKR by one mechanism or another is important for successful replication of many viruses.

The 2′,5′-OAS/RNase L System

Small oligonucleotides, 2′,5′-oligoadenylates (2–5As), were originally identified as low-molecular-weight inhibitors of viral replication induced in poly(I:C)-treated or virus-infected cells (56, 63, 103). 2–5As are a series of 5′-triphosphorylated oligoadenylates with 2′-5′ phosphodiester bonds rather than the 3′-5′ bonds normally seen in nucleic acids (17). IFN treatment induces expression of several 2–5A OAS genes. In the presence of dsRNA these proteins produce 2–5A from ATP (101). 2–5As can then bind to RNase L, resulting in its activation. In turn, activated RNase L can cleave ssRNA molecules, an event that usually occurs at UpU or UpA sequences (34, 127). As such, RNA degradation by RNase L prevents viral infection by degrading RNA replication intermediates of viral proteins (3).

Three structurally related classes of 2–5A OAS genes exist, and these enzymes are present at low levels in host cells but induced upon viral stress. Three human OAS genes, *OAS1*, *OAS2*, and *OAS3*, encode active proteins, while *OASL* encodes an inactive form of the protein (OAS-related protein or p59). *OAS1* encodes two splice variants of 40 and 46 kDa, differing only in the presence of the carboxyl terminus. *OAS2* encodes two splice variants of 69 and 71 kDa, which share a common amino terminus, and produces four different mRNAs upon IFN induction, of which one encodes the 71-kDa protein and the remaining three encode the 69-kDa protein and have different noncoding 3′ termini. *OAS3* encodes a 100-kDa protein and is produced from a unique mRNA (reviewed in reference 101). Oligomerization of some OAS subunits occurs and is essential for efficient catalytic activity, with OAS1 existing as tetramers and OAS2 existing as dimers (39). OAS3 remains a monomer. A crisscross enzymatic mechanism in which one monomer phosphorylates the substrate bound to the second monomer in the dimeric complex has been identified for OAS2 (109). The crystal structure of OAS1 suggested a similar crisscross enzymatic mechanism and identified a putative dsRNA activation site (48). In addition to structural differences, different OAS enzymes generate oligonucleotides of different lengths. OAS1 enzymes generate only up to hexamers of 2–5A, while OAS2 enzymes can generate up to 30-mers and OAS3 can only generate dimers (109). Homologs of human OAS1 have been found in vertebrate species (57, 118). In addition to the OAS enzymes, two OAS-like proteins induced by IFN were identified in human cells; however, these lack the catalytic activity of OAS proteins but may have other functions unrelated to nucleotide oligomerization (101). The OAS1 E I7 isozyme acquires a Bcl-2 homology 3 domain though alternative splicing and can interact with Bcl-2 family members through this domain (40). The enzymatically inactive p59 OAS-like protein can confer antiviral activity through ubiquitin-like sequences in its carboxyl terminus (49). In mice the OAS-like protein OASL, which is catalytically inactive, confers resistance to West Nile virus by an unknown mechanism (81).

Structurally, RNase L contains an amino-terminal regulatory domain containing nine ankyrin repeats. These ankyrin repeats contain two Walker A motifs required for ATP or GTP fixation and are required for 2–5A binding. Binding of 2–5A to the amino terminus induces a conformational change, exposing the carboxyl-terminal domain. The carboxyl terminus contains the nuclease domain required for RNA degradation. Efficient kinase activity of RNase L requires dimerization in addition to 2–5A binding (reviewed in reference 8). 2–5A molecules have been shown to be induced in cells infected with EMCV, vaccinia virus, or reovirus (reviewed in reference 8). Studies

in RNase L-deficient mice have shown an important role for this system in vivo in protection against EMCV, coxsackievirus B4, West Nile virus, and HSV-1 (107, 132). Recently, 2–5A molecules have been implicated in activating the RIG-I/MDA-5 system and amplifying the response to IFN or dsRNA (78). However, the OAS/RNase L system has also been implicated in a range of biological processes in addition to protection against viral infection. An RNase L-binding protein has recently been identified as a factor regulating interaction between RNase L and the translation termination release factor eRF3, thereby regulating mRNA translation (reviewed in reference 8). RNase L has also been shown in microrarray studies to regulate cellular RNA stability (79), including stability of mRNA for the ISGs *pkr*, *isg43*, and *isg15*. The ability to control stability of mRNA for ISGs implies that RNase L might have a role in limiting the IFN response. Studies have also implicated the RNase L pathway in control of apoptosis in response to stress or genotoxic stimuli (10, 106, 133). Mutations in RNase L have also been linked with susceptibility to human cancers, particularly prostate cancer. More recently an association between defects in RNase L activity, susceptibility to the novel gammaretrovirus xenotropic murine leukemia-related virus, and prostate cancer has been established (119).

ADAR

ADAR genes were first identified in *Xenopus laevis* as DNA-unwinding enzymes, with human and mouse homologues subsequently identified. In vertebrates, three *Adar* genes, *Adar1*, *Adar2*, and *Adar3*, are present (12). *Adar1* is constitutively expressed; however, its promoter contains an ISRE that mediates increased mRNA levels by IFN-α or IFN-γ treatment (62). In contrast, ADAR2 expression is controlled by CREB-binding protein elements present in the promoter (94), and the mechanism regulating ADAR3 expression is currently unclear.

The common structural features of ADAR family proteins include a dsRNA-binding domain and a conserved cytidine deaminase domain at the carboxyl terminus that contains highly conserved residues thought to be involved in catalysis. Functionally, ADAR proteins are involved in RNA editing, where they convert adenosine to inosine through hydrolytic deamination of adenine (5). Inosine residues are read as guanosine residues by translation machinery, leading to the introduction of missense mutations into mRNAs. Depending on the location of the A-to-I modification, this editing can lead to a codon change and alteration of protein function. Major targets of ADAR-mediated RNA editing include mammalian glutamate receptor R2 (GluR2), serotonin, and serotonin receptor mRNAs (53, 124). In the case of the glutamate receptor, eight adenosine editing sites have been identified, with editing responsible for converting a glycine residue to arginine at a site located in the channel pore loop domain of the protein and rendering the glutamate receptor impermeable to Ca^{2+}. In the case of the serotonin receptor, asparagine and isoleucine codons can be converted to six different amino acids through mRNA editing, resulting in up to 24 isoforms of the protein with altered G protein-coupling activity and ligand responsiveness. mRNA secondary structure appears to dictate whether or not nucleic acid can be edited by ADAR family members, with both intermolecular and intramolecular dsRNA molecules longer than 20 bp forming substrates for these enzymes (88). In shorter molecules, mismatched bases, bulges, and loops appear to be edited preferentially (70). Different ADAR family members appear to bind particular targets, with specificity possibly mediated by differences in the number of and spacing between dsRNA-binding domains between family members (115).

Expression of ADAR1 and ADAR2 has been identified in many tissues including human heart, brain, lung, liver, skeletal muscle, kidney, pancreas, and placenta, while expression of ADAR3 is restricted to the brain (12). Two alternative splice variants of ADAR1 (ADAR1S and ADAR1L) generated from four different transcription start sites have been identified, with two ADAR1S isoforms constitutively expressed and ADAR1L induced upon IFN treatment (92). Initially, four splice variants of human ADAR2—ADAR2a, ADAR2b, ADAR2c, and ADAR2d—have been identified, which vary based on the presence or absence of an in-frame *alu* cassette (present in ADAR2b and ADAR2c) and the length of the carboxyl terminus (long in ADAR2a and ADAR2b, short in ADAR2c and ADAR2d). More recent analysis has identified two additional human splice variants, one of which lacks the dsRNA-binding domains and another which results in an alternative 3'-untranslated region (61). Splice variants of ADAR proteins appear to have different subcellular localization, with ADAR1L predominantly found in the cytoplasm while ADAR1S and ADAR2 isoforms are located in the nucleus and nucleolus (25, 108).

Studies in mice lacking ADAR family members have revealed important developmental roles for these proteins. *Adar2$^{-/-}$* mice die several weeks after birth due to excessive neuronal Ca^{2+} influx and resulting cell death as a consequence of incorrect editing of *gluR2* mRNA (52). *Adar1* deficiency results in embryonic lethality caused by defective erythropoiesis and fetal hepatocyte apoptosis (50, 123). ADAR enzymes have also been found to edit transcripts of certain viral genomes, such as hepatitis delta virus (99). Escape mutants of human respiratory syncytial virus have also been proposed to arise through point mutations generated by ADAR-mediated RNA editing (80).

p56

The p56 family of proteins is the product of the viral stress-induced genes that are most strongly induced by dsRNA, viruses, or IFN (38). In mice, there are three members of the family—p56, p54, and p49—encoded by the *Ifit1*, *Ifit2*, and *Ifit3* genes, respectively (9). In humans, four p56 family members exist—p56, p54, p58, and p60—again encoded by the genes *Ifit1*, *Ifit2*, *Ifit5*, and *Ifit4* (26, 87, 129). While there are sequence similarities between family members, corresponding family members from different species are more structurally related than different family members from the same species. For example, human and mouse p54 have 63% sequence identity and human and mouse p56 have 53% sequence identity. However, human p54 and human p56 only have 46% identical residues (116).

The defining structural feature of the p56 family is the presence of one or more tetratricopeptide repeat (TPR) motifs involved in modulating protein-protein interactions (18). Human and mouse p56 both contain six TPR motifs located at similar positions in the linear protein sequence. Similarly, both human and mouse p54 contain four TPR motifs (Fig. 3). Through mediating

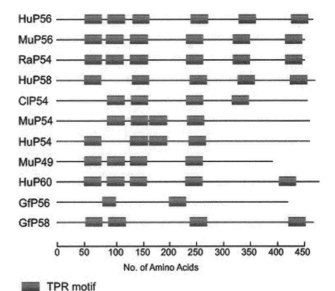

Figure 3 Arrangement of TPR motifs in p56 proteins. Corresponding members of the p56 family in different species have similar numbers of TPR motifs located in similar locations in the linear amino acid sequence of the protein (110).

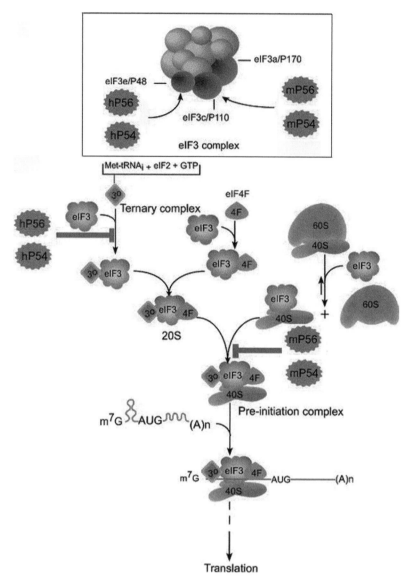

Figure 4 Inhibition of protein translation by p56 family members. Human and mouse p56 and p54 bind to different subunits of eIF3, inhibiting different points of the protein translation initiation process. Human p56 and human p54 bind to eIF3e and prevent formation of the ternary complex required for translation, while murine p54 and p56 bind to eIF3c and prevent assembly of the 48S preinitiation complex. Human p54 can bind to both eIF3c and eIF3e. (Adapted from reference 110.)

protein-protein interactions, TPR motifs allow binding of p56 family members to protein complexes, with the best-studied binding partner being members of the eIF3 protein complex. p56 family members interact with specific subunits of eIF3 and inhibit its function by different mechanisms, thereby inhibiting protein translation from mRNA (116). Different members of the family block different translation initiation functions of eIF3. Human p54 and p56 bind the "e" subunit of eIF3 and stop it from stabilizing the ternary complex of

eIF2.GTP.tRNA$_i$Met. Mouse p54 and p56 bind to the "c" subunit of eIF3 and prevent it facilitating formation of the 48S preinitiation complex. This preinitiation complex comprises the 40S ribosomal subunit and a 20S complex of eIF3, eIF4F, ternary complex, and mRNA (116) (Fig. 4). As well as binding to eIF3e, human p54 can also bind eIF3c. As with murine p54 and p56, human p54 can also disrupt formation of the 48S preinitiation complex but allow formation of the 20S complex. The human p56/eIF3e interaction occurs through binding of the TPR motif of p56 and the proteasome/COP9 signalosome/initiation factor (PCI) motif present at the carboxyl terminus of eIF3e (54). However, human p54 interacts with either the amino or carboxyl terminus of eIF3c. Mouse p54 is more specific in its interaction site, only interacting with the PCI domain of eIF3c, while mouse p56 interacts with the amino terminus of eIF3c (116).

Different members of the p56 family are generally coordinately induced; however, in human fibrosarcoma cells there are differences in the half-life of p56 and p54 mRNA following IFN-β treatment, with p54 mRNA being degraded much more quickly than p56 mRNA (116). More recently, inducer-specific and cell-type-specific induction patterns of murine p54 and p56 have been elucidated (117). p54 and p56 were induced in most tissues following tail vein injection of poly(I:C), IFN-α, or VSV, although the magnitude of induction tended to vary between different tissues tested. In liver, following IFN-α or VSV injection only p56 was induced, while IFN-β and poly(I:C) injection induced p54 equally well at both the protein and RNA levels. Also, in splenic B cells p54 but not p56 was induced following injection of IFN-α, poly(I:C), or VSV (117). Additional studies to determine the molecular basis for these different gene expression patterns may provide insight into additional regulatory mechanisms controlling expression of these genes.

In the absence of mouse models lacking expression of p56 family members, information regarding the role of these proteins in antiviral defense comes from in vitro systems. In the case of HCV and EMCV, studies have investigated the role of human p56 in inhibiting cap-dependent and internal ribosome entry site (IRES)-mediated translation of viral mRNA. Human p56 has been shown to significantly inhibit IRES-mediated translation of HCV but not EMCV mRNA (121). However, translation of 5′-capped mRNA was inhibited in both viruses, with this inhibition more pronounced for EMCV mRNA than HCV mRNA (121). This raises the possibility that different IRES elements may use distinct mechanisms to initiate translation and that these mechanisms may alter in their sensitivity to p56-mediated inhibition.

ISG20

Similar to the OAS/RNase L system, ISG20 is an IFN-induced nuclease. Human ISG20 was first independently identified through its upregulation in response to IFN (28) or estrogen (95). Sequence analysis has shown ISG20 to be a member of the DEDDh subgroup of DEDD exonucleases (135), which have 3′-to-5′ exonuclease

activity. The DEDDh family contains RNase enzymes and DNase enzymes; however, biochemical analysis has confirmed that ISG20 has a preference for RNA compared with DNA as a substrate (86). More recent studies have identified ISG20 induction not only in response to type I or type II IFNs, but also to TLR ligands (reviewed in reference 24).

Within the cell, ISG20 is located in both the nucleolus and the Cajal bodies of the nucleus. As well as its nuclear localization, ISG20 is also homologous with yeast proteins of the DEDD subfamily that are involved in processing of many RNA species including rRNAs and U4 and U5 small nuclear RNAs (120), suggesting a potential role in posttranscriptional processing of cellular RNAs.

ISG20 appears to have roles in antiviral defense as well as cellular physiology. ISG20 overexpression in HeLa cells reduces susceptibility to VSV, influenza virus, and EMCV infection, although the degree of protection conferred appears to vary between viruses (29). The exonuclease activity is also required for this antiviral activity (29). ISG20 is also involved in defense against HIV-1, with overexpression of ISG20 delaying replication of the virus and enhancing apoptosis of infected cells (30).

Mx

Mx proteins were first identified through their ability to protect against influenza virus infection in an inbred strain of mice (55, 75). The first member of this family, Mx1, was found to be an ISG whose expression was induced by type I and type III IFNs (44).

Mx proteins are high-molecular-weight GTPases and share a range of structural and functional features of dynamin. Structurally, Mx proteins consist of a highly conserved N-terminal GTPase domain, a "central interacting domain" (CID), and a C-terminal leucine zipper motif (46). Folding back of the leucine zipper domain onto the CID appears to increase GTPase activity (112). Assembly of Mx proteins into higher-order oligomers appears to be mediated by binding of the leucine zipper of one Mx molecule to the CID of another molecule (111), and is critical for GTPase activity, protein stability, and recognition of viral targets (46).

Subcellular localization of Mx proteins varies between species and appears to dictate their antiviral specificities. In rodents, Mx1 is found in the nucleus and confers resistance to viruses that replicate in the nucleus, such as influenza A virus and Thogoto virus (45). The rodent Mx2 is found in the cytoplasm and inhibits replication of viruses such as VSV in this compartment. In human cells, two Mx family members, MxA and

MxB, are found in the cytoplasm. While MxB has not been found to have antiviral functions, MxA is effective against a broad range of viruses, including some bunyaviruses, orthomyxoviruses, paramyxoviruses, rhabdoviruses, togaviruses, picornaviruses, and hepatitis B viruses (14, 42). The mechanism by which Mx proteins disrupt viral replication appears to vary between viruses and has not been clearly identified. A physical association between MxA and the nucleoprotein components of the nucleocapsid has been demonstrated for Thogoto virus and LaCrosse virus (66). Mx proteins may also redirect viral proteins to compartments where they cannot participate in viral RNA synthesis and are likely to be degraded. In bunyavirus-infected cells, MxA has been found to bind to the viral nucleocapsid protein and form a copolymer (67). These copolymers appeared to accumulate near the nuclear membrane and recruit more Mx molecules, eventually immobilizing the viral nucleocapsid proteins, resulting ultimately in their missorting (102).

Functional roles for Mx family members in antiviral defense have been shown in mice. Many inbred strains of laboratory mice lack functional Mx1 genes and show susceptibility to influenza A virus and Thogoto virus, which can be overcome by overexpression of Mx1 (45). A role for the Mx1 but not the MxA protein has also been shown in conferring resistance against Dhori virus in mice (45, 68). An antiviral role for MxA, independent of other ISGs, came from studies where the human MxA gene was expressed in mice lacking a functional type I IFN receptor ($Ifnar1^{-/-}$). Survival of the MxA transgenic mice was increased compared with $Ifnar1^{-/-}$ mice following Thogoto virus, LaCrosse virus, or Semliki Forest virus infection and reduced viral titers in organs from infected mice (93).

ISG15

ISG15 was first identified as an ISG in Ehrlich ascites tumor cells (31) and later identified as being immunologically related to ubiquitin (43). ISG15 cDNA encodes a 17-kDa pro-protein that is processed into its mature, 15-kDa form immediately upon translation by removal of a C-terminal peptide of eight amino acids (65). The mature protein consists of two ubiquitin-like domains bridged by a proline residue (43). Analogous to the ubiquitin system, ISG15 must undergo posttranslational modification by a pathway consisting of activation (E1), conjugating (E2), and ligating (E3) components (98). UBE1L has been identified as the E1 of the ISGylation system (130) and UbcH8, which is an established E2 ubiquitin ligase, was subsequently identified as the E2 component (64). Two candidates have been identified

for the E3 component of the pathway—estrogen-responsive finger protein (Efp) and Herc5 (19, 134). Multiple components of the ISGylation pathway are coordinately induced, as UBE1L and UbcH8 are also induced by IFN (26). Over 100 proteins involved in a broad range of cellular processes have been identified as targets of ISGylation in human cells (131). An independent study has identified 22 protein targets of ISGylation in both human and mouse cells (41).

IFN stimulation leads to increased intracellular ISG15 and unconjugated ISG15, which can be detected in cell culture supernatants or human serum (23). Intracellular ISG15 is required for ubiquitination and degradation of viral proteins. The importance of this pathway in resolution of some viral infections has been shown in mice lacking ISG15, which are more susceptible to influenza A and influenza B, HSV-1 infection, and Sindbis virus infection than wild-type mice (73). This is in contrast to earlier studies showing that $isg15^{-/-}$ mice did not show greater susceptibility to VSV and lymphocytic choriomeningitis virus (90). Extracellular ISG15 functions as a cytokine. Recombinant ISG15 has been shown to stimulate IFN-γ production in CD3$^+$ T cells and proliferation of CD56$^+$ natural killer cells, which in the presence of CD3$^+$ T cells results in enhanced target cell lysis. This killing is not major histocompatibility complex restricted and may represent a pathway for amplification of IFN-α and IFN-β signals (22).

p200

The p200 gene family consists of five members in mice (p202a, p202b, p203, p204, and p205) (76) and four members in humans: myeloid nuclear differentiation antigen (MNDA), absent in melanoma 2 (AIM2), interferon-inducible protein (IFI) 16, and IFIX (76). In addition to these genes, a murine pseudogene, $p202c$, and six splice isoforms of human IFIX protein (α1/2, β1/2, γ1/2) have been identified (77). Sequence analysis has also predicted the presence of novel family members p207, p208, and p209 (77). To date, members of the family have not been found in nonmammalian vertebrates. The structural feature common to these proteins is a long, partially conserved segment of at least 200 amino acids that is either of type a or b, with proteins containing one or two of these p200 domains. Among the conserved regions of p200 domains are the very highly conserved FM/LHATVAT/S sequence of amino acids, which may be involved in protein-protein interactions (60), and the LXCXE motif, involved in binding retinoblastoma protein (pRb) (77). This latter sequence distinguishes type a and b p200 domains, with the a-type domains containing a putative Rb binding site and b-type domains containing a TPKI motif (69). A more recent analysis has suggested that a third, *c* class of p200 domains exists (77). With the exception of p202, these proteins also contain a conserved N-terminal region with a nuclear localization sequence, a putative DNA-binding motif, and a 90-amino-acid domain in apoptosis and IFN response (DAPIN) domain (60). The five α-helices formed by the DAPIN domain may also function in protein-protein interactions (114).

p202 is the most extensively studied p200 family member, and is induced following treatment with IFN-α, IFN-β, IFN-γ, LPS, or poly(I:C) (37). Unlike most p200 family members, which are localized to the nucleus, p202 is initially expressed in cytoplasm and translocates to the nucleus (16). Once in the nucleus, p202 inhibits the activity of transcription factors including c-Fos, c-Jun, EdF1, EdF4, MyoD, myogenin, p53, and the p50 or p65 subunits of NF-κB (15). Patterns of expression vary between p200 family members (2); however, in mice, expression is consistently found in hematopoietic organs and myeloid cell lines. Expression of human family members has consistently been identified in myeloid and lymphoid cell lines (2). However, in both species there is developmental stage and hematopoietic lineage specificity of expression of each family member, with IFI16 expressed in CD34$^+$ hematopoietic stem cells and monocyte and lymphoid progenitors but not mature macrophages or granulocytes (20). In contrast, MNDA and p204 have similar expression patterns, being constitutively expressed in mature monocytes, granulocytes, and activated macrophages but not in lymphoid cells (84), while p205 is predominantly expressed during differentiation of monocytes and granulocytes (125). This finely tuned expression pattern suggests tight regulation of p200 family members and a possible role in development. Outside the hematopoietic system, proteins for all murine p200 family members have been detected in heart and skeletal muscle, suggesting a role in development. Consistent with these IFN-independent roles, family members have been shown to be induced by stress or differentiation stimuli such as platelet-derived growth factor (33), interleukin-6 (100), LPS and tumor necrosis factor (47), or dihydroxyvitamin D$_3$ (21).

More data exist to show a role for p200 family members in development and regulation of cellular differentiation, senescence, and apoptosis than in antiviral defense. Trangenic mice overexpressing p204 were not viable, with development blocked at the four-cell stage (72). Mice lacking *Ifi202a* were viable and did not show any distinct phenotype, but expression of p202b was upregulated in these animals to compensate for the loss

of p202a expression (122). The best evidence implicating p200 family proteins in controlling viral regulation comes from studies of mouse cytomegalovirus (MCMV)-infected MEFs. MCMV infection upregulated p204 expression (104), and in cells containing p204 with a mutant pRb-binding motif, MCMV replication was inhibited. However, a role for p204 in regulating viral replication did not extend to VSV, EMCV, ectromelia virus, or HSV-1 in this system (51). Expression of this mutant pRb-binding motif of p204 delayed expression of critical viral genes and prevented accumulation of dividing cells at the G$_1$-S checkpoint normally seen in MCMV-infected fibroblasts (51).

CONCLUSIONS

The evolutionary arms race between viruses and their hosts has produced a network of ISGs that detect an assortment of viral structural features, target various points of viral replication cycles, and have multiple layers of redundancy. While less well studied than their roles in viral defense, evidence is also emerging that many of these proteins also have important roles in cellular homeostasis or development in multicellular organisms, again through a variety of mechanisms. The potency and potentially wide scope of action of these ISGs has seen an equally complex regulatory network evolve to tightly control their transcription, translation, activation, and degradation. As the precise details regarding ISG function and regulation are elucidated, this allows greater understanding of questions such as which ISGs are most effective against a particular virus and how weaknesses in antiviral defenses are most readily exploited by viruses. Answering both of these questions, along with gaining a clear understanding of the complex homeostatic or developmental roles of ISGs, will allow them to be most effectively manipulated therapeutically.

Research in the authors' laboratory is supported by National Institutes of Health grants CA68782 and CA62220.

References

1. **Abraham, N., D. F. Stojdl, P. I. Duncan, N. Methot, T. Ishii, M. Dube, B. C. Vanderhyden, H. L. Atkins, D. A. Gray, M. W. McBurney, A. E. Koromilas, E. G. Brown, N. Sonenberg, and J. C. Bell.** 1999. Characterization of transgenic mice with targeted disruption of the catalytic domain of the double-stranded RNA-dependent protein kinase, PKR. *J. Biol. Chem.* **274:**5953–5962.
2. **Asefa, B., K. D. Klarmann, N. G. Copeland, D. J. Gilbert, N. A. Jenkins, and J. R. Keller.** 2004. The interferon-inducible p200 family of proteins: a perspective on their roles in cell cycle regulation and differentiation. *Blood Cells Mol. Dis.* **32:**155–167.

3. **Baglioni, C., M. A. Minks, and P. A. Maroney.** 1978. Interferon action may be mediated by activation of a nuclease by pppA2′p5′A2′p5′A. *Nature* **273:**684–687.

4. **Baltzis, D., L. K. Qu, S. Papadopoulou, J. D. Blais, J. C. Bell, N. Sonenberg, and A. E. Koromilas.** 2004. Resistance to vesicular stomatitis virus infection requires a functional cross talk between the eukaryotic translation initiation factor 2alpha kinases PERK and PKR. *J. Virol.* **78:**12747–12761.

5. **Bass, B. L., and H. Weintraub.** 1987. A developmentally regulated activity that unwinds RNA duplexes. *Cell* **48:** 607–613.

6. **Bennett, R. L., W. L. Blalock, D. M. Abtahi, Y. Pan, S. A. Moyer, and W. S. May.** 2006. RAX, the PKR activator, sensitizes cells to inflammatory cytokines, serum withdrawal, chemotherapy and viral infection. *Blood* **108:** 821–829.

7. **Beretta, L., M. Gabbay, R. Berger, S. M. Hanash, and N. Sonenberg.** 1996. Expression of the protein kinase PKR in modulated by IRF-1 and is reduced in 5q-associated leukemias. *Oncogene* **12:**1593–1596.

8. **Bisbal, C., and R. H. Silverman.** 2007. Diverse functions of RNase L and implications in pathology. *Biochimie* **89:**789–798.

9. **Bluyssen, H. A., R. J. Vlietstra, P. W. Faber, E. M. Smit, A. Hagemeijer, and J. Trapman.** 1994. Structure, chromosome localization, and regulation of expression of the interferon-regulated mouse *Ifi54/Ifi56* gene family. *Genomics* **24:**137–148.

10. **Castelli, J. C., B. A. Hassel, A. Maran, J. Paranjape, J. A. Hewitt, X. L. Li, Y. T. Hsu, R. H. Silverman, and R. J. Youle.** 1998. The role of 2′-5′ oligoadenylate-activated ribonuclease L in apoptosis. *Cell Death Differ.* **5:**313–320.

11. **Chang, K. S., Z. Cai, C. Zhang, G. C. Sen, B. R. Williams, and G. Luo.** 2006. Replication of hepatitis C virus (HCV) RNA in mouse embryonic fibroblasts: protein kinase R (PKR)-dependent and PKR-independent mechanisms for controlling HCV RNA replication and mediating interferon activities. *J. Virol.* **80:**7364–7374.

12. **Chen, C. X., D. S. Cho, Q. Wang, F. Lai, K. C. Carter, and K. Nishikura.** 2000. A third member of the RNA-specific adenosine deaminase gene family, ADAR3, contains both single- and double-stranded RNA binding domains. *RNA* **6:**755–767.

13. **Chen, G., C. Ma, K. A. Bower, Z. Ke, and J. Luo.** 2006. Interaction between RAX and PKR modulates the effect of ethanol on protein synthesis and survival of neurons. *J. Biol. Chem.* **281:**15909–15915.

14. **Chieux, V., W. Chehadeh, J. Harvey, O. Haller, P. Wattre, and D. Hober.** 2001. Inhibition of coxsackievirus B4 replication in stably transfected cells expressing human MxA protein. *Virology* **283:**84–92.

15. **Choubey, D., and J. U. Gutterman.** 1997. Inhibition of E2F-4/DP-1-stimulated transcription by p202. *Oncogene* **15:**291–301.

16. **Choubey, D., and P. Lengyel.** 1993. Interferon action: cytoplasmic and nuclear localization of the interferon-inducible 52-kD protein that is encoded by the Ifi 200 gene from the gene 200 cluster. *J. Interferon Res.* **13:**43–52.

17. **Clemens, M. J., and C. M. Vaquero.** 1978. Inhibition of protein synthesis by double-stranded RNA in reticulocyte

lysates: evidence for activation of an endoribonuclease. *Biochem. Biophys. Res. Commun.* **83:**59–68.

18. **D'Andrea, L. D., and L. Regan.** 2003. TPR proteins: the versatile helix. *Trends Biochem. Sci.* **28:**655–662.

19. **Dastur, A., S. Beaudenon, M. Kelley, R. M. Krug, and J. M. Huibregtse.** 2006. Herc5, an interferon-induced HECT E3 enzyme, is required for conjugation of ISG15 in human cells. *J. Biol. Chem.* **281:**4334–4338.

20. **Dawson, M. J., N. J. Elwood, R. W. Johnstone, and J. A. Trapani.** 1998. The IFN-inducible nucleoprotein IFI 16 is expressed in cells of the monocyte lineage, but is rapidly and markedly down-regulated in other myeloid precursor populations. *J. Leukoc. Biol.* **64:**546–554.

21. **Dawson, M. J., and J. A. Trapani.** 1995. IFI 16 gene encodes a nuclear protein whose expression is induced by interferons in human myeloid leukaemia cell lines. *J. Cell. Biochem.* **57:**39–51.

22. **D'Cunha, J., E. Knight, Jr., A. L. Haas, R. L. Truitt, and E. C. Borden.** 1996. Immunoregulatory properties of ISG15, an interferon-induced cytokine. *Proc. Natl. Acad. Sci. USA* **93:**211–215.

23. **D'Cunha, J., S. Ramanujam, R. J. Wagner, P. L. Witt, E. Knight, Jr., and E. C. Borden.** 1996. In vitro and in vivo secretion of human ISG15, an IFN-induced immunomodulatory cytokine. *J. Immunol.* **157:**4100–4108.

24. **Degols, G., P. Eldin, and N. Mechti.** 2007. ISG20, an actor of the innate immune response. *Biochimie* **89:**831–835.

25. **Desterro, J. M., L. P. Keegan, M. Lafarga, M. T. Berciano, M. O'Connell, and M. Carmo-Fonseca.** 2003. Dynamic association of RNA-editing enzymes with the nucleolus. *J. Cell Sci.* **116:**1805–1818.

26. **de Veer, M. J., M. Holko, M. Frevel, E. Walker, S. Der, J. M. Paranjape, R. H. Silverman, and B. R. Williams.** 2001. Functional classification of interferon-stimulated genes identified using microarrays. *J. Leukoc. Biol.* **69:**912–920.

27. **Durbin, R. K., S. E. Mertz, A. E. Koromilas, and J. E. Durbin.** 2002. PKR protection against intranasal vesicular stomatitis virus infection is mouse strain dependent. *Viral Immunol.* **15:**41–51.

28. **Eckert, M., S. E. Meek, and K. L. Ball.** 2006. A novel repressor domain is required for maximal growth inhibition by the IRF-1 tumor suppressor. *J. Biol. Chem.* **281:**23092–23102.

29. **Espert, L., G. Degols, C. Gongora, D. Blondel, B. R. Williams, R. H. Silverman, and N. Mechti.** 2003. ISG20, a new interferon-induced RNase specific for single-stranded RNA, defines an alternative antiviral pathway against RNA genomic viruses. *J. Biol. Chem.* **278:** 16151–16158.

30. **Espert, L., G. Degols, Y. L. Lin, T. Vincent, M. Benkirane, and N. Mechti.** 2005. Interferon-induced exonuclease ISG20 exhibits an antiviral activity against human immunodeficiency virus type 1. *J. Gen. Virol.* **86:**2221–2229.

31. **Farrell, P. J., R. J. Broeze, and P. Lengyel.** 1979. Accumulation of an mRNA and protein in interferon-treated Ehrlich ascites tumour cells. *Nature* **279:**523–525.

32. **Fasciano, S., A. Kaufman, and R. C. Patel.** 2007. Expression of PACT is regulated by Sp1 transcription factor. *Gene* **388:**74–82.

33. Flati, V., L. Frati, A. Gulino, S. Martinotti, and E. Toniato. 2001. The murine p202 protein, an IFN-inducible modulator of transcription, is activated by the mitogen platelet-derived growth factor. *J. Interferon Cytokine Res.* **21:**99–103.

34. Floyd-Smith, G., E. Slattery, and P. Lengyel. 1981. Interferon action: RNA cleavage pattern of a (2'-5')oligoadenylate-dependent endonuclease. *Science* **212:** 1030–1032.

35. Gabel, F., D. Wang, D. Madern, A. Sadler, K. Dayie, M. Z. Daryoush, D. Schwahn, G. Zaccai, X. Lee, and B. R. Williams. 2006. Dynamic flexibility of double-stranded RNA activated PKR in solution. *J. Mol. Biol.* **359:** 610–623.

36. Garcia, M. A., E. F. Meurs, and M. Esteban. 2007. The dsRNA protein kinase PKR: virus and cell control. *Biochimie* **89:**799–811.

37. Gariglio, M., E. Cinato, S. Panico, G. Cavallo, and S. Landolfo. 1991. Activation of interferon-inducible genes in mice by poly rI:rC or alloantigens. *J. Immunother.* **10:**20–27.

38. Geiss, G., G. Jin, J. Guo, R. Bumgarner, M. G. Katze, and G. C. Sen. 2001. A comprehensive view of regulation of gene expression by double-stranded RNA-mediated cell signaling. *J. Biol. Chem.* **276:**30178–30182.

39. Ghosh, A., S. N. Sarkar, W. Guo, S. Bandyopadhyay, and G. C. Sen. 1997. Enzymatic activity of 2'-5'-oligoadenylate synthetase is impaired by specific mutations that affect oligomerization of the protein. *J. Biol. Chem.* **272:**33220–33226.

40. Ghosh, A., S. N. Sarkar, T. M. Rowe, and G. C. Sen. 2001. A specific isozyme of 2'-5' oligoadenylate synthetase is a dual function proapoptotic protein of the Bcl-2 family. *J. Biol. Chem.* **276:**25447–25455.

41. Giannakopoulos, N. V., J. K. Luo, V. Papov, W. Zou, D. J. Lenschow, B. S. Jacobs, E. C. Borden, J. Li, H. W. Virgin, and D. E. Zhang. 2005. Proteomic identification of proteins conjugated to ISG15 in mouse and human cells. *Biochem. Biophys. Res. Commun.* **336:**496–506.

42. Gordien, E., O. Rosmorduc, C. Peltekian, F. Garreau, C. Brechot, and D. Kremsdorf. 2001. Inhibition of hepatitis B virus replication by the interferon-inducible MxA protein. *J. Virol.* **75:**2684–2691.

43. Haas, A. L., P. Ahrens, P. M. Bright, and H. Ankel. 1987. Interferon induces a 15-kilodalton protein exhibiting marked homology to ubiquitin. *J. Biol. Chem.* **262:**11315–11323.

44. Haller, O., H. Arnheiter, J. Lindenmann, and I. Gresser. 1980. Host gene influences sensitivity to interferon action selectively for influenza virus. *Nature* **283:**660–662.

45. Haller, O., M. Frese, D. Rost, P. A. Nuttall, and G. Kochs. 1995. Tick-borne Thogoto virus infection in mice is inhibited by the orthomyxovirus resistance gene product Mx1. *J. Virol.* **69:**2596–2601.

46. Haller, O., P. Staeheli, and G. Kochs. 2007. Interferon-induced Mx proteins in antiviral host defense. *Biochimie* **89:**812–818.

47. Hamilton, T. A., N. Bredon, Y. Ohmori, and C. S. Tannenbaum. 1989. IFN-gamma and IFN-beta independently stimulate the expression of lipopolysaccharide-inducible genes in murine peritoneal macrophages. *J. Immunol.* **142:**2325–2331.

48. Hartmann, R., J. Justesen, S. N. Sarkar, G. C. Sen, and V. C. Yee. 2003. Crystal structure of the 2'-specific and double-stranded RNA-activated interferon-induced antiviral protein 2'-5'-oligoadenylate synthetase. *Mol. Cell* **12:**1173–1185.

49. Hartmann, R., H. S. Olsen, S. Widder, R. Jorgensen, and J. Justesen. 1998. p59OASL, a 2'-5' oligoadenylate synthetase like protein: a novel human gene related to the 2'-5' oligoadenylate synthetase family. *Nucleic Acids Res.* **26:**4121–4128.

50. Hartner, J. C., C. Schmittwolf, A. Kispert, A. M. Muller, M. Higuchi, and P. H. Seeburg. 2004. Liver disintegration in the mouse embryo caused by deficiency in the RNA-editing enzyme ADAR1. *J. Biol. Chem.* **279:** 4894–4902.

51. Hertel, L., M. De Andrea, B. Azzimonti, A. Rolle, M. Gariglio, and S. Landolfo. 1999. The interferon-inducible 204 gene, a member of the Ifi 200 family, is not involved in the antiviral state induction by IFN-alpha, but is required by the mouse cytomegalovirus for its replication. *Virology* **262:**1–8.

52. Higuchi, M., S. Maas, F. N. Single, J. Hartner, A. Rozov, N. Burnashev, D. Feldmeyer, R. Sprengel, and P. H. Seeburg. 2000. Point mutation in an AMPA receptor gene rescues lethality in mice deficient in the RNA-editing enzyme ADAR2. *Nature* **406:**78–81.

53. Higuchi, M., F. N. Single, M. Kohler, B. Sommer, R. Sprengel, and P. H. Seeburg. 1993. RNA editing of AMPA receptor subunit GluR-B: a base-paired intron-exon structure determines position and efficiency. *Cell* **75:**1361–1370.

54. Hofmann, K., and P. Bucher. 1998. The PCI domain: a common theme in three multiprotein complexes. *Trends Biochem. Sci.* **23:**204–205.

55. Horisberger, M. A., P. Staeheli, and O. Haller. 1983. Interferon induces a unique protein in mouse cells bearing a gene for resistance to influenza virus. *Proc. Natl. Acad. Sci. USA* **80:**1910–1914.

56. Hovanessian, A. G., R. E. Brown, and I. M. Kerr. 1977. Synthesis of low molecular weight inhibitor of protein synthesis with enzyme from interferon-treated cells. *Nature* **268:**537–540.

57. Hovnanian, A., D. Rebouillat, M. G. Mattei, E. R. Levy, I. Marie, A. P. Monaco, and A. G. Hovanessian. 1998. The human 2',5'-oligoadenylate synthetase locus is composed of three distinct genes clustered on chromosome 12q24.2 encoding the 100-, 69-, and 40-kDa forms. *Genomics* **52:**267–277.

58. Huang, X., B. Hutchins, and R. C. Patel. 2002. The C-terminal, third conserved motif of the protein activator PACT plays an essential role in the activation of double-stranded-RNA-dependent protein kinase (PKR). *Biochem. J.* **366:**175–186.

59. Ito, T., M. Yang, and W. S. May. 1999. RAX, a cellular activator for double-stranded RNA-dependent protein kinase during stress signaling. *J. Biol. Chem.* **274:** 15427–15432.

60. Johnstone, R. W., and J. A. Trapani. 1999. Transcription and growth regulatory functions of the HIN-200 family of proteins. *Mol. Cell. Biol.* **19:**5833–5838.

61. Kawahara, Y., K. Ito, M. Ito, S. Tsuji, and S. Kwak. 2005. Novel splice variants of human ADAR2 mRNA: skipping

of the exon encoding the dsRNA-binding domains, and multiple C-terminal splice sites. *Gene* 363:193–201.

62. **Kawakubo, K., and C. E. Samuel.** 2000. Human RNA-specific adenosine deaminase (ADAR1) gene specifies transcripts that initiate from a constitutively active alternative promoter. *Gene* 258:165–172.

63. **Kerr, I. M., R. E. Brown, and A. G. Hovanessian.** 1977. Nature of inhibitor of cell-free protein synthesis formed in response to interferon and double-stranded RNA. *Nature* 268:540–542.

64. **Kim, K. I., N. V. Giannakopoulos, H. W. Virgin, and D. E. Zhang.** 2004. Interferon-inducible ubiquitin E2, Ubc8, is a conjugating enzyme for protein ISGylation. *Mol. Cell. Biol.* 24:9592–9600.

65. **Knight, E., Jr., D. Fahey, B. Cordova, M. Hillman, R. Kutny, N. Reich, and D. Blomstrom.** 1988. A 15-kDa interferon-induced protein is derived by COOH-terminal processing of a 17-kDa precursor. *J. Biol. Chem.* 263:4520–4522.

66. **Kochs, G., and O. Haller.** 1999. Interferon-induced human MxA GTPase blocks nuclear import of Thogoto virus nucleocapsids. *Proc. Natl. Acad. Sci. USA* 96:2082–2086.

67. **Kochs, G., C. Janzen, H. Hohenberg, and O. Haller.** 2002. Antivirally active MxA protein sequesters La Crosse virus nucleocapsid protein into perinuclear complexes. *Proc. Natl. Acad. Sci. USA* 99:3153–3158.

68. **Kolb, E., E. Laine, D. Strehler, and P. Staeheli.** 1992. Resistance to influenza virus infection of Mx transgenic mice expressing Mx protein under the control of two constitutive promoters. *J. Virol.* 66:1709–1716.

69. **Landolfo, S., M. Gariglio, G. Gribaudo, and D. Lembo.** 1998. The Ifi 200 genes: an emerging family of IFN-inducible genes. *Biochimie* 80:721–728.

70. **Lehmann, K. A., and B. L. Bass.** 1999. The importance of internal loops within RNA substrates of ADAR1. *J. Mol. Biol.* 291:1–13.

71. **Lemaire, P. A., J. Lary, and J. L. Cole.** 2005. Mechanism of PKR activation: dimerization and kinase activation in the absence of double-stranded RNA. *J. Mol. Biol.* 345:81–90.

72. **Lembo, M., C. Sacchi, C. Zappador, G. Bellomo, M. Gaboli, P. P. Pandolfi, M. Gariglio, and S. Landolfo.** 1998. Inhibition of cell proliferation by the interferon-inducible 204 gene, a member of the Ifi 200 cluster. *Oncogene* 16:1543–1551.

73. **Lenschow, D. J., C. Lai, N. Frias-Staheli, N. V. Giannakopoulos, A. Lutz, T. Wolff, A. Osiak, B. Levine, R. E. Schmidt, A. Garcia-Sastre, D. A. Leib, A. Pekosz, K. P. Knobeloch, I. Horak, and H. W. Virgin IV.** 2007. IFN-stimulated gene 15 functions as a critical antiviral molecule against influenza, herpes, and Sindbis viruses. *Proc. Natl. Acad. Sci. USA* 104:1371–1376.

74. **Li, S., G. A. Peters, K. Ding, X. Zhang, J. Qin, and G. C. Sen.** 2006. Molecular basis for PKR activation by PACT or dsRNA. *Proc. Natl. Acad. Sci. USA* 103:10005–10010.

75. **Lindenmann, J.** 1964. Inheritance of resistance to influenza virus in mice. *Proc. Soc. Exp. Biol. Med.* 116:506–509.

76. **Liu, C. J., H. Wang, and P. Lengyel.** 1999. The interferon-inducible nucleolar p204 protein binds the ribosomal RNA-specific UBF1 transcription factor and inhibits ribosomal RNA transcription. *EMBO J.* 18:2845–2854.

77. **Ludlow, L. E., R. W. Johnstone, and C. J. Clarke.** 2005. The HIN-200 family: more than interferon-inducible genes? *Exp. Cell Res.* 308:1–17.

78. **Malathi, K., B. Dong, M. Gale, Jr., and R. H. Silverman.** 2007. Small self-RNA generated by RNase L amplifies antiviral innate immunity. *Nature* 448:816–819.

79. **Malathi, K., J. M. Paranjape, E. Bulanova, M. Shim, J. M. Guenther-Johnson, P. W. Faber, T. E. Eling, B. R. Williams, and R. H. Silverman.** 2005. A transcriptional signaling pathway in the IFN system mediated by 2'-5'-oligoadenylate activation of RNase L. *Proc. Natl. Acad. Sci. USA* 102:14533–14538.

80. **Martinez, I., and J. A. Melero.** 2002. A model for the generation of multiple A to G transitions in the human respiratory syncytial virus genome: predicted RNA secondary structures as substrates for adenosine deaminases that act on RNA. *J. Gen. Virol.* 83:1445–1455.

81. **Mashimo, T., M. Lucas, D. Simon-Chazottes, M. P. Frenkiel, X. Montagutelli, P. E. Ceccaldi, V. Deubel, J. L. Guenet, and P. Despres.** 2002. A nonsense mutation in the gene encoding 2'-5'-oligoadenylate synthetase/L1 isoform is associated with West Nile virus susceptibility in laboratory mice. *Proc. Natl. Acad. Sci. USA* 99:11311–11316.

82. **McKenna, S. A., D. A. Lindhout, I. Kim, C. W. Liu, V. M. Gelev, G. Wagner, and J. D. Puglisi.** 2007. Molecular framework for the activation of RNA-dependent protein kinase. *J. Biol. Chem.* 282:11474–11486.

83. **Meurs, E., K. Chong, J. Galabru, N. S. Thomas, I. M. Kerr, B. R. Williams, and A. G. Hovanessian.** 1990. Molecular cloning and characterization of the human double-stranded RNA-activated protein kinase induced by interferon. *Cell* 62:379–390.

84. **Miranda, R. N., R. C. Briggs, K. Shults, M. C. Kinney, R. A. Jensen, and J. B. Cousar.** 1999. Immunocytochemical analysis of MNDA in tissue sections and sorted normal bone marrow cells documents expression only in maturing normal and neoplastic myelomonocytic cells and a subset of normal and neoplastic B lymphocytes. *Hum. Pathol.* 30:1040–1049.

85. **Nanduri, S., F. Rahman, B. R. Williams, and J. Qin.** 2000. A dynamically tuned double-stranded RNA binding mechanism for the activation of antiviral kinase PKR. *EMBO J.* 19:5567–5574.

86. **Nguyen, L. H., L. Espert, N. Mechti, and D. M. Wilson III.** 2001. The human interferon- and estrogen-regulated ISG20/HEM45 gene product degrades single-stranded RNA and DNA in vitro. *Biochemistry* 40:7174–7179.

87. **Niikura, T., R. Hirata, and S. C. Weil.** 1997. A novel interferon-inducible gene expressed during myeloid differentiation. *Blood Cells Mol. Dis.* 23:337–349.

88. **Nishikura, K., C. Yoo, U. Kim, J. M. Murray, P. A. Estes, F. E. Cash, and S. A. Liebhaber.** 1991. Substrate specificity of the dsRNA unwinding/modifying activity. *EMBO J.* 10:3523–3532.

89. **Onuki, R., Y. Bando, E. Suyama, T. Katayama, H. Kawasaki, T. Baba, M. Tohyama, and K. Taira.** 2004. An RNA-dependent protein kinase is involved in tunicamycin-induced apoptosis and Alzheimer's disease. *EMBO J.* 23:959–968.

90. Osiak, A., O. Utermohlen, S. Niendorf, I. Horak, and K. P. Knobeloch. 2005. ISG15, an interferon-stimulated ubiquitin-like protein, is not essential for STAT1 signaling and responses against vesicular stomatitis and lymphocytic choriomeningitis virus. *Mol. Cell. Biol.* **25:**6338–6345.

91. Patel, C. V., I. Handy, T. Goldsmith, and R. C. Patel. 2000. PACT, a stress-modulated cellular activator of interferon-induced double-stranded RNA-activated protein kinase, PKR. *J. Biol. Chem.* **275:**37993–37998.

92. Patterson, J. B., and C. E. Samuel. 1995. Expression and regulation by interferon of a double-stranded-RNA-specific adenosine deaminase from human cells: evidence for two forms of the deaminase. *Mol. Cell. Biol.* **15:** 5376–5388.

93. Pavlovic, J., H. A. Arzet, H. P. Hefti, M. Frese, D. Rost, B. Ernst, E. Kolb, P. Staeheli, and O. Haller. 1995. Enhanced virus resistance of transgenic mice expressing the human MxA protein. *J. Virol.* **69:**4506–4510.

94. Peng, P. L., X. Zhong, W. Tu, M. M. Soundarapandian, P. Molner, D. Zhu, L. Lau, S. Liu, F. Liu, and Y. Lu. 2006. ADAR2-dependent RNA editing of AMPA receptor subunit GluR2 determines vulnerability of neurons in forebrain ischemia. *Neuron* **49:**719–733.

95. Pentecost, B. T. 1998. Expression and estrogen regulation of the HEM45 MRNA in human tumor lines and in the rat uterus. *J. Steroid Biochem. Mol. Biol.* **64:**25–33.

96. Peters, G. A., R. Hartmann, J. Qin, and G. C. Sen. 2001. Modular structure of PACT: distinct domains for binding and activating PKR. *Mol. Cell. Biol.* **21:**1908–1920.

97. Peters, G. A., S. Li, and G. C. Sen. 2006. Phosphorylation of specific serine residues in the PKR activation domain of PACT is essential for its ability to mediate apoptosis. *J. Biol. Chem.* **281:**35129–35136.

98. Pickart, C. M. 2001. Mechanisms underlying ubiquitination. *Annu. Rev. Biochem.* **70:**503–533.

99. Polson, A. G., B. L. Bass, and J. L. Casey. 1996. RNA editing of hepatitis delta virus antigenome by dsRNA-adenosine deaminase. *Nature* **380:**454–456.

100. Pramanik, R., T. N. Jorgensen, H. Xin, B. L. Kotzin, and D. Choubey. 2004. Interleukin-6 induces expression of *Ifi202*, an interferon-inducible candidate gene for lupus susceptibility. *J. Biol. Chem.* **279:**16121–16127.

101. Rebouillat, D., and A. G. Hovanessian. 1999. The human 2′,5′-oligoadenylate synthetase family: interferon-induced proteins with unique enzymatic properties. *J. Interferon Cytokine Res.* **19:**295–308.

102. Reichelt, M., S. Stertz, J. Krijnse-Locker, O. Haller, and G. Kochs. 2004. Missorting of LaCrosse virus nucleocapsid protein by the interferon-induced MxA GTPase involves smooth ER membranes. *Traffic* **5:**772–784.

103. Roberts, W. K., A. Hovanessian, R. E. Brown, M. J. Clemens, and I. M. Kerr. 1976. Interferon-mediated protein kinase and low-molecular-weight inhibitor of protein synthesis. *Nature* **264:**477–480.

104. Rolle, S., M. De Andrea, D. Gioia, D. Lembo, L. Hertel, S. Landolfo, and M. Gariglio. 2001. The interferon-inducible 204 gene is transcriptionally activated by mouse cytomegalovirus and is required for its replication. *Virology* **286:**249–255.

105. Rowe, T. M., M. Rizzi, K. Hirose, G. A. Peters, and G. C. Sen. 2006. A role of the double-stranded RNA-binding protein PACT in mouse ear development and hearing. *Proc. Natl. Acad. Sci. USA* **103:**5823–5828.

106. Rusch, L., A. Zhou, and R. H. Silverman. 2000. Caspase-dependent apoptosis by 2′,5′-oligoadenylate activation of RNase L is enhanced by IFN-beta. *J. Interferon Cytokine Res.* **20:**1091–1100.

107. Samuel, M. A., K. Whitby, B. C. Keller, A. Marri, W. Barchet, B. R. Williams, R. H. Silverman, M. Gale, Jr., and M. S. Diamond. 2006. PKR and RNase L contribute to protection against lethal West Nile virus infection by controlling early viral spread in the periphery and replication in neurons. *J. Virol.* **80:**7009–7019.

108. Sansam, C. L., K. S. Wells, and R. B. Emeson. 2003. Modulation of RNA editing by functional nucleolar sequestration of ADAR2. *Proc. Natl. Acad. Sci. USA* **100:**14018–14023.

109. Sarkar, S. N., S. Pal, and G. C. Sen. 2002. Crisscross enzymatic reaction between the two molecules in the active dimeric P69 form of the 2′-5′ oligodenylate synthetase. *J. Biol. Chem.* **277:**44760–44764.

110. Sarkar, S. N., and G. C. Sen. 2004. Novel functions of proteins encoded by viral stress-inducible genes. *Pharmacol. Ther.* **103:**245–259.

111. Schumacher, B., and P. Staeheli. 1998. Domains mediating intramolecular folding and oligomerization of MxA GTPase. *J. Biol. Chem.* **273:**28365–28370.

112. Schwemmle, M., M. F. Richter, C. Herrmann, N. Nassar, and P. Staeheli. 1995. Unexpected structural requirements for GTPase activity of the interferon-induced MxA protein. *J. Biol. Chem.* **270:**13518–13523.

113. Sen, G. C., and G. A. Peters. 2007. Viral stress-inducible genes. *Adv. Virus Res.* **70:**233–263.

114. Staub, E., E. Dahl, and A. Rosenthal. 2001. The DAPIN family: a novel domain links apoptotic and interferon response proteins. *Trends Biochem. Sci.* **26:**83–85.

115. Stefl, R., M. Xu, L. Skrisovska, R. B. Emeson, and F. H. Allain. 2006. Structure and specific RNA binding of ADAR2 double-stranded RNA binding motifs. *Structure* **14:**345–355.

116. Terenzi, F., D. J. Hui, W. C. Merrick, and G. C. Sen. 2006. Distinct induction patterns and functions of two closely related interferon-inducible human genes, ISG54 and ISG56. *J. Biol. Chem.* **281:**34064–34071.

117. Terenzi, F., C. White, S. Pal, B. R. Williams, and G. C. Sen. 2007. Tissue-specific and inducer-specific differential induction of ISG56 and ISG54 in mice. *J. Virol.* **81:** 8656–8665.

118. Truve, E., M. Kelve, A. Aaspollu, H. C. Schroder, and W. E. Muller. 1994. Homologies between different forms of 2–5A synthetases. *Prog. Mol. Subcell. Biol.* **14:** 139–149.

119. Urisman, A., R. J. Molinaro, N. Fischer, S. J. Plummer, G. Casey, E. A. Klein, K. Malathi, C. Magi-Galluzzi, R. R. Tubbs, D. Ganem, R. H. Silverman, and J. L. DeRisi. 2006. Identification of a novel gammaretrovirus in prostate tumors of patients homozygous for R462Q *RNASEL* variant. *PLoS Pathog.* **2:**e25.

120. van Hoof, A., P. Lennertz, and R. Parker. 2000. Three conserved members of the RNase D family have unique and overlapping functions in the processing of 5S, 5.8S, U4, U5, RNase MRP and RNase P RNAs in yeast. *EMBO J.* **19:**1357–1365.

121. Wang, C., J. Pflugheber, R. Sumpter, Jr., D. L. Sodora, D. Hui, G. C. Sen, and M. Gale, Jr. 2003. Alpha interferon induces distinct translational control programs to suppress hepatitis C virus RNA replication. *J. Virol.* 77:3898–3912.

122. Wang, H., G. Chatterjee, J. J. Meyer, C. J. Liu, N. A. Manjunath, P. Bray-Ward, and P. Lengyel. 1999. Characteristics of three homologous *202* genes (*Ifi202a*, *Ifi202b*, and *Ifi202c*) from the murine interferon-activatable gene *200* cluster. *Genomics* 60:281–294.

123. Wang, Q., M. Miyakoda, W. Yang, J. Khillan, D. L. Stachura, M. J. Weiss, and K. Nishikura. 2004. Stress-induced apoptosis associated with null mutation of *ADAR1* RNA editing deaminase gene. *J. Biol. Chem.* 279:4952–4961.

124. Wang, Q., P. J. O'Brien, C. X. Chen, D. S. Cho, J. M. Murray, and K. Nishikura. 2000. Altered G protein-coupling functions of RNA editing isoform and splicing variant serotonin$_{2C}$ receptors. *J. Neurochem.* 74: 1290–1300.

125. Weiler, S. R., J. M. Gooya, M. Ortiz, S. Tsai, S. J. Collins, and J. R. Keller. 1999. D3: a gene induced during myeloid cell differentiation of Lin^{lo} c-Kit$^+$ Sca-1$^+$ progenitor cells. *Blood* 93:527–536.

126. Williams, B. R. 1999. PKR; a sentinel kinase for cellular stress. *Oncogene* 18:6112–6120.

127. Wreschner, D. H., J. W. McCauley, J. J. Skehel, and I. M. Kerr. 1981. Interferon action—sequence specificity of the ppp(A2'p)nA-dependent ribonuclease. *Nature* 289: 414–417.

128. Yang, Y. L., L. F. Reis, J. Pavlovic, A. Aguzzi, R. Schafer, A. Kumar, B. R. Williams, M. Aguet, and C. Weissmann. 1995. Deficient signaling in mice devoid of double-stranded RNA-dependent protein kinase. *EMBO J.* 14:6095–6106.

129. Yu, M., J. H. Tong, M. Mao, L. X. Kan, M. M. Liu, Y. W. Sun, G. Fu, Y. K. Jing, L. Yu, D. Lepaslier, M. Lanotte, Z. Y. Wang, Z. Chen, S. Waxman, Y. X. Wang, J. Z. Tan, and S. J. Chen. 1997. Cloning of a gene (*RIG-G*) associated with retinoic acid-induced differentiation of acute promyelocytic leukemia cells and representing a new member of a family of interferon-stimulated genes. *Proc. Natl. Acad. Sci. USA* 94: 7406–7411.

130. Yuan, W., and R. M. Krug. 2001. Influenza B virus NS1 protein inhibits conjugation of the interferon (IFN)-induced ubiquitin-like ISG15 protein. *EMBO J.* 20: 362–371.

131. Zhao, C., C. Denison, J. M. Huibregtse, S. Gygi, and R. M. Krug. 2005. Human ISG15 conjugation targets both IFN-induced and constitutively expressed proteins functioning in diverse cellular pathways. *Proc. Natl. Acad. Sci. USA* 102:10200–10205.

132. Zheng, X., R. H. Silverman, A. Zhou, T. Goto, B. S. Kwon, H. E. Kaufman, and J. M. Hill. 2001. Increased severity of HSV-1 keratitis and mortality in mice lacking the 2-5A-dependent RNase L gene. *Invest. Ophthalmol. Vis. Sci.* 42:120–126.

133. Zhou, A., J. M. Paranjape, B. A. Hassel, H. Nie, S. Shah, B. Galinski, and R. H. Silverman. 1998. Impact of RNase L overexpression on viral and cellular growth and death. *J. Interferon Cytokine Res.* 18:953–961.

134. Zou, W., and D. E. Zhang. 2006. The interferon-inducible ubiquitin-protein isopeptide ligase (E3) EFP also functions as an ISG15 E3 ligase. *J. Biol. Chem.* 281:3989–3994.

135. Zuo, Y., and M. P. Deutscher. 2001. Exoribonuclease superfamilies: structural analysis and phylogenetic distribution. *Nucleic Acids Res.* 29:1017–1026.

Cellular Signaling and Innate Immune
Responses to RNA Virus Infections
Edited by A. R. Brasier et al.
© 2009 ASM Press, Washington, D.C.

Amer A. Beg
Xingyu Wang

8

The Nuclear Factor-κB Transcription Factor Pathway

The nuclear factor-κB (NF-κB) transcription factor was discovered more than 20 years ago as a potential regulator of the mouse immunoglobulin κ light-chain gene (79, 80). It would have been difficult to imagine then that this DNA-binding protein would turn out to have such wide-ranging functions in so many different cell types and signaling pathways. Almost since their initial discovery, NF-κB transcription factors have served as a paradigm for understanding how extracellular signals modulate nuclear gene expression patterns. It is also remarkable that early models proposed to explain the mechanism of NF-κB activation have stood the test of time, albeit with some refinements. Despite the multitude of functions that NF-κB transcription factors perform, their role in orchestrating host gene expression responses to infectious agents is among the most crucial. Key general aspects of the NF-κB signaling pathway will be discussed first, followed by their role in mediating host responses to RNA virus infection.

OVERVIEW OF THE NF-κB SIGNALING PATHWAY

"Classic" NF-κB comprises a heterodimer of a 50-kDa protein called p50/NF-κB1 and a 65-kDa protein called p65/RelA (32) (Fig. 1). Indeed, this is the most prevalent form of NF-κB in most cell types. p50 is generated by processing of a 105-kDa precursor (p105), which can occur both constitutively and inducibly. The molecular cloning of the p50 and RelA genes revealed striking sequence homology to the cRel protein, the oncogenic form of which (vRel) had been known for many years to cause tumors in birds (26, 65). An additional member of the NF-κB family, p100/p52 (NF-κB2), is most similar to p105/p50 and has also been implicated in tumorigenesis (64). The fifth member of the NF-κB family is RelB, which together with RelA and cRel comprise the three most important transcription-activating subunits (32). All members of the NF-κB family have a conserved Rel homology domain, which mediates DNA binding and dimerization (Fig. 1). While different NF-κB subunits can homo- and heterodimerize in virtually any combination, it appears that certain combinations are preferred. Thus, in addition to p50/RelA, complexes of p50/cRel and p52/RelB can be readily detected in many cell types. It is likely that different dimers have different transcriptional activation potential, thus allowing fine tuning of gene expression responses.

Soon after the initial discovery of NF-κB, its unique mode of regulation also became evident. NF-κB complexes typically reside in the cytoplasm in a complex with

Amer A. Beg and Xingyu Wang, Department of Interdisciplinary Oncology, H. Lee Moffitt Cancer Center and Research Institute, University of South Florida, 12902 Magnolia Dr., Tampa, FL 33612.

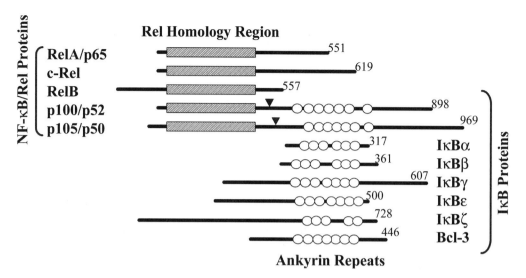

Figure 1 Mammalian NF-κB/Rel and IκB family proteins. The mammalian NF-κB family comprises five members: RelA/p65, cRel, RelB, p100/p52, and p105/p50. They have a structurally conserved amino-terminal Rel homology region, which contains functional domains responsible for dimerization and DNA binding. RelA/p65, cRel, and RelB also have nonhomologous carboxy-terminal transactivation domains. p100 and p105 have carboxy-terminal ankyrin repeats that are homologous to IκB family proteins. p50 and p52 proteins are generated by proteolytic processing of p105 and p100, respectively. IκB family proteins, which are featured by ankyrin repeats, contain IκB-α, IκB-β, IκB-γ, IκB-ε, IκB-ζ (also known as MAIL and INAP), Bcl-3, p105, and p100. IκB-α, IκB-β, and IκB-ε are prototypical IκBs that retain NF-κB proteins in the cytoplasm. Unprocessed p100 and p105 also function as inhibitors of NF-κB. IκB-γ is identical to the carboxy-terminal region of p105 and specifically inhibits p50-containing NF-κB dimers. p105, IκB-γ, and prototypical IκBs (IκB-α and IκB-β) use a similar mechanism to bind but a different mechanism to regulate the subcellular localization of NF-κB. In contrast, Bcl-3 interacts specifically with p50 and p52 homodimers and can induce expression of NF-κB-regulated genes. Similar to Bcl-3, IκB-ζ also functions as an activator of gene expression by specific association with p50.

inhibitory IκB proteins (Fig. 2). NF-κB activating signals, which are discussed in more detail below, induce phosphorylation of IκB proteins, leading to their ubiquitination and proteasome-mediated degradation (42). Free NF-κB then translocates to the nucleus and activates target gene expression. The key enzyme responsible for IκB phosphorylation is IκB kinase β (IKK-β), activation of which is regulated by signaling mediators involved in many distinct pathways (32). Among the most important activators of IKK-β are proinflammatory cytokines and Toll-like receptor (TLR) ligands (1, 2). Three members of the NF-κB family, RelA, cRel, and RelB, contain transcriptional activation domains. Nuclear translocation of these subunits, however, may not be sufficient for transcriptional activation. Thus, NF-κB subunits can be modified by posttranslational modifications, which enhances their transcriptional activation potential. Modifications of the RelA subunit have been studied most extensively. Phosphorylation at a single key residue (Ser-276) by protein kinase A enhances RelA interaction with the coactivator CREB-binding protein (CBP), resulting in enhanced transcriptional activation potential (102). Interestingly, CBP/p300 can also induce acetylation of RelA, resulting in enhanced RelA activity (14, 28). RelA can also be directly phosphorylated by IKK-β, leading to enhancement of transcription activation (75, 97). Finally, NF-κB-induced transcriptional activation can be potentiated by several additional proteins, including IKK-α, IκB-ζ, GSK-3β (glycogen synthase kinase 3β), and ELKS proteins (4, 21, 86, 95, 96). In addition to the canonical IKK-β-dependent pathway, NF-κB subunits can also be activated by the noncanonical or alternative IKK-α-dependent pathway (81). The alternative pathway leads to processing of p100/RelB complexes in the cytoplasm, leading to release of p52/RelB heterodimers (8, 81). Interestingly, this pathway responds to a distinct subset of NF-κB signals, including lymphotoxin, BAFF (B-cell-activating factor of the TNF family), and CD40 (8, 18). Together, the combined functions of different NF-κB subunits allow regulation of a wide array of genes with many

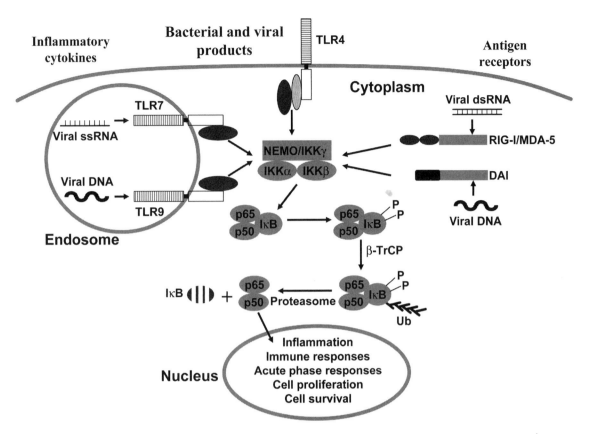

Figure 2   NF-κB activation pathways during viral infection. NF-κB can be activated by many stimuli, including bacterial and viral products, inflammatory cytokines, and antigen receptors. NF-κB plays vital roles in mediating intracellular responses to pathogens that are recognized by pattern recognition receptors (PRRs) such as TLRs. Bacterial product LPS or RSV F protein is recognized by the cell surface-expressed TLR4. Viral single-stranded RNA (ssRNA) and viral DNA can be recognized by TLR7 and TLR9, respectively, which are expressed in the endosomal compartment. Active virus replication-generated viral double-stranded RNA (dsRNA) and viral DNA are sensed by intracellular PRRs RIG-I/MDA-5 and DAI (DNA-dependent activator of IFN regulatory factors), respectively, in the cytoplasm. PRR engagement by its ligand results in activation of IKK complex, composed of kinases IKK-α and IKK-β and the regulatory subunit IKK-γ (also known as NEMO). IκB-α and IκB-β proteins are phosphorylated by the activated IKK complex. Phosphorylated IκBs are then ubiquitinated by β-TrCP (transducin repeat-containing protein), which targets it for degradation by the proteasome. This releases NF-κB dimers from cytoplasmic NF-κB/IκB complexes and allows them to translocate to the nucleus and activate target gene expression.

distinct functions. However, a significant subset of NF-κB targets are genes involved in inflammatory and immune responses.

THE NF-κB PATHWAY IN HOST RESPONSE TO INFECTION—MECHANISMS OF ACTIVATION

NF-κB is a key mediator of inflammatory cytokine-induced responses. Indeed, among the best-known activators of NF-κB is tumor necrosis factor-α (TNF-α)

(32). This proinflammatory cytokine plays a crucial role in mediating host responses to bacterial and viral infection, but is also important in autoimmune disease-induced tissue damage. It is also one of the key mediators of septic or endotoxic shock response. The pathway through which TNF-α induces NF-κB activation is among the most well studied in higher eukaryotes (13, 32). It involves a series of kinases, ubiquitination enzymes, and adaptor molecules, which together mediate IKK-β activation, leading to phosphorylation of IκB proteins (13). Nuclear NF-κB complexes then induce expression of key TNF-α target genes involved in the inflammatory responses, including leukocyte adhesion molecules and chemokines. In fact, it is probably safe to state that NF-κB is the most crucial mediator of the TNF-α-induced inflammatory response.

Studies performed over the last several years have led to the discovery of mammalian TLRs as key mediators of host responses to microbial agents (2, 41). Different TLRs recognize distinct microbial structures: e.g., TLR4 and TLR7 specifically detect and induce host responses to bacterial lipopolysaccharide (LPS) and viral single-stranded RNA, respectively (20, 41, 60). TLR4 is the most well studied member of the TLR family. Although many cell types express TLR4, dendritic cells (DCs) and macrophages play an especially important role in inducing host responses to LPS, which is manifested through induction of expression of proinflammatory cytokines and immunomodulatory molecules (41). Two families of transcription factors are known to play an especially important role in TLR-induced transcriptional responses: NF-κB and interferon regulatory factor (IRF) (1, 2, 36) (Fig. 2). TLR induction of NF-κB is mediated by IKK-β (42). As in the TNF-α-induced NF-κB activation pathway, TLRs also utilize a combination of enzymes and adaptor molecules for NF-κB activation (details of these are provided elsewhere in this volume). In addition, the virus-induced retinoic acid-inducible gene I (RIG-I)/melanoma, differentiation-associated gene 5 (MDA-5) pathways also induce NF-κB activation (2). While it is now well known that NF-κB can be activated by all known microbe-induced pathways, overall NF-κB function and the specific function of individual NF-κB subunits in these pathways are far from clear.

NF-κB FUNCTION—LESSONS FROM KNOCKOUT MOUSE STUDIES

Much of our current understanding of the biological function of NF-κB is obtained from studies of mice deficient in NF-κB subunits or signaling mediators involved in NF-κB activation (24). Overall, these studies have led to the identification of crucial functions for NF-κB not only in innate and adaptive immune responses, but also in tumor development and progression and in the function of the nervous system (32, 43, 53). Interestingly, as will become clear in the ensuing discussion, several broad themes concerning the biological function of NF-κB are now evident. A brief description of knockout mouse phenotypes follows, after which the major biological functions of NF-κB relevant to the topic of this publication are discussed. Due to

space limitations, the phenotypes of knockouts of IκB family members and additional key regulators of NF-κB activation (NIK, IKK-γ/NEMO) are not discussed.

p105/p50⁻/⁻

Absence of p50 does not result in any developmental abnormalities (82). However, B-cell function, and to a lesser extent T-cell function, is impaired. These mice are more susceptible to *Listeria monocytogenes* and *Streptococcus pneumoniae*, but interestingly, more resistant to infection with murine encephalomyocarditis virus (EMCV) (82). The latter findings are discussed in more detail in the section on the role of NF-κB in virus infection.

cRel⁻/⁻, p50⁻/⁻cRel⁻/⁻

cRel⁻/⁻ mice show a phenotype that resembles that of p50⁻/⁻ mice (45). Thus, both B-cell and T-cell function is impaired (45). Additional studies have established significant redundancy in the function of p50 and cRel subunits in lymphocytes and DCs (67, 73, 101). Thus, lymphocyte activation and survival is significantly more impaired in p50⁻/⁻cRel⁻/⁻ mice than in either single knockout (73, 101).

RelA⁻/⁻

The most striking developmental phenotype was noticed in RelA⁻/⁻ mice. Absence of RelA results in massive hepatocyte apoptosis during embryogenesis, resulting in embryonic lethality (6). These findings first established a definitive function for NF-κB in preventing apoptosis. Subsequent studies, which are described below, have now established antiapoptosis as one of the most important biological functions of NF-κB.

p100/p52⁻/⁻

Studies of p52⁻/⁻ mice have identified functions of this subunit in lymphoid tissue architecture (11, 22), in antigen-presenting cell function (85), and quite interestingly, in central tolerance through regulation of Aire (103). As an additional indication of redundancy in function of different subunits, combined absence of p52 and p50 results in impaired osteoclast and B-cell development (23). Also worth mentioning is that combined function of p52/Bcl-3 (an IκB-like protein) seen in vitro has been extended to in vivo systems. Thus, in another case of functional redundancy, only p52⁻/⁻Bcl-3⁻/⁻ mice show breakdown of central tolerance (99) A second important lesson from studies of p52⁻/⁻ mice is how the absence of one subunit results in manifestation of abnormalities through aberrant activation of other subunits. Thus, p52 deficiency results in enhanced DC activation as a consequence of increased nuclear presence of RelB (85) and enhanced activation of naïve T cells because of enhanced activation of RelA (40). At least some of these activities are likely mediated by the IκB-like activity of the p52 precursor, p100 (Fig. 1).

RelB$^{-/-}$

RelB is required for normal homeostasis of the hematopoietic compartment (93), regulation of innate and adaptive immunity to infectious agents (10), secondary lymphoid tissue development (92), and tolerance (62). As discussed above, p52 and RelB are components of the alternative NF-κB activation pathway. Thus, it is likely that similarities in at least some phenotypes seen in p52$^{-/-}$ and RelB$^{-/-}$ mice are indicative of their unique combined functions in the alternative NF-κB pathway.

IKK-α$^{-/-}$ and IKK-β$^{-/-}$

Generation of mice deficient in IKK-α has led to identification of its unique role in activation of the alternative NF-κB pathway as well as potential functions outside of the NF-κB pathway (38, 39, 81). Studies of IKK-β$^{-/-}$ mice established it as the main activator of the canonical NF-κB pathway (55, 56). The phenotype of IKK-β$^{-/-}$ mice resembles that of p50$^{-/-}$RelA$^{-/-}$ mice (55, 56). The most novel findings have been obtained from conditional knockout studies of IKK-β in different tissues, which have led to identification of unique roles for NF-κB proteins that were not evident in NF-κB subunit knockout mice. These include a critical role for IKK-β/NF-κB in linking inflammation and cancer (43). Additional studies have demonstrated both NF-κB-dependent and -independent functions for IKK-α and IKK-β in the skin (54, 71, 84). The use of IKK-α and IKK-β knockout mice in studies of host responses to viral infection are discussed later in this chapter.

TRANSCRIPTIONAL REGULATION BY NF-κB TRANSCRIPTION FACTORS

Knockouts of individual NF-κB subunits, for example, RelA and cRel, reveal distinct phenotypes, which likely indicate their ability to regulate different target genes. A key mechanism by which individual NF-κB subunits or specific dimers show target gene specificity is the sequence of κB sites (51, 76). Interestingly, the sequence of κB sites does not determine binding but rather modulates coactivator interaction through an as yet undefined allosteric mechanism (51). Additional mechanisms of temporal control are provided by change in NF-κB dimers bound to a specific κB sites, as well as TNF-α-dependent autocrine signaling pathways (9, 17, 63, 74, 94). In

the case of cRel, a distinct region of the DNA-binding domain is responsible for cRel-induced expression of the IL-12p40 gene in macrophages (76). In the case of RelA, the sequence of the κB site is not the primary determinant of target gene specificity (90). Instead, in contrast to cRel, only RelA can physically and functionally interact with CBP (90). Indeed, this interaction plays a crucial role in determining RelA subunit-specific interleukin-6 (IL-6) gene expression and explains a possible mechanism by which RelA and cRel may regulate different target genes (90). p52/RelB activation through IKK-α plays a crucial role in regulating expression of chemokines involved in establishing lymphoid tissue architecture (7). These findings are also consistent with observed phenotypes of mice lacking p52 or RelB. Interestingly, p52/RelB target genes have κB sites different from consensus sites bound by p50/RelA (7). Thus, multiple mechanisms can together define target gene specificity of different NF-κB dimers.

BIOLOGICAL ROLES OF NF-κB TRANSCRIPTION FACTORS

As is evident from the above discussion, a wide range of phenotypes result from deficiency of different NF-κB or IKK subunits (24). However, certain recurrent themes are also obvious. For example, individual absence of p50, cRel, and RelA can impact cell survival (24). Additionally, all five NF-κB subunits influence aspects of the innate and adaptive immunity. These functions of NF-κB, which are also crucial in the NF-κB-mediated response to virus infection, are discussed next.

Apoptosis

Studies of mice deficient in the NF-κB RelA subunit provided the first direct evidence of a role for NF-κB in inhibiting apoptosis (6). Additional studies demonstrated that NF-κB inhibits TNF-α-induced apoptosis, and most interestingly, embryonic lethality of RelA$^{-/-}$ mice can be rescued in a TNF receptor-1- or TNF-α-deficient background (3, 5). Thus, NF-κB regulates both TNF-α-induced inflammatory and cell death pathways. Furthermore, NF-κB also prevents apoptosis in cells treated with bacterial or viral constituents (52, 70). The antiapoptotic function of NF-κB is critical for lymphocyte activation and is discussed in the section on adaptive immunity (24).

The mechanisms by which NF-κB inhibits apoptosis are also many. A widely accepted mechanism, which is especially important in lymphocytes, is through the induction of expression of antiapoptotic members of the Bcl-2 family, including Bcl-2, Bcl-xL, and A-1 (24). The best-studied antiapoptotic mechanism regulated by NF-κB is the one utilized to inhibit TNF-α-induced cell death. TNF concomitantly initiates a cell death pathway and a survival pathway dependent on NF-κB. In the absence of RelA or following inhibition of NF-κB, the proapoptotic pathway dominates (5). Interestingly, NF-κB inhibition results in enhanced generation of reactive oxygen species and hyperactivation of proapoptotic JNK (12, 19, 72, 87). Although numerous NF-κB-regulated genes have been purported to inhibit TNF-induced cell death, a key target gene is the one encoding ferritin heavy chain, expression of which reduces reactive oxygen species and suppresses JNK (Jun kinase) signaling (72). Interestingly, NF-κB has also been shown to play a key role in inducing expression of proapoptotic Fas/CD95, thereby promoting Fas-induced cell death (46, 68). Thus, NF-κB can also be proapoptotic in certain instances.

Innate Immunity and Inflammation

NF-κB is activated by all known TLRs. Upon activation, NF-κB can enhance expression of a multitude of genes involved in inflammatory and immune responses. In DCs, it was shown that distinct sets of genes are regulated by different NF-κB subunits. Thus, the p50 and cRel NF-κB subunits were found to be crucial for regulating genes important for T-cell responses (e.g., CD40, IL-12, and IL-18) but not for genes encoding inflammatory cytokines (e.g., TNF-α, IL-1α, and IL-6) (90). In contrast, the RelA subunit was crucial for expression of inflammatory cytokine genes but not T-cell stimulatory genes (90). An especially important role for NF-κB is in regulating chemokine expression, with different subunits regulating distinct functional classes of chemokines (3, 7, 59, 68). In vivo studies have also demonstrated the importance of NF-κB in chemokine-induced neutrophil recruitment (3). Interestingly, while NF-κB is required for induction of proinflammatory cytokine expression by TLR ligands, it also plays a key role in mediating responses to these same cytokines. Thus, cytokines such as TNF-α and IL-1α/β are both regulated by NF-κB and potent NF-κB activators. These dual roles may help sustain TLR-induced NF-κB activation by autocrine signaling (17, 94). As will be discussed later, NF-κB plays a critically important role in regulating cytokine and chemokine expression during viral infection.

Adaptive Immunity

Although a role for NF-κB in regulating lymphocyte function has been suspected from its initial discovery, definitive proof for this was obtained from studies of p50 and cRel knockout mice (33, 77, 83). In particular, these studies underscored the crucial role of NF-κB in

B-cell activation and antibody responses. Importantly, the requirement for NF-κB in B-cell activation is, at least in part, because of its prosurvival functions (25). NF-κB has a similarly vital role during T-cell activation (77). The combined absence of p50 and cRel subunits, in particular, severely impairs T-cell survival and in vivo function (101). In both B and T cells, NF-κB enhances expression of antiapoptotic Bcl-2 family members (24, 25). In addition, RelB is essential for the development of both innate and adaptive T-cell responses to *Toxoplasma gondii* (10). Given these key cell intrinsic roles in lymphocytes, NF-κB transcription factors are likely to be crucial for both antiviral cell-mediated and humoral immune responses.

FUNCTION OF THE NF-κB PATHWAY IN THE HOST RESPONSE TO RNA VIRUSES

The NF-κB transcription factors play key functions in many aspects of the antiviral response (35). It should therefore come as no surprise that many viruses have devised strategies for modulating this pathway (35). Discussed below are some of the key roles of NF-κB during RNA virus infection.

Regulation of Apoptosis

As is evident from the above discussion, a key function of NF-κB in many cell types and in response to many different stimuli is to prevent apoptosis. A similar function for NF-κB has also been identified in the context of RNA virus infection. Initial studies showed that absence of p50 allows murine EMCV-infected mice to survive (82). Later studies with type I interferon (IFN) receptor knockout mice showed that resistance was not due to IFN-β (78). Instead, $p50^{-/-}$ fibroblasts infected with EMCV underwent premature apoptosis before significant levels of viral particles could be released from infected cells (78). A similar observation was made in $RelA^{-/-}$ fibroblasts (78). In vivo, premature apoptosis of infected $p50^{-/-}$ cells resulted in reduced viral burden, allowing infected mice to survive (78). The precise mechanism by which apoptosis is induced in infected knockout cells, and how p50 or RelA normally prevent apoptosis induction during EMCV infection, are not clear. Interestingly, $RelA^{-/-}$ cells are more sensitive to double-stranded RNA-induced cell death, which is produced during the course of viral infection, than WT cells (52). Thus, double-stranded RNA may represent at least one of the signals by which virus infection can trigger apoptosis. As is obvious from the above discussion, overall NF-κB activity generally promotes inflammatory and adaptive immune responses, both of which will limit replication of infectious agents. In the above example, however, it is apparent that NF-κB can enhance EMCV virulence. Thus, NF-κB activation during EMCV infection promotes infected cell survival, allowing more efficient viral replication.

Studies of coxsackievirus also indicate that NF-κB can prevent apoptosis of infected cells (98). A rather interesting mechanism has been proposed for NF-κB regulation during coxsackievirus infection, which is worth discussing further. The viral protease 3C^{pro} was shown to cleave IκB-α to generate a stable proteolytic fragment, which then associates with NF-κB and prevents its activation. This sensitizes infected cells to apoptosis, resulting in decreased overall viral replication (98). This mechanism therefore represents a novel way for the host to limit infection. It is interesting to note that during both EMCV and coxsackievirus infection, NF-κB promotes survival of infected cells. However, by preventing NF-κB activation through the 3C^{pro} protease, coxsackievirus-infected cells become susceptible to apoptosis. It is also worth noting that in vivo consequences of NF-κB inhibition during coxsackievirus infection have not been determined. Thus, NF-κB inhibition by coxsackievirus infection may also limit expression of inflammatory mediators and type I IFNs, which could allow viral replication to occur more robustly. Thus, the eventual outcome of NF-κB inhibition will be determined by a balance between increased cell death and an attenuated host inflammatory/type I IFN response. Similar to the above examples of RNA virus infection, NF-κB also prevents apoptosis after infection with the DNA virus herpes simplex virus type 1 (27). Thus, NF-κB-mediated antiapoptosis can be considered a relatively general mechanism for promoting cell survival during viral infection.

Unlike in the above examples, reovirus induces apoptosis in an NF-κB-dependent manner (16). Similar proapoptotic functions for NF-κB have been identified during dengue virus infection (61) and following infection with a DNA virus (46). Reovirus infection induces NF-κB and apoptosis, and fibroblasts deficient in either the p50 or RelA subunits are resistant to reovirus-induced apoptosis (16, 31). Surprisingly, IKK-β, which generally induces p50 and RelA, is not required for reovirus-induced apoptosis. Instead, the alternative NF-κB pathway promoting IKK-α is required for preventing cell death (31). Precisely how canonical NF-κB subunits (p50 and RelA) and components of the alternative NF-κB pathway together orchestrate cell survival following reovirus infection remains to be determined. Additionally, how NF-κB induces apoptosis during reovirus infection is also not clear. The impact of apoptosis regulation by NF-κB during reovirus infection has

also been investigated in vivo (66). Interestingly, opposing effects of the absence of p50 were noticed in different organs. In the central nervous system, apoptosis was significantly reduced in p50$^{-/-}$ mice compared with WT mice. These findings therefore validate prior in vitro findings by further demonstrating the key proapoptotic role of NF-κB in reovirus infection. However, greatly increased apoptosis was noticed in the hearts of p50$^{-/-}$ mice (66). This was likely due to reduced generation of IFN-β in the absence of p50, which resulted in uncontrolled virus replication in the heart (66). These findings therefore indicate the opposing detrimental and beneficial functions performed by NF-κB in the context of virus infection.

Regulation of Inflammatory and Chemokine Gene Expression

As discussed above, NF-κB plays a key role in regulating proinflammatory gene expression. This key function of NF-κB has been investigated following respiratory syncytial virus (RSV) infection. RSV is a member of the paramyxovirus family that causes acute lower respiratory tract infections, which are characterized by high expression of proinflammatory cytokines and chemokines. Importantly, NF-κB is strongly activated in lungs after infection with RSV, concomitant with increased expression of chemokines including RANTES, macrophage inflammatory protein (MIP)-1α/β, and MIP-2 (29, 89). The use of a specific blocker of IKK-β function showed that expression of these genes following RSV infection is controlled by NF-κB (29). Importantly, IKK-β inhibition also led to significant reduction in airway inflammation (29). Thus, specific blockade of NF-κB may be used to reduce inflammation during acute infection of the lung. However, one must also consider the possibility that inhibition of NF-κB may impair generation of immunity to infectious agents by compromising antigen-presenting cell function and/or lymphocyte function.

The mechanism of activation of NF-κB by RSV has also been subject to detailed study. Typically, RNA virus infection, including RSV, of most cell types induces IRF and NF-κB activation and type I IFN expression through the RIG-1/MDA5 pathway (57). Thus, the RIG-1 pathway is required for type I IFN expression in both conventional DCs and macrophages (44, 47). In contrast, the TLR pathway is dispensable for type I IFN induction in these cell types (44). In this respect, it is important to note that the F protein of RSV has been shown to be a ligand for TLR4, which is a main mediator of RSV-induced proinflammatory cytokines by macrophages (30, 34, 48). Using TLR4-deficient C3H/HeJ mice, it was shown that TLR4 is required for NF-κB activation by RSV (30). These findings suggest that selective activation of TLR signaling in macrophages by some viruses may provide a unique mechanism for viral virulence.

Regulation of Type I IFN Expression

Among the first described functions of NF-κB in the antiviral response was its role in regulating IFN-β mRNA expression (50). Virus induction of the IFN-β gene is one of the most studied transcriptional systems in mammals (36). Virus infection induces formation of the IFN-β enhanceosome, which comprises transcription factors activating transcription factor (ATF)-2/c-Jun, IRF3/IRF7, and NF-κB p50/RelA (36, 88). Each of these factors is thought to be crucial for enhanceosome formation and for high-level IFN-β expression. Studies of IRF3 and IRF7 knockout mice have provided clear evidence of their crucial roles in IFN-β and IFN-α expression in multiple cell types (36). In contrast, Sendai and Newcastle disease virus-induced IFN-β (and IFN-α) expression can still take place in the combined absence of the NF-κB p50 and RelA subunits in fibroblasts (91). Previously, IKK-β was reported to control virus-induced IFN-β expression (15). However, consistent with studies of p50/RelA knockout cells (91), we have found that IKK-β is dispensable for virus-induced IFN-β expression (X. Wang and A. A. Beg, in preparation). Thus, while IRF3/IRF7 are essential for IFN-β expression, NF-κB p50/RelA (and IKK-β) appear to be dispensable. Recent crystallographic studies of the IFN-β enhanceosome may provide possible explanations for these findings (69). A surprising aspect of this structure is the lack of protein-protein interactions between individual transcription factors (69). The 5′ side of the enhanceosome consists of ATF-2/c-Jun and IRF3/IRF7 complexes, which form a composite element since the individual DNA-binding sites overlap. These transcription factors may therefore jointly regulate IFN-β expression despite lack of protein-protein interactions. The NF-κB site, on the other hand, does not directly overlap, and therefore a "sub"-enhanceosome could form in the absence of NF-κB. In addition, the fact that IRFs and ATF-2/c-Jun can independently interact with the CBP coactivator suggests that NF-κB may not be absolutely required for CBP recruitment. Nonetheless, the evolutionary conservation of the NF-κB site suggests that it must play a role in regulating IFN-β expression. Another key area that needs further investigation is the role of NF-κB in virus-induced expression of type I IFNs in plasmacytoid DCs, which are likely the most important IFN-producing cell type in vivo (58). Based on different published findings, it therefore appears likely that NF-κB factors, unlike IRFs, are not universally important in regulating type I IFN expression, but instead participate in a tissue-specific (66) and perhaps viral agent-specific manner (49).

NF-κB Independent Functions of IKK-α and IKK-γ/NEMO in the Antiviral Response

TBK1 (TANK-binding kinase 1) and IKK-*i* (inducible IKK) are two IKK-related kinases that are essential for virus-induced type I IFN expression (2). However, these kinases are involved in phosphorylation and activation of IRF3 and IRF7 rather than NF-κB factors. Two additional proteins, IKK-α and IKK-γ, on the other hand, are known to be crucial for NF-κB activation. Recent studies indicate that these proteins may also be important in regulating type I IFN expression, but through non-NF-κB pathways. Thus, IKK-α was shown to participate in regulating type I IFN expression by activating IRF7 (37), while IKK-γ (also known as NF-κB essential modulator, or NEMO) was found to be essential for TBK1 and IKK-*i* activation, which in turn mediate IRF3 and IRF7 activation (100). Thus, key components of the NF-κB activation pathway also perform vital functions in the antiviral response through regulation of non-NF-κB targets.

Targeting the NF-κB Pathway during Virus Infection

The complexity of the NF-κB pathway and its many paradoxical functions during virus infection should be evident from the above discussion. Thus, during EMCV infection, NF-κB promotes infected cell survival and therefore enhances viral replication. In reovirus infection, NF-κB functions in a proapoptotic capacity. However, this is not necessarily beneficial for the host because reovirus induction of NF-κB causes damage to the central nervous system. In the heart, however, NF-κB is important for inducing IFN-β expression and therefore in limiting viral replication. Similarly disparate functions for NF-κB also come into play during induction of virus-induced inflammatory and adaptive immune responses. Thus, global inhibition of NF-κB is unlikely to be therapeutically efficacious during virus infection. However, short-term and tissue-specific NF-κB inhibition may provide therapeutic benefit in certain instances. For example, NF-κB inhibition in the lung during acute RSV infection may significantly dampen inflammation. In any case, substantially more work needs to be done before we can identify specific situations where inhibition of this key transcription factor may be therapeutically efficacious.

A.A.B. and X.W. are supported by National Institutes of Health grant R01 AI059715 and institutional funds from the H. Lee Moffitt Cancer Center to A.A.B.

References

1. Akira, S., and K. Takeda. 2004. Toll-like receptor signalling. *Nat. Rev. Immunol.* 4:499–511.
2. Akira, S., S. Uematsu, and O. Takeuchi. 2006. Pathogen recognition and innate immunity. *Cell* 124:783–801.
3. Alcamo, E., J. P. Mizgerd, B. H. Horwitz, R. Bronson, A. A. Beg, M. Scott, C. M. Doerschuk, R. O. Hynes, and D. Baltimore. 2001. Targeted mutation of TNF receptor I rescues the RelA-deficient mouse and reveals a critical role for NF-κB in leukocyte recruitment. *J. Immunol.* 167:1592–1600.
4. Anest, V., J. L. Hanson, P. C. Cogswell, K. A. Steinbrecher, B. D. Strahl, and A. S. Baldwin. 2003. A nucleosomal function for IkappaB kinase-alpha in NF-kappaB-dependent gene expression. *Nature* 423:659–663.
5. Beg, A. A., and D. Baltimore. 1996. An essential role for NF-kappaB in preventing TNF-alpha-induced cell death. *Science* 274:782–784.
6. Beg, A. A., W. C. Sha, R. T. Bronson, S. Ghosh, and D. Baltimore. 1995. Embryonic lethality and liver degeneration in mice lacking the RelA component of NF-κB. *Nature* 376:167–170.
7. Bonizzi, G., M. Bebien, D. C. Otero, K. E. Johnson-Vroom, Y. Cao, D. Vu, A. G. Jegga, B. J. Aronow, G. Ghosh, R. C. Rickert, and M. Karin. 2004. Activation of IKKalpha target genes depends on recognition of specific kappaB binding sites by RelB:p52 dimers. *EMBO J.* 23:4202–4210.
8. Bonizzi, G., and M. Karin. 2004. The two NF-kappaB activation pathways and their role in innate and adaptive immunity. *Trends Immunol.* 25:280–288.
9. Bosisio, D., I. Marazzi, A. Agresti, N. Shimizu, M. E. Bianchi, and G. Natoli. 2006. A hyper-dynamic equilibrium between promoter-bound and nucleoplasmic dimers controls NF-kappaB-dependent gene activity. *EMBO J.* 25:798–810.
10. Caamano, J., J. Alexander, L. Craig, R. Bravo, and C. A. Hunter. 1999. The NF-κB family member RelB is required for innate and adaptive immunity to *Toxoplasma gondii*. *J. Immunol* 163:4453–4461.
11. Caamano, J. H., C. A. Rizzo, S. K. Durham, D. S. Barton, C. Raventos-Suarez, C. M. Snapper, and R. Bravo. 1998. Nuclear factor (NF)-kappa B2 (p100/p52) is required for normal splenic microarchitecture and B cell-mediated immune responses. *J. Exp. Med.* 187:185–196.
12. Chang, L., H. Kamata, G. Solinas, J. L. Luo, S. Maeda, K. Venuprasad, Y. C. Liu, and M. Karin. 2006. The E3 ubiquitin ligase itch couples JNK activation to TNFα-induced cell death by inducing c-FLIP(L) turnover. *Cell* 124:601–613.
13. Chen, G., and D. V. Goeddel. 2002. TNF-R1 signaling: a beautiful pathway. *Science* 296:1634–1635.
14. Chen, L. F., Y. Mu, and W. C. Greene. 2002. Acetylation of RelA at discrete sites regulates distinct nuclear functions of NF-kappaB. *Embo J.* 21:6539–6548.
15. Chu, W. M., D. Ostertag, Z. W. Li, L. Chang, Y. Chen, Y. Hu, B. Williams, J. Perrault, and M. Karin. 1999. JNK2 and IKKbeta are required for activating the innate response to viral infection. *Immunity* 11:721–731.
16. Connolly, J. L., S. E. Rodgers, P. Clarke, D. W. Ballard, L. D. Kerr, K. L. Tyler, and T. S. Dermody. 2000. Reovirus-induced apoptosis requires activation of transcription factor NF-kappaB. *J. Virol.* 74:2981–2989.
17. Covert, M. W., T. H. Leung, J. E. Gaston, and D. Baltimore. 2005. Achieving stability of lipopolysaccharide-induced NF-kappaB activation. *Science* 309:1854–1857.
18. Dejardin, E., N. M. Droin, M. Delhase, E. Haas, Y. Cao, C. Makris, Z. W. Li, M. Karin, C. F. Ware, and D. R. Green. 2002. The lymphotoxin-beta receptor induces different patterns of gene expression via two NF-kappaB pathways. *Immunity* 17:525–535.
19. De Smaele, E., F. Zazzeroni, S. Papa, D. U. Nguyen, R. Jin, J. Jones, R. Cong, and G. Franzoso. 2001. Induction of gadd45beta by NF-kappaB downregulates pro-apoptotic JNK signalling. *Nature* 414:308–313.
20. Diebold, S. S., T. Kaisho, H. Hemmi, S. Akira, and C. Reis e Sousa. 2004. Innate antiviral responses by means of TLR7-mediated recognition of single-stranded RNA. *Science* 303:1529–1531.
21. Ducut Sigala, J. L., V. Bottero, D. B. Young, A. Shevchenko, F. Mercurio, and I. M. Verma. 2004. Activation of transcription factor NF-kappaB requires ELKS, an IkappaB kinase regulatory subunit. *Science* 304:1963–1967.
22. Franzoso, G., L. Carlson, L. Poljak, E. W. Shores, S. Epstein, A. Leonardi, A. Grinberg, T. Tran, T. Scharton-Kersten, M. Anver, P. Love, K. Brown, and U. Siebenlist. 1998. Mice deficient in nuclear factor (NF)-kappa B/p52 present with defects in humoral responses, germinal center reactions, and splenic microarchitecture. *J. Exp. Med.* 187:147–159.
23. Franzoso, G., L. Carlson, L. Xing, L. Poljak, E. W. Shores, K. D. Brown, A. Leonardi, T. Tran, B. F. Boyce, and U. Siebenlist. 1997. Requirement for NF-kappaB in osteoclast and B-cell development. *Genes Dev.* 11:3482–3496.
24. Gerondakis, S., R. Grumont, R. Gugasyan, L. Wong, I. Isomura, W. Ho, and A. Banerjee. 2006. Unravelling the complexities of the NF-kappaB signalling pathway using mouse knockout and transgenic models. *Oncogene* 25:6781–6799.
25. Gerondakis, S., and A. Strasser. 2003. The role of Rel/NF-kappaB transcription factors in B lymphocyte survival. *Semin. Immunol.* 15:159–166.
26. Ghosh, S., A. M. Gifford, L. R. Rivere, P. Tempest, G. P. Nolan, and D. Baltimore. 1990. Cloning of the p50 DNA binding subunit of NF-kB: homology to *rel* and *dorsal*. *Cell* 62:1019–1029.
27. Goodkin, M. L., A. T. Ting, and J. A. Blaho. 2003. NF-kappaB is required for apoptosis prevention during herpes simplex virus type 1 infection. *J. Virol.* 77:7261–7280.
28. Greene, W. C., and L. F. Chen. 2004. Regulation of NF-kappaB action by reversible acetylation. *Novartis Found. Symp.* 259:208–217; discussion 218–225.
29. Haeberle, H. A., A. Casola, Z. Gatalica, S. Petronella, H. J. Dieterich, P. B. Ernst, A. R. Brasier, and R. P. Garofalo. 2004. IkappaB kinase is a critical regulator of chemokine expression and lung inflammation in respiratory syncytial virus infection. *J. Virol.* 78:2232–2241.
30. Haeberle, H. A., R. Takizawa, A. Casola, A. R. Brasier, H. J. Dieterich, N. Van Rooijen, Z. Gatalica, and R. P.

Garofalo. 2002. Respiratory syncytial virus-induced activation of nuclear factor-kappaB in the lung involves alveolar macrophages and Toll-like receptor 4-dependent pathways. *J. Infect. Dis.* 186:1199–1206.

31. Hansberger, M. W., J. A. Campbell, P. Danthi, P. Arrate, K. N. Pennington, K. B. Marcu, D. W. Ballard, and T. S. Dermody. 2007. IkappaB kinase subunits alpha and gamma are required for activation of NF-kappaB and induction of apoptosis by mammalian reovirus. *J. Virol.* 81:1360–1371.

32. Hayden, M. S., and S. Ghosh. 2004. Signaling to NF-kappaB. *Genes Dev.* 18:2195–2224.

33. Hayden, M. S., A. P. West, and S. Ghosh. 2006. NF-kappaB and the immune response. *Oncogene* 25:6758–6780.

34. Haynes, L. M., D. D. Moore, E. A. Kurt-Jones, R. W. Finberg, L. J. Anderson, and R. A. Tripp. 2001. Involvement of Toll-like receptor 4 in innate immunity to respiratory syncytial virus. *J. Virol.* 75:10730–10737.

35. Hiscott, J., T. L. Nguyen, M. Arguello, P. Nakhaei, and S. Paz. 2006. Manipulation of the nuclear factor-kappaB pathway and the innate immune response by viruses. *Oncogene* 25:6844–6867.

36. Honda, K., A. Takaoka, and T. Taniguchi. 2006. Type I interferon gene induction by the interferon regulatory factor family of transcription factors. *Immunity* 25:349–360.

37. Hoshino, K., T. Sugiyama, M. Matsumoto, T. Tanaka, M. Saito, H. Hemmi, O. Ohara, S. Akira, and T. Kaisho. 2006. IkappaB kinase-alpha is critical for interferon-alpha production induced by Toll-like receptors 7 and 9. *Nature* 440:949–953.

38. Hu, Y., V. Baud, M. Delhase, P. Zhang, T. Deerinck, M. Ellisman, R. Johnson, and M. Karin. 1999. Abnormal morphogenesis but intact IKK activation in mice lacking the IKKalpha subunit of IkappaB kinase. *Science* 284:316–320.

39. Hu, Y., V. Baud, T. Oga, K. I. Kim, K. Yoshida, and M. Karin. 2001. IKKalpha controls formation of the epidermis independently of NF-kappaB. *Nature* 410:710–714.

40. Ishimaru, N., H. Kishimoto, Y. Hayashi, and J. Sprent. 2006. Regulation of naive T cell function by the NF-kappaB2 pathway. *Nat. Immunol.* 7:763–772.

41. Iwasaki, A., and R. Medzhitov. 2004. Toll-like receptor control of the adaptive immune responses. *Nat. Immunol.* 5:987–995.

42. Karin, M., and Y. Ben-Neriah. 2000. Phosphorylation meets ubiquitination: the control of NF-κB activity. *Annu. Rev. Immunol.* 18:621–663.

43. Karin, M., and F. R. Greten. 2005. NF-kappaB: linking inflammation and immunity to cancer development and progression. *Nat. Rev. Immunol.* 5:749–759.

44. Kato, H., S. Sato, M. Yoneyama, M. Yamamoto, S. Uematsu, K. Matsui, T. Tsujimura, K. Takeda, T. Fujita, O. Takeuchi, and S. Akira. 2005. Cell type-specific involvement of RIG-I in antiviral response. *Immunity* 23:19–28.

45. Kontgen, F., R. J. Grumont, A. Strasser, D. Metcalf, R. Li, D. Tarlinton, and S. Gerondakis. 1995. Mice lacking the c-rel proto-oncogene exhibit defects in lymphocyte proliferation, humoral immunity, and interleukin-2 expression. *Genes Dev.* 9:1965–1977.

46. Kuhnel, F., L. Zender, Y. Paul, M. K. Tietze, C. Trautwein, M. Manns, and S. Kubicka. 2000. NFkappaB mediates apoptosis through transcriptional activation of Fas (CD95) in adenoviral hepatitis. *J. Biol. Chem.* 275:6421–6427.

47. Kumagai, Y., O. Takeuchi, H. Kato, H. Kumar, K. Matsui, E. Morii, K. Aozasa, T. Kawai, and S. Akira. 2007. Alveolar macrophages are the primary interferon-alpha producer in pulmonary infection with RNA viruses. *Immunity* 27:240–252.

48. Kurt-Jones, E. A., L. Popova, L. Kwinn, L. M. Haynes, L. P. Jones, R. A. Tripp, E. E. Walsh, M. W. Freeman, D. T. Golenbock, L. J. Anderson, and R. W. Finberg. 2000. Pattern recognition receptors TLR4 and CD14 mediate response to respiratory syncytial virus. *Nat. Immunol* 1:398–401.

49. Le Bon, A., M. Montoya, M. J. Edwards, C. Thompson, S. A. Burke, M. Ashton, D. Lo, D. F. Tough, and P. Borrow. 2006. A role for the transcription factor RelB in IFN-alpha production and in IFN-alpha-stimulated cross-priming. *Eur. J. Immunol* 36:2085–2093.

50. Lenardo, M. J., C. M. Fan, T. Maniatis, and D. Baltimore. 1989. The involvement of NF-κB in β-interferon gene regulation reveals its role as widely inducible mediator of signal transduction. *Cell* 57:287–294.

51. Leung, T. H., A. Hoffmann, and D. Baltimore. 2004. One nucleotide in a kappaB site can determine cofactor specificity for NF-kappaB dimers. *Cell* 118:453–464.

52. Li, M., W. Shillinglaw, W. J. Henzel, and A. A. Beg. 2001. The RelA(p65) subunit of NF-kappa B is essential for inhibiting double-stranded RNA-induced cytotoxicity. *J. Biol. Chem.* 276:1185–1194.

53. Li, Q., G. Estepa, S. Memet, A. Israel, and I. M. Verma. 2000. Complete lack of NF-kappaB activity in IKK1 and IKK2 double-deficient mice: additional defect in neurulation. *Genes Dev.* 14:1729–1733.

54. Li, Q., Q. Lu, J. Y. Hwang, D. Buscher, K. F. Lee, J. C. Izpisua-Belmonte, and I. M. Verma. 1999. IKK1-deficient mice exhibit abnormal development of skin and skeleton. *Genes Dev.* 13:1322–1328.

55. Li, Q., D. Van Antwerp, F. Mercurio, K. F. Lee, and I. M. Verma. 1999. Severe liver degeneration in mice lacking the IkappaB kinase 2 gene. *Science* 284:321–325.

56. Li, Z. W., W. Chu, Y. Hu, M. Delhase, T. Deerinck, M. Ellisman, R. Johnson, and M. Karin. 1999. The IKKbeta subunit of IkappaB kinase (IKK) is essential for nuclear factor kappaB activation and prevention of apoptosis. *J. Exp. Med.* 189:1839–1845.

57. Liu, P., M. Jamaluddin, K. Li, R. P. Garofalo, A. Casola, and A. R. Brasier. 2007. Retinoic acid-inducible gene I mediates early antiviral response and Toll-like receptor 3 expression in respiratory syncytial virus-infected airway epithelial cells. *J. Virol.* 81:1401–1411.

58. Liu, Y. J. 2005. IPC: professional type 1 interferon-producing cells and plasmacytoid dendritic cell precursors. *Annu. Rev. Immunol.* 23:275–306.

59. Lo, D., L. Feng, L. Li, M. J. Carson, M. Crowley, M. Pauza, A. Nguyen, and C. R. Reilly. 1999. Integrating innate and adaptive immunity in the whole animal. *Immunol. Rev.* 169:225–239.

60. Lund, J. M., L. Alexopoulou, A. Sato, M. Karow, N. C. Adams, N. W. Gale, A. Iwasaki, and R. A. Flavell. 2004.

Recognition of single-stranded RNA viruses by Toll-like receptor 7. *Proc. Natl. Acad. Sci. USA* **101**:5598–5603.

61. Marianneau, P., A. Cardona, L. Edelman, V. Deubel, and P. Despres. 1997. Dengue virus replication in human hepatoma cells activates NF-kappaB which in turn induces apoptotic cell death. *J. Virol.* **71**:3244–3249.

62. Martin, E., B. O'Sullivan, P. Low, and R. Thomas. 2003. Antigen-specific suppression of a primed immune response by dendritic cells mediated by regulatory T cells secreting interleukin-10. *Immunity* **18**:155–167.

63. Natoli, G., S. Saccani, D. Bosisio, and I. Marazzi. 2005. Interactions of NF-kappaB with chromatin: the art of being at the right place at the right time. *Nat. Immunol.* **6**:439–445.

64. Neri, A., C.-C. Chang, L. Lombardi, M. Salina, P. Corradini, A. T. Maiolo, R. S. Chaganti, and R. Dalla-Favera. 1991. B cell lymphoma-associated chromosomal translocation involves candidate oncogene, lyt-10, homologous to NF-κB p50. *Cell* **67**:1075–1087.

65. Nolan, G. P., S. Ghosh, H.-C. Liou, P. Tempst, and D. Baltimore. 1991. DNA binding and IkB inhibition of the cloned p65 subunit of NF-kB, a rel-related polypeptide. *Cell* **64**:961–969.

66. O'Donnell, S. M., M. W. Hansberger, J. L. Connolly, J. D. Chappell, M. J. Watson, J. M. Pierce, J. D. Wetzel, W. Han, E. S. Barton, J. C. Forrest, T. Valyi-Nagy, F. E. Yull, T. S. Blackwell, J. N. Rottman, B. Sherry, and T. S. Dermody. 2005. Organ-specific roles for transcription factor NF-kappaB in reovirus-induced apoptosis and disease. *J. Clin. Invest.* **115**:2341–2350.

67. Ouaaz, F., J. Arron, Y. Zheng, Y. Choi, and A. A. Beg. 2002. Dendritic cell development and survival require distinct NF-kappaB subunits. *Immunity* **16**:257–270.

68. Ouaaz, F., M. Li, and A. A. Beg. 1999. A critical role for the RelA subunit of nuclear factor kappaB in regulation of multiple immune-response genes and in Fas-induced cell death. *J. Exp. Med.* **189**:999–1004.

69. Panne, D., T. Maniatis, and S. C. Harrison. 2007. An atomic model of the interferon-beta enhanceosome. *Cell* **129**:1111–1123.

70. Park, J. M., F. R. Greten, A. Wong, R. J. Westrick, J. S. Arthur, K. Otsu, A. Hoffmann, M. Montminy, and M. Karin. 2005. Signaling pathways and genes that inhibit pathogen-induced macrophage apoptosis—CREB and NF-kappaB as key regulators. *Immunity* **23**:319–329.

71. Pasparakis, M., G. Courtois, M. Hafner, M. Schmidt-Supprian, A. Nenci, A. Toksoy, M. Krampert, M. Goebeler, R. Gillitzer, A. Israel, T. Krieg, K. Rajewsky, and I. Haase. 2002. TNF-mediated inflammatory skin disease in mice with epidermis-specific deletion of IKK2. *Nature* **417**:861–866.

72. Pham, C. G., C. Bubici, F. Zazzeroni, S. Papa, J. Jones, K. Alvarez, S. Jayawardena, E. De Smaele, R. Cong, C. Beaumont, F. M. Torti, S. V. Torti, and G. Franzoso. 2004. Ferritin heavy chain upregulation by NF-kappaB inhibits TNFalpha-induced apoptosis by suppressing reactive oxygen species. *Cell* **119**:529–542.

73. Pohl, T., R. Gugasyan, R. J. Grumont, A. Strasser, D. Metcalf, D. Tarlinton, W. Sha, D. Baltimore, and S. Gerondakis. 2002. The combined absence of NF-kappa B1 and c-Rel reveals that overlapping roles for these transcription factors in the B cell lineage are restricted to the

activation and function of mature cells. *Proc. Natl. Acad. Sci. USA* **99**:4514–4519.

74. Saccani, S., S. Pantano, and G. Natoli. 2003. Modulation of NF-kappaB activity by exchange of dimers. *Mol. Cell* **11**:1563–1574.

75. Sakurai, H., H. Chiba, H. Miyoshi, T. Sugita, and W. Toriumi. 1999. IkappaB kinases phosphorylate NF-kappaB p65 subunit on serine 536 in the transactivation domain. *J. Biol. Chem.* **274**:30353–30356.

76. Sanjabi, S., K. J. Williams, S. Saccani, L. Zhou, A. Hoffmann, G. Ghosh, S. Gerondakis, G. Natoli, and S. T. Smale. 2005. A c-Rel subdomain responsible for enhanced DNA-binding affinity and selective gene activation. *Genes Dev.* **19**:2138–2151.

77. Schulze-Luehrmann, J., and S. Ghosh. 2006. Antigen-receptor signaling to nuclear factor kappa B. *Immunity* **25**:701–715.

78. Schwarz, E. M., C. Badorff, T. S. Hiura, R. Wessely, A. Badorff, I. M. Verma, and K. U. Knowlton. 1998. NF-kappaB-mediated inhibition of apoptosis is required for encephalomyocarditis virus virulence: a mechanism of resistance in p50 knockout mice. *J. Virol.* **72**:5654–5660.

79. Sen, R., and D. Baltimore. 1986. Inducibility of κ immunoglobulin enhancer-binding protein NF-kB by a posttranslational mechanism. *Cell* **47**:921–928.

80. Sen, R., and D. Baltimore. 1986. Multiple nuclear factors interact with the immunoglobulin enhancers. *Cell* **46**:705–716.

81. Senftleben, U., Y. Cao, G. Xiao, F. R. Greten, G. Krahn, G. Bonizzi, Y. Chen, Y. Hu, A. Fong, S. C. Sun, and M. Karin. 2001. Activation by IKKalpha of a second, evolutionary conserved, NF-kappa B signaling pathway. *Science* **293**:1495–1499.

82. Sha, W. C., H. C. Liou, E. I. Tuomanen, and D. Baltimore. 1995. Targeted disruption of the p50 subunit of NF-kappa B leads to multifocal defects in immune responses. *Cell* **80**:321–330.

83. Siebenlist, U., K. Brown, and E. Claudio. 2005. Control of lymphocyte development by nuclear factor-kappaB. *Nat. Rev. Immunol.* **5**:435–445.

84. Sil, A. K., S. Maeda, Y. Sano, D. R. Roop, and M. Karin. 2004. IkappaB kinase-alpha acts in the epidermis to control skeletal and craniofacial morphogenesis. *Nature* **428**:660–664.

85. Speirs, K., L. Lieberman, J. Caamano, C. A. Hunter, and P. Scott. 2004. Cutting edge: NF-kappaB2 is a negative regulator of dendritic cell function. *J. Immunol* **172**:752–756.

86. Steinbrecher, K. A., W. Wilson, III, P. C. Cogswell, and A. S. Baldwin. 2005. Glycogen synthase kinase 3β functions to specify gene-specific, NF-κB-dependent transcription. *Mol. Cell. Biol.* **25**:8444–8455.

87. Tang, G., Y. Minemoto, B. Dibling, N. H. Purcell, Z. Li, M. Karin, and A. Lin. 2001. Inhibition of JNK activation through NF-kappaB target genes. *Nature* **414**:313–317.

88. Thanos, D., and T. Maniatis. 1995. Virus induction of human IFN beta gene expression requires the assembly of an enhanceosome. *Cell* **83**:1091–1100.

89. Tian, B., Y. Zhang, B. A. Luxon, R. P. Garofalo, A. Casola, M. Sinha, and A. R. Brasier. 2002. Identification of NF-kappaB-dependent gene networks in respiratory syncytial virus-infected cells. *J. Virol.* **76**:6800–6814.

90. Wang, J., X. Wang, S. Hussain, Y. Zheng, S. Sanjabi, F. Ouaaz, and A. A. Beg. 2007. Distinct roles of different NF-kappa B subunits in regulating inflammatory and T cell stimulatory gene expression in dendritic cells. *J. Immunol.* **178:**6777–6788.

91. Wang, X., S. Hussain, E. J. Wang, X. Wang, M. O. Li, A. Garcia-Sastre, and A. A. Beg. 2007. Lack of essential role of NF-kappa B p50, RelA, and cRel subunits in virus-induced type 1 IFN expression. *J. Immunol.* **178:**6770–6776.

92. Weih, F., and J. Caamano. 2003. Regulation of secondary lymphoid organ development by the nuclear factor-kappaB signal transduction pathway. *Immunol. Rev.* **195:**91–105.

93. Weih, F., D. Carrasco, S. K. Durham, D. S. Barton, C. A. Rizzo, R.-P. Ryseck, S. A. Lira, and R. Bravo. 1995. Multiorgan inflammation and hematopoietic abnormalities in mice with a targeted disruption of RelB, a member of the NF-κB/Rel family. *Cell* **80:**331–340.

94. Werner, S. L., D. Barken, and A. Hoffmann. 2005. Stimulus specificity of gene expression programs determined by temporal control of IKK activity. *Science* **309:**1857–1861.

95. Yamamoto, M., S. Yamazaki, S. Uematsu, S. Sato, H. Hemmi, K. Hoshino, T. Kaisho, H. Kuwata, O. Takeuchi, K. Takeshige, T. Saitoh, S. Yamaoka, N. Yamamoto, S. Yamamoto, T. Muta, K. Takeda, and S. Akira. 2004. Regulation of Toll/IL-1-receptor-mediated gene expression by the inducible nuclear protein IkappaBzeta. *Nature* **430:**218–222.

96. Yamamoto, Y., U. N. Verma, S. Prajapati, Y. T. Kwak, and R. B. Gaynor. 2003. Histone H3 phosphorylation by IKK-alpha is critical for cytokine-induced gene expression. *Nature* **423:**655–659.

97. Yang, F., E. Tang, K. Guan, and C. Y. Wang. 2003. IKK beta plays an essential role in the phosphorylation of RelA/p65 on serine 536 induced by lipopolysaccharide. *J. Immunol.* **170:**5630–5635.

98. Zaragoza, C., M. Saura, E. Y. Padalko, E. Lopez-Rivera, T. R. Lizarbe, S. Lamas, and C. J. Lowenstein. 2006. Viral protease cleavage of inhibitor of kappaBalpha triggers host cell apoptosis. *Proc. Natl. Acad. Sci. USA* **103:** 19051–19056.

99. Zhang, X., H. Wang, E. Claudio, K. Brown, and U. Siebenlist. 2007. A role for the IkappaB family member Bcl-3 in the control of central immunologic tolerance. *Immunity* **27:**438–452.

100. Zhao, T., L. Yang, Q. Sun, M. Arguello, D. W. Ballard, J. Hiscott, and R. Lin. 2007. The NEMO adaptor bridges the nuclear factor-kappaB and interferon regulatory factor signaling pathways. *Nat. Immunol.* **8:**592–600.

101. Zheng, Y., M. Vig, J. Lyons, L. Van Parijs, and A. A. Beg. 2003. Combined deficiency of p50 and cRel in CD4+ T cells reveals an essential requirement for NF-κB in regulating mature T cell survival and in vivo function. *J. Exp. Med.* **197:**861–874.

102. Zhong, H., R. E. Voll, and S. Ghosh. 1998. Phosphorylation of NF-kappa B p65 by PKA stimulates transcriptional activity by promoting a novel bivalent interaction with the coactivator CBP/p300. *Mol. Cell.* **1:** 661–671.

103. Zhu, M., R. K. Chin, P. A. Christiansen, J. C. Lo, X. Liu, C. Ware, U. Siebenlist, and Y. X. Fu. 2006. NF-kappaB2 is required for the establishment of central tolerance through an Aire-dependent pathway. *J. Clin. Invest.* **116:**2964–2971.

Cellular Signaling and Innate Immune
Responses to RNA Virus Infections
Edited by A. R. Brasier et al.
© 2009 ASM Press, Washington, D.C.

Allan R. Brasier

9

The Nuclear Factor-κB Signaling Network: Insights from Systems Approaches

Highly conserved across species, the innate immune system is designed to recognize the presence of RNA viral infections and evoke protective antiviral responses via specific signaling pathways. Of these, the nuclear factor-κB (NF-κB) pathway plays a central role in the host response by mediating expression of chemokines, cell adhesion molecules, and antiapoptotic proteins that play key roles in mucosal inflammation (60, 75, 77). Not only is NF-κB important in mucosal antiviral response, but in certain cases it also mediates viral pathogenesis. For example, in mouse models of respiratory syncytial virus (RSV) infection, NF-κB inhibition attenuates monocyte recruitment, airway hypercellularity, pulmonary cytokine expression, and clinical disease, despite enhanced levels of viral replication (17).

NF-κB is a family of inducible transcription factors activated by a number of discrete pathways in response to inflammatory mediators produced during RNA virus infections, such as double-stranded (ds) RNA, cytokines, and reactive oxygen species (ROS). Described by Amer Beg and Xingyu Wang in chapter 8 of this volume, the NF-κB transcriptional activators are cytoplasmic proteins. A major pathway for activation involves liberation from sequestered cytoplasmic complexes and subsequent nuclear translocation. Recent work has also indicated the presence of independent, parallel signaling pathways whose function is required for transcriptional function. This chapter will review the nature of these bipartite NF-κB regulatory pathways activated in response to mono-kines produced as a consequence of viral infection, and those involved in mediating RNA viral infection. Recent work elucidating the dynamics of nucleo-cytoplasmic NF-κB shuttling and transient interactions with target chromatin in cellulo will be discussed.

Systematic ("unbiased") approaches have been applied to determine the structure of the NF-κB protein interaction network and determine the downstream genes NF-κB controls. These studies hold the promise of being able to comprehensively understand the mechanisms governing NF-κB signal transduction, signal cross talk, and diverse cellular responses to inflammation. In this vein, the basal and inducible protein interaction networks elucidated by affinity tagging methodologies have been determined and are presented. Finally, work using mRNA profiling studies to determine the genetic networks downstream of NF-κB in response to mono-kines and RNA virus replication will be discussed and integrated. Overall, these studies indicate that the NF-κB

Allan R. Brasier, Department of Medicine and the Sealy Center for Molecular Medicine, The University of Texas Medical Branch, Galveston, TX 77555–1060.

network is highly dynamic, capable of responding to diverse signals as a consequence of the pleiotropic functions of its regulatory kinases, and elicits stimulus type- and time-dependent expression of gene networks that control inflammation or pathogenesis. These insights will form the basis for more specific interruption or augmentation of NF-κB actions in viral infections.

NF-κB REGULATORY PATHWAYS AND MODULES

NF-κB is a family of inducible transcription factors related by a common NH_2-terminal Rel homology domain (RHD). This family includes the proteolytically processed DNA-binding subunits NF-κB1 and NF-κB2 and the transcriptional activators RelA, cRel, and RelB (see chapter 8). NF-κB is complexed and inactivated in the cellular cytoplasm by binding inhibitors of NF-κBs: IκBs, which are ankyrin repeat domain-containing proteins that bind the RHD and block nuclear translocation and DNA-binding activity (3). The major IκBs include the IκB isoforms α/β/ε and the 100-kDa NF-κB2 precursor.

NF-κB is controlled by distinct regulatory pathways, termed the "canonical," "noncanonical," and "crosstalk" pathways, that are activated by certain stimuli produced during viral-host cell interaction. The canonical pathway (36) is induced by cytokines, Toll-like receptor (TLR) ligands, and viral dsRNA. The noncanonical pathway (12, 70) is induced by RNA virus infection or lymphokines (B-cell activating factor [BAFF], stromal cell-derived factor [SDF]). The cross-talk module (2) is more recently described to be activated by lymphotoxin-β (LT-β). Mechanistic studies of these pathways indicate that each is composed of "modules"; modules are signaling subnetworks whose actions are to control either nuclear translocation or transcriptional activation. For example, the canonical pathway controls release of RelA sequestered by IκB-α via an IκB kinase (IKK)–IκB-α module, and transcriptional activation of RelA by Ser-276 phosphorylation via an ROS-catalytic subunit of protein kinase A (PKAc) module. These pathways and component modules are schematically diagrammed in Fig. 1 to 3 and described in more detail below.

The Canonical Pathway

The most-studied NF-κB signaling pathway is initiated by activation of the tumor necrosis factor receptor (TNFR) superfamily and TLRs. The TLR pathway was discussed in chapter 8 and will not be elaborated here. TNF-α is a potent monokine that is released from

tissue-resident macrophages upon encountering viral products during infection. TNF-α binds to two classes of ubiquitously expressed receptors, TNFRs 1 and 2, with TNFR1 playing the primary role in innate immune responses. Lacking intrinsic kinase activity, TNFR1 signals by inducing the formation of a submembranous complex containing adaptor signaling proteins that interact with the TNFR death domain (DD). These adaptor proteins include TNFR-associated DD (TRADD), FAS-associated protein with DD (FADD), TNFR-associated factors (TRAFs) 2 and 6, and receptor-interacting protein (RIP) (27, 28). Downstream, the mitogen-activated protein kinase kinase kinase (MAP3K) transforming growth factor-β-associated kinase 1 (TAK1) is activated by TRAF6. Here, TRAF6 is involved in a novel Lys-63-linked polyubiquitination step involving the UBC13-UBE2V1 ligase (14, 52, 56, 69, 74). This submembranous complex activates IKK, the first committed step in RelA·NF-κB1 translocation, by TAK1-initiated phosphorylation (Fig. 1), (34, 52, 56, 69, 74).

IKK is a multisubunit kinase complex (49) whose core is composed of two highly homologous serine/threonine kinases, IKK-α and -β (51, 83), and a regulatory subunit, IKK-γ (37), that exist in a precise stoichiometric relationship of two catalytic subunits to two regulatory subunits (16, 50). In spite of significant sequence similarity between the IKK-α/β catalytic subunits, IKK-β has ~30-fold higher activity toward IκB-α (30, 83), and gene knockout studies have shown that IKK-β is the major IκB-α kinase (29). IKK-γ, also known as the NF-κB essential modulator (NEMO) (43, 50, 89), is essential for inducible IKK activity, as IKK-γ-deficient cells are unable to activate IKK in response to all stimuli tested (65, 66, 89). IKK-γ plays multiple roles in IKK activation through its ability to organize the assembly of IKKs into the activated high-molecular-weight complex (62, 88); bind ubiquitinated signaling adaptors (84, 91); recruit the IκB-α inhibitor into the activated IKK complex, where it becomes phosphorylated, marking IκB-α for degradation (88); and serve as an adaptor molecule to recruit upstream kinases that phosphorylate the catalytic subunits (65, 90, 91). Through these activities, IKK-γ forms a molecular bridge between IKK, its upstream activators, and its substrate.

A common pathway for IKK activation is serine phosphorylation at residues 170/172 in the activation loop (13). Activated IKK, in turn, phosphorylates IκB-α on serine residues 32 and 36 in its NH_2-terminal regulatory domain (36), making it a substrate for proteolysis through the 26S proteasome and calpain pathways (21, 49). Liberated from its IκB-α inhibitor, the heterodimeric RelA·NF-κB1 rapidly enters the nucleus.

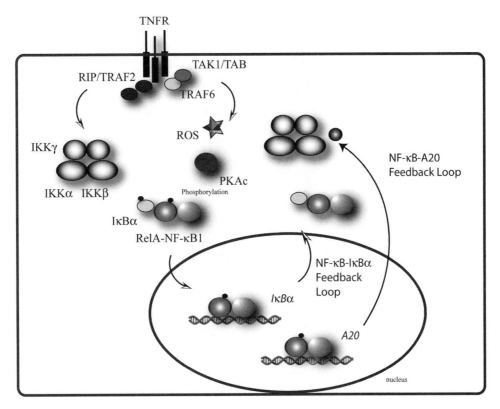

Figure 1 The canonical NF-κB activation pathway induced by the TNF-α monokine. A schematic view of the canonical NF-κB signaling pathway coupled to the TNFR. Ligation with TNF-α induces receptor trimerization and recruitment of signaling adaptors. The two modules controlling NF-κB translocation (IKK–IκB-α) and NF-κB activation/phosphorylation (ROS-PKAc) are shown. The sites of the NF-κB–IκB-α and NF-κB–TNFAIP3/A20 negative autoregulatory "feedback loops" are diagrammed.

RelA Ser-276 Phosphorylation Is an Independent Signaling Module Necessary for Transcriptional Competency

Although most studies have focused on NF-κB translocation as a major mechanism for its activation, NF-κB is a target for stimulus-inducible phosphorylation. In fact, RelA is Ser-phosphorylated at multiple sites, including Ser residues 276 and 311 in its NH_2-terminal RHD and Ser residues 529 and 536 in its COOH-terminal transactivation domain. Although the functional consequences of RelA site-specific phosphorylation have not been completely dissected, Ser-276 and -536 phosphorylation are thought to be the most important in regulating transcriptional activity (68, 94). We have defined a TNF-α-induced regulatory pathway mediated by intracellular ROS as a second messenger that is required for NF-κB-dependent gene expression independent of NF-κB translocation. Here, inhibition of ROS formation blocks Ser-276

phosphorylation and NF-κB-dependent gene expression without affecting NF-κB translocation (32, 78). This pathway is dependent on the catalytic subunit of PKAc, PKA, a kinase that copurifies with IκB-α (94) and is activated by TNF-α. Because inhibition of ROS formation or small interfering RNA (siRNA)-mediated PKAc knockdown prevents TNF-α-induced RelA Ser-276 phosphorylation without affecting its nuclear translocation, the ROS-PKAc transcriptional activating module is a separate pathway whose function is also necessary for NF-κB-dependent gene expression. In separate studies, a phosphoinositide-3 kinase (PI3K) pathway controlling RelA phosphorylation was described for interleukin-1 (IL-1)-inducible RelA activation (72). Like the ROS-PKAc pathway in TNF activation, the PI3K signaling pathway was independent of IL-1-induced RelA nuclear translocation.

The consequences of RelA Ser-276 phosphorylation are partially known. Phosphorylation at Ser-276 reduces

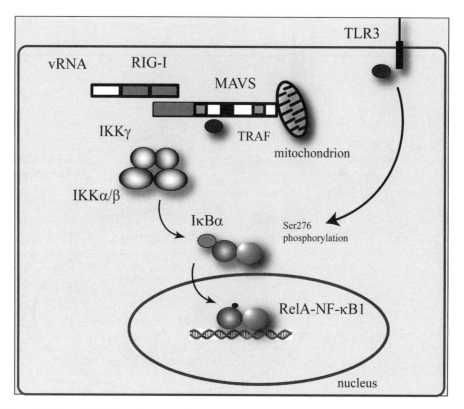

Figure 2   The viral-induced canonical NF-κB activation pathway downstream of the RIG-I/
MAVS complex. Schematic view of the RIG-I/MAVS-induced NF-κB activation pathway and
the nuclear translocation (IKK–IκB-α) and activation (TLR3-RelA Ser-276 kinase) modules.
Caspase recruitment domains are shown in dark gray; MAVS-TRAF interaction domains are
shaded light gray. vRNA, viral (ds) RNA.

intermolecular NH$_2$- and COOH-terminal interactions,
allowing the Ser-276-phosphorylated RelA to complex
with p300/CREB binding protein (CBP) coactivators (95),
resulting in RelA acetylation (11). Phosphorylation of
RelA also stabilizes its association with endogenous gene
targets in chromatin immunoprecipitation assays (32). We
have also found that RelA Ser-276 phosphorylation
allows complex formation with the cyclin-dependent
kinases CDK-9/Ccn T1, known as positive transcription
elongation factor-b (PTEF-b) (57). CDK-9/Ccn T1 are involved in phosphorylating the COOH-terminal domain
of RNA polymerase II, licensing Pol II to produce fully
elongated transcripts (63). We interpret these findings to
indicate that IKK–IκB-α and the PKAc-RelA Ser-276 kinase modules are two independent signaling arms whose
activation are both required for the canonical pathway.

Viral-Induced Activation of the Canonical NF-κB Pathway by RIG-I/MAVS

The presence of RNA virus replication is sensed by
cytoplasmic pattern recognition receptors known as

retinoic acid-inducible gene I (RIG-I) and melanoma differentiation-associated gene 5 (MDA-5) (1, 40). These
proteins both contain NH$_2$-terminal caspase activation
and recruitment domains (CARDs) and a COOH-
terminal DExD/H box RNA helicase. Of these two,
RIG-I has emerged as a major sensor of RNA virus
infection contributing to antiviral interferon (IFN) signaling, whereas MDA-5 is activated in response to
picornavirus infection (15, 39). Upon RIG-I binding
to single-stranded 5′-phosphorylated RNA, the RIG-I
CARD domains associate with an adaptor protein termed
the mitochondrial activiral signaling protein (MAVS),
also known as IFN-β promoter stimulator 1 (IPS-1) or
virus-induced signaling adaptor (VISA). MAVS is a mi-
tochondrial-localized protein also containing an NH$_2$-
terminal CARD domain as well as multiple TRAF2, 3,
and 6 interaction motifs (TIMS) and a COOH-terminal
transmembrane sequence (67, 71). The NH$_2$-terminal
MAVS CARD domain initiates RIG-I recruitment, forming an activated signaling complex. Importantly, MAVS
is required for RIG-I activation and is downstream of

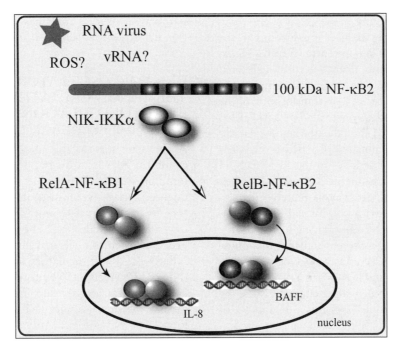

Figure 3 The viral-induced noncanonical NF-κB activation pathway. Schematic view of the noncanonical and cross-talk pathways leading to RelB·NF-κB2 and RelA·NF-κB1 nuclear translocation. The mechanisms of viral replication coupling to NIK–IKK-α and the role of RelA/RelB phosphorylation have not yet been elucidated.

RIG-I because inhibition of MAVS completely blocks RIG-I-induced IFN production (42, 71).

RIG-I recruitment to MAVS produces bifurcating downstream signals to activate IFN regulatory factor (IRF) 3 and the canonical NF-κB activation pathway (87). IRF3 is a latent cytoplasmic protein whose COOH-terminal phosphorylation, mediated by the "atypical" IKKs, TANK binding kinase-1 (TBK1)/IKK-*i*, induces its nuclear translocation and acquisition of transcriptional activator function. The IRF3 signaling pathway is initiated by TRAF3 recruitment to TIM domains on activated (RIG-I-complexed) MAVS (67). Interestingly, IKK-γ, one of the key regulatory kinases in the canonical NF-κB activation pathway, is required for atypical IKK recruitment and activation mediated through TRAF-associated NF-κB activator (TANK) (93). Conversely, the mechanisms by which activated MAVS initiates NF-κB activation are not completely understood. Although TRAF3 is capable of activating NF-κB, and TRAF3 is recruited to the MAVS complex mediating IRF3 activation, TRAF3 is not required for RIG-I/MAVS activation of IKK. Cells genetically deficient in TRAF3 or expressing a dominant negative TRAF3 lack MAVS-induced IRF signaling, but the canonical NF-κB pathway is not affected (93). These data suggest the role of other TRAF isoforms in activating the IKK–IκB-α module, whose identity will require further exploration.

Phosphorylation Control of RelA Activation Downstream of the RIG-I/MAVS Regulatory Module

The canonical NF-κB pathway downstream of the RIG-I/MAVS complex also induces RelA phosphorylation at Ser-276 and Ser-536 residues and is important in acquisition of full transactivating function. We have recently found that RelA phosphorylation on Ser-276 is important in RSV-induced RelA activation (47). In this pathway, TLR3 expression is upregulated downstream of the IRF pathway. siRNA-mediated knockdown of TLR3 inhibits RSV-induced RelA Ser-276 formation without affecting Ser-536 formation (47). TLR3 upregulation is mediated in a "feed-forward" mechanism mediated by the production and signaling of IFN-β (47). Like the cytokine-induced canonical pathway, inhibition of Ser-276 formation has a dramatic inhibitory action on RSV-induced NF-κB transcriptional activity, without affecting NF-κB translocation (47). These data indicate the existence of parallel signaling modules controlling NF-κB nuclear translocation and transcriptional activation downstream of the RIG-I/MAVS complex.

The Noncanonical Pathway

The noncanonical pathway induces processing of the 100-kDa NF-κB2 precursor into its mature 52-kDa DNA-binding form (NF-κB2) and liberation of sequestered p100-associated complexes. The original descriptions of the noncanonical pathway focused on translocation of the Rel·NF-κB2 complex, a pathway activated by BAFF, LT-β (5, 6, 53, 54, 70), and RSV (12). In the noncanonical pathway, p100 processing is initiated by IKK-α-mediated phosphorylation at two Ser residues in the p100 COOH terminus (86). Interestingly, neither IKK-β nor IKK-γ, key regulators of the canonical pathway, is required for signaling in the noncanonical pathway. Instead, IKK-α activation is controlled by the NF-κB-inducing kinase (NIK). NIK serves as a rate-limiting upstream activator that phosphorylates IKK-α and also serves as a docking protein to recruit both p100 NF-κB2 and IKK-α into a complex (86). By virtue of its different DNA-binding characteristics, the activated NF-κB2·RelB complex controls a distinct set of downstream genes important in the immune response, including Epstein-Barr virus-induced molecule 1 chemokine ligand 19 (ELC/CCL19), SDF, and BAFF (5).

Although the focus of the noncanonical pathway has been on the RelB·NF-κB2 complexes, recent work has indicated that this may be an oversimplification. Using cells deficient in the major IκB-α/β/ε inhibitors, the existence of a p100-sequestered RelA·NF-κB1 complex has been discovered whose liberation can be induced by LT-β (26). This latter represents a "cross-talk" pathway where prototypical RelA·NF-κB1 DNA-binding complex is released as a consequence of activating the noncanonical NIK–IKK-α kinases.

The role of the noncanonical pathway has been largely investigated in lymphoid cells and osteoclasts. We have found that this pathway is also activated in epithelial cells in response to RSV infection mediated by NIK (12). The RSV-induced noncanonical pathway appears to be more rapidly induced than the more potent canonical pathway, where it amplifies expression of NF-κB-dependent genes, such as IκB and IL-8 (12). The contribution of the cross-talk pathway in viral response has not yet been determined.

Although it is likely that separate signaling modules controlling transcriptional activation of translocated RelA·NF-κB1 and RelB·NF-κB2 complexes will play an important role in the noncanonical pathway, there is currently little work in this area. However, one study indicates that LT-β stimulation appears to induce RelA phosphorylation at Ser-536, a site important in LT-β-induced gene activation (35). This intriguing observation suggests that RelA may have distinct phosphorylation patterns when activated by the noncanonical pathway, a process that may affect its protein interaction network or spectrum of target genes.

DYNAMIC OSCILLATORY BEHAVIOR OF THE CANONICAL PATHWAY PRODUCED BY AUTOREGULATION: THE ROLE OF NEGATIVE-FEEDBACK LOOPS

A characteristic finding of time course studies of the canonical NF-κB pathway is that the complex is actively terminated. NF-κB pathway termination is the consequence of negative autoregulatory control, autoregulatory "loops," that represent NF-κB-dependent induction of pathway inhibitors. These feedback loops affect processes such as nuclear translocation (NF-κB–IκB and NF-κB–Bcl-3 autoregulatory loops), IKK kinase activity (NF-κB–TNFα-inducible protein 3 [TNFAIP3]/A20 and NF-κB–TRAF loops), and, finally, control of TNF-α mRNA message stability (NF-κB–tristetraprolein [TTP] loop). The physiological role of these negative-feedback loops is confirmed in genetic knockout models, where mice deficient in IκB-α, TNFAIP3/A20, and TTP exhibit phenotypes of chronic inflammation in response to a transient stimulus (8). Interestingly, the noncanonical pathway does not strongly activate IκB or A20 negative-feedback loops (33) and appears not to be subject to any known negative autoregulation.

The NF-κB–IκB feedback loop begins when IκB is proteolyzed, allowing activated NF-κB to enter the nucleus and stimulate transcription of the ankyrin repeat-containing proteins IκB-α IκB-ε Bcl-3, and NF-κB1 and 2. As a result, IκB-α is rapidly resynthesized to levels seen in untreated cells (<1 h after TNF-α stimulation), where it binds to RelA, exporting it from the nucleus via its nuclear export sequence (10, 23). IκB-ε has functions similar to those of IκB-α, binding RelA and cRel, but under normal conditions is thought to play a relatively minor role relative to that of IκB-α (26, 82). Similarly, Bcl-3 is an NF-κB inhibitor that selectively binds to 50-kDa NF-κB1. Bcl-3 is more slowly resynthesized (3 to 9 h) than IκB-α/ε (9, 20). This finding suggests that inactivation of NF-κB isoforms by the NF-κB–IκB autoregulatory loop is highly temporally orchestrated.

The NF-κB–TNFAIP3/A20 feedback loop is a second negative autoregulatory loop mediated by NF-κB-dependent activation of the TNFAIP3/A20 ubiquitin ligase (41). Newly synthesized TNFAIP3/A20 associates with IKK-γ to inhibit IKK directly and also induces RIP degradation through the ubiquitin-proteasome pathway (25, 81), thereby uncoupling IKK from the upstream activated TNFR. This pathway is required to prevent newly resynthesized IκB-α from being proteolyzed due to ongoing IKK activity.

A third level of negative autoregulation is mediated by the TTP zinc-finger protein (61). TTP is a prototypic member of the CCCH motif-containing zinc-finger family that promotes the degradation of mRNA transcripts containing AU-rich elements in their 3′-untranslated region, including TNF-α mRNA itself. Destruction of TNF-α mRNA inhibits further generation of NF-κB activating signals. The presence of negative-feedback loops allows for a cell to rapidly respond to NF-κB signals and then promptly terminate the response after the actions of NF-κB have occurred.

Dynamics of the Canonical NF-κB Pathway

The NF-κB pathway is surprisingly dynamic in terms of its nucleo-cytoplasmic shuttling and chromatin interactions. The observations supporting this statement have been produced using real-time imaging of NF-κB tagged to intrinsic fluorescent proteins.

For many years, it has been observed that canonical pathway activators induce biphasic NF-κB binding in gel mobility shift assays of nuclear extracts (19). Here, a rapid phase of NF-κB binding was followed by its loss, then reappearance in the nucleoplasm. Using a fluorescent protein fused to NF-κB- and IκB-α (the latter being driven by an NF-κB-inducible promoter to recreate the NF-κB–IκBα autoregulatory loop), Nelson et al. demonstrated that NF-κB translocation is oscillatory in cells continuously stimulated with TNF-α, but not in those only briefly stimulated (55). The dynamics of the NF-κB pathway arising from its negative autoregulation is schematically diagrammed in Fig. 4. These studies further indicated an emergent property of the system, in that there was significant heterogeneity in the magnitude and periodicity of the oscillatory response. Although the initial spike in NF-κB activation was observed in most cells, the timing and magnitude of subsequent oscillations were highly variable from cell to cell.

In separate studies, the dynamics of RelA-DNA interaction were examined in cellulo using macroscopic "arrays" of multimeric RelA-binding sites that could be visualized by real-time confocal microscopy (7). RelA binding and off-rates were measured on these arrays using fluorescence recovery after photobleaching (FRAP) and fluorescence lifetime measurements (FLIP). Activated RelA in the nucleoplasm is highly mobile and more rapid than the ability of the FRAP technique to measure its recovery time (7). Although RelA binding resulted in a transient immobilization relative to the uncomplexed NF-κB in the nucleoplasm, interaction with active chromatin was surprisingly transient, with an estimated turnover of bound RelA of less than 30 s (7). These data challenge the notion that NF-κB forms stable complexes on target genes and suggests instead that it binds and dissociates rapidly. This physical description indicates that the NF-κB interaction with target genes is governed by stochastic processes.

MATHEMATICAL MODELING TO UNDERSTAND PROPERTIES OF THE NF-κB PATHWAY

Mathematical analysis and simulations have enriched our understanding of the IKK–IκB-α regulatory module. Most of these approaches involve deterministic models that describe changes in nuclear and cytoplasmic concentrations of the major regulatory proteins based on ordinary differential equations (ODE). Hoffmann et al. used this approach of deterministic mathematical modeling by ODE to simulate the three NF-κB–IκB autoregulatory loops, showing the redundant function of IκB-ε (26). A separate model generated by Lipniacki et al. extended the modeling to the NF-κB–TNFAIP3/A20 feedback loop and corrected errors in nuclear transport and concentration (44). Much of the observed behavior of the canonical pathway could be simplified by considering all the diverse IκB species as a single IκB-α. This latter two-component model accurately predicted IKK activation profiles. From these simulations, these authors were able to infer the effects of deficiency in the TNFAIP3/A20 gene on persistent IKK activation, thereby explaining the prolonged NF-κB activation observed in the TNFAIP3/A20 knockout mouse (44). From this analysis, we have come to understand that nuclear RelA oscillatory behavior is critically dependent on the ratio of nuclear to cytoplasmic volume, a parameter that determines how rapidly the initial concentration of NF-κB rises, thereby controlling the initial rate of IκB-α resynthesis. Lipniacki et al. then adapted the deterministic ODE model to incorporate probabilistic response at the level of target gene expression. This analysis allowed study of how single cells respond to a given stimulation and suggested that the population-averaged cellular response does not, in fact, exist in any individual cell (46). This insight has profound implications for design and interpretation of signaling network studies using biochemical and population averaging techniques.

While the models discussed above have focused on TNF-α-induced IKK–IκB-α modules, Werner et al. recently extended this analysis to lipopolysaccharide (LPS)-induced TLR4 signaling (80). In contrast to the

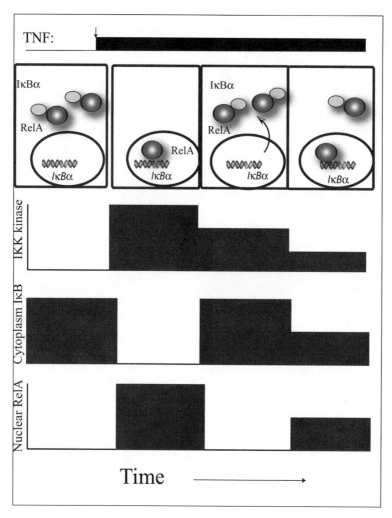

Figure 4   Dynamics of NF-κB: "oscillations." In certain cells with appropriate cytoplasmic:nuclear volumes, chronic stimulation with TNF-α induces NF-κB oscillatory behavior, where RelA cycles in and out of the nucleus. This diagram shows the control of NF-κB subcellular localization, IKK kinase activity, RelA nuclear abundance, and cytoplasmic IκB-α levels during a single oscillation of the IKK–IκB-α module. (Top) Timing of the TNF-α stimulus.

monophasic IKK activation profile induced by TNF-α (Fig. 4), LPS induces a biphasic activation profile; the second phase of IKK activation was found to depend on paracrine TNF-α secretion. Although specific aspects of this model will require further experimental verification, this analysis suggested that LPS induces NF-κB via a cytokine-mediated feed-forward mechanism (80), somewhat analogous to the TLR3-mediated "feed-forward" mechanism for RelA phosphorylation downstream of the RIG-I/MAVS complex.

The prediction that the IκB inhibitors had primary nonredundant activity to sequester RelA in the cytoplasm was challenged by the experimental finding that RelA was still cytoplasmic and productively inducible in triple-knockout (IκB-α/β/ε-deficient) cells. Here, cytoplasmic RelA was discovered to be primarily complexed to the 100-kDa NF-κB2 precursor. In response to LT-β, 100-kDa NF-κB2 was processed to liberate the canonical RelA·NF-κB1 complex. The liberation of RelA·NF-κB1 was independent of IKK-γ, but required the NIK–IKK-α pathway instead, defining this NIK–NF-κB2 regulatory module as a "cross-talk" module. The behavior of the NF-κB pathway with the NIK-NF-κB2 module was then mathematically examined (2). Computational simulations indicated that genetic deficiencies of the IκB proteins alone and in various combinations were compensated by enhanced expression of the 100-kDa NF-κB2 precursor (2). This phenomenon is understandable in terms of the NF-κB–NF-κB2 autoregulatory loop, where due to the genetic deficiency of IκB-α, NF-κB2 synthesis would be induced until cellular homeostasis has been reached.

Lipniacki et al. continued work in stochastic modeling by incorporating two classes of stochastic switches in the IKK–IκB-α module—one at the TNFR and the second at target gene expression (45). Both of these stochastic switches are associated with key signal amplification nodes, where, for example, activated TNFR can produce many activated IKK molecules, and activated nuclear NF-κB can produce many copies of target gene mRNA molecules. One of the interesting findings from this study was single-cell behavior in response to various TNF-α doses. At low TNF-α concentrations only a fraction of cells are activated, but in these activated cells the amplification mechanisms assure that the NF-κB nuclear translocation remains above a threshold. The lower nuclear NF-κB concentration only reduces the probability of gene activation, but does not reduce gene expression of those responding.

These two effects provide stochastic "robustness" in cell regulation, allowing cells to respond differently to the same stimuli but causing their individual responses to be unequivocal. Stated differently, these experiments indicate that a single cell is either fully activated ("on") or not ("off"). The dose of TNF-α primarily affects the probability that the cell is in the "on" state. In this way the computational analysis of NF-κB regulatory modules has enriched our understanding of the dynamics and stochastic processes governing its behavior at the single-cell level.

SYSTEM-BASED STUDIES OF THE NF-κB PROTEIN INTERACTION NETWORK

NF-κB is a prototypical intracellular signaling network, forming dynamic protein interactions induced by reversible posttranslational modifications that affect a cellular response. Although a number of protein-protein interactions have been described using hypothesis-driven studies, the possibility for systematic (unbiased) determination of protein interaction networks in cellulo has arisen with development of affinity isolation and mass spectrometry. In this technique, a bait protein with tandem affinity tags is expressed and protein complexes are allowed to form (64). The bait protein is then purified under nondenaturing conditions, and the protein complex is identified using highly sensitive mass spectrometers. In a tour de force study by the Superti-Furga group at Cellzome, 32 major regulators of the NF-κB pathway were used as bait and their complexes identified

using liquid chromatography-tandem mass spectrometry (8). These workers identified 680 nonredundant interacting proteins and confirmed 171 of 241 protein interactions that had been previously reported in the literature. By using a stringent bioinformatic filter to eliminate nonspecific and high-abundance contaminants, 131 proteins were identified as "high-confidence" interactors. The physical interaction map from these studies is reproduced in Color Plate 2 according to the convention established by the Alliance for Cellular Signaling. Consistent with the known canonical and noncanonical pathway connections, RelA is found to associate with IκB-α/β/γ and 100-kDa NF-κB2. NIK associates with IKK-α and 100-kDa NF-κB2. Ten novel interactors were further investigated using siRNA-mediated gene knockdown, which revealed the previously unknown role of the TRAF7 ubiquitin ligase in mitogen-ERK kinase kinase (MEKK3) activation upstream of the Jun NH$_2$-terminal kinase (JNK) and IKK pathways (Color Plate 2). Surprisingly few interactions of nuclear coreceptors were identified, perhaps because these protein interactions are either transient or low abundance. Further method development in proteomic affinity techniques will be required to identify nuclear protein interactions and those influenced by site-specific phosphorylation.

NF-κB-DEPENDENT GENE NETWORKS

As a transcriptional activator, the primary action of NF-κB is to induce expression of noncontiguous groups of genes important in the inflammatory and antiviral response. Understanding the biological functions encoded by this genetic network will provide important insights into the myriad homeostatic and pathological roles NF-κB plays in cytokine cascades and viral infection. Strategies for elucidating the identity of the NF-κB-controlled genetic network have employed the candidate gene approach, where the analysis of the regulation of a gene results in the identification that it is under NF-κB control. In the literature, over 150 different genes have been identified to be NF-κB responsive in a wide variety of cell types and in response to a vast collection of activating agents (60). From these studies, NF-κB has been found to control inducible cytokine/chemokine expression, immunoreceptor presentation, cell adhesion molecule expression, antigen presentation, and induction of the hepatic acute-phase response (60), implicating its actions as a mediator of inflammation. Additionally, NF-κB controls expression of negative regulators of the canonical pathway producing the IκB-α and TNFAIP3/A20 negative regulatory loops discussed previously (4, 24, 73). Other NF-κB actions include expression of antiapoptotic proteins (immediate early inducible gene-long [IEX-L], cellular inhibitor of apoptosis protein [CIAP]) (38, 79, 85) to prevent inducible cell death in responsive cells. NF-κB also controls inducible expression of diverse transcription factor families, such as activating protein-1 (AP-1) and inducible IRF1 and IRF7 isoforms (48, 76), actions that confer distinct signaling properties to the stimulated cell population. Although this information has been extremely useful to dissect details of NF-κB binding and transcription, these studies were conducted in isolation in a wide variety of cell types and used different stimulation protocols. Thus, it is difficult to infer the spectrum of NF-κB-controlled genes and their temporal coordination of expression from these data.

Systematic Identification of NF-κB Dependent Gene Networks by mRNA Profiling

mRNA expression profiling using high-density DNA microarrays has enabled the systematic analysis of NF-κB-induced transcriptional responses (31). It bears emphasis that TNF-α and viral infections are complex biological stimuli that activate many signaling pathways, including JNK–AP-1, Jak-Stat (Janus kinase–signal transducer and activator of transcription) MAP-ERK (and others), in addition to NF-κB. So, although it is trivial to systematically measure stimulus-induced changes in gene expression, one challenge lies in inferring which genes or groups of genes are controlled by a specific pathway. One approach is to perform mRNA profiling in wild-type cells and those containing an inactivated signaling pathway and compare the differences in gene expression patterns between the two. Validation experiments are then performed to determine which genes are directly controlled by NF-κB versus those indirectly regulated (e.g., by a secondary transcription factor mediator).

NF-κB-Dependent Genes in Response to the TNF-α Monokine

In a series of studies from our laboratory, we have used a clonal tetracycline-regulated cell line (Tet-Off) expressing a nondegradable IκB-α mutant to investigate the genes downstream of the TNF-α- and viral-induced pathways (59, 75–77). We chose this method for inhibition because we could uniformly induce high levels of dominant negative inhibitor expression and could isolate cells without chronically exposing them to the proapoptotic effect of NF-κB blockade. Briefly, when doxycycline is present in tissue culture medium, tetracycline transactivator (tTA) is inactivated, and IκB-α Mut levels are barely detectable by Western immunoblot, resulting in a wild-type phenotype, with normal levels of activated NF-κB in the nucleus being produced after stimulation (76, 77). Conversely,

upon doxycycline withdrawal, tTA is activated, and IκB-α Mut expression occurs at similar levels to endogenous IκB-α (76). These levels of IκB-α Mut expression are sufficient to completely inhibit NF-κB translocation and target gene expression (76, 77).

A critical issue in fully exploiting mRNA profiling data was our experimental design incorporating a time series of stimulation, where cells were TNF-α stimulated for 1, 3, or 6 h, times where primary effects on NF-κB-dependent gene regulation occurred and where distinct gene expression patterns could be resolved. Because four replicate experiments were performed for each time point, we used a rigorous statistical analysis to identify genes that both were regulated by TNF-α and demonstrated a ±3-fold change in signal intensity relative to control at any time during the stimulation. Of 50 genes that were TNF-α regulated, 28 were NF-κB dependent, indicating that NF-κB signaling is responsible for mediating a significant fraction of the TNF-α-induced genomic response. Moreover, all of the NF-κB-dependent genes were upregulated by TNF-α, indicating that activated NF-κB does not play an identifiable role in gene repression in these cell types.

The biological functions of these genes were analyzed by Gene Ontology, Ingenuity pathway analysis (see below), and expert classification. Major biological pathways under NF-κB control that were previously known included cytokines/chemokines (IL-6, IL-8, growth-related oncogene [GRO]α/β/γ, and CCL20), and the NF-κB negative autoregulators (IκB-α, Bcl-3, and NF-κB1 and 2). In addition, we identified NF-κB members that were positively autoregulated, such as RelB and cRel, indicating that NF-κB is also under positive autoregulatory control. Surprising biological functions included metabolism (GTP hydrolase) and receptor/cell surface signaling (CD83, syndecan-4, and IL-7R). These data suggest that NF-κB action influences cellular nucleoside metabolism as well as the cell's ability to respond to different signaling pathways. For example, the upregulation of IL-7R would allow the cell to respond to subsequent IL-7 ligand. These cellular effects may play significant, yet not fully resolved roles in the inflammatory response to monokine activation.

Gene Targets of the NF-κB Pathway Are Activated in Distinct Expression "Waves"

Kinetic analysis of the NF-κB-dependent genes further revealed significant information about the homeostatic response to TNF-α stimulation. As seen in the heat map shown in Color Plate 3, NF-κB-dependent genes are induced in an orchestrated cascade of distinct groups of genes, which we term early, middle, and late based on their relative expression profiles. We have validated that these three groups are directly under NF-κB control, because: (i) the expression of all three groups is completely blocked by overexpression of the nondegradable IκB-α Mut inhibitor (Color Plate 3); (ii) these genes contain high-affinity NF-κB-binding sites in their promoters that haave been experimentally verified to bind NF-κB using chromatin immunoprecipitation assays (58, 77); and (iii) expression of the middle and late gene group is induced in the presence of translational inhibitors, excluding the possibility that their expression requires intermediate protein synthesis (77).

Inflammatory Implications for the "Waves" of NF-κB-Dependent Gene Expression

Bioinformatic analysis of the biological functions of the early, middle, and late genes showed that these groups were statistically enriched for specific (and distinct) molecular functions.

The early genes are enriched in two major molecular functions: cytokines and NF-κB autoregulatory loop proteins. An important biological property of TNF-α in viral infections is to initiate and amplify the cytokine cascade by inducing expression of secondary (downstream) cytokines, thus enhancing leukocyte chemotaxis and activation. A major part of the early gene group is composed of members of the CXC chemokine family (CXCL1, 2, and 3 and IL-8). CXC chemokines are numerically the largest of the chemokine families, responsible for inducing migration of neutrophilic leukocytes, stimulating wound healing, initiating angiogenesis, and promoting tumorigenesis. In addition, NF-κB rapidly induces the CC chemokine CCL 20, which is responsible for stimulating monocytes and dendritic cells, and the cytokine mediator of the acute phase, IL-6, which is important in systemic inflammatory response. Therefore, part of the early genetic response to NF-κB is the rapid induction of secondary cytokine cascades that control leukocyte trafficking, wound healing, angiogenesis, and systemic inflammation. The other important members of the early genes encode IκB-α and TNFAIP3/A20 autoregulatory loops. These observations indicate that another function of the early NF-κB response is to terminate the TNFR–IKK–NF-κB signaling pathway to restore cellular homeostasis.

By contrast, the late genes encode adhesion molecules (intercellular adhesion molecule [ICAM], killer cell lectin receptor subfamily C, member 2 [KLRC2]) and proteins important in major histocompatibility class I antigen processing/presentation (TAP, TAPBP). The late genes therefore play important roles in lymphocyte recognition, recruitment, and cytotoxic T-cell-mediated cytolysis. As a result, chronically TNF-α-stimulated cells, such as those

produced in the context of persistent infection, would be targeted for enhanced immune recognition and clearance. Also in this group are the TRAF signal adaptor molecules that normally couple TNFR to intracellular responses (Fig. 1). TRAF1 is distinct from other TRAF isoforms in that it apparently serves to protect cells from apoptosis and plays a role in the negative-feedback regulation of receptor signaling.

NF-κB-Dependent Gene Networks in Response to RNA Virus Infection

Two additional stimuli were studied systematically using the tTA-IκB-α Mut cell system to identify NF-κB-dependent genes in response to RNA virus infection. In the first study, we investigated early response genes induced by infection with the paramyxovirus RSV. RSV is a major human pathogen important in inducing airway inflammation and responsible for epidemic bronchiolitis in children (18). RSV is a potent inducer of global changes in gene expression in airway epithelial cells including 17 distinct CC (I-309, Exodus-1, TARC, RANTES, MCP-1, MDC, and MIP-1α and -1β), CXC (GRO-α , -β , and -γ , ENA-78, IL-8, and I-TAC), and CX$_3$C (fractalkine) chemokines in three distinct expression profiles (92). Using a pairwise analysis in the experimental design, we identified 380 genes that were affected by NF-κB inhibition.

Strikingly, the spectrum of genes under NF-κB control in response to RSV infection had little resemblance to those NF-κB-dependent genes that were identified in response to TNF-α stimulation. Although chemokines, metabolic functions, and NF-κB autoregulation were identified as common biological functions shared by the network controlled by TNF-α, new functions were identified. These included IFN signaling (IRF1 and 7B, Stat1), cholesterol synthesis (cholesterol 25-hydroxylase), protein synthesis/turnover (ubiquitin-conjugating enzymes), tyrosine kinases, cadherins, and others (IL-15Rα, complement B). Because these mRNA profiling experiments were performed in the same cells as those for the TNF-α stimulation, these findings indicate that the spectrum of genes downstream of NF-κB is highly stimulus dependent.

We have further expanded this analysis to understand the NF-κB-dependent gene network activated in response to mammalian reovirus infection (59). Reovirus is a segmented dsRNA virus that induces cellular apoptosis in an NF-κB-dependent manner (22). In this study, the tTA-IκB-α Mut cell system was exposed to reovirus for various times from 0 to 10 h in the presence or absence of functional NF-κB signaling. Strikingly, 176 genes were NF-κB dependent, and their expression also occurred in "waves," as we had observed for the TNF-dependent

TNF

RSV

REO

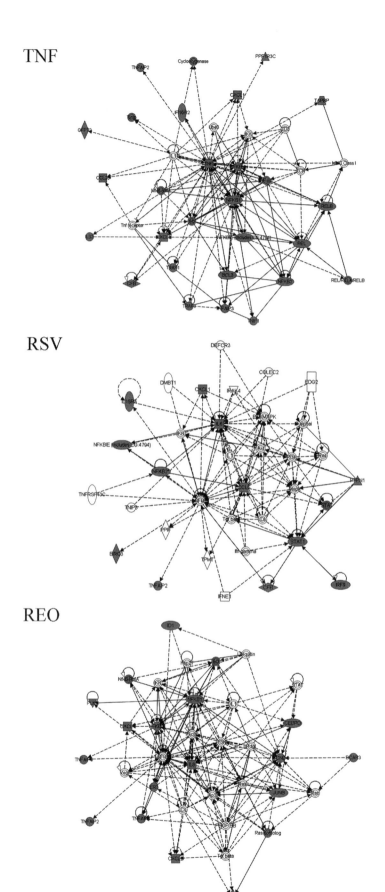

A) IRF-3

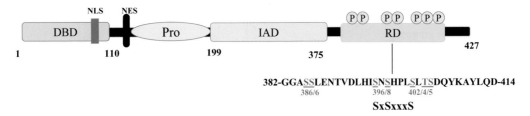

382-GGA<u>SS</u>LENTVDLHI<u>S</u>N<u>S</u>HPL<u>S</u>L<u>TS</u>DQYKAYLQD-414
386/6   396/8   402/4/5

SxSxxxS

B)

C)

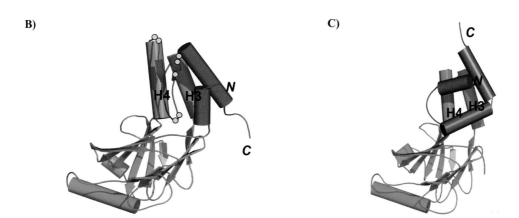

Color Plate 1 (chapter 5) Schematic representation of IRF3 and crystal structure. (A) Schematic representation of IRF3. TBK1/IKK-ε phosphorylation consensus site is also shown at the top of the figure as SxSxxxS. Pro, proline-rich region; RD, signal response domain; P, phosphorylation site. (B and C) The ribbon diagrams illustrate the interactions between the IAD of IRF3 and the IBiD region of CBP. In panel B, the intramolecular interactions between the IAD of IRF3 (in green) and the flanking autoinhibitory structures (in red) are shown. Phosphorylation sites are in yellow. In panel C, the intermolecular interactions between the IAD of IRF3 (in green) and the IBiD region of CBP (in blue) are shown. Figure graciously provided by Kai Lin, courtesy of *Structure* (127).

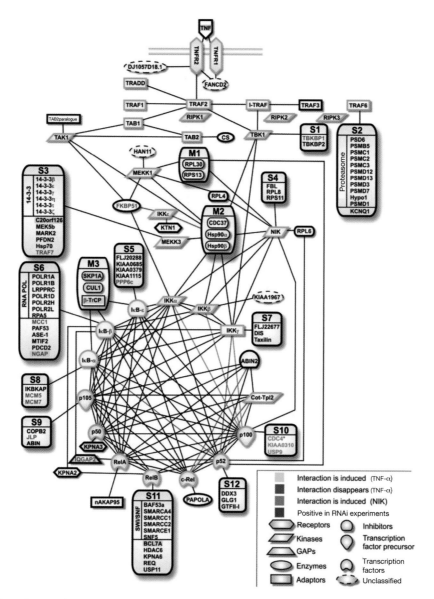

Color Plate 2 (chapter 9) NF-κB protein interaction network. Shown is the NF-κB interaction network determined by affinity isolation and liquid chromatography-tandem mass spectrometry. Network structure and connections are drawn according to the Alliance for Cellular Signaling convention, where interactions are indicated by a line and protein functions are indicated by the shape of the symbol, according to the legend. The Tagged components are indicated by green lines. (Reprinted with permission from reference 8.)

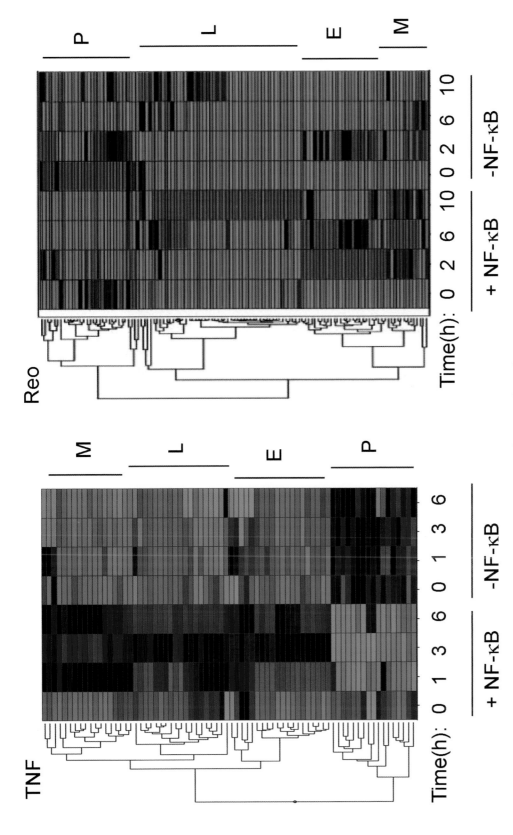

Color Plate 3 (chapter 9) Waves of NF-κB-dependent gene expression in response to monokines and RNA virus infection. Heat maps of mRNA profiling studies of NF-κB-dependent genes activated in response to TNF-α stimulation (left) and reovirus infection (right). (Bottom) Gene expression is coded by the following colors: green, low expression; black, middle level of expression; red, high expression. Groupings according to the kinetics of expression are indicated to the right of each heat map. E, early; M, middle; L, late; P, paradoxical group. (Data are from references 75 and 59.)

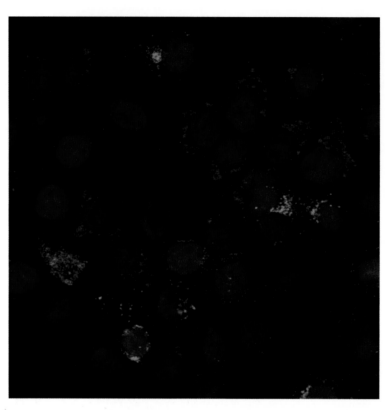

Color Plate 4 (chapter 21) Laser-scanning confocal microscopic image of HeLa cells infected with HAV. HAV capsid proteins are labeled (green) with a monoclonal antibody (K2-4F2) directed against a conformational neutralizing epitope. The cells were also labeled with a rabbit polyclonal antibody specific for MAVS/IPS-1 (red). MAVS/IPS-1 is found only in uninfected cells and is absent in cells infected with HAV due to degradation mediated by the viral 3ABC protease (96).

Figure 5 Ingenuity Pathways networks of NF-κB-dependent gene networks. The NF-κB-dependent genes determined from mRNA profiling studies in response to TNF-α, RSV, and reovirus (REO) were mapped onto the IPKB (http://www.ingenuity.com/). Shown are labeled nodes representing individual protein functions and their relationship represented by edges. Nodes are colored by identification, with gray shapes indicating upregulation of the gene in the input dataset. Squares indicate cytokines, circles indicate chemokines, and ovals indicate transcription factors. For the edges, an arrow indicates "acts on"; straight lines indicate binding interactions.

---

genes (Color Plate 3, right panel). Although the "early" inducible genes included some cytokines (IL-6, CXCL1 and 2, and IL-8) and NF-κB negative autoregulatory loop proteins (TNFAIP3/A20), other novel biological functions were also found within this group. These novel functions included cellular proliferation, proapoptotic, and DNA damage/repair genes, such as *GADD45*, *DDB2*, *ERCC4*, and *FUS* (59).

A comparison of these biological functions of the NF-κB-dependent gene network was performed using network analysis by the Ingenuity Pathways Knowledge Base (IPKB). The IPKB is a manually curated database of published protein-protein and protein-gene interactions that allows inference of the major biological functions controlled by gene networks. We mapped the NF-κB-dependent genes downstream of TNF, RSV, and reovirus onto the master IPKB; the primary biological network involving each NF-κB-dependent network is shown in Fig. 5. These studies indicate a striking complexity in the biological functions controlled by the NF-κB-dependent gene network as a function of the stimulus. More systematic studies of its network in response to diverse stimuli will be required to fully understand its pleiotropic role in monokine- and viral-induced inflammation.

SUMMARY AND FUTURE DIRECTIONS

The NF-κB transcription factor plays an important role in monokine- and viral-induced inflammation. This protein network functions to integrate multiple signaling pathways using common regulatory kinases that play pleiotropic roles. The major pathways are now understood to be composed of bipartite signaling modules whose actions affect nuclear translocation and, separately, transcription factor activation. In this chapter, I have reviewed systematic approaches to understanding this signaling pathway, including proteomics, mathematical modeling, real-time imaging, and gene network analyses. These approaches have yielded significant insights into the underlying protein interaction network and dynamic control of its subcellular distribution via negative autoregulatory feedback loops. Major questions remain in the analysis and function of this

pathway: How exactly does phosphorylation influence transcriptional activation, and what are the consequences of phosphorylation at different sites in the transactivator molecules? What is the identity of the nuclear protein interaction network and how is this different in response to different stimulation pathways? How does NF-κB induce these "waves" of gene expression, and is it possible to modulate the expression of the biologically distinct groups of genes? This understanding will allow for the development of subnetwork modulators to produce selective anti-inflammatory drugs or to modulate viral pathogenesis in vivo.

Work in our laboratory is supported by National Institute of Allergy and Infectious Diseases grants R01 AI40218 and P01 AI062885. Core Laboratory support was from National Institute of Environmental Health Science, grant P30 ES06676 (to J. Halpert, University of Texas Medical Branch) and BAA-HL-02–04 (A. Kurosky, University of Texas Medical Branch).

REFERENCES

1. **Akira, S., S. Uematsu, and O. Takeuchi.** 2006. Pathogen recognition and innate immunity. *Cell* **124:**783–801.
2. **Basak, S., H. Kim, J. D. Kearns, V. Tergaonkar, E. O'Dea, S. L. Werner, C. A. Benedict, C. F. Ware, G. Ghosh, I. M. Verma, and A. Hoffmann.** 2007. A fourth IκB protein within the NF-κB signaling module. *Cell* **128:** 369–381.
3. **Beg, A. A. and A. S. Baldwin, Jr.** 1993. The IκB proteins: multifunctional regulators of Rel/NF-κB transcription factors. *Genes Dev.* 7:2064–2070.
4. **Beyaert, R., K. Heyninck, and S. VanHuffel.** 2000. A20 and A20-binding proteins as cellular inhibitors of nuclear factor-κB-dependent gene expression and apoptosis. *Biochem. Pharmacol.* 60:1143–1151.
5. **Bonizzi, G., M. Bebien, D. C. Otero, K. Johnson-Vroom, Y. Cao, D. Vu, A. G. Jegga, B. Aronow, G. Ghosh, R. Rickert, and M. Karin.** 2004. Activation of IKKalpha target genes depends on recognition of specific kappaB binding sites by RelB:p52 dimers. *EMBO J.* 23: 4202–4210.
6. **Bonizzi, G., and M. Karin.** 2004. The two NF-κB activation pathways and their role in innate and adaptive immunity. *Trends Immunol.* 25:280–288.
7. **Bosisio, D., I. Marazzi, A. Agresti, N. Shimizu, M. E. Bianchi, and G. Natoli.** 2006. A hyper-dynamic equilibrium between promoter-bound and nucleoplasmic dimers

controls NF-kappaB-dependent gene activity. *EMBO J.* **25:**798–810.

8. Bouwmeester, T., A. Bauch, H. Ruffner, P. O. Angrand, G. Bergamini, K. Croughton, C. Cruciat, D. Eberhard, J. Gagneur, S. Ghidelli, C. Hopf, B. Huhse, R. Mangano, A. M. Michon, M. Schirle, J. Schlegl, M. Schwab, M. A. Stein, A. Bauer, G. Casari, G. Drewes, A. C. Gavin, D. B. Jackson, G. Joberty, G. Neubauer, J. Rick, B. Kuster, and G. Superti-Furga. 2004. A physical and functional map of the human TNF-α/NF-κB signal transduction pathway. *Nat. Cell. Biol.* **6:**97–105.

9. Brasier, A. R., M. Lu, T. Hai, Y. Lu, and I. Boldogh. 2001. NF-κB-inducible BCL-3 expression is an autoregulatory loop controlling nuclear p50/NF-κB1 residence. *J. Biol. Chem.* **276:**32080–32093.

10. Carlotti, F., S. K. Dower, and E. E. Qwarnstrom. 2000. Dynamic shuttling of nuclear factor kappa B between the nucleus and cytoplasm as a consequence of inhibitor dissociation. *J. Biol. Chem.* **275:**41028–41034.

11. Chen, L. F., S. A. Williams, Y. Mu, H. Nakano, J. M. Duerr, L. Buckbinder, and W. C. Greene. 2005. NF-κB RelA phosphorylation regulates RelA acetylation. *Mol. Cell. Biol.* **25:**7966–7975.

12. Choudhary, S., S. Boldogh, R. P. Garofalo, M. Jamaluddin, and A. R. Brasier. 2005. RSV influences NF-κB dependent gene expression through a novel pathway involving MAP3K14/NIK expression and nuclear complex formation with NF-κB2. *J. Virol.* **79:**8948–8959.

13. Delhase, M., M. Hayakawa, Y. Chen, and M. Karin. 1999. Positive and negative regulation of IκB kinase activity through IKKβ subunit phosphorylation. *Science* **284:**309–313.

14. Deng, L., C. Wang, E. Spencer, L. Yang, A. Braun, J. You, C. Slaughter, C. Pickart, and Z. J. Cheng. 2000. Activation of the IκB kinase complex by TRAF6 requires a dimeric ubiquitin-conjugating enzyme complex and a unique polyubiquitin chain. *Cell* **103:**351–361.

15. Gitlin, L., W. Barchet, S. Gilfillan, M. Cella, B. Beutler, R. A. Flavell, M. S. Diamond, and M. Colonna. 2006. Essential role of mda-5 in type I IFN responses to polyriboinosinic:polyribocytidylic acid and encephalomyocarditis picornavirus. *Proc. Nat. Acad. Sci. USA* **103:** 8459–8464.

16. Goshe, M. B., T. P. Conrads, E. A. Panisko, N. H. Angell, T. D. Veenstra, and R. D. Smith. 2001. Phosphoprotein isotope-coded affinity tag approach for isolating and quantitating phosphopeptides in proteome-wide analysis. *Anal. Chem.* **73:**2578–2586.

17. Haeberle, H., A. Casola, Z. Gatalica, S. Petronella, H.-J. Dieterich, P. B. Ernst, A. R. Brasier, and R. P. Garofalo. 2004. IκB kinase is a critical regulator of chemokine expression and lung inflammation in respiratory syncytial virus infection. *J. Virol.* **78:**2232–2241.

18. Hall, C. B. 2001. Respiratory syncytial virus and parainfluenza virus. *N. Engl. J. Med.* **344:**1917–1928.

19. Han, Y., and A. R. Brasier. 1997. Mechanism for biphasic Rel A:NF-κB1 nuclear translocation in tumor necrosis factor α-stimulated hepatocytes. *J. Biol. Chem.* **272:** 9823–9830.

20. Han, Y., T. Meng, N. R. Murray, A. P. Fields, and A. R. Brasier. 1999. IL-1 Induced NF-κB-IκBα autoregulatory feedback loop in hepatocytes. A role for PKCα in post-

21. Han, Y., S. Weinman, I. Boldogh, R. K. Walker, and A. R. Brasier. 1999. Tumor necrosis factor-α-inducible IκBα proteolysis mediated by cytosolic m-calpain. A mechanism parallel to the ubiquitin-proteasome pathway for nuclear factor-κB activation. *J. Biol. Chem.* **274:**787–794.

22. Hansberger, M. W., J. A. Campbell, P. Danthi, P. Arrate, K. N. Pennington, K. B. Marcu, D. W. Ballard, and T. S. Dermody. 2007. IκB kinase subunits α and γ are required for activation of NF-κB and induction of apoptosis by mammalian reovirus. *J. Virol.* **81:**1360–1371.

23. Harhaj, E. W., and S.-C. Sun. 1999. Regulation of RelA subcellular localization by a putative nuclear export signal and p50. *Mol. Cell. Biol.* **19:**7088–7095.

24. He, K.-L., and A. T. Ting. 2002. A20 inhibits tumor necrosis factor (TNF) alpha-induced apoptosis by disrupting recruitment of TRADD and RIP to the TNF receptor 1 complex in Jurkat T cells. *Mol. Cell. Biol.* **22:**6034–6045.

25. Heyninck, K., D. De Valck, W. Vanden Berghe, W. Van Criekinge, R. Contreras, W. Fiers, G. Haegeman, and R. Beyaert. 1999. The zinc finger protein A20 inhibits TNF-induced NF-κB dependent gene expression by interfering with an RIP or TRAF2-mediated transactivation signal and directly binds to a novel NF-κB inhibiting protein ABIN. *J. Cell. Biol.* **145:**1471–1482.

26. Hoffmann, A., A. Levchenko, M. L. Scott, and D. Baltimore. 2002. The Ikappa B-NF-kappa B signaling module: temporal control and selective gene activation. *Science* **298:**1241–1245.

27. Hsu, H., J. Huang, H. B. Shu, V. Baichwal, and D. V. Goeddel. 1996. TNF-dependent recruitment of the protein kinase RIP to the TNF receptor-1 signaling complex. *Immunity* **4:**387–396.

28. Hsu, H., H.-B. Shu, M.-G. Pan, and D. V. Goeddel. 1996. TRADD-TRAF2 and TRADD-FADD interactions define two distinct TNF receptor 1 signal transduction pathways. *Cell* **84:**299–308.

29. Hu, Y., V. Baud, M. Delhase, P. Zhang, T. Deerinck, M. Ellisman, R. Johnson, and M. Karin. 1999. Abnormal morphogenesis but intact IKK activation in mice lacking the IKKalpha subunit of IkappaB kinase. *Science* **284:** 316–320.

30. Huynh, Q. K., H. Boddupalli, S. A. Rouw, C. M. Koboldt, T. Hall, C. Sommers, S. D. Hauser, J. L. Pierce, R. G. Combs, B. A. Reitz, J. A. Diaz-Collier, R. A. Weinberg, B. L. Hood, B. F. Kilpatrick, and C. S. Tripp. 2000. Characterization of the recombinant IKK1/IKK2 heterodimer. Mechanisms regulating kinase activity. *J. Biol. Chem.* **275:**25883–25891.

31. Iyer, V. R., M. B. Eisen, D. T. Ross, G. Schuler, T. Moore, J. C. F. Lee, J. M. Trent, L. M. Staudt, J. Hudson, M. S. Boguski, D. Lashkari, D. Shalon, D. Botstein, and P. O. Brown. 1999. The transcriptional program in the response of human fibroblasts to serum. *Science* **283:** 83–87.

32. Jamaluddin, M., S. Wang, I. Boldogh, B. Tian, and A. R. Brasier. 2007. TNF-α-induced NF-κB/Rel A Ser 276 phosphorylation and enhanceosome formation on the IL-8 promoter is mediated by a reactive oxygen species (ROS)-dependent pathway. *Cell. Signal.* **9:**1419–1433.

transcriptional regulation of IkBa resynthesis. *J. Biol. Chem.* **274:**939–947.

33. Jamaluddin, M., A. Casola, R. P. Garofalo, Y. Han, T. Elliott, P. L. Ogra, and A. R. Brasier. 1998. The major component of IκBα proteolysis occurs independently of the proteasome pathway in respiratory syncytial virus-infected pulmonary epithelial cells. *J. Virol.* **72:**4849–4857.

34. Jian, Z., J. Ninomiya-Tsuji, Y. Qian, K. Matsumoto, and X. Li. 2002. Interleukin-1 receptor-associated kinase-dependent IL-1 induced signaling complexes phosphorylate TAK1 and TAB2 at the plasma membrane and activate TAK1 in the cytosol. *Mol. Cell. Biol.* **22:**7158–7167.

35. Jiang, X., N. Takahashi, N. Matsui, T. Tetsuka, and T. Okamoto. 2003. The NF-kappa B activation in lymphotoxin beta receptor signaling depends on the phosphorylation of p65 at serine 536. *J. Biol. Chem.* **278:**919–926.

36. Karin, M. 1999. The beginning of the end: IκB kinase (IKK) and NF-κB activation. *J. Biol. Chem.* **274:** 27339–27342.

37. Karin, M., and Y. Ben Neriah. 2000. Phosphorylation meets ubiquitination: the control of NF-κB activity. *Annu. Rev. Immunol.* **18:**621–663.

38. Karin, M., and A. Lin. 2005. NF-κB at the crossroads of life and death. *Nat. Immunol.* **3:**221–227.

39. Kato, H., S. Sato, M. Yoneyama, M. Yamamoto, S. Uematsu, K. Matsui, T. Tsujimura, K. Takeda, T. Fujita, O. Takeuchi, and S. Akira. 2005. Cell type-specific involvement of RIG-I in antiviral response. *Immunity* **23:**19–28.

40. Kato, H., O. Takeuchi, S. Sato, M. Yoneyama, M. Yamamoto, K. Matsui, S. Uematsu, A. Jung, T. Kawai, K. J. Ishii, O. Yamaguchi, K. Otsu, T. Tsujimura, C. S. Koh, C. Reis e Sousa, Y. Matsuura, T. Fujita, and S. Akira. 2006. Differential roles of MDA5 and RIG-I helicases in the recognition of RNA viruses. *Nature* **441:**101–105.

41. Krikos, A., C. D. Laherty, and V. M. Dixit. 1992. Transcriptional activation of the tumor necrosis factor alpha-inducible zinc finger protein, A20, is mediated by kappa B elements. *J. Biol. Chem.* **267:**17971–17976.

42. Kumar, H., T. Kawai, H. Kato, S. Sato, K. Takahashi, C. Coban, M. Yamamoto, S. Uematsu, K. J. Ishii, O. Takeuchi, and S. Akira. 2006. Essential role of IPS-1 in innate immune responses against RNA viruses. *J. Exp. Med.* **203:**1795–1803.

43. Li, Y., J. Kang, J. Frieman, L. Tarassishin, J. Ye, A. Kovalenko, D. Wallach, and M. S. Horwitz. 1999. Identification of a cell protein (FIP-3) as a modulator of NF-κB activity and as a target of an adenovirus inhibitor of tumor necrosis factor α-induced apoptosis. *Proc. Natl. Acad. Sci. USA* **96:**1042–1047.

44. Lipniacki, T., P. Paszek, A. R. Brasier, B. Luxon, and M. Kimmel. 2004. Mathematical model of NF-κB regulatory module. *J. Theor. Biol.* **228:**195–215.

45. Lipniacki, T., K. Puszynski, P. Paszek, A. R. Brasier, and M. Kimmel. 2006. Stochastic robustness of NF-kappaB signalling: low dose TNF stimulation. *Bioinformatics* **8:**376.

46. Lipniacki, T., P. Paszek, A. R. Brasier, B. A. Luxon, and M. Kimmel. 2006. Stochastic regulation in early immune response. *Biophys. J.* **90:**725–742.

47. Liu, P., S. Choudhary, M. Jamaluddin, K. Li, R. Garofalo, A. Casola, and A. R. Brasier. 2006. Respiratory syncytial virus activates interferon regulatory factor-3 in airway epithelial cells by upregulating a RIG-I-Toll like receptor-3 pathway. *J. Virol.* **81:**1401–1411.

48. Lu, R., P. A. Moore, and P. M. Pitha. 2002. Stimulation of IRF-7 gene expression by tumor necrosis factor alpha: requirement for NF-κB transcription factor and gene accessibility. *J. Biol. Chem.* **277:**16592–16598.

49. Maniatis, T. 1997. Catalysis by a multiprotein IκB kinase complex. *Science* **278:**818–819.

50. Mercurio, F., B. W. Murray, A. Shevchenko, B. L. Bennett, D. B. Young, J. W. Li, G. Pascual, A. Motiwala, H. Zhu, M. Mann, and A. M. Manning. 1999. IκB kinase (IKK)-associated protein 1, a common component of the heterogeneous IKK complex. *Mol. & Cell. Biol.* **19:**1526–1538.

51. Mercurio, F., H. Zhu, B. W. Murray, A. Shevchenko, B. L. Bennett, J. Li, D. B. Young, M. Barbosa, M. Mann, A. Manning, and A. Rao. 1997. IKK-1 and IKK-2: cytokine-activated IκB kinases essential for NF-κB activation. *Science* **278:**818–819.

52. Mizukami, J., G. Takaesu, H. Akatsuka, H. Sakurai, J. Ninomiya-Tsuji, K. Matsumoto, and N. Sakurai. 2002. Receptor activator of NF-κB ligand (RANKL) activates TAK1 mitogen-activated protein kinase kinase kinase through a signaling complex containing RANK, TAB2, and TRAF6. *Mol. Cell. Biol.* **22:**992–1000.

53. Mordmuller, B., D. Krappmann, M. Esen, M. Wegener, and C. Scheidereit. 2003. Lymphotoxin and lipopolysaccharide induce NF-kappaB-p52 generation by a co-translational mechanism. *EMBO Rep.* **4:**82–87.

54. Morrison, M. D., W. Reiley, M. Zhang, and S. C. Sun. 2005. An atypical tumor necrosis factor (TNF) receptor-associated factor-binding motif of B cell-activating factor belonging to the TNF family (BAFF) receptor mediates induction of the noncanonical NF-κB signaling pathway. *J. Biol. Chem.* **280:**10018–10024.

55. Nelson, D. E., A. E. C. Ihekwaba, M. Elliott, J. R. Johnson, C. A. Gibney, B. E. Foreman, G. Nelson, V. See, C. A. Horton, D. G. Spiller, S. W. Edwards, H. P. McDowell, J. F. Unitt, E. Sullivan, R. Grimley, N. Benson, D. Broomhead, D. B. Kell, and M. R. H. White. 2004. Oscillations in NF-κB signaling control the dynamics of gene expression. *Science* **306:**704–708.

56. Ninomiya-Tsuji, J., K. Kishimoto, A. Hiyama, J.-I. Inoue, Z. Cao, and K. Matsumoto. 1999. The kinase TAK1 can activate the NIK-IκB as well as the MAP kinase cascade in the IL–1 signalling pathway. *Nature* **398:** 252–256.

57. Nowak, D., B. Tian, M. Jamaluddin, I. Boldogh, L. Vergara, J. S. Choudhary, and A. R. Brasier. 2008. RelA Ser276 phosphorylation is required for activation of a subset of NF-κB dependent genes by recruiting Cdk-9/cyclin T1 complexes. *Mol. Cell. Biol.* **28:**3623–3638.

58. Nowak, D. E., B. Tian, and A. R. Brasier. 2005. Two-step cross-linking method for identification of NF-κB gene network by chromatin immunoprecipitation. *Biotechniques* **39:**715–725.

59. O'Donnell, S. M., G. Holm, J. M. Pierce, B. Tian, R. S. Chari, D. W. Ballard, A. R. Brasier, and T. S. Dermody. 2006. Identification of an NF-κB-dependent gene network in cells infected by mammalian reovirus. *J. Virol.* **80:**1077–1086.

60. Pahl, H. 1999. Activators and target genes of Rel/NF-κB transcription factors. *Oncogene* **18:**6853–6866.

61. Phillips, K., N. Kedersha, L. Shen, P. J. Blackshear, and P. Anderson. 2004. Arthritis suppressor genes TIA-1 and

TTP dampen the expression of tumor necrosis factor α, cyclooxygenase 2, and inflammatory arthritis. *Proc. Nat. Acad. Sci. USA* **101**:2011–2016.

62. **Poyet, J.-L., S. M. Srinivasula, J.-H. Lin, T. Fernandes-Alnemri, S. Yamaoka, P. N. Tsichlis, and E. S. Alnemri.** 2000. Activation of the IκB kinases by RIP via IKK-γ/NEMO-mediated oligomerization. *J. Biol. Chem.* **275**:37966–37977.

63. **Price, D. H.** 2000. P-TEFb, a cyclin-dependent kinase controlling elongation by RNA polymerase II. *Mol. Cell. Biol.* **20**:2629–2634.

64. **Puig, O., F. Caspary, G. Riguat, B. Rutz, E. Bouveret, E. Bragado-Nilsson, M. Wilm, and B. Seraphin.** 2001. The tandem affinity purification (TAP) method: a general procedure of protein complex purification. *Methods* **24**:218–229.

65. **Rothwarf, D. M., E. Zandi, G. Natoli, and M. Karin.** 1998. IKK-gamma is an essential regulatory subunit of the IkappaB kinase complex. *Nature* **395**:297–300.

66. **Rudolph, D., W. C. Yeh, A. Wakeham, B. Rudolph, D. Nallainathan, J. Potter, A. J. Elia, and T. W. Mak.** 2000. Severe liver degeneration and lack of NF-kappaB activation in NEMO/IKKgamma-deficient mice. *Genes Dev.* **14**:854–862.

67. **Saha, S. K., E. M. Pietras, J. Q. He, J. R. Kang, S.-Y. Liu, G. Oganesyan, A. Shahangian, B. Zarnegar, T. L. Shiba, Y. Wang, and G. Cheng.** 2006. Regulation of antiviral responses by a direct and specific interaction between TRAF3 and Cardif. *EMBO J.* **25**:3257–3263.

68. **Sakurai, H., H. Chiba, H. Miyoshi, T. Sugita, and W. Toriumi.** 1999. IkappaB kinases phosphorylate NF-kappaB p65 subunit on serine 536 in the transactivation domain. *J. Biol. Chem.* **274**:30353–30366.

69. **Sakurai, H., H. Miyoshi, W. Toriumi, and T. Sugita.** 1999. Functional interactions of transforming growth factor β-activated kinase 1 with IκB kinases to stimulate NF-κB activation. *J. Biol. Chem.* **274**:10641–10648.

70. **Senftleben, U., Y. Cao, G. Xiao, F. R. Greten, G. Krahn, G. Bonizzi, Y. Chen, Y. Hu, A. Fong, S. C. Sun, and M. Karin.** 2001. Activation by IKKalpha of a second, evolutionary conserved, NF-kappa B signaling pathway. *Science* **293**:1495–1499.

71. **Seth, R. B., L. Sun, C. K. Ea, and Z. J. Chen.** 2005. Identification and characterization of MAVS, a mitochondrial antiviral signaling protein that activates NF-κB and IRF3. *Cell* **122**:669–682.

72. **Sizemore, N., S. Leung, and G. R. Stark.** 1999. Activation of phosphatidylinositol 3-kinase in response to interleukin-1 leads to phosphorylation and activation of the NF-kappa B p65/RelA subunit. *Mol. Cell. Biol.* **19**:4798–4805.

73. **Sun, S. C., P. A. Ganchi, C. Beraud, D. W. Ballard, and W. C. Greene.** 1994. Autoregulation of the NF-kappa B transactivator RelA (p65) by multiple cytoplasmic inhibitors containing ankyrin motifs. *Proc. Natl. Acad. Sci. USA* **91**:1346–1350.

74. **Takaesu, G., R. M. Surabhi, K. J. Park, J. Ninomiya-Tsuji, K. Matsumoto, and R. B. Gaynor.** 2003. TAK1 is critical for IkappaB kinase-mediated activation of the NF-kappaB pathway. *J. Mol. Biol.* **326**:105–115.

75. **Tian, B., D. Nowak, and A. R. Brasier.** 2005. A TNF induced gene expression program under oscillatory NF-κB control. *BMC Genomics* **6**:137.

76. **Tian, B., Y. Zhang, B. A. Luxon, R. P. Garofalo, A. Casola, M. Sinha, and A. R. Brasier.** 2002. Identification of NF-κB dependent gene networks in respiratory syncytial virus-infected cells. *J. Virol.* **76**:6800–6814.

77. **Tian, B., D. E. Nowak, M. Jamaluddin, S. Wang, and A. R. Brasier.** 2005. Identification of direct genomic targets downstream of the NF-kappa B transcription factor mediating TNF signaling. *J. Biol. Chem.* **280**:17435–17448.

78. **Vlahopoulos, S., I. Boldogh, and A. R. Brasier.** 1999. NF-κB dependent induction of interleukin-8 gene expression by tumor necrosis factor α: evidence for an antioxidant sensitive activating pathway distinct from nuclear translocation. *Blood* **94**:1878–1889.

79. **Wang, C.-Y., M. W. Mayo, R. G. Korneluk, D. B. Goeddel, and A. S. Baldwin, Jr.** 1998. NF-κB antiapoptosis: induction of TRAF1 and TRAF2 and C-IAP1 and C-IAP2 to suppress caspase-8 activation. *Science* **281**:1680–1683.

80. **Werner, S. L., D. Barken, and A. Hoffmann.** 2005. Stimulus specificity of gene expression programs determined by temporal control of IKK activity. *Science* **309**:1857–1861.

81. **Wertz, I., K. M. O'Rourke, H. Zhou, M. Eby, L. Aravind, S. Seshagiri, P. Wu, C. Wiesmann, R. Baker, D. Boone, A. Ma, E. V. Koonin, and V. M. Dixit.** 2004. De-ubiquitination and ubiquitin ligase domains of A20 downregulate NF-κB signalling. *Nature* **430**:694–699.

82. **Whiteside, S. T., J.-C. Epinat, N. R. Rice, and A. Israel.** 1997. IκB epsilon, a novel member of the IκB family, controls RelA and cRel NF-κB activity. *EMBO J.* **16**:1413–1426.

83. **Woronicz, J., X. Gao, Z. Cao, M. Rothe, and D. V. Goeddel.** 1997. IκB kinase-β: NF-κB activation and complex formation with IκB kinase-α and NIK. *Science* **278**:866–869.

84. **Wu, C. J., D. B. Conze, T. Li, S. M. Srinivasula, and J. D. Ashwell.** 2006. Sensing of Lys 63-linked polyubiquitination by NEMO is a key event in NF-κB activation. *Nat. Cell. Biol.* **8**:398–406.

85. **Wu, M. X., Z. Ao, K. V. Prasad, R. Wu, and S. F. Schlossman.** 1998. IEX-1L, an apoptosis inhibitor involved in NF-kappaB-mediated cell survival. *Science* **281**:998–1001.

86. **Xiao, G., A. Fong, and S. C. Sun.** 2004. Induction of p100 processing by NF-κB-inducing kinase involves docking IκB kinase α (IKKα) to p100 and IKKα-mediated phosphorylation. *J. Biol. Chem.* **279**:30099–30105.

87. **Xu, L. G., Y. Y. Wang, K. J. Han, L. Y. Li, Z. Zhai, and H. B. Shu.** 2005. VISA is an adapter protein required for virus-triggered IFN-β signaling. *Mol. Cell* **19**:727–740.

88. **Yamamoto, Y., D. W. Kim, Y. T. Kwak, S. Prajapati, U. Verma, and R. B. Gaynor.** 2001. IKKgamma/NEMO facilitates the recruitment of the IkappaB proteins into the IkappaB kinase complex. *J. Biol. Chem.* **276**:36327–36336.

89. **Yamaoka, S., G. Courtois, C. Bessia, S. T. Whiteside, R. Weil, F. Agou, H. E. Kirk, R. J. Kay, and A. Israel.** 1998. Complementation cloning of NEMO, a component of the IκB kinase complex essential for NF-κB activation. *Cell* **93**:1231–1240.

90. **Ye, J., X. Xie, L. Tarassishin, and M. S. Horwitz.** 2000. Regulation of the NF-kappaB activation pathway by isolated domains of FIP3/IKKgamma, a component of the

IkappaB-alpha kinase complex. *J. Biol. Chem.* **275:** 9882–9889.

91. **Zhang, S. Q., A. Kovalenko, G. Cantarella, and D. Wallach.** 2000. Recruitment of the IKK signalosome to the p55 TNF receptor: RIP and A20 bind to NEMO (IKKgamma) upon receptor stimulation. *Immunity* **12:**301–311.

92. **Zhang, Y., B. A. Luxon, A. Casola, R. P. Garofalo, M. Jamaluddin, and A. R. Brasier.** 2001. Expression of RSV-induced chemokine gene networks in lower airway epithelial cells revealed by cDNA microarrays. *J. Virol.* **75:**9044–9058.

93. **Zhao, T., L. Yang, Q. Sun, M. Arguello, D. W. Ballard, J. Hiscott, and R. Lin.** 2007. The NEMO adaptor bridges the nuclear factor-κB and interferon regulatory factor signaling pathways. *Nat. Immunol.* **8:**592–600.

94. **Zhong, H., H. Su Yang, H. Erdjument-Bromage, P. Tempst, and S. Ghosh.** 1997. The transcriptional activity of NF-κB is regulated by the IκB-associated PKAc subunit through a cyclic AMP-independent mechanism. *Cell* **89:**413–424.

95. **Zhong, H., R. E. Voll, and S. Ghosh.** 1998. Phosphorylation of NF-κB p65 by PKA stimulates transcriptional activity by promoting a novel bivalent interaction with the coactivator CBP/p300. *Mol. Cell* **1:**661–671.

Cellular Signaling and Innate Immune
Responses to RNA Virus Infections
Edited by A. R. Brasier et al.
© 2009 ASM Press, Washington, D.C.

Christine A. Biron

10

Type I Interferon Signaling in Shaping Cellular Innate and Adaptive Immunity to Viral Infection

The immune system is broadly composed of components identified as innate (in place prior to infection or activated early after infection) and adaptive (dependent upon expansion and activation at later times after primary infections). A variety of innate cytokines can be produced in response to infections, and these can stimulate numerous signaling pathways to induce antimicrobial states in an infected individual (5, 8). The factors help promote the expression of direct defense mechanisms delivered within infected cells, but also have immunoregulatory effects to shape other innate and adaptive immune responses. These innate cytokines include members of the proinflammatory cytokine cascade: the type I interferons (IFNs), tumor necrosis factor-α, interleukin-12 (IL-12), and IFN-γ produced by natural killer (NK) cells. The best-studied cytokines in terms of the understanding of their signaling for immunoregulatory effects on innate and adaptive immunity during viral infections are the type I IFNs, and certain viruses appear to preferentially induce these innate factors to high and sustained levels.

As reviewed in other chapters, the type I IFNs, including a single β gene and multiple α genes (IFN-α/β), are elicited in response to a variety of stimuli including exposure to certain viral products (37). They can induce a surprisingly wide range of biological functions. Although other innate cytokines can have certain overlapping effects, the type I IFNs appear to be particularly potent at signaling for expression of direct antiviral defense mechanisms and a range of important immunoregulatory functions shaping endogenous responses to viral infections (Fig. 1) (7, 8, 37). Because the cytokines can unleash many potent antiviral effects, viruses have incorporated machinery, delivered within infected cells through interactions between viral products and cell molecules, to inhibit type I IFN induction, signaling, and direct antiviral effects (37). To protect individuals against these viral avoidance mechanisms, flexibility in pathways for the induction and functions of the type I IFNs have evolved. In addition to families of sensors for detection of infection within all cells, specialized cells have families of sensors for detection of viral products to elicit responses prior to their own infection (2, 19, 27, 37). Likewise, in addition to the stimulation of multiple direct defense mechanisms within infected cells, the immunoregulatory effects delivered by type I IFNs lead to independent effector mechanisms, mediated by uninfected immune cells, to block viral replication.

Christine A. Biron, Department of Molecular Microbiology and Immunology, Division of Biology and Medicine, Brown University, Providence, RI 02912.

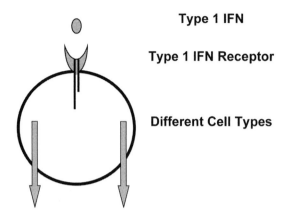

Type 1 IFN

Type 1 IFN Receptor

Different Cell Types

Direct Antiviral Mechanisms	*Immunoregulatory Effects*
-PKR	**Innate**
-2-5-OAS synthetase	-NK cell cytotoxicity
-Mx	-pro-proliferative
-ADAR	-IRF-7
	-Cytokine Expression
	Type 1 IFNs ↑
	IL-12 ↑ ↓
	IL-15 ↑
	IFN-γ ↓
	-DC
	pDC accumulation ↑
	cDC accumulation ↓
	DC maturation ↑
	Class I MHC
	Co-stimulatory molecules
	Adaptive
	-class I MHC
	-T cell responses
	-anti-proliferation
	-IFN-γ ↑

Figure 1   The biological effects of type I IFNs. The innate type I IFN cytokines, IFN-α/β, have a wide range of biological effects. These are important in promoting antiviral defense mechanisms. The cytokines induce expression of a number of biochemical pathways for cellular resistance to viral infection, including the Mx, PKR, and OAS enzymes. In addition, type I IFNs have a number of immunoregulatory effects, and these can also enhance antiviral states within an infected individual. In terms of innate cell responses, the cytokines activate NK-cell cytotoxicity and promote the maturation and accumulation of different DC populations. They also have effects on the expression of other innate cytokines, including induction of IL-15, concentration-dependent enhancement or inhibition of IL-12 expression, and inhibition of NK-cell IFN-γ production. In contrast to the effects on NK cells, the type I IFNs enhance T-cell IFN-γ production, and certain of their effects on DCs can have downstream consequences for adaptive immune responses. The antiproliferative effects of the cytokines present a challenge for the expansion of antigen-specific adaptive responses. MHC, major histocompatibility complex.

The type I IFN immunoregulatory effects on innate and adaptive immune system components are the focus of this chapter. These biological functions are remarkably broad; are often paradoxical, with positive or negative consequences depending on the cell types and conditions examined; and make important contributions to protection against infection in vivo. The type I IFNs act on other cytokines as well as directly on immune cells to regulate innate and adaptive responses. The most extensively studied cellular effects are those shaping responses of NK cells and dendritic cells (DCs) of the innate and T cells of the adaptive immune systems. Insights into how the downstream consequences of type I IFN exposure are regulated, to access such different biological effects when needed, are only now being developed. It is becoming clear that regulation of access to intracellular signaling pathways plays an important role in shaping type I IFN functions. The cytokines can be induced in response to different microbial infections including infections with either DNA or RNA viruses (8, 81), but they appear to be uniquely induced to high levels for extended periods of time during certain viral challenges, exemplified during infections of mice with the RNA virus lymphocytic choriomeningitis virus (LCMV) (6, 10, 70, 73). Moreover, although many viruses induce type I IFNs as part of a complex, innate, proinflammatory cytokine cascade, the extended production of these cytokines during LCMV infection is a more dominant innate response with lower-level production of other proinflammatory cytokines. Because of this strong and more focused innate cytokine response to LCMV, characterization of the type I IFN effects on innate and adaptive immune cells—as well as regulation of the signaling pathways to control these—have been largely advanced in studies examining LCMV infections of mice, and these are highlighted here.

BIOLOGICAL EFFECTS OF TYPE I IFNs

The different forms of type I IFNs bind to a common heterodimeric receptor comprising the IFN-α receptor-1 (IFNAR1) and IFNAR2 subunits (74, 76, 83, 95), but the biological effects required under different conditions of exposure are extremely divergent. Specific mediators of direct antiviral functions are targets of type I IFN action, and the cytokines have effects on both innate and adaptive immune functions (Fig. 1). The dissection of their actions on individual components is reviewed here.

Direct Antiviral Functions

The known biochemical mechanisms activated by the cytokines for direct defense in infected cells include the 2′,5′-oligoadenylate synthase (OAS)/RNase L pathways to degrade RNA, the protein kinase R (PKR) system to inhibit protein synthesis, and the guanosine triphosphatase (GTPase) products of the Mx genes sequestering viral ribonucleoproteins to specific subcellular compartments (8, 37). The double-stranded RNA-specific adenosine deaminase, ADAR, is also induced by type I IFNs to result in deamination of adenosine to inosine for RNA editing (8, 88). Additional pathways are likely to be identified in the future and to include effects mediated as a result of induction of the IFN-stimulated gene 15 (*isg15*) (54). Through interactions with other cytokines, type I IFNs can also elicit expression of the inducible nitric oxide synthase, and this enzyme can react with oxygen to chemically modify proteins essential for viral replication (8). Immunoregulatory effects of the type I IFNs at these earliest times after infection result from induction of a variety of IFN regulatory transcription factors (IRFs) (53, 60), with one of these, IRF7, promoting expression of a wider range of IFN-α/β genes to result in higher-level production of the cytokines for longer periods of time (60). These first targets of type I IFN action are likely to be accessible in all cell types, including immune cells, to protect them against viral infection. The effects are part of the earliest responses to infections and can be demonstrated both in vitro in culture and in vivo during infections.

Innate Cell Functions

In terms of their immunoregulatory effects on innate immune cells, the cytokines activate NK-cell cytotoxicity at early times during viral infections (10, 67, 71). This effect has been appreciated for some time, is seen in both human and mouse systems, and has been observed during any viral infection inducing type I IFNs (10). By providing an innate mechanism for lysing virus-infected cells, the activation of NK cell-killing functions is thought to contribute in vivo to defense against certain viral infections such as murine cytomegalovirus, an agent expressing a ligand for one of the NK-cell-activating receptors (8, 92). In addition, however, NK-cell cytotoxicity may also have other downstream immunoregulatory functions because viral infections can trigger profound diseases characterized by hyperactivation of macrophages and cytokine production, such as hemophagocytic lymphohistiocytosis, in mice and men with defects in the molecular mechanisms required for delivery of killing function (84).

The type I IFNs also have a number of effects on different DC populations. The cytokines help support the accumulation of plasmacytoid DCs (pDCs), major contributors to type I IFN production during certain viral infections (26). In addition, they have effects on other DC populations thought to be important in mediating antigen-presenting cell functions for the activation of adaptive immunity, although some of these are paradoxical (26, 42, 79). Type I IFNs act to limit the overall accumulation of conventional DCs (cDCs) (42, 79), and this is likely to be a consequence of their anti-proliferative effects on precursors of these cells (42). A consequence of limiting cDCs is an accompanying delay in the induction of T-cell responses (79), and this may benefit viruses by providing an extended period without defense mediated by adaptive immunity (42, 79). The negative effect on cDC accumulation may be controlled indirectly by NK cells when they act to limit overall viral burdens early in infection and, as a result, control the overall levels of type I IFNs induced (79). In contrast, however, the type I IFNs can promote the expression of the histocompatibility molecules used to present antigen to T cells (8), and the co-stimulatory molecules used to drive the activation of naive T cells are induced as a result of DC maturation (26, 64). Thus, type I IFNs produced during early viral infections link the immediate earliest responses to induction of innate defense mechanisms delivered by NK cells and shape the DC responses for type I IFN production and induction of adaptive immunity. The effects, however, are complex, interwoven, and sometimes paradoxical.

Cytokine Expression by Innate and Adaptive Cells

As noted above, type I IFNs can enhance their own expression. They also regulate expression of other cytokines, but do so in both negative and positive directions. As an example, they have confounding effects on the expression of IL-12, an innate factor known to be a potent inducer of IFN-γ. At low concentrations, type I IFNs can enhance IL-12 expression (38). At high but physiologically relevant concentrations, however, type I IFNs inhibit the production of biologically active IL-12 (22, 47, 61). Certain of these effects are likely to be influenced by the cell type and conditions examined. IFN-γ is both an innate and adaptive factor because it can be produced by either NK or T cells if they are appropriately stimulated. Remarkably, the responsiveness of NK cells to IL-12 for IFN-γ production is also negatively regulated by type I IFNs (66). Moreover, although they are not potent inducers of NK-cell IFN-γ responses, type I IFNs can induce IFN-γ by human CD4 T cells

(14, 59, 80) and enhance CD8 T-cell IFN-γ production in the mouse (23, 73). Thus, the effects of type I IFNs on the expression of other cytokines are diverse and paradoxical, with differential effects on IL-12 expression depending on doses and cell types, and inhibition of NK- but enhancement of T-cell IFN-γ responses.

Cell Proliferation in Innate and Adaptive Responses

The strong antiproliferative effects of type I IFNs have long been appreciated (8). There are cell types, however, that are relatively resistant to proliferation inhibition mediated by the cytokines, and type I IFNs can even act to promote division under particular conditions (9–11, 67). The mechanisms induced by type I IFNs to enhance antiviral states are likely to contribute to the antiproliferative effects because they shut down cell functions, such as protein synthesis, needed for both viral and cell replication. In terms of normal immune function, however, proliferation inhibition presents a serious challenge to the adaptive immune system because T- and B-lymphocyte subsets recognizing a pathogen through their antigen-specific receptors have to be selected from within a diverse repertoire and expanded.

In the case of effects on cellular proliferation, examination of the functions of type I IFNs during infections of mice have yielded a number of surprising and important findings. Under these conditions, it has been shown that type I IFNs induce IL-15 to drive NK- and memory-T-cell proliferation at early times after infections (67, 97). This effect appears to help sustain NK-cell numbers during acute periods of activation and memory-T-cell populations following exposure to their specific antigens. At intermediate times of infection when type I IFN expression overlaps with the selection of antigen-specific T cells for expansion, a population of CD8 T cells can be identified, in an ex vivo sensitivity assay, with increased resistance to the antiproliferative effects of type I IFNs (40). Thus, although the cytokines can profoundly inhibit cell proliferation, type I IFNs also induce IL-15, a factor that supports the expansion of NK and memory T cells, and certain T cells acquire resistance to the growth-inhibitory effects of these factors. These observations indicate that type I IFN effects on proliferation are modified by intermediary factors and apparent cellular conditioning during the development of endogenous immune responses.

PATHWAYS FOR TYPE I IFN SIGNALING

The mechanisms controlling the many diverse and conflicting biological effects of type I IFNs are poorly understood, but regulation of access to intracellular

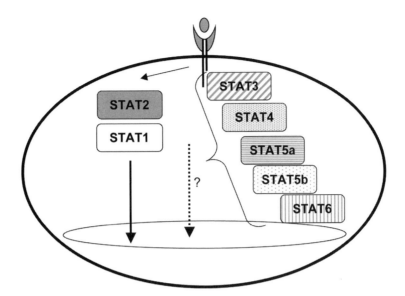

Figure 2 Signaling pathways used by type I IFNs. The classic signaling pathway stimulated by type I IFNs results in the stimulation of Jak1 and Tyk2 to activate, by phosphorylation, Stat2 and Stat1. These, in association with IRF9, translocate to the nucleus to stimulate the expression of gene targets expressing appropriate promoter elements. The cytokines, however, can also activate Stat1/Stat1 homodimers to stimulate expression of gene targets with promoter elements for these. Furthermore, there are a total of seven Stats, Stat1 through Stat6 (including Stat5a and Stat5b), and type I IFNs have been reported to conditionally activate all of these (13, 37, 90).

signaling pathways may contribute. As a consequence of binding to the type I IFN receptor, the cytokines activate the tyrosine kinases Tyk2 and Jak1 (74, 95). Studies in limited culture systems indicate that the activated enzymes in turn phosphorylate the receptor complex to enable recruitment and activation, by phosphorylation, of the signal transducers and activators of transcription (Stats). The classic signaling pathway results in the activation of Stat1 and Stat2, but other Stats can also be conditionally activated by type I IFNs (Fig. 2) (13, 37, 90). A general discussion of these pathways and the developing understanding of their functions in shaping type I IFN effects is presented here. More detailed aspects of their how their access is regulated during infections in vivo are provided below.

Classic Pathway

The best-characterized signaling pathway activated by type I IFNs is dependent on Stats 1 and 2. Heterodimers of the phosphorylated Stats, pStat1 and pStat2, in association with IRF9 to form an IFN-stimulated gene factor (ISGF), translocate to the nucleus and bind to the IFN-stimulated response elements (ISREs) in promoters of sensitive genes to activate transcription. Stat1 homodimers can also be activated (74). IFN-γ uses a different receptor to independently result in the activation of

Stat1 homodimers. Once activated, these function as a transcription factor complex for the IFN-γ-activated sequences (GAS) in sensitive gene promoters (28, 74). The activation of Stat1 homodimers by both type I IFNs and IFN-γ explains how these different cytokines, using different receptor systems, can mediate overlapping biological effects. Studies in a limited number of cell types indicate that the expression of hundreds of genes is stimulated by type I IFNs (28, 29). Many of the biochemical mediators of the antiviral effects induced by type I IFNs are Stat1 dependent, including the Mx and OAS genes. In fact, this classic signaling pathway was elucidated in part by following the antiviral effects induced by the cytokines.

Alternate Pathways

There are a total of seven different Stat molecules, Stat1 through Stat6 (with Stat5a and 5b), and type I IFNs have been reported to conditionally activate all of these Stats, including Stat4 (13, 18, 35, 37, 45, 68, 90). Remarkably, the Stats have a high degree of homology, including within the SH2 domains used for dimer formation (17, 43), and a wide range of homo- and heterodimers have been reported (43). The potential to activate virtually all Stats, along with the potential for variable interactions between Stats, may provide mechanisms to extend type I IFN functions from direct antiviral to immunoregulatory effects.

The understanding of Stat interactions with each other and the type I IFN receptor, as well as the consequences of these interactions, remains largely limited to information derived in select cells and culture conditions. This is also true for the known negative regulators of Stat activation, such as the suppressors of cytokine signaling (SOCS), protein inhibitors of activated Stats (PIAS), and SLIM (a nuclear protein containing both PDZ and LIM domains, acting as a ubiquitin $\varepsilon 3$ ligase on Stats) (1, 34, 55, 82, 87, 96). The flexibility of cellular responses to type I IFNs and mechanisms evolved to control these, however, are not easily predicted without examination in an in vivo context. In addition to providing an opportunity for unanticipated discoveries, in vivo experiments also ensure that the results are physiologically significant. The importance of evaluating the signaling pathways under biologically relevant conditions is made clear by the controversy concerning Stat4 activation by type I IFN. Based on culture studies, sequences present in human but absent in mouse Stat2 were proposed to be required for type I IFN activation of Stat4 (33). However, the species difference and a putative required role for Stat2 in type I IFN activation of Stat4 have not held up (3, 63, 68, 93). As will be seen below, type I IFNs have other pathways to the signaling molecule, and these are important under physiological conditions.

SIGNALING IN THE REGULATION OF IMMUNITY

Because of the multidisciplinary expertise required and the complexities of the regulation of individual immune system components, unraveling the mechanisms controlling type I IFN activation of different Stats and the biological consequences of regulating these have been challenging tasks. Nevertheless, information is accumulating. Studies of infections in mice (32, 62) and humans (31) with genetic Stat1 deficiencies have shown that this signaling molecule is important for defense against viral replication and virus-induced disease. It has been generally assumed that Stat1 mediates these beneficial effects through direct antiviral defense mechanisms, but the immunoregulatory effects mediated through Stat1 have not been well defined. Clearly, Stat1 is important for the antiproliferative effects of type I IFNs (15), and regulation of lymphocyte proliferation is important to the development of endogenous immune responses (40). The signaling molecule also contributes to type I IFN induction of NK-cell cytotoxicity (9, 52, 67), IL-15 (67), and IRF7 (58), as well as the accumulation of pDCs, during viral infections (26).

In contrast to the demonstrated effects of Stat1, Stat4 has the anticipated role of contributing to IL-12 signaling for IFN-γ production during innate responses to infections eliciting this cytokine (46, 89). It also contributes to a type I IFN-dependent CD8 T-cell IFN-γ response at times with overlapping innate and adaptive immune responses during viral infections (68) (see below). There are reports on the effects, through undefined mechanisms, of various Stat deficiencies on T-cell expansion and activation (25, 40, 41). The consequences of eliminating one Stat molecule for access to other signaling pathways and the importance of this in shaping cytokine responses, however, have only begun to be appreciated as a result of three independent lines of evidence using Stat-deficient cells or mice. Studies of Stat1-deficient mice demonstrated that although Stat1 is required for a number of innate immune responses activated by type I IFNs, infections in the absence of Stat1 are accompanied by a disregulation of the immune response. This is characterized by the appearance of high IFN-γ levels early in infection, and by type I IFN induction, in culture, of IFN-γ production by Stat1-deficient but not immunocompetent cells (66). Other studies examining IFN-γ induction of target genes in Stat1-deficient as compared with competent cells in culture showed that gene responses can be separated into three categories: Stat1 dependent, Stat1 independent, and inhibited by Stat1 (39, 77). Evaluation of the effects of IL-6, a cytokine that uses Stat3 and Stat1 to signal, demonstrated that in the absence of Stat3, this cytokine behaves like IFN-γ and induces antiviral effects (20). In addition, there is evidence that Stat1 is important in negatively regulating nonspecific CD8 T-cell proliferation at early times after infection (40). Taken together, the studies argue that changes in access to intracellular signaling pathways not only control individual biological effects of cytokine exposure but also help shape them. Moreover, as they also show that the presence of Stat1 negatively regulates a potential type I IFN-mediated biological effect, i.e., induction of IFN-γ, the observations suggest that Stat1 access may provide a mechanism to explain the paradox of divergent type I IFN effects.

LCMV INFECTION SYSTEM

Endogenous immune responses to acute infections with the RNA virus LCMV provide an excellent and flexible system for studying the immunoregulatory effects of type I IFNs (Fig. 3) (9, 69). The virus is relatively noncytopathic. When present, disease is a result of immune responses to infection. NK and T cells are activated to respond, but CD8 T and not NK cells are key in viral

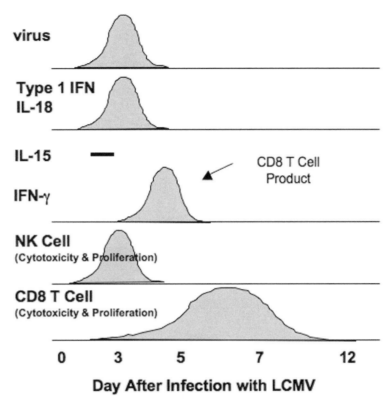

Figure 3 Schematic representation of endogenous immune responses to LCMV infection. At early times after infection with LCMV, immune-competent mice have high type I IFNs with IL-18 induction. The type I IFNs promote NK-cell cytotoxicity and elicit induction of IL-15 to result in NK-cell proliferation. There is, however, little biologically active IL-12. NK-cell IFN-γ production is also at low to undetectable levels, and NK cells become refractory to IL-12 for IFN-γ induction. At intermediate times after infection, there is an endogenous IFN-γ response produced by antigen-specific CD8 T cells, and this is dependent on type I IFNs and enhanced by IL-18. A dramatic expansion of the antigen-specific CD8 T cells is observed at late times after infection. (Presentation derived from compiled results in references 9, 48, 65, 70, 73, and 85.)

clearance (10, 69). Peak innate responses to LCMV occur in a period through day 3 of infection. During these times, immune-competent mice have high type I IFNs, resulting in induction of NK cell killing and IL-15 expression to support NK-cell proliferation (9, 67), but little biologically active IL-12 (70). There is an intermediate period between days 4 and 6 with overlapping waning innate but increasing adaptive responses. It is during this period that CD8 T-cell subsets with increasing resistance to type I IFN-mediated inhibition of proliferation, as evaluated ex vivo, can be identified (40), and antigen-specific CD8 T-cell production of IFN-γ dependent on type I IFNs is detected in vivo (56, 73). IL-18, a cytokine enhancing IFN-γ production in response to type I IFNs, is induced at times overlapping with the type I IFN response to LCMV, and contributes to the CD8 T-cell IFN-γ

response (73). Proliferation of antigen-specific T cells can first be detected after day 5 of infection, but maximal numbers are observed on day 8 (16, 40, 65). The expanded cells are primed for IFN-γ and IL-2 expression (16, 48, 49, 65, 85). Endogenous IL-2 contributes to overall CD8 T-cell expansion (21). There is evidence that responsiveness to either type I IFNs or IFN-γ promotes the long-term maintenance of CD8 T cells (51, 94), but type I IFN responsiveness is not required for CD8 T-cell proliferation during LCMV infection (23).

The LCMV infection system has been used to show that type I IFNs, through Stat1-dependent pathways, induce innate NK cell killing, IL-15, and IRF7 (9, 58, 67), but inhibit NK-cell IFN-γ responses in vivo (66), and that the intermediate IFN-γ production by CD8 T cells depends on Stat4 as well as type I IFNs (68, 73).

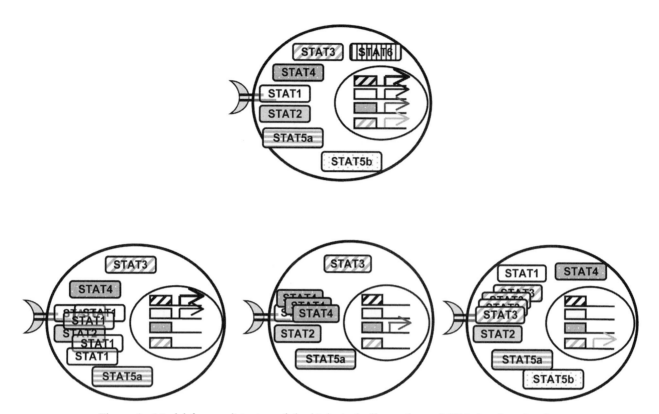

Figure 4 Model for conditioning of the biological effects of type I IFNs by changing Stat concentrations. Taken together with the experimental evidence demonstrating that the type I IFNs can conditionally activate all of the Stats, the newer results showing that relative Stat concentrations can be regulated suggest a model for changing biological effects of the cytokines by differentially regulating Stat levels. The model implies that access to different signaling pathways is a result of total and relative concentrations of different Stats.

STAT 1 AND 4 LEVELS AT DIFFERENT TIMES AFTER, AND IN DIFFERENT CELL POPULATIONS DURING, LCMV INFECTION

The studies evaluating the role of different Stats in shaping immune cell responses to type I IFNs, the ability of type I IFNs to contextually activate all of the Stats, and the demonstration of different effects of type I IFNs using cells rendered deficient in different Stats by genetic mutation suggest a model for type I IFN activation of different Stat molecules and their downstream targets controlled by relative concentrations of particular Stats (Fig. 4). Evolution of this kind of regulation would extend value to a limited number of genes and help explain the different and paradoxical effects of the cytokines. Most importantly, it would provide an infected individual with the flexibility to use type I IFNs in mediating a wide range of important immunoregulatory effects and in enhancing the development of different defense mechanisms delivered through uninfected immune cells. Experiments in cell cultures have shown that changes in total Stat1 levels can be induced by

exposure to type I IFNs or IFN-γ (28). To test the model of differential regulation of Stat concentrations under immunocompetent conditions, experiments carried out by my laboratory and our collaborators have been measuring the relative concentrations of Stat1 and Stat4 at different times after, and in different cell populations during, infections of mice with LCMV. The results indicate that there are mechanisms in place to differentially regulate Stat1 and Stat4 expression at different times during infection and within different cell types.

Stat1 Levels

Western blot analyses of lysates from total splenic leukocytes have revealed that Stat1 protein levels are dramatically elevated on day 1.5 through 5 of LCMV infection (63, 68). To extend the characterization of Stat1 levels to cell subsets, cellular levels of total Stat1 protein have been analyzed by Western blotting of total and purified cell subsets and by fluorescence-activated cell sorting (FACS) of mixed populations using surface staining for markers of cell subsets along with intracellular staining

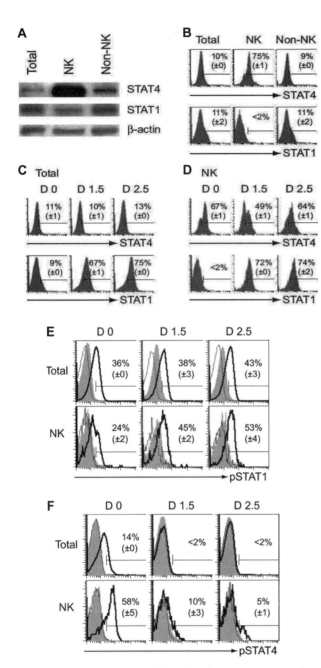

Figure 5 Total and NK-cell Stat levels—consequences for responsiveness to type I IFNs. The levels of Stat4 and Stat1 were evaluated by Western blot (A) and FACS analyses (B–D) in total, NK, and non-NK cells isolated from the spleens of uninfected mice and from mice at day 1.5 and 2.5 of LCMV infection, as indicated. The responsiveness of the total and NK-cell populations to type I IFNs for activation of pStat1 (E) or pStat4 (F) was examined using cells from uninfected (day 0) or LCMV-infected (day 1.5 or 2.5) mice. After treatment with type I IFN for 90 min in culture, the populations were stained intracellularly for the pStats. Gray areas represent results from untreated cells, solid lines represent results with IFN-treated cells, and the broken lines represent isotype control staining of treated cells. (Reproduced, in modified form, from reference 63 with permission of the Rockefeller University Press.)

for total Stat1 protein. NK (CD49b[+] and TCR[−]) and T (CD8[+]) cells were contrasted because of their different functions and kinetics of activation. Remarkably, all tested cell subsets showed increased Stat1 levels during the course of infection (40, 63). The change is a direct consequence of exposure to the earliest induced type I IFNs, but can also be elicited by IFN-γ if this factor is produced during the infection (63). The Stat1 molecule is important for the enhanced induction of Stat1 in response to type I IFNs or IFN-γ, while Stat2 is required for the induction of Stat1 in response to type I IFNs but not IFN-γ. As a result, LCMV infection of Stat2-deficient mice elicits a delayed induction of Stat1, whereas blocking responses to both type I IFNs and IFN-γ profoundly inhibits the response (63). Although all populations show an increase in Stat1 at early times during infection of immunocompetent mice, there are differences in the kinetics of the return of basal expression of Stat1 at later times during infection. In particular, subsets of CD8 T cells appear with lower concentrations of Stat1 as compared with the total cell populations (40). Thus, Stat1 levels are dynamically regulated during viral infections. This change is a consequence of exposure to type I IFNs themselves, and there are differences between cell subsets in the kinetics of return to basal levels.

Stat4 Levels

To extend the characterization to Stat4 levels, intracellular levels of Stat4 protein have been analyzed by Western blotting of total and purified cell subsets and by FACS. Interestingly, NK cells are unique in having high basal Stat4 levels, and their concentrations of the molecule are maintained during early times of infection when Stat1 levels are being induced (Fig. 5A–D) (63). The high levels of Stat4 found in populations of freshly isolated cells are surprising, but two other reports have shown innate cell populations derived in culture or from inflammatory sites with high Stat4 concentrations (36, 57). Thus, there are cell-lineage-specific differences in basal concentrations of Stat4, and different basal levels of Stats may be a characteristic of particular cell types.

CHANGES IN TYPE I IFN ACCESS TO STATS AND CYTOKINE-INDUCED BIOLOGICAL EFFECTS ASSOCIATED WITH DIFFERENCES IN STAT LEVELS

To evaluate the consequences of different Stat levels on cellular responses to type I IFNs and on the endogenous immune responses to viral infections, experiments have been carried out in uninfected and LCMV-infected immunocompetent mice and mice deficient in Stat1 or Stat2. The cellular responses of NK and T cells have been examined and compared to total cells in vivo or ex vivo. The results to date indicate that the basal and induced levels of Stats act to regulate the responses of these populations to type I IFNs, and that the effects are important to eliciting a beneficial immune response.

NK Cells

NK cells can be potent producers of IFN-γ, but this response must be regulated to protect against cytokine-mediated disease (4, 30, 44, 72). It is clear that Stat4 is an important mediator of IFN-γ induction in response to cytokines (46, 68). Under immunocompetent conditions, NK cells are not elicited to produce high levels of IFN-γ at times of type I IFN production during LCMV infections (9, 70, 71). To define type I IFN responsiveness of total and NK cells at different times after infection, intracellular FACS analyses have been carried out examining the activation by phosphorylation of Stat1 (pStat1) or Stat4 (pStat4) (63). The total populations have low Stat4 and increasing Stat1 levels during infection. They respond to ex vivo type I IFN exposure cytokines with activation of pStat1 but not pStat4 when they are isolated from uninfected mice or mice on days 1.5 and 2.5 after LCMV infection (Fig. 5E and F). The NK cells are less than 5% of total splenic populations. Their responsiveness to the cytokines changes when using cells from uninfected mice with high Stat4 and low Stat1 as compared with those from days 1.5 and 2.5 with high Stat4 and high Stat1. Under basal conditions, NK cells have a profound activation of pStat4 in response to type I IFN exposure, but this is turned off once the total Stat1 levels are increased. Detectable pStat1 is induced in all NK-cell populations, but the proportions are increased once Stat1 levels are induced during infection. Thus, there are profound differences in the signaling pathways activated by type I IFNs, and Stat4 access is associated with the presence of high concentrations of the molecule prior to Stat1 induction.

There are significant biological consequences linked to the expression of Stat4 and Stat1 within NK cells. The effects of changes in access to Stat1 within NK cells have been examined in Stat1-deficient mice and Stat2-deficient mice with delayed Stat1 induction during infection. Although minimal in immunocompetent mice, NK-cell IFN-γ responses, are strong in Stat1- and Stat2-deficient mice, with circulating levels of the cytokine on day 1.5 of infection (63, 66). The response is sustained through day 2.5 in Stat1- but not Stat2-deficient mice (63). Multiparameter FACS analysis has shown that only low frequencies of NK cells are induced to express IFN-γ or pStat4 during infection in immunocompetent mice. In contrast, the NK cells on day 1.5 from Stat1- or

Stat2-deficient mice have 13 to 19% of the populations expressing the cytokine and pStat4. As the Stat1 levels are increased in the Stat2-deficient mice on day 2.5, the responses subside in the NK cells. Thus, Stat2-deficient

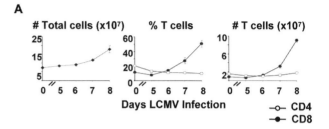

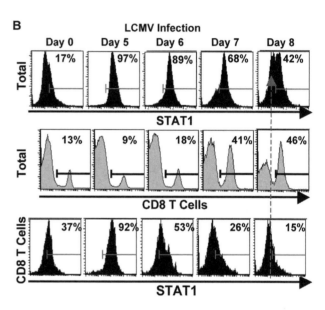

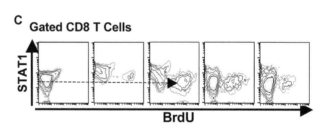

Figure 6 Changing Stat1 levels in CD8 T cells responding to LCMV infections. Cells were prepared from uninfected mice or mice infected with LCMV for the indicated times. (A) Spleen cell yields were measured and the number of CD8 and CD4 T cells determined using FACS analysis of subsets expressing cell markers. (B) Cytoplasmic staining of total Stat1 protein was determined in total cells and in the T-cell subset identified by cell surface staining with CD8. (C) To identify the cells proliferating in vivo, BrdU was administered for 2 hours prior to harvest, and the CD8 T-cell subsets were examined for expression of Stat1 along with BrdU. (Reproduced, in modified form, with permission from research originally published in reference 40.)

NK cells respond to infection in a manner reflecting Stat1-deficient cells when Stat1 levels are low, but immunocompetent cells when Stat1 levels are induced. Remarkably, IFN-γ production is accompanied by a strong endogenous induction of IFN-γ and pStat4 in the NK cells. All of these responses are dependent on in vivo exposure to type I IFNs because they are blocked by treatments neutralizing the function of the cytokines (63). Hence, NK cells are poised to respond to type I IFNs with Stat4 activation and IFN-γ induction, but inhibition of access to the pathway accompanies induction of high Stat1 levels.

In terms of the consequences for health, deficiencies in either Stat1 or Stat2 result in increased susceptibility to LCMV infection, with viral burdens greater than 2 log higher as compared with those observed under immunocompetent conditions (63). LCMV, however, is a relatively noncytopathic virus, and disease as measured by weight loss is only present in Stat1-deficient and not Stat2-deficient or immunocompenent mice. Neutralization of IFN-γ or depletion of NK cells results in some protection against the wasting disease. Thus, type I IFNs can activate Stat4 in NK cells, but this is tightly regulated by the induction of Stat1 during infection. This regulation is critical to the health of the infected individual.

CD8 T Cells

In contrast to the thorough understanding of the biological consequences of Stat expression in NK cells, an understanding of the role played by Stat concentrations in shaping CD8 T-cell responses is still under development. As noted above, the absence of Stat1 at early times after LCMV infection is associated with a disregulation of CD8 T-cell proliferation, as measured by the endogenous incorporation of the DNA analogue bromodeoxyuridine (BrdU) in cell subsets failing to demonstrate specificity for viral antigens (40). The Stat1-dependent antiproliferative effects of type I IFNs, however, present a challenge to the requirement for expansion of antigen-specific CD8 T cells during infections under immunocompetent conditions. In the case of LCMV infection, there is a profound expansion of CD8 but only a modest expansion of CD4 T cells (Fig. 6A). Studies examining CD8 T cells at intermediate to late times of LCMV infection have shown that subsets are becoming more resistant to type I IFN-mediated inhibition of proliferation (40), and that higher proportions of the CD8 T cells are returning to low-level Stat1 expression with a faster kinetics than other splenic populations (Fig. 6B) (40). Moreover, the CD8 T cells that are proliferating, as evidenced by in vivo

incorporation of BrdU, are preferentially found within the cells expressing lower levels of Stat1 (Fig. 6C), and the antigen-specific subsets identified by expression of receptors binding to tetramers of major histocompatibility molecules presenting immunodominant LCMV peptides are found in these subsets (40). Thus, although Stat1 may play an important role in helping to control nonspecific CD8 T-cell expansion at early times after infection, there are mechanisms in place to help the antigen-specific cells overcome type I IFN-mediated inhibition of proliferation, and lower Stat1 levels are associated with the differential responses of these cells.

REGULATION OF STAT1 AND STAT4 ACCESS AT THE LEVEL OF THE TYPE I IFN RECEPTOR

The studies reviewed above demonstrate that NK cells have high basal Stat4 and sustain this through the early innate response to infection when Stat1 levels are elevated. The higher levels of Stat4 are accompanied by a predisposition to activate Stat4, but this is lost once Stat1 is induced. To define the physical interactions between the Stats and the type I IFN receptor, immunoprecipitation studies have been carried out using purified cell subsets and a monoclonal antibody directed against the α chain of the type I IFN receptor. The results show that Stat4 is associated with the receptor when it is expressed at high concentrations in NK cells from uninfected mice, but that the highest levels of Stat1 induced during infection displace the Stat4 association in NK cells (Fig. 7). In

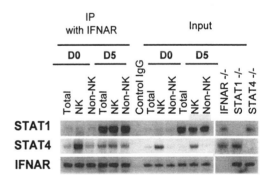

Figure 7 Stat associations with the type I IFN receptors in total, NK, and non-NK cells. Total splenic cells, NK cells, and non-NK cells were prepared from uninfected and day 5 LCMV-infected mice. The association of Stat1 or Stat4 with their type I IFN receptor was determined by coimmunoprecipation (IP) using antibodies specific for the receptor (lanes 1 to 6). Input samples were also examined (lanes 8 to 13). The specificity of the association was confirmed by detection of immunoreactivity in cell lysates from IFNAR-, Stat1-, or Stat4-deficient cells (lanes 14 to 16). (Reproduced from reference 63 with permission.)

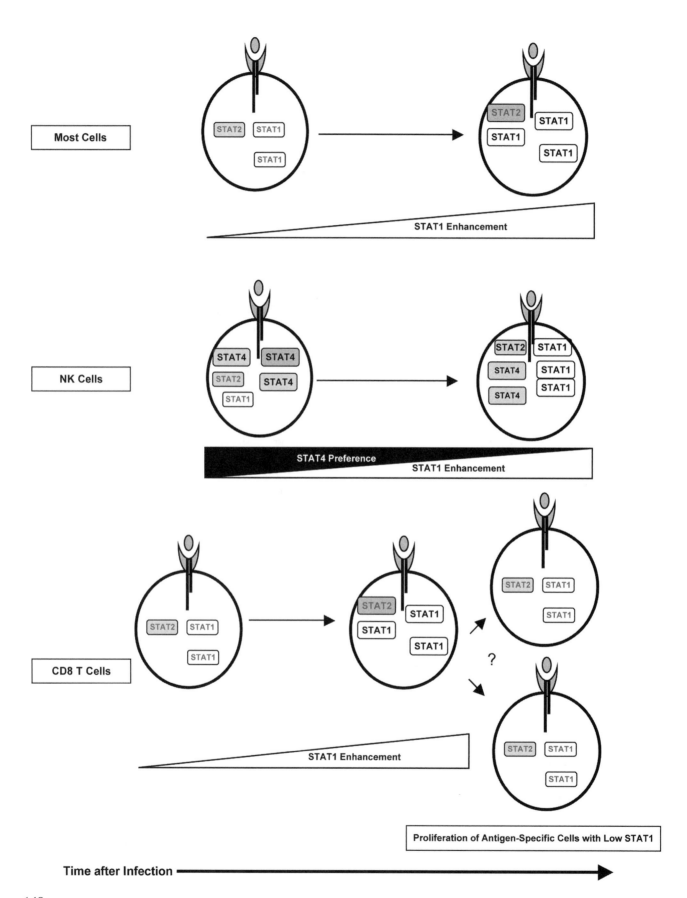

Most Cells

STAT1 Enhancement

NK Cells

STAT4 Preference
STAT1 Enhancement

CD8 T Cells

STAT1 Enhancement

Proliferation of Antigen-Specific Cells with Low STAT1

Time after Infection

148

addition to the infection-induced Stat1 association with the receptor in NK cells, total and non-NK cells have an infection-induced increase in the Stat1 association, but these cells do not demonstrate a strong Stat4 association under any of the conditions examined. Hence, a fundamental and apparently unique characteristic of NK cells is high basal Stat4 bound to the type I IFN receptor, but conditions of infection result in Stat1 induction and Stat4 displacement.

SUMMARY OF UNDERSTANDING OF THE REGULATION OF SIGNALING PATHWAYS IN SHAPING THE BIOLOGICAL EFFECTS OF TYPE I IFNs

In summary, after an extended period of accumulating information on many different and sometimes conflicting biological effects of type I IFNs (8) as well as their promiscuous activation of all the Stats (13, 37, 90), careful dissection of events at different times after infections and in different cell types is leading to the conclusion that everything fits together (Fig. 8). The original observations indicated that type I IFNs could activate important antimicrobial defense pathways within infected cells through signaling pathways dependent on Stat1 and Stat2, and that a number of the immune-enhancing effects of type I IFNs, particularly on innate components, are also dependent on Stat1. The observations of disregulated or surprising new responses revealed in Stat-deficient conditions (20, 39, 66) opened up an appreciation of the possibility that the Stats cross-regulate potential downstream consequences of exposure. The new studies in NK cells show that Stat4 can be preassociated with the type I IFN receptor when it is expressed at high concentrations in these cells, but that the induction of elevated Stat1 results in its displacement (63). In regard to T cells, the evidence is that Stat1 acts to limit nonspecific expansion during infections, but that mechanisms are in place to allow the proliferation of antigen-specific cells with lower Stat1 concentrations (40). Clearly, much remains to be learned, but the results to date indicate that (i) regulation of access to

Stat4 as compared with Stat1 is a consequence of basal and induced concentrations (63); (ii) without induced Stat1 levels, there are detrimental consequences resulting from an unregulated immune response (40, 63); and (iii) mechanisms are in place to help antigen-specific T cells overcome IFN-induced inhibition of proliferation, and these are associated with expansion of populations expressing lower levels of Stat1 (40). Thus, the picture emerging is that the apparent promiscuity of access that allows activation of different Stats is likely to be required to provide an infected individual with the opportunity to use these cytokines to regulate a wide range of endogenous immune responses and deliver a wide range of antimicrobial defense mechanisms, and that this is controlled in part by the different concentrations of the signaling molecules.

It is likely that the themes emerging from this work will have counterparts in signaling from other cytokines leading to Stat activation. In particular, it is worth noting that although stimulation of the IFN-γ receptor leads to activation of Stat1 homodimers, Stat3 can also be activated. Careful biochemical studies in cell culture systems have demonstrated an apparent competition between Stat1 and Stat3 for binding to the cytoplasmic regions of the IFN-γ receptor (75). It also will be interesting to extend the work on type I IFN activation of Stat1 and Stat4 to other Stats, including Stat3, to determine whether or not the range of biological effects are all mediated by Stat interactions with the type I IFN receptor (75). Given the wide potential for interactions between different activated Stat molecules (43), and the recently identified reduction of type I IFN signaling through Stat1 as a result of activated Stat1 binding by activated Stat3 (45), there are likely to be other mechanisms for Stat-Stat cross-regulation of biological functions.

BROADER PERSPECTIVES AND THERAPEUTIC IMPLICATIONS

A thorough understanding of the biochemical mechanisms regulating the broad range of biological effects of type I IFNs is critical to the understanding of endogenous

Figure 8 Summary of the current understanding of known Stat levels and use in NK and CD8 T cells during LCMV infections. NK cells begin with high Stat4 levels pre-associated with the type I IFN receptor. In contrast, they have little Stat1 associated with the receptor, and other cell types have little association of either Stat1 or Stat4. After infection, however, high Stat1 levels are induced in all populations, and the molecule is now associated with the receptor in NK cells as well as other cells. In the case of CD8 T cells, the receptor associations remain to be determined. Stat1 is required, however, to control nonspecific CD8 T-cell proliferation at early times after infection, and there is a preferential expansion of antigen-specific cells from within a subset expressing low Stat1 levels.

immune responses to infection. It is also advancing the broad conceptual understanding of how an individual can take advantage of a limited number of genes to access a wide range of cytokine functions. In addition, it is key to a number of therapeutic intervention strategies in place or under development. Type I IFNs are currently being used in the treatment of chronic infections with hepatitis C virus, cancers, and multiple sclerosis (12, 47, 86, 91). Their immunoregulatory functions are likely to be important in all of these applications. If variability between responders and nonresponders is a result of polymorphisms in genes unique to accessing particular required type I IFN immunoregulatory functions, identifying potential responsiveness would facilitate development of individual treatment strategies. Conversely, the expression in, and contribution of, circulating type I IFNs during autoimmune disease, particularly systemic lupus erythematosus, is now clear (24, 50), and polymorphisms in Stat4 genes have been linked to autoimmune susceptibility (78). Ablation protocols that utilize antibodies blocking the receptor and/or receptor antagonists are being developed to protect from the detrimental effects of endogenous cytokine expression. Because blocking all type I IFN functions may result in a profoundly increased susceptibility to infection, fine characterization of the signaling pathways leading to beneficial, as compared with problematic, effects would suggest more selective targets for intervention. Thus, although there has been a good deal of progress, additional future advances in understanding the regulation of particular type I IFN effects are important to a wide range of disease conditions and treatment protocols, and promise to contribute to strategies for promoting health.

References

1. **Alexander, W. S., and D. J. Hilton.** 2004. The role of suppressors of cytokine signaling (SOCS) proteins in regulation of the immune response. *Annu. Rev. Immunol.* 22:503–529.
2. **Asselin-Paturel, C., and G. Trinchieri.** 2005. Production of type I interferons: plasmacytoid dendritic cells and beyond. *J. Exp. Med.* 202:461–465.
3. **Berenson, L. S., M. Gavrieli, J. D. Farrar, T. L. Murphy, and K. M. Murphy.** 2006. Distinct characteristics of murine STAT4 activation in response to IL-12 and IFN-alpha. *J. Immunol.* 177:5195–5203.
4. **Billiau, A., H. Heremans, F. Vandekerckhove, and C. Dillen.** 1987. Anti-interferon-gamma antibody protects mice against the generalized Shwartzman reaction. *Eur. J. Immunol.* 17:1851–1854.
5. **Biron, C., M. Dalod, and T. Salazar-Mather.** 2002. Innate immunity and viral infection, p. 139–160. In S. H. E. Kaufman, A. Sher, and R. Ahmed (ed.), *Immunology of Infectious Diseases*. ASM Press, Washington, DC.
6. **Biron, C. A.** 1999. Initial and innate responses to viral infections—pattern setting in immunity or disease. *Curr. Opin. Microbiol.* 2:374–381.
7. **Biron, C. A.** 2001. Interferons alpha and beta as immune regulators—a new look. *Immunity* 14:661–664.
8. **Biron, C. A., and G. C. Sen.** 2007. Innate immune responses to viral infection, p. 249–278. In D. M. Knipe and P. M. Howley (ed.), *Fields Virology*, 5th ed. Walter Kluwer/Lippincott, Williams & Wilkins, Philadelphia, PA.
9. **Biron, C. A., K. B. Nguyen, and G. C. Pien.** 2002. Innate immune responses to LCMV infections: natural killer cells and cytokines. *Curr. Top. Microbiol. Immunol.* 263:7–27.
10. **Biron, C. A., K. B. Nguyen, G. C. Pien, L. P. Cousens, and T. P. Salazar-Mather.** 1999. Natural killer cells in antiviral defense: function and regulation by innate cytokines. *Annu. Rev. Immunol.* 17:189–220.
11. **Biron, C. A., G. Sonnenfeld, and R. M. Welsh.** 1984. Interferon induces natural killer cell blastogenesis in vivo. *J. Leukoc. Biol.* 35:31–37.
12. **Bretner, M.** 2005. Existing and future therapeutic options for hepatitis C virus infection. *Acta Biochim. Pol.* 52:57–70.
13. **Brierley, M. M., and E. N. Fish.** 2002. Review: IFN-alpha/beta receptor interactions to biologic outcomes: understanding the circuitry. *J. Interferon Cytokine Res.* 22:835–845.
14. **Brinkmann, V., T. Geiger, S. Alkan, and C. H. Heusser.** 1993. Interferon alpha increases the frequency of interferon gamma-producing human CD4+ T cells. *J. Exp. Med.* 178:1655–1663.
15. **Bromberg, J. F., C. M. Horvath, Z. Wen, R. D. Schreiber, and J. E. Darnell, Jr.** 1996. Transcriptionally active Stat1 is required for the antiproliferative effects of both interferon alpha and interferon gamma. *Proc. Natl. Acad. Sci. USA* 93:7673–7678.
16. **Butz, E. A., and M. J. Bevan.** 1998. Massive expansion of antigen-specific CD8+ T cells during an acute virus infection. *Immunity* 8:167–175.
17. **Chen, X., U. Vinkemeier, Y. Zhao, D. Jeruzalmi, J. E. Darnell, Jr., and J. Kuriyan.** 1998. Crystal structure of a tyrosine phosphorylated STAT-1 dimer bound to DNA. *Cell* 93:827–839.
18. **Cho, S. S., C. M. Bacon, C. Sudarshan, R. C. Rees, D. Finbloom, R. Pine, and J. J. O'Shea.** 1996. Activation of Stat4 by IL-12 and IFN-alpha: evidence for the involvement of ligand-induced tyrosine and serine phosphorylation. *J. Immunol.* 157:4781–4789.
19. **Colonna, M., G. Trinchieri, and Y. J. Liu.** 2004. Plasmacytoid dendritic cells in immunity. *Nat. Immunol.* 5:1219–1226.
20. **Costa-Pereira, A. P., S. Tininini, B. Strobl, T. Alonzi, J. F. Schlaak, H. Is'harc, I. Gesualdo, S. J. Newman, I. M. Kerr, and V. Poli.** 2002. Mutational switch of an IL-6 response to an interferon-gamma-like response. *Proc. Natl. Acad. Sci. USA* 99:8043–8047.
21. **Cousens, L. P., J. S. Orange, and C. A. Biron.** 1995. Endogenous IL-2 contributes to T cell expansion and IFN-gamma production during lymphocytic choriomeningitis virus infection. *J. Immunol.* 155:5690–5699.
22. **Cousens, L. P., J. S. Orange, H. C. Su, and C. A. Biron.** 1997. Interferon-alpha/beta inhibition of interleukin 12

and interferon-gamma production in vitro and endogenously during viral infection. *Proc. Natl. Acad. Sci. USA* 94:634–639.

23. Cousens, L. P., R. Peterson, S. Hsu, A. Dorner, J. D. Altman, R. Ahmed, and C. A. Biron. 1999. Two roads diverged: interferon alpha/beta- and interleukin 12-mediated pathways in promoting T cell interferon gamma responses during viral infection. *J. Exp. Med.* 189:1315–1328.

24. Crow, M. K. 2005. Interferon pathway activation in systemic lupus erythematosus. *Curr. Rheumatol. Rep.* 7:463–468.

25. Curtsinger, J. M., J. O. Valenzuela, P. Agarwal, D. Lins, and M. F. Mescher. 2005. Type I IFNs provide a third signal to CD8 T cells to stimulate clonal expansion and differentiation. *J. Immunol.* 174:4465–4469.

26. Dalod, M., T. Hamilton, R. Salomon, T. P. Salazar-Mather, S. C. Henry, J. D. Hamilton, and C. A. Biron. 2003. Dendritic cell responses to early murine cytomegalovirus infection: subset functional specialization and differential regulation by interferon alpha/beta. *J. Exp. Med.* 197:885–898.

27. Dalod, M., T. P. Salazar-Mather, L. Malmgaard, C. Lewis, C. Asselin-Paturel, F. Briere, G. Trinchieri, and C. A. Biron. 2002. Interferon alpha/beta and interleukin 12 responses to viral infections: pathways regulating dendritic cell cytokine expression in vivo. *J. Exp. Med.* 195:517–528.

28. Der, S. D., A. Zhou, B. R. Williams, and R. H. Silverman. 1998. Identification of genes differentially regulated by interferon alpha, beta, or gamma using oligonucleotide arrays. *Proc. Natl. Acad. Sci. USA* 95:15623–15628.

29. de Veer, M. J., M. Holko, M. Frevel, E. Walker, S. Der, J. M. Paranjape, R. H. Silverman, and B. R. Williams. 2001. Functional classification of interferon-stimulated genes identified using microarrays. *J. Leukoc. Biol.* 69:912–920.

30. Doherty, G. M., J. R. Lange, H. N. Langstein, H. R. Alexander, C. M. Buresh, and J. A. Norton. 1992. Evidence for IFN-gamma as a mediator of the lethality of endotoxin and tumor necrosis factor-alpha. *J. Immunol.* 149:1666–1670.

31. Dupuis, S., E. Jouanguy, S. Al-Hajjar, C. Fieschi, I. Z. Al-Mohsen, S. Al-Jumaah, K. Yang, A. Chapgier, C. Eidenschenk, P. Eid, A. Al Ghonaium, H. Tufenkeji, H. Frayha, S. Al-Gazlan, H. Al-Rayes, R. D. Schreiber, I. Gresser, and J. L. Casanova. 2003. Impaired response to interferon-alpha/beta and lethal viral disease in human Stat1 deficiency. *Nat. Genet.* 33:388–391.

32. Durbin, J. E., R. Hackenmiller, M. C. Simon, and D. E. Levy. 1996. Targeted disruption of the mouse Stat1 gene results in compromised innate immunity to viral disease. *Cell* 84:443–450.

33. Farrar, J. D., J. D. Smith, T. L. Murphy, S. Leung, G. R. Stark, and K. M. Murphy. 2000. Selective loss of type I interferon-induced Stat4 activation caused by a minisatellite insertion in mouse Stat2. *Nat. Immunol.* 1:65–69.

34. Fenner, J. E., R. Starr, A. L. Cornish, J. G. Zhang, D. Metcalf, R. D. Schreiber, K. Sheehan, D. J. Hilton, W. S. Alexander, and P. J. Hertzog. 2006. Suppressor of cytokine signaling 1 regulates the immune response to

infection by a unique inhibition of type I interferon activity. *Nat. Immunol.* 7:33–39.

35. Freudenberg, M. A., T. Merlin, C. Kalis, Y. Chvatchko, H. Stubig, and C. Galanos. 2002. Cutting edge: a murine, IL-12-independent pathway of IFN-gamma induction by gram-negative bacteria based on STAT4 activation by type I IFN and IL-18 signaling. *J. Immunol.* 169:1665–1668.

36. Frucht, D. M., M. Aringer, J. Galon, C. Danning, M. Brown, S. Fan, M. Centola, C. Y. Wu, N. Yamada, H. El Gabalawy, and J. J. O'Shea. 2000. Stat4 is expressed in activated peripheral blood monocytes, dendritic cells, and macrophages at sites of Th1-mediated inflammation. *J. Immunol.* 164:4659–4664.

37. Garcia-Sastre, A., and C. A. Biron. 2006. Type I interferons and the virus-host relationship: a lesson in detente. *Science* 312:879–882.

38. Gautier, G., M. Humbert, F. Deauvieau, M. Scuiller, J. Hiscott, E. E. Bates, G. Trinchieri, C. Caux, and P. Garrone. 2005. A type I interferon autocrine-paracrine loop is involved in Toll-like receptor-induced interleukin-12p70 secretion by dendritic cells. *J. Exp. Med.* 201:1435–1446.

39. Gil, M. P., E. Bohn, A. K. O'Guin, C. V. Ramana, B. Levine, G. R. Stark, H. W. Virgin, and R. D. Schreiber. 2001. Biologic consequences of Stat1-independent IFN signaling. *Proc. Natl. Acad. Sci. USA* 98:6680–6685.

40. Gil, M. P., R. Salomon, J. Louten, and C. A. Biron. 2006. Modulation of Stat1 protein levels: a mechanism shaping CD8 T-cell responses in vivo. *Blood* 107:987–993.

41. Gimeno, R., C. K. Lee, C. Schindler, and D. E. Levy. 2005. Stat1 and Stat2 but not Stat3 arbitrate contradictory growth signals elicited by alpha/beta interferon in T lymphocytes. *Mol. Cell. Biol.* 25:5456–5465.

42. Hahm, B., M. J. Trifilo, E. I. Zuniga, and M. B. Oldstone. 2005. Viruses evade the immune system through type I interferon-mediated Stat2-dependent, but Stat1-independent, signaling. *Immunity* 22:247–257.

43. Heim, M. 2003. The Stat protein family, p. 11–26. *In* P. B. Segal, D. E. Levy, and T. Hirano (ed.), *Signal Transducers and Activators of Transcription (STATS)*. Kluwer Academic Publishers, Dordrecht, Germany.

44. Heinzel, F. P. 1990. The role of IFN-gamma in the pathology of experimental endotoxemia. *J. Immunol.* 145:2920–2924.

45. Ho, H. H., and L. B. Ivashkiv. 2006. Role of Stat3 in type I interferon responses. Negative regulation of Stat1-dependent inflammatory gene activation. *J. Biol. Chem.* 281:14111–14118.

46. Kaplan, M. H., Y. L. Sun, T. Hoey, and M. J. Grusby. 1996. Impaired IL-12 responses and enhanced development of Th2 cells in Stat4-deficient mice. *Nature* 382:174–177.

47. Karp, C. L., C. A. Biron, and D. N. Irani. 2000. Interferon beta in multiple sclerosis: is IL-12 suppression the key? *Immunol. Today* 21:24–28.

48. Kasaian, M. T., and C. A. Biron. 1989. The activation of IL-2 transcription in L3T4+ and Lyt-2+ lymphocytes during virus infection in vivo. *J. Immunol.* 142:1287–1292.

49. Kasaian, M. T., and C. A. Biron. 1990. Effects of cyclosporin A on IL-2 production and lymphocyte proliferation during infection of mice with lymphocytic choriomeningitis virus. *J. Immunol.* 144:299–306.

50. Kirou, K. A., C. Lee, S. George, K. Louca, M. G. Peterson, and M. K. Crow. 2005. Activation of the interferon-alpha pathway identifies a subgroup of systemic lupus erythematosus patients with distinct serologic features and active disease. *Arthritis Rheum.* 52:1491–1503.

51. Kolumam, G. A., S. Thomas, L. J. Thompson, J. Sprent, and K. Murali-Krishna. 2005. Type I interferons act directly on CD8 T cells to allow clonal expansion and memory formation in response to viral infection. *J. Exp. Med,* 202:637–650.

52. Lee, C. K., D. T. Rao, R. Gertner, R. Gimeno, A. B. Frey, and D. E. Levy. 2000. Distinct requirements for IFNs and Stat1 in NK cell function. *J. Immunol.* 165:3571–3577.

53. Lehtonen, A., R. Lund, R. Lahesmaa, I. Julkunen, T. Sareneva, and S. Matikainen. 2003. IFN-alpha and IL-12 activate IFN regulatory factor 1 (IRF-1), IRF-4, and IRF-8 gene expression in human NK and T cells. *Cytokine* 24:81–90.

54. Lenschow, D. J., C. Lai, N. Frias-Staheli, N. V. Giannakopoulos, A. Lutz, T. Wolff, A. Osiak, B. Levine, R. E. Schmidt, A. Garcia-Sastre, D. A. Leib, A. Pekosz, K. P. Knobeloch, I. Horak, and H. W. T. Virgin. 2007. IFN-stimulated gene 15 functions as a critical antiviral molecule against influenza, herpes, and Sindbis viruses. *Proc. Natl. Acad. Sci. USA* 104:1371–1376.

55. Liu, B., S. Mink, K. A. Wong, N. Stein, C. Getman, P. W. Dempsey, H. Wu, and K. Shuai. 2004. PIAS1 selectively inhibits interferon-inducible genes and is important in innate immunity. *Nat. Immunol.* 5:891–898.

56. Liu, F., and J. L. Whitton. 2005. Cutting edge: re-evaluating the in vivo cytokine responses of CD8+ T cells during primary and secondary viral infections. *J. Immunol.* 174:5936–5940.

57. Longman, R. S., D. Braun, S. Pellegrini, C. M. Rice, R. B. Darnell, and M. L. Albert. 2007. Dendritic-cell maturation alters intracellular signaling networks, enabling differential effects of IFN-alpha/beta on antigen cross-presentation. *Blood* 109:1113–1122.

58. Malmgaard, L., T. P. Salazar-Mather, C. A. Lewis, and C. A. Biron. 2002. Promotion of alpha/beta interferon induction during in vivo viral infection through alpha/beta interferon receptor/Stat1 system-dependent and -independent pathways. *J. Virol.* 76:4520–4525.

59. Manetti, R., F. Annunziato, L. Tomasevic, V. Gianno, P. Parronchi, S. Romagnani, and E. Maggi. 1995. Polyinosinic acid: polycytidylic acid promotes T helper type I-specific immune responses by stimulating macrophage production of interferon-alpha and interleukin-12. *Eur. J. Immunol.* 25:2656–2660.

60. Marie, I., J. E. Durbin, and D. E. Levy. 1998. Differential viral induction of distinct interferon-alpha genes by positive feedback through interferon regulatory factor-7. *EMBO J.* 17:6660–6669.

61. McRae, B. L., R. T. Semnani, M. P. Hayes, and G. A. van Seventer. 1998. Type I IFNs inhibit human dendritic cell IL-12 production and Th1 cell development. *J. Immunol.* 160:4298–4304.

62. Meraz, M. A., J. M. White, K. C. Sheehan, E. A. Bach, S. J. Rodig, A. S. Dighe, D. H. Kaplan, J. K. Riley, A. C. Greenlund, D. Campbell, K. Carver-Moore, R. N. DuBois, R. Clark, M. Aguet, and R. D. Schreiber. 1996. Targeted disruption of the Stat1 gene in mice reveals unexpected physiologic specificity in the JAK-Stat signaling pathway. *Cell* 84:431–442.

63. Miyagi, T., M. P. Gil, X. Wang, J. Louten, W. M. Chu, and C. A. Biron. 2007. High basal Stat4 balanced by Stat1 induction to control type I interferon effects in natural killer cells. *J. Exp. Med.* 204:2383–2396.

64. Montoya, M., G. Schiavoni, F. Mattei, I. Gresser, F. Belardelli, P. Borrow, and D. F. Tough. 2002. Type I interferons produced by dendritic cells promote their phenotypic and functional activation. *Blood* 99:3263–3271.

65. Murali-Krishna, K., J. D. Altman, M. Suresh, D. J. Sourdive, A. J. Zajac, J. D. Miller, J. Slansky, and R. Ahmed. 1998. Counting antigen-specific CD8 T cells: a reevaluation of bystander activation during viral infection. *Immunity* 8:177–187.

66. Nguyen, K. B., L. P. Cousens, L. A. Doughty, G. C. Pien, J. E. Durbin, and C. A. Biron. 2000. Interferon alpha/beta-mediated inhibition and promotion of interferon gamma: Stat1 resolves a paradox. *Nat. Immunol.* 1:70–76.

67. Nguyen, K. B., T. P. Salazar-Mather, M. Y. Dalod, J. B. Van Deusen, X. Q. Wei, F. Y. Liew, M. A. Caligiuri, J. E. Durbin, and C. A. Biron. 2002. Coordinated and distinct roles for IFN-alpha beta, IL-12, and IL-15 regulation of NK cell responses to viral infection. *J. Immunol.* 169: 4279–4287.

68. Nguyen, K. B., W. T. Watford, R. Salomon, S. R. Hofmann, G. C. Pien, A. Morinobu, M. Gadina, J. J. O'Shea, and C. A. Biron. 2002. Critical role for Stat4 activation by type I interferons in the interferon-gamma response to viral infection. *Science* 297:2063–2066.

69. Oldstone, M. B. 2002. Biology and pathogenesis of lymphocytic choriomeningitis virus infection. *Curr. Top. Microbiol. Immunol.* 263:83–117.

70. Orange, J. S., and C. A. Biron. 1996. An absolute and restricted requirement for IL-12 in natural killer cell IFN-gamma production and antiviral defense. Studies of natural killer and T cell responses in contrasting viral infections. *J. Immunol.* 156:1138–1142.

71. Orange, J. S., and C. A. Biron. 1996. Characterization of early IL-12, IFN-alphabeta, and TNF effects on antiviral state and NK cell responses during murine cytomegalovirus infection. *J. Immunol.* 156:4746–4756.

72. Ozmen, L., M. Pericin, J. Hakimi, R. A. Chizzonite, M. Wysocka, G. Trinchieri, M. Gately, and G. Garotta. 1994. Interleukin 12, interferon gamma, and tumor necrosis factor alpha are the key cytokines of the generalized Shwartzman reaction. *J. Exp. Med.* 180:907–915.

73. Pien, G. C., K. B. Nguyen, L. Malmgaard, A. R. Satoskar, and C. A. Biron. 2002. A unique mechanism for innate cytokine promotion of T cell responses to viral infections. *J. Immunol.* 169:5827–5837.

74. Platanias, L. C. 2005. Mechanisms of type-I- and type-II-interferon-mediated signalling. *Nat. Rev. Immunol.* 5:375–386.

75. Qing, Y., and G. R. Stark. 2004. Alternative activation of Stat1 and Stat3 in response to interferon-gamma. *J. Biol. Chem.* 279:41679–41685.

76. Qureshi, S. A., M. Salditt-Georgieff, and J. E. Darnell, Jr. 1995. Tyrosine-phosphorylated Stat1 and Stat2 plus a 48-kDa protein all contact DNA in forming interferon-stimulated-gene factor 3. *Proc. Natl. Acad. Sci. USA* 92:3829–3833.

77. Ramana, C. V., M. P. Gil, R. D. Schreiber, and G. R. Stark. 2002. Stat1-dependent and -independent pathways in IFN-gamma-dependent signaling. *Trends Immunol.* 23:96–101.

78. Remmers, E. F., R. M. Plenge, A. T. Lee, R. R. Graham, G. Hom, T. W. Behrens, P. I. de Bakker, J. M. Le, H. S. Lee, F. Batliwalla, W. Li, S. L. Masters, M. G. Booty, J. P. Carulli, L. Padyukov, L. Alfredsson, L. Klareskog, W. V. Chen, C. I. Amos, L. A. Criswell, M. F. Seldin, D. L. Kastner, and P. K. Gregersen. 2007. STAT4 and the risk of rheumatoid arthritis and systemic lupus erythematosus. *N. Engl. J. Med.* 357:977–986.

79. Robbins, S. H., G. Bessou, A. Cornillon, N. Zucchini, B. Rupp, Z. Ruzsics, T. Sacher, E. Tomasello, E. Vivier, U. H. Koszinowski, and M. Dalod. 2007. Natural killer cells promote early CD8 T cell responses against cytomegalovirus. *PLoS Pathog.* 3:e123.

80. Sareneva, T., S. Matikainen, M. Kurimoto, and I. Julkunen. 1998. Influenza A virus-induced IFN-alpha/beta and IL-18 synergistically enhance IFN-gamma gene expression in human T cells. *J. Immunol.* 160:6032–6038.

81. Severa, M., and K. A. Fitzgerald. 2007. TLR-mediated activation of type I IFN during antiviral immune responses: fighting the battle to win the war. *Curr. Top. Microbiol. Immunol.* 316:167–192.

82. Shuai, K., and B. Liu. 2003. Regulation of JAK-Stat signalling in the immune system. *Nat. Rev. Immunol.* 3:900–911.

83. Stark, G. R., I. M. Kerr, B. R. Williams, R. H. Silverman, and R. D. Schreiber. 1998. How cells respond to interferons. *Annu. Rev. Biochem.* 67:227–264.

84. Stepp, S. E., R. Dufourcq-Lagelouse, F. Le Deist, S. Bhawan, S. Certain, P. A. Mathew, J. I. Henter, M. Bennett, A. Fischer, G. de Saint Basile, and V. Kumar. 1999. Perforin gene defects in familial hemophagocytic lymphohistiocytosis. *Science* 286:1957–1959.

85. Su, H. C., L. P. Cousens, L. D. Fast, M. K. Slifka, R. D. Bungiro, R. Ahmed, and C. A. Biron. 1998. CD4+ and CD8+ T cell interactions in IFN-gamma and IL-4 responses to viral infections: requirements for IL-2. *J. Immunol.* 160:5007–5017.

86. Tagliaferri, P., M. Caraglia, A. Budillon, M. Marra, G. Vitale, C. Viscomi, S. Masciari, P. Tassone, A. Abbruzzese, and S. Venuta. 2005. New pharmacokinetic and pharmacodynamic tools for interferon-alpha (IFN-alpha) treatment of human cancer. *Cancer Immunol. Immunother.* 54:1–10.

87. Tanaka, T., M. A. Soriano, and M. J. Grusby. 2005. SLIM is a nuclear ubiquitin E3 ligase that negatively regulates STAT signaling. *Immunity* 22:729–736.

88. tenOever, B. R., S. L. Ng, M. A. Chua, S. M. McWhirter, A. Garcia-Sastre, and T. Maniatis. 2007. Multiple functions of the IKK-related kinase IKKepsilon in interferon-mediated antiviral immunity. *Science* 315:1274–1278.

89. Thierfelder, W. E., J. M. van Deursen, K. Yamamoto, R. A. Tripp, S. R. Sarawar, R. T. Carson, M. Y. Sangster, D. A. Vignali, P. C. Doherty, G. C. Grosveld, and J. N. Ihle. 1996. Requirement for Stat4 in interleukin-12-mediated responses of natural killer and T cells. *Nature* 382:171–174.

90. van Boxel-Dezaire, A. H., M. R. Rani, and G. R. Stark. 2006. Complex modulation of cell type-specific signaling in response to type I interferons. *Immunity* 25:361–372.

91. Vannucchi, S., M. V. Chiantore, G. Mangino, Z. A. Percario, E. Affabris, G. Fiorucci, and G. Romeo. 2007. Perspectives in biomolecular therapeutic intervention in cancer: from the early to the new strategies with type I interferons. *Curr. Med. Chem.* 14:667–679.

92. Vidal, S. M., and L. L. Lanier. 2006. NK cell recognition of mouse cytomegalovirus-infected cells. *Curr. Top. Microbiol. Immunol.* 298:183–206.

93. Wang, J., N. Pham-Mitchell, C. Schindler, and I. L. Campbell. 2003. Dysregulated Sonic hedgehog signaling and medulloblastoma consequent to IFN-alpha-stimulated STAT2-independent production of IFN-gamma in the brain. *J. Clin. Invest.* 112:535–543.

94. Whitmire, J. K., J. T. Tan, and J. L. Whitton. 2005. Interferon-gamma acts directly on CD8+ T cells to increase their abundance during virus infection. *J. Exp. Med.* 201:1053–1059.

95. Yan, H., K. Krishnan, A. C. Greenlund, S. Gupta, J. T. Lim, R. D. Schreiber, C. W. Schindler, and J. J. Krolewski. 1996. Phosphorylated interferon-alpha receptor 1 subunit (IFNaR1) acts as a docking site for the latent form of the 113 kDa STAT2 protein. *EMBO J.* 15:1064–1074.

96. Yoshimura, A., T. Naka, and M. Kubo. 2007. SOCS proteins, cytokine signalling and immune regulation. *Nat. Rev. Immunol.* 7:454–465.

97. Zhang, X., S. Sun, I. Hwang, D. F. Tough, and J. Sprent. 1998. Potent and selective stimulation of memory-phenotype CD8+ T cells in vivo by IL-15. *Immunity* 8:591–599.

Cellular Signaling and Innate Immune
Responses to RNA Virus Infections
Edited by A. R. Brasier et al.
© 2009 ASM Press, Washington, D.C.

Bahram Razani, Arash Shahangian,
Beichu Guo, Genhong Cheng

Biological Impact of Type I Interferon Induction Pathways beyond Their Antivirus Activity

11

When we consider the host response to viral infection, we immediately begin asking questions such as how viruses initiate this response and how the activated signaling pathways result in an orchestrated defense against viral infection. Indeed, the other chapters in this section deal with just that, the cellular sensors of viral infection and the signaling pathways they subsequently activate. It is important to consider, however, that this powerful host response occurs within the context of an entire organism composed of multiple overlapping systems and cellular signaling pathways. And while antiviral signaling may be a physiological response to a pathological stimulus, it is becoming increasingly clear that these signaling pathways can modulate systems having very little to do with protection from viral infection and in doing so can themselves potentiate or protect from significant pathologies.

In this chapter we explore three categories in which antiviral signaling, either activated by virus or active within a noninfectious context, can significantly modulate pathology. And although at first glance it might seem surprising that diseases as diverse as lupus, atherosclerosis, or postviral pneumonia might possess within their complex etiologies the common thread of antiviral signaling, the lesson to draw would be an enhanced appreciation of how powerful the host response to virus is and how much mammalian physiology is invested in fighting off infection. After briefly reviewing antiviral signaling, we summarize recent findings that investigate the connection between the host antiviral and antibacterial responses and in so doing have begun to explain long-known clinical observations that viral infections can affect a patient's susceptibility to bacteria. In the next section we discuss how antiviral signaling can modulate the adaptive immune system and thereby result in enhanced, or in certain cases reduced, autoimmunity and inflammation. Finally, we explore the profound metabolic consequences of the host response to viral infection and specifically elaborate how such signaling can cross talk with nuclear receptor (NR) pathways critical for metabolism of lipids and xenobiotics, leading to atherosclerosis and hepatic pathology, respectively.

TYPE I INTERFERONS AS ANTIVIRAL AGENTS

The innate immune system is the first line of defense against microbes in vertebrates (87). It also has powerful influence over resultant adaptive immune responses.

Bahram Razani, Arash Shahangian, Beichu Guo, and Genhong Cheng, Department of Microbiology, Immunology and Molecular Genetics, University of California, Los Angeles, Los Angeles, CA 90095.

Induction of innate immunity is triggered by microbial components termed pathogen-associated molecular patterns (PAMPs), which can be viewed as a molecular "signature" of the invading pathogens (3, 5, 16, 79, 99, 183, 191, 195). Invading pathogens are recognized by a large and diverse family of pattern recognition receptors (PRRs) that survey extracellular, phagolysosomal, and intracellular milieus. Toll-like receptors (TLRs) are a major family of PRRs that are mainly expressed by cells of the immune system. TLRs sense invading pathogens by recognizing pathogen-associated PAMPs extracellularly or following phagocytosis in membrane-bound compartments of the cell (4, 16, 99, 195). Many pathogens, such as viruses and some bacteria, are not limited to these locations and spend a major aspect of their life cycle inside the cytoplasm of infected cells. Most if not all nucleated cells are capable of recognizing and responding to cytoplasmic infection following recognition by a diverse family of caspase activation and recruitment domain (CARD)-containing proteins that include nucleotide oligomerization domain (NOD) and retinoic acid-inducible gene I (RIG-I) family helicases (35, 95, 116, 125, 150, 160, 172, 183, 191, 214).

Induction of type I interferon (IFN) genes is achieved by a smaller subset of these PRRs that employ two IFN regulatory factor (IRF) family members, IRF3 and 7 (79, 94, 159, 181, 183, 196). Type I IFNs consist of a large family of closely related genes and include one gene for IFN-β, -ω, -κ, and -ε and several genes for IFN-α proteins (57, 121, 161, 182, 191, 196, 198). Activation of TLR 7/8 and 9 by single-stranded RNA and hypomethylated DNA, respectively, is one of the type I IFN-inducing pathways that is mainly employed by conventional dendritic cells (DCs) and plasmacytoid DCs (pDCs). In these cells, activation of the TLRs leads to recruitment of myeloid differentiation factor 88 (MyD88), which results in activation of IRF7 and nuclear factor-κB (NF-κB) transcription factors and subsequent induction of type I IFNs. Macrophages, however, rely on a completely different subset of TLRs for production of type I IFNs. Lipopolysaccharide (LPS) and double-stranded RNA, respectively, bind to and activate TLR3 and TLR4, culminating in recruitment of another MyD88 family adaptor molecule, Toll–IL-1 receptor domain-containing adaptor inducing IFN-β (TRIF), which results in activation of the type I IFN response through activation of IRF3 (16, 45, 81, 94, 183, 195). Of the cytoplasmic receptors described, only RIG-I and melanoma differentiation-associated gene 5 (MDA-5) are known to induce type I IFNs in response to viral RNA products (30, 78, 82, 97, 98, 124, 213–215). This pathway, which possibly exists in all cells of the host, results in production of type I IFNs following recruitment of the adaptor molecule IPS-1, also termed

Cardif/MAVS/VISA, with subsequent downstream activation of IRF3 (80, 85, 91, 100, 109, 135, 184, 211, 213). While NF-κB pathway activation requires the presence of tumor necrosis factor (TNF) receptor-associated factor 6 (TRAF6), recently it has become clear that the IRF/IFN arm of each of these pathways requires TRAF3 as well (70, 72, 147, 174). While rapid progress has been made on the function and signaling of TLRs along with intracellular RNA detection receptors, the mechanisms by which cells detect cytoplasmic DNA remain unclear. Cytoplasmic bacterial, viral, and possibly endogenous DNAs do indeed induce type I IFNs and other immune responses during infection or tissue damage. A recent report by Takaoka et al. indicates that DAI (DNA-dependent activator of IFN regulatory factors), a Z-DNA-binding protein, might be a cytosolic DNA sensor to initiate induction of type I IFN and innate immunity responses (194). However, further studies are needed to verify DAI's function in vivo and to identify other cytosolic DNA sensors.

Viral replication is potently inhibited by actions of type I IFNs in a cell-autonomous manner. Upon binding to the receptor, the type I IFN receptor 1 and 2 (IFNAR1:IFNAR2), Janus kinase (Jak)/signal transducer and activator of transcription (Stat) signaling pathway is activated, resulting in phosphorylation of Stat1 and Stat2. Stat1 can form homodimers, and upon translocation into the nucleus it can bind promoter IFN-γ-activated site (GAS) elements. Stat2, which does not seem to homodimerize effectively, heterodimerizes with Stat1. The final complex, which also includes IRF9, is dubbed IFN-stimulated gene factor 3 (ISGF3) and translocates to the nucleus and specifically binds IFN-stimulated respsonse element (ISRE) elements in the promoter of target genes (18, 57, 89, 121, 161, 183, 190, 191, 196, 198, 219). MX1, protein kinase R (PKR), RNase L, and IFN-stimulated gene 15 (ISG 15) are some of the IFN gene targets that are known to directly inhibit viral replication by various means such as inhibition of translation and viral RNA degradation. However, the antiviral activity of the vast majority of antiviral genes remains unknown (18, 27, 73, 121, 125, 161, 182, 190, 191, 196).

TYPE I IFNs AND BACTERIAL INFECTIONS

Viral infections are not unique in their ability to induce type I IFNs. TLR4 and 9, which respectively recognize LPS and bacterial hypomethylated DNA, culminate in production of type I IFNs when activated. Furthermore, bacterial DNA has been shown to induce the type I IFN response in an IPS-1-dependent manner. While for many years it had been thought that the effects of type I IFNs were purely antiviral, they are now deemed an integral

Table 1 Effects of type I interferons on host response to bacterial infection

Agent	Life cycle	Described effect of type I IFN (reference)	Comments
L. monocytogenes	Facultative intracellular (enteroinvasive)	Sensitizing in systemic infections (12, 36, 146a)	Mechanism attributed to increased apoptosis of effector cells
S. pneumoniae	Extracellular (lungs)	Protective in systemic and intrathecal infections (127)	Unpublished data (A. Shahangian et al.) suggest that type I IFN may sensitize host to secondary pneumococcal pneumonia
M. tuberculosis	Facultative intracellular (lungs)	Protective (68) and sensitizing (126)	Type I IFN successfully used as therapeutic
Salmonella enterica serovar Typhimurium	Facultative intracellular (enteroinvasive)	Protective (32, 64)	Direct in vivo data lacking
L. pneumophila	Facultative intracellular (lungs)	Protective (151, 180)	In vivo data lacking
Chlamydia spp.	Obligate intracellular (lungs/mucosal membranes)	Protective (33, 34, 180, 185)	In vivo data lacking
LPS		Sensitizing (96)	Data based on mice lacking IFN-β

part of immune responses that link innate and adaptive immunity and hence play a crucial role in clearance of many bacterial infections as well. Type I IFNs induce expression of several well-studied proteins that have been shown to possess antibacterial activity. These include inducible nitric oxide synthase and indoleamine 2,3-dioxygenase (IDO), which target bacteria by production of reactive oxygen species and deplete local stores of tryptophan. Type I IFNs also aid in bacterial clearance by augmenting natural killer- and T-cell activity, B-cell maturation, and isotype switching (55, 107, 112, 113, 115, 129, 148, 149, 202, 212). Despite this, it has been shown that type I IFNs may inhibit certain aspects of innate and adaptive immune responses by inducing apoptosis of effector cells (36, 217). While mostly protective, it seems that in certain specific examples, type I IFNs may paradoxically sensitize the host to bacterial infections (Table 1). Therefore, the role played by type I IFNs in the context of bacterial infections remains debatable. This section will summarize the literature for some of the most clinically pertinent bacterial pathogens that have been studied in the context of type I IFNs.

Gram-Positive Bacterial Infections

Listeria monocytogenes

Listeria is a gram-positive facultative intracellular pathogen. It is unusual for containing LPS, although its unique LPS is not recognized by TLR4. While commonly found in dairy products and undercooked meats, it is often of little clinical consequence to normal healthy individuals. However, it can cause fulminant and life-threatening infections in immunocompromised individuals, and infection during pregnancy can result in septic abortions. Once phagocytosed, *L. monocytogenes* expresses listeriolysin O (LLO), a pore-forming protein, which allows it to escape the phagolysosome and enter the cytoplasm (47). While the cellular receptor responsible for recognition of *L. monocytogenes* is not known, it is interesting to note that LLO mutants, which are incapable of escaping the phagosome, do not induce type I IFNs. Cytoplasmic escape of *L. monocytogenes*, nonetheless, results in activation of an unknown receptor that exclusively relies on the TANK binding kinase (TBK)/IRF3 axis to induce type I IFNs. Induction of type I IFNs, however, seems to have a detrimental effect on the host's ability to clear systemic infection. Intravenous or intraperitoneal infections of *Ifnar*[-/-] mice with *L. monocytogenes* are associated with increased survival and lower bacterial burdens in spleens and livers as compared with similarly infected wild-type mice (12, 36, 43). Further evidence for the detrimental role of type I IFNs during *L. monocytogenes* infections is provided by experiments showing that treatment of mice with the TLR3 ligand poly(I:C) increases sensitivity of wild-type mice to *L. monocytogenes* infection in a manner completely dependent on the presence of the type I IFN receptor (43). The mechanisms by which this enhanced sensitivity occurs have been attributed to the increased apoptosis of immune effector cells possibly brought on by IFN-mediated up-regulation of TNF-related apoptosis-inducting ligand (TRAIL). Indeed, TRAIL-deficient mice have increased resistance to *L. monocytogenes* infection (217). However, it is unclear to what extent the enhanced resistance of IFNAR-deficient mice to *L. monocytogenes* depends on the reduced apoptosis observed in these mice.

Mycobacterium tuberculosis

Mycobacterium tuberculosis is another facultative intracellular bacteria and is the causative agent for tuberculosis. *M. tuberculosis* infects millions of individuals worldwide but only becomes symptomatic in a small fraction of these individuals. However, active infections are often difficult to treat, requiring antibiotic therapy for several months which is further complicated by the emergence of antibiotic-resistant strains. *M. tuberculosis* mainly infects macrophages of the host, where upon uptake it prevents acidification and can persist chronically. Infection with *M. tuberculosis* does result in production of type I IFNs in infected cells. However, potential effects of type I IFN production on host immunity are complex. Previous studies have shown that certain hypervirulent strains of *M. tuberculosis* do induce higher levels of type I IFNs, and others have shown than treatment of mice with type I IFNs favored bacterial growth (126). However, studies with *Ifnar*$^{-/-}$ mice suggest that type I IFN signaling is not as detrimental to outcome as other studies suggest (46). These findings are further complicated by clinical studies that have successfully used type I IFNs to treat fulminant tuberculosis (68). These studies paint a complex picture about the role of type I IFNs in *M. tuberculosis* infections, and the variable results are perhaps confounded by temporal and system- and strain-specific variables.

Streptococcus pneumoniae

Streptococcus pneumoniae, which is a common causative agent for bacterial pneumonia, accounts for significant morbidity and mortality worldwide. Interestingly, bacterial pneumonias are very common in the context of a primary influenza infection; in fact, the overwhelming majority of morbidities during an influenza infection are attributable to secondary bacterial pneumonia, most commonly due to infection with *Sp. pneumoniae.* Type I IFNs have been shown to play a protective role against this common pulmonary bacterial pathogen. Exogenous administration of type I IFNs reduces bacterial burdens in the lungs of infected animals, and *Ifnar*$^{-/-}$ mice were found to be more sensitive to intravenous and intrathecal pneumococcal infections (127, 209). Nevertheless, *Ifnar*$^{-/-}$ mice appear to be equally resistant to pulmonary

challenges with *S. pneumoniae* as their wild-type counterparts, suggesting that type I IFN signaling may be required for host survival once the infection reaches the blood or crosses the blood-brain barrier. To complicate matters even further, mouse models of influenza and secondary pneumococcal pneumonia suggest that maximal susceptibility to pneumococcus occurs on days that the type I IFN response to the primary viral infection is maximal. Studies done in our laboratory suggest that in the context of a primary influenza infection, type I IFNs strongly inhibit antibacterial responses, leading to markedly enhanced sensitivity to bacterial infections. However, it remains unclear why on the one hand type I IFNs can protect mice and on the other hand they lead to enhanced susceptibility. Less is known about the role of type I IFNs in the context of other secondary bacterial pneumonias such as those caused by *Staphylococcus aureus* and *Klebsiella pneumoniae.*

Gram-Negative Bacterial Infections

Salmonella and Shigella

Similar to *L. monocytogenes,* discussed above, both of these pathogens interact with the host initially through the digestive tract following consumption of contaminated materials. Both of these pathogens cause dysentery and are highly invasive. Type I IFNs are thought to play a protective role against infection with both of these pathogens, by reducing epithelial cell invasion and augmenting IFN-γ production in splenocytes (32, 64, 143). It remains unclear, however, whether inhibition of invasion would also negatively alter clearance of these pathogens by innate immune effector cells such as macrophages and neutrophils that phagocytose bacteria before killing. No in vivo studies have been performed to date examining the effects of absence of type I IFN signaling on the outcome of enteric infections.

Other Gram-Negative Infections

Legionella pneumophila, the causative agent of Legionnaires' disease, can cause fulminant pneumonias and ultimately death of the patient. *L. pneumophila* is an environmental pathogen that replicates within amoebas. Once an individual is infected through inhalation of contaminated water aerosols, *L. pneumophila* invades its preferred host, macrophages, where it reproduces in vacuoles. Infection of macrophages leads to production of type I IFNs through an IPS-1-mediated pathway which in turn protects cells from infections (151). Pulmonary clearance of *L. pneumophila* is augmented by exogenous administration of type I IFNs independent of IFN-γ expression (180).

Chlamydia was one of the first bacterial species that was shown to be inhibited by type I IFNs. Similar to most gram-negative bacteria, various species of this pathogen induce type I IFNs through a TLR4/MyD88-dependent pathway. Studies with this bacterium using in vitro models have suggested a protective role for type I IFNs. The only in vivo models with *Chlamydia* infections have focused on mice deficient for type II IFN, and experiments on the effects of absence of type I IFN signaling on the outcome of *Chlamydia* infections are lacking. With this said, at least one prior study has shown that administration of the interferon-inducing agent poly(I:C) delayed mortality of mice infected with *C. trachomatis*, and it has been suggested that type I IFNs inhibit *Chlamydia* by upregulating IDO and degrading the amino acid tryptophan, thereby inhibiting bacterial replication (33, 34, 185).

Septic Shock

Gram-negative sepsis is a common cause of morbidity and mortality in the United States. The transition of this disease from a localized infection to bacteremia and sepsis occurs rather rapidly, and multiorgan failure due to the ensuing cytokine storm is not uncommon. Type I IFNs seem to play a major role in sensitization of animals to septic shock. Indeed, IFN-β-deficient mice are more resistant to LPS-induced septic shock despite normal production of inflammatory molecules such as tumor necrosis factor-α and IFN-γ (96). This finding is supported by another report showing that IRF3-deficient mice are more resistant to challenge with bacterial LPS (176). However, since induction of the type I IFN response is an early feature of sepsis, it remains to be seen whether neutralization of type I IFNs after initiation of septic shock can be of any therapeutic value.

Type I IFNs have traditionally been thought of as antiviral immune modulators. By induction of a wide variety of antiviral genes, they inhibit viral replication at various stages and induce proapoptotic genes that limit the spread of the virus in infected cells. However, over the past few years it has become exceedingly clear that type I IFNs are indeed more versatile than once thought and that they control various aspects of innate and adaptive immune responses. As an example, type I IFNs result in upregulation of the IDO and nitric oxide synthetase, both of which have been shown to be important for innate immune-mediated killing of phagocytosed bacteria. Type I IFNs have been shown to induce expression of IFN-γ in an interleukin-12 (IL-12)-independent manner. Cross presentation by professional antigen-presenting cells, which is an important aspect of developing cytotoxic-T-cell responses against intracellular bacterial pathogens, is only induced by activation of

type I IFN signaling but not IL-12. Furthermore, it has been shown that adjuvant-mediated class switching requires intact type I IFN signaling.

Given the complex and sometimes contradictory effects of type I IFN signaling on generation of immune responses, it is not surprising that the literature is often inconclusive in determining the importance of type I IFNs for bacterial infections. Elucidation of these complex cross-regulatory pathways, however, will perhaps allow investigators to ask more pertinent questions with regard to the effects of type I IFNs on bacterial responses.

TYPE I IFNs AND AUTOIMMUNITY

Innate immunity is the first line of defense against infection; however, uncontrolled activation of the innate immune system may facilitate or initiate harmful inflammatory or autoimmune responses. TLR signaling is essential for the induction of autoimmune diseases in a number of experimental models. For example, normal and lupus-prone mice can produce autoantibodies when repeatedly challenged with TLR ligands such as LPS, poly(I:C), and CpG (8, 29, 156). While TLRs and intracellular recognition receptors play a critical role in detecting pathogens and initiating the innate immune response, their function in recognizing endogenous ligands, and the very identity of these endogenous ligands, is an ongoing debate. Nevertheless, accumulating evidence supports the idea that TLRs also can sense endogenous danger signals generated from tissue injury or apoptotic cells, although the mechanisms remain unclear (130, 170). A well-studied example is that of immune complexes containing endogenous DNA/RNA and autoantibodies that can be recognized by TLR9 in DCs and B cells (28, 56, 77, 119). The interaction of these immune complexes with TLR9 has been shown to promote systemic autoimmune diseases such as systemic lupus erythematosus (SLE). In this section we will focus on the role of type I IFNs in promoting and inhibiting autoimmunity.

In addition to their antiviral functions, type I IFNs are capable of exerting immunomodulatory effects on both innate and adaptive immune cells. Type I IFNs have been shown to promote DC maturation, expression of major histocompatibility complex molecules, immunoglobulin (Ig) production from B cells, and induction of cytokines (48, 49, 120, 200). Since they are pleiotropic cytokines, studies have shown that they can promote or suppress autoimmune diseases, depending on the nature of immune disease and timing of IFN administration or induction (Fig. 1). Because of their ability to induce Ig isotype switching, type I IFNs are associated with autoantibody production and contribute to the pathogenesis of SLE

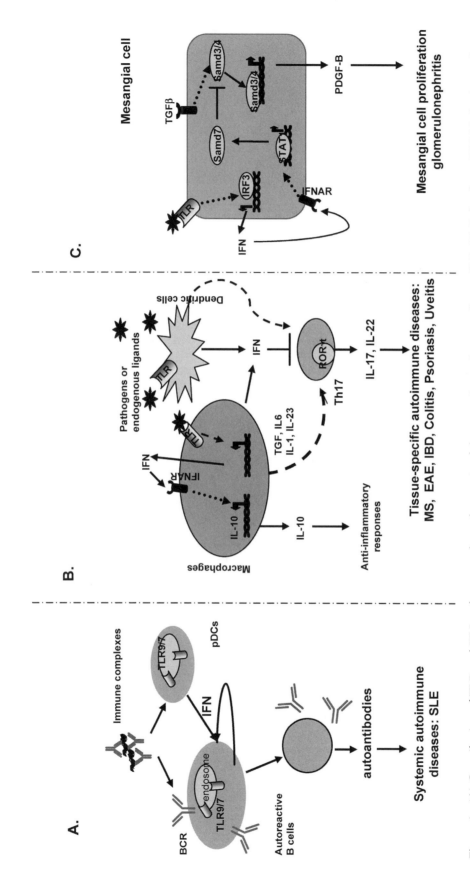

Figure 1 (A) Contribution of TLR and IFN pathways to the pathogenesis of the systemic autoimmune disease SLE. (B) Role of type I IFN induction pathway in the anti-inflammatory response and in suppression of Th17-mediated, tissue-specific autoimmune diseases. (C) Cross talk between type I IFN and TGF-β in regulating mesangial cell proliferation and glomerulonephritis. IBD, inflammatory bowel disease.

(48, 49, 130, 200). On the other hand, types I IFNs have been shown to function as potent anti-inflammatory cytokines. Clinically, systemic administration of IFN-β has been used to treat patients with multiple sclerosis (MS) (62, 158, 203). In this regard, SLE and MS may represent two opposite examples among a spectrum of autoimmune diseases affected by type I IFNs. Therefore, this section will discuss recent progress in understanding the function of type I IFNs in SLE, MS, and other autoimmune diseases.

IFN-Mediated Autoimmunity and SLE

Type I IFNs have been reported to exacerbate or attenuate several autoimmune diseases. The contribution of IFNs to the pathogenesis of certain autoimmune diseases has been well demonstrated in SLE. Early studies found that serum levels of IFNs correlated with disease activity in lupus patients. In addition, repeated type I IFN treatment leads to development of autoantibodies and in some cases SLE-like symptoms in certain patients with unrelated diseases (48, 49, 169). Strikingly, global gene expression analysis revealed increased expression of IFN-α/β-induced genes, or an "IFN signature," in cells from patients with active lupus (13, 23, 50). The detrimental effects of IFN-α/β on SLE also have been demonstrated in several animal models of lupus. For instance, treatment with IFN-α/β, as well as with TLR ligands, promotes lupus-like autoimmune diseases in susceptible animals such as New Zealand and BL/6-Faslpr mice. In contrast, deficiency of IFNAR in mice leads to reduced manifestation of lupus disease, including reduced levels of autoantibody immune complexes and decreased numbers of autoantibody-producing B cells. Furthermore, IFNAR-deficient mice also exhibit reduced frequencies of IFN-γ-producing T cells, as well as decreased maturation of DCs. These studies demonstrate the critical role of IFN-α/β in the pathogenesis of lupus (51, 56, 76, 104, 140, 156, 178).

Recent studies have begun to shed light on the mechanisms behind the increased type I IFN levels found in SLE patients. It has long been observed that the sera of patients with SLE frequently have nucleic acid-containing complexes constituted of cellular RNA, DNA, or nuclear ribonucleoproteins, as well as high titers of autoantibodies of the IgG isotype specific for them. These immune complexes have been demonstrated to activate innate immune cells. Particularly nucleic acid-containing immune complexes are potent stimuli for IFN-α production from pDCs (21, 49, 130, 156, 162). pDCs are referred to as natural IFN-producing cells because they produce large amounts of type I IFNs in response to viral or bacterial infection. One of the underlying mechanisms for efficient IFN production by pDCs is the constitutive expression of TLR9 and TLR7, which recognize hypomethylated CpG DNA and single-stranded RNA, respectively. Recent studies have shown that lupus IgG mixed with apoptotic or necrotic cell debris can activate pDCs to produce type I IFNs, and this response depends on TLR9/7 (21, 110, 111, 162, 171). Although the mechanism by which pDCs recognize endogenous nucleic acid-containing immune complexes is not completely clear, experimental evidence suggests that FcR expressed on the surface of pDCs binds and internalizes IgG/nucleic acid immune complexes, thereby delivering nucleic acid molecules to TLR9 and TLR7 in intracellular compartments (8, 14, 21, 132, 162).

Recent studies have demonstrated that autoreactive B cells can be activated by chromatin-antibody immune complexes through dual engagement of B-cell receptor (BCR) and TLR9 (110, 111). Similar to pDCs, B cells also express TLR9 and TLR7/8. Since TLR9 and TLR7 are sequestered in cytoplasmic compartments, endogenous ligands gain access to these compartments through the action of autoreactive BCR. Self-antigens, such as cellular DNA/RNA/nuclear proteins, are recognized by BCR expressed on self-reactive B cells and transported to endosome compartments that contain TLR9 and TLR7, which leads to the activation of self-reactive B cells. Endogenous immune complexes containing autoantibodies and RNA or DNA, possibly released by apoptotic cells, are commonly found in the serum of SLE patients. In normal conditions, mammalian DNA and RNA are very weak stimuli for TLRs; however, nucleic acids in immune complexes become potent endogenous TLR7 and 9 ligands, probably through uncharacterized modifications during apoptotic processes or via interaction with some protein components.

IFN-α/β have various effects on B-cell function. Experimental evidence revealed that IFN-α/β increase B-cell activation upon B-cell antigen receptor stimulation, enhance B-cell survival, and promote Ig isotype switching and antibody production (49, 77, 110, 148, 165, 200). The lupus-promoting effects of IFN-α/β were also verified in susceptible mice lacking IFNAR, which have reduced autoantibody-secreting B cells and IgG levels and decreased immune complex deposits (104, 156, 177). These studies indicate that IFN-α/β play a pivotal role in the pathogenesis of SLE. Interestingly, our results also show that B cells are able to produce type I IFNs in response to TLR stimulation (148). Unlike in DCs, B-cell-mediated type I IFN induction depended on the transcription factor IRF3. Utilizing type I IFN receptor-deficient mice, this study demonstrated that the IFN pathway enhanced B-cell differentiation and antibody production and was required for IgG2a production following CpG-DNA stimulation. Therefore, CpG-DNA and possibly cellular RNA/DNA immune complexes can directly activate B cells in an IFN-α/β-dependent manner and induce them to differentiate into plasma cells, producing autoantibodies. The unique type I IFN induction pathway in B cells may play an important role in mediating B-cell function and potentially in autoantibody production and the pathogenesis of autoimmune diseases like SLE.

IFN-Induced IL-10 Production in the Anti-Inflammatory Response

While type I IFNs have been suggested to contribute to the pathogenesis of SLE, both experimental and clinical studies show that they play protective roles in MS (158, 199, 203). In fact, IFN-β administration is an effective therapy for the treatment of MS patients. In addition to MS, a number of other human autoimmune diseases, including Sjögren syndrome, inflammatory bowel disease, chronic recurrent multifocal osteomyelitis, thrombocytopenic purpura, Behcet's disease, and rheumatoid arthritis, have been reported to benefit from treatment with type I IFNs. Although multiple mechanisms may be responsible for the beneficial or therapeutic effects of type I IFNs, induction of anti-inflammatory cytokines may represent a potential mechanism to limit inflammation and suppress autoimmune responses.

Innate immune cells respond to bacterial or viral infection by the rapid activation of proinflammatory cytokines as host defense mechanisms against microbial invasion. However, proinflammatory cytokines also contribute to disease pathogenesis by inducing inflammation and immune injury that are harmful to the host. Very likely, endogenous TLR ligands could induce similar consequences. To avert these deleterious effects, TLRs also initiate negative-feedback mechanisms to suppress inflammatory responses by producing anti-inflammatory cytokines such as IL-10. The essential role for anti-inflammatory cytokines in maintaining immune homeostasis is evident in IL-10-deficient mice, which have increased endotoxemia-associated mortality during microbial infection and increased susceptibility to autoimmune disorders such as inflammatory bowel disease and autoimmune encephalomyelitis (84, 193).

Recent studies in our lab provide the connection between type I IFNs (IFN-α/β) and IL-10 production upon TLR activation (131). We found that the type I IFN (IFN-α/β) induction pathway is required for LPS-induced IL-10 upregulation, and IFN-β was able to induce IL-10 production via the IFN-α/β receptor. The results showed that IFN-β-blocking antibody can inhibit IL-10 mRNA upregulation after LPS stimulation. In addition, IFNAR- or IRF3-deficient cells exhibited reduced IL-10 production

in response to TLR4 activation. Furthermore, defects in type I IFN production and induction signaling pathways led to significant upregulation of LPS-induced proinflammatory chemokines and cytokines in macrophages. These results indicate that, in addition to its antiviral functions, the type I IFN pathway may serve a novel anti-inflammatory role in regulating the TLR-mediated innate immune response.

Type I IFNs, Th17 Cells, and MS

Because type I IFN signaling can induce potent anti-inflammatory cytokines, systemic administration of IFN-β has been used to treat patients with a number of autoimmune diseases, particularly MS. MS is a chronic autoimmune disease characterized by the infiltration of immune cells, including macrophages and T cells, into the central nervous system (CNS) that results in the destruction of neuronal myelin sheaths (62, 158, 203). Currently, there are no curative treatments for MS. However, IFN-β has proven to be effective for the treatment of MS, as demonstrated by decreased inflammatory lesion formation in the CNS and lower relapse rates. Although the mechanisms responsible for its therapeutic effects remain unclear, these clinical data implicate the potential involvement of endogenous IFN-α/β in regulating inflammatory responses in the CNS.

MS and experimental allergic encephalomyelitis (EAE), an animal model of MS, were previously thought to be mediated by T-helper type 1 (Th1) cells. However, a number of recent studies provide strong evidence that IL-17-producing T cells play a dominant role in the pathogenesis of EAE (24, 65, 103, 106). Recently, this new T-helper subset that produces IL-17 was identified and termed Th17; it has a distinct phenotype, transcriptional control, and function (65, 75, 103, 105, 146, 153, 189, 192). Th17 cells have been shown to play a critical role in various human autoimmune diseases, such as adjuvant-induced arthritis and MS. The differentiation of Th17 cells is promoted by transforming growth factor-β (TGF-β), IL-6, and IL-21 in mice and by IL-6 and IL-1 in humans. While a number of cytokines have been shown to promote Th17 differentiation or expansion in combination with IL-6/TGF-β, studies from other groups and ours demonstrated that IFN-α/β can inhibit expansion of Th17 cells. Using EAE as a model, we also demonstrated that the type I IFN induction pathway constrains the differentiation of Th17 cells both in vitro and in vivo, as well as development of EAE.

Cross Talk between IFNs and TGF in Glomerulonephritis

Glomerulonephritis is characterized by inflammation in the glomeruli of the kidney. Immune-mediated glomerulonephritis, such as lupus nephritides and IgA nephropathy, is triggered by immune complexes deposited from the circulation or generated from the interaction of antibodies with glomerular antigens or antigens trapped in the glomerular area (2, 9). The presence of immune complexes leads to infiltration of inflammatory leukocytes into the glomerulus or tubulointerstitium. These events also trigger resident mesangial cell activation, proliferation, and release of mediators that further promote inflammation and sclerosis. Platelet-derived growth factor (PDGF) and TGF-β are important mediators of mesangial cell proliferation and the accumulation of extracellular matrix, ultimately leading to sclerosis (142, 173, 208). However, the role of type I IFNs in this process is less clear.

Recent data in our lab reveal a novel mechanism by which TLR agonists utilize TGF and type I IFN autocrine/paracrine loops to differentially regulate the induction of PDGF-B, and possibly glomerulonephritis (43). These results show that CpG induces mesangial cell proliferation through its induction of PDGF-B. Further, CpG acting through TGF-β autocrine/paracrine signaling leads to activation of Smad3/4, which, in combination with NF-κB, synergistically induces the transcription of PDGF-B. These data complement in vivo studies that demonstrate activation of the innate immune system with specific nucleic acid agonists of TLR9, leading to induction of lupus nephritis in genetically predisposed mice (8, 14, 76).

In contrast to CpG, poly(I:C) inhibits CpG-induced Sma- and Mad-related protein 3/4 (Smad3/4) activation, PDGF-B upregulation, and proliferation of mesangial cells (43). Type I IFNs were found to be critical for negative regulation of TGF-β signaling, PDGF-B production, and mesangial cell proliferation by poly(I:C). Indeed, CpG-induced PDGF-B upregulation in macrophages from IFNAR1-deficient mice is not affected by poly(I:C). Moreover, the repression of TGF-β-mediated Smad3/4 activation occurs through IFN/Stat-dependent induction of Smad7, which is one of the inhibitory Smads that can repress TGF signaling. These results demonstrate that the inhibition of TGF-β signaling by poly(I:C) depends upon the TLR3-mediated type I IFN autocrine/paracrine loop. In summary, these results demonstrate the importance of cross talk between TGF-β and type I IFNs in determining gene expression critical for mesangial cell proliferation. The results from this study may also suggest the possibility of IFN-mediated immune therapy for glomerulonephritis. We speculate that while systemic induction of or administration of type I IFNs leads to lupus-like symptoms, tissue- or cell-specific induction of type I IFNs may represent an attractive and effective therapy to limit inflammation and glomerulonephritis.

IFNs AND THE METABOLIC SYSTEM

Multiple lines of evidence have now established that the metabolic system is profoundly affected during the mammalian host response to infectious challenge (15, 19, 20, 31, 44, 101). While it is unclear whether such modulation of metabolism is ultimately beneficial to the host, a number of recent studies have suggested that this cross talk between the immune and metabolic systems contributes to the etiological basis of several human diseases (39, 43, 164). In this section we will specifically cover examples in which IFN activation represses important metabolic functions of macrophages and hepatocytes, leading to atherosclerosis and significant liver pathology, respectively, in mouse models. Before delving into such examples, we will briefly review nuclear hormone receptors, critical modulators of mammalian metabolic function, and the main targets of innate immune-mediated repression of the metabolic system.

Nuclear hormone receptors represent one of the largest families of mammalian transcription factors and are the molecular regulators of a wide variety of metabolic functions. They are basally held in an inactive conformation by various heat shock proteins (HSPs) and often sequestered in the cytosol, and binding of their hydrophobic ligands results in shedding of associated HSPs, dimerization, and transport to the nucleus, where active dimers bind particular DNA consensus sequences. Binding of these so-called nuclear receptor response elements results in the recruitment of a wide variety of transcriptional co-activators and ultimately the basic transcriptional machinery, leading to strong transactivation of target genes (167). Nuclear receptors (NRs) are categorized into a number of subfamilies largely on the basis of dimer partner preference and knowledge of the activating hormones. Interestingly, those NRs which have so far been shown to be modulated by the innate immune system are those of the type II NR subfamily, all of which effect their transactivation solely through obligatory heterodimerization with the promiscuous NR retinoic X receptor (RXR) and tend to be activated by low-affinity ligands frequently endogenous in most cells. These type II NRs are contrasted by type I NRs, which homodimerize upon binding by high-affinity ligands such as estrogen, testosterone, or glucocorticoids that are frequently produced by only particular organs (41).

While NRs are primarily thought of in their role as transactivators, nontranscriptional roles for these proteins are increasingly being appreciated. Interestingly, NR modulation of the immune system appears to be primarily by means of these nontranscriptional mechanisms and is generically termed trans-repression. While the anti-inflammatory effects of corticosteroids, acting through glucocorticoid receptor (GR), have long

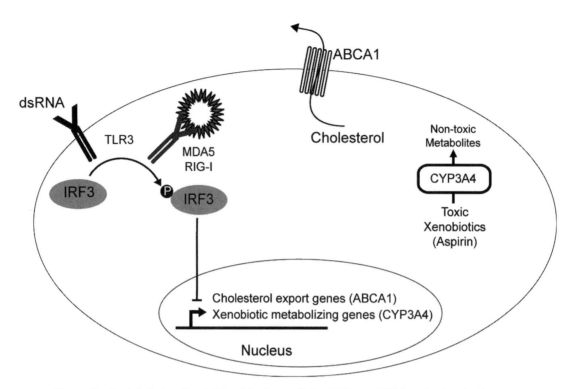

Figure 2   Antiviral signaling initiated by intracellular PRRs or TLR3 activation leads to an IRF3-mediated inhibition of LXR- and RXR-dependent gene transcription. Important targets of these NRs include cholesterol exporters in macrophages and aspirin-metabolizing enzymes in hepatocytes, respectively, which can enhance atherosclerosis or the aspirin-associated postviral hepatitis characteristic of Reye's syndrome. dsRNA, double-stranded RNA.

been known, recent work has also demonstrated similar roles for the RXR-partnering peroxisome proliferator-activated receptors (PPARs), liver X receptors (LXRs), and vitamin D receptor (VDR) (90, 92, 141, 168, 187, 216). A diverse variety of trans-repressive mechanisms have been identified that repress NF-κB-, activator protein 1 (AP-1)-, and IRF3-mediated transactivation of inflammatory target genes and are reviewed extensively elsewhere (155). Given the profound effect of NRs on the immune system, it is perhaps not surprising that cross talk in the reverse direction, that of the immune system on NR signaling, has begun to be appreciated.

That the IFN system might modulate metabolic signaling is suggested by a variety of clinical reports correlating viral or bacterial infection with metabolic dysfunction (6, 17, 54, 136, 164). Recent work in mice has now provided molecular evidence for a potential causal connection in the cases of atherosclerosis and Reye's syndrome. Interestingly, the common signaling denominator by which this cross talk occurs appears to be through IRF3 and its repression of critical metabolic functions in macrophages and hepatocytes (Fig. 2).

Atherosclerosis

The past 2 decades have seen significant revision of our etiological understanding of atherosclerosis. Although long thought to be largely a result of aberrant plasma lipid levels, work beginning in the mid-1980s began to point to an immunological component to the disease (11, 139). These immunohistochemical studies demonstrated the presence of a large variety of immune cells in atherosclerotic plaques, including macrophages, mast cells, and T cells. Several ideas have been put forth to explain the presence of this inflammatory infiltrate in the vascular wall, and while much attention has been directed toward endogenously produced, oxidized lipids as the primary culprit, a number of studies have advanced the hypothesis that viral and bacterial infection of atherosclerotic plaques may play an etiological role as well (74).

Evidence for an infectious etiology of atherosclerosis began with correlative studies demonstrating that a higher percentage of patients with atherosclerotic disease possess serological markers of infection with *Chlamydia pneumoniae* or cytomegalovirus (CMV) (26, 144, 175). Additional work using immunohistochemical staining, in situ hybridization, and PCR-based approaches demonstrated the presence of these pathogens in atherosclerotic plaques (69, 133, 134, 186). While some have cast doubt on the significance of these human studies, more recent work with mouse models has allowed for an experimental approach and has supported the notion of a causative link between infection and the development of atherosclerosis (52, 53). In these models, inoculation of hypercholesterolemic mice or rabbits with CMV or *C. pneumoniae* has been found to accelerate atherosclerotic lesion formation (138, 207). Further studies demonstrating reduced lesion development in *C. pneumoniae*-infected mice treated with antibiotics inspired a number of clinical trials, although the results from these human studies have been largely mixed (71, 137, 138).

The molecular mechanisms by which pathogen infection might enhance atherosclerotic lesion development have only recently begun to be uncovered and have focused on the ability of IRF3 to repress NR signaling critical for maintaining macrophage lipid homeostasis. That NRs are important mediators of atherosclerotic development has been confirmed through a large number of mouse studies demonstrating the disease-modifying effects of genetic NR deficiency or exogenous ligand administration. Thus far, two classes of NRs, the PPARs and the LXRs, have been implicated.

The PPAR NRs consist of three subtypes, PPAR-γ, PPAR-α, and PPAR-δ, and appear to be physiologically activated by endogenously present eicosanoids or polyunsaturated fatty acids (63, 102). Treatment of genetically disease-prone mice with synthetic high-affinity ligands for these receptors has been shown to reduce lesion development when PPAR-γ and PPAR-α have been targeted (59, 122). Furthermore, reconstitution of disease-prone mice with bone marrow deficient in each of these three receptors has been shown to impact lesion progression (40, 114, 201). A large variety of mechanisms have been proposed by which PPARs might affect atherosclerotic development, including modulation of cellular lipid uptake and efflux and changes in blood lipid levels, as well as regulation of inflammation (123). Importantly, PPAR-α and PPAR-γ have also been shown to directly transactivate expression of the LXRs, which are themselves important regulators of macrophage cholesterol homeostasis (40, 42).

Existing in two subtypes, LXR-α and LXR-β, the LXRs appear to be physiologically activated by modified forms of cholesterol known as oxysterols (88, 117). These oxysterols build up in macrophages as they take up the low-density lipoprotein (LDL) and ox-LDL present in atherosclerotic plaques and thus activate LXR-dependent gene transcription, which serves to enhance macrophage cholesterol efflux. In terms of cholesterol efflux, important targets of LXR gene transactivation appear to be a number of ABC cholesterol transporters including ABCA1, ABCG1, AGCG5, and ABCG8, which serve to transport intracellular cholesterol stores to plasma lipoproteins (166, 205, 206). Given the importance of LXR gene transcription in macrophage cholesterol efflux, it is easy to imagine how repression of LXR-dependent transactivation might lead to a pathological increase in macrophage cholesterol levels, ultimately leading to foam cell formation and thus contributing to atherosclerotic lesion development. This model has been confirmed by studies in mice using exogenous high-affinity ligand administration, which serves to decrease lesion formation, as well as reconstitution with bone marrow genetically deficient in LXRs, which leads to substantially increased atherosclerotic burden (93, 197).

It now appears that the molecular connection between pathogen infection and increased atherosclerotic burden may occur via repression of LXR signaling by IFN-activating signaling. Castrillo et al. first demonstrated that both bacterial and viral infection, along with specific TLR3 and TLR4 activation, leads to significant inhibition of LXR transactivation and thus repression of target genes, including ABC transporters, in macrophages (39). They ultimately demonstrated that this repression is dependent on IRF3 and results in decreased macrophage cholesterol efflux and increased foam cell formation. This study pointed to the possibility of competition for common coactivators as the primary mechanism by which this IRF3-dependent repression of LXR transactivation might occur. However, it appears that IRF3 activation may also serve to globally repress signaling from all RXR-heterodimerizing NRs through inhibition of RXR levels, and this seems to contribute to defects in liver metabolism following viral infection (43).

Liver Metabolism

Proper functioning of the liver's metabolic machinery is critical not only for processing of endogenously produced chemicals but also for detoxification of exogenous substances. As a result, suppressed liver metabolism can severely reduce the threshold at which exogenous toxins produce illness in mammals (10). This appears to be the case for the clinical entity known as Reye's syndrome, typified by an acute, febrile illness leading to encephalopathy and fatty degeneration of the liver, with characteristic ultrastructural damage to mitochondria. A progressive metabolic dysfunction of the liver occurs, and blood ammonia and liver enzyme levels increase in parallel with continuing hepatic damage (22). A number of clinical studies have correlated two major insults to patient physiology with the development of the syndrome (58, 67, 86, 152). They appear to be the occurrence of a viral infection concomitant with administration of aspirin, or acetylsalicylic acid. Several viruses have been implicated, including chickenpox, influenza A or B, adenoviruses, hepatitis A virus, paramyxovirus, picornaviruses, reoviruses, herpesviruses, measles, and varicella-zoster virus. Given the varied nature of these pathogens, this might suggest that the general host response to viral infection may be more contributory to the pathology than any virus-specific mechanisms. While subsequent work has questioned the validity of this clinical correlation, a sizable public health campaign in the United States during the 1980s designed to enhance awareness of the dangers of using aspirin to treat viral infections, especially among children, has been cited as a cause of reduced case reports of Reye's syndrome since then (22, 37, 38).

Recent work has lent credence to the notion that aspirin administration in the context of viral infection may lead to significant liver pathology. Metabolic by-products of aspirin, especially salicylic acid, are highly toxic to mitochondria, and their impaired processing by liver enzymes can lead to their buildup and subsequently to hepatotoxicity (131, 154, 204). Gene products important for salicylic acid clearance appear to be CYP3A4, CYP2C9, and UGT1A6, and interestingly, proper expression of these genes depends on the RXR-heterodimerizing NR pregnane X receptor (PXR) (7, 25, 60, 108, 118). Repression of PXR/RXR transcriptional activity might then be expected to severely compromise the liver's ability to metabolize salicylic acid, ultimately leading to the hepatoxicity and mitochondrial damage characteristic of Reye's syndrome.

Indeed, work by Chow et al. has shown that the murine host response to viral infection can potently inhibit PXR/RXR transactivation, resulting in downregulation of CYP3A4 and UGT1A6, ultimately leading to significant liver pathology and metabolic dysfunction. Once again, this repression of PXR/RXR signaling was shown to be dependent on the presence of IRF3, as IRF3-deficient mice were found to be highly protected against disease (43). However, the responsible mechanism appeared to be through suppression of the promiscuously heterodimerizing RXR-α. It was shown that viral infection or activation of TLR3 by cognate ligand led to a potent decrease in RXR-α mRNA levels by an IRF3-dependent mechanism. Subsequent studies demonstrated that IRF3 directly transactivated expression of the transcriptional repressor HES1, which subsequently bound the RXR-α promoter and presumably repressed RXR-α transcription. Interestingly, this led to an inhibition of not only PXR/RXR signaling but also

signaling through a number of other type II NRs, including LXR, PPAR-α, VDR, and farnesoid X receptor. This generalized repression of signaling through RXR-heterodimerizing NRs may ultimately be found to result in additional examples of viral infections leading to dysfunction of metabolic pathways, as discussed in the next section.

Before moving on, it is interesting to consider why the viral host response would repress NR signaling on such a generic scale, and indeed, work by others has shown that the acute-phase response in rodents leads to transcriptional downregulation of not only RXR-α but also other type II NRs (19, 61). Although the default hypothesis might be that decreased metabolism may be beneficial to a host during a viral illness, it would be an interesting possibility if repression of type II NR signaling directly contributed to an enhanced host immune response. These and other questions are clearly brought forth by any type of cross talk between two seemingly disparate systems.

Future Studies in Metabolic-Immune Cross Talk

Examples of cross talk between the immune system and metabolic signaling will undoubtedly continue to multiply in the coming years. Future investigations will likely begin either by pursuing the molecular mechanisms behind clinical diseases that appear to involve both systems or by identifying the functional significance of interactions between molecular components thought to be involved in these seemingly distinct signaling networks. An etiological role for immune-metabolic cross talk would seem to be a natural hypothesis for clinical entities that involve metabolic dysfunction caused by pathogen infection. To this end, several diseases readily come to mind, including hepatitis C virus (HCV)-induced liver steatosis, Paget's disease of the bone, and neonatal developmental defects caused by gestational viral infection.

Nearly 1% of the world's population is chronically infected with HCV, and 30% of these patients develop sufficiently severe liver fibrosis to be considered cirrhotic within 20 years of inoculation with the virus (163). Liver fibrosis leads to a large variety of clinical complications and ultimately results in death if liver transplantation is not feasible. Interestingly, HCV infection is accompanied by an accumulation of fatty droplets in liver cells known as hepatic steatosis, possibly indicating decompensation of the liver's homeostatic mechanisms (179). Although it is not known whether this steatosis is directly contributory to liver fibrosis, several reports have demonstrated a positive correlation between hepatic steatosis and fibrosis of increased severity (1, 83, 157). While it is possible that the steatosis seen in the context of HCV

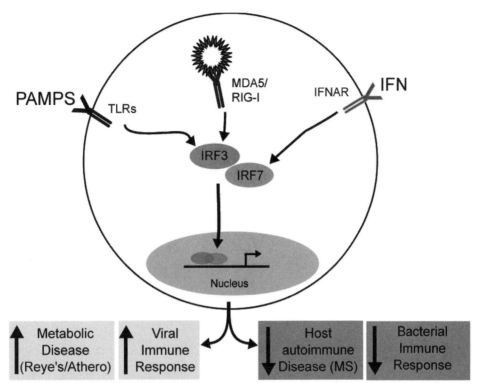

Figure 3 Outcomes of antiviral signaling initiated by PRRs or IFNAR affect host physiology in multiple ways in addition to defense against viral infection. These include modulation of antibacterial defense, adaptive immunity, and metabolic signaling.

infection is simply due to generic viral damage to the organ, it is also possible that the hepatic host response to viral infection may be dysregulating important metabolic functions of the liver such as processing of triglycerides, fatty acids, or bile. If so, a role for viral-induced downregulation of RXR-α would be an interesting hypothesis to investigate, given the importance of RXR/PXR heterodimers in regulating bile acid metabolism (210).

Chronic viral infection also seems to be at play in Paget's disease of the bone along with several other bone metabolic disorders linked to chronic HCV or human immunodeficiency virus infection (17, 66, 128, 164). As VDR plays a critical role in both the maintenance of blood calcium levels and cellular control of bone turnover, perturbation of VDR signaling could lead to significant dysfunction of bone metabolism (145). It is interesting to speculate whether IRF3-mediated downregulation of RXR-α might be playing a role as well in this context, as several VDR/RXR-α target genes are repressed when IRF3 is activated.

Finally, it has long been known that several pathogens known as the TORCH (toxoplasma, rubella, cytomegalovirus, and herpes simplex virus) organisms can cause severe developmental defects during gestation (188).

These organisms include rubella virus, CMV, and herpesviruses, and their infection could potentially affect retinoic acid receptor (RAR) signaling, which is critical for proper development of the fetus (218). It would be interesting to model such infections in wild-type and IRF3-deficient mice to determine whether IRF3-mediated downregulation of RAR/RXR signaling might be at play in this context as well.

CONCLUSION

In Fig. 3 we provide a basic model of the studies discussed in this chapter. As shown, multiple stimuli indicating the presence of a viral infection, either from direct sensing of viral components from PRRs or from the secondary IFN response acting through the IFN receptor, activate IRF3 and IRF7 to translocate to the nucleus and impact gene transactivation both positively and negatively. This modulation of gene transcription serves to generate an antiviral state in the host, but as detailed in this chapter, a number of other systems are also profoundly affected. These include other functions of the immune system related to its ability to defend against subsequent bacterial infections in addition to modulation of adaptive immunity, resulting in increased or decreased susceptibility to autoimmune disease. Additionally, certain components of the metabolic system are strongly repressed, including metabolic functions of both macrophages and hepatocytes.

These studies offer a more holistic view of the antiviral response and force us to consider that the host response to virus does come at a cost to other aspects of organismal physiology. Importantly, in the cases of autoimmune disease it appears that the IFN system can affect the development and progress of pathology in the absence of any viral infection. And although antibacterial defense and the metabolic system so far appear to be modulated by antiviral signaling only during an actual infection, given the fact that many patients suffer from chronic viral infections, these connections can only become increasingly appreciated in the years to come.

References

1. Adinolfi, L. E., M. Gambardella, A. Andreana, M. F. Tripodi, R. Utili, and G. Ruggiero. 2001. Steatosis accelerates the progression of liver damage of chronic hepatitis C patients and correlates with specific HCV genotype and visceral obesity. *Hepatology* 33:1358–1364.

2. Agrawal, N., L. K. Chiang, and I. R. Rifkin. 2006. Lupus nephritis. *Semin. Nephrol.* 26:95–104.

3. Akira, S. 2003. Toll-like receptor signaling. *J. Biol. Chem.* 278:38105–38108.

4. Akira, S., and K. Takeda. 2004. Toll-like receptor signalling. *Nat. Rev. Immunol.* 4:499–511.

5. Akira, S., M. Yamamoto, and K. Takeda. 2003. Role of adapters in Toll-like receptor signalling. *Biochem. Soc. Trans.* 31:637–642.

6. Alber, D. G., K. L. Powell, P. Vallance, D. A. Goodwin, and C. Grahame-Clarke. 2000. Herpesvirus infection accelerates atherosclerosis in the apolipoprotein E-deficient mouse. *Circulation* 102:779–785.

7. Al-Dosari, M. S., J. E. Knapp, and D. Liu. 2006. Activation of human CYP2C9 promoter and regulation by CAR and PXR in mouse liver. *Mol. Pharm.* 3:322–328.

8. Anders, H. J., B. Banas, Y. Linde, L. Weller, C. D. Cohen, M. Kretzler, S. Martin, V. Vielhauer, D. Schlondorff, and H. J. Grone. 2003. Bacterial CpG-DNA aggravates immune complex glomerulonephritis: role of TLR9-mediated expression of chemokines and chemokine receptors. *J. Am. Soc. Nephrol.* 14:317–326.

9. Anders, H. J., and D. Schlondorff. 2007. Toll-like receptors: emerging concepts in kidney disease. *Curr. Opin. Nephrol. Hypertens.* 16:177–183.

10. Anzenbacher, P., and E. Anzenbacherova. 2001. Cytochromes P450 and metabolism of xenobiotics. *Cell. Mol. Life Sci.* 58:737–747.

11. Aqel, N. M., R. Y. Ball, H. Waldmann, and M. J. Mitchinson. 1985. Identification of macrophages and smooth muscle cells in human atherosclerosis using monoclonal antibodies. *J. Pathol.* 146:197–204.

12. Auerbuch, V., D. G. Brockstedt, N. Meyer-Morse, M. O'Riordan, and D. A. Portnoy. 2004. Mice lacking the type I interferon receptor are resistant to *Listeria monocytogenes*. *J. Exp. Med.* 200:527–533.

13. Baechler, E. C., F. M. Batliwalla, G. Karypis, P. M. Gaffney, W. A. Ortmann, K. J. Espe, K. B. Shark, W. J. Grande, K. M. Hughes, V. Kapur, P. K. Gregersen, and T. W. Behrens. 2003. Interferon-inducible gene expression signature in peripheral blood cells of patients with severe lupus. *Proc. Natl. Acad. Sci. USA* 100:2610–2615.

14. Barrat, F. J., T. Meeker, J. H. Chan, C. Guiducci, and R. L. Coffman. 2007. Treatment of lupus-prone mice with a dual inhibitor of TLR7 and TLR9 leads to reduction of autoantibody production and amelioration of disease symptoms. *Eur. J. Immunol.* 37:3582–3586.

15. Bartalena, L., F. Bogazzi, S. Brogioni, L. Grasso, and E. Martino. 1998. Role of cytokines in the pathogenesis of the euthyroid sick syndrome. *Eur. J. Endocrinol.* 138:603–614.

16. Barton, G. M., and R. Medzhitov. 2003. Toll-like receptor signaling pathways. *Science* 300:1524–1525.

17. Basle, M. F., J. G. Fournier, S. Rozenblatt, A. Rebel, and M. Bouteille. 1986. Measles virus RNA detected in Paget's disease bone tissue by in situ hybridization. *J. Gen. Virol.* 67(Pt. 5):907–913.

18. Basler, C. F., and A. Garcia-Sastre. 2002. Viruses and the type I interferon antiviral system: induction and evasion. *Int. Rev. Immunol.* 21:305–337.

19. Beigneux, A. P., A. H. Moser, J. K. Shigenaga, C. Grunfeld, and K. R. Feingold. 2000. The acute phase response is associated with retinoid X receptor repression in rodent liver. *J. Biol. Chem.* 275:16390–16399.

20. Beigneux, A. P., A. H. Moser, J. K. Shigenaga, C. Grunfeld, and K. R. Feingold. 2003. Sick euthyroid syndrome is associated with decreased TR expression and DNA binding in mouse liver. *Am. J. Physiol. Endocrinol. Metab.* 284:E228–E236.

21. Bekeredjian-Ding, I. B., M. Wagner, V. Hornung, T. Giese, M. Schnurr, S. Endres, and G. Hartmann. 2005. Plasmacytoid dendritic cells control TLR7 sensitivity of naive B cells via type I IFN. *J. Immunol.* 174:4043–4050.

22. Belay, E. D., J. S. Bresee, R. C. Holman, A. S. Khan, A. Shahriari, and L. B. Schonberger. 1999. Reye's syndrome in the United States from 1981 through 1997. *N. Engl. J. Med.* 340:1377–1382.

23. Bennett, L., A. K. Palucka, E. Arce, V. Cantrell, J. Borvak, J. Banchereau, and V. Pascual. 2003. Interferon and granulopoiesis signatures in systemic lupus erythematosus blood. *J. Exp. Med.* 197:711–723.

24. Bettelli, E., M. Oukka, and V. K. Kuchroo. 2007. T$_H$-17 cells in the circle of immunity and autoimmunity. *Nat. Immunol.* 8:345–350.

25. Bigler, J., J. Whitton, J. W. Lampe, L. Fosdick, R. M. Bostick, and J. D. Potter. 2001. CYP2C9 and UGT1A6 genotypes modulate the protective effect of aspirin on colon adenoma risk. *Cancer Res.* 61:3566–3569.

26. Blum, A., M. Giladi, M. Weinberg, G. Kaplan, H. Pasternack, S. Laniado, and H. Miller. 1998. High anti-cytomegalovirus (CMV) IgG antibody titer is associated with coronary artery disease and may predict postcoronary balloon angioplasty restenosis. *Am. J. Cardiol.* 81:866–868.

27. Borden, E. C., G. C. Sen, G. Uze, R. H. Silverman, R. M. Ransohoff, G. R. Foster, and G. R. Stark. 2007. Interferons at age 50: past, current and future impact on biomedicine. *Nat. Rev. Drug Discov.* **6:**975–990.

28. Boule, M. W., C. Broughton, F. Mackay, S. Akira, A. Marshak-Rothstein, and I. R. Rifkin. 2004. Toll-like receptor 9-dependent and -independent dendritic cell activation by chromatin-immunoglobulin G complexes. *J. Exp. Med.* **199:**1631–1640.

29. Braun, D., P. Geraldes, and J. Demengeot. 2003. Type I interferon controls the onset and severity of autoimmune manifestations in *lpr* mice. *J. Autoimmun.* **20:**15–25.

30. Breiman, A., N. Grandvaux, R. Lin, C. Ottone, S. Akira, M. Yoneyama, T. Fujita, J. Hiscott, and E. F. Meurs. 2005. Inhibition of RIG-I-dependent signaling to the interferon pathway during hepatitis C virus expression and restoration of signaling by IKKepsilon. *J. Virol.* **79:**3969–3978.

31. Brierre, S., R. Kumari, and B. P. Deboisblanc. 2004. The endocrine system during sepsis. *Am. J. Med. Sci.* **328:**238–247.

32. Bukholm, G., B. P. Berdal, C. Haug, and M. Degre. 1984. Mouse fibroblast interferon modifies *Salmonella typhimurium* infection in infant mice. *Infect. Immun.* **45:**62–66.

33. Carlin, J. M., E. C. Borden, and G. I. Byrne. 1989. Interferon-induced indoleamine 2,3-dioxygenase activity inhibits *Chlamydia psittaci* replication in human macrophages. *J. Interferon. Res.* **9:**329–337.

34. Carlin, J. M., and J. B. Weller. 1995. Potentiation of interferon-mediated inhibition of *Chlamydia* infection by interleukin-1 in human macrophage cultures. *Infect. Immun.* **63:**1870–1875.

35. Carneiro, L. A., L. H. Travassos, and S. E. Girardin. 2007. Nod-like receptors in innate immunity and inflammatory diseases. *Ann. Med.* **39:**581–593.

36. Carrero, J. A., B. Calderon, and E. R. Unanue. 2004. Type I interferon sensitizes lymphocytes to apoptosis and reduces resistance to *Listeria* infection. *J. Exp. Med.* **200:**535–540.

37. Casteels-Van Daele, M., C. Van Geet, C. Wouters, and E. Eggermont. 2000. Reye syndrome revisited: a descriptive term covering a group of heterogeneous disorders. *Eur. J. Pediatr.* **159:**641–648.

38. Casteels-Van Daele, M., C. Wouters, C. Van Geet, M. C. McGovern, J. F. Glasgow, and M. C. Stewart. 2002. Reye's syndrome revisited. Outdated concept of Reye's syndrome was used. *BMJ* **324:**546.

39. Castrillo, A., S. B. Joseph, S. A. Vaidya, M. Haberland, A. M. Fogelman, G. Cheng, and P. Tontonoz. 2003. Crosstalk between LXR and Toll-like receptor signaling mediates bacterial and viral antagonism of cholesterol metabolism. *Mol. Cell* **12:**805–816.

40. Chawla, A., W. A. Boisvert, C. H. Lee, B. A. Laffitte, Y. Barak, S. B. Joseph, D. Liao, L. Nagy, P. A. Edwards, L. K. Curtiss, R. M. Evans, and P. Tontonoz. 2001. A PPAR gamma-LXR-ABCA1 pathway in macrophages is involved in cholesterol efflux and atherogenesis. *Mol. Cell* **7:**161–171.

41. Chawla, A., J. J. Repa, R. M. Evans, and D. J. Mangelsdorf. 2001. NRs and lipid physiology: opening the X-files. *Science* **294:**1866–1870.

42. Chinetti, G., S. Lestavel, V. Bocher, A. T. Remaley, B. Neve, I. P. Torra, E. Teissier, A. Minnich, M. Jaye, N. Duverger, H. B. Brewer, J. C. Fruchart, V. Clavey, and B. Staels. 2001. PPAR-alpha and PPAR-gamma activators induce cholesterol removal from human macrophage foam cells through stimulation of the ABCA1 pathway. *Nat. Med.* **7:**53–58.

43. Chow, E. K., A. Castrillo, A. Shahangian, L. Pei, R. M. O'Connell, R. L. Modlin, P. Tontonoz, and G. Cheng. 2006. A role for IRF3-dependent RXRalpha repression in hepatotoxicity associated with viral infections. *J. Exp. Med.* **203:**2589–2602.

44. Christeff, N., S. Gharakhanian, N. Thobie, W. Rozenbaum, and E. A. Nunez. 1992. Evidence for changes in adrenal and testicular steroids during HIV infection. *J. Acquir. Immune Defic. Syndr.* **5:**841–846.

45. Colonna, M. 2006. Toll-like receptors and IFN-alpha: partners in autoimmunity. *J. Clin. Invest.* **116:**2319–2322.

46. Cooper, A. M., J. E. Pearl, J. V. Brooks, S. Ehlers, and I. M. Orme. 2000. Expression of the nitric oxide synthase 2 gene is not essential for early control of *Mycobacterium tuberculosis* in the murine lung. *Infect. Immun.* **68:**6879–6882.

47. Cossart, P., and M. Lecuit. 1998. Interactions of *Listeria monocytogenes* with mammalian cells during entry and actin-based movement: bacterial factors, cellular ligands and signaling. *EMBO J.* **17:**3797–3806.

48. Crow, M. K. 2005. Interferon pathway activation in systemic lupus erythematosus. *Curr. Rheumatol Rep.* **7:**463–468.

49. Crow, M. K. 2007. Type I interferon in systemic lupus erythematosus. *Curr. Top. Microbiol. Immunol.* **316:**359–386.

50. Crow, M. K., K. A. Kirou, and J. Wohlgemuth. 2003. Microarray analysis of interferon-regulated genes in SLE. *Autoimmunity* **36:**481–490.

51. Dall'era, M. C., P. M. Cardarelli, B. T. Preston, A. Witte, and J. C. Davis, Jr. 2005. Type I interferon correlates with serological and clinical manifestations of SLE. *Ann. Rheum. Dis.* **64:**1692–1697.

52. Danesh, J., R. Collins, and R. Peto. 1997. Chronic infections and coronary heart disease: is there a link? *Lancet* **350:**430–436.

53. Danesh, J., P. Whincup, M. Walker, L. Lennon, A. Thomson, P. Appleby, Y. Wong, M. Bernardes-Silva, and M. Ward. 2000. *Chlamydia pneumoniae* IgG titres and coronary heart disease: prospective study and meta-analysis. *BMJ* **321:**208–213.

54. Davis, L. E., B. M. Woodfin, T. Q. Tran, L. S. Caskey, J. M. Wallace, O. U. Scremin, and K. S. Blisard. 1993. The influenza B virus mouse model of Reye's syndrome: pathogenesis of the hypoglycaemia. *Int. J. Exp. Pathol.* **74:**251–258.

55. Decker, T., M. Muller, and S. Stockinger. 2005. The Yin and Yang of type I interferon activity in bacterial infection. *Nat. Rev. Immunol.* **5:**675–687.

56. Ding, C., L. Wang, H. Al-Ghawi, J. Marroquin, M. Mamula, and J. Yan. 2006. Toll-like receptor engagement stimulates anti-snRNP autoreactive B cells for activation. *Eur. J. Immunol.* **36:**2013–2024.

57. Doly, J., A. Civas, S. Navarro, and G. Uze. 1998. Type I interferons: expression and signalization. *Cell. Mol. Life Sci.* **54:**1109–1121.

58. Duerksen, D. R., L. D. Jewell, A. L. Mason, and V. G. Bain. 1997. Co-existence of hepatitis A and adult Reye's syndrome. *Gut* **41**:121–124.

59. Duez, H., Y. S. Chao, M. Hernandez, G. Torpier, P. Poulain, S. Mundt, Z. Mallat, E. Teissier, C. A. Burton, A. Tedgui, J. C. Fruchart, C. Fievet, S. D. Wright, and B. Staels. 2002. Reduction of atherosclerosis by the peroxisome proliferator-activated receptor alpha agonist fenofibrate in mice. *J. Biol. Chem.* **277**:48051–48057.

60. Dupont, I., F. Berthou, P. Bodenez, L. Bardou, C. Guirriec, N. Stephan, Y. Dreano, and D. Lucas. 1999. Involvement of cytochromes P-450 2E1 and 3A4 in the 5-hydroxylation of salicylate in humans. *Drug. Metab. Dispos.* **27**:322–326.

61. Feingold, K., M. S. Kim, J. Shigenaga, A. Moser, and C. Grunfeld. 2004. Altered expression of nuclear hormone receptors and coactivators in mouse heart during the acute-phase response. *Am. J. Physiol. Endocrinol. Metab.* **286**:E201–E207.

62. Ford, H., and R. Nicholas. 2005. Multiple sclerosis. *Clin. Evid.* Dec:1637–1651.

63. Forman, B. M., J. Chen, and R. M. Evans. 1997. Hypolipidemic drugs, polyunsaturated fatty acids, and eicosanoids are ligands for peroxisome proliferator-activated receptors alpha and delta. *Proc. Natl. Acad. Sci. USA* **94**:4312–4317.

64. Freudenberg, M. A., T. Merlin, C. Kalis, Y. Chvatchko, H. Stubig, and C. Galanos. 2002. Cutting edge: a murine, IL-12-independent pathway of IFN-gamma induction by gram-negative bacteria based on STAT4 activation by type I IFN and IL-18 signaling. *J. Immunol.* **169**:1665–1668.

65. Furuzawa-Carballeda, J., M. I. Vargas-Rojas, and A. R. Cabral. 2007. Autoimmune inflammation from the Th17 perspective. *Autoimmun. Rev.* **6**:169–175.

66. Garcia Aparicio, A. M., S. Munoz Fernandez, J. Gonzalez, J. R. Arribas, J. M. Pena, J. J. Vazquez, M. E. Martinez, J. Coya, and E. Martin Mola. 2006. Abnormalities in the bone mineral metabolism in HIV-infected patients. *Clin. Rheumatol.* **25**:537–539.

67. Ghosh, D., D. Dhadwal, A. Aggarwal, S. Mitra, S. K. Garg, R. Kumar, and B. Kaur. 1999. Investigation of an epidemic of Reye's syndrome in northern region of India. *Indian Pediatr.* **36**:1097–1106.

68. Giosue, S., M. Casarini, L. Alemanno, G. Galluccio, P. Mattia, G. Pedicelli, L. Rebek, A. Bisetti, and F. Ameglio. 1998. Effects of aerosolized interferon-alpha in patients with pulmonary tuberculosis. *Am. J. Respir. Crit. Care Med.* **158**:1156–1162.

69. Grayston, J. T., C. C. Kuo, A. S. Coulson, L. A. Campbell, R. D. Lawrence, M. J. Lee, E. D. Strandness, and S. P. Wang. 1995. *Chlamydia pneumoniae* (TWAR) in atherosclerosis of the carotid artery. *Circulation* **92**:3397–400.

70. Guo, B., and G. Cheng. 2007. Modulation of the interferon antiviral response by the TBK1/IKKi adaptor protein TANK. *J. Biol. Chem.* **282**:11817–11826.

71. Gurfinkel, E., G. Bozovich, A. Daroca, E. Beck, and B. Mautner. 1997. Randomised trial of roxithromycin in non-Q-wave coronary syndromes: ROXIS Pilot Study. ROXIS Study Group. *Lancet* **350**:404–407.

72. Hacker, H., V. Redecke, B. Blagoev, I. Kratchmarova, L. C. Hsu, G. G. Wang, M. P. Kamps, E. Raz, H. Wagner, G. Hacker, M. Mann, and M. Karin. 2006. Specificity in Toll-like receptor signalling through distinct effector functions of TRAF3 and TRAF6. *Nature* **439**:204–207.

73. Haller, O., G. Kochs, and F. Weber. 2007. Interferon, Mx, and viral countermeasures. *Cytokine Growth Factor Rev.* **18**:425–433.

74. Hansson, G. K. 2005. Inflammation, atherosclerosis, and coronary artery disease. *N. Engl. J. Med.* **352**:1685–1695.

75. Harrington, L. E., R. D. Hatton, P. R. Mangan, H. Turner, T. L. Murphy, K. M. Murphy, and C. T. Weaver. 2005. Interleukin 17-producing CD4+ effector T cells develop via a lineage distinct from the T helper type 1 and 2 lineages. *Nat. Immunol.* **6**:1123–1132.

76. Hasegawa, K., and T. Hayashi. 2003. Synthetic CpG oligodeoxynucleotides accelerate the development of lupus nephritis during preactive phase in NZB x NZWF1 mice. *Lupus* **12**:838–845.

77. He, B., X. Qiao, and A. Cerutti. 2004. CpG DNA induces IgG class switch DNA recombination by activating human B cells through an innate pathway that requires TLR9 and cooperates with IL-10. *J. Immunol.* **173**:4479–4491.

78. Heim, M. H. 2005. RIG-I: an essential regulator of virus-induced interferon production. *J. Hepatol.* **42**:431–433.

79. Hertzog, P. J., L. A. O'Neill, and J. A. Hamilton. 2003. The interferon in TLR signaling: more than just antiviral. *Trends Immunol.* **24**:534–539.

80. Hiscott, J., R. Lin, P. Nakhaei, and S. Paz. 2006. MasterCARD: a priceless link to innate immunity. *Trends Mol. Med.* **12**:53–56.

81. Honda, K., Y. Ohba, H. Yanai, H. Negishi, T. Mizutani, A. Takaoka, C. Taya, and T. Taniguchi. 2005. Spatiotemporal regulation of MyD88-IRF-7 signalling for robust type-I interferon induction. *Nature* **434**:1035–1040.

82. Hornung, V., J. Ellegast, S. Kim, K. Brzozka, A. Jung, H. Kato, H. Poeck, S. Akira, K. K. Conzelmann, M. Schlee, S. Endres, and G. Hartmann. 2006. 5′-Triphosphate RNA is the ligand for RIG-I. *Science* **314**:994–997.

83. Hourigan, L. F., G. A. Macdonald, D. Purdie, V. H. Whitehall, C. Shorthouse, A. Clouston, and E. E. Powell. 1999. Fibrosis in chronic hepatitis C correlates significantly with body mass index and steatosis. *Hepatology* **29**:1215–1219.

84. Howard, M., T. Muchamuel, S. Andrade, and S. Menon. 1993. Interleukin 10 protects mice from lethal endotoxemia. *J. Exp. Med.* **177**:1205–1208.

85. Huang, J., T. Liu, L. G. Xu, D. Chen, Z. Zhai, and H. B. Shu. 2005. SIKE is an IKKepsilon/TBK1-associated suppressor of TLR3- and virus-triggered IRF-3 activation pathways. *EMBO J.* **24**:4018–4028.

86. Hurwitz, E. S., M. J. Barrett, D. Bregman, W. J. Gunn, P. Pinsky, L. B. Schonberger, J. S. Drage, R. A. Kaslow, D. B. Burlington, G. V. Quinnan, et al. 1987. Public Health Service study of Reye's syndrome and medications. Report of the main study. *JAMA* **257**:1905–1911.

87. Janeway, C. A., Jr., and R. Medzhitov. 2002. Innate immune recognition. *Annu. Rev. Immunol.* **20**:197–216.

88. Janowski, B. A., P. J. Willy, T. R. Devi, J. R. Falck, and D. J. Mangelsdorf. 1996. An oxysterol signalling pathway mediated by the NR LXR alpha. *Nature* **383**:728–731.

89. Jefferies, C. A., and K. A. Fitzgerald. 2005. Interferon gene regulation: not all roads lead to Tolls. *Trends Mol. Med.* **11**:403–411.

90. Jiang, C., A. T. Ting, and B. Seed. 1998. PPAR-gamma agonists inhibit production of monocyte inflammatory cytokines. *Nature* **391:**82–86.

91. Johnson, C. L., and M. Gale, Jr. 2006. CARD games between virus and host get a new player. *Trends Immunol.* **27:**1–4.

92. Joseph, S. B., A. Castrillo, B. A. Laffitte, D. J. Mangelsdorf, and P. Tontonoz. 2003. Reciprocal regulation of inflammation and lipid metabolism by liver X receptors. *Nat. Med.* **9:**213–219.

93. Joseph, S. B., E. McKilligin, L. Pei, M. A. Watson, A. R. Collins, B. A. Laffitte, M. Chen, G. Noh, J. Goodman, G. N. Hagger, J. Tran, T. K. Tippin, X. Wang, A. J. Lusis, W. A. Hsueh, R. E. Law, J. L. Collins, T. M. Willson, and P. Tontonoz. 2002. Synthetic LXR ligand inhibits the development of atherosclerosis in mice. *Proc. Natl. Acad. Sci. USA* **99:**7604–7609.

94. Kaisho, T., and S. Akira. 2006. Toll-like receptor function and signaling. *J. Allergy. Clin. Immunol.* **117:**979–987; quiz 988.

95. Kanneganti, T. D., M. Lamkanfi, and G. Nunez. 2007. Intracellular NOD-like receptors in host defense and disease. *Immunity* **27:**549–559.

96. Karaghiosoff, M., R. Steinborn, P. Kovarik, G. Kriegshauser, M. Baccarini, B. Donabauer, U. Reichart, T. Kolbe, C. Bogdan, T. Leanderson, D. Levy, T. Decker, and M. Muller. 2003. Central role for type I interferons and Tyk2 in lipopolysaccharide-induced endotoxin shock. *Nat. Immunol.* **4:**471–477.

97. Kato, H., S. Sato, M. Yoneyama, M. Yamamoto, S. Uematsu, K. Matsui, T. Tsujimura, K. Takeda, T. Fujita, O. Takeuchi, and S. Akira. 2005. Cell type-specific involvement of RIG-I in antiviral response. *Immunity* **23:**19–28.

98. Kato, H., O. Takeuchi, S. Sato, M. Yoneyama, M. Yamamoto, K. Matsui, S. Uematsu, A. Jung, T. Kawai, K. J. Ishii, O. Yamaguchi, K. Otsu, T. Tsujimura, C. S. Koh, C. Reis e Sousa, Y. Matsuura, T. Fujita, and S. Akira. 2006. Differential roles of MDA5 and RIG-I helicases in the recognition of RNA viruses. *Nature* **441:**101–105.

99. Kawai, T., and S. Akira. 2006. TLR signaling. *Cell Death Differ.* **13:**816–825.

100. Kawai, T., K. Takahashi, S. Sato, C. Coban, H. Kumar, H. Kato, K. J. Ishii, O. Takeuchi, and S. Akira. 2005. IPS-1, an adaptor triggering RIG-I- and Mda5-mediated type I interferon induction. *Nat. Immunol.* **6:**981–988.

101. Kim, M. S., J. Shigenaga, A. Moser, K. Feingold, and C. Grunfeld. 2003. Repression of farnesoid X receptor during the acute phase response. *J. Biol. Chem.* **278:**8988–8995.

102. Kliewer, S. A., S. S. Sundseth, S. A. Jones, P. J. Brown, G. B. Wisely, C. S. Koble, P. Devchand, W. Wahli, T. M. Willson, J. M. Lenhard, and J. M. Lehmann. 1997. Fatty acids and eicosanoids regulate gene expression through direct interactions with peroxisome proliferator-activated receptors alpha and gamma. *Proc. Natl. Acad. Sci. USA* **94:**4318–4323.

103. Komiyama, Y., S. Nakae, T. Matsuki, A. Nambu, H. Ishigame, S. Kakuta, K. Sudo, and Y. Iwakura. 2006. IL-17 plays an important role in the development of experimental autoimmune encephalomyelitis. *J. Immunol.* **177:**566–573.

104. Kono, D. H., R. Baccala, and A. N. Theofilopoulos. 2003. Inhibition of lupus by genetic alteration of the interferon-alpha/beta receptor. *Autoimmunity* **36:**503–510.

105. Korn, T., M. Oukka, V. Kuchroo, and E. Bettelli. 2007. Th17 cells: effector T cells with inflammatory properties. *Semin Immunol.* **19:**362–371.

106. Kramer, J. M., and S. L. Gaffen. 2007. Interleukin-17: a new paradigm in inflammation, autoimmunity, and therapy. *J. Periodontol.* **78:**1083–1093.

107. Krug, A., A. R. French, W. Barchet, J. A. Fischer, A. Dzionek, J. T. Pingel, M. M. Orihuela, S. Akira, W. M. Yokoyama, and M. Colonna. 2004. TLR9-dependent recognition of MCMV by IPC and DC generates coordinated cytokine responses that activate antiviral NK cell function. *Immunity* **21:**107–119.

108. Kuehl, G. E., J. Bigler, J. D. Potter, and J. W. Lampe. 2006. Glucuronidation of the aspirin metabolite salicylic acid by expressed UDP-glucuronosyltransferases and human liver microsomes. *Drug Metab. Dispos.* **34:**199–202.

109. Kumar, H., T. Kawai, H. Kato, S. Sato, K. Takahashi, C. Coban, M. Yamamoto, S. Uematsu, K. J. Ishii, O. Takeuchi, and S. Akira. 2006. Essential role of IPS-1 in innate immune responses against RNA viruses. *J. Exp. Med.* **203:**1795–1803.

110. Lau, C. M., C. Broughton, A. S. Tabor, S. Akira, R. A. Flavell, M. J. Mamula, S. R. Christensen, M. J. Shlomchik, G. A. Viglianti, I. R. Rifkin, and A. Marshak-Rothstein. 2005. RNA-associated autoantigens activate B cells by combined B cell antigen receptor/Toll-like receptor 7 engagement. *J. Exp. Med.* **202:**1171–1177.

111. Leadbetter, E. A., I. R. Rifkin, A. M. Hohlbaum, B. C. Beaudette, M. J. Shlomchik, and A. Marshak-Rothstein. 2002. Chromatin-IgG complexes activate B cells by dual engagement of IgM and Toll-like receptors. *Nature* **416:**603–607.

112. Le Bon, A., G. Schiavoni, G. D'Agostino, I. Gresser, F. Belardelli, and D. F. Tough. 2001. Type I interferons potently enhance humoral immunity and can promote isotype switching by stimulating dendritic cells in vivo. *Immunity* **14:**461–470.

113. Le Bon, A., C. Thompson, E. Kamphuis, V. Durand, C. Rossmann, U. Kalinke, and D. F. Tough. 2006. Cutting edge: enhancement of antibody responses through direct stimulation of B and T cells by type I IFN. *J. Immunol.* **176:**2074–2078.

114. Lee, C. H., A. Chawla, N. Urbiztondo, D. Liao, W. A. Boisvert, R. M. Evans, and L. K. Curtiss. 2003. Transcriptional repression of atherogenic inflammation: modulation by PPARdelta. *Science* **302:**453–457.

115. Lee, C. K., D. T. Rao, R. Gertner, R. Gimeno, A. B. Frey, and D. E. Levy. 2000. Distinct requirements for IFNs and Stat1 in NK cell function. *J. Immunol.* **165:**3571–3577.

116. Lee, M. S., and Y. J. Kim. 2007. Pattern-recognition receptor signaling initiated from extracellular, membrane, and cytoplasmic space. *Mol. Cells* **23:**1–10.

117. Lehmann, J. M., S. A. Kliewer, L. B. Moore, T. A. Smith-Oliver, B. B. Oliver, J. L. Su, S. S. Sundseth, D. A. Winegar, D. E. Blanchard, T. A. Spencer, and T. M. Willson. 1997. Activation of the NR LXR by oxysterols defines a new hormone response pathway. *J. Biol. Chem.* **272:**3137–3140.

118. Lehmann, J. M., D. D. McKee, M. A. Watson, T. M. Willson, J. T. Moore, and S. A. Kliewer. 1998. The human orphan NR PXR is activated by compounds that regulate CYP3A4 gene expression and cause drug interactions. *J. Clin. Invest.* **102:**1016–1023.

119. Lenert, P., R. Brummel, E. H. Field, and R. F. Ashman. 2005. TLR-9 activation of marginal zone B cells in lupus mice regulates immunity through increased IL-10 production. *J. Clin. Immunol.* **25:**29–40.

120. Le Page, C., P. Genin, M. G. Baines, and J. Hiscott. 2000. Interferon activation and innate immunity. *Rev. Immunogenet.* **2:**374–386.

121. Levy, D. E., and A. Garcia-Sastre. 2001. The virus battles: IFN induction of the antiviral state and mechanisms of viral evasion. *Cytokine Growth Factor Rev.* **12:**143–156.

122. Li, A. C., K. K. Brown, M. J. Silvestre, T. M. Willson, W. Palinski, and C. K. Glass. 2000. Peroxisome proliferator-activated receptor gamma ligands inhibit development of atherosclerosis in LDL receptor-deficient mice. *J. Clin. Invest.* **106:**523–531.

123. Li, A. C., and W. Palinski. 2006. Peroxisome proliferator-activated receptors: how their effects on macrophages can lead to the development of a new drug therapy against atherosclerosis. *Annu. Rev. Pharmacol. Toxicol.* **46:**1–39.

124. Loo, Y. M., J. Fornek, N. Crochet, G. Bajwa, O. Perwitasari, L. Martinez-Sobrido, S. Akira, M. A. Gill, A. Garcia-Sastre, M. G. Katze, and M. Gale, Jr. 2008. Distinct RIG-I and MDA5 signaling by RNA viruses in innate immunity. *J. Virol.* **82:**335–345.

125. Loo, Y. M., and M. Gale, Jr. 2007. Viral regulation and evasion of the host response. *Curr. Top. Microbiol. Immunol.* **316:**295–313.

126. Manca, C., L. Tsenova, A. Bergtold, S. Freeman, M. Tovey, J. M. Musser, C. E. Barry III, V. H. Freedman, and G. Kaplan. 2001. Virulence of a *Mycobacterium tuberculosis* clinical isolate in mice is determined by failure to induce Th1 type immunity and is associated with induction of IFN-alpha/beta. *Proc. Natl. Acad. Sci. USA* **98:**5752–5757.

127. Mancuso, G., A. Midiri, C. Biondo, C. Beninati, S. Zummo, R. Galbo, F. Tomasello, M. Gambuzza, G. Macri, A. Ruggeri, T. Leanderson, and G. Teti. 2007. Type I IFN signaling is crucial for host resistance against different species of pathogenic bacteria. *J. Immunol.* **178:**3126–3133.

128. Manganelli, P., N. Giuliani, P. Fietta, C. Mancini, M. Lazzaretti, A. Pollini, F. Quaini, and M. Pedrazzoni. 2005. OPG/RANKL system imbalance in a case of hepatitis C-associated osteosclerosis: the pathogenetic key? *Clin. Rheumatol.* **24:**296–300.

129. Marrack, P., J. Kappler, and T. Mitchell. 1999. Type I interferons keep activated T cells alive. *J. Exp. Med.* **189:**521–530.

130. Marshak-Rothstein, A. 2006. Toll-like receptors in systemic autoimmune disease. *Nat. Rev. Immunol.* **6:**823–835.

131. Martens, M. E., C. H. Chang, and C. P. Lee. 1986. Reye's syndrome: mitochondrial swelling and Ca^{2+} release induced by Reye's plasma, allantoin, and salicylate. *Arch. Biochem. Biophys.* **244:**773–786.

132. Means, T. K., E. Latz, F. Hayashi, M. R. Murali, D. T. Golenbock, and A. D. Luster. 2005. Human lupus autoantibody-DNA complexes activate DCs through cooperation of CD32 and TLR9. *J. Clin. Invest.* **115:**407–417.

133. Melnick, J. L., C. Hu, J. Burek, E. Adam, and M. E. DeBakey. 1994. Cytomegalovirus DNA in arterial walls of patients with atherosclerosis. *J. Med. Virol.* **42:**170–174.

134. Melnick, J. L., B. L. Petrie, G. R. Dreesman, J. Burek, C. H. McCollum, and M. E. DeBakey. 1983. Cytomegalovirus antigen within human arterial smooth muscle cells. *Lancet* **2:**644–647.

135. Meylan, E., J. Curran, K. Hofmann, D. Moradpour, M. Binder, R. Bartenschlager, and J. Tschopp. 2005. Cardif is an adaptor protein in the RIG-I antiviral pathway and is targeted by hepatitis C virus. *Nature* **437:**1167–1172.

136. Michitaka, K., N. Horiike, Y. Chen, T. N. Duong, I. Konishi, T. Mashiba, Y. Tokumoto, Y. Hiasa, Y. Tanaka, M. Mizokami, and M. Onji. 2004. Gianotti-Crosti syndrome caused by acute hepatitis B virus genotype D infection. *Intern. Med.* **43:**696–699.

137. Muhlestein, J. B., J. L. Anderson, J. F. Carlquist, K. Salunkhe, B. D. Horne, R. R. Pearson, T. J. Bunch, A. Allen, S. Trehan, and C. Nielson. 2000. Randomized secondary prevention trial of azithromycin in patients with coronary artery disease: primary clinical results of the ACADEMIC study. *Circulation* **102:**1755–1760.

138. Muhlestein, J. B., J. L. Anderson, E. H. Hammond, L. Zhao, S. Trehan, E. P. Schwobe, and J. F. Carlquist. 1998. Infection with *Chlamydia pneumoniae* accelerates the development of atherosclerosis and treatment with azithromycin prevents it in a rabbit model. *Circulation* **97:**633–636.

139. Munro, J. M., J. D. van der Walt, C. S. Munro, J. A. Chalmers, and E. L. Cox. 1987. An immunohistochemical analysis of human aortic fatty streaks. *Hum. Pathol.* **18:**375–380.

140. Nacionales, D. C., K. M. Kelly-Scumpia, P. Y. Lee, J. S. Weinstein, R. Lyons, E. Sobel, M. Satoh, and W. H. Reeves. 2007. Deficiency of the type I interferon receptor protects mice from experimental lupus. *Arthritis Rheum.* **56:**3770–3783.

141. Nagpal, S., J. Lu, and M. F. Boehm. 2001. Vitamin D analogs: mechanism of action and therapeutic applications. *Curr. Med. Chem.* **8:**1661–1679.

142. Niemir, Z. I., H. Stein, I. L. Noronha, C. Kruger, K. Andrassy, E. Ritz, and R. Waldherr. 1995. PDGF and TGF-beta contribute to the natural course of human IgA glomerulonephritis. *Kidney Int.* **48:**1530–1541.

143. Niesel, D. W., C. B. Hess, Y. J. Cho, K. D. Klimpel, and G. R. Klimpel. 1986. Natural and recombinant interferons inhibit epithelial cell invasion by *Shigella* spp. *Infect. Immun.* **52:**828–833.

144. Nieto, F. J., E. Adam, P. Sorlie, H. Farzadegan, J. L. Melnick, G. W. Comstock, and M. Szklo. 1996. Cohort study of cytomegalovirus infection as a risk factor for carotid intimal-medial thickening, a measure of subclinical atherosclerosis. *Circulation* **94:**922–927.

145. Norman, A. W. 2006. Minireview: vitamin D receptor: new assignments for an already busy receptor. *Endocrinology* **147:**5542–5548.

146. Nurieva, R., X. O. Yang, G. Martinez, Y. Zhang, A. D. Panopoulos, L. Ma, K. Schluns, Q. Tian, S. S. Watowich, A. M. Jetten, and C. Dong. 2007. Essential autocrine

regulation by IL-21 in the generation of inflammatory T cells. *Nature* **448**:480–483.

146a. O' Connell, R. M., S. K. Saha, S.A. Vaidya, K. W. Bruhn, G.A. Miranda, B. Zarnegar, A. K. Perry, B.O. Nguyen, T. F. Lane, T. Taniguchi, J. F. Miller, and G. Cheng. 2004. Type I interferon production enhances susceptibility to *Listeria monocytogenes* infection. *J. Exp. Med.* **200**:437–445.

147. Oganesyan, G., S. K. Saha, B. Guo, J. Q. He, A. Shahangian, B. Zarnegar, A. Perry, and G. Cheng. 2006. Critical role of TRAF3 in the Toll-like receptor-dependent and -independent antiviral response. *Nature* **439**:208–211.

148. Oganesyan, G., S. K. Saha, E. M. Pietras, B. Guo, A. K. Miyahira, B. Zarnegar, and G. Cheng. 2007. IRF3-dependent type I interferon response in B cells regulates CpG-mediated antibody production. *J. Biol. Chem.* **283**: 802–808.

149. Ogasawara, K., S. Hida, Y. Weng, A. Saiura, K. Sato, H. Takayanagi, S. Sakaguchi, T. Yokochi, T. Kodama, M. Naitoh, J. A. De Martino, and T. Taniguchi. 2002. Requirement of the IFN-alpha/beta-induced CXCR3 chemokine signalling for CD8$^+$ T cell activation. *Genes Cells* **7**:309–320.

150. Onomoto, K., M. Yoneyama, and T. Fujita. 2007. Regulation of antiviral innate immune responses by RIG-I family of RNA helicases. *Curr. Top. Microbiol. Immunol.* **316**:193–205.

151. Opitz, B., M. Vinzing, V. van Laak, B. Schmeck, G. Heine, S. Gunther, R. Preissner, H. Slevogt, P. D. N'Guessan, J. Eitel, T. Goldmann, A. Flieger, N. Suttorp, and S. Hippenstiel. 2006. *Legionella pneumophila* induces IFNbeta in lung epithelial cells via IPS-1 and IRF3, which also control bacterial replication. *J. Biol. Chem.* **281**: 36173–36179.

152. Orlowski, J. P., P. Campbell, and S. Goldstein. 1990. Reye's syndrome: a case control study of medication use and associated viruses in Australia. *Cleve. Clin. J. Med.* **57**:323–329.

153. Park, H., Z. Li, X. O. Yang, S. H. Chang, R. Nurieva, Y. H. Wang, Y. Wang, L. Hood, Z. Zhu, Q. Tian, and C. Dong. 2005. A distinct lineage of CD4 T cells regulates tissue inflammation by producing interleukin 17. *Nat. Immunol.* **6**:1133–1141.

154. Partin, J. C., W. K. Schubert, and J. S. Partin. 1971. Mitochondrial ultrastructure in Reye's syndrome (encephalopathy and fatty degeneration of the viscera). *N. Engl. J. Med.* **285**:1339–1343.

155. Pascual, G., and C. K. Glass. 2006. NRs versus inflammation: mechanisms of transrepression. *Trends Endocrinol. Metab.* **17**:321–327.

156. Pascual, V., L. Farkas, and J. Banchereau. 2006. Systemic lupus erythematosus: all roads lead to type I interferons. *Curr. Opin. Immunol.* **18**:676–682.

157. Patton, H. M., K. Patel, C. Behling, D. Bylund, L. M. Blatt, M. Vallee, S. Heaton, A. Conrad, P. J. Pockros, and J. G. McHutchison. 2004. The impact of steatosis on disease progression and early and sustained treatment response in chronic hepatitis C patients. *J. Hepatol.* **40**:484–490.

158. Paul, S., C. Ricour, C. Sommereyns, F. Sorgeloos, and T. Michiels. 2007. Type I interferon response in the central nervous system. *Biochimie* **89**:770–778.

159. Perry, A. K., G. Chen, D. Zheng, H. Tang, and G. Cheng. 2005. The host type I interferon response to viral and bacterial infections. *Cell Res.* **15**:407–422.

160. Petrilli, V., C. Dostert, D. A. Muruve, and J. Tschopp. 2007. The inflammasome: a danger sensing complex triggering innate immunity. *Curr. Opin. Immunol.* **19**: 615–622.

161. Pitha, P. M., and M. S. Kunzi. 2007. Type I interferon: the ever unfolding story. *Curr. Top. Microbiol. Immunol.* **316**:41–70.

162. Poeck, H., M. Wagner, J. Battiany, S. Rothenfusser, D. Wellisch, V. Hornung, B. Jahrsdorfer, T. Giese, S. Endres, and G. Hartmann. 2004. Plasmacytoid dendritic cells, antigen, and CpG-C license human B cells for plasma cell differentiation and immunoglobulin production in the absence of T-cell help. *Blood* **103**:3058–3064.

163. Ratziu, V., A. Heurtier, L. Bonyhay, T. Poynard, and P. Giral. 2005. Review article: an unexpected virus-host interaction—the hepatitis C virus-diabetes link. *Aliment. Pharmacol. Ther.* **22** (Suppl. 2):56–60.

164. Rebel, A., M. Basle, A. Pouplard, K. Malkani, R. Filmon, and A. Lepatezour. 1981. Towards a viral etiology for Paget's disease of bone. *Metab. Bone. Dis. Relat. Res.* **3**:235–238.

165. Renaudineau, Y., J. O. Pers, B. Bendaoud, C. Jamin, and P. Youinou. 2004. Dysfunctional B cells in systemic lupus erythematosus. *Autoimmun. Rev.* **3**:516–523.

166. Repa, J. J., K. E. Berge, C. Pomajzl, J. A. Richardson, H. Hobbs, and D. J. Mangelsdorf. 2002. Regulation of ATP-binding cassette sterol transporters ABCG5 and ABCG8 by the liver X receptors alpha and beta. *J. Biol. Chem.* **277**:18793–18800.

167. Ribeiro, R. C., P. J. Kushner, and J. D. Baxter. 1995. The nuclear hormone receptor gene superfamily. *Annu. Rev. Med.* **46**:443–453.

168. Ricote, M., A. C. Li, T. M. Willson, C. J. Kelly, and C. K. Glass. 1998. The peroxisome proliferator-activated receptor-gamma is a negative regulator of macrophage activation. *Nature* **391**:79–82.

169. Rifkin, I. R., E. A. Leadbetter, B. C. Beaudette, C. Kiani, M. Monestier, M. J. Shlomchik, and A. Marshak-Rothstein. 2000. Immune complexes present in the sera of autoimmune mice activate rheumatoid factor B cells. *J. Immunol.* **165**:1626–1633.

170. Rifkin, I. R., E. A. Leadbetter, L. Busconi, G. Viglianti, and A. Marshak-Rothstein. 2005. Toll-like receptors, endogenous ligands, and systemic autoimmune disease. *Immunol. Rev.* **204**:27–42.

171. Ronnblom, L., M. L. Eloranta, and G. V. Alm. 2003. Role of natural interferon-alpha producing cells (plasmacytoid dendritic cells) in autoimmunity. *Autoimmunity* **36**:463–472.

172. Rosenstiel, P., A. Till, and S. Schreiber. 2007. NOD-like receptors and human diseases. *Microbes Infect.* **9**:648–657.

173. Sadlier, D. M., X. Ouyang, B. McMahon, W. Mu, R. Ohashi, K. Rodgers, D. Murray, T. Nakagawa, C. Godson, P. Doran, H. R. Brady, and R. J. Johnson. 2005. Microarray and bioinformatic detection of novel and established genes expressed in experimental anti-Thy1 nephritis. *Kidney Int.* **68**:2542–2561.

174. Saha, S. K., E. M. Pietras, J. Q. He, J. R. Kang, S. Y. Liu, G. Oganesyan, A. Shahangian, B. Zarnegar, T. L. Shiba,

Y. Wang, and G. Cheng. 2006. Regulation of antiviral responses by a direct and specific interaction between TRAF3 and Cardif. *EMBO J.* **25:**3257–3263.

175. Saikku, P., M. Leinonen, K. Mattila, M. R. Ekman, M. S. Nieminen, P. H. Makela, J. K. Huttunen, and V. Valtonen. 1988. Serological evidence of an association of a novel Chlamydia, TWAR, with chronic coronary heart disease and acute myocardial infarction. *Lancet* **2:**983–986.

176. Sakaguchi, S., H. Negishi, M. Asagiri, C. Nakajima, T. Mizutani, A. Takaoka, K. Honda, and T. Taniguchi. 2003. Essential role of IRF-3 in lipopolysaccharide-induced interferon-β gene expression and endotoxin shock. *Biochem. Biophys. Res. Commun.* **306:**860–866.

177. Santiago-Raber, M. L., R. Baccala, K. M. Haraldsson, D. Choubey, T. A. Stewart, D. H. Kono, and A. N. Theofilopoulos. 2003. Type-I interferon receptor deficiency reduces lupus-like disease in NZB mice. *J. Exp. Med.* **197:**777–788.

178. Schaefer, C., T. R. Hidalgo, L. Cashion, H. Petry, A. Brooks, P. Szymanski, H. S. Qian, C. Gross, P. Wang, P. Liu, C. Goldman, G. M. Rubanyi, and R. N. Harkins. 2006. Gene-based delivery of IFN-beta is efficacious in a murine model of experimental allergic encephalomyelitis. *J. Interferon Cytokine Res.* **26:**449–454.

179. Scheuer, P. J., P. Ashrafzadeh, S. Sherlock, D. Brown, and G. M. Dusheiko. 1992. The pathology of hepatitis C. *Hepatology* **15:**567–571.

180. Schiavoni, G., C. Mauri, D. Carlei, F. Belardelli, M. Castellani Pastoris, and E. Proietti. 2004. Type I IFN protects permissive macrophages from *Legionella pneumophila* infection through an IFN-γ-independent pathway. *J. Immunol.* **173:**1266–1275.

181. Schneider, K., C. A. Benedict, and C. F. Ware. 2006. A TRAFfic cop for host defense. *Nat. Immunol.* **7:**15–16.

182. Sen, G. C., and G. A. Peters. 2007. Viral stress-inducible genes. *Adv. Virus Res.* **70:**233–263.

183. Seth, R. B., L. Sun, and Z. J. Chen. 2006. Antiviral innate immunity pathways. *Cell Res.* **16:**141–147.

184. Seth, R. B., L. Sun, C. K. Ea, and Z. J. Chen. 2005. Identification and characterization of MAVS, a mitochondrial antiviral signaling protein that activates NF-kappaB and IRF 3. *Cell* **122:**669–682.

185. Shemer-Avni, Y., D. Wallach, and I. Sarov. 1989. Reversion of the antichlamydial effect of tumor necrosis factor by tryptophan and antibodies to beta interferon. *Infect. Immun.* **57:**3484–3490.

186. Shi, Y., and O. Tokunaga. 2002. *Chlamydia pneumoniae* and multiple infections in the aorta contribute to atherosclerosis. *Pathol. Int.* **52:**755–763.

187. Smoak, K. A., and J. A. Cidlowski. 2004. Mechanisms of glucocorticoid receptor signaling during inflammation. *Mech. Ageing Dev.* **125:**697–706.

188. Stegmann, B. J., and J. C. Carey. 2002. TORCH Infections. Toxoplasmosis, other (syphilis, varicella-zoster, parvovirus B19), rubella, cytomegalovirus (CMV), and herpes infections. *Curr. Womens Health Rep.* **2:**253–258.

189. Steinman, L. 2007. A brief history of T_H17, the first major revision in the T_H1/T_H2 hypothesis of T cell-mediated tissue damage. *Nat. Med.* **13:**139–145.

190. Stetson, D. B., and R. Medzhitov. 2006. Antiviral defense: interferons and beyond. *J. Exp. Med.* **203:**1837–1841.

191. Stetson, D. B., and R. Medzhitov. 2006. Type I interferons in host defense. *Immunity* **25:**373–381.

192. Stockinger, B., and M. Veldhoen. 2007. Differentiation and function of Th17 T cells. *Curr. Opin. Immunol.* **19:**281–286.

193. Strober, W., I. J. Fuss, and R. S. Blumberg. 2002. The immunology of mucosal models of inflammation. *Annu. Rev. Immunol.* **20:**495–549.

194. Takaoka, A., Z. Wang, M. K. Choi, H. Yanai, H. Negishi, T. Ban, Y. Lu, M. Miyagishi, T. Kodama, K. Honda, Y. Ohba, and T. Taniguchi. 2007. DAI (DLM-1/ZBP1) is a cytosolic DNA sensor and an activator of innate immune response. *Nature* **448:**501–505.

195. Takeda, K., and S. Akira. 2005. Toll-like receptors in innate immunity. *Int. Immunol.* **17:**1–14.

196. Takeuchi, O., H. Hemmi, and S. Akira. 2004. Interferon response induced by Toll-like receptor signaling. *J. Endotoxin Res.* **10:**252–256.

197. Tangirala, R. K., E. D. Bischoff, S. B. Joseph, B. L. Wagner, R. Walczak, B. A. Laffitte, C. L. Daige, D. Thomas, R. A. Heyman, D. J. Mangelsdorf, X. Wang, A. J. Lusis, P. Tontonoz, and I. G. Schulman. 2002. Identification of macrophage liver X receptors as inhibitors of atherosclerosis. *Proc. Natl. Acad. Sci. USA* **99:**11896–11901.

198. Taniguchi, T., and A. Takaoka. 2002. The interferon-alpha/beta system in antiviral responses: a multimodal machinery of gene regulation by the IRF family of transcription factors. *Curr. Opin. Immunol.* **14:**111–116.

199. Teige, I., A. Treschow, A. Teige, R. Mattsson, V. Navikas, T. Leanderson, R. Holmdahl, and S. Issazadeh-Navikas. 2003. IFN-beta gene deletion leads to augmented and chronic demyelinating experimental autoimmune encephalomyelitis. *J. Immunol.* **170:**4776–4784.

200. Theofilopoulos, A. N., R. Baccala, B. Beutler, and D. H. Kono. 2005. Type I interferons (alpha/beta) in immunity and autoimmunity. *Annu. Rev. Immunol.* **23:**307–336.

201. Tordjman, K., C. Bernal-Mizrachi, L. Zemany, S. Weng, C. Feng, F. Zhang, T. C. Leone, T. Coleman, D. P. Kelly, and C. F. Semenkovich. 2001. PPARalpha deficiency reduces insulin resistance and atherosclerosis in apoE-null mice. *J. Clin. Invest.* **107:**1025–1034.

202. Tough, D. F., P. Borrow, and J. Sprent. 1996. Induction of bystander T cell proliferation by viruses and type I interferon in vivo. *Science* **272:**1947–1950.

203. Tourbah, A., and O. Lyon-Caen. 2007. Interferons in multiple sclerosis: ten years' experience. *Biochimie* **89:** 899–902.

204. Trost, L. C., and J. J. Lemasters. 1997. Role of the mitochondrial permeability transition in salicylate toxicity to cultured rat hepatocytes: implications for the pathogenesis of Reye's syndrome. *Toxicol. Appl. Pharmacol.* **147:**431–441.

205. Venkateswaran, A., B. A. Laffitte, S. B. Joseph, P. A. Mak, D. C. Wilpitz, P. A. Edwards, and P. Tontonoz. 2000. Control of cellular cholesterol efflux by the nuclear oxysterol receptor LXR alpha. *Proc. Natl. Acad. Sci. USA* **97:**12097–12102.

206. Venkateswaran, A., J. J. Repa, J. M. Lobaccaro, A. Bronson, D. J. Mangelsdorf, and P. A. Edwards. 2000. Human white/murine ABC8 mRNA levels are highly induced in lipid-loaded macrophages. A transcriptional

role for specific oxysterols. *J. Biol. Chem.* **275:** 14700–14707.

207. **Vliegen, I., S. B. Herngreen, G. E. Grauls, C. A. Bruggeman, and F. R. Stassen.** 2005. Mouse cytomegalovirus antigenic immune stimulation is sufficient to aggravate atherosclerosis in hypercholesterolemic mice. *Atherosclerosis* **181:**39–44.

208. **Wada, J., H. Sugiyama, and H. Makino.** 2003. Pathogenesis of IgA nephropathy. *Semin. Nephrol.* **23:**556–563.

209. **Weigent, D. A., T. L. Huff, J. W. Peterson, G. J. Stanton, and S. Baron.** 1986. Role of interferon in streptococcal infection in the mouse. *Microb. Pathog.* **1:**399–407.

210. **Xie, W., A. Radominska-Pandya, Y. Shi, C. M. Simon, M. C. Nelson, E. S. Ong, D. J. Waxman, and R. M. Evans.** 2001. An essential role for NRs SXR/PXR in detoxification of choleStatic bile acids. *Proc. Natl. Acad. Sci. USA* **98:**3375–3380.

211. **Xu, L. G., Y. Y. Wang, K. J. Han, L. Y. Li, Z. Zhai, and H. B. Shu.** 2005. VISA is an adapter protein required for virus-triggered IFN-beta signaling. *Mol. Cell.* **19:**727–740.

212. **Yoneyama, H., K. Matsuno, E. Toda, T. Nishiwaki, N. Matsuo, A. Nakano, S. Narumi, B. Lu, C. Gerard, S. Ishikawa, and K. Matsushima.** 2005. Plasmacytoid DCs help lymph node DCs to induce anti-HSV CTLs. *J. Exp. Med.* **202:**425–435.

213. **Yoneyama, M., and T. Fujita.** 2007. Function of RIG-I-like receptors in antiviral innate immunity. *J. Biol. Chem.* **282:**15315–15318.

214. **Yoneyama, M., and T. Fujita.** 2007. RIG-I family RNA helicases: cytoplasmic sensor for antiviral innate immunity. *Cytokine Growth Factor Rev.* **18:**545–551.

215. **Yoneyama, M., M. Kikuchi, T. Natsukawa, N. Shinobu, T. Imaizumi, M. Miyagishi, K. Taira, S. Akira, and T. Fujita.** 2004. The RNA helicase RIG-I has an essential function in double-stranded RNA-induced innate antiviral responses. *Nat. Immunol.* **5:**730–737.

216. **Zelcer, N., and P. Tontonoz.** 2006. Liver X receptors as integrators of metabolic and inflammatory signaling. *J. Clin. Invest.* **116:**607–614.

217. **Zheng, S. J., J. Jiang, H. Shen, and Y. H. Chen.** 2004. Reduced apoptosis and ameliorated listeriosis in TRAIL-null mice. *J. Immunol.* **173:**5652–5658.

218. **Zile, M. H.** 2001. Function of vitamin A in vertebrate embryonic development. *J. Nutr.* **131:**705–708.

219. **Zuniga, E. I., B. Hahm, and M. B. Oldstone.** 2007. Type I interferon during viral infections: multiple triggers for a multifunctional mediator. *Curr. Top. Microbiol. Immunol.* **316:**337–357.

Cellular Signaling and Innate Immune
Responses to RNA Virus Infections
Edited by A. R. Brasier et al.
© 2009 ASM Press, Washington, D.C.

Lennart Svensson
Elin Kindberg

12

Human Genetic Factors Involved in Viral Pathogenesis

The observation that certain individuals appear to have an inherited factor making them more susceptible or resistant to common infections is not new. Indeed, most of us know individuals who are more frequently ill than others or, the opposite, who are seldom ill. However, not long ago it was assumed that the clinical outcome of an infectious disease was due mainly to virulence factors associated with the microorganism, and little attention was given to host genetics. Today we are starting to reveal a complex interplay between environmental (microbial and nonmicrobial) and human (genetic and nongenetic) factors that determine immunity to infection or the resulting outcome of an infection. The behavior of pathogenic microorganisms can vary so much between strains that the effects of individual variation are best seen when the same strain of a microbe simultaneously infects previously unexposed individuals. This has occurred after inoculation or through accidents. In the early 1980s, hemophiliacs received human immunodeficiency virus (HIV)-infected blood products by mistake, and it soon became clear that the rate at which they progressed to AIDS differed between the infected individuals. While differences in susceptibility are more frequently observed at the individual level, differences can also be observed at the level of population. An example of this is that the Fulani tribe in West Africa appears more resistant to severe malaria than neighboring ethnic groups.

While mouse models have been explored to study pathogenesis and immunity of human viruses, surprisingly little information has been obtained about the factors that correlate with protection from disease, which is not necessarily the same as protection from infection. One reason is that certain human viruses are not naturally permissive for animals, and another is that administration of the virus sometimes is different from the natural route in humans. The principal advantage of the human model over animal models is that infection and immunity occur in natural as opposed to experimental conditions. Most human viruses are human-tropic and endure natural selection and coevolution with our own species. It is rare that virus-infected mice are used for breeding that have endured natural selection. Furthermore, several laboratory-inbred mice are not only immunodeficient but also show an age-dependent susceptibility to clinical disease. One such example is rotavirus. Rotavirus is the major cause of acute gastroenteritis in young children but can also symptomatically infect adults and the elderly. However, in contrast to humans, rotavirus-naïve mice

Lennart Svensson and Elin Kindberg, Division of Molecular Virology, Department of Clinical and Experimental Medicine, Medical Faculty, University of Linköping, 58185 Linköping, Sweden.

older than 2 weeks are resistant to symptomatic infection. More important, while no mortality is observed in rotavirus-infected mice, 600,000 children die due to rotavirus infections each year. Another contradiction between findings in mice and humans is that there is Mx-mediated natural resistance to influenza virus in some strains of mice (44), while humans are still highly susceptible to influenza virus infections despite the presence of Mx. Also, mice deficient in TAP1 (transporter associated with antigen processing-1) are not able to raise CD8$^+$ T-cell responses against a variety of viruses, including influenza virus, suggesting a strong dependency on antiviral CD8$^+$ cells together with a functional TAP complex. Over the last few years, several patients with defects in the HLA class I presentation pathway have been described (12). Analysis of clinical symptoms and immunological parameters of these TAP-deficient patients has revealed unexpected differences between TAP-deficient mice and TAP-deficient patients. Most surprisingly, lack of functional TAP in humans increases susceptibility to bacterial rather than viral infections (27). These findings illustrate the importance of comparing the phenotypes of mouse models with clinical features of patients.

Inherited factors modulating the outcome of a viral infection or disease could be host genes encoding cellular receptors for a virus or genes modulating innate or adaptive immune responses. While host genes involved in immune responses typically would affect the clinical pattern and clearance of an infection, absence of a functional cellular receptor for virus would prevent infection per se. A striking example of receptor dependence is the complete resistance to parvovirus B19 infection among individuals lacking the erythrocyte P antigen (8), the cellular receptor for parvovirus B19. Parvovirus B19 causes fifth disease or erythema infectiosum, a mild childhood illness characterized by rash. Furthermore, almost complete resistance is observed against winter vomiting disease caused by norovirus among individuals who are so-called nonsecretors. Nonsecretors are homozygous carriers of a nonsense mutation (G428A) in the *FUT2* gene encoding a fucosyltransferase and are protected against symptomatic norovirus disease (60). Strong protection against infection with HIV has also been found in individuals with a 32-bp deletion in chemokine receptor 5 (CCR5Δ32) (20).

The relationship between human genetics and infectious diseases has been studied following three different pathways: inborn errors of immunity that affect a single gene (Mendelian), resulting in predisposition for an infectious agent; multigenic (non-Mendelian) predisposition or resistance to common infectious agents; and finally, common Mendelian resistance to infection (Table 1). At least

five Mendelian susceptibility traits to a single pathogen have been elucidated and three Mendelian resistance traits to a specific pathogen have been identified (89). The interesting observation with the last group is that individuals homozygous for the wild-type alleles are susceptible to each pathogen, while homozygous mutant individuals display resistance (Table 1). It is reasonable to believe that many more mutations that confer Mendelian resistance or susceptibility will be discovered.

In this article we review genetic traits associated with susceptibility and resistance to several RNA viruses.

HIV

HIV is an enveloped retrovirus that has caused more than 25 million deaths worldwide. The devastating effect of HIV is caused by infection of immune cells, which debilitates the immune system, making the host more prone to catching opportunistic infections that can eventually lead to death without treatment.

For HIV to enter a cell, the glycoprotein gp120 has to interact with CD4 on the host cell. This causes conformational changes of the viral glycoprotein, and otherwise hidden domains of gp120 can interact with the coreceptor, most often CCR5 or CXCR4. CCR5-using (R5) strains are often found in the primary and early stages of infection, while CXCR4-using (X4) viral strains are found in the later stages of the disease. Strains that can use both coreceptors are termed R5X4. Binding of the coreceptor stabilizes host cell-virus binding and triggers changes in the conformation of gp41, which eventually cause fusion of the virus membrane with the host cell.

It is well known that our genes determine both susceptibility to HIV infection and disease progression of an already acquired infection. Host susceptibility studies on HIV have revealed why some individuals stay uninfected despite being exposed to the virus, while others develop AIDS shortly after the primary infection (3, 82) (Table 2). These new findings may not only help us understand and personalize antiretroviral treatment of HIV patients but may also lead to the development of new drugs, such as the CCR5 antagonist maraviroc, recently introduced on the market.

It was noted early on that some individuals stayed HIV seronegative despite high-risk behavior. A study of 424 prostitutes in Kenya, seronegative at the beginning of the study, revealed that persistent seronegativity could not be associated with differences in risk factors such as use of condoms, infection with other sexually transmitted diseases, or number of sex partners, but rather that some individuals have a natural protection against HIV-1 (26).

Table 1 Genetic factors associated with resistance or predisposition to specific viral infections[a]

Virus	Clinical phenotype	Infection/immunological phenotype	Species	Locus or gene	Reference(s)
Norovirus	Natural resistance	Lack of virus receptor	Human	*FUT2*	52, 61, 112
WNV	Susceptible	Impaired leukocyte migration	Human and mice	*CCR5*	30, 31
WNV	Susceptible	Impaired immune response to WNV	Mice	OAS L1	70
TBEV	Increased risk of encephalitis	Impaired leukocyte migration to CNS	Human	CCR5Δ32	54
HIV-1	Delayed progression to AIDS/resistance to infection	Unknown	Human	*FUT2*	1, 6, 53
HIV-1	Natural resistance/delayed progression	Lack of virus receptor	Human	*CCR5*	20
HSV-1	Increased risk of encephalitis	Impaired IFN response through TLR	Human	UNC-93B	11
HSV-1	Increased risk of encephalitis	TLR3 deficiency	Human	TLR3	125
HPV	Epidermodysplasia verruciformis	Unknown	Human	*EVER1, EVER2*	95, 107, 109
EBV	X-linked lymphoproliferative disease	Impaired host control of EBV	Human	SAP	78
Mouse hepatitis virus	Susceptible	Impaired immune response	Mice	Ceacam	43
Influenza A virus	Susceptible	Lack of Mx	Mice	*Mx*	37

[a] HSV, herpes simplex virus; TLR, Toll-like receptor; HPV, human papillomavirus; EBV, Epstein-Barr virus.

By 1995 it had been noted that RANTES, macrophage inflammatory protein (MIP)-1-α/CCL3, and MIP-1-β/CCL4, produced by CD8$^+$ cells, have an HIV-suppressive effect (18). These chemokines are the natural ligands of CCR5, and the presence of the ligand naturally inhibits binding of HIV to its coreceptor. At the time of the study, however, CCR5 was not known to be part of the viral entry process, and it was speculated that the protective effect was due to a chemoattractive effect on T cells and monocytes or that the chemokines had a direct antiviral effect (18). More recently, polymorphisms that reduce RANTES expression have been shown to protect from death in HIV-seropositive Ugandans with advanced disease (19).

The information that CCR5 ligands had a suppressive effect on HIV infection in vitro focused attention on the role of chemokine receptors in the natural process of HIV infection. Fusin, the chemokine receptor today

known as CXCR4, was found to be the coreceptor for T-cell-tropic HIV-1 strains (24), and a few months later Alkhatib et al. (2) and Choe et al. (16) showed that CCR5 was the coreceptor used by macrophage-tropic HIV-1 strains. The importance of CCR5 for infection of macrophages was later proved by the fact that a mutant *CCR5* allele, CCR5Δ32, giving a nonfunctional chemokine receptor was protective against HIV infection when carried homozygously (20). However, there have been reports of infection in homozygous carriers of this mutation (83), showing that the resistance against disease is not total.

The CCR5Δ32 mutation consists of a 32-bp deletion in the open reading frame, which causes a frame shift in amino acid 185 (20) corresponding to the second extracellular loop (64, 97). This allelic variant encodes a truncated protein, which cannot be expressed on the cell surface. Homozygous carriers hence lack the functional

Table 2 Influence of genetic variability on susceptibility to HIV infection and AIDS

Gene	Allele	Effect	Mechanism of action	Reference(s)
CCL3L1	Gene duplication	Prevents infection	CCR5 agonist	32
CCR5	Δ32 homozygosity	Prevents infection	Knock out CCR5 expression	20
CCR5	Δ32 heterozygosity	Delays disease progression	Decrease CCR5 expression	20
CCR5	P1	Accelerates disease progression	Increase CCR5 expression	68
CCR2	64I	Delays disease progression	Reduce CXCR4 availability	106, 115
SDF1	3′A	Delays disease progression	Unknown, may be due to gene upregulation	118
HLA	A, B, C homozygosity	Accelerates disease progression	Decrease breadth of antigen recognition	113
HLA	Supertype B7 (common)	Accelerates disease progression	High rate of HIV escape mutants	113
HLA	Supertypes B27 and B57 (uncommon)	Delays disease progression	Effective recognition of HIV	113
HLA	Bw4	Delays disease progression	Effective recognition of HIV	25
KIR	3DS1	Delays AIDS	Epistatic effect with specific HLA alleles	67, 69
FUT2	428A	Prevents infection	Unknown	1, 6
FUT2	428A	Delays disease progression	Unknown	53

CCR5 protein, and macrophage-tropic HIV-1, normally using CCR5 as a coreceptor, is not able to infect their cells. Heterozygous carriers of the CCR5Δ32 mutation express only low levels of CCR5 and often experience slow disease progression compared with CCR5 wild-type individuals (20).

Another *CCR5* allele variant, termed P1 (promoter allele 1), has been associated with accelerated progression toward AIDS (68). This promoter variant yields high expression of the *CCR5* gene and high CCR5 levels on the cell surface compared with other promoter alleles.

A mutation in *CCR2*, encoding another chemokine receptor, has also been associated with HIV, and this allele is thought to delay AIDS progression (106, 115). CCR2 is a chemokine receptor seldom used as a coreceptor for HIV-1, and the mode for protection must therefore be different than that of CCR5Δ32. The mutant allele variant CCR2 V64I is, in contrast to the wild type, thought to be able to dimerize with CXCR4 polypeptides, hindering the ability of X4 HIV-1 to interact with its coreceptor (72). In contrast to CCR5Δ32, the *CCR2* allele variation seems only to affect disease progression and not the risk of being infected with HIV-1 (106). However, not all studies have been able to show an association between the *CCR2* 64I allele and delay of disease progression (105), and further studies are needed on this possible association.

Not only the presence or absence of the receptor but also the levels of the natural ligands for CCR5 and CXCR4 may affect disease progression, since these chemokines compete with the HIV particle for binding of the receptor. Several studies have been performed on allelic variants of RANTES, MIP-1-α, and MIP-1-β, as well as on stromal cell-derived factor 1 (SDF1), the only known natural ligand for CXCR4. So far the only clear association between allelic variants of genes encoding ligands of the coreceptors for HIV is the finding that a low gene copy number of *CCL3L1* is a susceptibility factor for HIV (32). *CCL3L1* is a duplicated variant of *CCL3* (*MIP-1-α*), and this chemokine is the most potent CCR5 agonist and hence the most effective competitive inhibitor of R5 HIV-1 entry. In 2005, Gonzalez et al. (32) showed that carriers of high copy numbers of CCL3L1 are less prone to getting infected by HIV-1. However, it was not the absolute number that was important but the number compared with the population-specific average. The average number of gene copies varied from 2 to 10 in populations with different ethnic backgrounds, meaning that a high (and hence protective against HIV) number in one population could be normal in another population (32).

A less clear association has been suggested between a polymorphism in the 3′-untranslated region of the *SDF1* allele and progression to AIDS. In 1998, Winkler et al. (117) showed that this mutation, termed SDF1–3′A, had a recessive protective effect, delaying the onset of AIDS. However, some studies have failed to show this association (63), and others have found that SDF1–3′A increases the disease progression rate of HIV-1 infection (105).

Apart from receptor mutations and variations in chemokine expression levels, *HLA* genotypes have been shown to affect HIV disease rate (75, 113). HLA molecules present endogenous (HLA class I; HLA-A, -B, and -C) and exogenous (HLA class II; HLA-DP, -DQ, and -DR) proteins to immune cells and thereby coordinate the immune response to various pathogens. HLA class I molecules are present on all nucleated cells and interact with receptors on $CD8^+$ cytotoxic T lymphocytes, while class II molecules are present only on antigen-presenting cells and interact with T-helper cells. The HLA system is highly polymorphic, and since different *HLA* alleles can recognize different antigenic peptides, the variation in the population ensures that one single pathogen cannot decimate the whole population. Also, at the individual level, it is important to carry a heterozygous genotype at the six *HLA* alleles, and this has been shown to affect the ability to delay disease progression of HIV-1 infection (113). Individuals homozygous at *HLA* alleles, specifically at *HLA* class I alleles, are at highest risk of rapid progression to AIDS. Apart from this, certain *HLA* alleles are beneficial in terms of handling HIV infection. It has been suggested that HIV adapts to the most common *HLA* alleles in a population, which would mean that carriers of uncommon alleles would have an advantage compared with carriers of the common alleles. The most common HLA supertype, HLA-B7, has been associated with more rapid HIV-1 disease progression compared with the uncommon HLA-B27 and -B58 supertypes, which seem to have a protective effect (113).

HLA-B alleles can be divided into two groups based on the expression of the epitopes Bw4 and Bw6. These epitopes are defined by the amino acid sequences at residues 79 to 83 at the carboxyl-terminal end of the $-\alpha_1$-helix of the HLA class I binding groove. Flores-Villanueva and coworkers (25) found that a significant proportion of so-called controllers of viremia in their study were homozygous for Bw4, while none of the noncontrollers carried this genotype, suggesting that Bw4 has a protective role in HIV infection.

Another factor studied is variants of the killer-cell immunoglobulin (Ig)-like receptor (KIR), expressed on natural killer cells. The ligands for KIR include, for example, the HLA class I molecules (mainly HLA-C), and binding transmits inhibitory or activating signals to the cell (7). Different *KIR/HLA* genotypes make the natural killer cell react differently, and some genotypes may be advantageous or disadvantageous for a specific pathogen. Also, combinations of different KIR haplotypes have different levels of inhibition or activation (7). These haplotypes have been categorized into two main groups, A and B, of which A is seen as more inhibitory and B as more activating (123). The *KIR* genes also include polymorphisms that affect the level of ligand-receptor interaction.

Martin et al. (69) found an association between the activating *KIR* allele KIR3DS1 and delayed progression to AIDS among 1,039 patients from five different cohorts when KIR3DS1 was present together with HLA-B Bw4–80Ile (*HLA-B* alleles with isoleucine at position 80). In the absence of HLA-B Bw4–80Ile alleles, however, KIR3DS1 was associated with more rapid progression, suggesting that it is the epistatic interaction between the HLA and KIR loci that is determining the effect. Lopez-Vazquez and coworkers (67) studied a Zambian population but did not find the same interaction with KIR3DS1, probably because this allele is present at a low frequency in the study population. Instead, they found a protective role of KIR3DL1 together with B*57 containing Bw4–80Ile (67).

Another factor that seems to affect HIV disease is secretor status (1, 6, 53), also known to determine susceptibility to norovirus infection (60, 61, 112), as described below. Secretor status describes the presence or absence of histo-blood group antigens on mucosa and in body fluids such as saliva (48). The secretor status of an individual is determined by the *FUT2* gene (Fig. 1). Homozygous or heterozygous carriers of the wild-type *FUT2* allele are so-called secretors or secretor positive, and homozygous carriers of a nonsense allele are nonsecretors, also called secretor negative. Many polymorphisms are known in the *FUT2* gene, and approximately 20% of the European and North American population are nonsecretors due to a polymorphism at nucleotide 428 (*FUT2* G428A) (Fig. 1) (56). In 1991, Blackwell et al. showed an association between secretor status and susceptibility to HIV transmitted via heterosexual intercourse (6). Their conclusion was that nonsecretors seemed less susceptible to acquiring HIV through heterosexual contacts. This report was later followed by a study showing that Senegalese commercial sex workers carrying the nonsecretor phenotype seemed less prone to being infected by HIV through heterosexual transmission (1). In 2006, Kindberg et al. showed that secretor status also affects disease progression of HIV infection (53). Fifteen so-called long-term nonprogressors—HIV-positive individuals with a slower than expected disease progression—were genotyped regarding their secretor status. Kindberg and coworkers found that the secretor-negative phenotype was associated with slow disease progression since 10 of 15 (67%) of the long-term nonprogressors were nonsecretors, compared with 20% of the healthy controls and HIV-infected subjects with a normal disease progression rate (53).

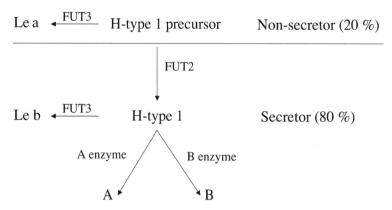

Figure 1   Secretor status, the ability to express histo-blood group antigens on mucosa and in secretions, is determined by the *FUT2* gene. Individuals lacking a functional allele are so-called nonsecretors, and carriers of at least one functional allele are termed secretors. Secretor status also indirectly determines the Lewis phenotype. Nonsecretors with a functional FUT3 enzyme are Lewis a (Le a) positive and secretors are Lewis b (Le b) positive. Individuals lacking the FUT3 enzyme are so-called Lewis negative.

NOROVIRUS

Noroviruses are the major cause of acute gastroenteritis world wide and are associated with the illness winter vomiting disease, characterized by a short incubation period (24 to 48 h) and significant vomiting and diarrhea. Noroviruses were earlier called Norwalk virus, Norwalk-agent, and "small round structured viruses." Norwalk virus was identified in an explosive outbreak of gastroenteritis in an elementary school in Norwalk, OH, in which 50% of teachers and students become ill. The virus (Fig. 2) was soon given to volunteers (21, 46, 85, 102, 121), and a general observation from the volunteer studies was that a subset of individuals never become symptomatically infected even after repeated inoculations. Parrino and coworkers reported that 6 of 12 individuals challenged with Norwalk virus developed clinical symptoms of infection. When rechallenged 27 to 42 months later, the six who become ill initially again had gastroenteritis, while the other six volunteers were resistant to two infections. Paradoxically, three of the volunteers who did not become ill following two challenges had little if any immune response. The authors made at that time a most provocative hypothesis and suggested host genetic control of susceptibility (85).

Schreiber and coworkers (102) inoculated six volunteers with another norovirus strain, the Hawaii agent, and two of the individuals developed gastroenteritis and intestinal mucosal lesion. The lesion was indistinguishable from the intestinal lesion demonstrated previously in Norwalk-agent-infected volunteers. Of interest was that two of the volunteers who remained asymptomatic developed the typical histological lesion characteristic of

infection with other strains. In another study from the same group (101), 12 of 15 volunteers acquired clinical gastroenteritis. In a multiple-challenge study of host susceptibility to Norwalk gastroenteritis, Johnson and coworkers (46) inoculated 40 subjects, of whom 22 were challenged 6 months later and 19 received a third challenge after 6 more months. Their major observation was that all 12 individuals with high ($\geq$1:200) prechallenge antibody titers but only 19 of 30 with low (<1:100) antibody titers responded with illness or a fourfold increase in antibody titer after the first challenge. Thus, those with high serum antibody titers were significantly more likely to experience disease. Furthermore, the authors

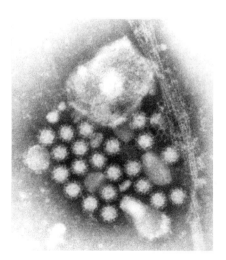

Figure 2   Electron micrograph of human norovirus in a clinical specimen.

concluded, "There appears to be a group in the USA who maintain low serum antibody levels yet are restricted to Norwalk infection, even after repeated challenge."

A more recent volunteer study investigated susceptibility to Norwalk virus with respect to secretor status, described earlier. Lindesmith and coworkers (62) inoculated 77 volunteers with Norwalk virus; of these, 71% were secretor positive and 29% were secretor negative. Norwalk virus-like particles (VLPs) bound to saliva from 41 of the 55 secretor-positive volunteers, but did not bind to any of the saliva samples from the 22 secretor-negative volunteers, showing that virus binding to saliva was strongly associated with secretor status. More important, none of the secretor-negative individuals responded with infection. An association between resistance to symptomatic Norwalk virus infection and nonsecretors was also found by Hutson and coworkers (45). Among 51 volunteers who participated in a Norwalk challenge study, all 8 nonsecretors were resistant to infection.

As mentioned earlier, the *FUT2* gene is responsible for the secretor phenotype. *FUT2* encodes an α1,2-fucosyltransferase that produces the carbohydrate H type 1 found on the surface of epithelial cells and mucosal secretions (Fig. 1), and this structure or a related structure probably functions as a receptor for the norovirus. The polymorphisms found in the *FUT2* gene show high ethnic specificity (56, 65), with the nonsense G428A mutation found in Caucasian populations (48, 56) and the C571T nonsense mutation found mainly in Pacific Islanders (40).

Since evidence for the role of secretor status in norovirus infection was first obtained with the less common genogroup I Norwalk virus that is rarely found in authentic outbreaks, it was important to investigate if the same phenomenon would also be valid in outbreaks associated with the globally dominant genotype II norovirus strains. To address this question, the role of the *FUT2* polymorphism and secretor status was investigated in norovirus outbreaks in Sweden (112). Allelic polymorphism characterization at nucleotide 428 from symptomatic ($n = 53$) and asymptomatic ($n = 62$) individuals associated with nosocomial and sporadic norovirus outbreaks revealed that a homozygous nonsense mutation (G428A) in *FUT2* segregated with complete resistance to the disease. Of all symptomatic individuals, 49% were homozygous ($FUT2^{+/+}$) and 51% heterozygous secretors ($FUT2^{+/-}$). None were secretor negative ($FUT2^{-/-}$), in contrast to 20% nonsecretors ($FUT2^{-/-}$) among Swedish blood donors (n = 104) (P < 0.0002) and 29% among asymptomatic individuals associated with nosocomial outbreaks (P < 0.00001). In addition, it was shown that the virus

strains from these outbreaks could bind to saliva mucins from *FUT2*-positive individuals who express the α1,2-linked fucose but not from $FUT2^{-/-}$ individuals who lack this carbohydrate, clearly linking their genetic resistance to the lack of ligand for the virus. Furthermore, Kindberg et al. (52) genotyped 61 individuals involved in five outbreaks in Denmark at alleles 428 and 571 of the *FUT2* gene and found complete resistance to symptomatic norovirus disease among individuals with the G428A nonsense mutation. Larsson et al. (58) studied the prevalence and titers of antibodies against norovirus and found that nonsecretors had significantly lower titers of norovirus-specific antibodies than secretors, which strengthens the hypothesis that secretor status determines susceptibility to norovirus infection. Since both the volunteer studies and the authentic outbreak studies revealed a strong correlation between secretor status and VLP binding to salivary mucin carbohydrates, it is reasonable to assume that carbohydrate-VLP binding analysis will provide information on norovirus–histo-blood group antigen interactions valid for authentic infections.

Most of the susceptibility studies of norovirus have been done in Caucasian and North American populations, and it is possible that susceptibility factors differ in other populations. It is possible that the pathogen adapts to its environment to be able to infect as many individuals in a population as possible. As mentioned, Lindesmith et al. (62) showed that the secretor-positive phenotype together with blood groups A and O was a susceptibility factor for norovirus in an American population, and it is an interesting note that secretor-positive individuals carrying the A or O blood group comprise the most common phenotype in the population studied.

In summary, volunteer challenge studies have concluded that (i) nonsecretors are resistant to infection, (ii) susceptibility occurs despite repeated challenge with identical virus, and (iii) short-term protective immunity can develop. Studies from authentic outbreaks show that protection from symptomatic norovirus infection is strongly associated with the G428A nonsense mutation in *FUT2*.

POLIOVIRUS

Poliovirus is a small single-standard RNA virus that belongs to the genus *Enterovirus* of the family *Picornaviridae*. Poliovirus infection is often asymptomatic, but about 4 to 8% of patients develop disease with mild fever and gastrointestinal signs (84). Poliovirus can also cause so-called aseptic meningitis, a nonparalytic illness with headache, fever, and meningeal signs but no signs of central nervous system (CNS) parenchymal involvement. About 1 in 200 poliovirus infections cause poliomyelitis (84), abbreviated polio, which is the outcome from infection of the motor neurons in the CNS (73). Many patients recover, at least partly, from the paralytic disease, while others have signs of paralysis for the rest of their lives. Poliovirus infection may in the worst case cause death due to respiratory arrest (80).

Vaccine-associated paralytic poliomyelitis (VAPP) is a disease caused by the Sabin live attenuated oral polio vaccine that affects about 1 child per 500,000 (73) who receives the first dose of the vaccine and, to a lesser extent, close contacts of vaccinees if not vaccinated.

Why some individuals develop poliomyelitis or VAPP is still not understood, but it has been demonstrated that individuals differ greatly in their response to a similar infectious dose of poliovirus, indicating that host factors play an important role in the outcome (80). A study by Morris et al. in 1973 found a significant difference in the HLA profile between a group of patients with paralytic poliomyelitis and a control group. The group of paralytic patients had a higher frequency of HLA-A*03 (79). A similar investigation during two outbreaks of paralytic poliomyelitis in Arab children showed that HLA-AW19 and HLA-B*7 were present more frequently in the children affected by the paralytic disease than in the healthy control group (59). Though they were not statistically significant, these differences suggested that HLA haplotypes lead to differential resistance to paralytic poliomyelitis. In a study of sibling pairs consisting of a patient with paralytic poliomyelitis and a nonparalytic control sibling, significantly less sharing of HLA haplotypes was observed than would be expected by chance (114). In a Russian study from 2002, which examined the immune status of children with VAPP, it was found that five out of six children had HLA-A*2 and -B*44, which considerably exceeded their average occurrence among the Belorussian population (22). However, a study conducted in Finland in 1986 during an outbreak of poliomyelitis found no distinct HLA association between patients with polio disease and those without (55).

The same Russian study also revealed that out of eight examined children with VAPP, five had IgA deficiency and one child had total variable immunodeficiency (22). In 1988, a Scottish child with low serum IgA levels and undetectable salivary IgA developed VAPP, confirming association to antibody deficiency (4). It is known that immunodeficient persons are at greater risk to develop VAPP than healthy individuals. VAPP has been observed in patients with agammaglobulinemia, HIV, and deficient cell-mediated immunity (10, 15, 41, 99, 120).

To infect a cell, poliovirus binds to the poliovirus receptor (PVR), CD155 (57, 74), which shows significant polymorphism. Saunderson et al. (100) demonstrated a significantly higher frequency of PVR polymorphism in patients with progressive muscular atrophy and other motor neuron diseases than in controls. Differences in the poliovirus CD155 receptor may result in slowly progressive viral cytopathic effects that lead to a mild motor neuron disease. Human neuroblastoma cells that are persistently infected with poliovirus select for mutants A67T, G39S, and R104Q in domain 1 of the PVR (86). The two latter mutants are suggested to be rare in humans, while the first exists as a heterozygote in 6.8% of the Caucasian population (100). Saunderson and coworkers also found that the A67T polymorphism, which is associated with persistent poliovirus infection in vitro, was more common in patients with motor neuron diseases and that the frequency of this polymorphism was significantly higher in patients with progressive muscular atrophy than in controls. The polymorphism is caused by a nucleotide change from G to A at position 199 in CD155, at exon 2, corresponding to an amino acid substitution at position 67, with an alanine being replaced by a threonine (100).

DENGUE VIRUS

Dengue virus is a mosquito-borne flavivirus estimated to cause more than 50 to 100 million cases of dengue fever and 250,000 to 500,000 cases of the severe dengue hemorrhagic fever (DHF) annually worldwide. There are four different dengue virus serotypes circulating (1 to 4), each being able to cause both asymptomatic and severe disease. Infection with one serotype induces lifelong immunity to the specific serotype but only temporary and partial protection against the other serotypes (119). Classic dengue fever produces a self-limited infection in humans with acute febrile illness (9). The more severe form of the disease, DHF, is characterized by four major clinical features: high fever and hemorrhagic phenomena, often with hepatomegaly and, in severe cases, signs of respiratory failure. This can in the worst cases be fatal, when it causes so-called dengue shock syndrome (DSS) (119). The mechanism behind DHF is not fully known, but it is known that it is more common in a second infection, suggesting an immunologically regulated pathogenesis (76). In 1988, Halstead (38) suggested that pre-existing non-neutralizing cross-reactive antibodies may facilitate virus entry on Fc receptor-bearing cells, which would increase the viral burden in a second dengue infection. DHF is mainly seen in young children experiencing their second dengue infection or in primary infection of infants born to dengue-immune mothers (36), who thus have maternal antibodies against dengue. However, it was also suggested early on that genetic factors of the host may increase or decrease the risk of developing DHF (38).

In 1981, an association between risk of developing DHF and certain HLA class I genotypes was suggested. Chiewsilp et al. (14) compared the genotypes of 87 unrelated Thai children who had been hospitalized because of DHF or DSS with those of 138 healthy controls who had no record of hospitalization due to dengue infection. They found that several *HLA-A* and *HLA-B* genotypes were related to increased risk of DSS and at least one case of an HLA-B antigen being protective. This study was followed by others, reviewed by Guzman et al. (36), of which several found associations with HLA class I genotypes and some also with an HLA class II antigen. In a recent study by Sierra et al. (104), HLA-A*31 and -B*15 were associated with higher risk of DHF, while HLA-DRB1*07 and DRB1*04 were found more frequently among controls than patients. It is an interesting annotation that the susceptibility genotypes are all, or at least mainly, HLA class I, while the protective genotypes are class II.

Dendritic cell-specific intercellular adhesion molecule-3-grabbing nonintegrin (DC-SIGN) is encoded by the *CD209* gene and expressed mainly on dendritic cells. DC-SIGN can function as a receptor for several different pathogens including HIV (29) and hepatitis C virus (HCV) (90), and recently it was also found to participate as an attachment receptor for dengue virus, essential for a productive infection of dendritic cells (81). This makes CD209 an interesting target for mutation studies. In 2003, Sakuntabhai et al. (96) showed an association between a promoter variant (DC-SIGN1–336) and an increased risk of dengue fever but a lower risk of severe DHF. The frequency of this mutant allele was similar among healthy controls and patients with DHF but was significantly increased among patients with dengue fever. The result was confirmed in three different Thai cohorts, indicating a different role for DC-SIGN in the pathogenesis of the two forms of the disease (96).

Loke et al. (66) reported an association between a variation in the vitamin D receptor (VDR) gene and a variation in the gene encoding the Fcγ receptor. They studied a Vietnamese population and found a variant allele, 352t, of the VDR gene associated with severity of dengue fever. Furthermore, they found that homozygosity for arginine instead of histidine at position 131 of the Fcγ receptor gene had a protective role for DHF. Both the VDR and Fcγ receptor have an immunoregulatory effect, but the exact mechanism behind the association to dengue virus infection is not known (66).

Another interesting finding was made by Kalayanarooj et al. (47), who found an association between the AB blood group and increased risk of severe dengue disease. Their study was performed on 399 patients with dengue fever who were stratified for severity of disease. ABO blood group frequencies did not differ between patients with primary and secondary dengue infection, but in the secondary infection, individuals with blood group AB were at higher risk of developing the most severe form of DHF (grade 3 of 3) than were those with blood group O, A, or B (47). More studies are needed to confirm the link between blood groups, VDR, and the Fcγ receptor and dengue virus disease.

WEST NILE VIRUS

West Nile virus (WNV) is a flavivirus that is usually transmitted to humans via mosquito bites. Symptomatic infections are typically biphasic, with an initial acute phase of fever, headache, myalgia, and fatigue lasting for about a week. Less than 1% of infected individuals develop severe neurological symptoms such as meningitis, encephalitis, or acute flaccid paralysis, but the risk of severe disease increases with age (reviewed in reference 88).

WNV appeared for the first time in 1937, in Uganda. Today it is known to cause disease in several places in the world. Why some individuals develop severe neurological disease while most infections are asymptomatic is not clear, but host genetic factors have been suggested. In 2002, Perelygin and coworkers (87) and Mashimo and coworkers (70) showed that an isoform of the 2′,5′-oligoadenylate synthetase (OAS) gene, L1, affects susceptibility to WNV in mice. Mashimo et al. (70) found a complete association between the OAS L1 isoform and susceptibility to WNV infection in mice, and the results from this study, as well as from the study by Perelygin et al., suggest that OAS is important for the host response to WNV (70, 87), at least in mice.

OAS is an interferon (IFN)-induced protein that binds viral double-stranded (ds) RNA, which leads to RNA degradation. Humans have four genes encoding proteins of the 2′,5′-OAS family (OAS1 to OAS3 and OASL). While OAS1 to OAS3 have enzymatic activity that is part of the antiviral response, OASL does not. In a study in patients hospitalized with severe WNV infection in the United States, it was shown that WNV-infected patients had a higher frequency than controls of the wild-type *OASL* allele C210T (122). The C210T mutation, which is more common among healthy controls, is thought to decrease the amount of full-length *OASL* gene transcripts, hence decreasing the inhibitory effect of OASL-binding dsRNA instead of the functional OAS. This would mean that carriers of the wild-type *OASL* allele have less effective degradation of dsRNA through the OAS pathway than carriers of the mutant allele, due to higher levels of OASL, and thus the mutation would be protective against severe WNV infection (122). Larger studies are needed on this and other mutations in the *OAS* gene family to confirm the association to WNV infections.

Two recent studies by Glass et al. (30, 31) have suggested that a mutation in the *CCR5* gene is associated with an increased risk of severe WNV infection. In the first study, Glass et al. showed that CCR5$^{-/-}$ mice had an increased viral burden and a decreased infiltration of CCR5-positive NK1.1$^+$ cells, macrophages, and CD4$^+$ and CD8$^+$ cells in the brain (30). The fatality rate was 100% among CCR5$^{-/-}$ mice and 40% among CCR5$^{+/+}$ mice. Transfer of splenocytes from infected CCR5$^{+/+}$ mice to infected CCR5$^{-/-}$ mice increased the survival of the CCR5$^{-/-}$ mice to the same level as the CCR5$^{+/+}$ mice (30). In the second study, Glass et al. showed that CCR5 deficiency also increases the risk of symptomatic WNV infection in humans (31). Two independent cohorts of Caucasian patients with laboratory-confirmed, symptomatic WNV infection were compared with a control population of healthy Caucasians regarding the deletion of 32 base pairs in the open reading frame of the *CCR5* gene, described earlier. A strong association was found between the *CCR5Δ32* allele and increased risk of developing symptomatic disease in both populations studied (P = 0.0013 and P < 0.0001, respectively). The association was also strong when comparing fatal cases of WNV infections with controls (P = 0.03) (31).

Several other studies of WNV infection of mice with different immunodeficiencies, such as B-cell, IFN, or complement deficiencies, have been done (reviewed in reference 98), but there is a need for more studies on human genetic susceptibility factors.

TICK-BORNE ENCEPHALITIS VIRUS

Another member of the flaviviruses is the tick-borne encephalitis virus (TBEV), which is endemic in many areas of Europe and Asia and is mainly transmitted to humans by *Ixodes ricinus* and *Ixodes persulcatus* ticks. The virus is maintained in nature in a cycle involving ticks and wild vertebrate hosts, often small mammals.

There are three main subtypes of TBEV species: Far Eastern, Siberian, and Western European. The Western European subtype causes a biphasic disease similar to that of WNV. Serological surveys have suggested that between 70 and 95% of infections with the European

subtype in endemic regions are asymptomatic (33). Symptomatic infections start with an influenza-like phase of fever, headache, malaise, and myalgia. About 20 to 30% of patients experience a second phase with neurological symptoms such as meningitis or meningoencephalitis. The fatality rate is about 1 to 2% and the disease is generally milder among children than adults. Human infection with the Far Eastern subtype results in the most severe form of CNS disorder, and case fatality rates of 20 to 60% have been recorded. The Siberian type has a fatality rate of 6 to 8% and often causes a milder acute disease and a high prevalence of the nonparalytic form of encephalitis.

There is limited information regarding human genetic susceptibility factors associated with TBEV. However, studies have been done to measure chemokine levels in serum and cerebrospinal fluid (CSF) of patients with TBEV, even though no genetic reason for differences was discussed. These studies may still provide important knowledge about the pathogenesis of TBEV infections, giving clues about genetic factors that may be of interest. In a study from 2006 (35), the levels of the chemokine MIP-1-α in serum and CSF were compared between TBEV-infected patients with varying degrees of symptoms and controls without infectious disease of the CNS. The synthesis of MIP-1-α was found to be increased among TBEV-infected patients, but the levels were lower in CSF than in serum, indicating that MIP-1-α does not seem to play a significant role as a chemoattractant to the CNS (35). The same group also studied the levels of RANTES in CSF and serum in TBEV-infected patients and patients with no neurological infection or infection with TBEV (34). The results showed that the concentration of RANTES in serum did not differ between the groups but the concentration in CSF was significantly higher among TBEV-infected patients than controls. The level of RANTES did not correlate with severity of disease, and the higher level was sustained after the disappearance of the symptoms of acute infection (34).

The only study on genetic differences and susceptibility to TBEV known to us is our study on TBEV and CCR5Δ32 (54). We compared 129 TBEV-infected patients with CNS symptoms with 76 patients with aseptic meningitis and 134 controls regarding the deletion of 32 base pairs in the *CCR5* gene, which is also associated with an increased risk of symptomatic WNV infection (31). Homozygous carriers of the CCR5Δ32 allele were only found among TBEV-infected patients (P = 0.026) and the allele frequency was higher among TBEV-infected patients than among other groups studied (P = 0.065), indicating that the deletion may increase the risk of developing TBEV. The allele frequency also increased with severity of disease, although no significant association was found (54). An interesting note is that the distribution of CCR5Δ32 and the Western European TBEV is similar.

HEPATITIS C VIRUS

HCV, a flavivirus earlier known as non-A, non-B hepatitis, was first identified as a causative agent of hepatitis in 1989. It was found to be the major cause of transfusion-acquired hepatitis before new methods to screen blood supplies were developed. Today, about 3% of the human population is infected with HCV.

The primary hepatitis C infection is often asymptomatic and is normally only identified if a known potential risk of infection has occurred. More than 70% of those infected with HCV develop a chronic infection, which predisposes the patient to develop chronic active hepatitis, liver cirrhosis, and hepatocellular carcinoma. Due to the high rate of chronic infections, HCV is today the most common cause of liver transplantations in the Western world. Progression from acute to chronic HCV is generally defined as persistence of increased levels of aminotransferases for 6 months or longer. Extrahepatic manifestations, such as mixed cryoglobulinemia or diabetes mellitus, are common among chronic carriers of HCV. Except for hepatic tropism, HCV shows a particular lymphotropism that causes many of the extrahepatic manifestations.

Apart from viral and environmental or behavioral factors, host genetic differences are also believed to contribute to the different outcomes of HCV infections. In 1994 it was revealed that batches of anti-rhesus D Ig given to Rh-negative women in Ireland during 1977 and 1978 were positive for HCV genotype 1 (92). Screening of 62,667 women who had received the preparations between 1970 and 1994 showed that 704 individuals (1.1%) had signs of past or current HCV infection. Of these, 390 individuals (55%) were positive for HCV RNA in serum (49). Since these women were all infected with the same strain of HCV, they are of great interest for studies of how host genetics affects the outcome of disease. Barrett and coworkers (5) found an association between HLA-DRB1*01 and spontaneous viral clearance in the above-mentioned Irish cohort, and these results were later confirmed by Fanning and coworkers (23). Thio et al. (110) found the same allele pattern in another cohort, including individuals with different ethnic backgrounds, but the association was stronger among white subjects. An American study of 142 individuals with a history of illicit drug abuse revealed that

individuals who were able to clear an HCV infection were more likely to be white than black (116), which is similar to observations by Thomas and coworkers (111). Furthermore, it has also been reported that women and individuals infected at a younger age have a lower rate of chronicity (124). Viral persistence has also been associated with certain *HLA* alleles. However, when looking at severity of HCV disease, the results have not been consistent. Furthermore, a trend of association between HLA-DRB1*11 and less severe disease, involving different disease definitions, has been observed (reviewed in reference 124).

In 1994, an outbreak of hepatitis C occurred among patients with hypogammaglobulinemia (17). Preparations of intravenous Ig were found to be contaminated, despite having been screened for the virus. Thirty of 36 patients known to have received the infected preparation were found to be HCV positive. Twenty-four patients started treatment with IFN-α, of whom 7 completed and responded to the treatment while 6 completed treatment but did not respond. Two patients spontaneously cleared the virus, while four of the patients experienced rapid disease progression with end-stage liver disease within 18 months (17). This is a good example of how infection caused by the same virus strain can cause different levels of disease in different patients.

Mannose-binding lectin (MBL) is an important molecule of the immune system with the ability to activate the complement system. MBL is believed to recognize the surface of HCV and may therefore affect the immune response to the virus. Mutations in three codons (52, 54, and 57) in the *MBL2* gene are known to lower the amount of MBL and to impair the biological function of the protein (28). The so-called *MBL2* 0 allele is known to have a dominant negative effect over the wild-type A allele. In a study including 111 Brazilian chronic HCV patients and 165 healthy HCV-negative controls, a possible association between the mutated *MBL2* 0 allele and risk of chronic HCV infection was investigated (103). The results showed a strong correlation between the mutated 0 allele and increased risk of chronic HCV infection. However, a similar study of Europeans with HCV did not find an association between HCV disease—neither in risk of infection nor progression—and these mutations in *MBL2* (51), concluding that further human studies are needed.

The deletion of the *CCR5* gene, earlier described to affect susceptibility to HIV, WNV, and TBEV, has also been investigated for a role in HCV pathogenesis. Woitas et al. (118) found the frequency of CCR5Δ32/Δ32 to be increased in HCV patients as well as an association between Δ32 homozygosity and higher viral load. The results from this study have, however, been greatly discussed, and other groups have found no difference in CCR5Δ32 frequency between HCV patients and controls (91, 94).

Promrat et al. (94) found an association between an altered RANTES promoter allele (RANTES -403-A) and less hepatic inflammation and a mutation in the promoter of the CCR5 gene (CCR5 59029-G/A) and treatment response to IFN. More studies, however, are needed to confirm these data.

As with HIV, mutations in KIR have also been studied for an association with the outcome of HCV infections. Khakoo et al. (50) found an association between the weak inhibitory HLA/KIR interaction (i.e., the lack of strong inhibitory interactions) and a protective effect on HCV infection. When patients were stratified for expected inocula (small inoculum size [drug injections] vs. large inocula [transfusion of blood/concentrated blood products]), the protective effect was only seen among those with expected small inoculum size. The suggested explanation for this is that when the exposure dose is too big, the innate immune system may be overwhelmed and unable to eliminate the virus (50). Montes-Cano and coworkers, however, found the KIR haplotype with stronger inhibitory interaction to be associated with HCV clearance (77). They were, however, not able to stratify their patients for inoculum size, which may explain the discrepancy. Also, differences in ethnicity of the populations studied may have an impact on the outcome.

Chen et al. studied the relationship between polymorphisms in different cytokine genes (interleukin-4 [IL-4], IL-10, and tumor necrosis factor-α) and the risk of chronic HCV infection (13). They found no difference in allele distribution between polymorphisms in the IL-4 and IL-10 genes, but they found an increased frequency of the mutant tumor necrosis factor-α allele, -308A, to be associated with risk of chronic HCV (13). In another study, a relationship was found between mutations in the gene encoding the IL-10 receptor IL-10RA and in the IL-22 gene and response to treatment as well as to grade of inflammation (39).

HCV is generally treated with IFN-α, often together with ribavirin. However, only approximately 40% of previously untreated patients experience a sustained virologic response to this very tough treatment (93). Several virologic and host factors, such as viral genotype and age of patient, affect the effect of this combination treatment, and host genetics may also play an important role. Matsushita and coworkers have studied the possible role of MBL in response to IFN therapy, and despite a relatively modest number of patients, they found a significant difference in the number of homozygous

mutants among sustained responders and nonresponders (71). The same group, as well as Hijikata et al., also studied a mutation in the promoter of the MxA gene (42, 108). MxA is induced by IFN stimulation, and a mutation at position −88 in the MxA gene is believed to cause more efficient expression of MxA (42). Both Suzuki et al. and Hijikata and coworkers found that carriers of at least one mutated allele at nucleotide −88 had a higher chance of sustained response to IFN therapy. Suzuki et al. also stratified for viral load and found that the mutation in MxA had an effect only in patients with low viral load (108).

CONCLUSION

A significant amount of robust data identifying host genetic determinants of viral diseases has now been recognized. However, translation of this information from bench to bedside may be problematic. An interesting example comes from the pharmacokinetics of an HIV medicine, in which abacavir, a nucleoside reverse transcriptase inhibitor, can induce hypersensitivity that limits its use. Investigators identified a particular HLA haplotype that greatly increases the risk of abacavir hypersensitivity. Another example is maraviroc, a chemokine receptor antagonist. It is designed to prevent HIV infection of CD4+ T cells by blocking the CCR5 receptor. When the CCR5 receptor is unavailable, R5-tropic HIV cannot engage with a CD4+ T cell to infect the cell. However, recent data indicate increased susceptibility to severe encephalitis by WNV and TBEV in humans homozygous for CCR5Δ32. The proposed mechanism is that homozygous CCR5Δ32 individuals have impaired leukocyte targeting to the CNS. Thus, maraviroc treatment can prevent HIV infection and progression but may facilitate severe viral encephalitis.

References

1. Ali, S., M. A. Niang, I. N'Doye, C. W. Critchlow, S. E. Hawes, A. V. Hill, and N. B. Kiviat. 2000. Secretor polymorphism and human immunodeficiency virus infection in Senegalese women. J. Infect. Dis. 181:737–739.
2. Alkhatib, G., C. Combadiere, C. C. Broder, Y. Feng, P. E. Kennedy, P. M. Murphy, and E. A. Berger. 1996. CC CKR5: a RANTES, MIP-1alpha, MIP-1beta receptor as a fusion cofactor for macrophage-tropic HIV-1. Science 272:1955–1958.
3. Arenzana-Seisdedos, F., and M. Parmentier. 2006. Genetics of resistance to HIV infection: role of co-receptors and co-receptor ligands. Semin. Immunol. 18:387–403.
4. Asindi, A. A., E. J. Bell, M. J. Browning, and J. B. Stephenson. 1988. Vaccine-induced polioencephalomyelitis in Scotland. Scott. Med. J. 33:306–307.

5. Barrett, S., E. Ryan, and J. Crowe. 1999. Association of the HLA-DRB1*01 allele with spontaneous viral clearance in an Irish cohort infected with hepatitis C virus via contaminated anti-D immunoglobulin. *J. Hepatol.* 30:979–983.

6. Blackwell, C. C., V. S. James, S. Davidson, R. Wyld, R. P. Brettle, R. J. Robertson, and D. M. Weir. 1991. Secretor status and heterosexual transmission of HIV. *BMJ* 303:825–826.

7. Boyton, R. J., and D. M. Altmann. 2007. Natural killer cells, killer immunoglobulin-like receptors and human leucocyte antigen class I in disease. *Clin. Exp. Immunol.* 149:1–8.

8. Brown, K. E., J. R. Hibbs, G. Gallinella, S. M. Anderson, E. D. Lehman, P. McCarthy, and N. S. Young. 1994. Resistance to parvovirus B19 infection due to lack of virus receptor (erythrocyte P antigen). *N. Engl. J. Med.* 330:1192–1196.

9. Burke, D. S., and T. P. Monath. 2001. Flaviviruses, p. 1043–1125. *In* D. M. Knipe and P. M. Howley (ed.), *Fields Virology*, 4th ed., vol. 1. Lippincott Williams & Wilkins, Philadelphia, PA.

10. Buttinelli, G., V. Donati, S. Fiore, J. Marturano, A. Plebani, P. Balestri, A. R. Soresina, R. Vivarelli, F. Delpeyroux, J. Martin, and L. Fiore. 2003. Nucleotide variation in Sabin type 2 poliovirus from an immunodeficient patient with poliomyelitis. *J. Gen. Virol.* 84: 1215–1221.

11. Casrouge, A., S. Y. Zhang, C. Eidenschenk, E. Jouanguy, A. Puel, K. Yang, A. Alcais, C. Picard, N. Mahfoufi, N. Nicolas, L. Lorenzo, S. Plancoulaine, B. Senechal, F. Geissmann, K. Tabeta, K. Hoebe, X. Du, R. L. Miller, B. Heron, C. Mignot, T. B. de Villemeur, P. Lebon, O. Dulac, F. Rozenberg, B. Beutler, M. Tardieu, L. Abel, and J. L. Casanova. 2006. Herpes simplex virus encephalitis in human UNC-93B deficiency. *Science* 314:308–312.

12. Cerundolo, V., and H. de la Salle. 2006. Description of HLA class I- and CD8-deficient patients: insights into the function of cytotoxic T lymphocytes and NK cells in host defense. *Semin. Immunol.* 18:330–336.

13. Chen, T. Y., Y. S. Hsieh, T. T. Wu, S. F. Yang, C. J. Wu, G. J. Tsay, and H. L. Chiou. 2007. Impact of serum levels and gene polymorphism of cytokines on chronic hepatitis C infection. *Transl. Res.* 150:116–121.

14. Chiewsilp, P., R. M. Scott, and N. Bhamarapravati. 1981. Histocompatibility antigens and dengue hemorrhagic fever. *Am. J. Trop. Med. Hyg.* 30:1100–1105.

15. Chitsike, I., and R. van Furth. 1999. Paralytic poliomyelitis associated with live oral poliomyelitis vaccine in child with HIV infection in Zimbabwe: case report. *BMJ* 318:841–843.

16. Choe, H., M. Farzan, Y. Sun, N. Sullivan, B. Rollins, P. D. Ponath, L. Wu, C. R. Mackay, G. LaRosa, W. Newman, N. Gerard, C. Gerard, and J. Sodroski. 1996. The beta-chemokine receptors CCR3 and CCR5 facilitate infection by primary HIV-1 isolates. *Cell* 85:1135–1148.

17. Christie, J. M., C. J. Healey, J. Watson, V. S. Wong, M. Duddridge, N. Snowden, W. M. Rosenberg, K. A. Fleming, H. Chapel, and R. W. Chapman. 1997. Clinical outcome of hypogammaglobulinaemic patients following outbreak of acute hepatitis C: 2 year follow up. *Clin. Exp. Immunol.* 110:4–8.

18. Cocchi, F., A. L. DeVico, A. Garzino-Demo, S. K. Arya, R. C. Gallo, and P. Lusso. 1995. Identification of RANTES, MIP-1 alpha, and MIP-1 beta as the major HIV-suppressive factors produced by CD8+ T cells. *Science* 270:1811–1815.

19. Cooke, G. S., K. Tosh, P. A. Ramaley, P. Kaleebu, J. Zhuang, J. S. Nakiyingi, C. Watera, C. F. Gilks, N. French, J. A. Whitworth, and A. V. Hill. 2006. A polymorphism that reduces RANTES expression is associated with protection from death in HIV-seropositive Ugandans with advanced disease. *J. Infect. Dis.* 194:666–669.

20. Dean, M., M. Carrington, C. Winkler, G. A. Huttley, M. W. Smith, R. Allikmets, J. J. Goedert, S. P. Buchbinder, E. Vittinghoff, E. Gomperts, S. Donfield, D. Vlahov, R. Kaslow, A. Saah, C. Rinaldo, R. Detels, and S. J. O'Brien. 1996. Genetic restriction of HIV-1 infection and progression to AIDS by a deletion allele of the CKR5 structural gene. Hemophilia Growth and Development Study, Multicenter AIDS Cohort Study, Multicenter Hemophilia Cohort Study, San Francisco City Cohort, ALIVE Study. *Science* 273:1856–1862.

21. Dolin, R., N. R. Blacklow, H. DuPont, S. Formal, R. F. Buscho, J. A. Kasel, R. P. Chames, R. Hornick, and R. M. Chanock. 1971. Transmission of acute infectious nonbacterial gastroenteritis to volunteers by oral administration of stool filtrates. *J. Infect. Dis.* 123:307–312.

22. Ermolovich, M. A., E. V. Fel'dman, E. O. Samoilovich, N. A. Kuzovkova, and V. I. Levin. 2002. [Characterization of the immune status of patients with vaccine-associated poliomyelitis]. *Zh. Mikrobiol. Epidemiol. Immunobiol.* 2:42–50.

23. Fanning, L. J., J. Levis, E. Kenny-Walsh, F. Wynne, M. Whelton, and F. Shanahan. 2000. Viral clearance in hepatitis C (1b) infection: relationship with human leukocyte antigen class II in a homogeneous population. *Hepatology* 31:1334–1337.

24. Feng, Y., C. C. Broder, P. E. Kennedy, and E. A. Berger. 1996. HIV-1 entry cofactor: functional cDNA cloning of a seven-transmembrane, G protein-coupled receptor. *Science* 272:872–877.

25. Flores-Villanueva, P. O., E. J. Yunis, J. C. Delgado, E. Vittinghoff, S. Buchbinder, J. Y. Leung, A. M. Uglialoro, O. P. Clavijo, E. S. Rosenberg, S. A. Kalams, J. D. Braun, S. L. Boswell, B. D. Walker, and A. E. Goldfeld. 2001. Control of HIV-1 viremia and protection from AIDS are associated with HLA-Bw4 homozygosity. *Proc. Natl. Acad. Sci. USA* 98:5140–5145.

26. Fowke, K. R., N. J. Nagelkerke, J. Kimani, J. N. Simonsen, A. O. Anzala, J. J. Bwayo, K. S. MacDonald, E. N. Ngugi, and F. A. Plummer. 1996. Resistance to HIV-1 infection among persistently seronegative prostitutes in Nairobi, Kenya. *Lancet* 348:1347–1351.

27. Garbi, N., S. Tanaka, M. van den Broek, F. Momburg, and G. J. Hammerling. 2005. Accessory molecules in the assembly of major histocompatibility complex class I/peptide complexes: how essential are they for CD8+T-cell immune responses? *Immunol. Rev.* 207:77–88.

28. Garred, P., H. O. Madsen, U. Balslev, B. Hofmann, C. Pedersen, J. Gerstoft, and A. Svejgaard. 1997. Susceptibility to HIV infection and progression of AIDS in relation to variant alleles of mannose-binding lectin. *Lancet* 349:236–240.

29. Geijtenbeek, T. B., D. S. Kwon, R. Torensma, S. J. van Vliet, G. C. van Duijnhoven, J. Middel, I. L. Cornelissen, H. S. Nottet, V. N. KewalRamani, D. R. Littman, C. G. Figdor, and Y. van Kooyk. 2000. DC-SIGN, a dendritic cell-specific HIV-1-binding protein that enhances trans-infection of T cells. *Cell* **100:**587–597.

30. Glass, W. G., J. K. Lim, R. Cholera, A. G. Pletnev, J. L. Gao, and P. M. Murphy. 2005. Chemokine receptor CCR5 promotes leukocyte trafficking to the brain and survival in West Nile virus infection. *J. Exp. Med.* **202:** 1087–1098.

31. Glass, W. G., D. H. McDermott, J. K. Lim, S. Lekhong, S. F. Yu, W. A. Frank, J. Pape, R. C. Cheshier, and P. M. Murphy. 2006. CCR5 deficiency increases risk of symptomatic West Nile virus infection. *J. Exp. Med.* **203:** 35–40.

32. Gonzalez, E., H. Kulkarni, H. Bolivar, A. Mangano, R. Sanchez, G. Catano, R. J. Nibbs, B. I. Freedman, M. P. Quinones, M. J. Bamshad, K. K. Murthy, B. H. Rovin, W. Bradley, R. A. Clark, S. A. Anderson, R. J. O'Connell, B. K. Agan, S. S. Ahuja, R. Bologna, L. Sen, M. J. Dolan, and S. K. Ahuja. 2005. The influence of *CCL3L1* gene-containing segmental duplications on HIV-1/AIDS susceptibility. *Science* **307:**1434–1440.

33. Gritsun, T. S., V. A. Lashkevich, and E. A. Gould. 2003. Tick-borne encephalitis. *Antiviral Res.* **57:**129–146.

34. Grygorczuk, S., J. Zajkowska, R. Swierzbinska, S. Pancewicz, M. Kondrusik, and T. Hermanowska-Szpakowicz. 2006. [Concentration of the beta-chemokine CCL5 (RANTES) in cerebrospinal fluid in patients with tick-borne encephalitis]. *Neurol. Neurochir. Pol.* **40:** 106–111.

35. Grygorczuk, S., J. Zajkowska, R. Swierzbinska, S. Pancewicz, M. Kondrusik, and T. Hermanowska-Szpakowicz. 2006. Elevated concentration of the chemokine CCL3 (MIP-1alpha) in cerebrospinal fluid and serum of patients with tick borne encephalitis. *Adv. Med. Sci.* **51:**340–344.

36. Guzman, M. G., and G. Kouri. 2003. Dengue and dengue hemorrhagic fever in the Americas: lessons and challenges. *J. Clin. Virol.* **27:**1–13.

37. Haller, O., M. Frese, and G. Kochs. 1998. Mx proteins: mediators of innate resistance to RNA viruses. *Rev. Sci. Tech.* **17:**220–230.

38. Halstead, S. B. 1988. Pathogenesis of dengue: challenges to molecular biology. *Science* **239:**476–481.

39. Hennig, B. J., A. J. Frodsham, S. Hellier, S. Knapp, L. J. Yee, M. Wright, L. Zhang, H. C. Thomas, M. Thursz, and A. V. Hill. 2007. Influence of IL-10RA and IL-22 polymorphisms on outcome of hepatitis C virus infection. *Liver Int.* **27:**1134–1143.

40. Henry, S., R. Mollicone, J. B. Lowe, B. Samuelsson, and G. Larson. 1996. A second nonsecretor allele of the blood group alpha(1,2)fucosyl-transferase gene (FUT2). *Vox Sang.* **70:**21–25.

41. Hidalgo, S., M. Garcia Erro, D. Cisterna, and M. C. Freire. 2003. Paralytic poliomyelitis caused by a vaccine-derived polio virus in an antibody-deficient Argentinean child. *Pediatr. Infect. Dis. J.* **22:**570–572.

42. Hijikata, M., Y. Ohta, and S. Mishiro. 2000. Identification of a single nucleotide polymorphism in the *MxA* gene promoter (G/T at nt -88) correlated with the response of hepatitis C patients to interferon. *Intervirology* **43:**124–127.

43. Holmes, K. V., and G. S. Dveksler. 1994. Specificity of coronavirus/receptor interactions, p. 403–443. *In* E. Wimmer (ed.), *Cellular Receptors for Animal Viruses.* Cold Spring Harbor Laboratory Press, Cold Spring Harbor, NY.

44. Horisberger, M. A., P. Staeheli, and O. Haller. 1983. Interferon induces a unique protein in mouse cells bearing a gene for resistance to influenza virus. *Proc. Natl. Acad. Sci. USA* **80:**1910–1914.

45. Hutson, A. M., F. Airaud, J. LePendu, M. K. Estes, and R. L. Atmar. 2005. Norwalk virus infection associates with secretor status genotyped from sera. *J. Med. Virol.* **77:**116–120.

46. Johnson, P. C., J. J. Mathewson, H. L. DuPont, and H. B. Greenberg. 1990. Multiple-challenge study of host susceptibility to Norwalk gastroenteritis in US adults. *J. Infect. Dis.* **161:**18–21.

47. Kalayanarooj, S., R. V. Gibbons, D. Vaughn, S. Green, A. Nisalak, R. G. Jarman, M. P. Mammen, Jr., and G. C. Perng. 2007. Blood group AB is associated with increased risk for severe dengue disease in secondary infections. *J. Infect. Dis.* **195:**1014–1017.

48. Kelly, R. J., S. Rouquier, D. Giorgi, G. G. Lennon, and J. B. Lowe. 1995. Sequence and expression of a candidate for the human *Secretor* blood group α(1,2)fucosyltransferase gene (*FUT2*). Homozygosity for an enzyme-inactivating nonsense mutation commonly correlates with the non-secretor phenotype. *J. Biol. Chem.* **270:**4640–4649.

49. Kenny-Walsh, E. 1999. Clinical outcomes after hepatitis C infection from contaminated anti-D immune globulin. Irish Hepatology Research Group. *N. Engl. J. Med.* **340:** 1228–1233.

50. Khakoo, S. I., C. L. Thio, M. P. Martin, C. R. Brooks, X. Gao, J. Astemborski, J. Cheng, J. J. Goedert, D. Vlahov, M. Hilgartner, S. Cox, A. M. Little, G. J. Alexander, M. E. Cramp, S. J. O'Brien, W. M. Rosenberg, D. L. Thomas, and M. Carrington. 2004. HLA and NK cell inhibitory receptor genes in resolving hepatitis C virus infection. *Science* **305:**872–874.

51. Kilpatrick, D. C., T. E. Delahooke, C. Koch, M. L. Turner, and P. C. Hayes. 2003. Mannan-binding lectin and hepatitis C infection. *Clin. Exp. Immunol.* **132:**92–95.

52. Kindberg, E., B. Akerlind, C. Johnsen, J. D. Knudsen, O. Heltberg, G. Larson, B. Bottiger, and L. Svensson. 2007. Host genetic resistance to symptomatic norovirus (GGII.4) infections in Denmark. *J. Clin. Microbiol.* **45:**2720–2722.

53. Kindberg, E., B. Hejdeman, G. Bratt, B. Wahren, B. Lindblom, J. Hinkula, and L. Svensson. 2006. A nonsense mutation (428G→A) in the fucosyltransferase FUT2 gene affects the progression of HIV-1 infection. *AIDS* **20:**685–689.

54. Kindberg, E., A. Mickiene, C. Ax, B. Åkerlind, S. Vene, L. Lindquist, A. Lundkvist, and L. Svensson. 2008. A deletion in the chemokine receptor 5 (CCR5) gene is associated with tick-borne encephalitis. *J. Infect. Dis.* **197:**266–269.

55. Kinnunen, E., T. Hovi, and S. Koskimies. 1986. Outbreak of poliomyelitis in Finland: no distinct HLA association. *Tissue Antigens* **28:**190–191.

56. Koda, Y., M. Soejima, and H. Kimura. 2001. The polymorphisms of fucosyltransferases. *Leg. Med. (Tokyo)* **3:**2–14.

57. Koike, S., H. Horie, I. Ise, A. Okitsu, M. Yoshida, N. Iizuka, K. Takeuchi, T. Takegami, and A. Nomoto. 1990. The poliovirus receptor protein is produced both as membrane-bound and secreted forms. *EMBO J.* **9:**3217–3224.

58. Larsson, M. M., G. E. Rydell, A. Grahn, J. Rodriguez-Diaz, B. Akerlind, A. M. Hutson, M. K. Estes, G. Larson, and L. Svensson. 2006. Antibody prevalence and titer to norovirus (genogroup II) correlate with secretor (FUT2) but not with ABO phenotype or Lewis (FUT3) genotype. *J. Infect. Dis.* **194:**1422–1427.

59. Lasch, E. E., H. Joshua, E. Gazit, M. El-Massri, O. Marcus, and R. Zamir. 1979. Study of the HLA antigen in Arab children with paralytic poliomyelitis. *Isr. J. Med. Sci.* **15:**12–13.

60. Le Pendu, J., N. Ruvoen-Clouet, E. Kindberg, and L. Svensson. 2006. Mendelian resistance to human norovirus infections. *Semin. Immunol.* **18:**375–386.

61. Lindesmith, L., C. Moe, J. Lependu, J. A. Frelinger, J. Treanor, and R. S. Baric. 2005. Cellular and humoral immunity following Snow Mountain virus challenge. *J. Virol.* **79:**2900–2909.

62. Lindesmith, L., C. Moe, S. Marionneau, N. Ruvoen, X. Jiang, L. Lindblad, P. Stewart, J. LePendu, and R. Baric. 2003. Human susceptibility and resistance to Norwalk virus infection. *Nat. Med.* **9:**548–553.

63. Liu, H., Y. Hwangbo, S. Holte, J. Lee, C. Wang, N. Kaupp, H. Zhu, C. Celum, L. Corey, M. J. McElrath, and T. Zhu. 2004. Analysis of genetic polymorphisms in CCR5, CCR2, stromal cell-derived factor-1, RANTES, and dendritic cell-specific intercellular adhesion molecule-3-grabbing nonintegrin in seronegative individuals repeatedly exposed to HIV-1. *J. Infect. Dis.* **190:**1055–1058.

64. Liu, R., W. A. Paxton, S. Choe, D. Ceradini, S. R. Martin, R. Horuk, M. E. MacDonald, H. Stuhlmann, R. A. Koup, and N. R. Landau. 1996. Homozygous defect in HIV-1 coreceptor accounts for resistance of some multiply-exposed individuals to HIV-1 infection. *Cell* **86:** 367–377.

65. Liu, Y., Y. Koda, M. Soejima, H. Pang, T. Schlaphoff, E. D. du Toit, and H. Kimura. 1998. Extensive polymorphism of the FUT2 gene in an African (Xhosa) population of South Africa. *Hum. Genet.* **103:**204–210.

66. Loke, H., D. Bethell, C. X. Phuong, N. Day, N. White, J. Farrar, and A. Hill. 2002. Susceptibility to dengue hemorrhagic fever in Vietnam: evidence of an association with variation in the vitamin D receptor and Fc gamma receptor IIa genes. *Am. J. Trop. Med. Hyg.* **67:**102–106.

67. Lopez-Vazquez, A., A. Mina-Blanco, J. Martinez-Borra, P. D. Njobvu, B. Suarez-Alvarez, M. A. Blanco-Gelaz, S. Gonzalez, L. Rodrigo, and C. Lopez-Larrea. 2005. Interaction between KIR3DL1 and HLA-B*57 supertype alleles influences the progression of HIV-1 infection in a Zambian population. *Hum. Immunol.* **66:**285–289.

68. Martin, M. P., M. Dean, M. W. Smith, C. Winkler, B. Gerrard, N. L. Michael, B. Lee, R. W. Doms, J. Margolick, S. Buchbinder, J. J. Goedert, T. R. O'Brien, M. W. Hilgartner, D. Vlahov, S. J. O'Brien, and M. Carrington. 1998. Genetic acceleration of AIDS progression by a promoter variant of CCR5. *Science* **282:**1907–1911.

69. Martin, M. P., X. Gao, J. H. Lee, G. W. Nelson, R. Detels, J. J. Goedert, S. Buchbinder, K. Hoots, D. Vlahov,

J. Trowsdale, M. Wilson, S. J. O'Brien, and M. Carrington. 2002. Epistatic interaction between KIR3DS1 and HLA-B delays the progression to AIDS. *Nat. Genet.* **31:**429–434.

70. Mashimo, T., M. Lucas, D. Simon-Chazottes, M. P. Frenkiel, X. Montagutelli, P. E. Ceccaldi, V. Deubel, J. L. Guenet, and P. Despres. 2002. A nonsense mutation in the gene encoding 2'-5'-oligoadenylate synthetase/L1 isoform is associated with West Nile virus susceptibility in laboratory mice. *Proc. Natl. Acad. Sci. USA* **99:** 11311–11316.

71. Matsushita, M., M. Hijikata, Y. Ohta, K. Iwata, M. Matsumoto, K. Nakao, K. Kanai, N. Yoshida, K. Baba, and S. Mishiro. 1998. Hepatitis C virus infection and mutations of mannose-binding lectin gene MBL. *Arch. Virol.* **143:**645–651.

72. Mellado, M., J. M. Rodriguez-Frade, A. J. Vila-Coro, A. M. de Ana, and A. C. Martinez. 1999. Chemokine control of HIV-1 infection. *Nature* **400:**723–724.

73. Melnick, J. L. 1996. Current status of poliovirus infections. *Clin. Microbiol. Rev.* **9:**293–300.

74. Mendelsohn, C. L., E. Wimmer, and V. R. Racaniello. 1989. Cellular receptor for poliovirus: molecular cloning, nucleotide sequence, and expression of a new member of the immunoglobulin superfamily. *Cell* **56:**855–865.

75. Migueles, S. A., M. S. Sabbaghian, W. L. Shupert, M. P. Bettinotti, F. M. Marincola, L. Martino, C. W. Hallahan, S. M. Selig, D. Schwartz, J. Sullivan, and M. Connors. 2000. HLA B*5701 is highly associated with restriction of virus replication in a subgroup of HIV-infected long term nonprogressors. *Proc. Natl. Acad. Sci. USA* **97:**2709–2714.

76. Monath, T. P. 1994. Dengue: the risk to developed and developing countries. *Proc. Natl. Acad. Sci. USA* **91:** 2395–2400.

77. Montes-Cano, M. A., J. L. Caro-Oleas, M. Romero-Gomez, M. Diago, R. Andrade, I. Carmona, J. Aguilar Reina, A. Nunez-Roldan, and M. F. Gonzalez-Escribano. 2005. HLA-C and KIR genes in hepatitis C virus infection. *Hum. Immunol.* **66:**1106–1109.

78. Morra, M., D. Howie, M. S. Grande, J. Sayos, N. Wang, C. Wu, P. Engel, and C. Terhorst. 2001. X-linked lymphoproliferative disease: a progressive immunodeficiency. *Annu. Rev. Immunol.* **19:**657–682.

79. Morris, P. J., and M. C. Pietsch. 1973. Letter: a possible association between paralytic poliomyelitis and multiple sclerosis. *Lancet* **2:**847–848.

80. Mueller, S., E. Wimmer, and J. Cello. 2005. Poliovirus and poliomyelitis: a tale of guts, brains, and an accidental event. *Virus Res.* **111:**175–193.

81. Navarro-Sanchez, E., R. Altmeyer, A. Amara, O. Schwartz, F. Fieschi, J. L. Virelizier, F. Arenzana-Seisdedos, and P. Despres. 2003. Dendritic-cell-specific ICAM3-grabbing non-integrin is essential for the productive infection of human dendritic cells by mosquito-cell-derived dengue viruses. *EMBO Rep.* **4:**723–728.

82. O'Brien, S. J., and G. W. Nelson. 2004. Human genes that limit AIDS. *Nat. Genet.* **36:**565–574.

83. O'Brien, T. R., C. Winkler, M. Dean, J. A. Nelson, M. Carrington, N. L. Michael, and G. C. White, II. 1997. HIV-1 infection in a man homozygous for CCR5 delta 32. *Lancet* **349:**1219.

84. Pallansch, M. A., and R. P. Roos. 2001. Enteroviruses: polioviruses, coxsackieviruses, echoviruses, and newer enteroviruses, p. 723–775. *In* H. P. Knipe et al. (ed.), *Fields Virology*, vol. 1. Lippincott Williams & Wilkins, Philadelphia, PA.

85. Parrino, T. A., D. S. Schreiber, J. S. Trier, A. Z. Kapikian, and N. R. Blacklow. 1977. Clinical immunity in acute gastroenteritis caused by Norwalk agent. *N. Engl. J. Med.* 297:86–89.

86. Pavio, N., T. Couderc, S. Girard, J. Y. Sgro, B. Blondel, and F. Colbere-Garapin. 2000. Expression of mutated poliovirus receptors in human neuroblastoma cells persistently infected with poliovirus. *Virology* 274:331–342.

87. Perelygin, A. A., S. V. Scherbik, I. B. Zhulin, B. M. Stockman, Y. Li, and M. A. Brinton. 2002. Positional cloning of the murine flavivirus resistance gene. *Proc. Natl. Acad. Sci. USA* 99:9322–9327.

88. Petersen, L. R., A. A. Marfin, and D. J. Gubler. 2003. West Nile virus. *JAMA* 290:524–528.

89. Picard, C., J. L. Casanova, and L. Abel. 2006. Mendelian traits that confer predisposition or resistance to specific infections in humans. *Curr. Opin. Immunol.* 18:383–390.

90. Pohlmann, S., J. Zhang, F. Baribaud, Z. Chen, G. J. Leslie, G. Lin, A. Granelli-Piperno, R. W. Doms, C. M. Rice, and J. A. McKeating. 2003. Hepatitis C virus glycoproteins interact with DC-SIGN and DC-SIGNR. *J. Virol.* 77:4070–4080.

91. Poljak, M., K. Seme, I. J. Marin, D. Z. Babic, M. Matcic, and J. Meglic. 2003. Frequency of the 32-base pair deletion in the chemokine receptor CCR5 gene is not increased in hepatitis C patients. *Gastroenterology* 124:1558–1560; author reply 1560–1561.

92. Power, J. P., E. Lawlor, F. Davidson, E. C. Holmes, P. L. Yap, and P. Simmonds. 1995. Molecular epidemiology of an outbreak of infection with hepatitis C virus in recipients of anti-D immunoglobulin. *Lancet* 345:1211–1213.

93. Poynard, T., P. Marcellin, S. S. Lee, C. Niederau, G. S. Minuk, G. Ideo, V. Bain, J. Heathcote, S. Zeuzem, C. Trepo, and J. Albrecht. 1998. Randomised trial of interferon alpha2b plus ribavirin for 48 weeks or for 24 weeks versus interferon alpha2b plus placebo for 48 weeks for treatment of chronic infection with hepatitis C virus. International Hepatitis Interventional Therapy Group (IHIT). *Lancet* 352:1426–1432.

94. Promrat, K., D. H. McDermott, C. M. Gonzalez, D. E. Kleiner, D. E. Koziol, M. Lessie, M. Merrell, A. Soza, T. Heller, M. Ghany, Y. Park, H. J. Alter, J. H. Hoofnagle, P. M. Murphy, and T. J. Liang. 2003. Associations of chemokine system polymorphisms with clinical outcomes and treatment responses of chronic hepatitis C. *Gastroenterology* 124:352–360.

95. Ramoz, N., L. A. Rueda, B. Bouadjar, L. S. Montoya, G. Orth, and M. Favre. 2002. Mutations in two adjacent novel genes are associated with epidermodysplasia verruciformis. *Nat. Genet.* 32:579–581.

96. Sakuntabhai, A., C. Turbpaiboon, I. Casademont, A. Chuansumrit, T. Lowhnoo, A. Kajaste-Rudnitski, S. M. Kalayanarooj, K. Tangnararatchakit, N. Tangthawornchaikul, S. Vasanawathana, W. Chaiyaratana, P. T. Yenchitsomanus, P. Suriyaphol, P. Avirutnan, K. Chokephaibulkit, F. Matsuda, S. Yoksan, Y. Jacob, G. M. Lathrop, P. Malasit, P. Despres, and C. Julier. 2005.

97. Samson, M., F. Libert, B. J. Doranz, J. Rucker, C. Liesnard, C. M. Farber, S. Saragosti, C. Lapoumeroulie, J. Cognaux, C. Forceille, G. Muyldermans, C. Verhofstede, G. Burtonboy, M. Georges, T. Imai, S. Rana, Y. Yi, R. J. Smyth, R. G. Collman, R. W. Doms, G. Vassart, and M. Parmentier. 1996. Resistance to HIV-1 infection in Caucasian individuals bearing mutant alleles of the CCR-5 chemokine receptor gene. *Nature* 382:722–725.

98. Samuel, M. A., and M. S. Diamond. 2006. Pathogenesis of West Nile virus infection: a balance between virulence, innate and adaptive immunity, and viral evasion. *J. Virol.* 80:9349–9360.

99. Saulsbury, F. T., J. A. Winkelstein, L. E. Davis, S. H. Hsu, B. J. D'Souza, G. R. Gutcher, and I. J. Butler. 1975. Combined immunodeficiency and vaccine-related poliomyelitis in a child with cartilage-hair hypoplasia. *J. Pediatr.* 86:868–872.

100. Saunderson, R., B. Yu, R. J. Trent, and R. Pamphlett. 2004. A polymorphism in the poliovirus receptor gene differs in motor neuron disease. *Neuroreport* 15:383–386.

101. Schreiber, D. S., N. R. Blacklow, and J. S. Trier. 1973. The mucosal lesion of the proximal small intestine in acute infectious nonbacterial gastroenteritis. *N. Engl. J. Med.* 288:1318–1323.

102. Schreiber, D. S., N. R. Blacklow, and J. S. Trier. 1974. The small intestinal lesion induced by Hawaii agent acute infectious nonbacterial gastroenteritis. *J. Infect. Dis.* 129:705–708.

103. Segat, L., L. R. Silva Vasconcelos, F. Montenegro de Melo, B. Santos Silva, L. C. Arraes, P. Moura, and S. Crovella. 2007. Association of polymorphisms in the first exon of mannose binding lectin gene (MBL2) in Brazilian patients with HCV infection. *Clin. Immunol.* 124:13–17.

104. Sierra, B., R. Alegre, A. B. Perez, G. Garcia, K. Sturn-Ramirez, O. Obasanjo, E. Aguirre, M. Alvarez, R. Rodriguez-Roche, L. Valdes, P. Kanki, and M. G. Guzman. 2007. HLA-A, -B, -C, and -DRB1 allele frequencies in Cuban individuals with antecedents of dengue 2 disease: advantages of the Cuban population for HLA studies of dengue virus infection. *Hum. Immunol.* 68:531–540.

105. Singh, K. K., C. F. Barroga, M. D. Hughes, J. Chen, C. Raskino, R. E. McKinney, and S. A. Spector. 2003. Genetic influence of CCR5, CCR2, and SDF1 variants on human immunodeficiency virus 1 (HIV-1)-related disease progression and neurological impairment, in children with symptomatic HIV-1 infection. *J. Infect. Dis.* 188:1461–1472.

106. Smith, M. W., M. Dean, M. Carrington, C. Winkler, G. A. Huttley, D. A. Lomb, J. J. Goedert, T. R. O'Brien, L. P. Jacobson, R. Kaslow, S. Buchbinder, E. Vittinghoff, D. Vlahov, K. Hoots, M. W. Hilgartner, and S. J. O'Brien. 1997. Contrasting genetic influence of CCR2 and CCR5 variants on HIV-1 infection and disease progression. Hemophilia Growth and Development Study (HGDS), Multicenter AIDS Cohort Study (MACS), Multicenter Hemophilia Cohort Study (MHCS), San Francisco City Cohort (SFCC), ALIVE Study. *Science* 277:959–965.

107. Sun, X. K., J. F. Chen, and A. E. Xu. 2005. A homozygous nonsense mutation in the EVER2 gene leads to

A variant in the CD209 promoter is associated with severity of dengue disease. *Nat. Genet.* 37:507–513.

epidermodysplasia verruciformis. *Clin. Exp. Dermatol.* 30:573–574.

108. Suzuki, F., Y. Arase, Y. Suzuki, A. Tsubota, N. Akuta, T. Hosaka, T. Someya, M. Kobayashi, S. Saitoh, K. Ikeda, M. Kobayashi, M. Matsuda, K. Takagi, J. Satoh, and H. Kumada. 2004. Single nucleotide polymorphism of the MxA gene promoter influences the response to interferon monotherapy in patients with hepatitis C viral infection. *J. Viral Hepat.* 11:271–276.

109. Tate, G., T. Suzuki, K. Kishimoto, and T. Mitsuya. 2004. Novel mutations of EVER1/TMC6 gene in a Japanese patient with epidermodysplasia verruciformis. *J. Hum. Genet.* 49:223–225.

110. Thio, C. L., D. L. Thomas, J. J. Goedert, D. Vlahov, K. E. Nelson, M. W. Hilgartner, S. J. O'Brien, P. Karacki, D. Marti, J. Astemborski, and M. Carrington. 2001. Racial differences in HLA class II associations with hepatitis C virus outcomes. *J. Infect. Dis.* 184:16–21.

111. Thomas, D. L., J. Astemborski, R. M. Rai, F. A. Anania, M. Schaeffer, N. Galai, K. Nolt, K. E. Nelson, S. A. Strathdee, L. Johnson, O. Laeyendecker, J. Boitnott, L. E. Wilson, and D. Vlahov. 2000. The natural history of hepatitis C virus infection: host, viral, and environmental factors. *JAMA* 284:450–456.

112. Thorven, M., A. Grahn, K. O. Hedlund, H. Johansson, C. Wahlfrid, G. Larson, and L. Svensson. 2005. A homozygous nonsense mutation (428G→A) in the human secretor (FUT2) gene provides resistance to symptomatic norovirus (GGII) infections. *J. Virol.* 79:15351–15355.

113. Trachtenberg, E., B. Korber, C. Sollars, T. B. Kepler, P. T. Hraber, E. Hayes, R. Funkhouser, M. Fugate, J. Theiler, Y. S. Hsu, K. Kunstman, S. Wu, J. Phair, H. Erlich, and S. Wolinsky. 2003. Advantage of rare HLA supertype in HIV disease progression. *Nat. Med.* 9:928–935.

114. van Eden, W., G. G. Persijn, H. Bijkerk, R. R. de Vries, R. K. Schuurman, and J. J. van Rood. 1983. Differential resistance to paralytic poliomyelitis controlled by histocompatibility leukocyte antigens. *J. Infect. Dis.* 147:422–426.

115. van Rij, R. P., A. M. de Roda Husman, M. Brouwer, J. Goudsmit, R. A. Coutinho, and H. Schuitemaker. 1998. Role of CCR2 genotype in the clinical course of syncytium-inducing (SI) or non-SI human immunodeficiency virus type 1 infection and in the time to conversion to SI virus variants. *J. Infect. Dis.* 178:1806–1811.

116. Villano, S. A., D. Vlahov, K. E. Nelson, S. Cohn, and D. L. Thomas. 1999. Persistence of viremia and the importance of long-term follow-up after acute hepatitis C infection. *Hepatology* 29:908–914.

117. Winkler, C., W. Modi, M. W. Smith, G. W. Nelson, X. Wu, M. Carrington, M. Dean, T. Honjo, K. Tashiro, D. Yabe, S. Buchbinder, E. Vittinghoff, J. J. Goedert, T. R. O'Brien, L. P. Jacobson, R. Detels, S. Donfield, A. Willoughby, E. Gomperts, D. Vlahov, J. Phair, and S. J. O'Brien. 1998. Genetic restriction of AIDS pathogenesis by an SDF-1 chemokine gene variant. ALIVE Study, Hemophilia Growth and Development Study (HGDS), Multicenter AIDS Cohort Study (MACS), Multicenter Hemophilia Cohort Study (MHCS), San Francisco City Cohort (SFCC). *Science* 279:389–393.

118. Woitas, R. P., G. Ahlenstiel, A. Iwan, J. K. Rockstroh, H. H. Brackmann, B. Kupfer, B. Matz, R. Offergeld, T. Sauerbruch, and U. Spengler. 2002. Frequency of the HIV-protective CC chemokine receptor 5-Delta32/Delta32 genotype is increased in hepatitis C. *Gastroenterology* 122:1721–1728.

119. World Health Organization. 1997. *Dengue Hemorrhagic Fever. Diagnosis, Treatment, Prevention and Control,* 2nd ed. World Health Organization, Geneva, Switzerland.

120. Wright, P. F., M. H. Hatch, A. G. Kasselberg, S. P. Lowry, W. B. Wadlington, and D. T. Karzon. 1977. Vaccine-associated poliomyelitis in a child with sex-linked agammaglobulinemia. *J. Pediatr.* 91:408–412.

121. Wyatt, R. G., R. Dolin, N. R. Blacklow, H. L. DuPont, R. F. Buscho, T. S. Thornhill, A. Z. Kapikian, and R. M. Chanock. 1974. Comparison of three agents of acute infectious nonbacterial gastroenteritis by cross-challenge in volunteers. *J. Infect. Dis.* 129:709–714.

122. Yakub, I., K. M. Lillibridge, A. Moran, O. Y. Gonzalez, J. Belmont, R. A. Gibbs, and D. J. Tweardy. 2005. Single nucleotide polymorphisms in genes for 2′-5′-oligoadenylate synthetase and RNase L in patients hospitalized with West Nile virus infection. *J. Infect. Dis.* 192:1741–1748.

123. Yawata, M., N. Yawata, M. Draghi, A. M. Little, F. Partheniou, and P. Parham. 2006. Roles for HLA and KIR polymorphisms in natural killer cell repertoire selection and modulation of effector function. *J. Exp. Med.* 203:633–645.

124. Yee, L. J. 2004. Host genetic determinants in hepatitis C virus infection. *Genes Immun.* 5:237–245.

125. Zhang, S. Y., E. Jouanguy, S. Ugolini, A. Smahi, G. Elain, P. Romero, D. Segal, V. Sancho-Shimizu, L. Lorenzo, A. Puel, C. Picard, A. Chapgier, S. Plancoulaine, M. Titeux, C. Cognet, H. von Bernuth, C. L. Ku, A. Casrouge, X. X. Zhang, L. Barreiro, J. Leonard, C. Hamilton, P. Lebon, B. Heron, L. Vallee, L. Quintana-Murci, A. Hovnanian, F. Rozenberg, E. Vivier, F. Geissmann, M. Tardieu, L. Abel, and J. L. Casanova. 2007. TLR3 deficiency in patients with herpes simplex encephalitis. *Science* 317:1522–1527.

Activation and Evasion of Host Antiviral Signaling Pathways by RNA Viruses

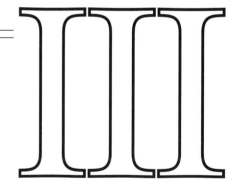

Cellular Signaling and Innate Immune
Responses to RNA Virus Infections
Edited by A. R. Brasier et al.
© 2009 ASM Press, Washington, D.C.

Peter W. Mason
Frank Scholle

13

RNA Virus Families: Distinguishing Characteristics, Differences, and Similarities

RNA viruses are unique. They are the only organisms known to use RNA to pass their genetic information to their progeny. Multiple lines of evidence exist to indicate that RNA viruses are a leftover from a world in which RNA was the only genetic material (24). Despite this "leftover" status, RNA is the most common genetic material of all known viruses, demonstrating the extraordinary fitness of this form of genetic material in a DNA world. Among the various families of RNA viruses, the single-stranded (ss) RNA viruses are the most numerous. In this chapter we will deal exclusively with RNA viruses that infect vertebrates, and we will highlight how differences between the biology of DNA and RNA viruses and differences among RNA viruses may help to determine how they can be detected by their hosts. By introducing and contrasting various aspects of the molecular biology of RNA virus families, we will provide a bridge between the chapters in section 1 of this volume, covering the host's antiviral signaling pathways, and those in section 2, describing how members of individual viral families evade or subvert these host responses.

While relationships among distantly related vertebrates can be easily calculated by evaluation of nucleic acid sequence data for conserved genes, such a luxury is not possible for RNA viruses. The only sequence that has been used to tie these viruses together is that of the RNA-dependent RNA polymerase (RdRP), which is shared by all RNA viruses and was first reported to be conserved in all known genera of positive-strand RNA viruses (23), strongly suggesting their evolution from a common ancestor. Diversity among the RNA viruses is manifest at both the structural and genetic levels. Many RNA viruses are encapsidated by a host cell-derived lipid bilayer containing viral glycoproteins, whereas others are encapsidated solely in a protein coat. At the genetic level, RNA viruses exist that have single-strand genomes that are positive sense, negative sense, or ambisense with respect to translational capacity. Other RNA viruses have double-strand genomes. Further complexity arises at the level of transcriptional control of gene expression, with some viruses using a single cistron to express all genes, whereas others have evolved complex mechanisms to create and regulate the

Peter W. Mason, Dept. of Pathology and Dept. of Microbiology & Immunology, and Sealy Center for Vaccine Development, University of Texas Medical Branch, 301 University Blvd., Galveston, TX 77555-0436. **Frank Scholle**, Dept. of Microbiology, North Carolina State University, 4512 Gardner Hall, Box 7615, Raleigh, NC 27695.

expression of their genes by producing multiple subgenomic RNAs.

RNA AND THE "OTHER" NUCLEIC ACID

The choice of selecting RNA (or retaining it, depending on your evolutionary bias) as the material for propagating genetic information has profound implications for viral biology. The use of a nucleic acid that is not actively propagated by the host cell has forced RNA viruses to encode their own RdRP. As indicated above, this polymerase contains the only sequence motif that has been used to group all RNA viruses together. Use of RNA as the genetic basis for passing information on to progeny produces differential constraints relative to the DNA viruses, resulting in several overarching differences in biology that distinguish the RNA viruses from DNA viruses. These include the fact that most (but not all, an exception being poxviruses) DNA viruses take advantage of the host cell's high-fidelity replicative machinery, reducing the need to expend viral genetic information to encode genome-replicating machinery. Nuclear localization of DNA virus genome replication may also provide convenient scaffolding for replication, whereas RNA viruses need to elaborate a purpose-built scaffold as part of the viral replication factories they produce in their host cells. Choice of genomic nucleic acid has been considered to be the driving force behind multiple DNA virus activities that force their host cells into S-phase, creating an environment favorable for viral DNA synthesis by inducing multiple DNA replication-associated enzymes and the production of additional deoxynucleotides. This acquisition of specific activities that are closely tied to the host's cell cycle could require a specialization among DNA viruses that tends to make them more host restricted than RNA viruses. In general, RNA viruses are not cell cycle dependent, but as expected, modulation of ribonucleoside triphosphates can have drastic effects on their replication.

The need to encode their own nucleic acid polymerase, and the resulting lack of fidelity of the simple replicative machine they can create with limited genetic material, contributes, in part, to the upper size limit of viable RNA virus genomes (17). Specifically, the expected lack of high-fidelity RdRPs would argue that if the genomes were extended beyond a certain size, the accumulation of errors in these genomes would produce only nonviable progeny. Although it is difficult to empirically determine how RdRP fidelity has influenced the size of the genomes of existing RNA viruses, an "inverse" experiment of testing the outcome of driving up error rates in genome replication has demonstrated that

error catastrophe is not too far away for the RNA viruses that exist today (7). These findings are consistent with the fact that the largest RNA virus genome identified to date is just over 30,000 bases in length, whereas DNA viruses exist with genomes 10 times as large. Limits on genome size likely contribute to the relatively simple transcription strategies described below, limiting transcriptional regulation of gene products and the number of gene products that RNA viruses can encode. This latter limitation produces several additional and interesting constraints, including the fact that acquisition of new functions by RNA viruses may require reshaping an existing protein product into one that can accomplish multiple activities, or employing various genetic conservation methods including expression of a new product from a secondary reading frame, or creation of an additional protein product by partial proteolytic processing or introduction of incomplete stop codons. One additional constraint imposed by the genome size that may be particularly important within the context of innate immune function is that RNA virus genomes are unlikely to be able to simply acquire whole genes (and their activities) from their hosts, a tactic that has been utilized repeatedly in the acquisition of immune function genes and decoys by the large DNA viruses.

Finally, genetic material restricts the way in which RNA viruses can be maintained within infected individuals. Many types of DNA viruses (including, of course, retroviruses, which appear to be RNA viruses that acquired a DNA lifestyle) can insert their genetic material into the chromosome of the host or have evolved mechanisms to maintain their genome extrachromosomally, permitting inter- or intragenerational persistence that has become part of their transmission strategy. In the context of immune evasion, gene expression from integrated DNA virus genomes or extrachromosomally maintained genomes is very limited and thus likely to be invisible to the host's pathogen-recognition apparatus, whereas RNA viruses that have acquired a transmission strategy that depends on long-term carriage by a host must establish a long-term persistent/chronic infection that may require more "active" host immune evasion strategies.

RNA VIRUSES HAVE EVOLVED TO FILL MANY NICHES; UNNATURAL NICHES MAY YIELD SPECTACULAR AND UNEXPECTED OUTCOMES

The concept of microbes evolving toward a state of benign commensalism has been debated for decades. Nevertheless, there are multiple examples that indicate that viruses can evolve into less pathogenic ones over time, including Fenner's groundbreaking work on myxoma virus evolution first reported in 1953 (12). Furthermore, fragments of nucleic acid found throughout the genomes of many vertebrate species have the hallmarks of remnants of microbes that have gone beyond a state of benign commensalism to become part of the genomes of their hosts. In the case of RNA viruses, speciation and adaptation to new hosts certainly represent a continuous process, in which viruses that spill over into new host species are only selected if they establish a useful relationship (in other words, effective transmission). Naturally, for many viruses where pathogenesis ensures effective shedding, for example, in the case of gastrointestinal and respiratory pathogens, virulence will undoubtedly be strongly selected for, as it may promote efficient transmission. However, the ability to produce disease is not necessarily a prerequisite for successful transmission (take, for example, murine norovirus transmission in mouse holding facilities). Thus, if disease is manifest after acquisition of a new host, evolutionary theory would predict that virulence could be lost if it is not essential for transmission. For all of these processes, the lack of fidelity of the RdRP can be expected to speed evolution, but other constraints might be expected to slow the evolution of RNA viruses, for example, the need to utilize the limited number of viral polypeptides for multiple purposes. Although we will not seek to provide definitive arguments for or against this "evolution toward avirulence" theory in this chapter, we will use the underpinnings of this idea, namely that effective viral transmission (which is not necessarily related to the ability to cause severe disease) is the driving force in viral evolution, and as such, viruses must evolve in a manner that permits them to be transmitted despite the innate and adaptive immune responses of their hosts.

To help us further establish context for the evolution of RNA viruses within the vertebrates they parasitize, we will point out that some of the most virulent infections within an individual host may be the least efficient in terms of transmission. Furthermore, multiple viruses, especially some well-known zoonotic agents, have a more or less benign parasitic relationship with their natural host(s) and only produce severe disease when they infect species outside of their normal host range. As further proof of the "unnatural" nature of these infections, most of these jumps into new hosts appear to be dead ends, with the highly pathogenic relationship not proving to be a suitable niche for viral transmission.

Key aspects of the inverse relationship between pathogenesis and effective transmission may also apply to experimental animal studies, which are often performed in inbred mice. These murine studies frequently utilize viruses that are important human pathogens and often utilize nonnatural routes of infection, resulting in diseases that do not recapitulate the outcome of infections in the natural host. In other cases, viruses are forcefully adapted to laboratory animals to produce pathogens capable of producing more profound disease, a type of adaptation that is certainly not the same as an adaptation selected in nature by the key driving factor for viral evolution, namely effective transmission. Thus, severe human infections (especially with zoonotic agents) and infections in some animal models may be limited in their usefulness for predicting how specific mechanisms of the innate immune response determine interactions of viruses with their host, and may miss key interactions that are important to the establishment of transmission cycles. However, despite the fact that many animal models and/or severe human infections may not yield useful information about the mechanisms at play during natural transmission, the disequilibrium situations they produce certainly provide useful information on virus-host interactions.

RNA VIRUS REPLICATION STRATEGIES

RNA viruses have evolved different ways to use their RNA genome to store their genetic information. RNA genomes can store genetic information in their viral particles as ssRNA of either positive, negative, or ambisense polarity, or as double-stranded (ds) RNA. Viral genomes can consist of a single RNA molecule or of multiple segments. Since host cells lack the enzymatic activity required to synthesize an RNA molecule from an RNA template, all RNA viruses need to encode their own RdRP to catalyze replication of their genomes. Based on the nature of their genomes, different strategies have to be employed in order to provide the RdRP in the host cell. In the context of immune recognition of the different virus families, described in more detail in subsequent chapters of this volume, it is important to take a look at the subcellular localization(s) where viral RNA replication occurs. With the exception of the *Orthomyxoviridae*, which replicate in the nucleus, RNA replication occurs in the cytoplasm of the host in association with molecular structures that presumably serve to anchor the RdRP and allow proper interaction among components of the viral replication machinery. Positive-strand RNA viruses are tightly associated with cellular membranes and often induce membrane proliferation in infected cells. Negative-sense and ambisense RNA viruses are tightly associated with their nucleocapsids, and neither genomes nor antigenomes are found as

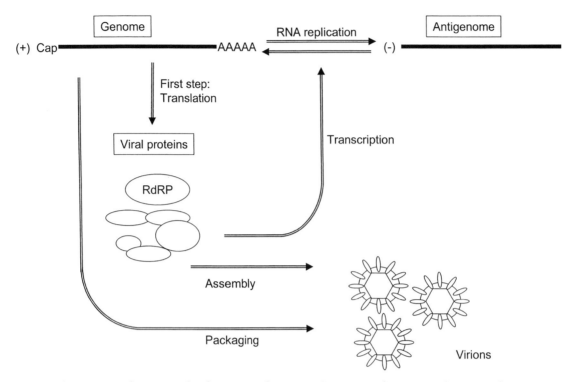

Figure 1 Replication cycle of viruses with an ssRNA genome of positive polarity. Viral genomic RNA is translated after uncoating to produce the viral proteins. Viral proteins then transcribe the genome into the minus-sense antigenome that, in turn, serves as a template for progeny genome synthesis. Progeny genomes and viral structural proteins assemble into progeny virions.

"naked" RNA molecules in the infected cells. As discussed below, dsRNA viruses are very tightly associated with their cores and newly synthesized subviral particles, in which all steps of RNA synthesis occur. Such sequestration of viral RNAs may help to limit their recognition as non-self by the cell.

In the following section of this chapter we will briefly describe general concepts of viral replication that depend on the nature of the RNA genome. The subsequent sections will then highlight the relationship between these different replication strategies and the ways in which different viruses can be detected by host cells.

Plus-Sense ssRNA Genomes

ssRNA genomes of positive polarity are infectious as naked RNA molecules. Naked plus-sense RNAs mimic mRNAs and are thus able to be translated into viral proteins as a first step of macromolecular synthesis following the uncoating of the viral genome. The genomes of many plus-strand RNA viruses contain 5′ cap structures that are recognized by translation initiation factors and the 40S ribosomal subunit to initiate viral gene expression. These caps are usually acquired through the action of virally

encoded guanylyl- and methyltransferases. Alternatively, other virus families have evolved a translation initiation mechanism independent of 5′ cap structures. Members of the *Picornaviridae* as well as some *Flaviviridae*, for example, contain RNA sequences that are able to fold into complex three-dimensional structures that serve as a landing pad for the 40S ribosomal subunit and translation initiation factors and assemble a translation initiation complex in the absence of a typical cellular mRNA cap structure. These complex dsRNA-rich structures have been termed internal ribosomal entry sites (IRESs) since they can direct translation initiation to AUG codons located at great distances from the 5' end of an RNA. Regardless of whether plus-strand viruses utilize a cap or an IRES, translation of the infecting copy of the viral genome leads to synthesis of the viral RdRP, among other viral proteins. Once synthesized this way, the RdRP together with other necessary proteins is able to transcribe the genome into its antigenomic negative-sense complement. Negative-sense antigenomes in turn serve as templates for more plus-strand synthesis. Progeny plus-sense genomes can either be translated to yield more proteins or are encapsidated into progeny virions (Fig. 1). Virus families discussed in the following chapters of section 3 of this volume that replicate following this general replication scheme are the *Picornaviridae, Flaviviridae, Togaviridae, Coronaviridae, Arteriviridae, Caliciviridae,* and *Astroviridae*. Aside from following this generalized scheme of RNA replication, viruses within the individual virus families have evolved varied mechanisms to express their proteins. While *Picornaviridae* and *Flaviviridae* only express a single open reading frame that yields a polyprotein that is proteolytically processed into mature individual viral proteins, other plus-strand RNA viruses such as the *Togaviridae, Caliciviridae, Coronaviridae,* and *Arteriviridae* generate subgenomic RNAs that specify additional open reading frames producing structural and/or accessory proteins; but in all cases, the translation of the genomic RNA produces a copy of the RdRP capable of copying the genome.

Negative-Sense ssRNA Genomes

Viruses with negative-sense RNA genomes require a different strategy to initiate their replication than described for positive-sense genomes. As pointed out above, the host cell does not provide for enzymatic activities that catalyze RNA-dependent RNA synthesis. Negative-sense RNA cannot be translated into proteins; thus, the RdRP cannot be produced this way and the naked RNA is not infectious. Negative-sense viruses therefore need to carry their RdRP with them into the host cell. Unlike positive-strand RNA viruses, whose genomes are thought to be initially "naked" within the cytoplasm of the cell following release from the infecting virion, negative-strand RNA genomes are tightly associated with many copies of nucleoproteins and, depending on the virus, one or more copies of the RdRP and accessory proteins. Following release of this ribonucleoprotein into the cell, the first step of macromolecular synthesis consists of RdRP-mediated transcription of mRNAs using the negative-sense genome as a template. 5′ caps for mRNAs are either synthesized by the polymerase complex or acquired by the action of a cap-specific endonuclease that cleaves caps off of cellular mRNAs and attaches them to viral mRNA. This latter mechanism, called "cap-snatching," is employed by members of the *Orthomyxoviridae* and *Bunyaviridae* (the latter of which utilize an ambisense strategy for genome organization; see below). Viral mRNAs are subsequently translated into various viral proteins. Minus-strand RNA is then transcribed into the plus-sense antigenomes that, in turn, serve as templates for more minus-strand synthesis (Fig. 2). Both genomic and antigenomic RNA are always associated with the nucleocapsid proteins. This overall replication scheme is valid for both viruses whose genomes consist of a single strand of negative-sense RNA, such as the *Paramyxoviridae, Rhabdoviridae,* and *Filoviridae,* and for viruses with segmented genomes, such as the *Orthomyxoviridae*. Viral protein expression is usually highly regulated to allow for high-level expression of the nucleocapsid protein needed for encapsidation of both genomic and antigenomic RNA and generally reduced expression of the catalytic enzymes required for RNA replication.

Ambisense Genomes

Ambisense viruses encode their genes both in plus- and minus-sense orientations. Genes located at the 3′ end of ambisense genomes are encoded in negative sense. In order for these latter genes to be expressed, genome-complementary mRNAs need to be transcribed. The genes at the 5′ end are encoded in plus (mRNA) sense although there is no evidence that they are directly translated from the input RNA. Viruses with ambisense genomes replicate similarly to negative-strand RNA viruses but further divide their protein synthesis into two stages, one of which provides the nucleocapsid protein and the RdRP, and the other of which leads to production of other viral proteins, including the structural proteins. As in the case of the negative-sense viruses, ambisense viruses begin their macromolecular synthesis by transcribing portions of their genome into mRNAs needed for the synthesis of the nucleocapsid protein and the RdRP. Nucleocapsid and RdRP are then responsible for RNA replication, resulting in synthesis of antigenomes and progeny genomes. Antigenomes serve as templates for the transcription of mRNAs for the other viral proteins, including the structural proteins that are encoded in the plus sense (Fig. 3). Nucleocapsid-associated progeny genomes and structural proteins then assemble into infectious progeny virions.

dsRNA Genomes

Replication of viruses with dsRNA genomes, such as the *Reoviridae,* occurs in close association with viral structures. The genomes of these viruses consist of multiple segments of dsRNA. Cell entry is followed by a limited degradation of the virion, leaving a subviral core particle in the cytoplasm of the host cell. The RdRP is part of the subviral core, and newly synthesized mRNAs are extruded from this apparatus and then translated into proteins. In contrast to other viruses, packaging of dsRNA viruses occurs at the level of the mRNAs, which are sorted into subviral particles assembled from newly synthesized viral proteins. Packaged mRNAs are then transcribed within the subviral particle by the RdRP to yield progeny dsRNA genomes. Thus, sorting of the

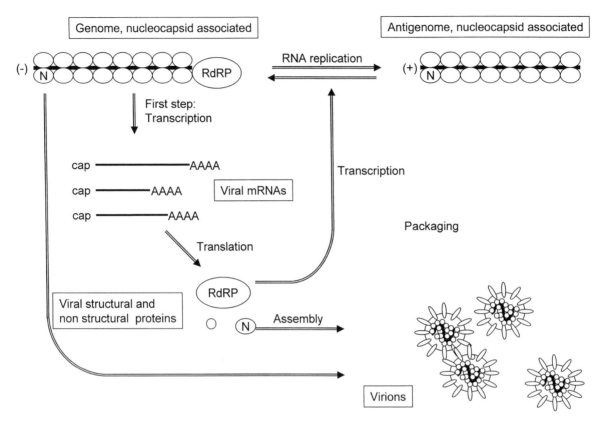

Figure 2 Replication cycle of viruses with an ssRNA genome of negative polarity. Viral genomes remain associated with the nucleocapsid protein (N) and are transcribed into mRNAs by the RdRP, which is carried into the host cell by the virion. Viral structural and nonstructural proteins are translated from the mRNAs. Nonstructural proteins serve to replicate the genome into its plus-sense antigenomic complement, which serves as a template for more genome synthesis. Nucleocapsid-associated genomes and viral structural proteins (including the RdRP) then assemble into progeny virions.

mRNAs into subviral particles needs to occur at the appropriate stoichiometry so that the full complement of the viral genome (all segments) is contained within the subviral particle following synthesis of the minus strand. These newly synthesized subviral particles then initiate a secondary round of transcription, translation, packaging, replication, and then assembly with the viral structural proteins to form complete virions (Fig. 4). It is this secondary transcription cycle followed by RNA replication that accounts for the bulk of viral amplification.

RECOGNITION OF THE NONSELF

Over 100 years ago, experiments demonstrated that sera obtained from convalescent animals contained factors that could prevent (or treat) viral infections. These findings revolutionized the prevention of diseases by providing the basis for production of vaccines that have helped to control many important viral diseases. The recognition of the specificity of this interaction led to the far-reaching paradigm that molecular recognition of an invading viral pathogen could be established by the clonal selection of pathogen-specific immunoglobulins and T cells. Likewise, the appreciation that the immune system has evolved an innate recognition system independent of clonal selection of lymphocytes has revolutionized our understanding of a variety of pathogenic states and produced new avenues to treat infectious diseases. The first chapters of this volume describe both the pathogen-associated molecular patterns (PAMPs) that are recognized by eukaryotic hosts experiencing microbial infections, as well as the receptors (pattern recognition receptors, or PRRs) that bind these PAMPs, permitting the host to recognize that it has been infected.

Logically, PAMPs that signal the presence of RNA viruses should constitute molecules that are associated with viruses themselves or molecules that are produced during their replicative cycle. Not surprisingly, specific

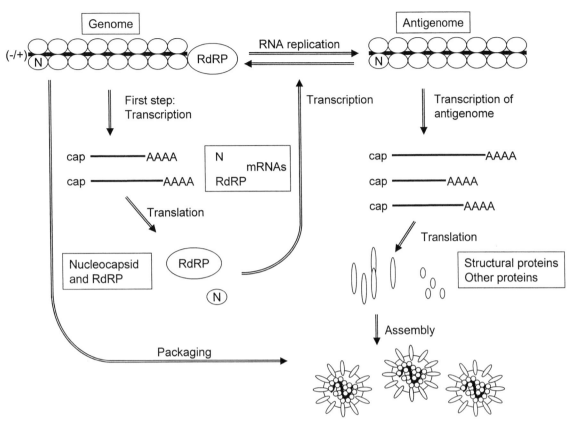

Figure 3 Replication cycle of viruses with an ssRNA genome of ambisense polarity. Nucleocapsid-associated genomes are transcribed into mRNAs for the nucleocapsid protein (N) and the RdRP. N and RdRP are involved in synthesis of the antigenome and progeny genomes. Antigenomes serve as templates for the transcription of mRNAs for structural and other viral proteins. Structural proteins and progeny genomes then assemble into progeny virions.

RNA structures, including dsRNAs and ssRNAs having a 5' triphosphate, have been shown to act as PAMPs. PAMPs recognized by PRRs can be as varied as viral glycoproteins, polysaccharides, or nucleic acids. PRRs are often expressed in specific cell types and tissues. While in several cases the same PAMP can be recognized by more than just one PRR, the availability of different PRRs with the capacity to recognize a variety of PAMPs ensures that the host is able to respond to a wide variety of different pathogens in an appropriate manner.

The major PRRs identified to date include Toll-like receptors (TLRs), nucleotide-binding oligomerization domain (NOD) family proteins, cytoplasmic RNA helicases, and the cytosolic protein kinase R (PKR). NOD family members and TLR2 and 4 participate mainly in the recognition of bacterial products, although these two TLRs can also recognize some viral glycoproteins. However, all of the remaining virus-specific PRRs appear to recognize distinct species of viral nucleic acids.

Among the TLRs, TLR9 recognizes unmethylated CpG-containing DNA found in many DNA viruses, TLR7 and 8 recognize ssRNA, and TLR3 recognizes dsRNA. dsRNA was also initially identified as a general activator of retinoic acid-inducible gene I (RIG-I) and melanoma differentiation-associated gene 5 (MDA-5) (the cytoplasmic helicases), as well as PKR. Much effort has recently been focused on further characterizing the nature of these nucleic acid ligands and how they are produced in the viral life cycle. In this section we will briefly review PRRs and their ligands and how they relate to the life cycle of different RNA virus families.

TLR7

TLR7 is expressed in many mononuclear leukocytes, notably plasmacytoid dendritic cells (pDCs), which are the major interferon (IFN)-α-producing cells in response to viral infection. TLR7 engagement activates the transcription factor IFN regulatory factor (IRF) 7, which is constitutively expressed to high levels in pDCs. TLR7

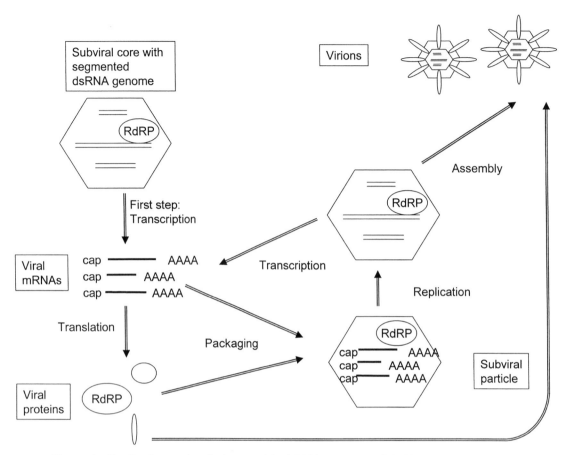

Figure 4 Replication cycle of viruses with dsRNA genomes. dsRNA genomes are transcribed into viral mRNAs within subviral core-particles by the core-associated RdRP. mRNAs are then extruded into the cytoplasm of the host cell and translated into viral proteins. mRNAs and viral proteins assemble into subviral particles, where synthesis of the minus-strand RNA occurs to yield replicated dsRNA. Subviral particles then initiate a secondary round of transcription, translation, packaging, and replication, producing newly formed cores that combine with structural proteins to form progeny virions.

was first demonstrated to act as a PRR in infection with influenza A virus (IAV) and vesicular stomatitis virus (VSV) (8, 32). PDCs from TLR7$^{-/-}$ mice were unable to respond to IAV or VSV infection with IFN-α synthesis (8), and TLR7$^{-/-}$ mice systemically infected with VSV showed a greatly reduced ability to mount an IFN-α response (32). TLR7 is localized to endosomes (1) and encounters its ligands in this cellular compartment. Examples of viruses known to activate TLR7 are dengue virus and IAV, which activate IFN-α production in human pDCs in a TLR7-dependent manner (43). TLR7 signaling requires viral entry via endosomes and viral fusion and uncoating in association with endosomal compartments (43). Several studies have addressed the nature of TLR7 ligands and demonstrated that guanosine- and uridine-rich ssRNA oligonucleotides as well as the ssRNA genome of IAV can trigger TLR7 signaling

(8, 16). Interestingly, for at least some viruses, viral replication does not seem to be required for IFN-α synthesis by pDCs. West Nile virus (WNV) does not productively infect human pDCs yet is able to elicit a vigorous IFN-α response from these cells (41). Dengue virus TLR7 engagement requires viral entry, but it was not reported whether dengue virus productively infects pDCs (43). Independence of TLR7 signaling from productive virus infection renders TLR7 a broader and potentially more sensitive receptor that is ideally matched to the role of pDCs as sentinel cells.

The role of TLR7 in viral recognition has so far only been addressed for a limited number of virus families, and one of the key questions is whether TLR7 indeed is a universal PRR for ssRNA viruses. Paramyxoviruses are generally considered to infect via direct fusion with the plasma membrane, independent of uptake into an endosome, a mode of entry that could circumvent virus detection by TLR7. Consistent with this hypothesis, Hornung et al. reported that paramyxovirus infection of pDCs results in efficient IFN-α production independent of TLR7/8 and endosomal acidification, and that viral replication was required for activation, suggesting that intracellular PRRs can also activate IFN-α production in this important cell type (19).

Another open question arises from the way RNA viral genomes come in contact with PRRs, especially those localized to endosomal compartments. For example, despite significant differences in the association of the genomes of positive- and negative-strand viruses with nucleocapsid proteins within the cytoplasm of infected cells (see above), members of each of these classes of viruses are efficiently recognized by TLR7. In the case of negative-sense and ambisense viruses, one could easily envision how a tight complex of nucleocapsid protein and genomic RNA might interfere with PRR recognition. However, this has not been observed and might be due to the possibility that independent of the "normal" state of their genomes, within the endosomes both ambisense and negative-sense RNA viruses may be degraded sufficiently to expose their genomes to PRRs.

TLR3

TLR3 is a PRR for dsRNA (2) that can be expected to be found in the secondary structures of the genomes of many RNA viruses and is produced during replication of all RNA virus genomes. TLR3 has been reported to be expressed on the cell surface of fibroblasts and in endosomal compartments in other cell types including DCs of the myeloid lineage (34, 35). As with TLR7, inhibitors of endosomal acidification inhibit TLR3 signaling. The exact mechanism by which TLR3 encounters its ligands is still not well understood. It has been hypothesized that dsRNA released from lysed virus-infected cells may bind TLR3 expressed in neighboring cells. Alternatively, dsRNA might be recognized by TLR3 in the endosomal compartment of a virus-infected cell. Purified reovirus RNA was shown to directly stimulate TLR3 (2), but reoviruses, as well as other dsRNA viruses, replicate within their cores or newly synthesized subviral particles. This compartmentalization of dsRNA synthesis could easily be expected to hide the genome from detection by PRRs.

Like other PRRs, TLR3 can also participate in the adaptive immune response. DCs that phagocytose virally infected cells can recognize the viral RNA within the phagocytosed cells in a TLR3-dependent manner. This results in DC activation and in an enhancement of their ability to present phagocytosed virus-derived peptides in the context of major histocompatibility molecules to cytotoxic T cells. Thus, TLR3 serves to increase the efficiency of "cross-priming," enabling the expansion of virus-specific cytotoxic T lymphocytes (39).

TLR3 ligation by the dsRNA mimetic poly(I:C) leads to activation of the transcription factors IRF3 and nuclear factor-κB (NF-κB), resulting in their translocation to the nucleus and promotion of transcription of the genes for type I IFNs and other proinflammatory cytokines. However, the role of TLR3 in the control of virus infections has been somewhat difficult to demonstrate and is controversial. Contrary to the expectation that abrogation of the function of a PRR would likely result in more severe disease, several pathogenesis studies using viruses of different families showed decreased severity of some aspects of viral pathogenesis in TLR3-deficient mice compared with wild-type mice. TLR3$^{-/-}$ mice infected with IAV (*Orthomyxoviridae*), respiratory syncytial virus (*Paramyxoviridae*), Punta Toro virus (*Bunyaviridae*), and WNV (*Flaviviridae*) showed reduced levels of proinflammatory cytokines (15, 28, 38, 44). In both IAV and WNV infection, higher levels of virus production could be detected in the absence of functional TLR3, and IAV-, WNV-, and Punta Toro virus-infected TLR3$^{-/-}$ mice were more resistant to lethal infection. TLR3 signaling in WNV-infected mice leads to increased permeability of the blood-brain barrier and therefore increased WNV invasion into the central nervous system, whereas TLR3-deficient mice showed less central nervous system invasion. This was attributed to TLR3-dependent production of tumor necrosis factor-α and possibly other proinflammatory cytokines and their effect on blood-brain barrier integrity. On the other hand, no differences in the adaptive immune response as a late outcome of infection were seen between TLR3-deficient and wild-type mice after infection with lymphocytic choriomeningitis virus, mouse cytomegalovirus, VSV, and reovirus (10). Thus, while these studies indicate that TLR3 mediates responses to a variety of different RNA viruses, no singular role for its exact involvement in the stimulation of the innate immune response has presented itself.

RNA Helicases

More information about the actual structure of viral PAMPs and the PRRs that they activate has come from studies on the intracellular RNA helicases. The putative RNA helicases RIG-I and MDA-5 contain two repeats of a caspase activation and recruitment domain (CARD) at the N-terminal region and a DExD/H box helicase homology region in their C-terminal region. Both helicases activate a novel adaptor protein, IFN-β promoter stimulator 1 (IPS-1, also known as MAVS,

Cardif, and VISA), that contains a single CARD (22, 25, 36, 40). IPS-1 activates TANK-binding kinase 1 (TBK1) and IκB kinase-related kinase (IKK-ε) through a mechanism not understood at present, thereby leading to activation of IRF3. These RNA helicases are expressed in the cytoplasm and are therefore poised to encounter PAMPs of cytosolically replicating viruses.

RIG-I was initially found to induce IFN-β in response to transfection of cells with the dsRNA mimetic poly(I:C) (46), and its role in the recognition of many different viruses was subsequently established in many different studies. Due to the difficulty of generating viable RIG-I knockout mice (20), it has been challenging to address the role of RIG-I in viral pathogenesis. Analysis of IFN production in various cell types, however, established that RIG-I can detect viruses in many different cell types while pDCs mostly rely on TLRs to detect pathogens (20).

MDA-5 was identified as a target for the V proteins of the *Paramyxoviridae*, which had been known to interfere with IRF3 activation by dsRNA (3). MDA-5 was further shown in knockout mouse studies to play a major role in poly(I:C)-induced IFN and proinflammatory cytokine responses (14). Examination of cytokine responses after infection with a variety of different viruses demonstrated a remarkable specificity of MDA-5 in the recognition of encephalomyocarditis virus (14). Besides being a target for paramyxovirus V proteins, MDA-5 has been described to be degraded in poliovirus-infected cells (4), and hepatitis A virus-infected cells degrade IPS-1, blocking MDA-5 signaling (45). Two recent surveys of the requirements for RIG-I and MDA-5 for viral recognition concluded that while the majority of viruses are recognized by RIG-I, some viruses such as reoviruses and flaviviruses can be recognized by either RIG-I or MDA-5 and the *Picornaviridae* are recognized by MDA-5 (21, 30).

A ligand other than dsRNA or its mimetic poly(I:C) has so far not been identified for MDA-5. However, RIG-I has been investigated more extensively, and these studies have shed light on the relationship between PAMP production and recognition of specific viruses by specific cytosolic PRRs. In 2006, it was demonstrated that 5′-triphosphate-containing RNA (3pRNA) is a ligand for RIG-I (18). Within the host cell, host RNA transcripts by all three RNA polymerases initially contain 5′ triphosphates, which are modified prior to the RNAs exiting the nucleus. RNA Pol II transcripts are modified by addition of a 7′-methyl guanosine cap, while RNA Pol I and Pol III transcripts, with some exceptions, become processed to contain a 5′ monophosphate. Thus, once they reach the cytoplasm, cellular RNAs are no longer thought to be able to be recognized by RIG-I.

In contrast, 3pRNA can be produced by several mechanisms during RNA virus infections. Although several virus families, such as the *Flaviviridae* and the *Togaviridae*, contain 5′-capped genomes which by themselves are not recognized by RIG-I, 3pRNA is still produced during viral replication. The RdRPs of many RNA viruses, irrespective of the nature of the viral genome (plus-stranded RNA, negative-sense ssRNA, or segmented ssRNA), are able to initiate de novo RNA synthesis without the use of a primer. Therefore, RdRP transcripts resulting in both antigenomes and newly synthesized genomes may contain 3pRNA at the 5′ end of the transcript and can therefore be recognized by RIG-I. The exception to the rule lies within the family *Picornaviridae*, which use a protein primer to initiate antigenome and progeny genome synthesis (27), but a cellular enzyme removes this protein from the RNA that enters the cell, producing a 1pRNA, and all actively translating picornavirus RNA is also missing the protein cap. On the other hand, the picornavirus IRES element is full of dsRNA structures, consistent with MDA-5-dependent detection (14), whereas RIG-I does not seem to be activated by picornavirus infection (21). However, some controversy over the exact nature(s) of the ligand for RIG-I still exists. A recent study described a role for RNase L in the amplification of IFN production (33). RNase L is an IFN-induced gene that generates small dsRNAs derived from cellular origins that were shown to be recognized in a RIG-I-dependent fashion. Interestingly, these small RNAs did not contain any 5′ phosphates, but a phosphate group at the 3′ end instead.

Viral gene expression is another process in the viral life cycle that leads to generation of 3pRNA, albeit only transiently. The mRNAs of viruses are translated by the host translation machinery and thus conform to the requirements for host translation. In some cases, viruses have subverted this process by encoding a functional replacement for the cap (namely the IRES that is encoded by all members of the *Picornaviridae* and some members of the *Flaviviridae*, as described above), whereas other RNA viruses utilize a 5′ cap, which should allow their mRNAs to escape recognition by RIG-I. Viral mRNAs acquire their caps either by their own encoded enzymatic action or by stealing caps from cellular mRNAs (cap-snatching). Thus, depending on the kinetics of genome synthesis and cap addition, there may be a transient production of viral mRNAs that contain 3pRNA prior to acquisition of their caps.

PKR

PKR has long been recognized as an IFN-induced gene that is activated by binding dsRNA (42) and exerts antiviral action by shutting down both cellular and viral protein translation. Activation of this serine/threonine kinase by dsRNA leads to phosphorylation of the eukaryotic translation initiation factor eIF2-α, resulting in a shutoff of translation initiation, inhibiting viral replication, and thus limiting virus spread in an infected host. After this key antitranslational activity was identified, many viruses were shown to have developed mechanisms to evade or counteract PKR activity. PKR has also been shown to activate transcription factors such as NF-κB and the mitogen-activated protein (MAP) kinases p38 and Jun N-terminal kinase (JNK), but despite this demonstration that PKR could participate in gene induction, it remained unclear if PKR could participate in stimulation of IFN synthesis. Following the discovery of other dsRNA-binding molecules such as TLR3, RIG-I, and MDA-5, a role for PKR in PAMP-induced IFN induction was not vigorously investigated. However, in recent years a role for PKR in induction of IFN has reemerged. Diebold et al. demonstrated that bone marrow-derived DCs require PKR for efficient IFN-α induction in response to poly(I:C) treatment (9). Using several human- and mouse-derived cells, a recent study demonstrated that IFN induction by WNV infection also requires PKR (13), and PKR is required for Theiler's murine encephalomyelitis virus-mediated induction of type I IFNs and other cytokines in murine astrocytes (5). Similarly, induction of proinflammatory cytokines in rhinovirus-infected bronchial epithelial cells was dependent on PKR and NF-κB activation (11). The mechanism of PKR-induced IFN production is not understood, but it seems likely that signal transduction events such as NF-κB activation by dsRNA-activated PKR will play important roles in this process.

PAMPs and PRRs: a Summary

In summary, PRRs can be divided into receptors that are ubiquitously expressed in many cell types, such as the intracellular RNA helicases PKR and TLR3, and receptors such as TLR7 and TLR9 that are expressed only in specific cell types such as DC subsets that perform a very specific function in the stimulation of an early systemic immune response. Furthermore, some PRRs recognize PAMPs that are commonly produced by many viruses, such as dsRNA, and others such as RIG-I are more specific as to the nature of the recognized molecules. The outcome of innate immune stimulation and the relevance of specific PRRs will of course depend on the tissues and cells in which they are expressed and the way these tissues are affected by viral infection. Furthermore, other cellular activities are likely to contribute to the ability of PRRs to come into contact with their ligands, as shown by recent work demonstrating

that autophagy can help to deliver RNA PAMPs to the intracellular compartments where they are recognized by TLRs (26), and studies showing that RNAse L-mediated cleavage of host or pathogen RNA molecules can produce PAMPs that trigger (or enhance) antiviral responses (33).

Studies on host recognition of PAMPs have revolutionized innate immunity research. However, other signals, including stress responses and changes in cellular physiology, may also contribute to the recognition of virus infection. These signals may include virion-induced membrane fusion, changes in the secretory pathway associated with virion morphogenesis, as well as virus-mediated membrane rearrangements associated with establishment of extranuclear replicative factories that are hallmarks of infection by many different RNA viruses. These signals, which share some properties with cellular changes that result from exposure to toxic environmental stimuli, may represent pathways that do not rely on the known PRRs to identify both pathogenic and nonpathogenic threats to the organism or tissue.

SEEK AND DESTROY OR EVADE AND SUBVERT?

The mode of transmission a particular virus has acquired has a profound influence on how it needs to interact with the innate immune system. Viruses that can effectively parasitize tissues at or near their portal of entry and/or exit will likely have evolved site-specific mechanisms for evading the host's antiviral responses. Likewise, viruses that establish a chronic infection will need to combat a sustained offense provided by the host's antiviral immune response. Here we will briefly review two remarkably different RNA virus pathogens, foot-and-mouth disease virus (FMDV) and hepatitis C virus (HCV), to contrast alternate and effective strategies to counteract the innate immune responses of their host.

FMDV was the first animal virus identified, likely due to a combination of its severe economic impact (providing research funding), ready transmission (making experimental studies simple), and presence of virus at very high titers in diseased tissues (facilitating filtration studies). FMDV can produce severe disease in animals within 24 h and can complete a replication cycle in cell culture within 5 h. The secret to this virus's competitive success appears to be an extremely effective shutoff of host translation that blocks the translation of most host proteins, including IFN, even though FMDV-infected cells recognize the infection and transcribe IFN mRNAs (6). Soon after infection, diseased animals shed tremendous amounts of virus, which can be maintained in fomites in infected areas for months, facilitating transmission. Furthermore, the virus has a very broad host range, permitting propagation of epidemics, even though infected animals are protected for life from reinfection with the same serotype. Viruses engineered to lack the protease (Lpro) that is responsible for shutoff of host mRNA synthesis are severely attenuated in natural hosts, but replicate well in selected in vitro and in vivo systems that lack IFN induction/recognition pathways (6, 37).

HCV, an important causative agent of human hepatitis, was not discovered until late in the 1980s, due in part to the lack of an in vitro cultivation system, the presence of low amounts of virus in tissues, and the absence of an inexpensive animal model. This virus establishes a long-term chronic infection of liver cells in humans. Acute infection produces a mild disease, and the majority of patients harbor a symptom-free disease state. However, in a fraction of patients the chronic liver insult results in cirrhosis that can lead to even more severe consequences, such as hepatocellular carcinoma. With this disease profile, it is not surprising that the key to HCV's success is the avoidance of innate immunity, facilitating a long-term chronic infection with low virus shedding, which promotes transmission (especially by routes that have only recently evolved, e.g., intravenous drug use and transfusion). Interestingly, the mechanism of innate immune evasion utilizes a virally encoded proteinase, in this case the NS3/4A proteinase, which cleaves IPS-1 (31) and TIR domain-containing adaptor inducing IFN-β (TRIF) (29), two key components of pathways that lead to the virus-induced induction of IFN. As expected from this lifestyle, HCV is sensitive to IFN, although treatments with IFN are not universally successful.

CLOSING THOUGHTS

Despite spectacular advances in deciphering how virally encoded PAMPs stimulate PRRs to induce antiviral responses in cells in vitro, identifying the in vivo biological significance of these pathways and activities will be a challenge. Interactions between different cell types in vivo are likely to greatly affect both the timing and the outcome of cytokine production during infection. Although in vivo models have helped to validate some in vitro observations, the nature of these in some cases "artificial" models of infection may obscure the interactions that are most important in human disease. This problem may also apply to responses to zoonotic viruses, where some of the virus-host interactions, especially those from "dead-end" host infections, produce outcomes that may not reflect the natural ecology of the infection. Thus, the interactions of the virus with this type of host may not expose the innate immune responses and viral countermeasures that have evolved to ensure the survival of both the virus and its natural host. Nevertheless, understanding how RNA viruses have evolved to combat immune responses in both natural and unnatural hosts, which are the subjects of the subsequent chapters of this volume, will undoubtedly help us to appreciate the host responses that are most important to limiting viral transmission.

References

1. Ahmad-Nejad, P., H. Hacker, M. Rutz, S. Bauer, R. M. Vabulas, and H. Wagner. 2002. Bacterial CpG-DNA and lipopolysaccharides activate Toll-like receptors at distinct cellular compartments. Eur. J. Immunol. 32:1958–1968.
2. Alexopoulou, L., A. C. Holt, R. Medzhitov, and R. A. Flavell. 2001. Recognition of double-stranded RNA and activation of NF-kappaB by Toll-like receptor 3. Nature 413:732–738.
3. Andrejeva, J., K. S. Childs, D. F. Young, T. S. Carlos, N. Stock, S. Goodbourn, and R. E. Randall. 2004. The V proteins of paramyxoviruses bind the IFN-inducible RNA helicase, mda-5, and inhibit its activation of the IFN-β promoter. Proc. Natl. Acad. Sci. USA 101: 17264–17269.
4. Barral, P. M., J. M. Morrison, J. Drahos, P. Gupta, D. Sarkar, P. B. Fisher, and V. R. Racaniello. 2007. MDA-5 is cleaved in poliovirus-infected cells. J. Virol. 81: 3677–3684.
5. Carpentier, P. A., B. R. Williams, and S. D. Miller. 2007. Distinct roles of protein kinase R and Toll-like receptor 3 in the activation of astrocytes by viral stimuli. Glia 55:239–252.
6. Chinsangaram, J., M. E. Piccone, and M. J. Grubman. 1999. Ability of foot-and-mouth disease virus to form plaques in cell culture is associated with suppression of alpha/beta interferon. J. Virol. 73:9891–9898.
7. Crotty, S., C. E. Cameron, and R. Andino. 2001. RNA virus error catastrophe: direct molecular test by using ribavirin. Proc. Natl. Acad. Sci. USA 98:6895–6900.
8. Diebold, S. S., T. Kaisho, H. Hemmi, S. Akira, and C. Reis e Sousa. 2004. Innate antiviral responses by means of TLR7-mediated recognition of single-stranded RNA. Science 303:1529–1531.
9. Diebold, S. S., M. Montoya, H. Unger, L. Alexopoulou, P. Roy, L. E. Haswell, A. Al-Shamkhani, R. Flavell, P. Borrow, and C. Reis e Sousa. 2003. Viral infection switches non-plasmacytoid dendritic cells into high interferon producers. Nature 424:324–328.
10. Edelmann, K. H., S. Richardson-Burns, L. Alexopoulou, K. L. Tyler, R. A. Flavell, and M. B. Oldstone. 2004. Does Toll-like receptor 3 play a biological role in virus infections? Virology 322:231–238.
11. Edwards, M. R., C. A. Hewson, V. Laza-Stanca, H. T. Lau, N. Mukaida, M. B. Hershenson, and S. L. Johnston. 2007. Protein kinase R, IkappaB kinase-beta and NF-kappaB are required for human rhinovirus induced proinflammatory cytokine production in bronchial epithelial cells. Mol. Immunol. 44:1587–1597.

12. Fenner, F. 1953. Changes in the mortality-rate due to myxomatosis in the Australian wild rabbit. *Nature* **172:** 228–230.

13. Gilfoy, F. D., and P. W. Mason. 2007. West Nile virus-induced interferon production is mediated by the double-stranded RNA-dependent protein kinase PKR. *J. Virol.* **81:**11148–11158.

14. Gitlin, L., W. Barchet, S. Gilfillan, M. Cella, B. Beutler, R. A. Flavell, M. S. Diamond, and M. Colonna. 2006. Essential role of mda-5 in type I IFN responses to polyriboinosinic:polyribocytidylic acid and encephalomyocarditis picornavirus. *Proc. Natl. Acad. Sci. USA* **103:** 8459–8464.

15. Gowen, B. B., J. D. Hoopes, M. H. Wong, K. H. Jung, K. C. Isakson, L. Alexopoulou, R. A. Flavell, and R. W. Sidwell. 2006. TLR3 deletion limits mortality and disease severity due to phlebovirus infection. *J. Immunol.* **177:** 6301–6307.

16. Heil, F., H. Hemmi, H. Hochrein, F. Ampenberger, C. Kirschning, S. Akira, G. Lipford, H. Wagner, and S. Bauer. 2004. Species-specific recognition of single-stranded RNA via Toll-like receptor 7 and 8. *Science* **303:**1526–1529.

17. Holmes, E. C. 2003. Error thresholds and the constraints to RNA virus evolution. *Trends Microbiol.* **11:**543–546.

18. Hornung, V., J. Ellegast, S. Kim, K. Brzozka, A. Jung, H. Kato, H. Poeck, S. Akira, K. K. Conzelmann, M. Schlee, S. Endres, and G. Hartmann. 2006. 5′-Triphosphate RNA is the ligand for RIG-I. *Science* **314:**994–997.

19. Hornung, V., J. Schlender, M. Guenthner-Biller, S. Rothenfusser, S. Endres, K. K. Conzelmann, and G. Hartmann. 2004. Replication-dependent potent IFN-alpha induction in human plasmacytoid dendritic cells by a single-stranded RNA virus. *J. Immunol.* **173:**5935–5943.

20. Kato, H., S. Sato, M. Yoneyama, M. Yamamoto, S. Uematsu, K. Matsui, T. Tsujimura, K. Takeda, T. Fujita, O. Takeuchi, and S. Akira. 2005. Cell type-specific involvement of RIG-I in antiviral response. *Immunity* **23:**19–28.

21. Kato, H., O. Takeuchi, S. Sato, M. Yoneyama, M. Yamamoto, K. Matsui, S. Uematsu, A. Jung, T. Kawai, K. J. Ishii, O. Yamaguchi, K. Otsu, T. Tsujimura, C. S. Koh, C. Reis e Sousa, Y. Matsuura, T. Fujita, and S. Akira. 2006. Differential roles of MDA5 and RIG-I helicases in the recognition of RNA viruses. *Nature* **441:**101–105.

22. Kawai, T., K. Takahashi, S. Sato, C. Coban, H. Kumar, H. Kato, K. J. Ishii, O. Takeuchi, and S. Akira. 2005. IPS-1, an adaptor triggering RIG-I- and Mda5-mediated type I interferon induction. *Nat. Immunol.* **6:**981–988.

23. Koonin, E. V. 1991. The phylogeny of RNA-dependent RNA polymerases of positive-strand RNA viruses. *J. Gen. Virol.* **72**(Pt. 9):2197–2206.

24. Koonin, E. V., T. G. Senkevich, and V. V. Dolja. 2006. The ancient Virus World and evolution of cells. *Biol. Direct.* **1:**29.

25. Kumar, H., T. Kawai, H. Kato, S. Sato, K. Takahashi, C. Coban, M. Yamamoto, S. Uematsu, K. J. Ishii, O. Takeuchi, and S. Akira. 2006. Essential role of IPS-1 in innate immune responses against RNA viruses. *J. Exp. Med.* **203:**1795–1803.

26. Lee, H. K., J. M. Lund, B. Ramanathan, N. Mizushima, and A. Iwasaki. 2007. Autophagy-dependent viral recognition by plasmacytoid dendritic cells. *Science* **315:** 1398–1401.

27. Lee, Y. F., A. Nomoto, B. M. Detjen, and E. Wimmer. 1977. A protein covalently linked to poliovirus genome RNA. *Proc. Natl. Acad. Sci. USA* **74:**59–63.

28. Le Goffic, R., V. Balloy, M. Lagranderie, L. Alexopoulou, N. Escriou, R. Flavell, M. Chignard, and M. Si-Tahar. 2006. Detrimental contribution of the Toll-like receptor (TLR)3 to influenza A virus-induced acute pneumonia. *PLoS Pathog.* **2:**e53.

29. Li, K., E. Foy, J. C. Ferreon, M. Nakamura, A. C. Ferreon, M. Ikeda, S. C. Ray, M. Gale, Jr., and S. M. Lemon. 2005. Immune evasion by hepatitis C virus NS3/4A protease-mediated cleavage of the Toll-like receptor 3 adaptor protein TRIF. *Proc. Natl. Acad. Sci. USA* **102:**2992–2997.

30. Loo, Y. M., J. Fornek, N. Crochet, G. Bajwa, O. Perwitasari, L. Martinez-Sobrido, S. Akira, M. A. Gill, A. Garcia-Sastre, M. G. Katze, and M. Gale, Jr. 2008. Distinct RIG-I and MDA5 signaling by RNA viruses in innate immunity. *J. Virol.* **82:**335–345.

31. Loo, Y. M., D. M. Owen, K. Li, A. K. Erickson, C. L. Johnson, P. M. Fish, D. S. Carney, T. Wang, H. Ishida, M. Yoneyama, T. Fujita, T. Saito, W. M. Lee, C. H. Hagedorn, D. T. Lau, S. A. Weinman, S. M. Lemon, and M. Gale, Jr. 2006. Viral and therapeutic control of IFN-beta promoter stimulator 1 during hepatitis C virus infection. *Proc. Natl. Acad. Sci. USA* **103:**6001–6006.

32. Lund, J. M., L. Alexopoulou, A. Sato, M. Karow, N. C. Adams, N. W. Gale, A. Iwasaki, and R. A. Flavell. 2004. Recognition of single-stranded RNA viruses by Toll-like receptor 7. *Proc. Natl. Acad. Sci. USA* **101:** 5598–5603.

33. Malathi, K., B. Dong, M. Gale, Jr., and R. H. Silverman. 2007. Small self-RNA generated by RNase L amplifies antiviral innate immunity. *Nature* **448:**816–819.

34. Matsumoto, M., K. Funami, M. Tanabe, H. Oshiumi, M. Shingai, Y. Seto, A. Yamamoto, and T. Seya. 2003. Subcellular localization of Toll-like receptor 3 in human dendritic cells. *J. Immunol.* **171:**3154–3162.

35. Matsumoto, M., S. Kikkawa, M. Kohase, K. Miyake, and T. Seya. 2002. Establishment of a monoclonal antibody against human Toll-like receptor 3 that blocks double-stranded RNA-mediated signaling. *Biochem. Biophys. Res. Commun.* **293:**1364–1369.

36. Meylan, E., J. Curran, K. Hofmann, D. Moradpour, M. Binder, R. Bartenschlager, and J. Tschopp. 2005. Cardif is an adaptor protein in the RIG-I antiviral pathway and is targeted by hepatitis C virus. *Nature* **437:**1167–1172.

37. Piccone, M. E., E. Rieder, P. W. Mason, and M. J. Grubman. 1995. The foot-and-mouth disease virus leader proteinase gene is not required for viral replication. *J. Virol.* **69:**5376–5382.

38. Rudd, B. D., J. J. Smit, R. A. Flavell, L. Alexopoulou, M. A. Schaller, A. Gruber, A. A. Berlin, and N. W. Lukacs. 2006. Deletion of TLR3 alters the pulmonary immune environment and mucus production during respiratory syncytial virus infection. *J. Immunol.* **176:**1937–1942.

39. Schulz, O., S. S. Diebold, M. Chen, T. I. Naslund, M. A. Nolte, L. Alexopoulou, Y. T. Azuma, R. A. Flavell, P. Liljestrom, and C. Reis e Sousa. 2005. Toll-like receptor 3 promotes cross-priming to virus-infected cells. *Nature* **433:**887–892.

40. Seth, R. B., L. Sun, C. K. Ea, and Z. J. Chen. 2005. Identification and characterization of MAVS, a mitochondrial antiviral signaling protein that activates NF-kappaB and IRF 3. *Cell* **122:**669–682.

41. Silva, M. C., A. Guerrero-Plata, F. D. Gilfoy, R. P. Garofalo, and P. W. Mason. 2007. Differential activation of human monocyte-derived and plasmacytoid dendritic cells by West Nile virus generated in different host cells. *J. Virol.* **81:**13640–13648.

42. Ucci, J. W., Y. Kobayashi, G. Choi, A. T. Alexandrescu, and J. L. Cole. 2007. Mechanism of interaction of the double-stranded RNA (dsRNA) binding domain of protein kinase R with short dsRNA sequences. *Biochemistry* **46:**55–65.

43. Wang, J. P., P. Liu, E. Latz, D. T. Golenbock, R. W. Finberg, and D. H. Libraty. 2006. Flavivirus activation of plasmacytoid dendritic cells delineates key elements of TLR7 signaling beyond endosomal recognition. *J. Immunol.* **177:**7114–7121.

44. Wang, T., T. Town, L. Alexopoulou, J. F. Anderson, E. Fikrig, and R. A. Flavell. 2004. Toll-like receptor 3 mediates West Nile virus entry into the brain causing lethal encephalitis. *Nat. Med.* **10:**1366–1373.

45. Yang, Y., Y. Liang, L. Qu, Z. Chen, M. Yi, K. Li, and S. M. Lemon. 2007. Disruption of innate immunity due to mitochondrial targeting of a picornaviral protease precursor. *Proc. Natl. Acad. Sci. USA* **104:**7253–7258.

46. Yoneyama, M., M. Kikuchi, T. Natsukawa, N. Shinobu, T. Imaizumi, M. Miyagishi, K. Taira, S. Akira, and T. Fujita. 2004. The RNA helicase RIG-I has an essential function in double-stranded RNA-induced innate antiviral responses. *Nat. Immunol.* **5:**730–737.

*Cellular Signaling and Innate Immune
Responses to RNA Virus Infections*
Edited by A. R. Brasier et al.
© 2009 ASM Press, Washington, D.C.

Krzysztof Brzózka
Karl-Klaus Conzelmann

14

Rhabdoviruses and Mechanisms of Type I Interferon Antagonism

RHABDOVIRIDAE

The first order to be established within the kingdom of viruses was the order *Mononegavirales*, also known as nonsegmented negative-strand RNA viruses, which comprises enveloped viruses with a single RNA of negative polarity (i.e., anti-mRNA sense) and which includes the families of *Rhabdoviridae*, *Paramyxoviridae*, *Filoviridae*, and *Bornaviridae*. Essential cis- and trans-acting elements for gene expression are conserved throughout the *Mononegavirales*, illustrating that they have originated from a common ancestor, probably a rhabdovirus-like one (for review, see reference 112). The *Mononegavirales* RNA is always contained in a highly stable, helical nucleocapsid (NC), and the information encoded is expressed by sequential and polar transcription of discrete subgenomic mRNAs from the NCs (for review, see reference 148).

Among the *Mononegavirales,* the members of the *Rhabdoviridae* family are characterized by a typical rod- or bullet-shaped morphology of the virions. They have by far the widest host range, including plant, invertebrate, fish, bird, and mammalian hosts (for review, see reference 48). Only a few of the approximately 160 described rhabdovirus species are assigned to the six genera established so far. Members of the *Lyssavirus* genus include the neurotropic rabies virus (RV) and rabies-related viruses from bats and other animals, for example, the Australian bat lyssavirus. Human rabies is a zoonosis; the main host reservoirs are foxes, raccoons, dogs, and bats, depending on the region. Most cases of human rabies are reported from India, accounting for nearly 50,000 deaths per year, though effective vaccines are available. The *Vesiculovirus* genus includes animal viruses like the prototypic vesicular stomatitis virus (VSV) and related viruses such as Cocal virus. Natural VSV may cause a severe disease in cattle and pigs that is clinically reminiscent of foot-and-mouth disease and which causes enormous economic loss. A closely related vesiculovirus, Chandipura virus, has recently emerged as a serious human pathogen associated with a number of outbreaks of acute encephalitis with a high fatality rate in different parts of India (for review, see reference 12). Most remarkably, as suggested by identical genome organization and sequence homologies of genes and gene junctions, a tentative member of the *Vesiculovirus* genus is even found among the fish rhabdoviruses: spring

Krzysztof Brzózka and Karl-Klaus Conzelmann, Max von Pettenkofer Institute & Gene Center, Ludwig-Maximilians-University Munich, Feodor-Lynen-Str. 25, 81377 Munich, Germany.

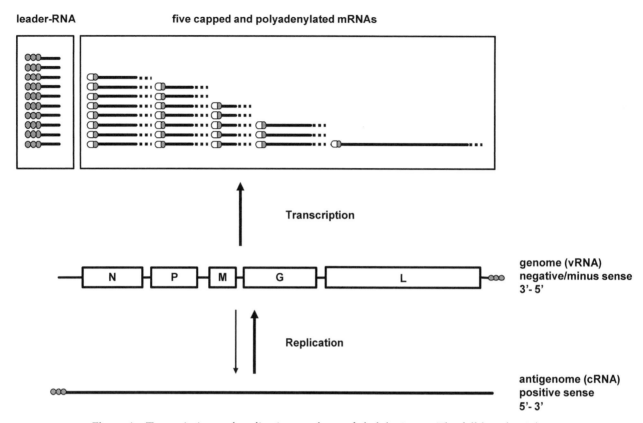

Figure 1 Transcription and replication products of rhabdoviruses. The full-length triphosphate genome and antigenome RNAs are present exclusively in an NC RNP. The genome NC is the template for production of a short triphosphate leader RNA and sequential transcription of five monocistronic capped and polyadenylated mRNAs. Since the polymerase complex P/L eventually dissociates from the template, a transcription gradient is generated. Replicative synthesis of full-length antigenomes involves concurrent packaging by N/P complexes into an NC.

viremia of carp virus (5). The few members of the *Ephemerovirus* genus, with the prototype bovine ephemeral fever virus, are pathogens of cattle and of water buffalo. In contrast to RV and other lyssaviruses, which are transmitted directly, most vesiculoviruses and ephemeroviruses are transmitted by insect vectors. The *Novirhabdovirus* genus includes typical rhabdoviruses from fish, such as viral hemorrhagic septicemia virus, which infects numerous marine and freshwater fish species (126). Finally, rhabdoviruses that infect plants are separated into the genera *Cytorhabdovirus* and *Nucleorhabdovirus*, based on their sites of replication and morphogenesis (63). They include important plant pathogens such as Lettuce necrotic yellows virus and Sonchus yellow net virus. Plant rhabdoviruses are also transmitted by insect vectors. This has led to the suggestion that the family evolved from an ancestral insect virus and that the natural host range is largely determined by the insect host (48).

Rhabdovirus Genomes and Gene Products

Rhabdoviruses of mammals may have the simplest genomes of all *Mononegavirales*, as exemplified by the RV and VSV genomes, which comprise only five genes in the order 3′-N-P-M-G-L-5′ (Fig. 1). In the case of RV, indeed only five proteins are encoded: (i) the nucleoprotein (N) encapsidates the viral RNA to form the NC; (ii) the phosphoprotein (P) is an essential cofactor for RNA synthesis and RNA encapsidation by N; (iii) the matrix protein (M) provides a structural lattice between the NC and the virus envelope, and thus has a critical role in virus assembly; (iv) the transmembrane spike glycoprotein (G) is responsible for attachment to target cells and membrane fusion; and (v) the "large" protein (L) is the catalytic subunit of the viral RNA polymerase. This minimal set of five genes, occupying less than 12 kb of RNA, as well as the order of these genes, are highly conserved in all *Mononegavirales*. The most conserved protein is the N, and L proteins of rhabdoviruses also share common sequence blocks with those from other *Mononegavirales* families, reflecting the conserved mode of gene expression from the NCs by an "NC-dependent" polymerase. In contrast, though also critically involved in RNA synthesis, the rhabdovirus P proteins show a relatively high variability, even within the genera.

However, the coding capacity of rhabdoviruses can be easily increased by expression of multiple products from a single RNA and by the addition of extra RNA and genes. The latter is facilitated by the helical nature of the NCs. In vesiculoviruses like VSV, a second open reading frame in the P gene codes for an additional small, nonessential and nonstructural C protein with unknown function. Up to six extra genes, including a nonstructural glycoprotein, have been identified between the G and L genes of ephemeroviruses. The fish novirhabdoviruses have a typical small protein-coding gene (nonviral, NV) with unknown function between the G and L genes, and sigma virus of *Drosophila* has three extra genes downstream of the N gene (48). All plant rhabdoviruses carry analogs of the five rhabdovirus core genes—N, P, M, G, and L—but encode extra genes, at least one of which is inserted between the P and M genes. While these extra genes are not similar among plant rhabdoviruses, two encode proteins with similarity to the 30K superfamily of plant virus movement proteins. The nucleorhabdoviral protein sequences of the N, P, M, and L proteins contain nuclear localization signals consistent with virus replication and morphogenesis of these viruses in the nucleus (114).

The Rhabdovirus Nucleocapsid and Rhabdovirus RNAs

The helical NCs of rhabdoviruses are extremely tight and stable (96). In the cytoplasm of host cells, they are present in the form of relaxed NCs, which represent the synthetically active forms. In the rod-shaped virion, a condensed NC superhelix is present, probably held together by M protein. Recent progress has revealed details at 2.9-Å resolution of the molecular structure of NCs of VSV and RV. Crystals could be obtained from ring complexes containing 10 or 11 N molecules, respectively (7, 53). The RNA is tightly sequestered in a cavity at the interface between the N- and C-terminal lobes of the N protein, which appears to clamp down onto the bound RNA. The characteristics of folding, RNA binding, and assembly are highly conserved in VSV and RV, despite their lack of significant homology in amino acid sequence (88). The complete occlusion of the RNA explains the excellent protection of the viral RNA against high salt, attack by RNases, and silencing by small interfering RNAs (13). It is suggested that for RNA synthesis, in which the

RNA has to be accessed by the viral polymerase, some conformational change in the N protein by the polymerase and/or the P protein is necessary. The easiest access to the RNA would indeed be at the ends of the NCs.

The NCs of rhabdoviruses are templates for two modes of RNA synthesis: sequential transcription of subgenomic RNAs and replication of full-length NCs (for a comprehensive review, see reference 148) (Fig. 1). Transcription is sequential and, according to the most widely accepted model, starts at the NC 3′ end with the synthesis of a short, approximately 50-nucleotide-long leader RNA that remains unmodified, i.e., begins with a 5′ triphosphate nucleotide, and in almost all rhabdoviruses with 5′-pppA(N)n (1, 28, 29). The 3′ end of the leader RNA is defined in the genome template by a conserved transcriptional restart sequence preceding the first (N) gene. At this leader/N junction the leader RNA is released and transcription of the first mRNA (N) is initiated. Restart involves the addition of a 5′ methylguanosine cap structure (18) by the L protein in an unconventional way (84, 100). Transcription proceeds until the polymerase encounters a stop/polyadenylation signal. This signal is characterized by a stretch of U residues on which the polymerase stutters back and forth to produce an approximately 50- to 150-nucleotide-long poly(A) tail. Transcriptional stop/polyadenylation signals are typically separated from the downstream restart signals by a few—probably nontranscribed—nucleotides, known as "intergenic region." The entry of the polymerase at the 3′ end and eventual dissociation of the polymerase at the gene junctions—probably influenced by the length of intergenic regions (45)—leads to the typical transcript gradient of *Mononegavirales* in which upstream RNAs are more abundantly produced than downstream mRNAs (Fig. 1). This is a simple and very efficient method to direct adequate expression levels of gene products. Shifting viral genes to other positions in the genome by reverse genetics (30) has become a versatile tool to study the impact of gene dosage on the rhabdovirus life cycle and pathogenesis (10, 20, 98).

Replication of NCs involves continuous synthesis of full-length RNA and concurrent encapsidation of the nascent RNA with N protein. In this process, the P protein plays important roles as a chaperone for N and for recruiting the L polymerase to the NC. Only N-P complexes have specificity for viral RNA, whereas N alone associates with other RNAs as well, or aggregates. Again, NC replication initiates at the 3′ end of the genome NC template and yields a full-length antigenome NC harboring cRNA. This serves only as a template for replication and amplification of genome NCs, although the production of a leader equivalent, the trailer RNA, has been suggested. Since the 3′-terminal sequences of the antigenome RNA acts as a very strong promoter compared with the genome promoter, a huge excess of genome over antigenome NCs is produced, in the case of RV and VSV approximately 50- and 10-fold, respectively (42, 43, 148).

The ends of the rhabdovirus genome and antigenome RNAs are identical in sequence to a certain extent and assumed to contain a specific N-encapsidation signal. Indeed, part of the leader RNA is found encapsidated by N protein as well (17), and it is assumed that the release of the encapsidation signal within leader RNA is important to prevent encapsidation of mRNAs. In view of the tight NC structure, which should allow easy access to RNA only at the ends, it appears plausible that initiation of transcription (leading to encapsidated leader RNA and MRNAs) and replication (leading to encapsidated full-length RNA) occurs exclusively at the ends. Although there is evidence for the existence of at least two different versions of the polymerase, i.e., a transcriptase and a replicase, the encapsidation of the leader RNA may suggest that initiation for all RNA synthesis is done with a single form of polymerase (the replicase). This is supported by the recent observation that the N protein of the Sendai paramyxovirus is critically required not only for replication but also for transcription (151, D. Kolakofsky, personal communication).

Replication in different hosts requires adaptation to the peculiar host defense mechanisms. Rhabdoviruses, with their wide host range, therefore represent an interesting group to study how this task is solved. In plant and insect hosts, viruses have to withstand RNA interference as a major antiviral strategy (32, 49, 144). Vertebrate rhabdoviruses have to replicate in the face of a powerful and sophisticated immune system in which innate recognition of pathogens, direct defense mechanisms, and adaptive humoral and cellular immune systems are effectively integrated. Notably, very closely related and similarly organized viruses, like VSV and RV, have developed quite distinct strategies to escape, reflecting their biology and host range.

RECOGNITION OF RHABDOVIRUSES BY INNATE IMMUNE RECEPTORS

The mammalian innate defense system has developed to reliably recognize virus infection, and to induce type I interferon (IFN) and other antiviral cytokines. Recognition of pathogens is based on receptors for conserved, invariant structures or molecular patterns (65, 66). In the case of bacteria, a distinction between self and nonself is straightforward, because of distinct metabolic

pathways and products, such as bacterial lipopolysaccharide (LPS), which is recognized by Toll-like receptor (TLR) 4. Virus proteins, however, are made by the host and may look like self. Rather, virus nucleic acids, the products of viral replication and transcription, represent a suitable means for discrimination, although in some cases even viral nucleic acids are made by the host cell. Indeed, almost exclusively, the pattern recognition receptors (PRRs) that induce IFN are activated by nucleic acids (for reviews, see references 108, 137, and 156 and chapters 2 and 3 in this volume). They include the cytoplasmic RNA helicases retinoic acid-inducible gene I (RIG-I) and melanoma differentiation-associated gene 5 (MDA-5) (RIG-like receptors, or RLRs), a cytoplasmic DNA receptor (DAI) (134), as well as some members of the transmembrane TLR family, which might sense viral nucleic acids outside the cells, but may also have access to cytoplasmic nucleic acids by the process of autophagy, as shown recently (81). Whereas expression of TLRs is mostly limited to specialized immune cells, the RLRs appear to be ubiquitous. Since the presence of viral RNA or DNA in the cytoplasm of any cell is a sign of utmost emergency, the latter are of utmost importance. Indeed, as shown by gene knockout mice, RLRs are critical for mounting an adequate innate antiviral defense against RNA viruses, including rhabdoviruses (70, 76).

Recognition of Rhabdoviruses by Cytoplasmic RLRs

Although double-stranded (ds) RNA, and synthetic dsRNA analogs like poly(I:C), have long been known to play a major role both in the induction of IFN and in stimulating antiviral responses (40, 73, 80), the first cytoplasmic receptors able to activate IFN gene transcription have been described only recently. The two DExD/H box RNA helicases, RIG-I and MDA-5, were found to be essential for type I IFN production in response to RNA virus infection (8, 157, 158) and also for production of IFN-λ (101). RIG-I and MDA-5 possess a unique domain structure among RNA helicases. They consist of two N-terminal caspase activation and recruitment domain (CARD) domains, a central subfamily 2 (SF2)-type DECH box ATPase domain, and a C-terminal extension of little sequence homology to other proteins. Downstream signaling to induce IFN- and nuclear factor-κB (NF-κB)-controlled cytokines occurs when the CARD domains bind to the CARD of the adaptor protein IFN-β promoter stimulator 1 (IPS-1) (also known as Cardif, MAVS, or VISA), which is located in the outer mitochondrial membrane (72, 94, 124, 153). This CARD-CARD interaction is thought to lead to the recruitment and activation of TANK-binding kinase 1

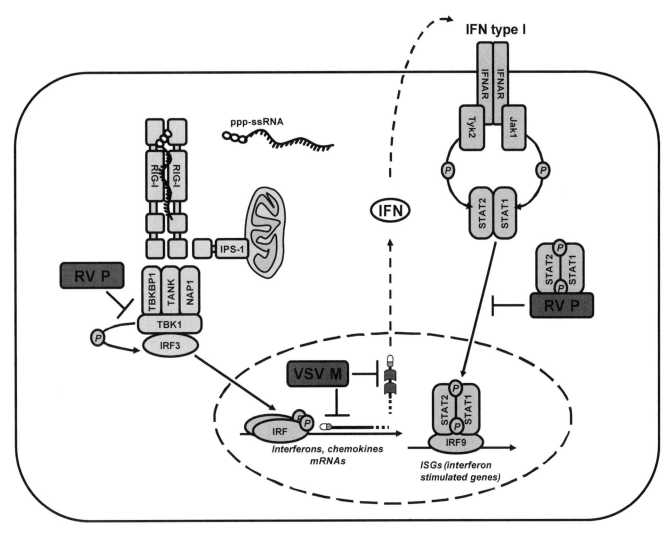

Figure 2 IFN antagonists of RV and VSV. Rhabdovirus triphosphate RNAs are recognized by RIG-I, resulting in association of the CARD domains of RIG-I and IPS-1 and recruitment of a complex in which IRF3 is phosphorylated by TBK1. Phosphorylation results in dimerization of IRF3, import into the nucleus, and transcriptional activation of the IFN-β gene. In the presence of RV P, IRF3 phosphorylation is prevented. RV P binds also to phosphorylated Stat1 and Stat2, preventing expression of ISGs after IFN stimulation of the IFN receptor (IFNAR). VSV M causes a shutdown of Pol II transcription, thereby preventing transcription of both IFN genes and ISGs. In addition, VSV M interferes with the nuclear export of mRNAs by blocking Nup98 (see text for details).

(TBK1), inhibitor of kappa-B kinase (IKK-*i*), and other IKK kinases that activate IFN regulatory factor (IRF) 3 and NF-κB, thereby stimulating the IFN-β promoter (Fig. 2). A third member of the RLR family, LGP2, binds RNA but lacks a CARD domain (116). Therefore, LGP2 was initially described as a negative regulator of RLR signaling, but recent studies from knockout mice suggest that this has to be revisited (139).

Intriguingly, the closely related RIG-I and MDA-5 respond to different RNA virus types. Whereas RIG-I recognizes a variety of positive- and negative-strand RNA viruses, including rhabdoviruses, MDA-5 is required for IFN induction by a picornavirus, encephalomyocarditis virus (51, 69, 70), and may contribute to recognition of reovirus and dengue virus (86). Differential activation of RLRs was also suggested by RNA transfection experiments. RIG-I was activated by in vitro-transcribed RNAs, whereas MDA-5 did respond to poly(I:C) (76, 133). Importantly, RNAs transcribed in vitro by the RNA polymerases of bacteriophages like T7 or Sp6 were previously found to trigger IFN expression (74). Like those, rhabdovirus polymerases initiate RNA synthesis with a triphosphate nucleotide at the 5' end of RNAs. In contrast, picornaviruses like encephalomyocarditis virus that are sensed by MDA-5 produce RNAs that lack a 5' triphosphate and instead have a protein (Vpg) linked covalently to the 5' terminus.

Recognition of Rhabdovirus Triphosphate RNA by RIG-I

The identity of 5' triphosphate RNA as the first molecularly defined ligand for RIG-I was confirmed recently (58, 109). As previously shown for VSV infection (70), IFN induction in cells infected with an RV mutant, SAD ΔPLP (see below), was found to depend on RIG-I. RNA from RV-infected cells (containing the 5' triphosphate leader RNA, vRNA, and cRNA), as well as RNA isolated from purified RV particles (containing vRNA and traces of cRNA), led to induction of IFN after transfection into cells. This activity was completely lost when the 5' triphosphate of the RNA was removed by alkaline phosphatase (58). Moreover, RIG-I directly bound single-stranded (ss)RNA oligonucleotides carrying a 5'-terminal triphosphate. IFN induction in response to triphosphate RNAs was independent of TLR recognition and was abolished in RIG-I knockout murine embryonic fibroblasts (MEFs), demonstrating a nonredundant function of RIG-I in triphosphate RNA-dependent IFN induction. As confirmed in independent work, influenza virus 5' triphosphate ssRNAs are also recognized by RIG-I (109). These authors could also show that transfected VSV vRNA induced an IFN response, which was completely abrogated by prior phosphatase treatment.

This suggested that, in addition to dsRNA, triphosphate ssRNA alone may be a signature of nonself viral RNA, though RIG-I also recognizes triphosphate dsRNA (K. Brzózka, unpublished data), and panhandle/ds structures in the RNA may even further stimulate RIG-I, as in case of the terminal nontranscribed hepatitis C virus RNAs (119). Triphosphate ssRNAs are also primary products of cellular polymerases, thus raising the question of how self RNAs may escape from RIG-I recognition. In cellular mRNAs, the triphosphates are perfectly obscured by the typical 5' methylguanosine cap structure added before export into the cytoplasm. In the case of tRNAs, triphosphate ends are removed by RNase P prior to export (67). Most small RNAs are in addition heavily modified by incorporation of uridine derivatives such as pseudouridine, 2-thiouridine, and 2-O-methyluridine (85). Intriguingly, both 5'-capping of in vitro-transcribed RNAs and incorporation of modified uridine were shown to abolish activation of RIG-I (58). These findings strongly suggest that the presence or the accessibility of cytoplasmic, unmodified

triphosphate ssRNA serves as a pathogen-associated molecular pattern (PAMP).

Structure/Function of RIG-I Signaling

The similar organization of RLRs and differential recognition of RNAs raises the question of how these proteins sense particular RNAs. In analogy to other SF2 helicases and the finding that mutations in the ATP-binding site or truncation of parts of the DECH box domain abolish RIG-I activity, it is suggested that ATP and nucleic acid substrate-dependent conformational changes are a central part of viral RNA sensing (58, 158). Additional experiments have also shown that the characteristic C-terminal domain is required for RIG-I activity and that its overexpression interferes with RIG-I signaling in response to Sendai virus infections (118). We recently demonstrated that it is the C-terminal regulatory domain (RD) of RIG-I that binds viral RNA in a 5'-triphosphate-dependent manner and activates the RIG-I ATPase by RNA-dependent dimerization (31) (Fig. 2). The crystal structure of RD could be determined and revealed a zinc-binding domain that is structurally related to GDP/GTP exchange factors of Rab-like GTPases. The zinc coordination site is essential for RIG-I signaling and also conserved in MDA-5 and LGP2, suggesting structurally related RD domains in all three enzymes. Structure-guided mutagenesis identified a positively charged groove as the likely 5' triphosphate binding site of RIG-I. This groove, however, is distinct in MDA-5 and LGP2, suggesting the possibility that the RD confers ligand specificity (31).

Leader RNA of Rhabdoviruses Is the Major IFN-Inducing Pattern

Triphosphate RNAs of rhabdoviruses include the leader RNA (and potentially trailer RNA transcribed from the antigenome) and the full-length genome and antigenome RNAs (see Fig. 1). The latter are probably completely enwrapped by N protein, though the details of the NC end structure are not known. From the above investigations it is clear that, in addition to a triphosphate, access of RIG-I to a short stretch of RNA is required for signaling (58). It is not clear yet whether the terminal triphosphate of RNA is so well covered by the N protein as to prevent access by RIG-I. In contrast to the full-length NC, part of the leader-RNA and leader-N mRNA read-through products are not or are only partially encapsidated by N protein. In addition, when synthesis of the leader RNA is considered essential for transcription initiation, a many-fold excess of leader-RNA over full-length replication products should be present in infected cells (42, 46). Thus, the leader RNA of rhabdoviruses is the prime target candidate for recognition by RIG-I. This is supported by experiments with a paramyxovirus, measles virus (111). During measles virus infection, IFN-β gene transcription was found to be paralleled by the synthesis of subgenomic RNAs rather than of full-length RNAs. As for RV and VSV, the leader RNAs of measles virus or Sendai virus synthesized in vitro or in vivo were efficient in inducing IFN-β expression, provided that it was delivered into the cytosol as a 5′-triphosphate-ended RNA (111, 132).

Role of Rhabdovirus DIs and dsRNA in IFN Induction

Defective interfering (DI) particle RNAs are generated notoriously in fast-replicating rhabdoviruses like VSV. DI RNAs originate from the viral polymerase jumping to other templates. Mostly, short noncoding copyback DIs are selected, which originate from switching of the polymerase to the newly generated antigenome RNA as a template and "back-transcription" to the 5′ end. Because these have the strong antigenome RNA promoter for replication, they can "interfere" with replication of the helper virus (79, 105). In infections with DI-containing virus stocks, there is therefore an overproduction of 5′ PPP-A ends, and probably of 5′ triphosphate trailer RNA, both of which could activate RIG-I. VSV DI particles containing copyback DI genomes were found to be very potent inducers of IFN, even in the absence of coinfecting nondefective helper virus (91, 122). For other viruses described as excellent IFN inducers, like the Sendai paramyxovirus Cantrell strain, the presence of DI RNA is responsible for IFN induction. The level of IFN-β activation was found to be proportional to that of DI genome replication (131).

Although it appears clear that IFN induction by rhabdoviruses is mainly via RIG-I and due to their triphosphate RNAs, a minor contribution of other potential PAMPs has not been excluded. Although dsRNA is not a major product of rhabdovirus RNA synthesis (146), it might be present in trace amounts, or abundantly in the presence of copyback DIs. Certain mutants of VSV (polR) overproduce abortive replication products, and indirect evidence has been provided that the cell-type-specific growth restriction of polR mutants may be due to enhanced production of dsRNA (103). However, there was no indication of secretion of antiviral factors from restricting cells, and inhibition of protein kinase R (PKR) did not show significant effects on restriction. Similarly, coinfecting cells with VSV recombinants expressing sense- and antisense-strand RNA of enhanced green fluorescent protein (EGFP), respectively, mimicked all aspects of polR restriction, and it was concluded that, for some cell types, overproduction of dsRNA during VSV infection triggers a host-cell antiviral effector response independent of IFN induction or signaling. In contrast, coinfections with corresponding EGFP⁺ and EGFP⁻ Sendai viruses increased IFN induction through RIG-I (132).

Notably, long dsRNA other than poly(I:C) has so far not been shown to activate RIG-I for IFN induction, but is able to moderately stimulate ATPase activity of full-length RIG-I and to considerably stimulate that of a CARD-less RIG-I. This suggests that in unstimulated RIG-I the CARD domain prevents access to the helicase, and it might be speculated that triphosphate RNA may serve as a first trigger for conformational changes in RIG-I and that subsequent binding of dsRNA to the RIG-I helicase domain may act synergistically on RIG-I activation. Also, a possible synergistic contribution of MDA-5 to RIG-I-dependent IFN induction in rhabdovirus-infected cells has not yet been excluded.

Recognition of Rhabdoviruses by TLRs

TLR7 and TLR9, recognizing artificial ssRNA and CpG DNA, respectively, are primarily expressed on myeloid dendritic cells (mDCs) and human plasmacytoid dendritic cells (pDCs) (59). pDCs are specialized sentinels for viruses and the major producers of IFN-α in humans. They also express high levels of IRF7, which otherwise is induced only upon IFN feedback. Human pDCs therefore can produce the entire "late" IFN-α repertoire immediately upon stimulation of TLR7 and TLR9 (11). Notably, this pathway is independent of the canonical IRF-activating kinases TBK1 and IKK-i and depends on the TLR adaptor myeloid differentiation primary response protein 88 (MyD88), which forms a complex with IRF7, TNF receptor-associated factor (TRAF), and the kinases IL-1 receptor-associated kinase 4 (IRAK-4), IRAK-1, and IKK-α (for review, see reference 137 and chapter 3 in this volume).

Recent work revealed IFN-α production in pDCs after incubation with a variety of inactivated and live RNA viruses, including the rhabdovirus VSV (11, 87) and the paramyxoviruses Sendai virus and respiratory syncytial virus (RSV) (60, 62). Notably, for VSV and RSV, entry into pDCs and replication was required for IFN-α production (11, 60), and recent evidence revealed that the recognition of VSV by TLR7 requires transport of cytosolic viral replication intermediates into the lysosome by the process of autophagy. This suggests that cytosolic replication intermediates of viruses rather than extracellular virus RNA or NCs may serve as pathogen signatures also for TLR7 (81). Notably, this work also revealed that components of the autophagic machinery like Atg5 are important for production of IFN (but not of interleukin-12p40) by exogenous TLR ligands. Therefore, the autophagosome sorting and transport machinery may be important for the quality of TLR downstream signaling.

Although TLR7 activation by viruses may lead to high systemic or local IFN levels, analyses of mice lacking components of either the TLR7/9 or RLR/IPS-1 pathway suggest that this may not be sufficient to mount the full antiviral capacity of hosts, and that cell-intrinsic recognition of viruses may be more important for virus elimination. In mice lacking TLR or MyD88, RLR/IPS-1 signaling was sufficient for mounting a protective antiviral inflammatory response (57), whereas IPS-1$^{-/-}$ mice and even heterozygous IPS-1$^{+/-}$ mice were highly susceptible to infection with VSV in spite of high IFN-α levels probably produced by pDCs via the TLR7-dependent pathway (6, 75, 133). This suggests that RLR/IPS-1 activation, but not TLR7 activation, leads to expression of signals labeling infected cells for destruction. In this regard, work toward integrating the expression profiles of IRF, Stat, NF-κB, AP-1, and MAPK target genes (52, 128) may lead to identification of the relevant components "beyond IFN" (127). However, in certain settings TLR-dependent IFN induction may also be critical, since mice lacking the TLR adaptor MyD88 were extremely sensitive to intranasal infection with VSV. This was correlated with impaired production of IFN-α and of neutralizing antibodies (160). Therefore, the route of infection, and an appropriate local host response, may be decisive for the outcome of infection.

Although TLR3 is activated by dsRNA and poly(I:C), a significant role of TLR3 for rhabdovirus sensing and the establishment of an antiviral state in vivo is questionable. In particular, TLR3$^{-/-}$ mice did not show increased susceptibility to VSV or changes in pathogenesis and adaptive antiviral response (34). The role of TLR3 may rather be in regulation of immunopathology associated with infection by certain neurotropic viruses and its effects on the integrity of the blood-brain barrier (78, 143).

Also unforeseen is the recent observation that TLR4 may play a role in recognizing VSV infection. TLR4 is expressed on the cell surface of many cell types, including epithelial cells, fibroblasts, and monocytes. TLR4 in conjunction with a coreceptor, CD14, can mediate IFN induction in response to bacterial LPS bound to the MD-2 protein (for review, see reference 71). Stimulation of TLR4 by LPS leads to the "classical" phosphorylation and activation of IRF3 and IRF7 by TBK1 and IKK-i and involves the Toll/interleukin-1 receptor (TIR) domain adaptors TRIF (TIR domain-containing adaptor inducing IFN-β; also known as TICAM-1) and TRAM (TRIF-related adaptor molecule) (47, 102, 154,

155). Interestingly, mutations of CD14 or TLR4 impaired IFN production and macrophage survival during infection with VSV. The VSV glycoprotein G was then reported to induce a previously unrecognized CD14/TLR4-dependent response pathway in which TRAM is involved and which leads to activation of IRF7 in mDCs and macrophages. Also noteworthy is that this pathway did not stimulate NF-κB (50).

RHABDOVIRUS COUNTERMEASURES TO PREVENT IFN INDUCTION AND IFN ACTION

Rhabdoviruses are known as IFN sensitive, a fact that is being widely used to assay the biological activity of IFNs (23). As outlined above, rhabdoviruses are being recognized by nucleic acid pattern receptors, although most of their RNA is shielded by N protein, and at the latest when they start gene expression by synthesis of leader RNA. The crucial point is therefore to limit the consequential IFN production and the expression of IFN-stimulated genes (ISGs) as long and as far as possible by expression of "IFN antagonists." The rhabdoviruses analyzed in some detail so far with respect to IFN antagonism are RV and VSV. Though closely related, these two viruses provide quite different examples of how to solve the problem. Whereas VSV is a fast and cytopathic virus, RV is slow and needs to conserve the integrity of host cells, in particular neurons, in order to reach the "immune-privileged" central nervous system, where it replicates best.

Shutdown of Host Gene Expression by VSV Matrix Protein

The major principle of how VSV counteracts IFN has been known for decades. Huang and Wagner found that the cytopathic infection of cells with VSV is related to the rapid inhibition of cellular RNA synthesis (61). These authors also reported that VSV, which inhibited cellular RNA synthesis very effectively in Krebs-2 cells, was incapable of stimulating production of IFN. Moreover, VSV inhibited IFN synthesis in cells previously infected with the avian Newcastle disease paramyxovirus, which is an efficient IFN stimulus in this system. They concluded that the inhibition of cellular RNA synthesis by VSV was responsible for the lack of IFN production (61). Furthermore, Wertz and Youngner correlated the efficiency of shutoff of different host cells with the ability of VSV to prevent IFN synthesis (147). In other words, if host gene expression is shut off by the virus before the required new RNA is transcribed, IFN production does not occur. These authors (147) formulated the idea of a contest between virus and host, which is still valid today (55).

Transfection of cDNA then implicated the VSV matrix (M) protein in cytopathogenicity (15) and inhibition of host-cell RNA synthesis by inactivation of the basal transcription factor TFIID (14, 159). The key role of M in inhibition of host gene expression and IFN production was further confirmed by analysis of IFN-inducing and -noninducing VSV isolates and identification of a critical mutation in M(M51R) (39). Recombinant VSV carrying the mutant protein M(M51R), was defective in its ability to inhibit both host RNA synthesis and IFN production (4).

Interestingly, in addition to preventing host transcription, VSV M is capable of inhibiting nuclear export of host mRNA and snRNA (107), which should greatly contribute to preventing IFN expression. VSV M was found to target the nucleoporin Nup98 and the export factor Rae1/mrnp41 (35, 38, 142). A mutant of M protein defective in Rae1 binding, and in which M residues 52 to 54 are changed to alanines, was unable to form a complex with Nup98 and was unable to inhibit mRNA nuclear export (142). Like M51R, this mutant has also lost the ability to shut down host transcription. Similar to VSV M, and despite only 28% amino acid sequence identity, the M protein of the human pathogen Chandipura virus is able to shut down host-cell gene expression (136) and to inhibit host nucleocytoplasmic transport (106).

The M-mediated inhibition of host gene expression of vesiculoviruses is therefore a means to suppress IFN response in mammalian cells. Most intriguingly, IFN-induced cellular mechanisms may again counteract this M function. The two targets of M, Nup98 and Rae1/mrnp41, were found to be type I and type II IFN inducible (35, 38), and the M-mediated mRNA nuclear export block was reversed when cells were treated with IFN-γ or transfected with a cDNA encoding Nup98 or Rae1/mrnp41. Thus, increased Nup98 expression constitutes an IFN-mediated mechanism to reverse M protein-mediated inhibition of IFN gene expression (35, 38).

VSV mutants like M(M51R), M(51Δ), and M(52-54AAA) are of special interest because they stimulate and cannot counteract host immune responses. Indeed, replication of an M(51Δ) VSV was controlled faster than wild-type VSV replication in the lungs of mice and did not spread systemically, which correlated with a more rapid induction of IFN in the lung (113). In contrast to wild-type VSV, an M(M51R) mutant VSV induced the maturation of mDC populations, which effectively activated naïve T cells in vitro. The mutant VSV therefore has the potential to promote effective T-cell responses to vector-expressed antigens or activate

DCs at tumor sites during therapy (2). Particular tumor cells are more susceptible to VSV infection than normal cells due to defects in their antiviral responses, in particular in the PKR and Stat pathways, which may be related to activated extracellular signal-related kinase (ERK) (97), making VSV a candidate for oncolytic virus therapy (33, 99, 130). The above M mutants should be attenuated in normal cells, resulting in a greater selectivity for such tumor cells. Indeed, in nude mouse experiments, an M51R-M virus was much less pathogenic than the recombinant wild-type virus from which it was derived but showed comparable therapeutic effects (3).

Inhibition of IFN Induction by RV Phosphoprotein

In contrast to VSV, RV infection is not associated with rapid cytopathic effects and host-cell shutdown. Rather, mechanisms appear to prevent destruction of host cells and to conserve, in particular, the integrity of the neuronal network, which is the basis for RV spread to the central nervous system (for reviews, see references 44 and 77) . Indeed, RV-infected neurons in brain slices appear to be healthy for weeks, which makes RV an ideal tracer for neurons, neuronal functions, and neuronal connectivity (149, 150). Expression of the RV M protein from plasmids does not result in a pronounced shutdown of host gene expression (46), and unlike VSV M, RV M is therefore not responsible for IFN-antagonistic activities.

We recently assigned the key role in counteracting innate immunity to the RV phosphoprotein P. RV P is known as a protein serving several essential functions during RNA synthesis and RNP formation, by association with other viral proteins. The N-terminal part of P is involved in binding P to the L polymerase (27). Two domains localized in the center (amino acids 69 to 177) and at the C terminus of P are essential for the interaction with the N protein (26, 64, 92, 93). P is phosphorylated by two cellular kinases, an as yet uncharacterized RV protein kinase (RVPK) and four isomers of protein kinase C (PKC) (54).

The first hint of the involvement of RV P in inhibition of host defense was the attenuation in some cell lines of a recombinant RV, SAD EGFP-P, expressing an EGFP-P fusion protein instead of wild-type (41). The attenuation of SAD EGFP-P virus was inversely correlated with production of IFN-β and of ISGs. To verify a role of wild-type P in controlling IFN in RV-infected cells, we made use of the possibility of shifting rhabdovirus genes to other positions and accordingly manipulating their expression. A recombinant RV expressing P from the most downstream position (SAD ΔPLP: 3′-N-M-G-L-P-5′; wild type: 3′-N-P-M-G-L-5′) expressed very small amounts of P, just enough to support RNA replication (20). SAD ΔPLP was only slightly attenuated in IFN-incompetent BSR cell lines, with a 10-fold reduction of maximum virus titer, but completely failed to replicate in IFN-competent HEp-2 cells. This effect was correlated with efficient transcription of IFN-β mRNA in SAD ΔPLP-infected cells, whereas in wild-type RV-infected cells IFN-β transcription was hardly detectable. In addition, SAD ΔPLP infection resulted in activation of the critical transcription factor IRF3, as demonstrated by Ser-386 phosphorylation and IRF3 dimerization. In contrast, in the presence of wild-type virus, or after ectopic expression of P from plasmids, IRF3 was not activated. Furthermore, P prevented IFN induction by transfection of TBK1 or of upstream components of the RIG-I and TLR3/4 pathways, such as IPS-1 or TRIF, but not by a constitutive active form of IRF3. This indicated that P directly targets the critical step of phosphorylation of IRF3 by TBK1. Although RV P inhibits IRF7 activation by TBK1, it is not able to interfere with the TLR7/9-MyD88-dependent activation of IRF7, in contrast to the paramyxoviruses measles virus and RSV (60, 121). By directed mutagenesis of RV P, we could identify a critical domain required for inhibition of IFN production by P, which is located between the two N-binding sites of the protein. Whereas transfection of the full-length P cDNA prevented IRF3 activation and IFN production, P mutants lacking this domain were significantly impaired in inhibition (Brzózka, unpublished). Intriguingly, direct binding of P to IRF and TBK1 is unlikely, as suggested by its failure in inhibiting constitutive active IRFs and by coprecipitation assays. Rather, P should target an adaptor protein required for the assembly of functional complexes of TBK1 and IRF3.

The importance of P functions in counteracting IFN for RV infection in vivo was confirmed by animal experiments. Severe attenuation of SAD ΔPLP was confirmed by intracranial injection of 10^2 or 10^5 focus-forming units into either wild-type mice or IFN-α receptor (IFNAR) knockout mice. SAD ΔPLP was completely apathogenic in wild-type mice, whereas all mice lacking the IFNAR receptor were killed, although death was delayed to 11 days postinfection, compared with wild-type RV, which killed mice on day 7 postinfection. These experiments provide evidence that SAD ΔPLP does replicate in mice but can kill mice only in the absence of an IFN response. In the wild-type host, the presence of sufficient amounts of P and the functions of RV P are therefore essential for virus survival in the host (Brzózka, unpublished).

Previous results indicated some IFN induction in RV-infected hosts (68, 89, 90). A correlation of the level of IFN induction and attenuation was suggested in a study comparing wild-type and laboratory-adapted RV. This revealed major differences in the cellular gene expression pattern after infection with the pathogenic silver-haired bat RV and the laboratory-adapted virus B2C. In contrast to the former, the latter induced extensive inflammation and expression of genes involved in innate immune responses (145).

Inhibition of Stat Signaling by RV Phosphoprotein P

The biochemistry of IFN signaling via Janus kinase (Jak)/Stat is well described in the literature (for recent reviews, see reference 110 and chapters 6 and 7 of this volume), but the understanding of the manifold biological effects of different IFNs in different cell types is still limited (19). Apart from stimulating expression of antiviral, antiproliferative, and immunomodulatory proteins, IFN signaling in most cells provides an important feedback loop for amplification of the RLR and TLR sensing machinery and the expression of the full set of late IFN-α by induction of IRF7. As probably no virus goes unrecognized, and some IFN is induced initially, viral targeting of IFN signaling is important not only to counteract the IFN-induced antiviral effects but also to confine further stimulation of IFN induction. Indeed, viruses have been shown to target virtually all steps of the classical Jak/Stat signaling pathway, including IFN binding to the receptor, recruitment and autophosphorylation of the Jaks, tyrosine phosphorylation of Stat1 and Stat2, and assembly of the heteromeric complex of Stat1, Stat2, and IRF9 (p48), known as IFN-stimulated gene factor 3 (ISGF3), which enters the nucleus to initiate transcription of ISGs defined by interferon-stimulated response element (ISRE) and gamma-activated sequence (GAS) promoter sequences (for review, see references 22 and 55).

In cells previously infected with RV, the induction of ISGs and ISRE- or GAS-controlled luciferase by IFN-α or IFN-γ treatment was found to be almost completely abolished, demonstrating the existence of a potent IFN signaling antagonist in RV (21, 140). Notably, the inhibitory effect was not observed in cells infected with the above-described SAD ΔPLP expressing little P. This again demonstrated a key role for the RV P protein. Interestingly, despite the lack of ISRE- and GAS-controlled gene expression in wild-type RV-infected cells, effective phosphorylation of both Stat1 and Stat2 at the critical tyrosine residues Y701 and Y689, respectively, was observed after IFN treatment. Moreover, the phosphorylated isoforms of Stats accumulated in RV-infected cells, whereas in mock-infected cells phospho-Stats were rapidly dephosphorylated. Immunofluorescence studies demonstrated Stat and RV P colocalization in the cytoplasm of cells infected with RV or transfected with P cDNA. Further experiments involving bleaching of fluorescent Stat1 revealed that RV P prevented the nuclear import of Stats, whereas Stat shuttling in nonstimulated cells was unaffected. Indeed, by coprecipitation experiments, efficient binding of P to phospho-Stat but not to nonphosphorylated Stats was observed. Such conditional, activation-dependent targeting of Stat1 and Stat2 signaling is unique among viruses. In view of the numerous tasks of P protein, including the roles in viral RNA synthesis, NC assembly, and IRF inhibition, the observed activity "on demand" may reflect a valuable specialization to maintain full capacity for other duties in nonalerted cells. Some residual binding of RV P to nonphosphorylated Stats, however, is indicated by the identification of the P protein of the RV CVS strain with Stat1 in the yeast two-hybrid system (140). In addition, coprecipitation of Stats with P from non-IFN-stimulated cells was reported, though it was not excluded that this represented phosphorylated isoforms of Stat. Intriguingly, and in contrast to the IRF-inhibitory function of P, inhibition of Jak/Stat signaling by P required the C-terminal part of the P protein (21, 140). A RV P mutant lacking 10 amino acids of the C terminus was shown to be deficient for inhibiting Stat nuclear import, while it was not affected in the capacity to prevent IRF phosphorylation. Thus, inhibition of IFN induction and inhibition of IFN signaling are two distinct, independent, and genetically dissociable roles of RV P (21, 22).

Although, in contrast to the P genes of VSV or of paramyxoviruses, the RV P gene is considered monocistronic, different N-terminally truncated P proteins are expressed from RV P mRNA in addition to full-length P (P1) by translation initiation at downstream in-frame AUG codons (P2, P3, P4, and P5) (25, 36). As all these forms share the C terminus, they should be able to bind to Stats (140). Whereas P1 and P2 of the RV CVS strain are cytoplasmic after expression from plasmids, P3 and P4 are also localized to the nucleus, because of the lack of one of the two nuclear export signals (104). Making use of the nuclear localization of P3, Vidy et al. showed that this protein was able to prevent binding of nuclear Stat to the host DNA (141). Although in RV-infected cells or in the presence of full-length P nuclear P products are not readily detectable, the nuclear P products may therefore provide a supplementary measure against activated Stat that escaped retention by P in the cytoplasm.

The role of truncated P proteins was also addressed by reverse genetics. In cells infected with the recombinant wild-type RV SAD L16, three truncated P forms are detectable, P2, P3, and P4, starting with methionines at

positions 20, 53, and 83, respectively. By exchanging methionine with isoleucine codons, recombinant viruses were generated lacking expression of individual or all truncated P proteins (P1xxx). Growth of the viruses was only slightly affected in the IFN-competent HEp-2 cell line, with the SAD P1xxx virus being attenuated most. From this experiment we concluded that full-length P (P1) is sufficient for both viral RNA synthesis and IFN escape and that individual P forms do not have particular essential functions. However, since the domains for IRF and Stat inhibition are present, we speculate that increased levels of the C-terminally nested forms of P may appreciably increase the capacity of the virus to suppress IFN activation and response in vivo.

Interplay with IFN-Induced Gene Products

To date, three major IFN-induced antiviral pathways have been firmly established. These are the dsRNA-dependent PKR, the 2′,5′-oligoadenylate synthetase (OAS)/RNase L system, and the Mx proteins (for reviews, see references 56, 125, and 152). PKR phosphorylates eukaryotic initiation factor 2 (eIF2) and inhibits translation initiation. PKR first needs to be activated by dsRNA (73, 80) or human PKR protein activator (PACT), another cellular protein (123). Intriguingly, PKR is a major component of IFN-mediated resistance to VSV infection (9, 82, 129), whereas OAS/RNase L appears not to have any effect on IFN-mediated protection from VSV or other rhabdoviruses (125). Mx proteins, which are dynamin-like GTPases, have broad antiviral activity against a range of diverse RNA viruses. Inhibition of VSV by one of the forms of rat Mx1, and of individual strains of RV by bovine but not human MxA, have been reported (83, 120). For a comprehensive review of Mx antiviral activities, see reference 56. Notably, several tumor suppressors that are induced by IFN or virus infection may have direct or indirect anti-VSV activities. These include, for example, p53, the alternative reading frame protein (ARF), and phospholipid scramblase 1 (PLSCR1) (for review, see reference 95), and this may contribute to the preferential replication of VSV in certain tumor cells.

Finally, promyelocytic leukemia (PML) protein was described as a component of the antiviral defense against VSV and RV (16, 24, 37, 115). Overexpression of PML in CHO cells induced resistance to infection by VSV (24), but the mechanism behind this remained unknown. In an elegant study from the group of Danielle Blondel (16), the RV phosphoprotein P was identified to be bound by cytoplasmic PML. In PML$^{-/-}$ MEFs, titers of RV CVS were 10- to 20-fold increased compared with wild-type MEFs. Virus infection was reported to be responsible for a reorganization of PML in PML-expressing CHO cells, and it was proposed that RV P is an antagonist of some PML function in antiviral defense. However, based on experiments with the recombinant SAD L16 and SAD ΔPLP, we would like to propose an opposite situation. In IFN-competent HEp2 cells, only infection with the IFN-inducing SAD ΔPLP, but not with the IFN-suppressing wild-type SAD L16, caused an increase in number and size of PML bodies. In Vero cells, which do not produce IFN, this phenotype was not observed after infection with either virus, but only after addition of exogenous IFN to SAD ΔPLP-infected or mock-infected cells. Thus, the reorganization of PML bodies is exclusively due to IFN and not directly to virus infection (Brzózka, unpublished). Interestingly, the domain required for the interaction of PML with P is localized in the RV P C terminus, which is also required for Stat binding (16). Binding of the IFN-induced PML to P in the cytoplasm could therefore prevent P from binding to activated Stat and may represent a means to restore Jak/Stat signaling. Following this line, PML may in this case act as an IFN-induced cellular antagonist of a viral antagonist of IFN rather than as a direct antiviral agent such as PKR or Mx. This is reminiscent of the situation with VSV M binding to Nup98, where the target of a viral antagonist is overexpressed in response to IFN to disarm the antagonist. Both examples may be a reflection of the coevolution of virus and host and the possibilities of both to accomplish a fine-balanced basis for coexistence.

CONCLUSIONS AND OUTLOOK

The first details of the interplay of rhabdoviruses and the host innate immune system are just emerging. Rhabdoviruses illustrate that in spite of their similarity in terms of genetic organization and morphology, they have developed a variety of possibilities to cope with the IFN network according to their requirements and niches. Further characterization of the mechanisms involved in IFN escape of rhabdoviruses will help to better explain how self and nonself are distinguished and how host signaling networks are connected and regulated and how they function. In addition, detailed knowledge of the viral antagonists and their mechanisms is required to generate attenuated viruses for purposes of immune stimulation, vaccination, gene therapy, and oncolytic therapy. Indeed, viruses with targeted deletions in IFN antagonist genes are excellent candidates for live virus vaccines, because they can be grown well in tissue culture and are highly attenuated in the host organism due to a robust IFN and immune response, as shown in pioneering

studies with influenza A virus (135) and RSV (138). Fine-tuning of IFN induction and IFN resistance capacity can be achieved in viruses having mutations in the separate genetic determinants of those functions, as exemplified by RV with mutations in either the IRF or Stat inhibitory domain of the P protein (K. Brzózka et al., unpublished data). Future use of cytopathic or apoptosis-inducing viruses like VSV will greatly profit from integrating knowledge on the relationship of tumor suppressors and oncogenes like *ras* with components of innate defense like PKR and IFN (117). Furthermore, studies on rhabdoviruses like VSV in its vector host, and of other rhabdoviruses of fish, insects, and plants, will undoubtedly provide further intriguing insights into the host defense mechanisms and the intricate relationship with viruses.

Work in the authors' laboratory was supported by the Deutsche Forschungsgemeinschaft through SFB 455, "Viral functions and immune modulation," and GK1202, "Oligonucleotides in cell biology and therapy."

References

1. **Abraham, G., and A. K. Banerjee.** 1976. Sequential transcription of the genes of vesicular stomatitis virus. *Proc. Natl. Acad. Sci. USA* 73:1504–1508.
2. **Ahmed, M., K. L. Brzoza, and E. M. Hiltbold.** 2006. Matrix protein mutant of vesicular stomatitis virus stimulates maturation of myeloid dendritic cells. *J. Virol.* 80:2194–2205.
3. **Ahmed, M., S. D. Cramer, and D. S. Lyles.** 2004. Sensitivity of prostate tumors to wild type and M protein mutant vesicular stomatitis viruses. *Virology* 330:34–49.
4. **Ahmed, M., M. O. McKenzie, S. Puckett, M. Hojnacki, L. Poliquin, and D. S. Lyles.** 2003. Ability of the matrix protein of vesicular stomatitis virus to suppress beta interferon gene expression is genetically correlated with the inhibition of host RNA and protein synthesis. *J. Virol.* 77:4646–4657.
5. **Ahne, W., H. V. Bjorklund, S. Essbauer, N. Fijan, G. Kurath, and J. R. Winton.** 2002. Spring viremia of carp (SVC). *Dis. Aquat. Organ.* 52:261–272.
6. **Akira, S., and K. Takeda.** 2004. Functions of Toll-like receptors: lessons from KO mice. *C. R. Biol.* 327:581–589.
7. **Albertini, A. A., A. K. Wernimont, T. Muziol, R. B. Ravelli, C. R. Clapier, G. Schoehn, W. Weissenhorn, and R. W. Ruigrok.** 2006. Crystal structure of the rabies virus nucleoprotein-RNA complex. *Science* 313:360–363.
8. **Andrejeva, J., K. S. Childs, D. F. Young, T. S. Carlos, N. Stock, S. Goodbourn, and R. E. Randall.** 2004. The V proteins of paramyxoviruses bind the IFN-inducible RNA helicase, mda-5, and inhibit its activation of the IFN-beta promoter. *Proc. Natl. Acad. Sci. USA* 101:17264–17269.
9. **Balachandran, S., P. C. Roberts, L. E. Brown, H. Truong, A. K. Pattnaik, D. R. Archer, and G. N. Barber.** 2000. Essential role for the dsRNA-dependent protein kinase PKR in innate immunity to viral infection. *Immunity* 13:129–141.

10. Ball, L. A., C. R. Pringle, B. Flanagan, V. P. Perepelitsa, and G. W. Wertz. 1999. Phenotypic consequences of rearranging the P, M, and G genes of vesicular stomatitis virus. *J. Virol.* **73:**4705–4712.

11. Barchet, W., M. Cella, B. Odermatt, C. Asselin-Paturel, M. Colonna, and U. Kalinke. 2002. Virus-induced interferon alpha production by a dendritic cell subset in the absence of feedback signaling in vivo. *J. Exp. Med.* **195:**507–516.

12. Basak, S., A. Mondal, S. Polley, S. Mukhopadhyay, and D. Chattopadhyay. 2007. Reviewing Chandipura: a vesiculovirus in human epidemics. *Biosci. Rep.* **27:**275–298.

13. Bitko, V., and S. Barik. 2001. Phenotypic silencing of cytoplasmic genes using sequence-specific double-stranded short interfering RNA and its application in the reverse genetics of wild type negative-strand RNA viruses. *BMC Microbiol.* **1:**34.

14. Black, B. L., and D. S. Lyles. 1992. Vesicular stomatitis virus matrix protein inhibits host cell-directed transcription of target genes in vivo. *J. Virol.* **66:**4058–4064.

15. Blondel, D., G. G. Harmison, and M. Schubert. 1990. Role of matrix protein in cytopathogenesis of vesicular stomatitis virus. *J. Virol.* **64:**1716–1725.

16. Blondel, D., T. Regad, N. Poisson, B. Pavie, F. Harper, P. P. Pandolfi, H. De The, and M. K. Chelbi-Alix. 2002. Rabies virus P and small P products interact directly with PML and reorganize PML nuclear bodies. *Oncogene* **21:**7957–7970.

17. Blumberg, B. M., and D. Kolakofsky. 1981. Intracellular vesicular stomatitis virus leader RNAs are found in nucleocapsid structures. *J. Virol.* **40:**568–576.

18. Both, G. W., A. K. Banerjee, and A. J. Shatkin. 1975. Methylation-dependent translation of viral messenger RNAs in vitro. *Proc. Natl. Acad. Sci. USA* **72:**1189–1193.

19. Boxel-Dezaire, A. H., M. R. Rani, and G. R. Stark. 2006. Complex modulation of cell type-specific signaling in response to type I interferons. *Immunity* **25:**361–372.

20. Brzózka, K., S. Finke, and K. K. Conzelmann. 2005. Identification of the rabies virus alpha/beta interferon antagonist: phosphoprotein P interferes with phosphorylation of interferon regulatory factor 3. *J. Virol.* **79:**7673–7681.

21. Brzózka, K., S. Finke, and K. K. Conzelmann. 2006. Inhibition of interferon signaling by rabies virus phosphoprotein P: activation-dependent binding of STAT1 and STAT2. *J. Virol.* **80:**2675–2683.

22. Brzózka, K., C. Pfaller, and K. K. Conzelmann. 2007. Signal transduction in the type I interferon system and viral countermeasures. *Signal Transduction* **7:**5–19.

23. Buckler, C. E., and S. Baron. 1966. Antiviral action of mouse interferon in heterologous cells. *J. Bacteriol.* **91:**231–235.

24. Chelbi-Alix, M. K., F. Quignon, L. Pelicano, M. H. Koken, and H. De The. 1998. Resistance to virus infection conferred by the interferon-induced promyelocytic leukemia protein. *J. Virol.* **72:**1043–1051.

25. Chenik, M., K. Chebli, and D. Blondel. 1995. Translation initiation at alternate in-frame AUG codons in the rabies virus phosphoprotein mRNA is mediated by a ribosomal leaky scanning mechanism. *J. Virol.* **69:**707–712.

26. Chenik, M., K. Chebli, Y. Gaudin, and D. Blondel. 1994. In vivo interaction of rabies virus phosphoprotein (P) and nucleoprotein (N): existence of two N-binding sites on P protein. *J. Gen. Virol.* **75**(Pt. 11)**:**2889–2896.

27. Chenik, M., M. Schnell, K. K. Conzelmann, and D. Blondel. 1998. Mapping the interacting domains between the rabies virus polymerase and phosphoprotein. *J. Virol.* **72:**1925–1930.

28. Colonno, R. J., and A. K. Banerjee. 1976. A unique RNA species involved in initiation of vesicular stomatitis virus RNA transcription in vitro. *Cell* **8:**197–204.

29. Colonno, R. J., and A. K. Banerjee. 1978. Complete nucleotide sequence of the leader RNA synthesized in vitro by vesicular stomatitis virus. *Cell* **15:**93–101.

30. Conzelmann, K. K. 2004. Reverse genetics of mononegavirales. *Curr. Top. Microbiol. Immunol.* **283:**1–41.

31. Cui, S., K. Eisenächer, A. Kirchhofer, K. Brzózka, A. Lammens, K. Lammens, T. Fujita, K. K. Conzelmann, A. Krug, and K. Hopfner. 2008. The C-terminal regulatory domain is the RNA 5′-triphosphate sensor of RIG-I. *Mol. Cell* **29:**169–179.

32. Deleris, A., J. Gallego-Bartolome, J. Bao, K. D. Kasschau, J. C. Carrington, and O. Voinnet. 2006. Hierarchical action and inhibition of plant Dicer-like proteins in antiviral defense. *Science* **313:**68–71.

33. Ebert, O., K. Shinozaki, T. G. Huang, M. J. Savontaus, A. Garcia-Sastre, and S. L. Woo. 2003. Oncolytic vesicular stomatitis virus for treatment of orthotopic hepatocellular carcinoma in immune-competent rats. *Cancer Res.* **63:**3605–3611.

34. Edelmann, K. H., S. Richardson-Burns, L. Alexopoulou, K. L. Tyler, R. A. Flavell, and M. B. Oldstone. 2004. Does Toll-like receptor 3 play a biological role in virus infections? *Virology* **322:**231–238.

35. Enninga, J., D. E. Levy, G. Blobel, and B. M. Fontoura. 2002. Role of nucleoporin induction in releasing an mRNA nuclear export block. *Science* **295:**1523–1525.

36. Eriguchi, Y., H. Toriumi, and A. Kawai. 2002. Studies on the rabies virus RNA polymerase: 3. Two-dimensional electrophoretic analysis of the multiplicity of non-catalytic subunit (P protein). *Microbiol. Immunol.* **46:**463–474.

37. Everett, R. D., and M. K. Chelbi-Alix. 2007. PML and PML nuclear bodies: implications in antiviral defence. *Biochimie* **89:**819–830.

38. Faria, P. A., P. Chakraborty, A. Levay, G. N. Barber, H. J. Ezelle, J. Enninga, C. Arana, J. van Deursen, and B. M. Fontoura. 2005. VSV disrupts the Rae1/mrnp41 mRNA nuclear export pathway. *Mol. Cell* **17:**93–102.

39. Ferran, M. C., and J. M. Lucas-Lenard. 1997. The vesicular stomatitis virus matrix protein inhibits transcription from the human beta interferon promoter. *J. Virol.* **71:**371–377.

40. Field, A. K., A. A. Tytell, G. P. Lampson, and M. R. Hilleman. 1967. Inducers of interferon and host resistance. II. Multistranded synthetic polynucleotide complexes. *Proc. Natl. Acad. Sci. USA* **58:**1004–1010.

41. Finke, S., K. Brzózka, and K. K. Conzelmann. 2004. Tracking fluorescence-labeled rabies virus: enhanced green fluorescent protein-tagged phosphoprotein P supports virus gene expression and formation of infectious particles. *J. Virol.* **78:**12333–12343.

42. Finke, S., and K. K. Conzelmann. 1997. Ambisense gene expression from recombinant rabies virus: random packaging of positive- and negative-strand ribonucleoprotein complexes into rabies virions. *J. Virol.* **71:**7281–7288.

43. Finke, S., and K. K. Conzelmann. 1999. Virus promoters determine interference by defective RNAs: selective amplification of mini-RNA vectors and rescue from cDNA by a 3′ copy-back ambisense rabies virus. *J. Virol.* 73:3818–3825.

44. Finke, S., and K. K. Conzelmann. 2005. Replication strategies of rabies virus. *Virus Res.* 111:120–131.

45. Finke, S., J. H. Cox, and K. K. Conzelmann. 2000. Differential transcription attenuation of rabies virus genes by intergenic regions: generation of recombinant viruses overexpressing the polymerase gene. *J. Virol.* 74:7261–7269.

46. Finke, S., R. Mueller-Waldeck, and K. K. Conzelmann. 2003. Rabies virus matrix protein regulates the balance of virus transcription and replication. *J. Gen. Virol.* 84:1613–1621.

47. Fitzgerald, K. A., D. C. Rowe, B. J. Barnes, D. R. Caffrey, A. Visintin, E. Latz, B. Monks, P. M. Pitha, and D. T. Golenbock. 2003. LPS-TLR4 signaling to IRF-3/7 and NF-kappaB involves the Toll adapters TRAM and TRIF. *J. Exp. Med.* 198:1043–1055.

48. Fu, Z. F. 2005. Genetic comparison of the rhabdoviruses from animals and plants. *Curr. Top. Microbiol. Immunol.* 292:1–24.

49. Galiana-Arnoux, D., C. Dostert, A. Schneemann, J. A. Hoffmann, and J. L. Imler. 2006. Essential function in vivo for Dicer-2 in host defense against RNA viruses in drosophila. *Nat. Immunol.* 7:590–597.

50. Georgel, P., Z. Jiang, S. Kunz, E. Janssen, J. Mols, K. Hoebe, S. Bahram, M. B. Oldstone, and B. Beutler. 2007. Vesicular stomatitis virus glycoprotein G activates a specific antiviral Toll-like receptor 4-dependent pathway. *Virology* 362:304–313.

51. Gitlin, L., W. Barchet, S. Gilfillan, M. Cella, B. Beutler, R. A. Flavell, M. S. Diamond, and M. Colonna. 2006. Essential role of mda-5 in type I IFN responses to polyriboinosinic:polyribocytidylic acid and encephalomyocarditis picornavirus. *Proc. Natl. Acad. Sci. USA* 103:8459–8464.

52. Grandvaux, N., M. J. Servant, B. tenOever, G. C. Sen, S. Balachandran, G. N. Barber, R. Lin, and J. Hiscott. 2002. Transcriptional profiling of interferon regulatory factor 3 target genes: direct involvement in the regulation of interferon-stimulated genes. *J. Virol.* 76:5532–5539.

53. Green, T. J., X. Zhang, G. W. Wertz, and M. Luo. 2006. Structure of the vesicular stomatitis virus nucleoprotein-RNA complex. *Science* 313:357–360.

54. Gupta, A. K., D. Blondel, S. Choudhary, and A. K. Banerjee. 2000. The phosphoprotein of rabies virus is phosphorylated by a unique cellular protein kinase and specific isomers of protein kinase C. *J. Virol.* 74:91–98.

55. Haller, O., G. Kochs, and F. Weber. 2007. Interferon, Mx, and viral countermeasures. *Cytokine Growth Factor Rev.* 18:425–433.

56. Haller, O., P. Staeheli, and G. Kochs. 2007. Interferon-induced Mx proteins in antiviral host defense. *Biochimie* 89:812–818.

57. Honda, K., and T. Taniguchi. 2006. IRFs: master regulators of signalling by Toll-like receptors and cytosolic pattern-recognition receptors. *Nat. Rev. Immunol.* 6:644–658.

58. Hornung, V., J. Ellegast, S. Kim, K. Brzózka, A. Jung, H. Kato, H. Poeck, S. Akira, K. K. Conzelmann, M. Schlee, S. Endres, and G. Hartmann. 2006. 5′-Triphosphate RNA is the ligand for RIG-I. *Science* 314:994–997.

59. Hornung, V., S. Rothenfusser, S. Britsch, A. Krug, B. Jahrsdorfer, T. Giese, S. Endres, and G. Hartmann. 2002. Quantitative expression of Toll-like receptor 1–10 mRNA in cellular subsets of human peripheral blood mononuclear cells and sensitivity to CpG oligodeoxynucleotides. *J. Immunol.* 168:4531–4537.

60. Hornung, V., J. Schlender, M. Guenthner-Biller, S. Rothenfusser, S. Endres, K. K. Conzelmann, and G. Hartmann. 2004. Replication-dependent potent IFN-alpha induction in human plasmacytoid dendritic cells by a single-stranded RNA virus. *J. Immunol.* 173:5935–5943.

61. Huang, A. S., and R. R. Wagner. 1965. Inhibition of cellular RNA synthesis by nonreplicating vesicular stomatitis virus. *Proc. Natl. Acad. Sci. USA* 54:1579–1584.

62. Izaguirre, A., B. J. Barnes, S. Amrute, W. S. Yeow, N. Megjugorac, J. Dai, D. Feng, E. Chung, P. M. Pitha, and P. Fitzgerald-Bocarsly. 2003. Comparative analysis of IRF and IFN-alpha expression in human plasmacytoid and monocyte-derived dendritic cells. *J. Leukoc. Biol.* 74:1125–1138.

63. Jackson, A. O., R. G. Dietzgen, M. M. Goodin, J. N. Bragg, and M. Deng. 2005. Biology of plant rhabdoviruses. *Annu. Rev. Phytopathol.* 43:623–660.

64. Jacob, Y., E. Real, and N. Tordo. 2001. Functional interaction map of lyssavirus phosphoprotein: identification of the minimal transcription domains. *J. Virol.* 75:9613–9622.

65. Janeway, C. A., Jr. 1989. Approaching the asymptote? Evolution and revolution in immunology. *Cold Spring Harb. Symp. Quant. Biol.* 54(Pt. 1):1–13.

66. Janeway, C. A., Jr., and R. Medzhitov. 2002. Innate immune recognition. *Annu. Rev. Immunol.* 20:197–216.

67. Jarrous, N., and R. Reiner. 2007. Human RNase P: a tRNA-processing enzyme and transcription factor. *Nucleic Acids Res.* 35:3519–3524.

68. Johnson, N., C. S. McKimmie, K. L. Mansfield, P. R. Wakeley, S. M. Brookes, J. K. Fazakerley, and A. R. Fooks. 2006. Lyssavirus infection activates interferon gene expression in the brain. *J. Gen. Virol.* 87:2663–2667.

69. Kato, H., S. Sato, M. Yoneyama, M. Yamamoto, S. Uematsu, K. Matsui, T. Tsujimura, K. Takeda, T. Fujita, O. Takeuchi, and S. Akira. 2005. Cell type-specific involvement of RIG-I in antiviral response. *Immunity* 23:19–28.

70. Kato, H., O. Takeuchi, S. Sato, M. Yoneyama, M. Yamamoto, K. Matsui, S. Uematsu, A. Jung, T. Kawai, K. J. Ishii, O. Yamaguchi, K. Otsu, T. Tsujimura, C. S. Koh, C. Reis e Sousa, Y. Matsuura, T. Fujita, and S. Akira. 2006. Differential roles of MDA5 and RIG-I helicases in the recognition of RNA viruses. *Nature* 441:101–105.

71. Kawai, T., and S. Akira. 2007. TLR signaling. *Semin. Immunol.* 19:24–32.

72. Kawai, T., K. Takahashi, S. Sato, C. Coban, H. Kumar, H. Kato, K. J. Ishii, O. Takeuchi, and S. Akira. 2005. IPS-1, an adaptor triggering RIG-I- and Mda5-mediated type I interferon induction. *Nat. Immunol.* 6:981–988.

73. Kerr, I. M., R. E. Brown, and L. A. Ball. 1974. Increased sensitivity of cell-free protein synthesis to double-stranded RNA after interferon treatment. *Nature* 250:57–59.

74. Kim, D. H., M. Longo, Y. Han, P. Lundberg, E. Cantin, and J. J. Rossi. 2004. Interferon induction by siRNAs and ssRNAs synthesized by phage polymerase. *Nat. Biotechnol.* **22:**321–325.

75. Kumar, H., T. Kawai, H. Kato, S. Sato, K. Takahashi, C. Coban, M. Yamamoto, S. Uematsu, K. J. Ishii, O. Takeuchi, and S. Akira. 2006. Essential role of IPS-1 in innate immune responses against RNA viruses. *J. Exp. Med.* **203:**1795–1803.

76. Kumar, H., T. Kawai, H. Kato, S. Sato, K. Takahashi, C. Coban, M. Yamamoto, S. Uematsu, K. J. Ishii, O. Takeuchi, and S. Akira. 2006. Essential role of IPS-1 in innate immune responses against RNA viruses. *J. Exp. Med.* **203:**1795–1803.

77. Lafon, M. 2005. Modulation of the immune response in the nervous system by rabies virus. *Curr. Top. Microbiol. Immunol.* **289:**239–258.

78. Lafon, M., F. Megret, M. Lafage, and C. Prehaud. 2006. The innate immune facet of brain: human neurons express TLR-3 and sense viral dsRNA. *J. Mol. Neurosci.* **29:**185–194.

79. Lazzarini, R. A., J. D. Keene, and M. Schubert. 1981. The origins of defective interfering particles of the negative-strand RNA viruses. *Cell* **26:**145–154.

80. Lebleu, B., G. C. Sen, S. Shaila, B. Cabrer, and P. Lengyel. 1976. Interferon, double-stranded RNA, and protein phosphorylation. *Proc. Natl. Acad. Sci. USA* **73:**3107–3111.

81. Lee, H. K., J. M. Lund, B. Ramanathan, N. Mizushima, and A. Iwasaki. 2007. Autophagy-dependent viral recognition by plasmacytoid dendritic cells. *Science* **315:**1398–1401.

82. Lee, S. B., R. Bablanian, and M. Esteban. 1996. Regulated expression of the interferon-induced protein kinase p68 (PKR) by vaccinia virus recombinants inhibits the replication of vesicular stomatitis virus but not that of poliovirus. *J. Interferon Cytokine Res.* **16:**1073–1078.

83. Leroy, M., G. Pire, E. Baise, and D. Desmecht. 2006. Expression of the interferon-alpha/beta-inducible bovine Mx1 dynamin interferes with replication of rabies virus. *Neurobiol. Dis.* **21:**515–521.

84. Li, J., J. T. Wang, and S. P. Whelan. 2006. A unique strategy for mRNA cap methylation used by vesicular stomatitis virus. *Proc. Natl. Acad. Sci. USA* **103:**8493–8498.

85. Limbach, P. A., P. F. Crain, and J. A. McCloskey. 1994. Summary: the modified nucleosides of RNA. *Nucleic Acids Res.* **22:**2183–2196.

86. Loo, Y. M., J. Fornek, N. Crochet, G. Bajwa, O. Perwitasari, L. Martinez-Sobrido, S. Akira, M. A. Gill, A. Garcia-Sastre, M. G. Katze, and M. Gale, Jr. 2007. Distinct RIG-I and MDA5 Signaling by RNA Viruses in innate Immunity. *J. Virol.* **82:**335–345.

87. Lund, J. M., L. Alexopoulou, A. Sato, M. Karow, N. C. Adams, N. W. Gale, A. Iwasaki, and R. A. Flavell. 2004. Recognition of single-stranded RNA viruses by Toll-like receptor 7. *Proc. Natl. Acad. Sci. USA* **101:**5598–5603.

88. Luo, M., T. J. Green, X. Zhang, J. Tsao, and S. Qiu. 2007. Conserved characteristics of the rhabdovirus nucleoprotein. *Virus Res.* **129:**246–251.

89. Marcovistz, R., J. Galabru, H. Tsiang, and A. G. Hovanessian. 1986. Neutralization of interferon produced early during rabies virus infection in mice. *J. Gen. Virol.* **67**(Pt. 2):387–390.

90. Marcovistz, R., E. C. Leal, D. C. Matos, and H. Tsiang. 1994. Interferon production and immune response induction in apathogenic rabies virus-infected mice. *Acta Virol.* **38:**193–197.

91. Marcus, P. I., and M. J. Sekellick. 1977. Defective interfering particles with covalently linked [+/−]RNA induce interferon. *Nature* **266:**815–819.

92. Mavrakis, M., A. A. McCarthy, S. Roche, D. Blondel, and R. W. Ruigrok. 2004. Structure and function of the C-terminal domain of the polymerase cofactor of rabies virus. *J. Mol. Biol.* **343:**819–831.

93. Mavrakis, M., S. Mehouas, E. Real, F. Iseni, D. Blondel, N. Tordo, and R. W. Ruigrok. 2006. Rabies virus chaperone: identification of the phosphoprotein peptide that keeps nucleoprotein soluble and free from non-specific RNA. *Virology* **349:**422–429.

94. Meylan, E., J. Curran, K. Hofmann, D. Moradpour, M. Binder, R. Bartenschlager, and J. Tschopp. 2005. Cardif is an adaptor protein in the RIG-I antiviral pathway and is targeted by hepatitis C virus. *Nature* **437:**1167–1172.

95. Munoz-Fontela, C., M. A. Garcia, M. Collado, L. Marcos-Villar, P. Gallego, M. Esteban, and C. Rivas. 2007. Control of virus infection by tumour suppressors. *Carcinogenesis* **28:**1140–1144.

96. Naeve, C. W., C. M. Kolakofsky, and D. F. Summers. 1980. Comparison of vesicular stomatitis virus intracellular and virion ribonucleoproteins. *J. Virol.* **33:**856–865.

97. Noser, J. A., A. A. Mael, R. Sakuma, S. Ohmine, P. Marcato, P. W. Lee, and Y. Ikeda. 2007. The RAS/Raf1/MEK/ERK signaling pathway facilitates VSV-mediated oncolysis: implication for the defective interferon response in cancer cells. *Mol. Ther.* **15:**1531–1536.

98. Novella, I. S., L. A. Ball, and G. W. Wertz. 2004. Fitness analyses of vesicular stomatitis strains with rearranged genomes reveal replicative disadvantages. *J. Virol.* **78:**9837–9841.

99. Obuchi, M., M. Fernandez, and G. N. Barber. 2003. Development of recombinant vesicular stomatitis viruses that exploit defects in host defense to augment specific oncolytic activity. *J. Virol.* **77:**8843–8856.

100. Ogino, T., and A. K. Banerjee. 2007. Unconventional mechanism of mRNA capping by the RNA-dependent RNA polymerase of vesicular stomatitis virus. *Mol. Cell* **25:**85–97.

101. Onoguchi, K., M. Yoneyama, A. Takemura, S. Akira, T. Taniguchi, H. Namiki, and T. Fujita. 2007. Viral infections activate types I and III interferon genes through a common mechanism. *J. Biol. Chem.* **282:**7576–7581.

102. Oshiumi, H., M. Matsumoto, K. Funami, T. Akazawa, and T. Seya. 2003. TICAM-1, an adaptor molecule that participates in Toll-like receptor 3-mediated interferon-beta induction. *Nat. Immunol.* **4:**161–167.

103. Ostertag, D., T. M. Hoblitzell-Ostertag, and J. Perrault. 2007. Cell-type-specific growth restriction of vesicular stomatitis virus polR mutants is linked to defective viral polymerase function. *J. Virol.* **81:**492–502.

104. Pasdeloup, D., N. Poisson, H. Raux, Y. Gaudin, R. W. Ruigrok, and D. Blondel. 2005. Nucleocytoplasmic shuttling of the rabies virus P protein requires a nuclear localization signal and a CRM1-dependent nuclear export signal. *Virology* **334:**284–293.

105. Perrault, J. 1981. Origin and replication of defective interfering particles. *Curr. Top. Microbiol. Immunol.* **93:**151–207.

106. Petersen, J. M., L. S. Her, and J. E. Dahlberg. 2001. Multiple vesiculoviral matrix proteins inhibit both nuclear export and import. *Proc. Natl. Acad. Sci. USA* **98:**8590–8595.

107. Petersen, J. M., L. S. Her, V. Varvel, E. Lund, and J. E. Dahlberg. 2000. The matrix protein of vesicular stomatitis virus inhibits nucleocytoplasmic transport when it is in the nucleus and associated with nuclear pore complexes. *Mol. Cell. Biol.* **20:**8590–8601.

108. Pichlmair, A., and C. Reis e Sousa. 2007. Innate recognition of viruses. *Immunity* **27:**370–383.

109. Pichlmair, A., O. Schulz, C. P. Tan, T. I. Naslund, P. Liljestrom, F. Weber, and C. Reis e Sousa. 2006. RIG-I-mediated antiviral responses to single-stranded RNA bearing 5′-phosphates. *Science* **314:**997–1001.

110. Platanias, L. C. 2005. Mechanisms of type-I- and type-II-interferon-mediated signalling. *Nat. Rev. Immunol.* **5:**375–386.

111. Plumet, S., F. Herschke, J. M. Bourhis, H. Valentin, S. Longhi, and D. Gerlier. 2007. Cytosolic 5′-triphosphate ended viral leader transcript of measles virus as activator of the RIG I-mediated interferon response. *PLoS ONE* **2:**e279.

112. Pringle, C. R. 1997. The order Mononegavirales—current Status. *Arch. Virol.* **142:**2321–2326.

113. Publicover, J., E. Ramsburg, M. Robek, and J. K. Rose. 2006. Rapid pathogenesis induced by a vesicular stomatitis virus matrix protein mutant: viral pathogenesis is linked to induction of tumor necrosis factor alpha. *J. Virol.* **80:**7028–7036.

114. Redinbaugh, M. G., and S. A. Hogenhout. 2005. Plant rhabdoviruses. *Curr. Top. Microbiol. Immunol.* **292:**143–163.

115. Regad, T., and M. K. Chelbi-Alix. 2001. Role and fate of PML nuclear bodies in response to interferon and viral infections. *Oncogene* **20:**7274–7286.

116. Rothenfusser, S., N. Goutagny, G. DiPerna, M. Gong, B. G. Monks, A. Schoenemeyer, M. Yamamoto, S. Akira, and K. A. Fitzgerald. 2005. The RNA helicase Lgp2 inhibits TLR-independent sensing of viral replication by retinoic acid-inducible gene-I. *J. Immunol.* **175:**5260–5268.

117. Russell, S. J., and K. W. Peng. 2007. Viruses as anticancer drugs. *Trends Pharmacol. Sci.* **28:**326–333.

118. Saito, T., R. Hirai, Y. M. Loo, D. Owen, C. L. Johnson, S. C. Sinha, S. Akira, T. Fujita, and M. Gale, Jr. 2007. Regulation of innate antiviral defenses through a shared repressor domain in RIG-I and LGP2. *Proc. Natl. Acad. Sci. USA* **104:**582–587.

119. Saitoh, T., A. Tun-Kyi, A. Ryo, M. Yamamoto, G. Finn, T. Fujita, S. Akira, N. Yamamoto, K. P. Lu, and S. Yamaoka. 2006. Negative regulation of interferon-regulatory factor 3-dependent innate antiviral response by the prolyl isomerase Pin1. *Nat. Immunol.* **7:**598–605.

120. Sandrock, M., M. Frese, O. Haller, and G. Kochs. 2001. Interferon-induced rat Mx proteins confer resistance to Rift Valley fever virus and other arthropod-borne viruses. *J. Interferon Cytokine Res.* **21:**663–668.

121. Schlender, J., V. Hornung, S. Finke, M. Gunthner-Biller, S. Marozin, K. Brzozka, S. Moghim, S. Endres, G.

Hartmann, and K. K. Conzelmann. 2005. Inhibition of Toll-like receptor 7- and 9-mediated alpha/beta interferon production in human plasmacytoid dendritic cells by respiratory syncytial virus and measles virus. *J. Virol.* **79:**5507–5515.

122. Sekellick, M. J., and P. I. Marcus. 1982. Interferon induction by viruses. VIII. Vesicular stomatitis virus: [+/−]DI-011 particles induce interferon in the absence of standard virions. *Virology* **117:**280–285.

123. Sen, G. C., and G. A. Peters. 2007. Viral stress-inducible genes. *Adv. Virus Res.* **70:**233–263.

124. Seth, R. B., L. Sun, C. K. Ea, and Z. J. Chen. 2005. Identification and characterization of MAVS, a mitochondrial antiviral signaling protein that activates NF-kappaB and IRF 3. *Cell* **122:**669–682.

125. Silverman, R. H. 2007. Viral encounters with 2′, 5′-oligodenylate synthetase and RNase L during the interferon antiviral response. *J. Virol.* **81:**12720–12729.

126. Skall, H. F., N. J. Olesen, and S. Mellergaard. 2005. Viral haemorrhagic septicaemia virus in marine fish and its implications for fish farming—a review. *J. Fish. Dis.* **28:**509–529.

127. Stetson, D. B., and R. Medzhitov. 2006. Antiviral defense: interferons and beyond. *J. Exp. Med.* **203:**1837–1841.

128. Stetson, D. B., and R. Medzhitov. 2006. Recognition of cytosolic DNA activates an IRF3-dependent innate immune response. *Immunity* **24:**93–103.

129. Stojdl, D. F., N. Abraham, S. Knowles, R. Marius, A. Brasey, B. D. Lichty, E. G. Brown, N. Sonenberg, and J. C. Bell. 2000. The murine double-stranded RNA-dependent protein kinase PKR is required for resistance to vesicular stomatitis virus. *J. Virol.* **74:**9580–9585.

130. Stojdl, D. F., B. Lichty, S. Knowles, R. Marius, H. Atkins, N. Sonenberg, and J. C. Bell. 2000. Exploiting tumor-specific defects in the interferon pathway with a previously unknown oncolytic virus. *Nat. Med.* **6:**821–825.

131. Strahle, L., D. Garcin, and D. Kolakofsky. 2006. Sendai virus defective-interfering genomes and the activation of interferon-beta. *Virology* **351:**101–111.

132. Strahle, L., J. B. Marq, A. Brini, S. Hausmann, D. Kolakofsky, and D. Garcin. 2007. Activation of the beta interferon promoter by unnatural Sendai virus infection requires RIG-I and is inhibited by viral C proteins. *J. Virol.* **81:**12227–12237.

133. Sun, Q., L. Sun, H. H. Liu, X. Chen, R. B. Seth, J. Forman, and Z. J. Chen. 2006. The specific and essential role of MAVS in antiviral innate immune responses. *Immunity* **24:**633–642.

134. Takaoka, A., Z. Wang, M. K. Choi, H. Yanai, H. Negishi, T. Ban, Y. Lu, M. Miyagishi, T. Kodama, K. Honda, Y. Ohba, and T. Taniguchi. 2007. DAI (DLM-1/ZBP1) is a cytosolic DNA sensor and an activator of innate immune response. *Nature* **448:**501–505.

135. Talon, J., M. Salvatore, R. E. O'Neill, Y. Nakaya, H. Zheng, T. Muster, A. Garcia-Sastre, and P. Palese. 2000. Influenza A and B viruses expressing altered NS1 proteins: a vaccine approach. *Proc. Natl. Acad. Sci. USA* **97:**4309–4314.

136. Taylor, A., A. J. Easton, and A. C. Marriott. 1999. Matrix protein of Chandipura virus inhibits transcription from an RNA polymerase II promoter. *Virus Genes* **19:**223–228.

137. Uematsu, S., and S. Akira. 2007. Toll-like receptors and type I interferons. *J. Biol. Chem.* **282:**15319–15323.

138. Valarcher, J. F., J. Furze, S. Wyld, R. Cook, K. K. Conzelmann, and G. Taylor. 2003. Role of alpha/beta interferons in the attenuation and immunogenicity of recombinant bovine respiratory syncytial viruses lacking NS proteins. *J. Virol.* **77:**8426–8439.

139. Venkataraman, T., M. Valdes, R. Elsby, S. Kakuta, G. Caceres, S. Saijo, Y. Iwakura, and G. N. Barber. 2007. Loss of DExD/H box RNA helicase LGP2 manifests disparate antiviral responses. *J. Immunol.* **178:**6444–6455.

140. Vidy, A., M. Chelbi-Alix, and D. Blondel. 2005. Rabies virus P protein interacts with Stat1 and inhibits interferon signal transduction pathways. *J. Virol.* **79:**14411–14420.

141. Vidy, A., J. El Bougrini, M. K. Chelbi-Alix, and D. Blondel. 2007. The nucleocytoplasmic rabies virus P protein counteracts interferon signaling by inhibiting both nuclear accumulation and DNA binding of Stat1. *J. Virol.* **81:**4255–4263.

142. von Kobbe, C., J. M. van Deursen, J. P. Rodrigues, D. Sitterlin, A. Bachi, X. Wu, M. Wilm, M. Carmo-Fonseca, and E. Izaurralde. 2000. Vesicular stomatitis virus matrix protein inhibits host cell gene expression by targeting the nucleoporin Nup98. *Mol. Cell* **6:**1243–1252.

143. Wang, T., T. Town, L. Alexopoulou, J. F. Anderson, E. Fikrig, and R. A. Flavell. 2004. Toll-like receptor 3 mediates West Nile virus entry into the brain causing lethal encephalitis. *Nat. Med.* **10:**1366–1373.

144. Wang, X. H., R. Aliyari, W. X. Li, H. W. Li, K. Kim, R. Carthew, P. Atkinson, and S. W. Ding. 2006. RNA interference directs innate immunity against viruses in adult *Drosophila*. *Science* **312:**452–454.

145. Wang, Z. W., L. Sarmento, Y. Wang, X. Q. Li, V. Dhingra, T. Tseggai, B. Jiang, and Z. F. Fu. 2005. Attenuated rabies virus activates, while pathogenic rabies virus evades, the host innate immune responses in the central nervous system. *J. Virol.* **79:**12554–12565.

146. Weber, F., V. Wagner, S. B. Rasmussen, R. Hartmann, and S. R. Paludan. 2006. Double-stranded RNA is produced by positive-strand RNA viruses and DNA viruses but not in detectable amounts by negative-strand RNA viruses. *J. Virol.* **80:**5059–5064.

147. Wertz, G. W., and J. S. Youngner. 1970. Interferon production and inhibition of host synthesis in cells infected with vesicular stomatitis virus. *J. Virol.* **6:**476–484.

148. Whelan, S. P., J. N. Barr, and G. W. Wertz. 2004. Transcription and replication of nonsegmented negative-strand RNA viruses. *Curr. Top. Microbiol. Immunol.* **283:**61–119.

149. Wickersham, I. R., S. Finke, K. K. Conzelmann, and E. M. Callaway. 2007. Retrograde neuronal tracing with a deletion-mutant rabies virus. *Nat. Methods* **4:**47–49.

150. Wickersham, I. R., D. C. Lyon, R. J. Barnard, T. Mori, S. Finke, K. K. Conzelmann, J. A. Young, and E. M. Callaway. 2007. Monosynaptic restriction of transsynaptic tracing from single, genetically targeted neurons. *Neuron* **53:**639–647.

151. Wiegand, M. A., S. Bossow, S. Schlecht, and W. J. Neubert. 2007. De novo synthesis of N and P proteins as a key step in Sendai viral gene expression. *J. Virol.* **81:**13835–1384

152. Williams, B. R. 1999. PKR; a sentinel kinase for cellular stress. *Oncogene* **18:**6112–6120.

153. Xu, L. G., Y. Y. Wang, K. J. Han, L. Y. Li, Z. Zhai, and H. B. Shu. 2005. VISA is an adapter protein required for virus-triggered IFN-beta signaling. *Mol. Cell* **19:**727–740.

154. Yamamoto, M., S. Sato, H. Hemmi, K. Hoshino, T. Kaisho, H. Sanjo, O. Takeuchi, M. Sugiyama, M. Okabe, K. Takeda, and S. Akira. 2003. Role of adaptor TRIF in the MyD88-independent Toll-like receptor signaling pathway. *Science* **301:**640–643.

155. Yamamoto, M., S. Sato, H. Hemmi, S. Uematsu, K. Hoshino, T. Kaisho, O. Takeuchi, K. Takeda, and S. Akira. 2003. TRAM is specifically involved in the Toll-like receptor 4-mediated MyD88-independent signaling pathway. *Nat. Immunol.* **4:**1144–1150.

156. Yoneyama, M., and T. Fujita. 2007. Function of RIG-I-like receptors in antiviral innate immunity. *J. Biol. Chem.* **282:**15315–15318.

157. Yoneyama, M., M. Kikuchi, K. Matsumoto, T. Imaizumi, M. Miyagishi, K. Taira, E. Foy, Y. M. Loo, M. Gale, Jr., S. Akira, S. Yonehara, A. Kato, and T. Fujita. 2005. Shared and unique functions of the DExD/H-box helicases RIG-I, MDA5, and LGP2 in antiviral innate immunity. *J. Immunol.* **175:**2851–2858.

158. Yoneyama, M., M. Kikuchi, T. Natsukawa, N. Shinobu, T. Imaizumi, M. Miyagishi, K. Taira, S. Akira, and T. Fujita. 2004. The RNA helicase RIG-I has an essential function in double-stranded RNA-induced innate antiviral responses. *Nat. Immunol.* **5:**730–737.

159. Yuan, H., B. K. Yoza, and D. S. Lyles. 1998. Inhibition of host RNA polymerase II-dependent transcription by vesicular stomatitis virus results from inactivation of TFIID. *Virology* **251:**383–392.

160. Zhou, S., E. A. Kurt-Jones, K. A. Fitzgerald, J. P. Wang, A. M. Cerny, M. Chan, and R. W. Finberg. 2007. Role of MyD88 in route-dependent susceptibility to vesicular stomatitis virus infection. *J. Immunol.* **178:**5173–5181.

Cellular Signaling and Innate Immune
Responses to RNA Virus Infections
Edited by A. R. Brasier et al.
© 2009 ASM Press, Washington, D.C.

Christopher F. Basler

Filoviruses

15

INTRODUCTION

Filoviruses—Ebola viruses (EBOVs) and Marburg viruses (MBGVs)—are infamous for their ability to cause a highly lethal viral hemorrhagic fever. A single species of MBGV and four species of EBOV—Zaire, Sudan, Ivory Coast (Côte d'Ivoire), and Reston—have been identified. MBGVs were identified when, in 1967, workers exposed to monkeys imported from Africa to European vaccine production facilities contracted a disease with a 23% case fatality rate. Researchers in Marburg, Germany, were able to isolate a novel virus, MBGV, that was the etiologic agent of this new disease. Later, in 1976, the first outbreaks of EBOV infection were described. One outbreak occurred in Zaire (now the Democratic Republic of Congo) and another in Sudan. The mortality rates, 88 and 53%, respectively, associated with these outbreaks, caused by related but distinct EBOVs, Zaire EBOV and Sudan EBOV, exceeded that seen in the Marburg outbreak (20). Because of the fatality rates associated with these and subsequent outbreaks, and because of the hemorrhagic manifestations commonly associated with filovirus disease, these viruses have acquired great notoriety and are viewed as potential bioweapons.

Not all human filovirus infections are lethal, however. The Reston Ebola virus is notable because, although it can cause lethal infections in nonhuman primates, the only known human infections, documented by seroconversion following exposure to infected monkeys from the Philippines, did not result in illness (25, 60, 89, 104). These data suggest that Reston Ebola virus may be avirulent in humans. It is also important to note that Ebola viruses are presumed zoonotic pathogens that occasionally find their way into human or nonhuman primates (107). Although bat species have been implicated, the natural reservoir host(s) for filoviruses remain to be identified definitively (72, 97, 119). Whatever the identity of the reservoir host, it is generally presumed that the natural reservoir will not succumb to filovirus infection at rates comparable to those in primates; otherwise the virus could not be maintained in nature (97). Accordingly, experimental infection of some species, such as mice or guinea pigs, does not result in disease unless the virus has been adapted to these hosts or the hosts have been somehow modified (as in interferon [IFN]-α/β receptor [IFNAR] knockout mice) to render them susceptible to disease (14, 22, 76, 123). These facts limit the use of small animals as models for human disease, but

Christopher F. Basler, Dept. of Microbiology, Box 1124, Mount Sinai School of Medicine, 1 Gustave L. Levy Pl., New York, NY 10029.

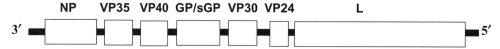

Figure 1 Structure of the EBOV genome. Each box represents an individual gene. The proteins produced from each gene are indicated above the boxes. (Note that MBGV has a similar genome but does not encode an sGP protein.) The viral proteins shown have the following functions: NP, RNA synthesis, structural role; VP35, RNA synthesis, inhibits IFNα/β production; VP40, viral budding, structural role; GP, viral attachment and entry, induces cell rounding; sGP, uncertain function, nonstructural secreted protein; VP30, viral transcription factor; VP24, inhibits IFN signaling, role in assembly; L, RNA polymerase.

the same small animal models can be exploited to explore virus and host determinants of species-specific virulence (13, 16, 27, 81).

Filovirus Structure and Replication

The structure and molecular biology of EBOV and MBGV have been reviewed in detail (107). Filoviruses are long, filamentous, enveloped viruses. They possess a nonsegmented, negative-strand RNA genome approximately 19 kilobases in length that is replicated in the cytoplasm of infected cells. The genomes of both Ebola viruses and Marburg viruses contain seven genes; in EBOVs these genes encode eight proteins, while the MBGV genome encodes seven proteins (Fig. 1). The gene products common to EBOVs and MBGVs are the nucleoprotein (NP), VP35, VP40, the glycoprotein (GP), VP30, VP24, and the large (L) protein (the viral RNA-dependent RNA polymerase). NP, VP35, VP30, and L are involved in viral RNA synthesis; specifically, NP, VP35, and L are required for genome replication of both EBOV and MBGV; additionally, VP30 strongly stimulates transcription (synthesis of mRNAs) in the EBOV system (93). VP40 is the viral matrix protein and is the driving force for budding of new viral particles (28, 51, 95, 118). GP is a type I transmembrane glycoprotein found on the surface of the viral envelope. GP mediates viral attachment and entry into host cells and mediates virus-host cell membrane fusion following endocytosis of the virus (30). In EBOV-infected cells, GP mRNA is produced by an RNA-editing mechanism where the viral polymerase inserts, at a specific site, nucleotides not encoded by the template RNA; the result is a frameshift yielding the GP protein (122). In contrast, faithful transcription of the EBOV GP gene results in an mRNA that encodes secretory GP (sGP). sGP is a truncated version of GP due to the presence of a stop codon found before the transmembrane domain sequence (122). This transmembrane domain anchors GP to cellular and viral membranes, but sGP, because it lacks a transmembrane domain, is secreted from infected cells and is not present in the virus particle. By comparison, MBGV does

not encode an sGP, and its GP mRNA does not undergo RNA editing. The function of the EBOV sGP remains uncertain, although several functions have been proposed, including modulation of host immune function through sGP's interaction with the FcγRIIIB receptor and suppression of inflammation (66, 126, 135). VP35 and VP24, in addition to their roles in virus replication and assembly, have roles in suppression of host IFN responses (6, 7, 102, 103). The GP protein also influences cellular processes at multiple levels (see, for examples, references 100, 113, 116, 124, 126, and 139). These functions, as they relate to filovirus disease, will be discussed in detail below.

Filovirus Disease in Humans and Nonhuman Primates

The best-characterized form of filovirus disease is Ebola hemorrhagic fever (for review, see reference 15). Human infections begin with an incubation period of a few days to around 2 weeks. This is followed by an illness of abrupt onset typically characterized by fever, myalgia, headache, and gastrointestinal symptoms including nausea, vomiting, and diarrhea (reviewed in reference 80). Roughly 50% of patients also develop a maculopapular rash. Changes are also seen in coagulation, although bleeding is not seen in all patients, and even when these symptoms are present, blood loss is not thought to typically contribute to the death of patients. Fatal outcome does seem to correlate with increasingly high viremia as the infection proceeds, and fatal cases progress to shock, convulsions, and diffuse coagulopathy (35). In contrast, nonfatal cases are characterized by an eventual decrease in viral titers in blood, which correlates with the presence of measurable antivirus antibody and clinical improvement. Nonhuman primates experimentally infected with Zaire EBOV exhibit a disease that resembles human EBOV infection, although these infections progress more rapidly and are almost invariably lethal. It is from these nonhuman primate studies that current views of EBOV pathogenesis have been derived (35).

Macaque studies suggest that early targets of infection are macrophages and dendritic cells (DCs), but the infection disseminates quite rapidly to other target tissues (38). The ability of filoviruses to elicit pathological host responses is likely coupled with the ability of these viruses to replicate systemically to high titers. Past reviews have identified several processes associated with filovirus pathogenesis (80). These include the early targeting of macrophages and DCs for infection, resulting ultimately in their destruction by virus. Additionally, the virus may cause direct tissue injury, perhaps in part because the viral glycoprotein is cytotoxic. The ability of the virus to suppress adaptive immunity both by impairing DC function and promoting lymphocyte apoptosis may also contribute to uncontrolled virus replication. Finally, the ability of filoviruses to elicit host production of cytokines, chemokines, and tissue factor likely contributes to several critical manifestations of filovirus disease, including vascular permeability, shock, and bleeding. Because substantial levels of viral replication are probably required for the virus to elicit the most severe manifestations of filovirus infection, the ability of EBOV and MBGV to overcome early innate responses to infection is expected to be central to their pathogenesis. Understanding how filoviruses overcome these early host responses may therefore lead to effective therapies targeting virus-encoded antagonists of innate immunity.

HOST-CELL IFN-α/β RESPONSES

The sustained, systemic infection in primates suggests that EBOV and MBGV possess an ability to overcome or evade host innate antiviral responses, such as the IFN response. The IFN-α/β response is a key component of host innate immunity to viral infection and also influences the development of adaptive immunity (9, 55). IFN-α and IFN-β are a family of secreted proteins expressed, in humans, from a single IFN-β gene and multiple IFN-α genes. IFN-β transcription is stimulated by the cellular transcription factors nuclear factor-κB (NF-κB), interferon regulatory factor 3 (IRF3) or IRF7, and AP-1. These transcription factors are activated by cellular kinases. Of note are the kinases IKK-ε and TBK1; these play a critical role in activation of IFN-α/β gene expression because they phosphorylate serine and threonine residues on IRF3 (and IRF7, if present), leading to the dimerization and nuclear accumulation of the IRF protein (54). Nuclear, activated IRF3 or IRF7 then participates in IFN gene expression (54).

In contrast to the case for IFN-β, most IFN-α gene promoters are activated specifically by IRF7, an IFN-induced protein. IFN-induced expression of IRF7 thus functions as a positive-feedback system that amplifies the IFN-α/β response by activating many IFN-α genes that would otherwise remain quiescent. It should be noted that some cell types, such as plasmacytoid DCs (pDCs), constitutively express IRF7 and, as a consequence, rapidly produce IFN-α in response to virus infection (83). This allows pDCs to rapidly produce large amounts of IFN-α in response to virus infection.

IFN-α/β production can be induced by a variety of stimuli including viral infection (Fig. 2) (83). Induction of the IFN-α/β response can occur via membrane-associated pattern recognition receptors (PRRs) of the Toll-like receptor (TLR) family or via intracellular PRRs (reviewed in reference 83). PRRs recognize and respond to a variety of pathogen-associated molecular patterns (PAMPs). Although the roles of specific PRRs in host response to filovirus infection have not been defined, those most likely to be relevant would include TLR3, 7, and 8 and the intracellular sensors of virus infection retinoic acid-inducible gene I (RIG-I) and melanoma differentiation-associated gene 5 (MDA-5). TLRs are type I transmembrane proteins with cytoplasmic tails that contain Toll/interleukin (IL)-1 receptor (TIR) domains (64). Upon activation by the appropriate PAMP, TLRs signal to transcriptionally activate the expression of a number of genes, and for TLR3, 4, 7, 8, and 9, IFN-β and IFN-α are among the genes turned on. TLR3, 7, and 8 are particularly relevant to RNA virus infection because they respond to nucleic acids that may be produced during virus infection. Thus, TLR3 responds to double-stranded (ds) RNA while TLR7 and TLR8 recognize viral single-stranded RNAs (63). Upon recognition of these PAMPs, TLRs signal in a manner that requires the TIR domain of the TLR as well as cellular adaptor molecules. Notably, dsRNA signaling through TLR3 requires the TIR-domain-containing adaptor inducing IFN-β (TRIF), whereas signaling through TLR7 and 8 requires the adaptor MyD88. For TLR3, signaling proceeds to activate several cellular kinases including IKK-ε and TBK1, which phosphorylate and activate IRF3; IKK-α/β, which phosphorylate I-κB to activate NF-κB; and MAP kinases, which activate AP-1. MyD88-dependent signaling through TLR7 and 8 involves the kinases IRAK-1, IRAK-4, and TRAF6 and leads to the activation of IRF7, NF-κB, and AP-1. The activation of IRF7, which is expressed constitutively in pDCs, permits the transcription of IFN-α as well as IFN-β genes and allows pDCs to produce copious amounts of IFN-α in response to virus infection.

TLR expression is limited to specific cell types, including macrophages and DCs, which, as noted above, are important targets of filovirus infection. However, because

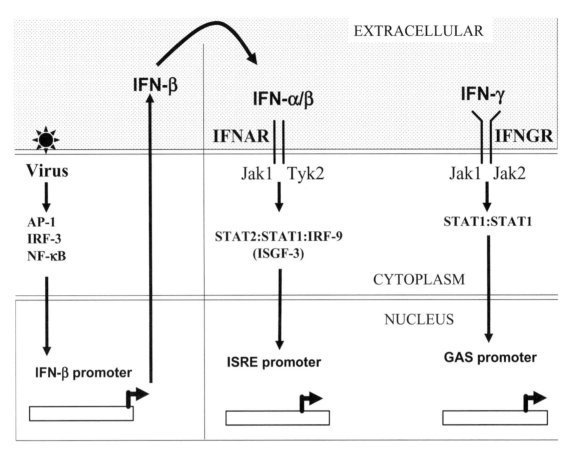

Figure 2   Viral induction of the host IFN system. A simplified schematic diagram of the signaling pathways leading to IFN-α/β synthesis following virus infection (left-side pathway) and the signaling pathways activated by IFN-α/β (center pathway) or by IFN-γ (right-side pathway). Virus infection can activate cellular transcription factors including the AP-1 transcription factor complex ATF-2/c-Jun, IRF3, and NF-κB. These transcription factors cooperate to activate transcription of the IFN-β gene. Expressed IFN-β is secreted and binds to the IFN-α/β receptor (IFNAR), triggering the activation of IFNAR-associated Jak family tyrosine kinases Jak1 and Tyk2. Tyrosine-phosphorylate Stat1 and Stat2, which form heterodimers, further interact with IRF9, forming the transcription factor complex ISGF3. ISGF3, when in the nucleus, activates transcription of genes with ISREs. IFN-γ binds to a distinct receptor, the IFN-γ receptor (IFNGR), leading to the activation of Jak1 and Jak2 and the formation of Stat1-Stat1 heterodimers, which move to the nucleus and activate promoters with gamma-activated sequence (GAS) elements.

TLR expression is restricted, the more ubiquitously expressed RIG-I and MDA-5 are expected to play critical roles in activating IFN responses in many of the cell types infected by EBOV and MBGV (138). RIG-I and MDA-5 are homologous to RNA helicases and possess amino-terminal protein-protein interaction domains referred to as caspase activation and recruitment domains (CARDs) (63). RIG-I recognizes RNAs possessing 5′ triphosphates, such as would be found on many RNA virus genomic RNAs, and MDA-5 recognizes the synthetic dsRNA molecule poly(I:C) (56, 98). These properties confer specificity on RIG-I and MDA-5, and in

apparent consequence allow these proteins to signal in response to specific virus families and to RNA molecules of different structure (62). Thus, RIG-I is essential for the production of IFN in response to paramyxoviruses, influenza viruses, flaviviruses, and transfected, in vitro-transcribed dsRNA. In contrast, MDA-5 is essential for production of IFN in response to picornaviruses or to transfected poly(I:C). RIG-I and MDA-5 signal via another CARD domain-containing protein, IPS-1 (IFN-β promotor stimulator 1; also called Cardif, MAVS, or VISA), which localizes to mitochondria (65, 87, 111, 133). Signaling through these IFN-α/β-inducing pathways ultimately results in the activation of either of two IRF3 kinases, IKK-ε and TBK1 (33, 112).

Upon production of IFN-α/β, these secreted cytokines bind to IFNAR, activating a Jak/Stat (Janus kinase/signal transducer and activator of transcription) signaling pathway, among other signaling pathways (99) (Fig. 2). Jak family tyrosine kinase activation leads to Stat1 and Stat2 tyrosine phosphorylation. These Stat proteins then heterodimerize via phosphotyrosine-SH2 domain interactions. This dimeric complex associates with IRF9 to create the transcription factor complex IFN-stimulated gene factor 3 (ISGF3). ISGF3 accumulates in the nucleus and activates expression of genes with promoters containing IFN-stimulated response elements (ISREs). Addition to cells of IFN-α/β induces an antiviral state such that cells become resistant to virus infection. Among the best-studied IFN-induced genes whose products exert antiviral effects are the dsRNA-activated protein kinase PKR, 2′,5′-oligoadenylate synthetases (OAS), and the MxA protein (called Mx1 in mouse cells), although other IFN-induced genes also contribute to IFN-induced antiviral effects (see, for example, references 70 and 127).

EBOV Evades the IFN-α/β Response

Infection of cells with filoviruses impairs the capacity of cells to either produce IFN-α/β or to respond to IFNs. For example, Zaire EBOV infection decreased IFN-α/β production by human umbilical vein endothelial cells (HUVECs) in response to poly(I:C) (46). In contrast, the uninfected HUVECs exhibited the expected responsiveness to poly(I:C) and expressed a number of IFN-induced proteins including class I major histocompatibility complex, IRF1, OAS, PKR, intercellular adhesion molecule-1, and IL-6 (46). A later study by Gupta and colleagues which examined human peripheral blood mononuclear cells (PBMCs) and macrophages infected with Zaire EBOV obtained similar results and found that infection elicits proinflammatory chemokines and cytokines but does not elicit IFN-α or IFN-β until 3 days postinfection, when only small amounts of IFN were seen (42). Additionally, Zaire EBOV infection suppressed IFN-α/β production induced by poly(I:C) (42). Separately, Zaire EBOV infection of HUVECs was also found to inhibit IFN signaling, as infected cells lacked normal cellular responses to exogenously added IFN-α or IFN-γ (47). Specifically, infection reduced formation of IFN-α/β or IFN-γ-induced transcription factor complexes, as assessed by electrophoretic mobility shift assays, and reduced the IFN-α- or IFN-γ-induced expression of IRF1 and OAS (47). This inhibition was IFN specific, as infection did not prevent formation of IL-1-β-induced NF-κB transcription factor complexes or IL-1β-induced gene expression (47).

Separately, microarray studies support the view that filoviruses suppress IFN responses. Specifically, the transcriptional responses of the human liver cell line Huh7 to infection with Zaire EBOV, Reston EBOV, and MBGV were studied. Classes of genes regulated by all three viruses included genes involved in immune response, IFN response, coagulation, and acute-phase response (61). Of note, gene expression profiles from Zaire EBOV- and MBGV-infected cells showed more similarity than did the profiles of Zaire EBOV- or MBGV- to Reston EBOV-infected cells (61). Specifically, Zaire EBOV and MBGV activated fewer IFN-inducible genes relative to Reston EBOV. Additionally, Zaire EBOV and MBGV each inhibited to a greater extent cellular responses to exogenously added IFN-α as compared with Reston EBOV (61). These observations demonstrate that different filoviruses influence cellular signaling pathways in similar but not identical ways. They also suggest that the ability of filoviruses to modulate innate immune response pathways, particularly IFN-related pathways, may influence pathogenesis.

IFN-α/β Produced during the Course of EBOV Infection

Despite the fact that filovirus infection impairs IFN-α/β responses in infected cells, IFN-α/β is produced at detectable levels during infections in vivo (42, 53, 120). Sera taken from EBOV-infected patients contained IFN-α and IFN-β as well as other cytokines, whereas pooled serum from uninfected individuals had undetectable levels of IFN-α/βs (42). Also, an IFN response was clearly evident in microarray analyses of the PBMCs of EBOV-infected nonhuman primates (105). Whether this systemic IFN response influences the outcome of infection remains unclear. Studies of human Zaire EBOV infections correlated high serum IFN-α levels with fatal outcome, whereas analysis of a larger number of Sudan EBOV-infected subjects correlated higher serum IFN-α with survival (59). Neutralization of the IFN-α/β produced during the course of experimental infections could help define the impact of serum IFN upon disease outcome. Which cell types produce this IFN-α/β and whether these cells are virus infected also remains to be determined.

INHIBITION OF IFN-α/β PRODUCTION BY THE VP35 PROTEIN

The Functions of Ebola VP35 Protein

Investigations into the molecular mechanisms by which EBOVs block host IFN responses have identified two proteins, VP35 and VP24, as filovirus-encoded inhibitors of IFN responses. The EBOV VP35 protein has three defined functions: (i) it is a component of the viral RNA-dependent RNA polymerase complex; (ii) it has a structural role as part of the viral nucleocapsid; and (iii) it inhibits production of IFN-α/β. Each of these VP35 functions is likely critical for the virus in vivo. For example, if the virus is unable to synthesize RNA, it will be unable to replicate its genome or make mRNA for viral protein expression. The essential role of VP35 for viral RNA synthesis has been demonstrated using a viral "minigenome" system. The minigenome is a replica of the viral genome that contains only a chloramphenicol acetyltransferase (CAT) reporter gene. Replication of the minigenome RNA and production of mRNA encoding CAT can only occur when VP35 is coexpressed with the viral NP, VP30, and L proteins (92, 93).

The structural role of VP35 has been less well defined. Like other negative-sense RNA viruses, filoviruses must package and deliver to newly infected cells a functional viral polymerase in order to initiate viral gene expression in the infected cell. Therefore, VP35 must be incorporated into viral particles, if the particles, are to be infectious. Additionally, EBOV virions contain long filamentous nucleocapsids that are composed primarily of NP and VP35 (57). Nucleocapsids could be reconstituted by expressing in 293T cells NP, VP35, and VP24, but did not form if any of these three proteins was absent (57). Whether such structures are absolutely required for virus assembly and replication is not fully clear; however, mutants of NP that are defective for nucleocapsid formation failed to support replication of a viral minigenome (129).

VP35 Inhibits IFN-α/β Production

The first indication that VP35 can antagonize IFN responses was the observation that the Zaire EBOV VP35 protein could functionally complement the impaired-growth phenotype of a mutant influenza virus, influenza ΔNS1 virus. This mutant lacks the influenza virus IFN-antagonist protein NS1 and grows poorly upon substrates that mount an effective IFN-α/β response (7). The ability of VP35 to complement, in *trans*, the influenza ΔNS1 virus growth defect correlated with the ability of VP35 to inhibit the virus- or dsRNA-induced activation of the IFN-β promoter (7). Follow-up studies demonstrated that VP35 blocks IFN-α/β production by preventing the phosphorylation and activation of IRF3 (6). In Sendai virus-infected cells, VP35 prevented IRF3-dependent gene expression, inhibited IRF3 nuclear accumulation, and blocked IRF3 hyperphosphorylation (6). The suppression of IRF3 hyperphosphorylation suggested that VP35 inhibits the virus-induced signaling pathways that activate IRF3 via the serine/threonine kinases IKK-ε and TBK1 (6). As noted above, in most cell

types, infection of cells with nonsegmented, negative-strand RNA viruses, such as EBOV and MBGV, is expected to trigger IFN-α/β synthesis by activating RIG-I-like sensors. These activate downstream signaling that leads to the activation of IRF3, NF-κB, and AP-1 (Fig. 3). Because transfected dsRNA can induce IFN-α/β production through RIG-I or MDA-5, it was of interest to observe that VP35 binds to dsRNA (18). When VP35 point mutants were generated that lacked detectable dsRNA-binding activity, they had partially impaired inhibition of IFN-β production (18). However, these mutants retained the ability to inhibit IFN-β production induced by overexpression of RIG-I, IPS-1, or the IRF3 kinases IKK-ε or TBK1. These data suggest that VP35's dsRNA-binding activity, while possibly contributing to inhibition of IRF3 activation, is probably not required for this function (18). This mirrors the situation of other dsRNA-binding proteins, such as the influenza virus NS1 protein, that inhibit IFN-α/β production. NS1 also can inhibit IFN production even when its dsRNA-binding activity is abrogated by mutation (26, 88). That VP35 inhibits IFN-β gene expression induced by overexpressed IKK-ε or TBK1, a situation where the upstream signaling components such as RIG-I or IPS-1 are not expected to contribute to activation of IFN-β gene expression, suggests that VP35 can act at some point proximal to the IRF3 kinases. However, additional effects upstream of the kinases, such as sequestering of dsRNA, cannot be excluded (Fig. 3). Defining the precise targets by which VP35 suppresses IRF3 activation and IFN-β production will be critical to fully understanding the role of VP35 during viral infection.

Effect of VP35 upon PKR and RNA Interference

In addition to its role as a suppressor of IFN-α/β production, VP35 can inhibit the activation of PKR. PKR is a well-studied IFN-induced, dsRNA-activated, cellular serine/threonine kinase that inhibits virus replication by phosphorylating the α subunit of the eukaryotic translation initiation factor eIF2, thereby inhibiting protein synthesis (34). PKR has long been known to mediate IFN-induced antiviral responses; and virus-encoded inhibitors of PKR, including the adenovirus VA₁ RNA, the vaccinia virus E3L protein, and the influenza A virus NS1 protein, were among the first viral products found to suppress the antiviral effects of IFNs (21, 67, 78).

Demonstration that VP35 can inhibit PKR came from studies in which VP35 was built into a mutant herpes simplex virus type 1 (HSV-1) that lacked the γ₁34.5 gene. The γ₁34.5 protein recruits cellular protein phosphatase 1 to dephosphorylate eIF2-α, and HSV-1 mutants impaired for this function exhibit increased sensitivity to the

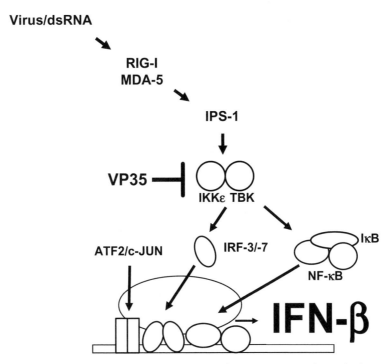

Figure 3 Model of EBOV VP35's IFN-antagonist function. Depicted are the basic components of the pathways that lead from detection of viral replication products by the cellular RNA helicases RIG-I or MDA-5 to the production of IFN-β. RIG-I and MDA-5 signal in an IPS-1-dependent manner and activate the IRF3 kinases IKK-ε or TBK1. These participate in the activation of transcription factors, including IRF3, required for IFN-β promoter activity. Present data suggest that VP35 inhibits these pathways at or near the level of the IRF3 kinases.

antiviral effects of IFN. When Vero cells were pretreated with IFN-α and then infected with either a parental HSV-1 mutant lacking $γ_1$34.5 or with a $γ_1$34.5 null virus expressing VP35, the VP35 virus was relatively resistant to the antiviral effects of IFN-α (31). This correlated with a relative suppression of PKR activation and eIF2-α phosphorylation in the VP35 virus-infected cells. Interestingly, an R312A mutation that impairs VP35's dsRNA-binding activity (18, 31) was still able to inhibit PKR, even though PKR is dsRNA activated. These data suggest that VP35 may counteract the antiviral effects of PKR in EBOV-infected cells by a mechanism independent of VP35 dsRNA-binding activity. However, this function has not been addressed in the context of an EBOV infection.

VP35 was also found to suppress RNA silencing by a mechanism that requires its dsRNA-binding activity (44). RNA silencing serves as an innate antiviral mechanism in plant and insect cells, and viruses that infect plants and insects frequently encode suppressors of RNA silencing (reviewed in reference 121). Several mammalian viruses have also been demonstrated to produce gene products that inhibit RNA silencing. Examples include the VA_I RNA of adenovirus, the NS1 protein of influenza viruses, the E3L protein of vaccinia virus, and the Tat protein of human immunodeficiency virus (8, 17, 74, 77). Similarly, wild-type VP35 inhibited short-hairpin RNA-mediated knockdown of a transfected luciferase reporter gene, but dsRNA binding-defective VP35 mutants lost this inhibitory activity (44). Additionally, VP35 could functionally substitute for the RNA silencing suppression function of the human immunodeficiency virus type 1 Tat protein (44). However, it remains unclear whether RNA silencing functions as an antiviral mechanism in mammalian cells (121). Therefore, the relevance of this function during EBOV infection remains unclear.

VP35s IFN-Antagonist Function in the Context of Virus Infection

Recombinant EBOVs bearing mutant VP35s have provided evidence that the IFN-antagonist function of VP35 is relevant to virus replication and pathogenesis. Recombinant EBOVs containing either of two VP35 mutations, R305A or R312A, were constructed (49). Based on in vitro studies, these mutations were predicted to impair VP35 inhibition of IRF3 activation (50). Growth of these mutants was compared to wild-type EBOV in Vero cells, U937 cells differentiated into macrophages, or the hepatocyte cell line Huh7, and the mutants consistently grew to lower titers (49). The growth attenuation in Vero cells was surprising given the inability of Vero cells to produce IFN-α/β; however, the mutant viruses did induce higher levels of the cytokine RANTES, which is expressed in an IRF3-responsive manner (49). Therefore, the mutant viruses were presumed to exhibit growth defects in Vero cells due to IFN-independent but IRF3-dependent cellular responses (49). Consistent with this model, the mutant virus activated IRF3 nuclear localization more strongly than did the wild-type virus (49). Additionally, microarray analyses comparing Huh7 cellular responses to infection with purified parental wild-type virus and purified R312A virus preparations revealed that the mutant virus induced the potent upregulation of the IFN response at late times in infection while the wild-type virus did not (49a). Finally, a recombinant Zaire EBOV engineered to contain mouse-adaptive mutations in NP and VP24 was further mutated to possess the VP35 R312A mutation. This mouse-adapted VP35 mutant virus was attenuated in mice relative to the parental (wild-type VP35) control virus, providing compelling evidence that VP35 modulates innate immune responses in the context of EBOV infection (48).

EBOV VP24 Inhibits IFN Signaling

Whereas the VP35 protein appears to block IFN-α/β production, the ability of EBOV to block IFN signaling appears to be accounted for by the VP24 protein. VP24 is a multifunctional protein that counters host IFN responses and influences EBOV pathogenesis (27, 102, 103). VP24 is also a virus structural protein and in the past was designated a "minor matrix protein" (28, 108). The EBOV matrix protein is VP40, and VP40 drives the budding of EBOV particles (28, 51, 95, 118). However, there is a possible role for VP24 in viral budding (45), although coexpression of VP24 with VP40 did not appear to alter VP40 budding efficiency (75). VP24 also plays an important role in assembly of viral nucleocapsids, as EBOV nucleocapsids could be reconstituted by coexpression of EBOV NP, VP35, and VP24 (57). In these experiments, VP24 could be immunoprecipitated with NP and VP35 (57), but VP24 did not cosediment in density gradients with NP and VP35, suggesting that while it promotes nucleocapsid formation, VP24 may not be an essential structural component of

the nucleocapsid (57). Separately, VP24 was found not to be required for either the formation or the infectivity of EBOV-like particles carrying an EBOV minigenome (130); however, a study employing small interfering RNAs targeting VP24 in MBGV-infected cells suggested a role for this protein in MBGV assembly and release from infected Vero cells (5).

As noted above, EBOV-infected HUVECs fail to respond to IFN-α or IFN-γ (47). However, expression of VP35 did not block IFN signaling, suggesting that EBOV encodes a different protein(s) to inhibit this arm of the IFN response (6). Zaire EBOV proteins were therefore screened for the ability to block IFN-β-induced transcription of a reporter gene, and VP24 protein was identified as an inhibitor of IFN-induced gene expression (102). Additionally, cells expressing VP24 were resistant to the antiviral effects of IFN-β (102). Because VP24 inhibited cellular transcription induced not only by IFN-α/β but also by IFN-γ, it was logical to expect that VP24 would target a cellular factor required for both the IFN-α/β and IFN-γ signaling pathways. For this reason, the impact of VP24 upon Stat1 was assessed. Interestingly, VP24 did not prevent the IFN-β- or IFN-γ-induced tyrosine phosphorylation of Stat1, and VP24 did not substantially alter levels of Stat1 within cells. Rather, despite the presence of tyrosine-phosphorylated Stat1 (PY-Stat1), Stat1 failed to accumulate in the nucleus. A similar block was seen in EBOV-infected cells; Stat1 became tyrosine phosphorylated following IFN treatment of EBOV-infected Vero cells; however, as in VP24-transfected cells, Stat1 failed to accumulate in the nucleus (102).

An explanation for the failure of tyrosine-phosphorylated Stat1 to accumulate in the nucleus came from the demonstration that VP24 can interact with a subset of human karyopherin α proteins (Fig. 4). Nuclear translocation of proteins above a certain size (~50 kDa) usually requires active transport through the nuclear pore, and this transport is mediated by nuclear localization signals (NLSs) (79). Many NLSs mediate binding of cargo proteins to the karyopherin α/β heterodimer (reviewed in reference 114). Karyopherin α functions as an adaptor by binding both the NLS and karyopherin β, and karyopherin β, in turn, docks the trimeric complex at the nuclear pore. The complex subsequently translocates into the nucleus (41, 132).

In humans, there are six karyopherin α's, which can be assigned, based on relative sequence similarity, into three subfamilies (68): the Rch1 subfamily (karyopherin α2) (24, 131), the Qip1 subfamily (karyopherin α3 and karyopherin α4) (68, 90, 94, 109), and the NPI-1 subfamily (karyopherin α1, karyopherin α5, and karyopherin α6)

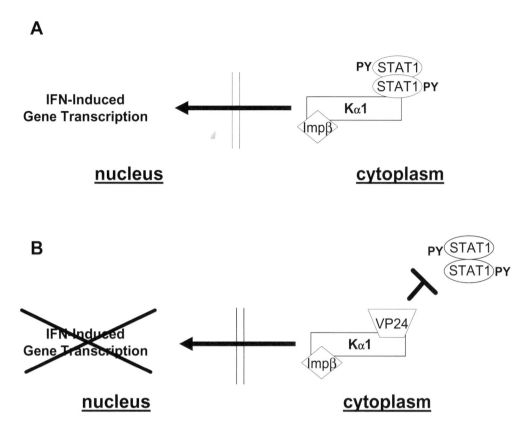

Figure 4 Model of EBOV VP24 inhibition of Stat1 nuclear accumulation. (A) Tyrosine phosphorylation of Stat1 by Jak family kinases results in dimerization of Stat1 with itself or other Stat proteins. In the case of Stat1-Stat1 homodimers, such as are activated by IFN-γ (depicted here) or Stat1-Stat2 heterodimers such as are activated by IFN-α/β, Stat1 nuclear accumulation is mediated by karyopherin α1 (K-α1). (B) VP24 binds to K-α1, preventing Stat1 from interacting with K-α1. This results in a failure of Stat1 to enter the nucleus and activate gene expression.

(23, 68, 69, 91, 96). Within the NPI-1 subfamily, karyopherin α1, karyopherin α5, and karyopherin α6 share greater than 80% amino acid identity (86).

As noted above, upon IFN-α/β or IFN-γ signaling, the Stat1-Stat2 heterodimer or the Stat1-Stat1 homodimer were demonstrated to interact specifically with karyopherin α1 (85, 86, 110). This interaction with karyopherin α1 can mediate the nuclear accumulation of the these Stat1-containing complexes (85, 86, 110). The consequence of the activation and nuclear accumulation of these complexes is the specific transcriptional regulation of numerous genes, some of which have antiviral properties (73).

Initially, VP24 was found to coimmunoprecipitate with karyopherin α1, but not with karyopherins α2, α3, or α4 (102). This observation correlated with the earlier reports that Stat1 nuclear import is mediated specifically by karyopherin α1. VP24 also prevented, in cells treated with IFN-β, the association of karyopherin α1

with PY-Stat1, thus explaining how VP24 prevents Stat1 nuclear import (19). Because of the homology among the NPI-1 karyopherin α subfamily members, the ability of both activated Stat1 and VP24 to interact with karyopherin α5 and α6 was also examined (103). By coimmunoprecipitation, PY-Stat1 interacted not only with karyopherin α1 but also with karyopherins α5 and α6. This was mirrored by the interaction of VP24 with the same subset of karyopherin αs (103). As was seen with karyopherin α1, VP24 prevented interaction of karyopherins α5 and α6 with PY-Stat1 (103). Mapping experiments concluded that VP24 and Stat1 bind to overlapping sites on karyopherin α1, and in vitro binding experiments suggest that VP24 can compete with PY-Stat1 for binding to karyopherin α1 (103). However, whether VP24 blocks IFN signaling simply by acting as a competitive inhibitor of Stat1-karyopherin α1 interaction or whether VP24 may also influence the trafficking of the NPI-1 subfamily of karyopherin α proteins remains unresolved. If the latter proves true, then VP24 may affect nuclear transport more broadly than might be the case if it only affects Stat1 interaction with karyopherins α1, α5, and α6.

Mouse Models Suggest a Role for VP24 in IFN Evasion and Pathogenesis

One major way in which EBOVs modulate host-cell signaling pathways is by antagonizing the IFN response, inhibiting both IFN-α/β production and IFN-α/β and IFN-γ-induced signaling. That inhibition of these pathways is relevant to filovirus pathogenesis is supported by data from experimental infections of mice. Specifically, filoviruses that have not previously been adapted to mice do not cause lethal disease in adult or weanling mice, but they can kill newborn mice and SCID mice (58). Nonadapted Zaire EBOV was also lethal to mice lacking either the IFN-α/β receptor (IFNAR$^{-/-}$ mice) or Stat1 (Stat1$^{-/-}$ mice) (13). These data suggest that the IFN-α/β response normally controls EBOV infection in mice unless the virus has been adapted to better evade mouse IFN responses.

Zaire EBOV was adapted by repeated mouse passage such that it became lethal to wild-type mice following intraperitoneal injection (14). In this system, induction of innate antiviral responses, including strong IFN-α responses, could prevent mouse-adapted EBOV lethality in mice. For example, 3-deazaneplanocin A, an adenosine analogue, was able to protect mice from an otherwise lethal challenge with mouse-adapted EBOV, and this protection correlated with induction of "massively increased IFN-α" due to the combination of drug and virus (16). Likewise, a subcutaneous inoculation of mice with mouse-adapted EBOV proved to be nonlethal, whereas an intraperitoneal inoculation was lethal (81). When mice were first subcutaneously administered the virus and then challenged at early time points following the subcutaneous dose, the mice were protected as early as the 48-h time point. This correlated with the appearance of detectable levels of IFN-α in liver, spleen, and serum (81). These data again suggest that IFN-α can, at least in the mouse system, influence EBOV pathogenesis (81).

Changes in VP24 have been associated with adaptation of Zaire EBOV from nonlethal to lethal in mouse and guinea pig models (27, 123), and at least in the mouse system, these changes appear to influence virus sensitivity to IFNs. When Zaire EBOV was adapted to mice, nine nucleotide changes occurred relative to the parental virus used to initiate the adaptation process (27). Five of these nucleotide changes caused amino acid changes. There were single changes in the NP, VP35,

and VP24 proteins and two in the L protein. Studies employing recombinant EBOVs engineered to contain different combinations of these changes demonstrated that the minimal set of the alterations required for lethal infection consisted of the one amino acid change in NP combined with the one amino acid change in VP24 (27). Virulence in this system also correlated with the replication capacity of the recombinant EBOVs in the mouse macrophage cell line RAW 264.7, and with the relative impact of exogenously added IFN-α/β upon virus replication in RAW 264.7 cells (27).

It is tempting to speculate, given these observations, that the mouse-adaptive change in VP24 resulted in an altered capacity of the virus to block IFN-induced Stat1 activation in infected cells. Therefore, the ability of the nonadapted and mouse-adapted Zaire EBOV VP24 to interact with human karyopherin α proteins was compared (103). In coimmunoprecipitation studies, the nonadapted Zaire EBOV and mouse-adapted Zaire EBOV VP24 coimmunoprecipitated with human karyopherins α1, α5, and α6 to similar extents (103). Similarly, the nonadapted Zaire EBOV and mouse-adapted Zaire EBOV VP24 did not differ in their capacity to coimmunoprecipitate with mouse karyopherin α's (unpublished observation). In reporter gene assays, nonadapted Zaire EBOV and mouse-adapted Zaire EBOV VP24 inhibited IFN-β-induced reporter gene activation to similar extents in two different mouse cell lines and in two human cell lines (103). Therefore, the extent to which mouse adaptation is related to VP24 modulation of Stat1 function is, at present, unclear. It should be recognized, however, that VP24 and NP interact; thus it is possible that NP may modulate VP24 function. Alternatively, it may be that in EBOV-infected cells the mouse-adaptive mutations influence VP24's IFN-antagonist function by influencing VP24 expression or VP24 stability. Future studies will be required to address these possibilities.

THE HOST RESPONSE TO FILOVIRUS INFECTION

Host Innate Responses Trigger Pathological Outcomes during Filovirus Infection

Filovirus infection of macrophages and DCs in vitro leads to significant proinflammatory cytokine production (42, 53, 115), and significant levels of proinflammatory cytokines are present in the serum of infected animals and humans (29, 53). Because these cytokines influence blood vessel permeability and also can recruit additional immune cells, the proinflammatory responses associated with infection are thought to make an important contribution to the manifestations of filovirus disease (80).

Early Innate Responses Influence the Outcome of EBOV Infection

A role for early innate response in determining the outcome of filovirus infection is supported both by animal studies and by studies on very limited numbers of human cases. Some human studies have compared fatal versus nonfatal human infections and suggest that "very early events in EBOV infection" determine the ultimate course of the disease. They also suggest that survival is associated with "orderly and well-regulated" immune responses (2, 3, 120). Supporting these hypotheses, 11 asymptomatic, EBOV-infected individuals had an early, transient period (2 to 3 days) in which proinflammatory cytokines and chemokines, including IL-6, IL-1-β, tumor necrosis factor-α (TNF-α), monocyte chemoattractant protein-1 (MCP-1), and macrophage inflammatory protein-1α and -1β, were detected (71). Detectable levels of EBOV RNA in asymptomatic patients remained low throughout the course of infection, although it is not clear to what extent the proinflammatory cytokine response helped control these infections (71). In contrast, cytokine production may also mediate the hemorrhagic manifestations of filovirus infection; for example, it was shown that MBGV infection of monocytes and macrophages induces sufficient TNF-α production to enhance endothelial permeability (29).

Profiling Host Responses to EBOV Infection in Cells and in Animals

Host responses to filovirus infection have been characterized by examining the sera and blood of infected human patients or experimentally infected nonhuman primates. Blood contains a complex mixture of cells, and the contents of serum may reflect cytokine responses not just of blood cells per se but also of other cell types. Such studies have, however, provided a characterization of filovirus disease and provide a broad picture as to how these viruses influence host innate responses to infection. For example, sera from several EBOV-infected patients were found to contain elevated levels of IFN-α/β, the chemokine MCP-1, and the cytokine TNF-α. Similarly, infection of human PBMCs with live Zaire EBOV resulted in production of TNF-α and MCP-1, although not in enhanced production of IFN-α/β, IL-1β, or IL-10. The infected PBMCs also became resistant to poly(I:C)-mediated induction of IFN-α/β expression (42). Because proinflammatory cytokine production may contribute significantly to filovirus pathogenesis, the capacity of EBOV to turn on TNF-α and MCP-1 expression without

simultaneously inducing antiviral IFN-α/β may be critical for disease progression. Similarly, separate studies on Sudan EBOV-infected individuals identified a number of upregulated cytokines in sera (59). Complementary studies have profiled global transcript levels in PBMCs from Zaire EBOV-infected nonhuman primates at multiple times postinfection (105). The nonhuman primate model of Zaire EBOV infection results in a fairly uniform pattern of clinical symptoms leading to death (38). Gene expression profiles of PBMCs from different animals at specific time points postinfection also showed a relatively uniform response to infection (105). This uniformity was perhaps surprising given the complexity of PBMC composition and the likelihood that this mixed cell population contains both infected and uninfected cells. Nonetheless, the profiles provided evidence for an early host response to infection in which IFN-induced gene expression was apparent (105). As infection progressed, a prominent proinflammatory response became evident and was evidenced by enhanced cytokine mRNA and protein levels and by an increase in genes related to TNF-α/NF-κB responses (105). Also detected were changes in expression of genes associated with apoptosis and genes related to the coagulation cascade (105). These changes in gene expression likely reflect, respectively, the bystander apoptosis of lymphocytes seen in infected animals and in lymphocytes cultured with EBOV-infected PBMCs and the coagulopathy associated with EBOV disease (2, 12, 32, 36–38, 43, 101). These observations provide a detailed and comprehensive view of the progression of EBOV infection in vivo and are largely consistent with the current view of EBOV pathogenesis (15, 36, 39, 53). These results also suggest that, in vivo, EBOV infection modulates the signaling pathways that regulate IFN responses, cytokine responses, and apoptosis. However, because this study was performed on a complex population of cells, gene expression changes might reflect a direct effect of virus upon a cell or a "bystander effect" occurring in uninfected cells.

Impact of EBOV Infection upon DCs and Macrophages

As early targets of productive infection, macrophages and DCs are thought to play an important role in dissemination of EBOV infection. However, these cells are also likely to be an important source of proinflammatory cytokines and other factors that influence blood vessel permeability and contribute to coagulopathy during the course of infection (15). Presumably, the same, as yet undefined filovirus replication products that trigger IFN responses will also trigger PAMPs in macrophages to activate cytokine production. Based on pathology studies, a model has emerged where cytokines and chemokines produced by EBOV-infected macrophages recruit additional cells to sites of infection (15). The cells would include immature neutrophils from vessel walls and bone marrow as well as inflammatory cells from the circulation. The cytokines and chemokines would recruit these cells by increasing endothelial permeability and increasing expression of adhesion molecules on the surface of the endothelium. When such a response is activated systemically, such as occurs in EBOV hemorrhagic fever, circulatory collapse would presumably result (15).

Filovirus Infection of Nonhuman Primates Resembles Disseminated Intravascular Coagulation

In EBOV-infected macaques, a syndrome similar to disseminated intravascular coagulation also occurs, and parallels have been drawn between Ebola hemorrhagic fever and severe sepsis (37, 39). Coagulopathy appears to be induced by production of cell surface tissue factor (TF) by virus-infected macrophages. Upon interaction with factors VIIa and X, the extrinsic coagulation cascade is activated. Direct evidence that TF plays a critical role in the pathogenesis of Ebola hemorrhagic fever in macaques was provided when administration of an inhibitor of the factor VIIa/TF complex, nematode anticoagulant protein c2, was found to reduce mortality in Zaire EBOV-infected monkeys (37, 39). Interestingly, the inhibitor also reduced IL-6 levels and reduced viral titers, and it has been suggested that the inhibitor might affect these parameters by modulating TF signaling (106). Another characteristic of Ebola hemorrhagic fever is a rapid decrease of plasma protein C levels, which is a characteristic of severe sepsis, and treatment with recombinant human activated protein C, a treatment for severe sepsis, delayed or prevented death. These data further support the view that Ebola hemorrhagic fever is a sepsis-like illness (52).

Filovirus Infection Induces Lymphocyte Apoptosis and Impairs DC Function

Zaire EBOV infections are also characterized by impairment of several components of the adaptive immune system (15, 108). For example, apoptotic loss of lymphocytes is seen during infection, despite the fact that these cells are not detectably infected, and experimental data suggest that this loss of lymphocytes is cytokine mediated (2, 4, 10, 36, 40, 82, 101). Additionally, infection of DCs, at least in vitro, results in impaired activation of these cells, which may in turn impair mobilization of T-cell responses to infection in vivo (10, 82). When immature monocyte-derived DCs were infected with

replication-competent Zaire EBOV, they secreted only limited cytokines and did not upregulate costimulatory molecules or class I major histocompatibility complex on their cell surface. These cells also failed to stimulate allogeneic T-cell responses (10, 82).

In contrast, several studies demonstrate that noninfectious EBOV virus-like particles (VLPs) can stimulate DCs to produce proinflammatory cytokines, upregulate cell surface markers, and stimulate allogeneic T-cell responses (10, 125, 128, 137). EBOV VLPs are readily produced by expression in mammalian cells of the viral matrix protein VP40 (95). When GP is coexpressed with VP40, GP becomes incorporated into the VLP membrane. The stimulatory activity of the VLPs toward DCs was shown to involve activation of NF-κB, and blocking the activity of NF-κB was sufficient block VLP-induced cytokine production (84). Interestingly, although expression of GP in cells inhibits ERK2 (extracellular signal-related kinase 2) activation (116), addition of VLPs with wild-type GP to DCs transiently activated ERK1/2 phosphorylation (84). How EBOV VLPs are recognized by macrophages and DCs to trigger their activation remains to be determined. However, the activating property of EBOV VLPs was largely abated by deletion of the mucin domain; despite the fact that VLPs containing GPΔmuc associated with DCs to the same extent as VLPs bearing wild-type GP, the GPΔmuc VLPs failed to activate the DCs (84). The differences in DC responses to live virus versus VLP raise a number of interesting questions. First, what host factor recognizes GP to trigger this signal remains to be defined. Additionally, although many manifestations of this signal are obviously masked by the virus, the initial signal may still occur, but its role in virus replication remains undefined. Third, the data suggest that during EBOV infection, a factor is produced that suppresses DC activation; however, the identity of this factor is not known.

GP Influences Cellular Signaling Pathways

In addition to the many pathological effects attributed to soluble factors produced by infected cells, evidence suggests that GP may contribute to viral pathogenesis. When expressed in mammalian cells, GP has several notable effects, including modulation of surface protein levels (113, 117, 125), induction of anoikis (1, 100), and stimulation of cytokine production from human macrophages (125) and DCs (11, 137). Further, its vascular cytotoxicity has been proposed to play an important role in virulence (136), and it may possess immune-suppressive properties (134), although the relevance of these functions to viral pathogenesis remains to be demonstrated.

The cell rounding and detachment phenotype of GP is closely tied to GP cytotoxicity and has been tied to changes in cellular MAP kinase signaling. Expression of GP in human embryonic kidney 293 cells was also found to reduce the amount of phosphorylated ERK1 and ERK2 MAP kinases, whereas a GP deleted of the mucin domain, a region of the protein modified with O-linked glycans (GPΔmuc), did not reduce phospho-ERK1 or -ERK2 levels (139). This inhibitory effect was more pronounced for ERK2 and correlated with the ability of GP to induce cell rounding and cell death in a mucin domain-dependent manner (139). As further evidence that inhibition of ERK2 phosphorylation contributes to GP-dependent cytotoxicity, knockdown of ERK2 with small interfering RNAs or expression of a dominant negative form of ERK1/2 enhanced the GP-induced downregulation of αV integrin, whereas expression of a constitutively activated ERK2 preserved αV integrin expression in GP-transfected cells (139). These data directly link the ability of GP to induce cell detachment and cytotoxicity with its ability to affect MAP kinase signaling. It remains to be determined how GP mediates the downregulation of ERK activity; however, because ERK1 and ERK2 phosphorylate numerous targets, it is reasonable to expect that GP expression will impact numerous cellular pathways, including those involved in innate immunity (139).

CONCLUSION

The severity of filovirus disease is associated with overwhelming inflammatory responses and immune suppression. Manifestations of disease may also be mediated by destruction of host cells by virus replication or be due to the effects of specific viral products, such as the viral GP. Systemic virus replication is most likely required to induce severe disease, whether this is through pathological host responses or through direct effects of virus replication. For this reason, the ability of these viruses to impair cellular IFN pathways is likely critical for pathogenesis, as suppression of IFN responses likely allows these viruses to first gain a foothold in macrophages and DCs and to then replicate systemically. However, while many aspects of the interactions between EBOV and host have been explored, many aspects of this model of disease remain to be verified experimentally. For example, many questions remain regarding the mechanisms by which the VP35 and VP24 proteins modulate host IFN responses, and how effectively these proteins function in different cell types during the course of EBOV infection. It will also be important to carefully assess whether the VP35 and VP24 proteins of other filoviruses function with similar efficiency as the Zaire EBOV VP35 and VP24. Additionally, the sensors of filovirus infection and the viral triggers that activate these sensors are not identified. A goal should be to identify the triggers of potent cytokine responses with the aim of devising strategies to mitigate damaging host responses to infection.

REFERENCES

1. **Alazard-Dany, N., M. Ottmann Terrangle, and V. Volchkov.** 2006. [Ebola and Marburg viruses: the humans strike back]. *Med. Sci. (Paris)* **22:**405–410.
2. **Baize, S., E. M. Leroy, M. C. Georges-Courbot, M. Capron, J. Lansoud-Soukate, P. Debre, S. P. Fisher-Hoch, J. B. McCormick, and A. J. Georges.** 1999. Defective humoral responses and extensive intravascular apoptosis are associated with fatal outcome in Ebola virus-infected patients. *Nat. Med.* **5:**423–426.
3. **Baize, S., E. M. Leroy, A. J. Georges, M. C. Georges-Courbot, M. Capron, I. Bedjabaga, J. Lansoud-Soukate, and E. Mavoungou.** 2002. Inflammatory responses in Ebola virus-infected patients. *Clin. Exp. Immunol.* **128:**163–168.
4. **Baize, S., E. M. Leroy, E. Mavoungou, and S. P. Fisher-Hoch.** 2000. Apoptosis in fatal Ebola infection. Does the virus toll the bell for immune system? *Apoptosis* **5:**5–7.
5. **Bamberg, S., L. Kolesnikova, P. Moller, H. D. Klenk, and S. Becker.** 2005. VP24 of Marburg virus influences formation of infectious particles. *J. Virol.* **79:**13421–13433.
6. **Basler, C. F., A. Mikulasova, L. Martinez-Sobrido, J. Paragas, E. Muhlberger, M. Bray, H. D. Klenk, P. Palese, and A. Garcia-Sastre.** 2003. The Ebola virus VP35 protein inhibits activation of interferon regulatory factor 3. *J. Virol.* **77:**7945–7956.
7. **Basler, C. F., X. Wang, E. Muhlberger, V. Volchkov, J. Paragas, H. D. Klenk, A. Garcia-Sastre, and P. Palese.** 2000. The Ebola virus VP35 protein functions as a type I IFN antagonist. *Proc. Natl. Acad. Sci. USA* **97:**12289–12294.
8. **Bennasser, Y., S. Y. Le, M. Benkirane, and K. T. Jeang.** 2005. Evidence that HIV-1 encodes an siRNA and a suppressor of RNA silencing. *Immunity* **22:**607–619.
9. **Biron, C. A.** 2001. Interferons alpha and beta as immune regulators—a new look. *Immunity* **14:**661–664.
10. **Bosio, C. M., M. J. Aman, C. Grogan, R. Hogan, G. Ruthel, D. Negley, M. Mohamadzadeh, S. Bavari, and A. Schmaljohn.** 2003. Ebola and Marburg viruses replicate in monocyte-derived dendritic cells without inducing the production of cytokines and full maturation. *J. Infect. Dis.* **188:**1630–1638.
11. **Bosio, C. M., B. D. Moore, K. L. Warfield, G. Ruthel, M. Mohamadzadeh, M. J. Aman, and S. Bavari.** 2004. Ebola and Marburg virus-like particles activate human myeloid dendritic cells. *Virology* **326:**280–287.
12. **Bradfute, S. B., D. R. Braun, J. D. Shamblin, J. B. Geisbert, J. Paragas, A. Garrison, L. E. Hensley, and T. W. Geisbert.** 2007. Lymphocyte death in a mouse model of Ebola virus infection. *J. Infect. Dis.* **196**(Suppl 2):S296–S304.
13. **Bray, M.** 2001. The role of the type I interferon response in the resistance of mice to filovirus infection. *J. Gen. Virol.* **82:**1365–1373.

14. Bray, M., K. Davis, T. Geisbert, C. Schmaljohn, and J. Huggins. 1998. A mouse model for evaluation of prophylaxis and therapy of Ebola hemorrhagic fever. *J. Infect. Dis.* **178:**651–661.

15. Bray, M., and T. W. Geisbert. 2005. Ebola virus: the role of macrophages and dendritic cells in the pathogenesis of Ebola hemorrhagic fever. *Int. J. Biochem. Cell. Biol.* **37:**1560–1566.

16. Bray, M., J. L. Raymond, T. Geisbert, and R. O. Baker. 2002. 3-Deazaneplanocin A induces massively increased interferon-alpha production in Ebola virus-infected mice. *Antiviral Res.* **55:**151–159.

17. Bucher, E., H. Hemmes, P. de Haan, R. Goldbach, and M. Prins. 2004. The influenza A virus NS1 protein binds small interfering RNAs and suppresses RNA silencing in plants. *J. Gen. Virol.* **85:**983–991.

18. Cardenas, W. B., Y. M. Loo, M. Gale, Jr., A. L. Hartman, C. R. Kimberlin, L. Martinez-Sobrido, E. O. Saphire, and C. F. Basler. 2006. Ebola virus VP35 protein binds double-stranded RNA and inhibits alpha/beta interferon production induced by RIG-I signaling. *J. Virol.* **80:**5168–5178.

19. Centers for Disease Control and Prevention. 2005. Outbreak of Marburg virus hemorrhagic fever—Angola, October 1, 2004-March 29, 2005. *Morb. Mortal. Wkly. Rep.* **54:**308–309.

20. Centers for Disease Control and Prevention, Special Pathogens Branch. 2002, posting date. Ebola hemorrhagic fever. http://www.cdc.gov/ncidod/dvrd/spb/mnpages/dispages/filoviruses.htm. [Online.]

21. Chang, H. W., J. C. Watson, and B. L. Jacobs. 1992. The E3L gene of vaccinia virus encodes an inhibitor of the interferon-induced, double-stranded RNA-dependent protein kinase. *Proc. Natl. Acad. Sci. USA* **89:**4825–4829.

22. Connolly, B. M., K. E. Steele, K. J. Davis, T. W. Geisbert, W. M. Kell, N. K. Jaax, and P. B. Jahrling. 1999. Pathogenesis of experimental Ebola virus infection in guinea pigs. *J. Infect. Dis.* **179**(Suppl. 1)**:**S203–S217.

23. Cortes, P., Z. S. Ye, and D. Baltimore. 1994. RAG-1 interacts with the repeated amino acid motif of the human homologue of the yeast protein SRP1. *Proc. Natl. Acad. Sci. USA* **91:**7633–7637.

24. Cuomo, C. A., S. A. Kirch, J. Gyuris, R. Brent, and M. A. Oettinger. 1994. Rch1, a protein that specifically interacts with the RAG-1 recombination-activating protein. *Proc. Natl. Acad. Sci. USA* **91:**6156–6160.

25. Dalgard, D. W., R. J. Hardy, S. L. Pearson, G. J. Pucak, R. V. Quander, P. M. Zack, C. J. Peters, and P. B. Jahrling. 1992. Combined simian hemorrhagic fever and Ebola virus infection in cynomolgus monkeys. *Lab. Anim. Sci.* **42:**152–157.

26. Donelan, N. R., C. F. Basler, and A. Garcia-Sastre. 2003. A recombinant influenza A virus expressing an RNA-binding-defective NS1 protein induces high levels of beta interferon and is attenuated in mice. *J. Virol.* **77:**13257–13266.

27. Ebihara, H., A. Takada, D. Kobasa, S. Jones, G. Neumann, S. Theriault, M. Bray, H. Feldmann, and Y. Kawaoka. 2006. Molecular determinants of Ebola virus virulence in mice. *PLoS Pathog* **2:**e73.

28. Elliott, L. H., M. P. Kiley, and J. B. McCormick. 1985. Descriptive analysis of Ebola virus proteins. *Virology* **147:**169–176.

29. Feldmann, H., H. Bugany, F. Mahner, H. D. Klenk, D. Drenckhahn, and H. J. Schnittler. 1996. Filovirus-induced endothelial leakage triggered by infected monocytes/macrophages. *J. Virol.* **70:**2208–2214.

30. Feldmann, H., V. E. Volchkov, V. A. Volchkova, U. Stroher, and H. D. Klenk. 2001. Biosynthesis and role of filoviral glycoproteins. *J. Gen. Virol.* **82:**2839–2848.

31. Feng, Z., M. Cerveny, Z. Yan, and B. He. 2007. The VP35 protein of Ebola virus inhibits the antiviral effect mediated by double-stranded RNA-dependent protein kinase PKR. *J. Virol.* **81:**182–192.

32. Fisher-Hoch, S. P., G. S. Platt, G. H. Neild, T. Southee, A. Baskerville, R. T. Raymond, G. Lloyd, and D. I. Simpson. 1985. Pathophysiology of shock and hemorrhage in a fulminating viral infection (Ebola). *J. Infect. Dis.* **152:**887–894.

33. Fitzgerald, K. A., S. M. McWhirter, K. L. Faia, D. C. Rowe, E. Latz, D. T. Golenbock, A. J. Coyle, S. M. Liao, and T. Maniatis. 2003. IKKepsilon and TBK1 are essential components of the IRF3 signaling pathway. *Nat. Immunol.* **4:**491–496.

34. Gale, M., Jr., and M. G. Katze. 1998. Molecular mechanisms of interferon resistance mediated by viral-directed inhibition of PKR, the interferon-induced protein kinase. *Pharmacol. Ther.* **78:**29–46.

35. Geisbert, T. W., and L. E. Hensley. 2004. Ebola virus: new insights into disease aetiopathology and possible therapeutic interventions. *Expert Rev. Mol. Med.* **6:**1–24.

36. Geisbert, T. W., L. E. Hensley, T. R. Gibb, K. E. Steele, N. K. Jaax, and P. B. Jahrling. 2000. Apoptosis induced in vitro and in vivo during infection by Ebola and Marburg viruses. *Lab. Invest.* **80:**171–186.

37. Geisbert, T. W., L. E. Hensley, P. B. Jahrling, T. Larsen, J. B. Geisbert, J. Paragas, H. A. Young, T. M. Fredeking, W. E. Rote, and G. P. Vlasuk. 2003. Treatment of Ebola virus infection with a recombinant inhibitor of factor VIIa/tissue factor: a study in rhesus monkeys. *Lancet* **362:**1953–1958.

38. Geisbert, T. W., L. E. Hensley, T. Larsen, H. A. Young, D. S. Reed, J. B. Geisbert, D. P. Scott, E. Kagan, P. B. Jahrling, and K. J. Davis. 2003. Pathogenesis of Ebola hemorrhagic fever in cynomolgus macaques: evidence that dendritic cells are early and sustained targets of infection. *Am. J. Pathol.* **163:**2347–2370.

39. Geisbert, T. W., H. A. Young, P. B. Jahrling, K. J. Davis, E. Kagan, and L. E. Hensley. 2003. Mechanisms underlying coagulation abnormalities in Ebola hemorrhagic fever: overexpression of tissue factor in primate monocytes/macrophages is a key event. *J. Infect. Dis.* **188:**1618–1629.

40. Gibb, T. R., M. Bray, T. W. Geisbert, K. E. Steele, W. M. Kell, K. J. Davis, and N. K. Jaax. 2001. Pathogenesis of experimental Ebola Zaire virus infection in BALB/c mice. *J. Comp. Pathol.* **125:**233–242.

41. Gorlich, D., P. Henklein, R. A. Laskey, and E. Hartmann. 1996. A 41 amino acid motif in importin-alpha confers binding to importin-beta and hence transit into the nucleus. *EMBO J.* **15:**1810–1817.

42. Gupta, M., S. Mahanty, R. Ahmed, and P. E. Rollin. 2001. Monocyte-derived human macrophages and peripheral blood mononuclear cells infected with Ebola virus secrete MIP-1alpha and TNF-alpha and inhibit poly-IC-induced IFN-alpha in vitro. *Virology* **284:**20–25.

43. Gupta, M., C. Spiropoulou, and P. E. Rollin. 2007. Ebola virus infection of human PBMCs causes massive death of macrophages, CD4 and CD8 T cell sub-populations in vitro. *Virology* 364:45–54.

44. Haasnoot, J., W. de Vries, E. J. Geutjes, M. Prins, P. de Haan, and B. Berkhout. 2007. The Ebola virus VP35 protein is a suppressor of RNA silencing. *PLoS Pathog.* 3:e86.

45. Han, Z., H. Boshra, J. O. Sunyer, S. H. Zwiers, J. Paragas, and R. N. Harty. 2003. Biochemical and functional characterization of the Ebola virus VP24 protein: implications for a role in virus assembly and budding. *J. Virol.* 77:1793–1800.

46. Harcourt, B. H., A. Sanchez, and M. K. Offermann. 1998. Ebola virus inhibits induction of genes by double-stranded RNA in endothelial cells. *Virology* 252:179–188.

47. Harcourt, B. H., A. Sanchez, and M. K. Offermann. 1999. Ebola virus selectively inhibits responses to interferons, but not to interleukin-1beta, in endothelial cells. *J. Virol.* 73:3491–3496.

48. Hartman, A. L., B. H. Bird, J. S. Towner, Z. A. Antoniadou, S. R. Zaki, and S. T. Nichol. 2008. Inhibition of IRF-3 activation by VP35 is critical for the high virulence of Ebola virus. *J. Virol.* 82:2699–2704.

49. Hartman, A. L., J. E. Dover, J. S. Towner, and S. T. Nichol. 2006. Reverse genetic generation of recombinant Zaire Ebola viruses containing disrupted IRF-3 inhibitory domains results in attenuated virus growth in vitro and higher levels of IRF-3 activation without inhibiting viral transcription or replication. *J. Virol.* 80:6430–6440.

49a. Hartman, A. L., L. Ling, S. T. Nichol, and M. L. Hibberd. 2008. Whole-genome profiling reveals that inhibition of host innate immune response pathways by Ebola virus can be reversed by a single amino acid change in the VP35 protein. *J. Virol.* 82:5348–5358.

50. Hartman, A. L., J. S. Towner, and S. T. Nichol. 2004. A C-terminal basic amino acid motif of Zaire ebolavirus VP35 is essential for type I interferon antagonism and displays high identity with the RNA-binding domain of another interferon antagonist, the NS1 protein of influenza A virus. *Virology* 328:177–184.

51. Harty, R. N., M. E. Brown, G. Wang, J. Huibregtse, and F. P. Hayes. 2000. A PPxY motif within the VP40 protein of Ebola virus interacts physically and functionally with a ubiquitin ligase: implications for filovirus budding. *Proc. Natl.Acad. Sci. USA* 97:13871–13876.

52. Hensley, L. E., E. L. Stevens, S. B. Yan, J. B. Geisbert, W. L. Macias, T. Larsen, K. M. Daddario-DiCaprio, G. H. Cassell, P. B. Jahrling, and T. W. Geisbert. 2007. Recombinant human activated protein C for the postexposure treatment of Ebola hemorrhagic fever. *J. Infect. Dis.* 196(Suppl. 2):S390–S399.

53. Hensley, L. E., H. A. Young, P. B. Jahrling, and T. W. Geisbert. 2002. Proinflammatory response during Ebola virus infection of primate models: possible involvement of the tumor necrosis factor receptor superfamily. *Immunol. Lett.* 80:169–179.

54. Hiscott, J. 2007. Triggering the innate antiviral response through IRF-3 activation. *J. Biol. Chem.* 282:15325–15329.

55. Hoebe, K., and B. Beutler. 2004. LPS, dsRNA and the interferon bridge to adaptive immune responses: Trif, Tram, and other TIR adaptor proteins. *J. Endotoxin Res.* 10:130–136.

56. Hornung, V., J. Ellegast, S. Kim, K. Brzozka, A. Jung, H. Kato, H. Poeck, S. Akira, K. K. Conzelmann, M. Schlee, S. Endres, and G. Hartmann. 2006. 5′-Triphosphate RNA is the ligand for RIG-I. *Science* 314:994–997.

57. Huang, Y., L. Xu, Y. Sun, and G. Nabel. 2002. The assembly of Ebola virus nucleocapsid requires virion-associated proteins 35 and 24 and posttranslational modification of nucleoprotein. *Mol. Cell* 10:307.

58. Huggins, J. W., Z. X. Zhang, and T. I. Monath. 1995. Inhibition of Ebola virus replication in vitro and in vivo in a SCID mouse model by S-adenosylhomocysteine hydrolase inhibitors. *Antiviral Research Suppl.* 1:122.

59. Hutchinson, K. L., and P. E. Rollin. 2007. Cytokine and chemokine expression in humans infected with Sudan Ebola virus. *J. Infect. Dis.* 196(Suppl. 2): S357–S363.

60. Jahrling, P. B., T. W. Geisbert, D. W. Dalgard, E. D. Johnson, T. G. Ksiazek, W. C. Hall, and C. J. Peters. 1990. Preliminary report: isolation of Ebola virus from monkeys imported to USA. *Lancet* 335:502–505.

61. Kash, J. C., T. M. Tumpey, S. C. Proll, V. Carter, O. Perwitasari, M. J. Thomas, C. F. Basler, P. Palese, J. K. Taubenberger, A. Garcia-Sastre, D. E. Swayne, and M. G. Katze. 2006. Genomic analysis of increased host immune and cell death responses induced by 1918 influenza virus. *Nature* 443:578–581.

62. Kato, H., O. Takeuchi, S. Sato, M. Yoneyama, M. Yamamoto, K. Matsui, S. Uematsu, A. Jung, T. Kawai, K. J. Ishii, O. Yamaguchi, K. Otsu, T. Tsujimura, C. S. Koh, C. Reis e Sousa, Y. Matsuura, T. Fujita, and S. Akira. 2006. Differential roles of MDA5 and RIG-I helicases in the recognition of RNA viruses. *Nature* 441:101–105.

63. Kawai, T., and S. Akira. 2007. Antiviral signaling through pattern recognition receptors. *J. Biochem. (Tokyo)* 141:137–145.

64. Kawai, T., and S. Akira. 2006. TLR signaling. *Cell. Death Differ.* 13:816–825.

65. Kawai, T., K. Takahashi, S. Sato, C. Coban, H. Kumar, H. Kato, K. J. Ishii, O. Takeuchi, and S. Akira. 2005. IPS-1, an adaptor triggering RIG-I- and Mda5-mediated type I interferon induction. *Nat. Immunol.* 6:981–988.

66. Kindzelskii, A. L., Z. Yang, G. J. Nabel, R. F. Todd III, and H. R. Petty. 2000. Ebola virus secretory glycoprotein (sGP) diminishes FcγRIIIB-to-CR3 proximity on neutrophils. *J. Immunol.* 164:953–958.

67. Kitajewski, J., R. J. Schneider, B. Safer, S. M. Munemitsu, C. E. Samuel, B. Thimmappaya, and T. Shenk. 1986. Adenovirus VAI RNA antagonizes the antiviral action of interferon by preventing activation of the interferon-induced eIF-2 alpha kinase. *Cell* 45:195–200.

68. Kohler, M., S. Ansieau, S. Prehn, A. Leutz, H. Haller, and E. Hartmann. 1997. Cloning of two novel human importin-alpha subunits and analysis of the expression pattern of the importin-alpha protein family. *FEBS Lett.* 417:104–108.

69. Kohler, M., C. Speck, M. Christiansen, F. R. Bischoff, S. Prehn, H. Haller, D. Gorlich, and E. Hartmann. 1999. Evidence for distinct substrate specificities of importin alpha family members in nuclear protein import. *Mol. Cell. Biol.* 19:7782–7791.

70. Lenschow, D. J., C. Lai, N. Frias-Staheli, N. V. Giannakopoulos, A. Lutz, T. Wolff, A. Osiak, B. Levine,

R. E. Schmidt, A. Garcia-Sastre, D. A. Leib, A. Pekosz, K. P. Knobeloch, I. Horak, and H. W. Virgin IV. 2007. IFN-stimulated gene 15 functions as a critical antiviral molecule against influenza, herpes, and Sindbis viruses. *Proc. Natl. Acad. Sci. USA* **104**:1371–1376.

71. Leroy, E. M., S. Baize, V. E. Volchkov, S. P. Fisher-Hoch, M. C. Georges-Courbot, J. Lansoud-Soukate, M. Capron, P. Debre, J. B. McCormick, and A. J. Georges. 2000. Human asymptomatic Ebola infection and strong inflammatory response. *Lancet* **355**:2210–2215.

72. Leroy, E. M., B. Kumulungui, X. Pourrut, P. Rouquet, A. Hassanin, P. Yaba, A. Delicat, J. T. Paweska, J. P. Gonzalez, and R. Swanepoel. 2005. Fruit bats as reservoirs of Ebola virus. *Nature* **438**:575–576.

73. Levy, D. E., and A. Garcia-Sastre. 2001. The virus battles: IFN induction of the antiviral state and mechanisms of viral evasion. *Cytokine Growth Factor Rev.* **12**:143–156.

74. Li, W. X., H. Li, R. Lu, F. Li, M. Dus, P. Atkinson, E. W. Brydon, K. L. Johnson, A. Garcia-Sastre, L. A. Ball, P. Palese, and S. W. Ding. 2004. Interferon antagonist proteins of influenza and vaccinia viruses are suppressors of RNA silencing. *Proc. Natl. Acad. Sci. USA* **101**: 1350–1355.

75. Licata, J. M., R. F. Johnson, Z. Han, and R. N. Harty. 2004. Contribution of Ebola virus glycoprotein, nucleoprotein, and VP24 to budding of VP40 virus-like particles. *J. Virol.* **78**:7344–7351.

76. Lofts, L. L., M. S. Ibrahim, D. L. Negley, M. C. Hevey, and A. L. Schmaljohn. 2007. Genomic differences between guinea pig lethal and nonlethal Marburg virus variants. *J. Infect. Dis.* **196**(Suppl. 2):S305–S312.

77. Lu, S., and B. R. Cullen. 2004. Adenovirus VA1 noncoding RNA can inhibit small interfering RNA and microRNA biogenesis. *J. Virol.* **78**:12868–12876.

78. Lu, Y., M. Wambach, M. G. Katze, and R. M. Krug. 1995. Binding of the influenza virus NS1 protein to double-stranded RNA inhibits the activation of the protein kinase that phosphorylates the eIF-2 translation initiation factor. *Virology* **214**:222–228.

79. Macara, I. G. 2001. Transport into and out of the nucleus. *Microbiol. Mol. Biol. Rev.* **65**:570–594.

80. Mahanty, S., and M. Bray. 2004. Pathogenesis of filoviral haemorrhagic fevers. *Lancet Infect. Dis.* **4**:487–498.

81. Mahanty, S., M. Gupta, J. Paragas, M. Bray, R. Ahmed, and P. E. Rollin. 2003. Protection from lethal infection is determined by innate immune responses in a mouse model of Ebola virus infection. *Virology* **312**:415–424.

82. Mahanty, S., K. Hutchinson, S. Agarwal, M. McRae, P. E. Rollin, and B. Pulendran. 2003. Cutting edge: impairment of dendritic cells and adaptive immunity by Ebola and Lassa viruses. *J. Immunol.* **170**:2797–2801.

83. Malmgaard, L. 2004. Induction and regulation of IFNs during viral infections. *J. Interferon Cytokine Res.* **24**: 439–454.

84. Martinez, O., C. Valmas, and C. F. Basler. 2007. Ebola virus-like particle-induced activation of NF-kappaB and Erk signaling in human dendritic cells requires the glycoprotein mucin domain. *Virology* **364**:342–354.

85. McBride, K. M., G. Banninger, C. McDonald, and N. C. Reich. 2002. Regulated nuclear import of the Stat1 transcription factor by direct binding of importin-alpha. *EMBO J.* **21**:1754–1763.

86. Melen, K., R. Fagerlund, J. Franke, M. Kohler, L. Kinnunen, and I. Julkunen. 2003. Importin alpha nuclear localization signal binding sites for Stat1, Stat2, and influenza A virus nucleoprotein. *J. Biol. Chem.* **278**: 28193–28200.

87. Meylan, E., J. Curran, K. Hofmann, D. Moradpour, M. Binder, R. Bartenschlager, and J. Tschopp. 2005. Cardif is an adaptor protein in the RIG-I antiviral pathway and is targeted by hepatitis C virus. *Nature* **437**:1167–1172.

88. Min, J. Y., and R. M. Krug. 2006. The primary function of RNA binding by the influenza A virus NS1 protein in infected cells: inhibiting the 2′-5′ oligo (A) synthetase/RNase L pathway. *Proc. Natl. Acad. Sci. USA* **103**: 7100–7105.

89. Miranda, M. E., T. G. Ksiazek, T. J. Retuya, A. S. Khan, A. Sanchez, C. F. Fulhorst, P. E. Rollin, A. B. Calaor, D. L. Manalo, M. C. Roces, M. M. Dayrit, and C. J. Peters. 1999. Epidemiology of Ebola (subtype Reston) virus in the Philippines, 1996. *J. Infect. Dis.* **179**(Suppl. 1): S115–S119.

90. Miyamoto, Y., N. Imamoto, T. Sekimoto, T. Tachibana, T. Seki, S. Tada, T. Enomoto, and Y. Yoneda. 1997. Differential modes of nuclear localization signal (NLS) recognition by three distinct classes of NLS receptors. *J. Biol. Chem.* **272**:26375–26381.

91. Moroianu, J., G. Blobel, and A. Radu. 1995. Previously identified protein of uncertain function is karyopherin alpha and together with karyopherin beta docks import substrate at nuclear pore complexes. *Proc. Natl. Acad. Sci. USA* **92**:2008–2011.

92. Muhlberger, E., B. Lotfering, H. D. Klenk, and S. Becker. 1998. Three of the four nucleocapsid proteins of Marburg virus, NP, VP35, and L, are sufficient to mediate replication and transcription of Marburg virus-specific monocistronic minigenomes. *J. Virol.* **72**:8756–8764.

93. Muhlberger, E., M. Weik, V. E. Volchkov, H. D. Klenk, and S. Becker. 1999. Comparison of the transcription and replication strategies of Marburg virus and Ebola virus by using artificial replication systems. *J. Virol.* **73**:2333–2342.

94. Nachury, M. V., U. W. Ryder, A. I. Lamond, and K. Weis. 1998. Cloning and characterization of hSRP1 gamma, a tissue-specific nuclear transport factor. *Proc. Natl. Acad. Sci. USA* **95**:582–587.

95. Noda, T., H. Sagara, E. Suzuki, A. Takada, H. Kida, and Y. Kawaoka. 2002. Ebola virus VP40 drives the formation of virus-like filamentous particles along with GP. *J. Virol.* **76**:4855–4865.

96. O'Neill, R. E., and P. Palese. 1995. NPI-1, the human homolog of SRP-1, interacts with influenza virus nucleoprotein. *Virology* **206**:116–125.

97. Peterson, A. T., D. S. Carroll, J. N. Mills, and K. M. Johnson. 2004. Potential mammalian filovirus reservoirs. *Emerg. Infect. Dis.* **10**:2073–2081.

98. Pichlmair, A., O. Schulz, C. P. Tan, T. I. Naslund, P. Liljestrom, F. Weber, and C. Reis e Sousa. 2006. RIG-I-mediated antiviral responses to single-stranded RNA bearing 5′-phosphates. *Science* **314**:997–1001.

99. Platanias, L. C. 2005. Mechanisms of type-I- and type-II-interferon-mediated signalling. *Nat. Rev. Immunol.* **5**:375–386.

100. Ray, R. B., A. Basu, R. Steele, A. Beyene, J. McHowat, K. Meyer, A. K. Ghosh, and R. Ray. 2004. Ebola virus

glycoprotein-mediated anoikis of primary human cardiac microvascular endothelial cells. *Virology* **321**:181–188.

101. **Reed, D. S., L. E. Hensley, J. B. Geisbert, P. B. Jahrling, and T. W. Geisbert.** 2004. Depletion of peripheral blood T lymphocytes and NK cells during the course of Ebola hemorrhagic fever in cynomolgus macaques. *Viral Immunol.* **17**:390–400.

102. **Reid, S. P., L. W. Leung, A. L. Hartman, O. Martinez, M. L. Shaw, C. Carbonnelle, V. E. Volchkov, S. T. Nichol, and C. F. Basler.** 2006. Ebola virus VP24 binds karyopherin alpha1 and blocks Stat1 nuclear accumulation. *J. Virol.* **80**:5156–5167.

103. **Reid, S. P., C. Valmas, O. Martinez, F. M. Sanchez, and C. F. Basler.** 2007. Ebola virus VP24 proteins inhibit the interaction of NPI-1 subfamily karyopherin alpha proteins with activated Stat1. *J. Virol.* **81**:13469–13477.

104. **Rollin, P. E., R. J. Williams, D. S. Bressler, S. Pearson, M. Cottingham, G. Pucak, A. Sanchez, S. G. Trappier, R. L. Peters, P. W. Greer, S. Zaki, T. Demarcus, K. Hendricks, M. Kelley, D. Simpson, T. W. Geisbert, P. B. Jahrling, C. J. Peters, and T. G. Ksiazek.** 1999. Ebola (subtype Reston) virus among quarantine nonhuman primates recently imported from the Philippines to the United States. *J. Infect. Dis.* **179** (Suppl. 1):S108–S114.

105. **Rubins, K. H., L. E. Hensley, V. Wahl-Jensen, K. M. Daddario Dicaprio, H. A. Young, D. S. Reed, P. B. Jahrling, P. O. Brown, D. A. Relman, and T. W. Geisbert.** 2007. The temporal program of peripheral blood gene expression in the response of nonhuman primates to Ebola hemorrhagic fever. *Genome Biol.* **8**:R174.

106. **Ruf, W.** 2004. Emerging roles of tissue factor in viral hemorrhagic fever. *Trends Immunol.* **25**:461–464.

107. **Sanchez, A., T. W. Geisbert, and H. Feldmann.** 2007. Filoviridae: Marburg and Ebola viruses, p. 1410–1448. *In* D. M. Knipe and P. M. Howley (ed.), *Fields Virology*, 5th ed. Lippincott Williams and Wilkins, Philadelphia, PA.

108. **Sanchez, A., A. S. Khan, S. R. Zaki, G. J. Nabel, T. G. Ksiazek, and C. J. Peters.** 2001. Filoviridae: Marburg and Ebola viruses, p. 1279–1304. *In* D. M. Knipe and P. M. Howley (ed.), *Fields Virology*, 4th ed., vol. 1. Lippincott Williams and Wilkins, Philadelphia, PA.

109. **Seki, T., S. Tada, T. Katada, and T. Enomoto.** 1997. Cloning of a cDNA encoding a novel importin-alpha homologue, Qip1: discrimination of Qip1 and Rch1 from hSrp1 by their ability to interact with DNA helicase Q1/RecQL. *Biochem. Biophys. Res. Commun.* **234**:48–53.

110. **Sekimoto, T., N. Imamoto, K. Nakajima, T. Hirano, and Y. Yoneda.** 1997. Extracellular signal-dependent nuclear import of Stat1 is mediated by nuclear pore-targeting complex formation with NPI-1, but not Rch1. *EMBO J.* **16**:7067–7077.

111. **Seth, R. B., L. Sun, C. K. Ea, and Z. J. Chen.** 2005. Identification and characterization of MAVS, a mitochondrial antiviral signaling protein that activates NF-kappaB and IRF 3. *Cell* **122**:669–682.

112. **Sharma, S., B. R. tenOever, N. Grandvaux, G. P. Zhou, R. Lin, and J. Hiscott.** 2003. Triggering the interferon antiviral response through an IKK-related pathway. *Science* **300**:1148–1151.

113. **Simmons, G., R. J. Wool-Lewis, F. Baribaud, R. C. Netter, and P. Bates.** 2002. Ebola virus glycoproteins induce global surface protein down-modulation and loss of cell adherence. *J. Virol.* **76**:2518–2528.

114. **Stewart, M.** 2007. Molecular mechanism of the nuclear protein import cycle. *Nat. Rev. Mol. Cell Biol.* **8**:195–208.

115. **Stroher, U., E. West, H. Bugany, H. D. Klenk, H. J. Schnittler, and H. Feldmann.** 2001. Infection and activation of monocytes by Marburg and Ebola viruses. *J. Virol.* **75**:11025–11033.

116. **Sullivan, N. J., M. Peterson, Z. Y. Yang, W. P. Kong, H. Duckers, E. Nabel, and G. J. Nabel.** 2005. Ebola virus glycoprotein toxicity is mediated by a dynamin-dependent protein-trafficking pathway. *J. Virol.* **79**:547–553.

117. **Takada, A., S. Watanabe, H. Ito, K. Okazaki, H. Kida, and Y. Kawaoka.** 2000. Downregulation of beta1 integrins by Ebola virus glycoprotein: implication for virus entry. *Virology* **278**:20–26.

118. **Timmins, J., S. Scianimanico, G. Schoehn, and W. Weissenhorn.** 2001. Vesicular release of Ebola virus matrix protein VP40. *Virology* **283**:1–6.

119. **Towner, J. S., X. Pourrut, C. G. Albarino, C. N. Nkogue, B. H. Bird, G. Grard, T. G. Ksiazek, J. P. Gonzalez, S. T. Nichol, and E. M. Leroy.** 2007. Marburg virus infection detected in a common African bat. *PLoS ONE* **2**:e764.

120. **Villinger, F., P. E. Rollin, S. S. Brar, N. F. Chikkala, J. Winter, J. B. Sundstrom, S. R. Zaki, R. Swanepoel, A. A. Ansari, and C. J. Peters.** 1999. Markedly elevated levels of interferon (IFN)-gamma, IFN-alpha, interleukin (IL)-2, IL-10, and tumor necrosis factor-alpha associated with fatal Ebola virus infection. *J. Infect. Dis.* **179** (Suppl. 1):S188–S191.

121. **Voinnet, O.** 2005. Induction and suppression of RNA silencing: insights from viral infections. *Nat. Rev. Genet.* **6**:206–220.

122. **Volchkov, V. E., S. Becker, V. A. Volchkova, V. A. Ternovoj, A. N. Kotov, S. V. Netesov, and H. D. Klenk.** 1995. GP mRNA of Ebola virus is edited by the Ebola virus polymerase and by T7 and vaccinia virus polymerases. *Virology* **214**:421–430.

123. **Volchkov, V. E., A. A. Chepurnov, V. A. Volchkova, V. A. Ternovoj, and H. D. Klenk.** 2000. Molecular characterization of guinea pig-adapted variants of Ebola virus. *Virology* **277**:147–155.

124. **Volchkov, V. E., V. A. Volchkova, E. Muhlberger, L. V. Kolesnikova, M. Weik, O. Dolnik, and H. D. Klenk.** 2001. Recovery of infectious Ebola virus from complementary DNA: RNA editing of the GP gene and viral cytotoxicity. *Science* **291**:1965–1969.

125. **Wahl-Jensen, V., S. K. Kurz, P. R. Hazelton, H. J. Schnittler, U. Stroher, D. R. Burton, and H. Feldmann.** 2005. Role of Ebola virus secreted glycoproteins and virus-like particles in activation of human macrophages. *J. Virol.* **79**:2413–2419.

126. **Wahl-Jensen, V. M., T. A. Afanasieva, J. Seebach, U. Stroher, H. Feldmann, and H. J. Schnittler.** 2005. Effects of Ebola virus glycoproteins on endothelial cell activation and barrier function. *J. Virol.* **79**:10442–10450.

127. **Wang, X., E. R. Hinson, and P. Cresswell.** 2007. The interferon-inducible protein viperin inhibits influenza virus release by perturbing lipid rafts. *Cell Host Microbe* **2**:96–105.

128. **Warfield, K. L., C. M. Bosio, B. C. Welcher, E. M. Deal, M. Mohamadzadeh, A. Schmaljohn, M. J. Aman, and**

S. Bavari. 2003. Ebola virus-like particles protect from lethal Ebola virus infection. *Proc. Natl. Acad. Sci. USA* **100:**15889–15894.

129. Watanabe, S., T. Noda, and Y. Kawaoka. 2006. Functional mapping of the nucleoprotein of Ebola virus. *J. Virol.* **80:**3743–3751.

130. Watanabe, S., T. Watanabe, T. Noda, A. Takada, H. Feldmann, L. D. Jasenosky, and Y. Kawaoka. 2004. Production of novel Ebola virus-like particles from cDNAs: an alternative to Ebola virus generation by reverse genetics. *J. Virol.* **78:**999–1005.

131. Weis, K., I. W. Mattaj, and A. I. Lamond. 1995. Identification of hSRP1 alpha as a functional receptor for nuclear localization sequences. *Science* **268:**1049–1053.

132. Weis, K., U. Ryder, and A. I. Lamond. 1996. The conserved amino-terminal domain of hSRP1 alpha is essential for nuclear protein import. *EMBO J.* **15:**1818–1825.

133. Xu, L. G., Y. Y. Wang, K. J. Han, L. Y. Li, Z. Zhai, and H. B. Shu. 2005. VISA is an adapter protein required for virus-triggered IFN-beta signaling. *Mol. Cell* **19:**727–740.

134. Yaddanapudi, K., G. Palacios, J. S. Towner, I. Chen, C. A. Sariol, S. T. Nichol, and W. I. Lipkin. 2006. Implication of a retrovirus-like glycoprotein peptide in the immunopathogenesis of Ebola and Marburg viruses. *FASEB J.* **20:**2519–2530.

135. Yang, Z., R. Delgado, L. Xu, R. F. Todd, E. G. Nabel, A. Sanchez, and G. J. Nabel. 1998. Distinct cellular interactions of secreted and transmembrane Ebola virus glycoproteins. *Science* **279:**1034–1037.

136. Yang, Z. Y., H. J. Duckers, N. J. Sullivan, A. Sanchez, E. G. Nabel, and G. J. Nabel. 2000. Identification of the Ebola virus glycoprotein as the main viral determinant of vascular cell cytotoxicity and injury. *Nat. Med.* **6:**886–889.

137. Ye, L., J. Lin, Y. Sun, S. Bennouna, M. Lo, Q. Wu, Z. Bu, B. Pulendran, R. W. Compans, and C. Yang. 2006. Ebola virus-like particles produced in insect cells exhibit dendritic cell stimulating activity and induce neutralizing antibodies. *Virology* **351:**260–270.

138. Yoneyama, M., M. Kikuchi, T. Natsukawa, N. Shinobu, T. Imaizumi, M. Miyagishi, K. Taira, S. Akira, and T. Fujita. 2004. The RNA helicase RIG-I has an essential function in double-stranded RNA-induced innate antiviral responses. *Nat. Immunol.* **5:**730–737.

139. Zampieri, C. A., J. F. Fortin, G. P. Nolan, and G. J. Nabel. 2007. The ERK mitogen-activated protein kinase pathway contributes to Ebola virus glycoprotein-induced cytotoxicity. *J. Virol.* **81:**1230–1240.

Cellular Signaling and Innate Immune
Responses to RNA Virus Infections
Edited by A. R. Brasier et al.
© 2009 ASM Press, Washington, D.C.

Antonella Casola, Xiaoyong Bao,
Allan R. Brasier, Roberto P. Garofalo

16

Inhibition of Antiviral Signaling Pathways by Paramyxovirus Proteins

CLASSIFICATION AND DISEASE BURDEN

The *Paramyxoviridae* family includes enveloped, negative-sense, single-stranded (ss) RNA viruses, that are major and ubiquitous disease-causing pathogens of humans and animals (83). Among them are important viruses that cause acute respiratory morbidity, particularly in infants, the elderly, and immunocompromised subjects of any age. The family is taxonomically divided into two subfamilies: the *Paramyxovirinae*, with five genera; and the *Pneumovirinae*, which includes two genera (Table 1). The classification of these viruses is based on their genome organization, morphological and biological characteristics, and sequence relationship of the encoded proteins. The complete genome sequence for all members of the *Paramyxoviridae* is available (www.ncbi.nlm.nih.gov/). The 15,000- to 19,000-nucleotide genomic RNAs contain a 3' extracistronic region of ~50 nucleotides known as the leader and a 5' extracistronic region of 50 to 161 nucleotides (trailer), which are critical for the processes of transcription and replication. Six to ten genes (depending on the specific virus), each flanked by intergenic regions of variable length, encode the viral proteins (Fig. 1).

Major etiologic agents of human respiratory infections are included in the *Paramyxovirinae* genera *Respirovirus* and *Rubulavirus*. These include parainfluenza viruses (PIV), genetically and antigenically divided into serotypes 1 to 4, which are second only to respiratory syncytial virus (RSV) as causative agents of lower respiratory tract infections in infancy. Since immunity is incomplete and of short duration, reinfections with PIV and RSV occur throughout life (52). The genus *Rubulavirus* also includes the mumps virus (MuV), the etiologic agent of parotitis, which infects the respiratory tract; clinical manifestations, however, usually relate to infection of the parotid gland, without respiratory tract illness. The genus *Morbillivirus* includes one human virus, measles virus (MeV). Infection with MeV is characterized by fever, cough, and conjunctivitis followed by the appearance of generalized maculopapular rash. Lower respiratory tract infections have been reported, but in countries where live attenuated vaccine is widely used, measles virus is no longer a major cause of respiratory morbidity. The *Avulavirus* genus contains only avian viruses. The genus *Henipavirus* includes the human viruses Nipah (NiV), identified in 1999, when it caused an outbreak of neurological and respiratory disease on pig farms in Malaysia; and Hendra (HeV), which was isolated in 1994 following an outbreak in horses in a training complex in Australia. In humans, the infections

Antonella Casola, Xiaoyong Bao, Allan R. Brasier, and Roberto P. Garofalo, University of Texas Medical Branch, Galveston, TX 77555.

present with fever, headache, drowsiness, and neurological symptoms consistent with encephalitis. Emergence of these viruses has been linked to an increase in contact between bats (which may harbor these viruses) and humans, sometimes involving an intermediate animal host (152).

The pneumoviruses can be distinguished from the *Paramyxovirinae* members morphologically because they contain narrower nucleocapsids. In addition, pneumoviruses have differences in genome organization and in the number of encoded proteins, and an attachment protein that is different from that of members of the subfamily *Paramyxovirinae*. There are two genera in the *Pneumovirinae* family: the *Pneumovirus* genus, which includes human and bovine RSV; and the *Metapneumovirus* genus, which includes human metapneumovirus (hMPV) and avian metapneumovirus. Human RSV encodes 11 separate proteins, while hMPV encodes 9 proteins that generally correspond to those of RSV, except that hMPV lacks NS1 and NS2 (Fig. 1) (12, 149). Human RSV has a single serotype with two antigenic subgroups, A and B (4). Sequence analysis of representatives of the two subgroups has shown ~80% nucleotide identity. The M2–2, SH, and G proteins are the most divergent among subgroups (26). As for hMPV, recent data indicate that there is one serotype with two genetic subgroups, A and B, which have extensive cross-reactivity and cross-protection (133). RSV is responsible for ~60% of all lower respiratory tract infections in infants and is a major cause of severe respiratory morbidity and mortality in the elderly (52). The World Health Organization

Table 1 Major members of the *Paramyxoviridae* family

Subfamily	Genus	Virus
Paramyxovirinae	*Henipavirus*	Hendra virus
		Nipah virus
	Morbillivirus	Measles virus
	Respirovirus	Sendai virus
		Human parainfluenza virus type 1
		Human parainfluenza virus type 3
		Bovine parainfluenza virus type 3
	Rubulavirus	Parainfluenza virus type 5
		Mumps virus
		Human parainfluenza virus type 2
Pneumovirinae	*Pneumovirus*	Human respiratory syncytial virus
		Bovine respiratory syncytial virus
	Metapneumovirus	Avian pneumovirus
		Human metapneumovirus

estimates that RSV is responsible for 64 million clinical infections and 160,000 deaths annually worldwide (37). In the United States, infection with RSV is the most frequent cause of hospitalization of infants (84). Since its discovery in 2001, hMPV has been isolated from

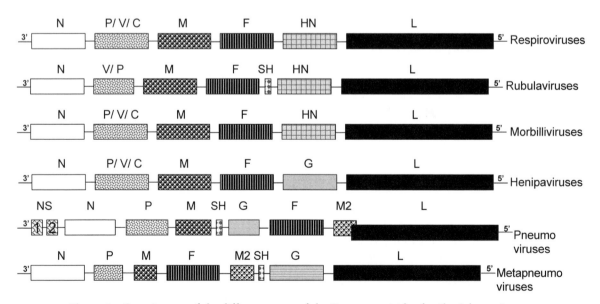

Figure 1 Genetic map of the different genes of the *Paramyxoviridae* family. Schematic representation of the genetic map of typical members of each genus of the *Paramyxoviridae*. (Adapted from reference 82, page 1453.)

Table 2 Viral proteins encoded by the P gene in members of the *Paramyxovirinae*[a]

| Genus | Virus | mRNA insertion | | | Overlapping ORFs |
		0	+ 1(or 4)G	+ 2G	
Henipavirus	NiV, HeV	P	V	W	C
Morbillivirus	MeV	P	V	W	C
Respirovirus	SeV	P	V	W	C′ C Y1 Y2
	bPIV3	P	V	D	C
	hPIV3	P	V	D	C
	hPIV1	P			C′ C
Rubulavirus	PIV5 (SV5), MuV, hPIV2, hPIV4	V	I	P	

[a] ORFs, open reading frames; b, bovine.

individuals of all ages with acute respiratory tract infection worldwide (110). Virtually all children older than 5 years show 100% serologic evidence of infection (150). Around 12% of all respiratory tract infections in children are caused by hMPV, second only to RSV (27, 70, 110, 154). hMPV also accounts for 10% of all hospitalizations of elderly patients with respiratory tract infections, and it has been isolated from respiratory samples in a single winter season as often as PIV (36).

PARAMYXOVIRUS PROTEINS ANTAGONIZING HOST INNATE SIGNALING PATHWAYS

To date, the paramyxovirus P gene-encoded proteins, the NS1 and NS2 proteins of pneumoviruses, and the envelope glycoproteins G and SH have been shown to play a major role in antagonizing type I interferon (IFN) signaling and other host innate immune responses. A discussion of the inhibitory function for each of these proteins is presented in this review.

The P gene is unique because it produces more than one polypeptide species by a process known as RNA editing or pseudotemplated addition of nucleotides (reviewed in reference 82). For example, the Sendai virus (SeV) P gene directs the expression of at least seven polypeptides, including the P, V, W, C′, C, Y1, and Y2 proteins. The P mRNA is an exact copy of the P gene, whereas the V or W mRNA is generated by the insertion of one or two G residues (+1G, +2G), respectively, at the editing site during viral mRNA synthesis. The proteins that are produced as a result of RNA editing share a common N-terminal region but differ in their C-terminal regions, starting at the G insertion site. Virtually all members of the *Paramyxovirinae* encode a characteristic editing site in the P gene, with a specific frequency and number of inserted G residues. The P protein is 400 to 600 amino acids long, is heavily phosphorylated, and is the only P gene product essential for viral RNA synthesis. The V protein is a ~25- to 30-kDa polypeptide that shares an N-terminal domain with the P protein but has a distinct C-terminal domain due to RNA editing. V protein plays a critical role in the virus replication cycle, as mutant viruses lacking the V protein cysteine-rich domain are severely attenuated for growth in vivo or are cleared more rapidly than wild-type viruses from the lung, suggesting a role of this protein in inhibition of host immune responses. The W/D/I proteins of respiroe-, morbilli-, and henipaviruses are expressed from mRNAs with two inserted G residues. The W protein is essentially a truncated P protein containing the N-terminal assembly module of the P protein alone and is abundantly expressed in SeV-infected cells. Overall, the role of the W/D/I proteins in viral growth has not been established. In addition to RNA editing, some paramyxoviruses use a second mechanism to express P gene proteins that involves the use of alternative translation initiation codons to yield the C proteins. The SeV C′, Y1, and Y2 proteins comprise a nested set of polypeptides that range in size from 175 to 215 residues. The C protein is abundantly expressed in infected cells at higher levels than C′, Y1, and Y2, but virions contain low levels of these polypeptides. As shown in Table 2, morbilliviruses and henipaviruses express one C protein, while respiroviruses can express all four polypeptides. Rubula- and avulaviruses do not express C proteins. C proteins, which are small basic polypeptides, have several different functions in the viral growth cycle, including viral RNA synthesis, counteracting host antiviral pathways, and release of virus from infected cells.

The RSV genome encodes two proteins considered to be nonstructural, NS1 (139 amino acids) and NS2 (124 amino acids). Only trace amounts of these proteins (probably due to contamination of the infected cells) are recovered from purified viral preparations (64). Deletion of NS1 or NS2 from recombinant human RSV results in reduction of viral growth both in cells that are IFN competent and in cells that lack IFN-α and -β genes, but neither protein is considered to be essential for transcription or replication in vitro or in vivo (139, 153). As we discuss later in this review, human RSV and bovine RSV NS1 and NS2 gene products have been shown to be important viral suppressors of type I IFN induction.

A small hydrophobic protein (64 amino acids), designated SH protein, is an integral membrane protein of RSV. SH accumulates in infected cells in multiple forms, including SH_0, the full-length, nonglycosylated species; SH_g, containing a single N-linked carbohydrate side chain; SH_p, with an additional polylactosaminoglycan side chain; and SH_t, a nonglycosylated species derived from the initiation of protein synthesis at an internal AUG codon (3). SH_0 and SH_p are the predominant forms in virions. The role of the SH protein in the RSV life cycle is not understood, as spontaneous deletion of the gene by extensive passages in vitro or in a recombinant strain results in only minor alterations in virus growth properties in tissue culture cells or the respiratory tract of mice or chimpanzees (21, 72). Also, hMPV expresses an SH protein that is nearly three times longer than the RSV SH and does not affect viral replication (13). The rubulaviruses PIV5 and MuV contain a small gene located between F and HN designated SH (Fig. 1) (57, 58). The PIV5 SH protein (44 amino acids) is a type II integral membrane protein that is expressed at the plasma membrane and is packaged in virions. The MuV virus SH protein is a 57-residue integral membrane protein orientated in the opposite direction from the PIV5 SH protein with a C-terminal cytoplasmic domain (33). PIV5 lacking SH grows as well as wild-type virus in tissue culture cells, but the virus is attenuated in vivo. MuV SH is not required for virus replication in vitro. The SH protein of rubulaviruses and *Pneumovirinae* appears to interfere with nuclear factor-κB (NF-κB)-mediated signaling pathways that lead to the production of antiviral cytokines and cell apoptosis (88).

The G protein of human RSV is a type II transmembrane glycoprotein of 282 to 319 amino acids in length and is considered the major attachment protein (85; reviewed in reference 25a). The G protein has a single hydrophobic region near the N-terminal end that serves as an uncleaved signal peptide and a membrane anchor, leaving the C terminal of the molecule oriented externally. In addition, a secreted form of G is also expressed as result of a translational initiation at the second AUG in the open reading frame, which is located within the signal/anchor

(56). The truncated species is trimmed proteolytically so that the final secreted form lacks the N-terminal 65 amino acids. The secreted form represents ~80% of the G protein produced in infected cells by 24 h, the remaining being virion associated (56). While the ectodomain of RSV G is highly divergent between strains, a central conserved domain of 13 residues (position 164 to 176 in A2 strain) and an overlapping segment containing 4 cysteine residues (positions 173, 176, 182, and 186) are fully conserved among RSV strains. This central domain contains a CX3C motif with limited sequence relatedness to the CX3C domain of the chemokine fractalkine (146) and has been shown to modulate NF-κB-mediated cytokine production (20). The G protein of hMPV does not reveal significant amino acid sequence identity with the G protein of RSV, but the overall organization and glycosylation pattern is consistent with that of an anchored type II transmembrane protein similar to the RSV G protein (149). The hMPV G protein amino acid sequence is also highly divergent between hMPV subgroups. One apparent difference between the RSV and hMPV G proteins is the fact that the latter lacks the conserved central domain and cysteine noose. A secreted form of hMPV G has not been identified so far and the protein appears to be nonessential for viral replication (13).

MAJOR CELLULAR SIGNALING ELICITED BY PARAMYXOVIRUS INFECTIONS

Paramyxovirus infection induces a myriad of host-cell signaling responses leading to the production of antiviral and inflammatory gene products. As a result, infected cells secrete type I IFNs, antimicrobial peptides, cytokines, chemokines, and other metabolites whose actions are to produce an antiviral state, coordinate the complex processes of inflammation (vascular permeability, paracrine activation, and leukocyte recruitment), and facilitate the development of adaptive immunity.

Both epithelial cells and monocytes/macrophages play important roles in initiating the primary innate immune response to paramyxovirus infection (59, 68). Due to differences in the expression of pattern recognition receptors (PRRs) and other factors not yet understood, the signaling pathways activated are cell type dependent (73). Currently, most evidence suggests that epithelial cells initially respond to paramyxovirus replication via cytoplasmic PRRs mediated by the retinoic acid-inducible gene I (RIG-I) and melanoma differentiation-associated gene 5 (MDA-5) RNA helicases, whereas monocytes/macrophages respond in a viral replication-independent manner via cell surface-associated PRRs, such as the Toll-like receptors (TLRs) 3 and 4. The downstream signaling pathways that play major roles in the cellular response are the interferon regulatory factor (IRF), NF-κB, and signal transducer and activator of transcription (Stat) pathways. The structures of these pathways were introduced earlier in this text; their mechanisms for activation are discussed in detail below.

The Cytoplasmic PRRs: Early Mediators of IRF and NF-κB Activation

In 2004, RIG-I, a gene encoding a DExD/H box motif-containing RNA helicase, was identified as an essential regulator of double-stranded (ds) RNA-induced signaling (159). Through its helicase domain, RIG-I binds and unwinds dsRNA produced during paramyxovirus infection (25, 89) and signals via its NH_2-terminal caspase activation and recruitment domains (CARDs) that interact with the downstream mitochondrial antiviral signaling (MAVS) adaptor, resulting in IRF and NF-κB activation. Of the more than 100 genomic RNA helicases, MDA-5 is highly related to RIG-I, containing a DExD/H box-containing helicase and two tandem CARD domains (71, 78). RIG-I and MDA-5 recognize different types of dsRNAs and are activated by different viruses: RIG-I is essential for the production of type I IFNs in response to most ssRNA viruses, including paramyxoviruses (89), influenza virus, and Japanese encephalitis virus; whereas MDA-5 is critical for picornavirus detection (74). LGP2 is a third RIG-I-like RNA helicase that has been identified but lacks an NH_2-terminal CARD domain. LGP2 inhibits Newcastle disease virus (NDV)- and SeV-induced IRF3 and NF-κB activation by competing for dsRNA binding, but lacking the CARD domain, is unable to complex with MAVS (116).

Activated RIG-I and MDA-5 converge on MAVS, a mitochondrial anchored protein with an NH_2-terminal CARD domain, mid-molecule tumor necrosis factor (TNF)-α receptor associated factor (TRAF) interaction motifs (TIMs), and a COOH-terminal transmembrane domain (75, 80, 95, 102). Upon activation, MAVS recruits the TRAF2, 3, and 6 isoforms, required for coupling with IRF and NF-κB (119, 156). The RIG-I/MAVS pathway is highly inducible by SeV (92, 119, 129) and has been implicated as an important early intracellular sensor for RSV in epithelial cells and mouse embryonic fibroblasts (MEFs) (89, 92). Using small interfering RNA (siRNA)-mediated knockdown of RIG-I, RSV-induced activation of NF-κB and IRF3 was significantly inhibited, as were downstream IFN-β, IP-10, CCL-5, and ISG15 genes during the early phase of infection (89). These findings were reproduced by Seya and colleagues, who proved that RSV-induced IFN-β production is initiated by RIG-I (and not the TLR3 pathway), and also showed

that NAK-associated protein 1 (NAP1) was involved in IFN-β expression downstream (121). Loo et al. have recently shown that RSV and SeV infection of MDA-5$^{-/-}$ MEFs was able to induce IFN and IFN-stimulated gene (ISG) expression, which was almost completely abolished in RIG-I$^{-/-}$ MEFs (92). Similarly, we have recently found that hMPV infection of airway epithelial cells induced the expression of both RIG-I and MDA-5 and that RIG-I, but not MDA-5, plays a fundamental role in hMPV-induced cellular signaling, as inhibition of RIG-I expression significantly decreased activation of IRF and NF-κB transcription factors and production of type I IFN and proinflammatory cytokines and chemokines. RIG-I-dependent signaling was necessary to induce a cellular antiviral state, as reduction of RIG-I expression resulted in enhanced hMPV replication (85a).

Measles virus has also been shown to induce the expression of RIG-I, MDA-5, and TLR3 in airway epithelial cells and endothelial cells (10). Increased expression of MDA-5, but not RIG-I or TLR3, led to enhanced IFN-β activation in MeV-infected airway epithelial cells, suggesting a role for this helicase in MeV-induced cellular signaling (10).

The IRF Pathway

Many members of the *Paramyxoviridae* family, including SeV, MeV, RSV, and hMPV, have been shown to be potent activators of IRF3 (9, 22, 76, 128). IRF3 is constitutively expressed and normally present in the cytoplasm; in response to viral replication, IRF3 is activated in a time-, dose (multiplicity of infection [MOI])-, and replication-dependent manner in a variety of epithelial cells (9, 22, 143). The process of IRF3 activation involves phosphorylation at multiple sites in its COOH terminus, a posttranslational modification that induces dimerization, nuclear translocation, and association with the p300/CREB-binding protein (CBP) coactivator (87, 122, 160). IRF3 phosphorylation is mediated by the viral-activated kinase, a heterodimer of atypical I-κB kinases (IKKs), now known as TRAF-associated NF-κB-activated (TANK)-binding kinase (TBK) 1/IKK-ε (107). TBK1/IKK-ε are recruited to the activated RIG-I/MAVS complex in a manner dependent on a complex formed with the regulatory IKK, IKK-γ and the TBK1 binding protein, TANK (48, 162). IRF3 activation is also dependent on the TRAF3 protein because IRF3 is not activated in TRAF3-deficient cells (119).

In addition to the PRR RIG-I-induced IRF3 activation mediated by TBK1/IKK-ε, a second pathway for IRF3 activation has also been described for MeV, mediated by its nucleocapsid (N) protein (143). Ectopic expression of MeV N activates IRF3 phosphorylation, dimerization, and activation of downstream genes. This latter pathway involves complex formation of MeV N with the COOH-terminal IRF association domain of IRF3 (143). These data suggest that several mechanisms are involved in MeV-induced IRF3 activation; at low MOIs, viral replication/dsRNA induce the RIG-I/MAVS pathway, whereas at higher MOIs, late in the course of infection, accumulated MeV N protein directly activates IRF3 by binding its COOH terminus. Whether other paramyxovirus N proteins induce IRF3 through this mechanism has not yet been determined. It is worth mentioning that the pneumoviruses of the *Paramyxoviridae* family (RSV) express nonstructural proteins that interfere with IRF3 activation late in the course of infection (19, 135). The subtle interplays of IRF activation/inactivation and relationship to viral replication are undoubtedly important in disease pathogenesis.

NF-κB Activation Pathways

Canonical Pathway Activation via the RIG-I/MAVS Complex

NF-κB is a family of highly inducible cytoplasmic transcription factors that play a central role as a mediator of inflammation in paramyxovirus infection (145). In a mouse model of RSV infection, we have shown that following intranasal administration of a specific cell-permeable IKK inhibitor, NF-κB DNA-binding activity, chemokine gene expression, and airway inflammation were markedly reduced in response to RSV infection (50), indicating that NF-κB plays a central role in viral-induced airway inflammation. Virtually all paramyxoviruses are known to activate NF-κB with various kinetics, including hPIV3 (17), SeV (43), RSV (42), hMPV (9), and MeV (55).

NF-κB activation is controlled by distinct pathways, including the canonical, noncanonical, and cross-talk pathways, which were discussed earlier (see chapters 8 and 9). The relative use of these pathways has not been rigorously dissected for many paramyxoviruses, RSV being the exception, but the canonical pathway appears to be a major pathway important in hPIV, RSV, and MeV infections. The canonical pathway is activated by RIG-I induction, TLR engagement, paracrine TNF production, or enhanced reactive oxygen species (ROS) production, resulting in proteolysis of the cytoplasmic IκB-α inhibitor, resulting in NF-κB translocation. Recent work has determined that a parallel pathway is required to activate latent transcriptional activity of the RelA transcription factor via Ser-276 phosphorylation. This modification is similar to the COOH-terminal phosphorylation of IRF3, and permits complex formation with p300/CBP coactivators (66, 89, 164).

In epithelial cells, RSV infection induces a time-, dose-, and replication-dependent induction of IκB-α/β degradation and RelA translocation that begins 12 h after viral adsorption and peaks 24 h later (42, 65). During the early phase of RSV infection (12 to 24 h), the RIG-I/MAVS complex is important in canonical pathway activation (89). At later times, other pathways participate, including TLR3; in addition, TLR3 is a major inducer of the RelA Ser-276 kinase, whose action is required for maximal transcriptional activation of the nuclear RelA induced by RIG-I/MAVS (89). The molecular mechanisms by which the RIG-I/MAVS complex mediates IKK activation are not completely understood. Current evidence suggests that IKK-γ is required as an adaptor between IKK and the RIG-I/MAVS complex. Surprisingly, studies using TRAF3-deficient MEFs have found that although TRAF3 is required for IRF3 activation, it is not required for the NF-κB pathway (119). The precise role of other TRAF isoforms in the IKK–IκB-α module will require further exploration.

The role of RSV NS proteins in NF-κB activation has been suggested through mechanisms not completely understood. Using recombinant RSV, Spann et al. found that NS2 deletion reduced the magnitude of NF-κB activation, suggesting that NS2 may play a role in induction of NF-κB in RSV infection (135). Another report described the findings that the transcriptional processing factor M2–1 induces NF-κB translocation and forms a nuclear complex with RelA (112). The interrelationships between the NS proteins and NF-κB activation pathways will require further study.

The Noncanonical NF-κB Activation Pathway

The noncanonical NF-κB activation pathway is a separate pathway that controls processing of the 100-kDa NF-κB2 precursors into its mature 52-kDa DNA-binding form (NF-κB2) and liberation of associated RelB complexes. Although the noncanonical pathway has largely been investigated in lymphocytes (15, 16, 99, 100, 127), our studies have shown that RSV replication is a potent noncanonical pathway activator (24). In time course studies, activation of the noncanonical pathway is more rapid than the canonical pathway and is associated with nuclear translocation of NF-κB-inducing kinase (NIK) (24). The demonstration that distinct groups of genes are activated by RelB complexes suggests that further work is required to fully understand the role of the noncanonical pathway in RSV infection (15).

TLR-Dependent Pathways

The TLR family members are transmembrane PRRs that recognize different types of pathogens by binding lipopolysaccharide (LPS), carbohydrate, peptide, dsRNA, ssRNA, and ssDNA. TLR1, 2, 4, 5, and 6 localize at the cell surface, whereas TLR7, 8, and 9 are located within the endosomal compartment. The localization of TLR3 is apparently more dynamic, with both cell surface-associated and endosomal-bound TLR3s having been reported (96, 101, 144).

TLRs are composed of an ectodomain of leucine-rich repeats involved in ligand binding and a cytoplasmic Toll/interleukin-1 (IL-1) receptor (TIR) domain that interacts with TIR-domain-containing adaptor molecules (1). Two types of downstream adaptors have been identified to associate with different TLRs. Myeloid differentiation primary-response gene 88 (MyD88) has been shown to interact with most of the TLRs, except TLR-3. MyD88 then binds IL-1 receptor-associated kinase-1 (IRAK-1), IRAK-2, and TRAF6 to activate IRF and NF-κB. The other TLR adaptor, TIR-domain-containing adaptor protein inducing IFN-β (TRIF), is only recruited by TLR3 and TLR4 during signaling transduction. In response to paramyxovirus infection, TLR3, 4, 7/8, and 9 play various roles in different cell types. In airway epithelial cells, TLR3 and 4 have been shown to interact with RSV, whereas TLR7/8 and 9 function in plasmacytoid dendritic cells (pDCs). Measles virus is the only member of the *Paramyxoviridae* family that has been shown to activate TLR2 via its hemmagglutinin (H) protein, resulting in production of inflammatory cytokines and expression of CD150 (measles virus receptor) (14).

The TLR4 Pathway

TLR4 has been reported to recognize bacterial endotoxin (LPS) and RSV fusion protein (F protein) (81). In unstimulated airway epithelial cells, however, TLR4 is expressed at very low levels on the cell surface and, as a result, is barely responsive to LPS. However, RSV infection increases expression and cell surface translocation of TLR4, suggesting that TLR4 may play a role in epithelial signaling later during the evolution of infection (163). Confirming this finding, increased TLR4 expression has been reported in monocytes of RSV-infected infants (40). One ramification of this finding is that after RSV exposure, the airway epithelium can be activated by subsequent environmental LPS exposure (98). Interestingly, TLR4 coding region D299G and T399I polymorphisms have been found to be associated with severe RSV bronchiolitis in human studies (137).

By contrast, studies of TLR4 function in RSV-infected animal models have revealed mixed results where the genetic background significantly influences the role of TLR4. Although one study showed impaired pulmonary CD14+ cell recruitment and reduced virus clearance in RSV-infected TLR4null mice, another study

in BALB/c background found no impact on RSV clearance and recruitment of inflammatory cells. The second study also indicated that the genetic background used in the previous study had an additional defect in the IL-12 receptor that may impair RSV control (32). In a separate study investigating NF-κB activation in the airways of RSV-infected BALB/c mice, the roles of alveolar macrophages and TLR4 were described in mediating two different patterns of NF-κB activation. The first NF-κB response occurs early after RSV inoculation, is alveolar macrophage and TLR4 dependent, and is viral replication independent, whereas the second response involves epithelial cells and/or inflammatory cells, is TLR4 independent, and requires viral replication (49). These data suggest that TLR4 plays a minor role early in the innate immune response to RSV.

Studies of TLR4 function in response to other paramyxoviruses have produced similar conclusions. In a mouse model of SeV infection, TLR4 mutant and wild-type mice showed no difference in pulmonary viral loads, histopathology, cytokine levels, and leukocyte influx (151). In a model of pneumovirus of mice, body weight, pulmonary function values, histopathology, and pulmonary viral loads were no different between TLR4 wild-type and TLR4-deficient mice (35). All these results suggest that TLR4 may not play as important a role as initially expected. More studies will be required to resolve the issue of whether TLR4 plays a role in human RSV infections, as suggested by recent data with certain TLR4 polymorphisms (8, 137).

Interestingly, as we discuss later in regard to the TLR7/9 pathway, paramyxoviruses are capable of interfering with TLR pathways triggered by specific agonists. In the case of TLR4, recent work has shown that measles virus, a known pathogen inducing immune suppression, inhibits IL-12 production in murine DCs in response to LPS but not TLR2, 3, 7, or 9 agonists (51). The P protein of MeV has been recently associated with the inhibition of the TLR4 signal via a mechanism that involves the upregulation of the ubiqitin-modifying enzyme A20, a negative-feedback regulator of NF-κB (157).

The TLR3 Pathway

Because TLR3 is a receptor for dsRNA, it may function as an intracellular sensor for paramyxovirus infections. TLR3 is relatively unique in the TLR family because its intracellular signaling is mediated by TRIF and functions independently of the MyD88 pathway. TLR3-TRIF recruits the phosphoinositide 3-kinase (PI3K) and associates with the TBK1 and TRAF6 adaptors to activate atypical TBK1/IKK-ε, producing IRF activation (38, 123). Alternatively, NF-κB activation appears to be mediated through two distinct domains of TRIF, with one domain being coupled to IKK activation via receptor-interacting protien-1 (RIP-1) (94). Like IRF activation, recruitment of PI3K is also apparently involved in NF-κB activation by mediating inducible phosphorylation and transcriptional activation (120).

The role of TLR3 in paramyxovirus infection has not been fully elucidated. In uninfected epithelial cells, our group and others have observed that most of the TLR3 is located in the endosomal compartment, whereas after viral infection, TLR3 expression is induced and the protein is redistributed to the cell surface (45, 89). siRNA-mediated inhibition of TLR3 expression decreases synthesis of CXCL10 and CCL5 but did not significantly reduce levels of IL-8 in response to RSV infection, but the effect on downstream signaling pathways was not described (117). Our study found that TLR3 in response to RSV infection is a paracrine effect, i.e., dependent on paracrine IFN-β signaling (89). In IFN-β-deficient Vero cells, RSV infection does not increase TLR3 expression, whereas after adding conditioned media from RSV-infected A549 cells into Vero cells, TLR3 expression was significantly induced (89). This result not only indicates the essential role of IFN-β for TLR3 induction, but also suggests that TLR3 is downstream of the RIG-I/IRF pathway in RSV infection.

Mouse models of TLR3 deficiency have suggested a role for this receptor in innate defense against RSV infection. One study showed that in TLR3$^{-/-}$ mice, but not wild-type mice, RSV induced enhanced eosinophil accumulation (118). TLR3$^{-/-}$ mice also produced significant increases in the Th2-type cytokines IL-5 and IL-13, as well as an increase of mucus production compared with wild-type mice after RSV infection. However, RSV clearance is apparently not reduced in TLR3$^{-/-}$ mice. These data suggested that TLR3 is necessary to maintain the proper immune environment in the lung and protect animals from developing further pathologic symptoms (118).

The TLR7/8/9 Pathway

It is well established that viral nucleic acids are the predominant trigger of type I IFNs and that ssRNA viruses such as influenza and vesicular stomatitis virus activate TLR7 in the endosomal compartment (30, 97). The role of TLR7/8/9-mediated IFN production by paramyxoviruses has not been conclusively characterized. Virus strain tropism, species- and strain-specific genetic factors (i.e., human, mouse, distinct mouse strains), and cell-specific signaling pathways (such as those in DC subsets) are likely to differentially regulate the complex cellular responses to infection by each pathogen. As such, SeV has been shown to activate antiviral (cytokine) responses

in human nonmyeloid cell lines by a mechanism entirely independent of TLR7/8, while in myeloid cell lines the virus seems to utilize both the ssRNA-sensing TLR7/8 and dsRNA-dependent mechanisms (93). Using mice with targeted deletions of TLR adaptors, including MyD88, MyD88 adaptor-like (Mal), TRIF, and TRIF-related adaptor molecule (TRAM), it has been shown that the recognition of SeV in Flt-3L-derived DCs (consisting of 40 to 50% pDCs and 50 to 60% myeloid DCs) occurs independently of the TLR system (116). In contrast to these data, elegant studies using MyD88-, TLR7-, TLR9-, or RIG-I-deficient mice have demonstrated that IFN-α and IL-6 production in response to the paramyxovirus NDV is entirely MyD88/TLR7 dependent in mouse pDCs (73). On the other hand, in conventional DCs (CD11c$^+$ B220$^-$) and fibroblasts, response to NDV was largely RIG-I dependent and TLR independent. The reason for the discrepancy between the latter studies is unclear, but type and purity of cells utilized and type of virus used (SeV or NDV) may explain in part the different findings.

Even when highly purified primary cells such as pDCs have been utilized, the results have been contrasting. For example, human pDCs have been reported to produce IFN-α in response to the RSV Long strain by a mechanism independent of the endosomal TLR pathway (60), while the RSV A2 strain clearly signals and interferes with the TLR7/9 pathway (125; R. P. Garofalo, personal observation). Experimental evidence suggests that both replication-competent RSV (A2 strain) and hMPV (A strain) recognize endosomal TLRs in human pDCs, leading to IFN-α production, while murine pDCs produce IFN-α only in response to hMPV (R. P. Garofalo, personal communication). In addition, infection of human DCs with RSV (A2 strain) or hMPV (A strain) effectively blocks production of type I IFNs in response to TLR agonists (47). In this regard, in human monocyte-derived DCs production of IFN-α in response to the TLR3 agonist poly(I:C) has been shown to be severely impaired by RSV or hMPV infection. Similarly, in human pDCs, production of IFN-α by the TLR9 agonist CpG-ODN is significantly blocked by RSV or hMPV infection. This inhibitory effect occurs also following stimulation of infected pDCs via the TLR7 pathway, involves production of IFN-α but not other cytokines such as IL-8, and is not restricted to members of the *Pneumovirinae* family, since MeV displays similar inhibitory activity on TLR-mediated signaling (125). The relevance of these in vitro observations is also supported by observations in vivo experimental models. Mice that were infected with RSV or hMPV showed a dramatic inhibition of IFN-α production in the lung following intranasal inoculation with TLR3 or TLR9 agonists compared with uninfected controls (46).

The Jak/Stat Pathway

Paramyxovirus replication in epithelial cells is a potent inducer of production of type I IFN (41, 67, 134), an antiviral cytokine that plays a central role in mucosal immunity (105). IFN production is thought to be mediated by two phases of IRF activation. The first phase involves the detection of virus by cytoplasmic RIG-I/MAVS, resulting in activation of IRF3 and initial production of IFN-β/IFN-α4, which induces IRF7 transcription. The second phase includes the activation of IRF7 and the induction of other type I IFNs. The IRF-controlled second phase of IFN secretion ensures a maximum antiviral response from host cells (130). SeV and RSV both induce IRF1 and 7 induction in response to viral replication (22, 143). Although paramyxovirus infections induce a number of type I IFNs (134), bioassays indicate that IFN-β has the primary antiviral biological activity (67, 89).

IFN-β induces gene expression programs to produce an antiviral state through several mechanisms. One involves enhancing peptide production from intracellular pathogens by inducing 26S proteasome catalytic activity by expression of large multifunctional proteasome subunit 2 (LMP2), and inducing cytosolic-to-endoplasmic reticulum transport by expression of TAP1 and TAP2. Together, coordinate expression of the LMP2/TAP genetic element results in the extracellular display of pathogen-derived peptides within the context of the major histocompatibility complex class I. As a result, cytotoxic CD8-expressing T lymphocytes can then recognize and clear the infected cells. The second mechanism is to induce expression of various ISGs, including MxA, oligoadenylate synthetase, protein kinase R, speckled protein-100, and others that produce an antiviral state by inhibition of viral translation and replication through largely unknown mechanisms (29).

The role and elements of IFN-induced signaling pathways are being intensively investigated (reviewed in references 28, 124, and 138). Type I IFNs signal cells by activating IFN-α/β receptor (IFNAR)-associated tyrosine kinases, known as Janus kinase 1 (Jak1) and Tyk2, followed by recruitment of the cytoplasmic Stat 1 and 2 isoforms and their phosphorylation on critical tyrosine residues 701 and 689, respectively. This modification produces intermolecular SH2-SH3 domain interactions, which results in Stat homo- and heterotypic association, forming distinct types of complexes. Newly created Stat1 homodimers, Stat1-Stat 2 heterodimers, and a complex of Stat1-Stat 2 heterodimers and IRF9, termed ISG factor-3 (ISGF3), are transported into the nucleus, where they bind high-affinity sequences in target genes, recruiting p300/CBP coactivators, producing chromatin and factor-induced acetylation, and inducing gene expression (61). Subsequently, Stats are dephosphorylated and exported from the nucleus (124).

Although the classic view of Stat activation involves kinase activation, the mechanisms by which RSV induces Stat activation appear more complex. RSV replication is a rapid inducer of ROS, an event associated with inactivation of cellular phosphatases and accumulation of tyrosine-phosphorylated Stat, without detectable changes in Jak/Tyk kinase activity (90). NADPH oxidase inhibitors successfully blocked the production of ROS, inhibited the phosphorylation of Stat, and consequently downregulated IRF1 and 7 expression (90). Together these findings indicate that RSV activates the Stat pathway in the IRF activation loop via intracellular ROS.

IFN EVASION AND MODULATION OF CELLULAR SIGNALING BY PARAMYXOVIRUS PROTEINS

IFN represents a major line of defense against virus infection, and in response, viruses have evolved countermeasures to inhibit IFN responses. Paramyxoviruses have been shown to directly suppress IFN production as well as to interfere with IFN-dependent cellular signaling. Although the strategies of IFN evasion are similar, the specific mechanisms by which paramyxovirus proteins inhibit IFN responses are quite diverse and are discussed in detail below.

V Protein

Many members of the *Paramyxovirinae* subfamily, including *Rubulavirus*, *Morbillivirus*, *Respirovirus*, and *Henipavirus*, encode a V protein, which is transcribed from a polycistronic gene also encoding the phosphoprotein P and in some species additional small proteins called C, W/I, X, or Y. V proteins have been shown to prevent both IFN synthesis and IFN-dependent signaling (reviewed in reference 63).

The V proteins of PIV5, hPIV2, and SeV have been shown to limit IFN-β induction and activation of both IRF3 and NF-κB by intracellular synthetic dsRNA (109), and this property was also shared by the SeV C protein (77). Recent studies have identified a possible mechanism for this inhibition of IFN-β production. MDA-5 was identified as an interaction partner for the PIV5 V protein as well as for the V proteins of hPIV2, SeV, MuV, and HeV (5, 23). MDA-5 interaction with the cysteine-rich C terminus of PIV5 V was dependent on

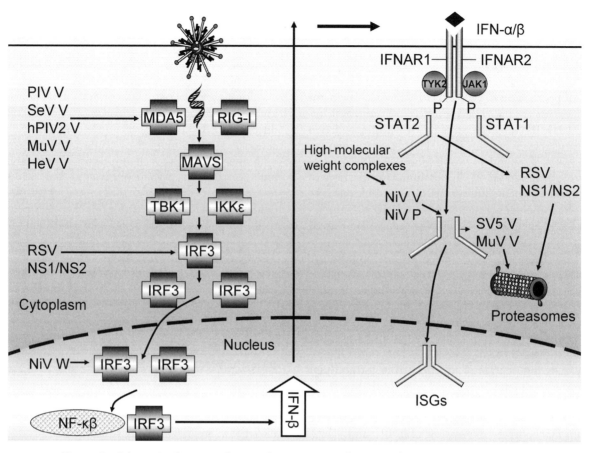

Figure 2 Schematic diagram of sites of antagonism of IFN production and signaling by paramyxovirus proteins.

the helicase domain and blocked the ability of MDA-5 to stimulate IRF3 and NF-κB activation both constitutively and in response to dsRNA (5). Surprisingly, there was no effect on RIG-I-induced activation of IFN-β gene transcription. Experiments with knockout mice have shown that MDA-5 is essential for IFN production in response to picornaviruses (74) but may play a less important role in responses to other virus types, in particular paramyxoviruses, where RIG-I is critical (74, 92); therefore, it is not clear whether interaction of V proteins with MDA-5 does play a meaningful role in limiting IFN production in infected cells.

In the henipaviruses, induction of IFN-β is inhibited not only by the V protein but also by the accessory W protein (131, 132). Whether NiV V protein has been shown to affect viral-induced IRF3 activation and subsequent type I IFN production, NiV W protein, which contains a unique nuclear localization signal in the C-terminal domain, was able to inhibit activation of IRF-responsive promoters in response to both intracellular RNA signaling and signaling through TLR3 (131).

IFNs display potent antiviral and immunomodulatory activities through interaction with their cognate cell surface receptors. They initiate a signaling cascade ultimately leading to the activation of the Stat family of transcription factors and to the subsequent expression of many genes with antiviral function. Many members of the *Paramyxovirinae* subfamily use one or more products of their P/V/C gene to antagonize IFN signaling by various mechanisms, including the targeted degradation of Stat proteins by the V proteins of rubulaviruses such as PIV5, hPIV2, and MuV; disruption of Stat phosphorylation by SeV C proteins; inhibition of Stat nuclear translocation by MeV V protein; and sequestration of Stats in high-molecular-weight complexes observed for the V, W, and P proteins of HeV and NiV (reviewed in references 62 and 63), as summarized in Fig. 2 (in which the RSV NS1/NS2 proteins are also represented as Stat antagonists).

Paramyxoviruses of the *Rubulavirus* genus effectively overcome IFN antiviral responses by promoting the degradation of Stat proteins. Specifically, PIV5 V protein causes the degradation of Stat1, whereas hPIV2 V protein targets Stat2 for degradation, and the V protein encoded by MuV eliminates both Stat1 and Stat3 (reviewed in references 62 and 63). PIV5 and hPIV2 V proteins have been shown to catalyze the transfer of ubiquitin in vitro when the ubiquitin-activating enzymes E1 and E2 are present, meeting the requirement for being defined as ubiquitin ligase enzymes (E3) (147). Expression of simian virus 5 (SV5), hPIV2, or MuV V protein in human cells by cDNA transfection results in Stat polyubiquitylation, which results in efficient proteasomal degradation (147, 148). SV5 and hPIV2 V proteins need to form multisubunit complexes with the nontarget Stat for their full E3 ubiquitin ligase activity toward Stat1 or Stat2, meaning that PIV5 can target Stat1 only in cells that express Stat2, while hPIV2-mediated Stat2 degradation requires the presence of Stat1. A number of additional cellular proteins have been identified in these complexes, including UV-damaged DNA-binding protein (DDB1) (6, 86), and several members of the cullin family of ubiquitin ligase subunits, prominently including cullin 4A (147, 148). Like other *Rubulavirus* species, MuV V protein induces Stat1 proteasomal degradation, but its expression also leads to degradation of cellular Stat3 instead of Stat2 (148). Stat1 targeting required the participation of Stat2, while the Stat3 targeting activity was Stat2 independent (148). Affinity purification of MuV V-interacting proteins revealed a pattern of associated proteins that resembles the SV5 and hPIV2 VDCs, with common components including Stat1, Stat2, DDB1, and cullin 4A (148).

Henipavirus species share V-dependent IFN signaling evasion properties with other paramyxoviruses (104), but unlike *Rubulavirus* species, their V proteins do not induce Stat degradation but rather sequester Stat1 and Stat2 in high-molecular-weight cytoplasmic complexes (114, 115). In addition to the ability to bind to both Stat1 and Stat2, the *Henipavirus* V proteins can shuttle between the nuclear and cytoplasmic compartments in a chromosome region maintenance 1 (CRM1) nuclear export-dependent fashion (114). This nuclear-cytoplasmic shuttling affects the subcellular distribution of the V protein, but it also alters the distribution of latent Stat1. Stat1 shuttling between the nucleus and cytoplasm results in a steady-state accumulation in both compartments (114). However, in cells expressing *Henipavirus* V protein, Stat1 was completely relocalized to the cytoplasm (113, 115), suggesting that the V protein enters the nucleus, binds to Stat1, and brings it back to the cytoplasm. All the three activities of nuclear export, Stat protein interaction, and IFN signaling inhibition of NiV V protein are associated with the N-terminal portion of the protein (113), while localization

of HeV V protein binding activity and anti-IFN signaling activity in general is not known yet.

In addition to the V protein, NiV W and C proteins have also been shown to have IFN-antagonist activity (104). While the target of C protein remains to be determined, W specifically inhibits the IFN signaling pathway. This is the first W protein of a paramyxovirus to display such a function. The observation that both V and W were IFN antagonists led to the identification of the common N-terminal domain of NiV V and W proteins as the one having anti-IFN activity (104), in contrast to the V proteins of hPIV2, PIV5, and MuV, for which it has been ascribed to the cysteine-rich domain present in the C-terminal domain. As said before, NiV W protein has a unique nuclear localization signal; and unlike the V protein, which inhibits Stat1 only in the cytoplasm, W acts also from the nuclear compartment (132).

MeV V protein expression effectively prevents both IFN-α/β- and IFN-γ-induced transcriptional responses. The MeV V protein does not degrade Stats or prevent IFN-induced Stat protein-activating tyrosine phosphorylation, but effectively blocks IFN-induced Stat1 and Stat2 nuclear import (103). Unlike the henipaviruses, MeV V does not shuttle between nucleus and cytoplasm and consequently does not alter the distribution pattern of latent Stat1. Affinity chromatography purification has demonstrated that the MeV V protein interacts with Stat1, Stat2, Stat3, and IRF9, but not the cellular components required for *Rubulavirus* VDC ubiquitin ligase function, in agreement with its distinct mechanism of action (103). In addition, MeV V protein can suppress Jak1 phosphorylation by associating with the type I IFN receptor (158). The MeV nucleocapsid protein, used as a marker for infected cells, was found to accumulate in specific cytoplasmic bodies (103). In measles-infected cells, a portion of the Stat1 and Stat2 proteins were also redistributed to cytoplasmic aggregates that colocalized with the nucleocapsid protein and were shown to contain nucleic acids (103), suggesting that this specific cellular compartment is important for viral replication and antagonism of host cellular responses.

NS1 and NS2

A unique characteristic that distinguishes RSV from the rest of the *Paramyxoviridae* family is the presence of two nonstructural (NS) proteins, NS1 and NS2. The proteins are so called because they are not packaged into mature virions (64). Both genes are synthesized abundantly during RSV infection, starting from the early phase of infection, as they represent the first two genes of the viral genome. Recombinant RSV lacking NS1 and/or NS2 is attenuated in vivo and in vitro (69, 141, 153). The mechanisms by which the NS proteins modulate viral replication have been recently revealed to be related to their ability to prevent both type I IFN synthesis and IFN-dependent signaling (18, 19, 135).

Spann et al. have recently shown that human RSV lacking NS1 and NS2 induced high levels of IFN-α/β in a human pulmonary epithelial cell line (A549) and in monocyte-derived macrophages, with NS1 having a greater independent role in inhibiting IFN-β secretion (135). This result is consistent with previous findings from human RSV mutants in chimpanzees and in BALB/c mice, where the ΔNS1 virus was significantly more attenuated than the ΔNS2 mutant (141, 153). The authors subsequently investigated the effect of RSV ΔNS1 and 2 on activation of the major transcription factors regulating IFN-β production, NF-κB and IRF3 (135). Wild-type RSV and NS mutants shared a very similar pattern of IRF3 nuclear translocation at early times postinfection; however, IRF3 activation was suppressed in wild-type virus-infected cells at late times postinfection, but not in cells infected with NS-deleted mutants. Once NS1 and NS2 were significantly expressed, they acted cooperatively to suppress activation and nuclear translocation of IRF3. NS1 and NS2 deletion did not enhance NF-κB activation in response to RSV infection; on the contrary, NS2 was found to be necessary for NF-κB activation (135).

Similarly, Bossert et al. provided evidence that bovine RSV has the capability to block IFN production by interfering with IRF3 activation, thereby preventing induction of IFN-β and the establishment of an antiviral state (19). While NF-κB and activator protein-1 (AP-1) activities were comparable in both wild-type virus- and ΔNS1/2-infected cells, phosphorylation and transcriptional activity of IRF3 were significantly enhanced after ΔNS1/2 infection (19).

Human RSV infection or NS2 protein overexpression in epithelial cells has also been shown to affect IFN-dependent signaling by decreasing cellular Stat2 levels (91, 111). Wild-type RSV-infected cells displayed less Stat2 expression and IFN responsiveness than ΔNS1/2-infected cells (91). Similarly, expression of A2 strain-derived NS2 protein in airway epithelial cells was sufficient to decrease Stat2 cellular abundance (111). Human recombinant RSV infection lacking NS2 did not result in decreased Stat2 level or loss of IFN-dependent signaling, indicating that NS2 was the protein modulating IFN-dependent gene expression via regulating cellular Stat2 levels (111).

The mechanism by which NS proteins regulate Stat2 abundance was recently investigated by Elliott et al. and found to be due to enhanced proteasomal degradation, similar to rubulavirus V proteins (34). NS1/2-dependent loss of Stat2 was inhibited by siRNA treatment targeting

E3 ligase components, suggesting that E3 ligase activity is crucial for the ability of RSV to degrade Stat2. NS1, but not NS2, was shown to be able to bind elongin C and cullin 2 in vitro, suggesting that NS1 has the potential to act as an elongin-cullin-SOCS (ECS)-like E3 ubiquitin ligase (34).

The role of NS protein in antagonizing IFN responses was also confirmed in vivo using siRNA nanoparticles targeting RSV NS1 gene (siNS1) expression (161). Administration of siNS1 nanoparticles to BALB/c mice, followed by RSV infection, significantly decreased viral replication in the lungs, as well as airway hyperresponsiveness and pulmonary inflammation, compared with untreated mice, suggesting that siNS1 could be used as a prophylactic and/or therapeutic agent against RSV infection in humans (161).

Recently, hMPV infection has been shown to inhibit IFN-α-mediated induction of the IFN-stimulated response element (ISRE) and upregulation of ISGs. hMPV infection prevented IFN-α-induced phosphorylation and nuclear translocation of Stat1 (31). This inhibitory effect on Stat1 phosphorylation and translocation was abolished by UV-inactivation. Regulation of Stat1 by hMPV was specific, as phosphorylation of Stat2, Tyk2, and Jak1 by IFN-α and the surface expression of the IFN-α receptor were unaltered by hMPV infection (31). As hMPV does not possess NS1 and NS2, future studies will hopefully identify which viral protein(s) is responsible for interfering with IFN signaling.

Attachment Glycoprotein G

G proteins are type II transmembrane glycoproteins with an N-terminal cytoplasmic tail and hydrophobic membrane anchor and a C-terminal domain located extracellularly (26, 149). RSV but not hMPV G is also made as an N-terminal truncated soluble form (56). The ectodomain of RSV and hMPV G protein consists of two mucin-like domains with divergent amino acid sequence between isolates, separated by a short central region highly conserved among antigenic subgroups (26, 149). RSV G protein also contains a cysteine-rich region (GCRR), while hMPV G protein does not (26, 149). Although both RSV and hMPV G proteins seem to be unnecessary for efficient viral replication in vitro (11, 13, 142), they are important for efficient infection in vivo (11, 13, 142).

The conservation of the RSV G protein central region and GCRR was initially thought to be important for its receptor-binding activity, but this has not been proven to be the case (140). On the other hand, RSV G and in particular the GCRR region has been recently shown to modulate cytokine and chemokine production both in vitro and in vivo (7, 108). Infection with a mutant RSV lacking G protein (RSV-ΔG) enhanced production of IL-6 and IL-1β in monocytes (108), as well as IL-8 and RANTES secretion and intercellular adhesion molecule expression in airway epithelial cells (7). RSV-ΔG also caused more lung inflammation, as well as decreased airway function, assessed by whole-body plethysmography, compared with RSV wild-type in a mouse model of infection (108, 126). Furthermore, soluble G protein (sG) was shown to decrease cytokine production in response to TLR stimulation in monocytes, indicating that RSV G inhibits innate immune responses (108). The RSV G GCRR region was found to be responsible for this inhibitory activity, as infection of monocytes with RSV mutants either lacking the whole GCRR region or containing single mutations affecting disulfide bridge formation within the GCRR also induced enhanced cytokine production (108). Both in airway epithelial cells and monocytes, infection with RSV-ΔG, compared with wild-type RSV, led to increased NF-κB activation, which likely mediates RSV-ΔG-induced enhanced proinflammatory cytokine and chemokine production (7, 108).

The role of hMPV G protein in modulation of hMPV-induced cellular responses was also recently investigated. Airway epithelial cells infected with a recombinant hMPV lacking G protein expression (rhMPV-ΔG) produced higher levels of cytokines, chemokines, and type I IFN compared with cells infected with wild-type rhMPV (8b). Similar results were observed in experimentally infected mice, in which rhMPV-ΔG induced increased cytokine, chemokine, and type I IFN secretion compared with wild-type rhMPV (A. Casola, unpublished data). Infection of airway epithelial cells with rhMPV-ΔG enhanced activation of transcription factors belonging to the NF-κB and IRF families, as demonstrated by increased nuclear translocation or phosphorylation of these transcription factors. Compared with wild-type rhMPV, rhMPV-ΔG also increased IRF- and NF-κB-dependent gene transcription, which was reversely inhibited by G protein overexpression.

The inhibitory effect of pneumovirus G proteins on cellular signaling could be a common feature of surface glycoproteins of enveloped, negative-sense ssRNA viruses. The G protein of pneumonia virus of mice, a murine relative of RSV, has also been recently identified as an important virulence factor, as viral replication of a recombinant mutant virus lacking G is severely restricted in a BALB/c mouse model of infection, but not in vitro (79). Similarly, the surface glycoproteins of hantaviruses have been shown to affect IRF3 activation and IFN production via interaction with RIG-I and TBK1, a kinase responsible for viral-induced IRF3 phosphorylation, as well as to inhibit IFN-mediated cellular responses (2, 44).

SH Protein

The small hydrophobic (SH) gene is not common to all members of the *Paramyxoviridae* family, as it has been identified only in RSV, hMPV, PIV5, and MuV genomes (82). SH proteins are either type I or II transmembrane proteins and do not seem to be necessary for viral replication in vitro, as recombinant RSV, hMPV, PIV5, and MuV viruses lacking the SH gene have been shown to grow as well as their wild-type counterparts (13, 21, 53, 136). However, SH proteins might play an important role in pathogenesis, as shown for RSV and PIV5 lacking SH protein expression, which are attenuated in vivo (21, 54). Furthermore, a recombinant PIV5 lacking the SH gene (rPIV5-ΔSH) induces apoptosis in tissue culture cells, compared with wild-type rPIV5 (54). Recent studies using rPIV5-ΔSH have shown that the SH protein inhibits TNF-α-induced apoptosis in L929 cells (88). Neutralizing antibodies against TNF-α and TNF-α receptor 1 blocked rPIV5-ΔSH-induced apoptosis, suggesting that PIV5 SH plays an essential role in blocking the TNF-α-mediated apoptosis pathway (88). SH protein of MuV, as well as SH of RSV, have been recently shown to function similarly to that of PIV5, even though there is no sequence homology among them (39, 155). Lack of PIV5 and MuV SH expression led to enhanced NF-κB activation, and ectopically expressed PIV5 SH and MuV SH blocked activation of NF-κB by TNF-α in a reporter gene assay, suggesting inhibition of NF-κB as a mechanism for the observed enhanced apoptosis in the absence of SH expression (39, 155). The role of hMPV SH protein was also recently investigated. RhMPV-ΔSH infection of airway epithelial cells or mice enhanced secretion of proinflammatory mediators, including IL-6 and IL-8, two NF-κB-dependent genes, compared with wild-type rhMPV (8a). RhMPV-ΔSH infection led to modification of NF-κB-dependent gene transcription by affecting nuclear levels of phosphorylated and acetylated NF-κB, identifying a different mechanism by which hMPV SH protein modulates NF-κB activation and cellular signaling (8a).

References

1. Akira, S., and K. Takeda. 2004. Toll-like receptor signalling. *Nat. Rev. Immunol.* 4:499–511.
2. Alff, P. J., I. N. Gavrilovskaya, E. Gorbunova, K. Endriss, Y. Chong, E. Geimonen, N. Sen, N. C. Reich, and E. R. Mackow. 2006. The pathogenic NY-1 hantavirus G1 cytoplasmic tail inhibits RIG-I- and TBK-1-directed interferon responses. *J. Virol.* 80:9676–9686.
3. Anderson, K., A. M. King, R. A. Lerch, and G. W. Wertz. 1992. Polylactosaminoglycan modification of the respiratory syncytial virus small hydrophobic (SH) protein: a conserved feature among human and bovine respiratory syncytial viruses. *Virology* 191:417–430.

4. Anderson, L. J., J. C. Hierholzer, C. Tsou, R. M. Hendry, B. F. Fernie, Y. Stone, and K. McIntosh. 1985. Antigenic characterization of respiratory syncytial virus strains with monoclonal antibodies. *J. Infect. Dis.* **151:**626–633.

5. Andrejeva, J., K. S. Childs, D. F. Young, T. S. Carlos, N. Stock, S. Goodbourn, and R. E. Randall. 2004. The V proteins of paramyxoviruses bind the IFN-inducible RNA helicase, mda-5, and inhibit its activation of the IFN-beta promoter. *Proc. Natl. Acad. Sci. USA* **101:**17264–17269.

6. Andrejeva, J., E. Poole, D. F. Young, S. Goodbourn, and R. E. Randall. 2002. The p127 subunit (DDB1) of the UV-DNA damage repair binding protein is essential for the targeted degradation of Stat1 by the V protein of the paramyxovirus simian virus 5. *J. Virol.* **76:**11379–11386.

7. Arnold, R., B. Konig, H. Werchau, and W. Konig. 2004. Respiratory syncytial virus deficient in soluble G protein induced an increased proinflammatory response in human lung epithelial cells. *Virology* **330:**384–397.

8. Awomoyi, A. A., P. Rallabhandi, T. I. Pollin, E. Lorenz, M. B. Sztein, M. S. Boukhvalova, V. G. Hemming, J. C. Blanco, and S. N. Vogel. 2007. Association of TLR4 polymorphisms with symptomatic respiratory syncytial virus infection in high-risk infants and young children. *J. Immunol.* **179:**3171–3177.

8a. Bao, X., D. Kolli, T. Liu, Y. shan, R. P. Garofalo, and A. Casola. 2008. Human metapneumovirus small hydrophobic protein inhibits NF-κB transcriptional activity. *J. Virol.* Jun 11. [Epub ahead of print.]

8b. Bao, X., T. Liu, Y. Shan, R. P. Garofalo, and A. Casola. 2008. Human metapneumovirus glycoprotein G inhibits innate immune responses. *PLoS Pathog.* **4:**e1000077.

9. Bao, X., T. Liu, L. Spetch, D. Kolli, R. P. Garofalo, and A. Casola. 2007. Airway epithelial cell response to human metapneumovirus infection. *Virology* **368:**91–101.

10. Berghall, H., J. Siren, D. Sarkar, I. Julkunen, P. B. Fisher, R. Vainionpaa, and S. Matikainen. 2006. The interferon-inducible RNA helicase, mda-5, is involved in measles virus-induced expression of antiviral cytokines. *Microbes Infect.* **8:**2138–2144.

11. Biacchesi, S., Q. N. Pham, M. H. Skiadopoulos, B. R. Murphy, P. L. Collins, and U. J. Buchholz. 2005. Infection of nonhuman primates with recombinant human metapneumovirus lacking the SH, G, or M2–2 protein categorizes each as a nonessential accessory protein and identifies vaccine candidates. *J. Virol.* **79:**12608–12613.

12. Biacchesi, S., M. H. Skiadopoulos, G. Boivin, C. T. Hanson, B. R. Murphy, P. L. Collins, and U. J. Buchholz. 2003. Genetic diversity between human metapneumovirus subgroups. *Virology* **315:**1–9.

13. Biacchesi, S., M. H. Skiadopoulos, L. Yang, E. W. Lamirande, K. C. Tran, B. R. Murphy, P. L. Collins, and U. J. Buchholz. 2004. Recombinant human metapneumovirus lacking the small hydrophobic SH and/or attachment G glycoprotein: deletion of G yields a promising vaccine candidate. *J. Virol.* **78:**12877–12887.

14. Bieback, K., E. Lien, I. M. Klagge, E. Avota, J. Schneider-Schaulies, W. P. Duprex, H. Wagner, C. J. Kirschning, V. ter Meulen, and S. Schneider-Schaulies. 2002. Hemagglutinin protein of wild-type measles virus activates Toll-like receptor 2 signaling. *J. Virol.* **76:**8729–8736.

15. Bonizzi, G., M. Bebien, D. C. Otero, K. E. Johnson-Vroom, Y. Cao, D. Vu, A. G. Jegga, B. J. Aronow, G. Ghosh,

R. C. Rickert, and M. Karin. 2004. Activation of IKKalpha target genes depends on recognition of specific kappaB binding sites by RelB:p52 dimers. *EMBO J.* **23:**4202–4210.

16. Bonizzi, G., and M. Karin. 2004. The two NF-kappaB activation pathways and their role in innate and adaptive immunity. *Trends Immunol.* **25:**280–288.

17. Bose, S., N. Kar, R. Maitra, J. A. DiDonato, and A. K. Banerjee. 2003. Temporal activation of NF-kappaB regulates an interferon-independent innate antiviral response against cytoplasmic RNA viruses. *Proc. Natl. Acad. Sci. USA* **100:**10890–10895.

18. Bossert, B., and K. K. Conzelmann. 2002. Respiratory syncytial virus (RSV) nonstructural (NS) proteins as host range determinants: a chimeric bovine RSV with NS genes from human RSV is attenuated in interferon-competent bovine cells. *J. Virol.* **76:**4287–4293.

19. Bossert, B., S. Marozin, and K. K. Conzelmann. 2003. Nonstructural proteins NS1 and NS2 of bovine respiratory syncytial virus block activation of interferon regulatory factor 3. *J. Virol.* **77:**8661–8668.

20. Bukreyev, A., M. E. Serra, F. R. Laham, G. A. Melendi, S. R. Kleeberger, P. L. Collins, and F. P. Polack. 2006. The cysteine-rich region and secreted form of the attachment G glycoprotein of respiratory syncytial virus enhance the cytotoxic T-lymphocyte response despite lacking major histocompatibility complex class I-restricted epitopes. *J. Virol.* **80:**5854–5861.

21. Bukreyev, A., S. S. Whitehead, B. R. Murphy, and P. L. Collins. 1997. Recombinant respiratory syncytial virus from which the entire SH gene has been deleted grows efficiently in cell culture and exhibits site-specific attenuation in the respiratory tract of the mouse. *J. Virol.* **71:**8973–8982.

22. Casola, A., N. Burger, T. Liu, M. Jamaluddin, A. R. Brasier, and R. P. Garofalo. 2001. Oxidant tone regulates RANTES gene transcription in airway epithelial cells infected with respiratory syncytial virus: role in viral-induced interferon regulatory factor activation. *J. Biol. Chem.* **276:**19715–19722.

23. Childs, K., N. Stock, C. Ross, J. Andrejeva, L. Hilton, M. Skinner, R. Randall, and S. Goodbourn. 2007. mda-5, but not RIG-I, is a common target for paramyxovirus V proteins. *Virology* **359:**190–200.

24. Choudhary, S., S. Boldogh, R. Garofalo, M. Jamaluddin, and A. R. Brasier. 2005. Respiratory syncytial virus influences NF-kappaB-dependent gene expression through a novel pathway involving MAP3K14/NIK expression and nuclear complex formation with NF-kappaB2. *J. Virol.* **79:**8948–8959.

25. Civas, A., P. Genin, P. Morin, R. Lin, and J. Hiscott. 2006. Promoter organization of the interferon-A genes differentially affects virus-induced expression and responsiveness to TBK1 and IKKepsilon. *J. Biol. Chem.* **281:**4856–4866.

25a. Collins, P. L., and J. E. Crowe, Jr. 2007. Respiratory syncytial virus and metapneumovirus, p. 1601–1646. *In* D. M. Knipe and P. M. Howley (ed.)., *Fields Virology*, 3rd ed. Lippincott, Williams & Wilkins, Philadelphia, PA.

26. Collins, P. L., K. McIntosh, and R. M. Chanock. 1996. Respiratory syncytial virus, p. 1313–1351. *In* B. N. Fields, D. M. Knipe, and P. M. Howley (ed.), *Fields Virology*, 5th ed. Lippincott, Williams & Wilkins, Philadelphia, PA.

27. Crowe, J. E., Jr. 2004. Human metapneumovirus as a major cause of human respiratory tract disease. *Pediatr. Infect. Dis. J.* **23:**S215-S221.

28. Darnell, J. E., Jr. 1997. Stats and gene regulation. *Science* **277:**1630–1635.

29. Der, S. D., A. Zhou, B. R. Williams, and R. H. Silverman. 1998. Identification of genes differentially regulated by interferon alpha, beta, or gamma using oligonucleotide arrays. *Proc. Natl. Acad. Sci. USA* **95:**15623–15628.

30. Diebold, S. S., T. Kaisho, H. Hemmi, S. Akira, and C. Reis e Sousa. 2004. Innate antiviral responses by means of TLR7-mediated recognition of single-stranded RNA. *Science* **303:**1529–1531.

31. Dinwiddie, D. L., and K. S. Harrod. 2008. Human metapneumovirus inhibits IFN-α signaling through inhibition of Stat1 phosphorylation. *Am. J. Respir. Cell Mol. Biol.* **38:**661–670.

32. Ehl, S., R. Bischoff, T. Ostler, S. Vallbracht, J. Schulte-Monting, A. Poltorak, and M. Freudenberg. 2004. The role of Toll-like receptor 4 versus interleukin-12 in immunity to respiratory syncytial virus. *Eur. J. Immunol.* **34:**1146–1153.

33. Elango, N., J. Kovamees, T. M. Varsanyi, and E. Norrby. 1989. mRNA sequence and deduced amino acid sequence of the mumps virus small hydrophobic protein gene. *J. Virol.* **63:**1413–1415.

34. Elliott, J., O. T. Lynch, Y. Suessmuth, P. Qian, C. R. Boyd, J. F. Burrows, R. Buick, N. J. Stevenson, O. Touzelet, M. Gadina, U. F. Power, and J. A. Johnston. 2007. Respiratory syncytial virus NS1 protein degrades Stat2 by using the Elongin-Cullin E3 ligase. *J. Virol.* **81:**3428–3436.

35. Faisca, P., D. B. Tran Anh, A. Thomas, and D. Desmecht. 2006. Suppression of pattern-recognition receptor TLR4 sensing does not alter lung responses to pneumovirus infection. *Microbes Infect.* **8:**621–627.

36. Falsey, A. R., D. Erdman, L. J. Anderson, and E. E. Walsh. 2003. Human metapneumovirus infections in young and elderly adults. *J. Infect. Dis.* **187:**785–790.

37. Falsey, A. R., P. A. Hennessey, M. A. Formica, C. Cox, and E. E. Walsh. 2005. Respiratory syncytial virus infection in elderly and high-risk adults. *N. Engl. J. Med.* **352:**1749–1759.

38. Fitzgerald, K. A., S. M. McWhirter, K. L. Faia, D. C. Rowe, E. Latz, D. T. Golenbock, A. J. Coyle, S. M. Liao, and T. Maniatis. 2003. IKKepsilon and TBK1 are essential components of the IRF3 signaling pathway. *Nat. Immunol.* **4:**491–496.

39. Fuentes, S., K. C. Tran, P. Luthra, M. N. Teng, and B. He. 2007. Function of the respiratory syncytial virus small hydrophobic protein. *J. Virol.* **81:**8361–8366.

40. Gagro, A., M. Tominac, V. Krsulovic-Hresic, A. Bace, M. Matic, V. Drazenovic, G. Mlinaric-Galinovic, E. Kosor, K. Gotovac, I. Bolanca, S. Batinica, and S. Rabatic. 2004. Increased Toll-like receptor 4 expression in infants with respiratory syncytial virus bronchiolitis. *Clin. Exp. Immunol.* **135:**267–272.

41. Garofalo, R. P., F. Mei, R. Espejo, G. Ye, H. Haeberle, S. Baron, P. L. Ogra, and V. E. Reyes. 1997. Respiratory syncytial virus infection of human respiratory epithelial cells up-regulates class I MHC expression through the induction of IFN-βand IL-1α. *J. Immunol.* **157:**2506–2513.

42. Garofalo, R. P., M. Sabry, M. Jamaluddin, R. K. Yu, A. Casola, P. L. Ogra, and A. R. Brasier. 1996. Transcriptional activation of the interleukin-8 gene by respiratory syncytial virus infection in alveolar epithelial cells: nuclear translocation of the RelA transcription factor as a mechanism producing airway mucosal inflammation. *J. Virol.* **70:**8773–8781.

43. Garoufalis, E., I. Kwan, R. Lin, A. Mustafa, N. Pepin, A. Roulston, J. Lacoste, and J. Hiscott. 1994. Viral induction of the human beta interferon promoter: modulation of transcription by NF-κB/rel proteins and interferon regulatory factors. *J. Virol.* **68:**4707–4715.

44. Geimonen, E., R. LaMonica, K. Springer, Y. Farooqui, I. N. Gavrilovskaya, and E. R. Mackow. 2003. Hantavirus pulmonary syndrome-associated hantaviruses contain conserved and functional ITAM signaling elements. *J. Virol.* **77:**1638–1643.

45. Groskreutz, D. J., M. M. Monick, L. S. Powers, T. O. Yarovinsky, D. C. Look, and G. W. Hunninghake. 2006. Respiratory syncytial virus induces TLR3 protein and protein kinase R, leading to increased double-stranded RNA responsiveness in airway epithelial cells. *J. Immunol.* **176:**1733–1740.

46. Guerrero-Plata, A., S. Baron, J. S. Poast, P. A. Adegboyega, A. Casola, and R. P. Garofalo. 2005. Activity and regulation of alpha interferon in respiratory syncytial virus and human metapneumovirus experimental infections. *J. Virol.* **79:**10190–10199.

47. Guerrero-Plata, A., A. Casola, G. Suarez, X. Yu, L. Spetch, M. E. Peeples, and R. P. Garofalo. 2006. Differential response of dendritic cells to human metapneumovirus and respiratory syncytial virus. *Am. J. Respir. Cell Mol. Biol.* **34:**320–329.

48. Guo, B., and G. Cheng. 2007. Modulation of the interferon antiviral response by the TBK1/IKKi adaptor protein TANK. *J. Biol. Chem.* **282:**11817–11826.

49. Haeberle, H., R. Takizawa, A. Casola, A. R. Brasier, H. J. Dieterich, N. van Rooijen, Z. Gatalica, and R. P. Garofalo. 2002. Respiratory syncytial virus-induced activation of NF-κB in the lung involves alveolar macrophages and Toll-like receptor 4-dependent pathways. *J. Infect. Dis.* **186:**1199–1206.

50. Haeberle, H. A., A. Casola, Z. Gatalica, S. Petronella, H. J. Dieterich, P. B. Ernst, A. R. Brasier, and R. P. Garofalo. 2004. IkappaB kinase is a critical regulator of chemokine expression and lung inflammation in respiratory syncytial virus infection. *J. Virol.* **78:**2232–2241.

51. Hahm, B., J. H. Cho, and M. B. Oldstone. 2007. Measles virus-dendritic cell interaction via SLAM inhibits innate immunity: selective signaling through TLR4 but not other TLRs mediates suppression of IL-12 synthesis. *Virology* **358:**251–257.

52. Hall, C. B. 2001. Respiratory syncytial virus and parainfluenza virus. *N. Engl. J. Med.* **344:**1917–1928.

53. He, B., G. P. Leser, R. G. Paterson, and R. A. Lamb. 1998. The paramyxovirus SV5 small hydrophobic (SH) protein is not essential for virus growth in tissue culture cells. *Virology* **250:**30–40.

54. He, B., G. Y. Lin, J. E. Durbin, R. K. Durbin, and R. A. Lamb. 2001. The SH integral membrane protein of the paramyxovirus simian virus 5 is required to block apoptosis in MDBK cells. *J. Virol.* **75:**4068–4079.

55. Helin, E., R. Vainionpaa, T. Hyypia, I. Julkunen, and S. Matikainen. 2001. Measles virus activates NF-kappa B and Stat transcription factors and production of IFN-alpha/beta and IL-6 in the human lung epithelial cell line A549. *Virology* **290:**1–10.

56. Hendricks, D. A., K. McIntosh, and J. L. Patterson. 1988. Further characterization of the soluble form of the G glycoprotein of respiratory syncytial virus. *J. Virol.* **62:**2228–2233.

57. Hiebert, S. W., R. G. Paterson, and R. A. Lamb. 1985. Hemagglutinin-neuraminidase protein of the paramyxovirus simian virus 5: nucleotide sequence of the mRNA predicts an N-terminal membrane anchor. *J. Virol.* **54:**1–6.

58. Hiebert, S. W., R. G. Paterson, and R. A. Lamb. 1985. Identification and predicted sequence of a previously unrecognized small hydrophobic protein, SH, of the paramyxovirus simian virus 5. *J. Virol.* **55:**744–751.

59. Hippenstiel, S., B. Opitz, B. Schmeck, and N. Suttorp. 2006. Lung epithelium as a sentinel and effector system in pneumonia—molecular mechanisms of pathogen recognition and signal transduction. *Respir. Res.* **7:**97.

60. Hornung, V., J. Schlender, M. Guenthner-Biller, S. Rothenfusser, S. Endres, K. K. Conzelmann, and G. Hartmann. 2004. Replication-dependent potent IFN-alpha induction in human plasmacytoid dendritic cells by a single-stranded RNA virus. *J. Immunol.* **173:**5935–5943.

61. Horvai, A. E., L. Xu, E. Korzus, G. Brard, D. Kalafus, T. M. Mullen, D. W. Rose, M. G. Rosenfeld, and C. K. Glass. 1997. Nuclear integration of JAK/Stat and Ras/AP-1 signaling by CBP and p300. *Proc. Natl. Acad. Sci. USA* **94:**1074–1079.

62. Horvath, C. M. 2004. Silencing Stats: lessons from paramyxovirus interferon evasion. *Cytokine Growth Factor Rev.* **15:**117–127.

63. Horvath, C. M. 2004. Weapons of Stat destruction. Interferon evasion by paramyxovirus V protein. *Eur. J. Biochem.* **271:**4621–4628.

64. Huang, Y. T., P. L. Collins, and G. W. Wertz. 1985. Characterization of the 10 proteins of human respiratory syncytial virus: identification of a fourth envelope-associated protein. *Virus Res.* **2:**157–173.

65. Jamaluddin, M., A. Casola, R. P. Garofalo, Y. Han, T. Elliott, P. L. Ogra, and A. R. Brasier. 1998. The major component of IκBα proteolysis occurs independently of the proteasome pathway in respiratory syncytial virus-infected pulmonary epithelial cells. *J. Virol.* **72:**4849–4857.

66. Jamaluddin, M., S. Wang, I. Boldogh, B. Tian, and A. R. Brasier. 2007. TNF-α-induced NF-κB/RelA Ser(276) phosphorylation and enhanceosome formation is mediated by an ROS-dependent PKAc pathway. *Cell. Signal.* **19:**1419–1433.

67. Jamaluddin, M., S. Wang, R. P. Garofalo, T. Elliott, A. Casola, S. Baron, and A. R. Brasier. 2001. IFN-beta mediates coordinate expression of antigen-processing genes in RSV-infected pulmonary epithelial cells. *Am. J. Physiol.* **280:**L248-L257.

68. Jewell, N. A., N. Vaghefi, S. E. Mertz, P. Akter, R. S. Peebles, Jr., L. O. Bakaletz, R. K. Durbin, E. Flano, and J. E. Durbin. 2007. Differential type I interferon induction by respiratory syncytial virus and influenza A virus in vivo. *J. Virol.* **81:**9790–9800.

69. Jin, H., H. Zhou, X. Cheng, R. Tang, M. Munoz, and N. Nguyen. 2000. Recombinant respiratory syncytial viruses with deletions in the NS1, NS2, SH, and M2–2 genes are attenuated in vitro and in vivo. *Virology* **273:**210–218.

70. Kahn, J. S. 2006. Epidemiology of human metapneumovirus. *Clin. Microbiol. Rev.* **19:**546–557.

71. Kang, D. C., R. V. Gopalkrishnan, Q. Wu, E. Jankowsky, A. M. Pyle, and P. B. Fisher. 2002. mda-5: an interferon-inducible putative RNA helicase with double-stranded RNA-dependent ATPase activity and melanoma growth-suppressive properties. *Proc. Natl. Acad. Sci. USA* **99:**637–642.

72. Karron, R. A., D. A. Buonagurio, A. F. Georgiu, S. S. Whitehead, J. E. Adamus, M. L. Clements-Mann, D. O. Harris, V. B. Randolph, S. A. Udem, B. R. Murphy, and M. S. Sidhu. 1997. Respiratory syncytial virus (RSV) SH and G proteins are not essential for viral replication in vitro: clinical evaluation and molecular characterization of a cold-passaged, attenuated RSV subgroup B mutant. *Proc. Natl. Acad. Sci. USA* **94:**13961–13966.

73. Kato, H., S. Sato, M. Yoneyama, M. Yamamoto, S. Uematsu, K. Matsui, T. Tsujimura, K. Takeda, T. Fujita, O. Takeuchi, and S. Akira. 2005. Cell type-specific involvement of RIG-I in antiviral response. *Immunity* **23:**19–28.

74. Kato, H., O. Takeuchi, S. Sato, M. Yoneyama, M. Yamamoto, K. Matsui, S. Uematsu, A. Jung, T. Kawai, K. J. Ishii, O. Yamaguchi, K. Otsu, T. Tsujimura, C. S. Koh, C. Reis e Sousa, Y. Matsuura, T. Fujita, and S. Akira. 2006. Differential roles of MDA5 and RIG-I helicases in the recognition of RNA viruses. *Nature* **441:**101–105.

75. Kawai, T., K. Takahashi, S. Sato, C. Coban, H. Kumar, H. Kato, K. J. Ishii, O. Takeuchi, and S. Akira. 2005. IPS-1, an adaptor triggering RIG-I- and Mda5-mediated type I interferon induction. *Nat. Immunol.* **6:**981–988.

76. Kim, T. K., and T. Maniatis. 1997. The mechanism of transcriptional synergy of an in vitro assembled interferon-beta enhanceosome. *Mol. Cell* **1:**119–129.

77. Komatsu, T., K. Takeuchi, J. Yokoo, and B. Gotoh. 2004. C and V proteins of Sendai virus target signaling pathways leading to IRF-3 activation for the negative regulation of interferon-beta production. *Virology* **325:**137–148.

78. Kovacsovics, M., F. Martinon, O. Micheau, J. L. Bodmer, K. Hofmann, and J. Tschopp. 2002. Overexpression of Helicard, a CARD-containing helicase cleaved during apoptosis, accelerates DNA degradation. *Curr. Biol.* **12:**838–843.

79. Krempl, C. D., A. Wnekowicz, E. W. Lamirande, G. Nayebagha, P. L. Collins, and U. J. Buchholz. 2007. Identification of a novel virulence factor in recombinant pneumonia virus of mice. *J. Virol.* **81:**9490–9501.

80. Kumar, H., T. Kawai, H. Kato, S. Sato, K. Takahashi, C. Coban, M. Yamamoto, S. Uematsu, K. J. Ishii, O. Takeuchi, and S. Akira. 2006. Essential role of IPS-1 in innate immune responses against RNA viruses. *J. Exp. Med.* **203:**1795–1803.

81. Kurt-Jones, E. A., L. Popova, L. Kwinn, L. M. Haynes, L. P. Jones, R. A. Tripp, E. E. Walsh, M. W. Freeman, D. T. Golenbock, L. J. Anderson, and R. W. Finberg. 2000. Pattern recognition receptors TLR4 and CD14 mediate response to respiratory syncytial virus. *Nat. Immunol.* **1:**398–401.

82. Lamb, R. A., and G. D. Parks. 2007. Paramyxoviridae: the viruses and their replication. *In* D. M. Knipe and

P. M. Howley (ed.), *Fields Virology*, 5th ed. Lippincott, Williams & Wilkins, Philadelphia, PA.

83. **Lamb, R. A., and D. Kolakofsky.** 2001. Paramyxoviridae: the viruses and their replication, p. 689–724. In D. M. Knipe and P. M. Howley (ed.), *Fundamental Virology*, 4th ed. Lippincott, Williams and Wilkins, Philadelphia, PA.

84. **Leader, S., and K. Kohlhase.** 2002. Respiratory syncytial virus-coded pediatric hospitalizations, 1997 to 1999. *Pediatr. Infect. Dis. J.* **21:**629–632.

85. **Levine, S., R. Klaiber-Franco, and P. R. Paradiso.** 1987. Demonstration that glycoprotein G is the attachment protein of respiratory syncytial virus. *J. Gen. Virol.* **68**(Pt. 9)**:**2521–2524.

85a. **Liao, S.-L., X. Bao, T. Liu, C. Hong, R. P. Garofalo, and A. Casola.** Role of RNA helicases in human metapneumovirus-induced cellular signaling. *J. Gen. Virol.*, in press.

86. **Lin, G. Y., R. G. Paterson, C. D. Richardson, and R. A. Lamb.** 1998. The V protein of the paramyxovirus SV5 interacts with damage-specific DNA binding protein. *Virology* **249:**189–200.

87. **Lin, R., C. Heylbroeck, P. M. Pitha, and J. Hiscott.** 1998. Virus-dependent phosphorylation of the IRF-3 transcription factor regulates nuclear translocation, transactivation potential, and proteasome-mediated degradation. *Mol. Cell. Biol.* **18:**2986–2996.

88. **Lin, Y., A. C. Bright, T. A. Rothermel, and B. He.** 2003. Induction of apoptosis by paramyxovirus simian virus 5 lacking a small hydrophobic gene. *J. Virol.* **77:**3371–3383.

89. **Liu, P., M. Jamaluddin, K. Li, R. P. Garofalo, A. Casola, and A. R. Brasier.** 2007. Retinoic acid-inducible gene I mediates early antiviral response and Toll-like receptor 3 expression in respiratory syncytial virus-infected airway epithelial cells. *J. Virol.* **81:**1401–1411.

90. **Liu, T., S. Castro, A. R. Brasier, M. Jamaluddin, R. P. Garofalo, and A. Casola.** 2004. Reactive oxygen species mediate virus-induced Stat activation: role of tyrosine phosphatases. *J. Biol. Chem.* **279:**2461–2469.

91. **Lo, M. S., R. M. Brazas, and M. J. Holtzman.** 2005. Respiratory syncytial virus nonstructural proteins NS1 and NS2 mediate inhibition of Stat2 expression and alpha/beta interferon responsiveness. *J. Virol.* **79:**9315–9319.

92. **Loo, Y. M., J. Fornek, N. Crochet, G. Bajwa, O. Perwitasari, L. Martinez-Sobrido, S. Akira, M. A. Gill, A. Garcia-Sastre, M. G. Katze, and M. Gale, Jr.** 2008. Distinct RIG-I and MDA5 signaling by RNA viruses in innate immunity. *J. Virol.* **82:**335–345.

93. **Melchjorsen, J., S. B. Jensen, L. Malmgaard, S. B. Rasmussen, F. Weber, A. G. Bowie, S. Matikainen, and S. R. Paludan.** 2005. Activation of innate defense against a paramyxovirus is mediated by RIG-I and TLR7 and TLR8 in a cell-type-specific manner. *J. Virol.* **79:**12944–12951.

94. **Meylan, E., K. Burns, K. Hofmann, V. Blancheteau, F. Martinon, M. Kelliher, and J. Tschopp.** 2004. RIP1 is an essential mediator of Toll-like receptor 3-induced NF-kappa B activation. *Nat. Immunol.* **5:**503–507.

95. **Meylan, E., J. Curran, K. Hofmann, D. Moradpour, M. Binder, R. Bartenschlager, and J. Tschopp.** 2005. Cardif is an adaptor protein in the RIG-I antiviral pathway and is targeted by hepatitis C virus. *Nature* **437:**1167–1172.

96. **Meylan, E., J. Tschopp, and M. Karin.** 2006. Intracellular pattern recognition receptors in the host response. *Nature* **442:**39–44.

97. **Mogensen, T. H., and S. R. Paludan.** 2005. Reading the viral signature by Toll-like receptors and other pattern recognition receptors. *J. Mol. Med.* **83:**180–192.

98. **Monick, M. M., T. O. Yarovinsky, L. S. Powers, N. S. Butler, A. B. Carter, G. Gudmundsson, and G. W. Hunninghake.** 2003. Respiratory syncytial virus up-regulates TLR4 and sensitizes airway epithelial cells to endotoxin. *J. Biol. Chem.* **278:**53035–53044.

99. **Mordmuller, B., D. Krappmann, M. Esen, E. Wegener, and C. Scheidereit.** 2003. Lymphotoxin and lipopolysaccharide induce NF-kappaB-p52 generation by a co-translational mechanism. *EMBO Rep.* **4:**82–87.

100. **Morrison, M. D., W. Reiley, M. Zhang, and S. C. Sun.** 2005. An atypical tumor necrosis factor (TNF) receptor-associated factor-binding motif of B cell-activating factor belonging to the TNF family (BAFF) receptor mediates induction of the noncanonical NF-kappaB signaling pathway. *J. Biol. Chem.* **280:**10018–10024.

101. **O'Neill, L. A., and A. G. Bowie.** 2007. The family of five: TIR-domain-containing adaptors in Toll-like receptor signalling. *Nat. Rev. Immunol.* **7:**353–364.

102. **Opitz, B., M. Vinzing, V. van Laak, B. Schmeck, G. Heine, S. Gunther, R. Preissner, H. Slevogt, P. D. N'Guessan, J. Eitel, T. Goldmann, A. Flieger, N. Suttorp, and S. Hippenstiel.** 2006. *Legionella pneumophila* induces IFNbeta in lung epithelial cells via IPS-1 and IRF3, which also control bacterial replication. *J. Biol. Chem.* **281:**36173–36179.

103. **Palosaari, H., J. P. Parisien, J. J. Rodriguez, C. M. Ulane, and C. M. Horvath.** 2003. Stat protein interference and suppression of cytokine signal transduction by measles virus V protein. *J. Virol.* **77:**7635–7644.

104. **Park, M. S., M. L. Shaw, J. Munoz-Jordan, J. F. Cros, T. Nakaya, N. Bouvier, P. Palese, A. Garcia-Sastre, and C. F. Basler.** 2003. Newcastle disease virus (NDV)-based assay demonstrates interferon-antagonist activity for the NDV V protein and the Nipah virus V, W, and C proteins. *J. Virol.* **77:**1501–1511.

105. **Pestka, S., C. D. Krause, and M. R. Walter.** 2004. Interferons, interferon-like cytokines, and their receptors. *Immunol. Rev.* **202:**8–32.

106. [Reference deleted.]

107. **Peters, R. T., and T. Maniatis.** 2001. A new family of IKK-related kinases may function as I kappa B kinase kinases. *Biochim. Biophys. Acta* **1471:**M57-M62.

108. **Polack, F. P., P. M. Irusta, S. J. Hoffman, M. P. Schiatti, G. A. Melendi, M. F. Delgado, F. R. Laham, B. Thumar, R. M. Hendry, J. A. Melero, R. A. Karron, P. L. Collins, and S. R. Kleeberger.** 2005. The cysteine-rich region of respiratory syncytial virus attachment protein inhibits innate immunity elicited by the virus and endotoxin. *Proc. Natl. Acad. Sci. USA* **102:**8996–9001.

109. **Poole, E., B. He, R. A. Lamb, R. E. Randall, and S. Goodbourn.** 2002. The V proteins of simian virus 5 and other paramyxoviruses inhibit induction of interferon-beta. *Virology* **303:**33–46.

110. **Principi, N., S. Bosis, and S. Esposito.** 2006. Human metapneumovirus in paediatric patients. *Clin. Microbiol. Infect.* **12:**301–308.

111. **Ramaswamy, M., L. Shi, S. M. Varga, S. Barik, M. A. Behlke, and D. C. Look.** 2006. Respiratory syncytial virus nonstructural protein 2 specifically inhibits type I interferon signal transduction. *Virology* **344:**328–339.

112. **Reimers, K., K. Buchholz, and H. Werchau.** 2005. Respiratory syncytial virus M2–1 protein induces the activation of nuclear factor kappa B. *Virology* **331:**260–268.

113. **Rodriguez, J. J., C. D. Cruz, and C. M. Horvath.** 2004. Identification of the nuclear export signal and Stat-binding domains of the Nipah virus V protein reveals mechanisms underlying interferon evasion. *J. Virol.* **78:**5358–5367.

114. **Rodriguez, J. J., J. P. Parisien, and C. M. Horvath.** 2002. Nipah virus V protein evades alpha and gamma interferons by preventing Stat1 and Stat2 activation and nuclear accumulation. *J. Virol.* **76:**11476–11483.

115. **Rodriguez, J. J., L. F. Wang, and C. M. Horvath.** 2003. Hendra virus V protein inhibits interferon signaling by preventing Stat1 and Stat2 nuclear accumulation. *J. Virol.* **77:**11842–11845.

116. **Rothenfusser, S., N. Goutagny, G. DiPerna, M. Gong, B. G. Monks, A. Schoenemeyer, M. Yamamoto, S. Akira, and K. A. Fitzgerald.** 2005. The RNA helicase Lgp2 inhibits TLR-independent sensing of viral replication by retinoic acid-inducible gene-I. *J. Immunol.* **175:**5260–5268.

117. **Rudd, B. D., E. Burstein, C. S. Duckett, X. Li, and N. W. Lukacs.** 2005. Differential role for TLR3 in respiratory syncytial virus-induced chemokine expression. *J. Virol.* **79:**3350–3357.

118. **Rudd, B. D., J. J. Smit, R. A. Flavell, L. Alexopoulou, M. A. Schaller, A. Gruber, A. A. Berlin, and N. W. Lukacs.** 2006. Deletion of TLR3 alters the pulmonary immune environment and mucus production during respiratory syncytial virus infection. *J. Immunol.* **176:**1937–1942.

119. **Saha, S. K., E. M. Pietras, J. Q. He, J. R. Kang, S. Y. Liu, G. Oganesyan, A. Shahangian, B. Zarnegar, T. L. Shiba, Y. Wang, and G. Cheng.** 2006. Regulation of antiviral responses by a direct and specific interaction between TRAF3 and Cardif. *EMBO J.* **25:**3257–3263.

120. **Sarkar, S. N., K. L. Peters, C. P. Elco, S. Sakamoto, S. Pal, and G. C. Sen.** 2004. Novel roles of TLR3 tyrosine phosphorylation and PI3 kinase in double-stranded RNA signaling. *Nat. Struct. Mol. Biol.* **11:**1060–1067.

121. **Sasai, M., M. Shingai, K. Funami, M. Yoneyama, T. Fujita, M. Matsumoto, and T. Seya.** 2006. NAK-associated protein 1 participates in both the TLR3 and the cytoplasmic pathways in type I IFN induction. *J. Immunol.* **177:**8676–8683.

122. **Sato, M., N. Tanaka, N. Hata, E. Oda, and T. Taniguchi.** 1998. Involvement of the IRF family transcription factor IRF-3 in virus-induced activation of the IFN-beta gene. *FEBS Lett.* **425:**112–116.

123. **Sato, S., M. Sugiyama, M. Yamamoto, Y. Watanabe, T. Kawai, K. Takeda, and S. Akira.** 2003. Toll/IL-1 receptor domain-containing adaptor inducing IFN-beta (TRIF) associates with TNF receptor-associated factor 6 and TANK-binding kinase 1, and activates two distinct transcription factors, NF-kappa B and IFN-regulatory factor-3, in the Toll-like receptor signaling. *J. Immunol.* **171:**4304–4310.

124. **Schindler, C., D. E. Levy, and T. Decker.** 2007. JAK-Stat signaling: from interferons to cytokines. *J. Biol. Chem.* **282:**20059–20063.

125. **Schlender, J., V. Hornung, S. Finke, M. Gunthner-Biller, S. Marozin, K. Brzozka, S. Moghim, S. Endres, G.**

Hartmann, and K. K. Conzelmann. 2005. Inhibition of Toll-like receptor 7- and 9-mediated alpha/beta interferon production in human plasmacytoid dendritic cells by respiratory syncytial virus and measles virus. *J. Virol.* **79:**5507–5515.

126. **Schwarze, J., and U. Schauer.** 2004. Enhanced virulence, airway inflammation and impaired lung function induced by respiratory syncytial virus deficient in secreted G protein. *Thorax* **59:**517–521.

127. **Senftleben, U., Y. Cao, G. Xiao, F. R. Greten, G. Krahn, G. Bonizzi, Y. Chen, Y. Hu, A. Fong, S. C. Sun, and M. Karin.** 2001. Activation by IKKalpha of a second, evolutionary conserved, NF-kappa B signaling pathway. *Science* **293:**1495–1499.

128. **Servant, M. J., B. tenOever, C. LePage, L. Conti, S. Gessani, I. Julkunen, R. Lin, and J. Hiscott.** 2001. Identification of distinct signaling pathways leading to the phosphorylation of interferon regulatory factor 3. *J. Biol. Chem.* **276:**355–363.

129. **Seth, R. B., L. Sun, C. K. Ea, and Z. J. Chen.** 2005. Identification and characterization of MAVS, a mitochondrial antiviral signaling protein that activates NF-kappaB and IRF 3. *Cell* **122:**669–682.

130. **Sharma, S., B. R. tenOever, N. Grandvaux, G. P. Zhou, R. Lin, and J. Hiscott.** 2003. Triggering the interferon antiviral response through an IKK-related pathway. *Science* **300:**1148–1151.

131. **Shaw, M. L., W. B. Cardenas, D. Zamarin, P. Palese, and C. F. Basler.** 2005. Nuclear localization of the Nipah virus W protein allows for inhibition of both virus- and Toll-like receptor 3-triggered signaling pathways. *J. Virol.* **79:**6078–6088.

132. **Shaw, M. L., A. Garcia-Sastre, P. Palese, and C. F. Basler.** 2004. Nipah virus V and W proteins have a common Stat1-binding domain yet inhibit Stat1 activation from the cytoplasmic and nuclear compartments, respectively. *J. Virol.* **78:**5633–5641.

133. **Skiadopoulos, M. H., S. Biacchesi, U. J. Buchholz, J. M. Riggs, S. R. Surman, E. Amaro-Carambot, J. M. McAuliffe, W. R. Elkins, M. St. Claire, P. L. Collins, and B. R. Murphy.** 2004. The two major human metapneumovirus genetic lineages are highly related antigenically, and the fusion (F) protein is a major contributor to this antigenic relatedness. *J. Virol.* **78:**6927–6937.

134. **Spann, K. M., K. C. Tran, B. Chi, R. L. Rabin, and P. L. Collins.** 2004. Suppression of the induction of alpha, beta, and lambda interferons by the NS1 and NS2 proteins of human respiratory syncytial virus in human epithelial cells and macrophages [corrected]. *J. Virol.* **78:**4363–4369.

135. **Spann, K. M., K. C. Tran, and P. L. Collins.** 2005. Effects of nonstructural proteins NS1 and NS2 of human respiratory syncytial virus on interferon regulatory factor 3, NF-kappaB, and proinflammatory cytokines. *J. Virol.* **79:**5353–5362.

136. **Takeuchi, K., K. Tanabayashi, M. Hishiyama, and A. Yamada.** 1996. The mumps virus SH protein is a membrane protein and not essential for virus growth. *Virology* **225:**156–162.

137. **Tal, G., A. Mandelberg, I. Dalal, K. Cesar, E. Somekh, A. Tal, A. Oron, S. Itskovich, A. Ballin, S. Houri, A. Beigelman, O. Lider, G. Rechavi, and N. Amariglio.**

2004. Association between common Toll-like receptor 4 mutations and severe respiratory syncytial virus disease. *J. Infect. Dis.* **189:**2057–2063.

138. Taniguchi, T., and A. Takaoka. 2002. The interferon-alpha/beta system in antiviral responses: a multimodal machinery of gene regulation by the IRF family of transcription factors. *Curr. Opin. Immunol.* **14:**111–116.

139. Teng, M. N., and P. L. Collins. 1999. Altered growth characteristics of recombinant respiratory syncytial viruses which do not produce NS2 protein. *J. Virol.* **73:**466–473.

140. Teng, M. N., and P. L. Collins. 2002. The central conserved cystine noose of the attachment G protein of human respiratory syncytial virus is not required for efficient viral infection in vitro or in vivo. *J. Virol.* **76:**6164–6171.

141. Teng, M. N., S. S. Whitehead, A. Bermingham, M. St. Claire, W. R. Elkins, B. R. Murphy, and P. L. Collins. 2000. Recombinant respiratory syncytial virus that does not express the NS1 or M2-2 protein is highly attenuated and immunogenic in chimpanzees. *J. Virol.* **74:**9317–9321.

142. Teng, M. N., S. S. Whitehead, and P. L. Collins. 2001. Contribution of the respiratory syncytial virus G glycoprotein and its secreted and membrane-bound forms to virus replication in vitro and in vivo. *Virology* **289:**283–296.

143. tenOever, B. R., M. J. Servant, N. Grandvaux, R. Lin, and J. Hiscott. 2002. Recognition of the measles virus nucleocapsid as a mechanism of IRF-3 activation. *J. Virol.* **76:**3659–3669.

144. Thompson, A. J., and S. A. Locarnini. 2007. Toll-like receptors, RIG-I-like RNA helicases and the antiviral innate immune response. *Immunol. Cell Biol.* **85:**435–445.

145. Tian, B. Y. Zhang, B. Luxon, R. P. Garofalo, A. Casola, M. Sinha, and A. R. Brasier. 2002. Identification of NF-κB dependent gene networks in respiratory syncytial virus-infected cells. *J. Virol.* **76:**6800–6814.

146. Tripp, R. A., L. P. Jones, L. M. Haynes, H. Zheng, P. M. Murphy, and L. J. Anderson. 2001. CX3C chemokine mimicry by respiratory syncytial virus G glycoprotein. *Nat. Immunol.* **2:**732–738.

147. Ulane, C. M., and C. M. Horvath. 2002. Paramyxoviruses SV5 and HPIV2 assemble Stat protein ubiquitin ligase complexes from cellular components. *Virology* **304:**160–166.

148. Ulane, C. M., J. J. Rodriguez, J. P. Parisien, and C. M. Horvath. 2003. Stat3 ubiquitylation and degradation by mumps virus suppress cytokine and oncogene signaling. *J. Virol.* **77:**6385–6393.

149. van den Hoogen, B. G., T. M. Bestebroer, A. D. Osterhaus, and R. A. Fouchier. 2002. Analysis of the genomic sequence of a human metapneumovirus. *Virology* **295:**119–132.

150. van den Hoogen, B. G., J. C. de Jong, J. Groen, T. Kuiken, R. de Groot, R. A. Fouchier, and A. D. Osterhaus. 2001. A newly discovered human pneumovirus isolated from young children with respiratory tract disease. *Nat. Med.* **7:**719–724.

151. van der Sluijs, K. F., L. van Elden, M. Nijhuis, R. Schuurman, S. Florquin, H. M. Jansen, R. Lutter, and T. van der Poll. 2003. Toll-like receptor 4 is not involved in host defense against respiratory tract infection with Sendai virus. *Immunol. Lett.* **89:**201–206.

152. Wacharapluesadee, S., B. Lumlertdacha, K. Boongird, S. Wanghongsa, L. Chanhome, P. Rollin, P. Stockton, C. E. Rupprecht, T. G. Ksiazek, and T. Hemachudha. 2005. Bat Nipah virus, Thailand. *Emerg. Infect. Dis.* **11:**1949–1951.

153. Whitehead, S. S., A. Bukreyev, M. N. Teng, C. Y. Firestone, M. St.Claire, W. R. Elkins, P. L. Collins, and B. R. Murphy. 1999. Recombinant respiratory syncytial virus bearing a deletion of either the NS2 or SH gene is attenuated in chimpanzees. *J. Virol.* **73:**3438–3442.

154. Williams, J. V., P. A. Harris, S. J. Tollefson, L. L. Halburnt-Rush, J. M. Pingsterhaus, K. M. Edwards, P. F. Wright, and J. E. Crowe, Jr. 2004. Human metapneumovirus and lower respiratory tract disease in otherwise healthy infants and children. *N. Engl. J. Med.* **350:**443–450.

155. Wilson, R. L., S. M. Fuentes, P. Wang, E. C. Taddeo, A. Klatt, A. J. Henderson, and B. He. 2006. Function of small hydrophobic proteins of paramyxovirus. *J. Virol.* **80:**1700–1709.

156. Xu, L. G., Y. Y. Wang, K. J. Han, L. Y. Li, Z. Zhai, and H. B. Shu. 2005. VISA is an adapter protein required for virus-triggered IFN-beta signaling. *Mol. Cell* **19:**727–740.

157. Yokota, S., T. Okabayashi, N. Yokosawa, and N. Fujii. 2008. Measles virus P protein suppresses Toll-like receptor signal through up-regulation of ubiquitin-modifying enzyme A20. *FASEB J.* **22:**74–83.

158. Yokota, S., H. Saito, T. Kubota, N. Yokosawa, K. Amano, and N. Fujii. 2003. Measles virus suppresses interferon-alpha signaling pathway: suppression of Jak1 phosphorylation and association of viral accessory proteins, C and V, with interferon-alpha receptor complex. *Virology* **306:**135–146.

159. Yoneyama, M., M. Kikuchi, T. Natsukawa, N. Shinobu, T. Imaizumi, M. Miyagishi, K. Taira, S. Akira, and T. Fujita. 2004. The RNA helicase RIG-I has an essential function in double-stranded RNA-induced innate antiviral responses. *Nat. Immunol.* **5:**730–737.

160. Yoneyama, M., W. Suhara, Y. Fukuhara, M. Fukuda, E. Nishida, and T. Fujita. 1998. Direct triggering of the type I interferon system by virus infection: activation of a transcription factor complex containing IRF-3 and CBP/p300. *EMBO. J.* **17:**1087–1095.

161. Zhang, W., H. Yang, X. Kong, S. Mohapatra, H. S. Juan-Vergara, G. Hellermann, S. Behera, R. Singam, R. F. Lockey, and S. S. Mohapatra. 2005. Inhibition of respiratory syncytial virus infection with intranasal siRNA nanoparticles targeting the viral NS1 gene. *Nat. Med.* **11:**56–62.

162. Zhao, T., L. Yang, Q. Sun, M. Arguello, D. W. Ballard, J. Hiscott, and R. Lin. 2007. The NEMO adaptor bridges the nuclear factor-kappaB and interferon regulatory factor signaling pathways. *Nat. Immunol.* **8:**592–600.

163. Zheng, S., B. P. De, S. Choudhary, S. A. Comhair, T. Goggans, R. Slee, B. R. Williams, J. Pilewski, S. J. Haque, and S. C. Erzurum. 2003. Impaired innate host defense causes susceptibility to respiratory virus infections in cystic fibrosis. *Immunity* **18:**619–630.

164. Zhong, H., R. E. Voll, and S. Ghosh. 1998. Phosphorylation of NF-kappa B p65 by PKA stimulates transcriptional activity by promoting a novel bivalent interaction with the coactivator CBP/p300. *Mol. Cell* **1:**661–671.

Cellular Signaling and Innate Immune
Responses to RNA Virus Infections
Edited by A. R. Brasier et al.
© 2009 ASM Press, Washington, D.C.

Randy A. Albrecht
Adolfo García-Sastre

17

Suppression of Innate Immunity by Orthomyxoviruses

Viruses have undeniably coevolved with their hosts. As hosts have evolved, they have acquired what has come to be known as innate and adaptive immune responses that provide the capability for the host to detect and defend itself against virus infection. Evidence supporting the evolution of the host's immune response is provided by comparison of the immune responses of diverse organisms, including mammalian and avian species and even the fruit fly (*Drosophila melanogaster*). These immune responses are sufficiently conserved to the extent that some observations of the *Drosophila* immune response can be applied to mammalian immune responses. Viruses have evolved to counteract the host's immune responses by acquiring virulence factors that actively antagonize components of the immune response. Although virus families differ in how they antagonize adaptive and innate immune responses, members within each family share similar strategies for antagonizing the host's immune response. For orthomyxoviruses, the nonstructural protein 1 (NS1) is the main virulence factor that most of the viruses in this family utilize to restrict the induction of innate immunity. It should be mentioned that the coevolution of viruses with their hosts has resulted in the development of species specificity in the ability to antagonize immunity for some viruses.

Infection with contemporary strains of human influenza viruses generally causes, in the absence of secondary infections, limited clinical pathology. However, there are a significant number of complicated cases of influenza virus infections that result in severe disease and even lethality in humans. In addition, rare cases of human infections with highly pathogenic avian influenza viruses can result in a severe innate immune response within the respiratory tract that is often referred to as a "cytokine storm" (87, 97, 102); some of these cytokines may contribute more to pathology than to the control of influenza virus replication (33, 156, 168, 183). Of the cytokines produced during viral infection, type I interferon (IFN) cytokines (IFN-β and IFN-α) are essential for directly controlling influenza virus infection and thus limiting morbidity and mortality. This chapter will discuss how influenza virus infection triggers the production of type I IFNs, the signaling cascades that these cytokines stimulate to establish a protective antiviral response, and the mechanisms that orthomyxoviruses employ to circumvent the host's innate immunity.

Randy A. Albrecht and Adolfo García-Sastre, Department of Microbiology, Mount Sinai School of Medicine, New York, NY 10029.

THE ORTHOMYXOVIRUSES

The *Orthomyxoviridae* family comprises viruses possessing segmented, single-stranded (ss) RNA genomes of negative-sense polarity. Whereas the biology of orthomyxoviruses has been extensively described elsewhere (23, 154), their regulation of innate immunity will be the focus of this chapter. Members of the *Orthomyxoviridae* family are divided into five genera: *Influenzavirus A*, *Influenzavirus B*, *Influenzavirus C*, *Thogotovirus*, and *Isavirus*. Influenza A and B viruses are currently circulating in the human population and cause significant morbidity and mortality in humans. A hallmark of influenza A viruses is their ability to infect many different animal species, including different bird species, pigs, horses, dogs, and humans. While specific viral strains appear to be adapted to a specific host, there are frequent jumps of viral strains between different hosts. These jumps between hosts are the basis of influenza A virus pandemics in the human population, which are based on the appearance of a new virus strain with antigens derived from an avian virus for which there is no preexisting immunity in humans. Influenza C viruses typically cause a mild febrile upper respiratory tract illness; however, complications can develop in children. Although human infections are extremely rare, the tick-borne *Thogotovirus* members can cause febrile illness and encephalitis in humans. *Isavirus*, or infectious salmon anemia virus, is a significant economic threat to fishing industries.

Orthomyxoviruses employ several strategies to circumvent the host's innate and adaptive immune responses; the effectiveness by which these viruses apply these strategies contributes to the clinical outcome of virus infections. For influenza A viruses, the viral hemagglutinin and neuraminidase glycoproteins substantially contribute to the pathogenesis of influenza virus infections. Antigenic drift due to polymerase infidelity and antigenic shift resulting from the exchange of genomic segments between viruses of differing subtypes (reassortment) provide a formidable challenge to the host's adaptive immune response. However, the influenza A virus NS1 protein is also a virulence factor whose ability to antagonize the type I IFN system contributes to morbidity and mortality. This chapter will review what has been detailed regarding the mechanisms by which this protein subverts the host innate immune response.

INDUCTION OF INNATE IMMUNITY BY INFLUENZA VIRUS

The host immune response against viral infection involves the concerted actions of innate and adaptive immunity. Innate immunity not only represents the host's initial response in restricting virus replication, but it also influences the generation and specificity of the adaptive immune response. Type I IFN constitutes one of the main components of innate immunity. The induction of signaling pathways involved in innate immunity is described extensively in other chapters of this book; therefore, only the most recent observations will be included in this brief description of the innate immune response as it pertains to replication of orthomyxoviruses.

The activation of the host's innate immunity by virus infection is mediated by two classes of pathogen recognition receptors (PRRs). The first class of PRR includes Toll-like receptors (TLRs), which are predominantly present in myeloid-derived antigen-presenting cells. Of the 13 known human and mouse TLRs, TLR3, TLR7, and TLR8 have been implicated in stimulating IFN-β production in response to influenza virus infection. Endosomal TLR3 recognizes double-stranded (ds) RNA, while endosomal TLR7 and TLR8 recognize uridine-rich sequences in ssRNA. The second class of PRR includes the cytoplasmic melanoma differentiation-associated gene 5 (MDA-5) and retinoic acid-inducible gene I (RIG-I), both of which belong to the DExD/H box family of RNA helicases.

Binding of TLR3 to dsRNA activates a TRIF (TIR domain-containing adaptor protein inducing IFN-β)-dependent signaling cascade that diverges into two arms. One arm of the TRIF-dependent pathway results in the activation of TANK-binding kinase 1 (TBK1) and IKK-ε, which phosphorylate IFN regulatory factor (IRF) 3, while the other arm results in activation of the canonical IκB kinases (IKKs) and nuclear factor-κB (NF-κB)-dependent transcription. Even though influenza virus replication has not been demonstrated to produce detectable levels of dsRNA (159, 202), virus infection of BEAS-2B bronchial epithelial cells expressing dominant negative TRIF revealed the TLR3-dependent activation of NF-κB in response to virus infection (67, 112). However, subsequent studies with TLR3$^{-/-}$ mice demonstrated that TLR3 signaling contributes to the pathological outcome of influenza virus infection (111); specifically, the absence of TLR3-dependent responses increased mouse survival rates and reduced pulmonary levels of several cytokines in a cytokine array blot. Binding of TLR7 to ssRNA activates a MyD88-dependent signaling cascade that involves IL-1 receptor-associated kinase 1 (IRAK-1), IRAK-4, and TNF receptor-associated factor 6 (TRAF6), which ultimately results in the phosphorylation of IRF7. Analysis of kinase-inactive IRAK-4 knockin mice revealed the essential role of the TLR7 signaling pathway in initiating IFN-β production in plasmacytoid dendritic cells (pDCs) (101). This essential role was recently confirmed by the absence of type I IFN production by TLR7$^{-/-}$ or MyD88$^{-/-}$ pDCs in response to influenza virus infection (107).

In vivo data obtained from RIG-I$^{-/-}$ mice revealed that this cytoplasmic PRR is essential for detecting influenza virus replication and initiating innate immune responses through IFN induction (98, 121). The RIG-I protein consists of two amino-terminal caspase activation and recruitment domains (CARDs), an internal ATP-dependent helicase domain, and a carboxy-terminal regulatory domain that holds the protein in an inactive state (29, 184). The ability of the RIG-I protein to initiate the transcriptional induction of type I IFN is regulated by its conformational state (61, 184). During influenza virus replication, new progeny viral RNAs (vRNAs) are exported into the cytoplasm as a viral ribonucleoprotein complex consisting of the polymerase subunits PB1, PB2, and PA and NP. Unlike cellular mRNA transcripts that contain a 7-methyl-guanosine cap structure, the vRNAs within the ribonucleoprotein complex possess 5′ triphosphates (88, 159). The specific binding of the RIG-I regulatory domain to triphosphorylated vRNA induces the prerequisite conformational change that exposes the CARDs, which are subsequently ubiquitinated on Lys-172 by tripartite motif protein 25 (TRIM25) (88), multimerize, and interact with the CARD domain of mitochondrial antiviral signaling adaptor (MAVS), which is also known as IPS-1/Cardif/VISA (99, 138, 176, 204). Once engaged by RIG-I, MAVS sequentially recruits the E3 ubiquitin ligase TRAF3 and NF-κB essential modulator (NEMO) (69, 167). NEMO either stimulates the canonical IκB kinases IKK-α and IKK-β to activate NF-κB (211), or it recruits TANK to stimulate the noncanonical I-κB kinases TBK1 and IKK-ε (18, 69, 149, 177), which then phosphorylate IRF3 and/or IRF7 (53, 84, 157, 177). The RIG-I/MAVS interaction has also been implicated in initiation of TRAF6-dependent signal transduction pathways leading to the activation of NF-κB and ATF-2/c-Jun (96, 176, 204). The result of this signaling pathway is the nucleation of the IFN-β enhanceosome complex consisting of ATF-2/c-Jun, IRF3 or IRF7, and NF-κB on the IFN-β promoter (95, 155, 172, 199, 200, 207).

Newly synthesized IFN-β is immediately secreted from the cell and engages the IFN-α/β receptor (IFN-α/β R) on the same cell or neighboring cell to stimulate autocrine or paracrine signaling, respectively. Binding of IFN-β to the receptor induces dimerization of the IFN-α/βR1 and IFN-α/βR2 subunits. The juxtaposition of these receptor subunits enables the cross-phosphorylation of the receptor-associated kinases Tyk2 and Jak1, which subsequently phosphorylate Stat1 and Stat2. Phosphorylated Stat1 either forms homodimers, designated as gamma-activated factor (GAF), or forms heterodimers with phosphorylated Stat2. The Stat1-Stat2 heterodimers then associate with IRF9 to form the IFN-stimulated gene factor-3 complex (ISGF3). Binding of GAF to gamma-activated sequences (GAS) and/or ISGF3 to IFN-stimulated response elements (ISREs) results in the transcription of over 100 genes (35, 36), including genes encoding transcription factors, signaling factors, sensors of virus replication, and effectors that directly antagonize virus replication.

An important consideration in the induction of IFN by influenza virus relates to the actual nature of the viral inducer recognized by the cellular sensors. Of the two possible RNA helicases that recognize viral RNAs, MDA-5 and RIG-I, the host's innate immune response appears to be completely dependent on RIG-I for IFN production in response to influenza virus infection. This is in contrast to infections with other negative-strand RNA viruses, in which both RIG-I and MDA-5 appear to contribute to IFN induction, although RIG-I appears to have the most prominent role (121). RIG-I is known to recognize both 5′-triphosphate-containing RNAs as well as dsRNA (61, 88, 159). As mentioned, there is no detectable dsRNA in influenza virus-infected cells, as determined by staining with a dsRNA-specific antibody, which points toward viral 5′-triphosphate-containing RNA as the molecule that most likely triggers RIG-I activation during influenza virus infection. However, it is unclear how accessible the 5′ triphosphates of the vRNA are when complexed within the viral RNP. A likely scenario is that most of the vRNA is sequestered within viral complexes, such as RNPs, and in this form is not recognized by RIG-I. However, uncomplexed vRNA or viral dsRNA, generated as by-products of viral RNA replication and transcription and likely existing at levels that are below the limit of detection by most biochemical techniques, are the actual molecules that activate this cellular sensor. This still remains to be determined.

Recent studies using mice deficient in TLR7, MyD88, or MAVS or mice deficient in both MyD88 and MAVS revealed that both the TLR7/MyD88-mediated pathway and the RIG-I/MAVS-mediated pathway contribute to IFN production during influenza virus infection (107). Only when both pathways are knocked out is a failure in the induction of IFN by influenza virus infection observed. Thus, it is likely that pulmonary epithelial cells that are TLR7 negative and are the target cells for virus replication use RIG-I to induce IFN, while pDCs that are TLR7 positive use TLR7 to induce IFN during influenza virus infection. In vitro results obtained from multiple cell types, including pDCs, support this concept (107).

SUBVERSION OF TYPE I IFN INDUCTION BY THE NS1 PROTEIN OF *INFLUENZAVIRUS A*

Given the recent concerns about past and future pandemics, influenza A viruses have been the most extensively studied of the orthomyxoviruses. Pandemics occurred in 1918, 1957, and 1968, with the appearance of the H1N1, H2N2, and H3N2 subtypes, respectively. While the "Spanish" influenza pandemic of 1918 resulted in an estimated 50 million deaths worldwide (93), the 1957 "Asian" influenza and 1968 "Hong Kong" influenza pandemics resulted in approximately 2 million and 1 million influenza-associated estimated deaths worldwide, respectively. More recently, avian influenza A viruses of the H5N1 subtype have captured the attention of both the public and scientific community as a potential candidate for the next pandemic influenza virus. Since their initial detection in 1997, highly pathogenic H5N1 influenza viruses have spread from Southeast Asia to countries in Eastern Europe. Human infections with H5N1 influenza viruses have resulted in 353 human cases and 221 deaths [World Health Organization, Cumulative Number of Confirmed Human Cases of Avian Influenza A/(H5N1) Reported to WHO, http://www.who.int/csr/disease/avian_influenza/country/cases_table_2008_01_24/en/index.html, accessed 24 January 2008].

The NS1 Protein Is the Master Viral Regulator of the Type I IFN Pathway

While detection of influenza virus infection resulting in induction of type I IFN is mediated in vivo by the viral sensors RIG-I and TLR7, influenza viruses are in general poor inducers of IFN (89). This would suggest that influenza virus encodes mechanisms that prevent a robust induction of type I IFNs. However, a few publications have reported that specific influenza virus strains are more efficient at inducing IFN production than other virus strains (83, 132). Variability in content of defective interfering particles could in part account for the production of IFN by different preparations of influenza A viruses. A series of studies with a genetically engineered influenza A/Puerto Rico/8/1934 H1N1 (PR8) virus that does not encode the NS1 protein (PR8ΔNS1) revealed that this protein is the main IFN antagonist encoded by influenza A viruses. Whereas infection with wild-type PR8 virus produces negligible amounts of IFN and is lethal in mice, the PR8ΔNS1 virus is a potent inducer of IFN and is extremely attenuated in mice (50, 60). Similar findings have also been reported with an NS1-deficient H7N7 avian influenza virus (104), indicating that this phenotype is not subtype specific. The contribution of IFN to restricting replication of PR8ΔNS1 in mice was confirmed by the restoration of virulence in Stat1-deficient mice (60). Additionally, the ability of the NS1

protein to inhibit IFN production is a prerequisite for systemic replication of mouse-adapted influenza viruses (59). Infection of IFN-$\gamma^{-/-}$ mice with influenza virus revealed that type II IFN does not play a major role in the control of viral replication in vivo (10, 15, 144, 161).

The interplay between influenza viruses and the host immune response has been described as an intracellular war. If this analogy is developed further, then the influenza A virus NS1 protein could be viewed as the agent that utilizes multiple approaches to disrupt the line of communication between the PRRs and the establishment of an effective antiviral response.

One of the strategies employed by the NS1 protein of influenza A viruses to prevent the induction of type I IFNs is based on its RNA-binding properties. The NS1 protein contains a noncanonical dsRNA-binding domain located within the first N-terminal 73 amino acids (79, 81, 162). The crystal structure of this domain revealed important structural features required for binding to RNA. The domain forms a dimer, each monomer being composed of three α-helixes separated by short loops, with the third α-helix from each monomer containing basic residues that are predicted to make contact with dsRNA. Both dimerization and these basic residues, especially Arg-38 and Lys-41, were found to be required for its dsRNA-binding activity (24, 25, 119, 195). Importantly, this RNA-binding activity is needed for optimal inhibition of type I IFN induction in virus-infected cells, as a recombinant influenza virus expressing an NS1 mutant in which Arg-38 and Lys-41 are mutated to Ala induces higher levels of IFN than wild-type virus and is attenuated with respect to replication and pathogenicity in mice (39, 50, 60). Repeated passaging of a recombinant influenza virus expressing an RNA-binding-defective NS1 protein in tissue culture yielded a mutant virus possessing an additional S42G substitution in the NS1 protein (39). Although the S42G substitution did not restore RNA-binding activity to the mutant NS1 protein, the S42G substitution restored virus replication to titers close to wild-type levels in vitro and increased virulence in mice (39). Intriguingly, analysis of genetically similar, recombinant H5N1 avian influenza viruses revealed that the presence of either Ser or Pro at amino acid position 42 in the NS1 protein could significantly influence the pathogenicity of the recombinant viruses in mice and IFN induction in vitro (92). These observations suggest that the amino acid sequence adjacent to the RNA-binding domain can influence the ability of the NS1 protein to antagonize IFN production. It is highly possible that binding of NS1 to dsRNA generated during influenza virus infection will shield these molecules from being recognized by the cytoplasmic sensor, RIG-I. Consistent with this, the NS1 protein also prevents the activation during viral infection of two other cellular proteins involved in antiviral responses that bind to and become activated by dsRNA, dsRNA-dependent protein kinase (PKR), and oligoadenylate synthetase (12, 80, 116, 125, 141, 186). However, one limitation to the applicability of this model to the regulation of IFN-β production during influenza virus infections is that influenza virus replication does not appear to generate dsRNA by-products (202). Therefore, the nature of the molecules produced during influenza virus replication that activates the RIG-I-dependent assembly of the IFN-β enhanceosome remains to be determined. Influenza virus genomic and antigenomic RNAs possess 5′ triphosphates, and these structures are known to trigger RIG-I activation (88, 159), but how accessible these molecules are during viral infection is unclear. Interestingly, the NS1 protein has also been reported to bind to the 5′ ends of the viral RNAs (81), which are also able to be recognized by RIG-I due to their terminal 5′ triphosphate. Experiments should be performed to characterize the RNA bound to RIG-I and to NS1 that has been isolated from influenza virus-infected cells to elucidate the main viral RNA species shielded from RIG-I recognition by the NS1 protein.

The second strategy employed by the NS1 protein of influenza A virus to prevent IFN induction involves its interaction with RIG-I (70, 139, 151); this interaction appears to occur independently of an RNA bridge (M. Mibayashi and A. García-Sastre, unpublished observations). Consistently, NS1 expression inhibits the IFN-inducing properties of a constitutively activated, truncated form of RIG-I that lacks the helicase domain and therefore does not bind to RNA. Also in agreement, initial studies on the regulation of IFN-β production by the NS1 protein indicated that this virulence factor inhibits the activation of the RIG-I-stimulated transcription factors NF-κB, IRF3, and ATF-2/c-Jun, components that constitute the IFN-β enhanceosome (126, 185, 198). The precise molecular events involved in the interaction between the NS1 protein and RIG-I and the domains that mediate this interaction have not yet been elucidated. In this regard, the recent crystal structures of the amino-terminal RNA-binding domain and carboxy-terminal effector domain of the NS1 protein are going to be of great help in these studies (14, 24, 119). The carboxy-terminal effector domain likely contributes to RIG-I binding, helps in the dimerization of the NS1 protein, is required for its RNA-binding properties, and contains a domain associated with general inhibition of cellular gene expression, an activity that also contributes to NS1's inhibitory effects on the IFN system.

The third strategy employed by the NS1 protein in inhibiting the induction of type I IFNs is based on regulation of cellular and viral gene expression via interactions with components involved in cellular transcription and the processing, transport, and translation of mRNA. While some of the NS1 protein interactions with components of the cellular translational machinery enhance viral gene expression (3, 16, 34, 45, 133), some of these interactions suppress cellular gene expression and, by extension, antiviral responses. Moreover, the NS1 protein inhibits the 3′ processing of cellular mRNA via its interaction with the cellular factors poly(A)-binding protein II (PABII) and the 30-kDa subunit of cleavage and polyadenylation specificity factor (CPSF-30) (22, 143). These factors are important for proper polyadenylation, processing, and export of cellular mRNAs, but not of viral mRNAs, which are polyadenylated by the viral RNA-dependent RNA polymerase (140). The NS1 protein, by virtue of its interaction with CPSF-30, selectively inhibits polyadenylation of cellular mRNAs, including the IFN mRNA. The interaction with CPSF-30 is mediated by two domains of the NS1 protein, both of which are located in the carboxy-terminal amino acids 73 to 230 of the effector domain. The first domain is a Gly-Leu-Glu-Trp-Asn motif centered on Trp-186 of the NS1 protein (143, 146). Mutation of this sequence motif in the A/Udorn/1972 influenza virus NS1 protein resulted in a mutant virus that induced more IFN than the parental virus, thus demonstrating the contribution of this domain to inhibition of IFN induction by influenza virus (146). The second domain, despite being located in a more central part of the amino acid sequence of the protein, is spatially adjacent to the first domain, as revealed by the crystal structure of the NS1 effector domain (14). In fact, the identity of the amino acids at positions 103 and 106 influences the interaction of NS1 with CPSF-30 (103). Whereas NS1 proteins containing Phe-103 and Met-106 interact with CPSF-30, the NS1 protein from the A/Puerto Rico/8/1934 strain with Ser-103 and Ile-106 fails to interact with CPSF-30 (103). As suggested by Bornholdt and Prasad, the region encompassing Phe-103 and Met-106 may therefore stabilize the interaction of the NS1 Gly-Leu-Glu-Trp-Asn motif with CPSF-30 (14). Interestingly, the PR8 virus strain, as already mentioned, as well as several other influenza virus strains, including some of the highly pathogenic H5N1 viruses, do not interact with CPSF-30, yet they efficiently inhibit IFN-β production. This might be explained by other viral proteins, such as the polymerase proteins, contributing to the NS1 protein binding to CPSF-30 (191), or by the remaining inhibitory properties of the more conserved RNA-binding domain of NS1. In fact, only the RNA-binding domain of the NS1 protein appears to be able to efficiently inhibit the induction of type I IFNs in the context of viral infections if provided with a strong carboxy-terminal dimerization domain (196).

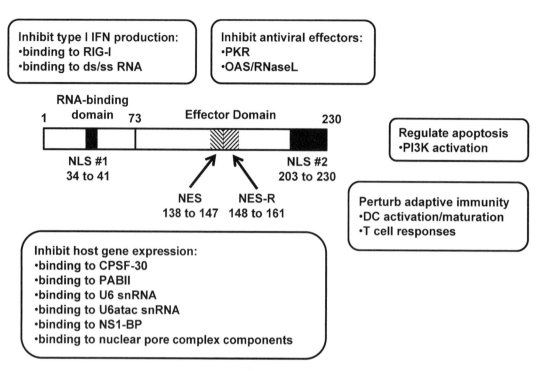

Figure 1   Regulation of innate and adaptive immune responses by the influenza A virus NS1 protein. Mutational analyses of the influenza A virus NS1 protein have identified several domains that control its intracellular localization or are essential for its regulation of general gene expression and antiviral immunity. The diverse cellular processes that are regulated by the NS1 protein and the molecules targeted within each process are indicated. The schematic depicts a hypothetical NS1 protein of 230 amino acids. Amino acid positions are identified for the indicated functional domains.

NS1 also inhibits splicing of pre-mRNA by binding to several cellular components (55, 124), including CPSF-30 (117), U6 snRNA (164), U6atac snRNA (194), a cellular protein of unknown function termed NS1-BP (203), and components of the nuclear pore complex (174). The major consequence of these interactions is that the NS1 protein effectively blocks the nuclear export of cellular and viral mRNA (2, 133, 163). The NS1 protein appears to indiscriminately block mRNA export, as it has been demonstrated that the NS1 protein also diminishes the nuclear export of its own mRNA (57). It is then intriguing that multiple strategies are utilized by the influenza A virus NS1 protein to inhibit the type I IFN system, which likely gives this protein high flexibility in different cell types and hosts, where one strategy might be less efficient than another. Consistent with these multiple activities (some of them occurring at the cytoplasm, such as RIG-I interaction, and some of them in the nucleus, such as CPSF-30 interaction), the cellular localization of the NS1 protein is regulated in a complex, not well understood manner, as multiple signals have been identified that regulate its import and export

to and from the nucleus (135, 153). As shown in Fig. 1, NS1 proteins, depending on the influenza virus strain, can possess two nuclear localization signals (NLSs) (64, 135): the first NLS was mapped to the amino-terminal amino acids 34 to 41, whereas the second NLS was mapped to the carboxy-terminal amino acids 203 to 230. The second NLS also serves as a nucleolar localization signal (135); however, the implications of the nucleolar targeting by the NS1 protein with respect to its regulation of innate immune responses remain to be determined. The cytoplasmic accumulation of the NS1 protein is mediated by a Leu-rich nuclear export sequence (NES) that has been mapped to amino acids 138 to 147. However, the effectiveness by which the NES mediates the nuclear export of the NS1 protein is regulated by an adjacent regulatory sequence, designated as NES-R, that was mapped to amino acids 148 to 161 (118).

NS1 Sequence Diversity and Host Specificity
Avian influenza viruses can possess two different types, or alleles, of the NS segment (5). In general, the type A NS segment is present in influenza viruses isolated from

mammalian and avian species, while the type B NS segment is present in influenza viruses isolated from avian species. The NS1 proteins encoded by the type A and B NS segments exhibit approximately 70% amino acid identity (5, 188). Despite the low conservation of identity, there seems to be no difference in the ability of the two types of NS1 proteins encoded by avian influenza viruses to antagonize the human type I IFN system (82). However, a comparison of the amino acid sequences of NS1 proteins encoded by influenza viruses isolated from mammalian and avian species has indicated that the identity of the amino acid at position 227 may confer species specificity to influenza virus infections (21). Of 704 avian influenza NS1 proteins examined, >98% contained Glu/Gly-227, while of 904 human influenza virus NS1 protein segments examined, >99% contained Arg/Lys-227. Amino acid 227 is located within a PDZ ligand motif of Glu-Ser-Glu-Val sequence that is present in NS1 proteins encoded by avian, but not human, influenza viruses (148). Considering that avian influenza virus NS1 proteins containing the consensus PDZ ligand motif efficiently antagonize the human IFN pathway, this domain most likely contributes an additional function to influenza biology. The concept of the NS1 protein conferring species specificity to influenza viruses is best exemplified by results with recombinants of the mouse-adapted influenza virus WSN (9). Intranasal infection of BALB/c mice with recombinant, wild-type WSN resulted in a 100% mortality rate. However, intranasal infection with recombinant WSN viruses expressing the NS1 protein of the human, non-mouse-adapted strain that caused the 1918 pandemic resulted in no lethality, and pulmonary viral titers were reduced by 1.5 log at days 2 and 4 postinfection (9).

NS1 proteins encoded by influenza viruses isolated from mammalian and avian species are typically 230 amino acids in length, but can range from 126 to 237 amino acids in length. The size of the NS1 protein does appear to influence the replication of influenza viruses; the greater the truncation, the less efficiently the virus replicates. For instance, the avian influenza A/turkey/Oregon/1971 virus encodes a truncated NS1 protein of 124 amino acids that efficiently replicated in embryonated hens eggs (147), but was apathogenic in mice, chickens, and ferrets (94). Later on it was found that the NS1 truncation spontaneously arose during in vitro passaging of the virus (147). These findings suggest that domains within the carboxyl terminus of the NS1 protein are required for influenza virus replication in the presence of a fully functional type I IFN response, as discussed above. This concept was further confirmed by reverse genetics experiments, in which carboxy-terminal

truncations of the NS1 proteins of mouse-adapted (48, 50), swine (179, 192), equine (165), and human (7) virus strain isolates resulted in viruses with enhanced IFN-inducing properties and reduced virulence in mice, pigs, horses, and macaques, respectively. Interestingly, these viruses have attenuated and immunogenic properties consistent with their potential use as live-attenuated virus vaccines, as discussed below; they are able to induce not only a neutralizing humoral response, but also strong cellular responses (7, 50).

ANTIVIRAL INHIBITORS OF INFLUENZA VIRUS REPLICATION INDUCED BY TYPE I IFNs

Many genes are known to be induced by type I IFNs. Of all of these, a few have been found to have antiviral activity against influenza viruses. We will review several of the genes encoding factors known to have clear anti-influenza virus activity. However, it is likely that many other understudied IFN-inducible genes also have antiviral activity against the influenza virus.

Role of Mx Homologues in the IFN-Mediated Antiviral Response against Influenza Viruses

IFN-inducible Mx homologues belong to the dynamin-like family of GTPases (86) that are normally associated with intracellular vesicle trafficking and organelle homeostasis. Mx homologues are also potent effectors of the host's antiviral responses to a range of viruses, including orthomyxoviruses (74–77, 106, 171, 181, 187, 189). Human MxA localizes to and establishes a unique membranous compartment at the smooth endoplasmic reticulum (1, 182). Mouse Mx1 localizes in the nucleus, and it is clearly a major factor in controlling influenza virus infection, as strains devoid in Mx1, including most of the commonly utilized laboratory mouse strains, are highly susceptible to influenza virus infection (169, 189). In vitro assays that involved an influenza virus minigenome reporter replicon and influenza infections of $Mx^{+/+}$ or $Mx^{-/-}$ cells revealed that the influenza virus NP is the viral target of Mx (38, 189, 190). In these in vitro studies, the polymerase complex of avian influenza viruses, including the highly pathogenic H5N1 influenza virus A/Vietnam/1203/2004, were more sensitive to murine Mx1 than the polymerase complex from human influenza viruses (38). Interestingly, the in vitro minigenome reporter replicon assays suggested that the amount of Mx-expression plasmid required to inhibit the 1918 influenza virus polymerase activity by 50% was fivefold more than was required to inhibit the A/Vietnam/1203/2004 polymerase activity by 50% (38). These findings are consistent with previous findings that intranasal infection of Mx$^{-/-}$ BALB/c mice with A/Vietnam/1203/2004 consistently results in 100% lethality; however, infection of Mx$^{+/+}$ BALB/c mice resulted in increased survival (169, 189). In addition, recombinant universal IFN-αB/D afforded 100% protection to transgenic Mx$^{+/+}$ BALB/c mice from lethal infection with 1,000 50% lethal doses of the highly pathogenic H5N1 influenza Vietnam/1203/2004 virus (189). Consistent with the infections with A/Vietnam/1203/2004, pretreatment with universal IFN-αB/D conferred 100% protection to transgenic Mx$^{+/+}$ BALB/c mice against intranasal infection with the complete 1918 influenza virus, a virus strain that is also highly pathogenic in mice (189). These results reveal that IFN produced during pulmonary infection with an influenza virus can result in expression of IFN-stimulated genes (ISGs) and the generation of an antiviral state that provides protection from influenza virus infection.

Role of PKR in the IFN-Mediated Antiviral Response against Influenza Viruses

PKR is an IFN-inducible serine/threonine kinase that is activated by dsRNA (27, 58). The antiviral function of PKR involves its phosphorylation of the cellular translation factor eukaryotic initiation factor 2α (eIF2α), which results in the shutoff of translation (37, 123, 136). PKR also appears to contribute to IFN-β transcription by stimulating the kinase activity of IKK-β, leading to the activation of NF-κB (26), a component of the IFN-β enhanceosome. This implies that PKR functions in both arms of the IFN pathway. However, antiviral responses were normal in PKR$^{-/-}$ mice infected with encephalomyocarditis virus (205), and PKR does not appear to be necessary for phosphorylation of either IRF3 or IRF7 or IFN-β transcription in response to infection with Newcastle disease virus (178), suggesting that PKR may not be essential for induction of IFN-β expression under some circumstances. These results also imply that the main function of PKR in innate antiviral immunity is to inhibit both cellular and viral protein synthesis. Infection of PKR$^{-/-}$ mice with a mouse-adapted influenza virus revealed that PKR does contribute to the control of influenza virus infection (6). Importantly, NS1-deficient influenza viruses gain replication in PKR$^{-/-}$ mice, demonstrating that PKR is one of the major effectors of antiviral responses to influenza virus infection (12, 104).

Interestingly, in addition to the already discussed inhibitory actions of the influenza A virus NS1 protein on induction of IFN, this protein is known to directly antagonize the kinase activity of PKR. While part of this inhibitory activity may involve sequestration by the NS1 protein of the activator of PKR, i.e., dsRNA (25, 79, 80,

116, 125), the NS1 protein is known to directly interact with PKR, resulting in PKR inhibition (186). Moreover, influenza A virus infection results in the activation of a cellular inhibitor of PKR, the tetratricopeptide family member p58^{IPK}, by disrupting the association of p58^{IPK} with its inhibitor Hsp40 (110, 137). As would be expected, influenza virus infection of mouse embryonic fibroblasts from p58$^{IPK-/-}$ mice revealed that influenza viruses benefit from stimulating p58^{IPK} regulation of PKR by decreased phosphorylation of eIF2α and increased translation of viral mRNA (63). These results suggest that influenza viruses employ a variety of approaches that contribute to the inhibition of PKR.

Role of OAS in the IFN-Mediated Antiviral Response against Influenza Viruses

A third class of IFN-inducible proteins with broad antiviral activity is the 2',5'-oligoadenylate synthetases (OAS). As is the case with PKR, these proteins are also activated by dsRNA. Once activated, OAS synthesizes 2',5'-oligoadenylates, which are coactivators of a latent cellular RNase, RNase L, which cleaves viral and cellular RNAs, thus preventing viral replication (41, 42, 166, 212). Although the contribution of OAS to inhibiting influenza virus replication is not well understood, once again, based on its ability to sequester dsRNA, the NS1 protein of influenza A virus prevents the activation of OAS enzymes and the induction of RNase L during infection (141). Interestingly, since RNase L is also known to produce RNA species that activate RIG-I and therefore amplify the induction of IFN by viruses (130, 131), one would predict that the OAS-inhibitory properties of NS1 would also contribute to inhibition of IFN production during influenza virus infection. In any case, the multiple activities of the NS1 protein of influenza A virus antagonize the type I IFN system at multiple levels, including IFN induction, IFN-induced gene expression, and the activities of important antiviral mediators induced by IFN, such as PKR and OAS, illustrating the complexity of IFN antagonism exerted by this multifunctional viral protein (Fig. 2).

Role of Other IFN-Induced Genes in the IFN-Mediated Antiviral Response against Influenza Viruses

It is now becoming clear that other IFN-induced genes also have antiviral activity against influenza viruses, and this underscores the need of this virus (as well as of many other viruses) to decrease IFN production during viral infection. Of special mention, viperin is an IFN-inducible protein that inhibits influenza virus budding by disrupting lipid-raft microdomains required for optimal virus release at the plasma membrane (197). Another

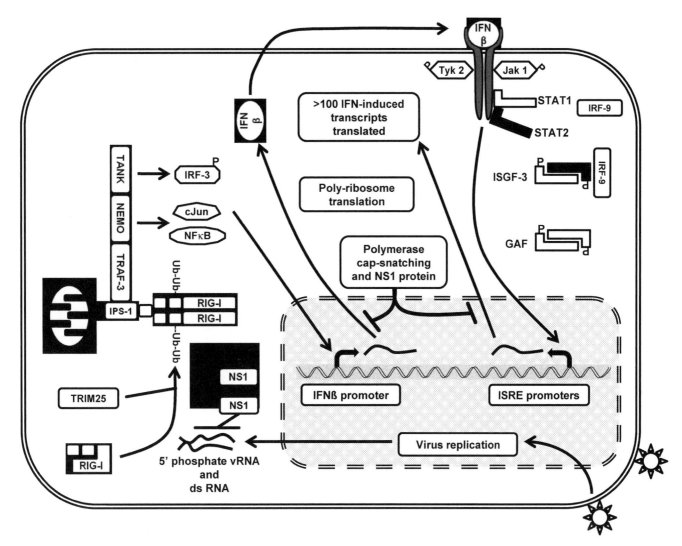

Figure 2   Regulation of IFN-β and IFN-mediated gene induction during influenza A virus infection. The influenza A virus NS1 protein utilizes several strategies to regulate the induction of type I IFNs, including (i) binding to and sequestration of dsRNA and ssRNA from the cellular sensor, RIG-I; (ii) interaction with RIG-I; and (iii) binding to components involved in mRNA processing, export, and translation. As a consequence of the first two strategies, the NS1 protein blocks the activation of the transcription factors IRF3, NF-κB, and ATF-2/c-Jun, thus preventing the formation of the IFN-β enhanceosome. The third strategy employed by the NS1 protein coupled with the cap-snatching function of the viral polymerase complex contributes to the general inhibition of IFN-β and IFN-induced gene expression.

interesting IFN-inducible gene with antiviral activities against many viruses, including influenza virus, is the ISG15 gene (113), which is discussed below in the context of influenza B virus infections. Promyelocytic leukemia (PML) protein (also known as TRIM19) is an additional IFN-inducible protein that accumulates within nuclear structures that have been referred to as nuclear dot 10 (ND10) or nuclear bodies (19, 66, 85, 180). Although PML exerts antiviral activity against the

mouse-adapted influenza virus A/WSN, this antiviral activity can be outcompeted by high multiplicity of infection (20). Influenza virus infections, as demonstrated with laboratory and human influenza viruses, influence the protein composition of PML-containing nuclear bodies (68, 173). PML-containing nuclear bodies also incorporate ISG20, an IFN-inducible 3′-to-5′ ssRNA exonuclease that, in stably expressing HeLa cells, mediated an approximately 2-log reduction in influenza virus titers (46, 62, 145). Therefore, PML-containing nuclear bodies may serve as a repository of IFN-inducible antiviral effectors (47), the activities of which are potentially regulated by one or more influenza virus proteins.

Sensitivity of Influenza A Viruses to the Antiviral Action of IFN

Differences in the sensitivity of different strains of influenza A viruses to the antiviral action of IFN have been described. In particular, Seo et al. have reported that, in cell culture studies, the highly pathogenic H5N1 influenza viruses that first emerged in Asia in 1997 are highly resistant to IFN and tumor necrosis factor-α pretreatment. This increased resistance appears to be conferred by a D92E substitution that is present in NS1 proteins of H5N1 influenza viruses first isolated in 1997 (175). In a different study, it was shown that binding of the effector domain of NS1 to CPSF-30, by virtue of inhibiting general gene expression, results in resistance to the action of IFN (103). Based on the crystal structure of the NS1 effector domain (amino acids 74 to 230), the D92E substitution could introduce a minor conformational change in the effector domain that could restrict the accessibility of two potential phosphorylation sites in the NS1 effector domain, namely Ser-195 and Thr-197 (14, 114). Whether or not these residues are phosphorylated and whether the phosphorylation state of these two residues regulates the stability of the NS1 protein are questions that remain to be addressed. A naturally occurring deletion of codons 79 to 83 of the NS1 proteins encoded by H5N1 influenza viruses isolated after 2002 has also been suggested to confer IFN resistance (175). Bornholdt and Prasad have proposed that based on their crystallographic data (14), the loss of amino acids 80 to 84 could induce a conformational change in the NS1 effector domain such that the reoriented effector domain enhances the affinity of the amino-terminal RNA-binding domain for dsRNA (14). This hypothesis remains to be confirmed in vitro by RNA-protein interaction assays.

ALTERNATIVE MECHANISMS OF GENE REGULATION AND IFN ESCAPE BY INFLUENZA A VIRUSES

It is known that infection with influenza viruses results in inhibition of cellular protein expression. As discussed, this activity can be mediated by NS1 proteins with the ability to interact with CPSF-30. However, an influenza A virus lacking the NS1 gene induces cellular protein synthesis shutoff, indicating that other viral factors contribute to inhibition of cellular gene expression during viral infection, which in turn might also inhibit

induction of the type I IFN response (170). Somewhat surprisingly, PKR is not essential to induce host cellular shutoff in virus-infected cells (213). By contrast, the cap-dependent endonuclease function of the virus polymerase complex could likely contribute, together with the NS1 protein, to antagonism of IFN production (54, 78, 90, 108, 115, 132). The endonuclease activity of the viral polymerase cleaves cellular pre-mRNA 10 to 15 nucleotides downstream of the 7-methyl cap structure (11, 160). The stolen cellular cap then serves as a primer for initiating transcription of the vRNA. The end result is the production of viral mRNA with leaders that contain, depending on the segment that has been transcribed, 20 to 50 nucleotides derived from host mRNA and the elimination of the cap from cellular mRNAs, which are rendered unable to be translated.

In addition, influenza virus infection also perturbs cellular gene expression by targeting eIF4F (49), which is composed of the eIF4E cap-binding protein, eIF4G, and the eIF4A helicase. Specifically, virus infection results in a significant dephosphorylation of the cellular cap-binding protein eIF4E; phosphorylation of Ser-209 increases its affinity for the 5′ cap structures of mRNA. The principal function of eIF4F is to recruit the 40S ribosomal subunit; however, the helicase activity of eIF4F is also necessary for translation of mRNAs with substantial secondary structure in the 5′ untranslated region. The 5′ untranslated regions of influenza virus mRNA, i.e., from the 7-methyl cap to the initiator methionine codon, appear to be devoid of secondary structure, which potentially allows their selective translation in the presence of diminished levels of active eIF4F. Recently, screening of type A influenza viruses in transgenic $Mx^{+/+}$ mice identified a highly virulent isolate of the A/Puerto Rico/8/1934 influenza virus isolate (65). The increased virulence of this isolate in transgenic $Mx^{+/+}$ mice was attributable to increased function of the polymerase complex that resulted in increased viral gene expression and replication in vitro (65). In fact, the fast polymerase activity associated with this virus isolate appeared to outrun the induction of the type I IFN response (65).

SUBVERSION OF INNATE IMMUNITY BY *INFLUENZAVIRUS B*

Influenza B viruses are responsible for seasonal epidemics of mild respiratory infections in humans and also infections of seals (152). In the absence of an avian pool of different hemagglutinin and neuraminidase subtypes, influenza B viruses do not cause pandemics. However, members of this genus can be divided into different lineages; the two lineages that circulated during the 2006–2007 influenza season belonged to the B/Yamagata/16/88 and B/Victoria/02/87 viruses. As with influenza A viruses, influenza B virus replication is sensed by RIG-I (121). In addition, the B/NS1 protein, like its cousin the A/NS1 protein, is essential for the influenza B virus to regulate RIG-I-dependent IFN-β production (31). The A/NS1 and B/NS1 proteins, although divergent in sequences, share some structural similarities (193), which allow these NS1 proteins to share most, but not all, functions. One example of the divergence in functions between these two virulence factors is the observation that the influenza A virus NS1 protein activates the phosphoinositide-3 kinase (PI3K) pathway, while the influenza B virus NS1 protein is lacking in this function (44). However, whether or not this A/NS1 protein-mediated activation contributes to IFN antagonism remains to be addressed. The amino-terminal 93 amino acids of B/NS1, similarly to the amino-terminal 73 amino acids of A/NS1, encompass an RNA-binding domain (193), the crystal structure of which was recently solved (206). The B/NS1 RNA-binding domain recognizes the same RNA species [e.g., poly(A), U6 snRNA, and dsRNA] that are also bound by the A/NS1 protein; however, this RNA-binding activity does not enable the B/NS1 protein to inhibit host-cell splicing (193). Interestingly, the main contribution of dsRNA binding by the B/NS1 protein in IFN antagonism appears to be inhibition of PKR during virus infection (32), while both the amino-terminal dsRNA-binding domain and an unidentified carboxy-terminal domain independently inhibit IRF3 and IFN-β promoter activation (40). It is unclear at this moment whether the carboxyl terminus of the B/NS1 protein stabilizes the amino-terminal RNA-binding domain, as has been modeled for the A/NS1 protein (196).

The B/NS1 protein mediates one function that is not shared by the A/NS1 protein: regulation of ISG15 conjugation to cellular proteins (208, 209). ISG15 is induced by type I IFN treatment, and, similarly to ubiquitin, with which ISG15 shares structural similarities, is conjugated to Lys residues of cellular proteins involved in diverse cellular processes, including innate immunity (i.e., MxA, PKR, and RIG-I) (71, 120, 210). This process, also known as ISGylation, is mediated by specific E1, E2, and E3 ligases, all of which are IFN inducible (30). Interestingly, a de-ISGylating, IFN-inducible cellular enzyme, UBP43, that deconjugates ISG15 has also been identified (17, 127–129). Therefore, the whole cycle of ISGylation/de-ISGylation is mediated by IFN-inducible proteins, suggesting that this process is an important IFN-regulated process. Genetically modified mice lacking ISG15 demonstrated the role of ISG15 in antiviral immunity, as these mice are more susceptible to severe disease caused by infection with influenza virus, herpesvirus, and alphaviruses (113). The mechanism by which ISG15 and/or ISGylation results in viral inhibition is still not clear. Modification by ISG15 conjugation negatively regulates RIG-I activation (100) but increases the stability and consequent steady-state levels of IRF3 (122), suggesting that ISGylation may regulate induction of IFN through negative- and positive-feedback loops. ISG15 may also inhibit ubiquitin-mediated proviral processes, such as viral budding (150). Inhibition of ISGylation by B/NS1 maps to its amino-terminal region, which binds to ISG15 (208). The recent identification of viral proteins encoded by bunyaviruses and arteriviruses that deconjugate ISG15 suggests that ISGylation is subverted by different virus families to evade IFN-mediated antiviral responses (4, 56).

SUBVERSION OF INNATE IMMUNITY BY *INFLUENZAVIRUS C*

To date, there have been no publications describing the ability of influenza C viruses to subvert innate immunity, but it is likely that the C/NS1 protein of this virus is responsible for IFN antagonism. The recent establishment of a plasmid rescue system for influenza C virus (28) is likely to facilitate the rational insertion of mutations in the NS1 gene of this virus in order to study its role in evading innate immune responses.

SUBVERSION OF INNATE IMMUNITY BY *THOGOTOVIRUS* AND *ISAVIRUS*

The *Thogotovirus* genus contains four different species, Thogoto virus, Dhori virus, Batken virus, and Araguari virus. Whereas Thogoto virus, Dhori virus, and Batken virus are tick-borne viruses, the presence of Araguari virus, a recently added member, in ticks has not been formally demonstrated. These viruses do not encode an NS1 homologue and therefore use a different mechanism to antagonize type I IFN-mediated antiviral responses. As observed with the influenza A virus polymerase complex, transcription by the Thogoto virus polymerase complex engages in cap-snatching activity (109). However, unlike influenza A virus 7-methyl-capped mRNAs, which contain 10 to 15 nucleotides of host sequence, the 5′ sequences of capped Thogoto virus mRNAs contain only 1 to 2 nucleotides of host sequence (201). The cap-snatching function of the Thogoto virus polymerase complex could contribute to the inhibition of host gene expression, in particular innate immune responses, as has been postulated for influenza A viruses.

In vitro studies have demonstrated that the segment 6-encoded ML protein of Thogoto virus suppresses IRF3-mediated gene expression and subsequent IFN induction by dsRNA (72, 91). The ML protein is specific for this group of viruses, and it is translated from an unspliced mRNA transcript encoded by the M gene, which also encodes the matrix M protein. The ML protein is identical to the M protein in its amino terminus, but contains a unique carboxy-terminal extension of 38 amino acids (72, 105). The requirement for ML suppression of IFN-β production was subsequently demonstrated in vivo by analysis of a rescued ML-deficient Thogoto virus (73). The inhibition of IFN induction by the ML protein is mediated by its unique carboxyl terminus (72); however, the precise mechanism by which the ML protein inhibits IRF3 activity remains to be determined. Analysis of an ML-deficient Thogoto virus in transgenic Mx$^{+/+}$ mice not only confirmed the importance of the ML protein in blocking IFN-β production, but also revealed that, as observed with influenza A viruses, the Mx protein is one of the main effector proteins of the antiviral state created by type I IFNs (158).

Very little is known about the regulation of innate immune responses by infectious salmon anemia virus, the only species in the *Isavirus* genus. The segment 7-encoded 7i protein has been reported to be an antagonist that putatively blocks poly(I:C)-induced production of IFN-β (134). Although the regulation of type I IFNs by influenza A viruses has been the most extensively studied, it appears that all orthomyxoviruses utilize multiple approaches, some of which are conserved among all members of this family, to control induction of type I IFNs.

CONCLUSIONS AND POTENTIAL APPLICATIONS

As indicated in the beginning of this chapter, orthomyxovirus infection can stimulate expression of type I IFNs and ISGs by several signaling pathways that can initiate from the cytoplasmic sensor RIG-I or the endosomal sensors TLR3, TLR7, and TLR8. Therefore, pulmonary infections by influenza viruses could result in robust IFN production by most, if not all, cell types present in the respiratory tract. The studies discussed in this chapter have concisely described the mechanisms by which, in the absence of the NS1 protein, the host detects virus replication, stimulates expression of the IFN-β gene and ISGs, and establishes an antiviral state that controls replication of and pathogenicity resulting from infection by orthomyxoviruses. In addition, the findings summarized in this chapter have not only elucidated the multiple

approaches employed by the NS1 protein to thwart the type I IFN pathway, but have also identified potential consequences resulting from the actions mediated by the NS1 protein.

Type I IFNs mediate antiviral, antitumoral, and immunomodulatory functions that also include the shaping of adaptive immune responses. The implication of this statement is that perturbation of type I IFNs by the NS1 protein not only limits the host's innate antiviral responses but also profoundly influences other biological functions of the host, including adaptive immunity. As one example of the ability of the NS1 protein to interfere with adaptive immunity, influenza A virus infection of myeloid dendritic cells (mDCs) results in the NS1 protein-dependent suppression of mDC maturation (52). As a consequence, influenza virus-infected mDCs secrete lower levels of cytokines that stimulate mDC maturation, including IFN-β and IFN-α, and inefficiently prime CD4 T cells, as revealed by diminished IFN-γ secretion (52). Therefore, the NS1 protein negatively regulates adaptive immune responses by inhibiting type I IFN production by numerous cell types and by suppressing mDC priming of T cells.

Because of the ability of the NS1 protein to suppress innate and adaptive immune responses (51), new research avenues have been initiated that have focused on the potential application of influenza viruses that either lack the NS1 gene (ΔNS1) or encode truncated forms of the NS1 protein as live-attenuated vaccines or as oncolytic agents for cancer therapy. The vaccine potential of modified influenza viruses is borne from observations that murine (48, 50), swine (179, 192), equine (165), and human (7) influenza viruses expressing NS1 proteins possessing carboxy-terminal truncations resulted in enhanced IFN production and reduced pathogenicity in mice, pigs, horses, and macaques, respectively. Furthermore, as compared with infection of mice with wild-type influenza virus A/Puerto Rico/8/1934, infection with influenza viruses encoding NS1 proteins possessing carboxy-terminal truncations resulted in increased humoral (immunoglobulins G and A) and NP-specific T-cell responses (50). It remains to be determined whether the immunogenicity and antigenicity of influenza viruses expressing modified forms of the NS1 protein are markedly improved compared with those of conventional cold-adapted influenza virus vaccines.

The therapeutic application of modified influenza viruses as oncolytic agents is borne on the observation that most, if not all, cancers are associated with cellular defects in the type I IFN pathway. Thus, influenza viruses expressing modified NS1 proteins are restricted to replication in type I IFN-deficient cancer cells, while virus replication is inhibited in normal cells (142). This concept is additionally supported by observations that influenza viruses that either lack the NS1 gene (i.e., ΔNS1) or express truncated NS1 polypeptides stimulate tumor cell-specific CD8 T-cell responses (43) and stimulate tumor regression in mice (13). Therefore, influenza viruses could represent a noninvasive therapeutic approach to specifically target cancer cells. However, it should be mentioned that prior vaccination with the U.S. Food and Drug Administration (FDA)-approved trivalent (inactivated) or live-attenuated (cold-adapted) influenza vaccines could limit the therapeutic efficacy of modified influenza viruses as oncolytic agents.

Current therapies for the treatment of influenza virus infections are limited to the FDA-approved neuraminidase inhibitors (oseltamivir and zanamivir) and M2 ion channel inhibitors (amantadine and rimantadine). The increasing occurrence of clinical isolates of oseltamivir- and amantadine-resistant influenza viruses and the high probability that supplies of oseltamivir will be limited during the next influenza virus pandemic necessitate the development of new therapeutic approaches and virus targets to expand existing treatment options and ensure that a larger percentage of the population can be protected during the next pandemic. In addition to current efforts to identify new compounds targeting the activities of the influenza virus hemagglutinin, neuraminidase, and M2 ion channel, the NS1 protein represents an intriguing yet problematic viral target, given its intracellular functions (8). Analysis of the available crystal structures of the NS1 protein RNA-binding domain (24, 119) and effector domain (14) should facilitate the rational design of and/or high-throughput screening of peptide, short hairpin RNA, small interfering RNA, and chemical libraries to identify novel antiviral compounds that specifically target the functions of the NS1 protein.

References

1. Accola, M. A., B. Huang, A. Al Masri, and M. A. McNiven. 2002. The antiviral dynamin family member, MxA, tubulates lipids and localizes to the smooth endoplasmic reticulum. *J. Biol. Chem.* 277:21829–21835.

2. Alonso-Caplen, F. V., M. E. Nemeroff, Y. Qiu, and R. M. Krug. 1992. Nucleocytoplasmic transport: the influenza virus NS1 protein regulates the transport of spliced NS2 mRNA and its precursor NS1 mRNA. *Genes Dev.* 6:255–267.

3. Aragon, T., S. de la Luna, I. Novoa, L. Carrasco, J. Ortin, and A. Nieto. 2000. Eukaryotic translation initiation factor 4GI is a cellular target for NS1 protein, a translational activator of influenza virus. *Mol. Cell. Biol.* 20:6259–6268.

4. Arguello, M. D., and J. Hiscott. 2007. Ub surprised: viral ovarian tumor domain proteases remove ubiquitin and ISG15 conjugates. *Cell Host Microbe* 2:367–369.

5. Baez, M., J. J. Zazra, R. M. Elliott, J. F. Young, and P. Palese. 1981. Nucleotide sequence of the influenza A/duck/Alberta/60/76 virus NS RNA: conservation of the NS1/NS2 overlapping gene structure in a divergent influenza virus RNA segment. *Virology* 113:397–402.

6. Balachandran, S., P. C. Roberts, L. E. Brown, H. Truong, A. K. Pattnaik, D. R. Archer, and G. N. Barber. 2000. Essential role for the dsRNA-dependent protein kinase PKR in innate immunity to viral infection. *Immunity* 13:129–141.

7. Baskin, C. R., H. Bielefeldt-Ohmann, A. Garcia-Sastre, T. M. Tumpey, N. Van Hoeven, V. S. Carter, M. J. Thomas, S. Proll, A. Solorzano, R. Billharz, J. L. Fornek, S. Thomas, C. H. Chen, E. A. Clark, K. Murali-Krishna, and M. G. Katze. 2007. Functional genomic and serological analysis of the protective immune response resulting from vaccination of macaques with an NS1-truncated influenza virus. *J. Virol.* 81:11817–11827.

8. Basler, C. F. 2007. Influenza viruses: basic biology and potential drug targets. *Infect. Disord. Drug Targets* 7:282–293.

9. Basler, C. F., A. H. Reid, J. K. Dybing, T. A. Janczewski, T. G. Fanning, H. Zheng, M. Salvatore, M. L. Perdue, D. E. Swayne, A. Garcia-Sastre, P. Palese, and J. K. Taubenberger. 2001. Sequence of the 1918 pandemic influenza virus nonstructural gene (NS) segment and characterization of recombinant viruses bearing the 1918 NS genes. *Proc. Natl. Acad. Sci. USA* 98:2746–2751.

10. Baumgarth, N., and A. Kelso. 1996. In vivo blockade of gamma interferon affects the influenza virus-induced humoral and the local cellular immune response in lung tissue. *J. Virol.* 70:4411–4418.

11. Beaton, A. R., and R. M. Krug. 1981. Selected host cell capped RNA fragments prime influenza viral RNA transcription in vivo. *Nucleic Acids Res.* 9:4423–4436.

12. Bergmann, M., A. Garcia-Sastre, E. Carnero, H. Pehamberger, K. Wolff, P. Palese, and T. Muster. 2000. Influenza virus NS1 protein counteracts PKR-mediated inhibition of replication. *J. Virol.* 74:6203–6206.

13. Bergmann, M., I. Romirer, M. Sachet, R. Fleischhacker, A. Garcia-Sastre, P. Palese, K. Wolff, H. Pehamberger, R. Jakesz, and T. Muster. 2001. A genetically engineered influenza A virus with ras-dependent oncolytic properties. *Cancer Res.* 61:8188–8193.

14. Bornholdt, Z. A., and B. V. Prasad. 2006. X-ray structure of influenza virus NS1 effector domain. *Nat. Struct. Mol. Biol.* 13:559–560.

15. Bot, A., S. Bot, and C. A. Bona. 1998. Protective role of gamma interferon during the recall response to influenza virus. *J. Virol.* 72:6637–6645.

16. Burgui, I., T. Aragon, J. Ortin, and A. Nieto. 2003. PABP1 and eIF4GI associate with influenza virus NS1 protein in viral mRNA translation initiation complexes. *J. Gen. Virol.* 84:3263–3274.

17. Catic, A., E. Fiebiger, G. A. Korbel, D. Blom, P. J. Galardy, and H. L. Ploegh. 2007. Screen for ISG15-crossreactive deubiquitinases. *PLoS ONE* 2:e679.

18. Chariot, A., A. Leonardi, J. Muller, M. Bonif, K. Brown, and U. Siebenlist. 2002. Association of the adaptor TANK

with the I kappa B kinase (IKK) regulator NEMO connects IKK complexes with IKK epsilon and TBK1 kinases. *J. Biol. Chem.* **277:**37029–37036.

19. Chelbi-Alix, M. K., L. Pelicano, F. Quignon, M. H. Koken, L. Venturini, M. Stadler, J. Pavlovic, L. Degos, and H. de The. 1995. Induction of the PML protein by interferons in normal and APL cells. *Leukemia* **9:**2027–2033.

20. Chelbi-Alix, M. K., F. Quignon, L. Pelicano, M. H. Koken, and H. de The. 1998. Resistance to virus infection conferred by the interferon-induced promyelocytic leukemia protein. *J. Virol.* **72:**1043–1051.

21. Chen, G. W., S. C. Chang, C. K. Mok, Y. L. Lo, Y. N. Kung, J. H. Huang, Y. H. Shih, J. Y. Wang, C. Chiang, C. J. Chen, and S. R. Shih. 2006. Genomic signatures of human versus avian influenza A viruses. *Emerg. Infect. Dis.* **12:**1353–1360.

22. Chen, Z., Y. Li, and R. M. Krug. 1999. Influenza A virus NS1 protein targets poly(A)-binding protein II of the cellular 3′-end processing machinery. *EMBO J.* **18:** 2273–2283.

23. Cheung, T. K., and L. L. Poon. 2007. Biology of influenza a virus. *Ann. N. Y. Acad. Sci.* **1102:**1–25.

24. Chien, C. Y., R. Tejero, Y. Huang, D. E. Zimmerman, C. B. Rios, R. M. Krug, and G. T. Montelione. 1997. A novel RNA-binding motif in influenza A virus non-structural protein 1. *Nat. Struct. Biol.* **4:**891–895.

25. Chien, C. Y., Y. Xu, R. Xiao, J. M. Aramini, P. V. Sahasrabudhe, R. M. Krug, and G. T. Montelione. 2004. Biophysical characterization of the complex between double-stranded RNA and the N-terminal domain of the NS1 protein from influenza A virus: evidence for a novel RNA-binding mode. *Biochemistry* **43:**1950–1962.

26. Chu, W. M., D. Ostertag, Z. W. Li, L. Chang, Y. Chen, Y. Hu, B. Williams, J. Perrault, and M. Karin. 1999. JNK2 and IKKbeta are required for activating the innate response to viral infection. *Immunity* **11:**721–731.

27. Cole, J. L. 2007. Activation of PKR: an open and shut case? *Trends Biochem. Sci.* **32:**57–62.

28. Crescenzo-Chaigne, B., and S. van der Werf. 2007. Rescue of influenza C virus from recombinant DNA. *J. Virol.* **81:**11282–11289.

29. Cui, S., K. Eisenacher, A. Kirchhofer, K. Brzozka, A. Lammens, K. Lammens, T. Fujita, K. K. Conzelmann, A. Krug, and K. P. Hopfner. 2008. The C-terminal regulatory domain is the RNA 5′-triphosphate sensor of RIG-I. *Mol. Cell* **29:**169–179.

30. Dao, C. T., and D. E. Zhang. 2005. ISG15: a ubiquitin-like enigma. *Front. Biosci.* **10:**2701–2722.

31. Dauber, B., G. Heins, and T. Wolff. 2004. The influenza B virus nonstructural NS1 protein is essential for efficient viral growth and antagonizes beta interferon induction. *J. Virol.* **78:**1865–1872.

32. Dauber, B., J. Schneider, and T. Wolff. 2006. Double-stranded RNA binding of influenza B virus nonstructural NS1 protein inhibits protein kinase R but is not essential to antagonize production of alpha/beta interferon. *J. Virol.* **80:**11667–11677.

33. de Jong, M. D., C. P. Simmons, T. T. Thanh, V. M. Hien, G. J. Smith, T. N. Chau, D. M. Hoang, N. V. Chau, T. H. Khanh, V. C. Dong, P. T. Qui, B. V. Cam, D. Q. Ha, Y. Guan, J. S. Peiris, N. T. Chinh, T. T. Hien, and J. Farrar. 2006. Fatal outcome of human influenza A (H5N1) is

associated with high viral load and hypercytokinemia. *Nat. Med.* **12:**1203–1207.

34. de la Luna, S., P. Fortes, A. Beloso, and J. Ortin. 1995. Influenza virus NS1 protein enhances the rate of translation initiation of viral mRNAs. *J. Virol.* **69:**2427–2433.

35. Der, S. D., A. Zhou, B. R. Williams, and R. H. Silverman. 1998. Identification of genes differentially regulated by interferon alpha, beta, or gamma using oligonucleotide arrays. *Proc. Natl. Acad. Sci. USA* **95:**15623–15628.

36. de Veer, M. J., M. Holko, M. Frevel, E. Walker, S. Der, J. M. Paranjape, R. H. Silverman, and B. R. Williams. 2001. Functional classification of interferon-stimulated genes identified using microarrays. *J. Leukoc. Biol.* **69:**912–920.

37. Dever, T. E., J. J. Chen, G. N. Barber, A. M. Cigan, L. Feng, T. F. Donahue, I. M. London, M. G. Katze, and A. G. Hinnebusch. 1993. Mammalian eukaryotic initiation factor 2 alpha kinases functionally substitute for GCN2 protein kinase in the GCN4 translational control mechanism of yeast. *Proc. Natl. Acad. Sci. USA* **90:**4616–4620.

38. Dittmann, J., S. Stertz, D. Grimm, J. Steel, A. Garcia-Sastre, O. Haller, and G. Kochs. 2008. Influenza A virus strains differ in sensitivity to the antiviral action of the Mx-GTPase. *J. Virol.* **82:**3624–3631.

39. Donelan, N. R., C. F. Basler, and A. Garcia-Sastre. 2003. A recombinant influenza A virus expressing an RNA-binding-defective NS1 protein induces high levels of beta interferon and is attenuated in mice. *J. Virol.* **77:**13257–13266.

40. Donelan, N. R., B. Dauber, X. Wang, C. F. Basler, T. Wolff, and A. Garcia-Sastre. 2004. The N- and C-terminal domains of the NS1 protein of influenza B virus can independently inhibit IRF-3 and beta interferon promoter activation. *J. Virol.* **78:**11574–11582.

41. Dong, B., M. Niwa, P. Walter, and R. H. Silverman. 2001. Basis for regulated RNA cleavage by functional analysis of RNase L and Ire1p. *RNA* **7:**361–373.

42. Dong, B., and R. H. Silverman. 1997. A bipartite model of 2–5A-dependent RNase L. *J. Biol. Chem.* **272:** 22236–22242.

43. Efferson, C. L., N. Tsuda, K. Kawano, E. Nistal-Villan, S. Sellappan, D. Yu, J. L. Murray, A. Garcia-Sastre, and C. G. Ioannides. 2006. Prostate tumor cells infected with a recombinant influenza virus expressing a truncated NS1 protein activate cytolytic CD8+ cells to recognize noninfected tumor cells. *J. Virol.* **80:**383–394.

44. Ehrhardt, C., T. Wolff, and S. Ludwig. 2007. Activation of phosphatidylinositol 3-kinase signaling by the nonstructural NS1 protein is not conserved among type A and B influenza viruses. *J. Virol.* **81:**12097–12100.

45. Enami, K., T. A. Sato, S. Nakada, and M. Enami. 1994. Influenza virus NS1 protein stimulates translation of the M1 protein. *J. Virol.* **68:**1432–1437.

46. Espert, L., G. Degols, C. Gongora, D. Blondel, B. R. Williams, R. H. Silverman, and N. Mechti. 2003. ISG20, a new interferon-induced RNase specific for single-stranded RNA, defines an alternative antiviral pathway against RNA genomic viruses. *J. Biol. Chem.* **278:**16151–16158.

47. Everett, R. D., and M. K. Chelbi-Alix. 2007. PML and PML nuclear bodies: implications in antiviral defence. *Biochimie* **89:**819–830.

48. Falcon, A. M., A. Fernandez-Sesma, Y. Nakaya, T. M. Moran, J. Ortin, and A. Garcia-Sastre. 2005. Attenuation and immunogenicity in mice of temperature-sensitive

influenza viruses expressing truncated NS1 proteins. *J. Gen. Virol.* **86**:2817–2821.

49. Feigenblum, D., and R. J. Schneider. 1993. Modification of eukaryotic initiation factor 4F during infection by influenza virus. *J. Virol.* **67**:3027–3035.

50. Ferko, B., J. Stasakova, J. Romanova, C. Kittel, S. Sereinig, H. Katinger, and A. Egorov. 2004. Immunogenicity and protection efficacy of replication-deficient influenza A viruses with altered NS1 genes. *J. Virol.* **78**:13037–13045.

51. Fernandez-Sesm, A. 2007. The influenza virus NS1 protein: inhibitor of innate and adaptive immunity. *Infect. Disord. Drug Targets* **7**:336–343.

52. Fernandez-Sesma, A., S. Marukian, B. J. Ebersole, D. Kaminski, M. S. Park, T. Yuen, S. C. Sealfon, A. Garcia-Sastre, and T. M. Moran. 2006. Influenza virus evades innate and adaptive immunity via the NS1 protein. *J. Virol.* **80**:6295–6304.

53. Fitzgerald, K. A., S. M. McWhirter, K. L. Faia, D. C. Rowe, E. Latz, D. T. Golenbock, A. J. Coyle, S. M. Liao, and T. Maniatis. 2003. IKKepsilon and TBK1 are essential components of the IRF3 signaling pathway. *Nat. Immunol.* **4**:491–496.

54. Fodor, E., M. Crow, L. J. Mingay, T. Deng, J. Sharps, P. Fechter, and G. G. Brownlee. 2002. A single amino acid mutation in the PA subunit of the influenza virus RNA polymerase inhibits endonucleolytic cleavage of capped RNAs. *J. Virol.* **76**:8989–9001.

55. Fortes, P., A. Beloso, and J. Ortin. 1994. Influenza virus NS1 protein inhibits pre-mRNA splicing and blocks mRNA nucleocytoplasmic transport. *EMBO J.* **13**:704–712.

56. Frias-Staheli, N., N. V. Giannakopoulos, M. Kikkert, S. L. Taylor, A. Bridgen, J. Paragas, J. A. Richt, R. R. Rowland, C. S. Schmaljohn, D. J. Lenschow, E. J. Snijder, A. Garcia-Sastre, and H. W. Virgin IV. 2007. Ovarian tumor domain-containing viral proteases evade ubiquitin- and ISG15-dependent innate immune responses. *Cell Host Microbe* **2**:404–416.

57. Garaigorta, U., and J. Ortin. 2007. Mutation analysis of a recombinant NS replicon shows that influenza virus NS1 protein blocks the splicing and nucleo-cytoplasmic transport of its own viral mRNA. *Nucleic Acids Res.* **35**:4573–4582.

58. Garcia, M. A., E. F. Meurs, and M. Esteban. 2007. The dsRNA protein kinase PKR: virus and cell control. *Biochimie* **89**:799–811.

59. Garcia-Sastre, A., R. K. Durbin, H. Zheng, P. Palese, R. Gertner, D. E. Levy, and J. E. Durbin. 1998. The role of interferon in influenza virus tissue tropism. *J. Virol.* **72**:8550–8558.

60. Garcia-Sastre, A., A. Egorov, D. Matassov, S. Brandt, D. E. Levy, J. E. Durbin, P. Palese, and T. Muster. 1998. Influenza A virus lacking the NS1 gene replicates in interferon-deficient systems. *Virology* **252**:324–330.

61. Gee, P., P. K. Chua, J. Gevorkyan, K. Klumpp, I. Najera, D. C. Swinney, and J. Deval. 2008. Essential role of the N-terminal domain in the regulation of RIG-I ATPase activity. *J. Biol. Chem.* **283**:9488–9496.

62. Gongora, C., G. David, L. Pintard, C. Tissot, T. D. Hua, A. Dejean, and N. Mechti. 1997. Molecular cloning of a new interferon-induced PML nuclear body-associated protein. *J. Biol. Chem.* **272**:19457–19463.

63. Goodman, A. G., J. A. Smith, S. Balachandran, O. Perwitasari, S. C. Proll, M. J. Thomas, M. J. Korth, G. N. Barber, L. A. Schiff, and M. G. Katze. 2007. The cellular protein P58[IPK] regulates influenza virus mRNA translation and replication through a PKR-mediated mechanism. *J. Virol.* **81**:2221–2230.

64. Greenspan, D., P. Palese, and M. Krystal. 1988. Two nuclear location signals in the influenza virus NS1 nonstructural protein. *J. Virol.* **62**:3020–3026.

65. Grimm, D., P. Staeheli, M. Hufbauer, I. Koerner, L. Martinez-Sobrido, A. Solorzano, A. Garcia-Sastre, O. Haller, and G. Kochs. 2007. Replication fitness determines high virulence of influenza A virus in mice carrying functional Mx1 resistance gene. *Proc. Natl. Acad. Sci. USA* **104**:6806–6811.

66. Grotzinger, T., K. Jensen, and H. Will. 1996. The interferon (IFN)-stimulated gene Sp100 promoter contains an IFN-gamma activation site and an imperfect IFN-stimulated response element which mediate type I IFN inducibility. *J. Biol. Chem.* **271**:25253–25260.

67. Guillot, L., R. Le Goffic, S. Bloch, N. Escriou, S. Akira, M. Chignard, and M. Si-Tahar. 2005. Involvement of Toll-like receptor 3 in the immune response of lung epithelial cells to double-stranded RNA and influenza A virus. *J. Biol. Chem.* **280**:5571–5580.

68. Guldner, H. H., C. Szostecki, T. Grotzinger, and H. Will. 1992. IFN enhance expression of Sp100, an autoantigen in primary biliary cirrhosis. *J. Immunol.* **149**:4067–4073.

69. Guo, B., and G. Cheng. 2007. Modulation of the interferon antiviral response by the TBK1/IKKi adaptor protein TANK. *J. Biol. Chem.* **282**:11817–11826.

70. Guo, Z., L. M. Chen, H. Zeng, J. A. Gomez, J. Plowden, T. Fujita, J. M. Katz, R. O. Donis, and S. Sambhara. 2007. NS1 protein of influenza A virus inhibits the function of intracytoplasmic pathogen sensor, RIG-I. *Am. J. Respir. Cell Mol. Biol.* **36**:263–269.

71. Haas, A. L., P. Ahrens, P. M. Bright, and H. Ankel. 1987. Interferon induces a 15-kilodalton protein exhibiting marked homology to ubiquitin. *J. Biol. Chem.* **262**:11315–11323.

72. Hagmaier, K., H. R. Gelderblom, and G. Kochs. 2004. Functional comparison of the two gene products of Thogoto virus segment 6. *J. Gen. Virol.* **85**:3699–3708.

73. Hagmaier, K., S. Jennings, J. Buse, F. Weber, and G. Kochs. 2003. Novel gene product of Thogoto virus segment 6 codes for an interferon antagonist. *J. Virol.* **77**:2747–2752.

74. Haller, O., and G. Kochs. 2002. Interferon-induced Mx proteins: dynamin-like GTPases with antiviral activity. *Traffic* **3**:710–717.

75. Haller, O., G. Kochs, and F. Weber. 2007. Interferon, Mx, and viral countermeasures. *Cytokine Growth Factor Rev.* **18**:425–433.

76. Haller, O., P. Staeheli, and G. Kochs. 2007. Interferon-induced Mx proteins in antiviral host defense. *Biochimie* **89**:812–818.

77. Haller, O., S. Stertz, and G. Kochs. 2007. The Mx GTPase family of interferon-induced antiviral proteins. *Microbes Infect.* **9**:1636–1643.

78. Hara, K., F. I. Schmidt, M. Crow, and G. G. Brownlee. 2006. Amino acid residues in the N-terminal region of the PA subunit of influenza A virus RNA polymerase play

a critical role in protein stability, endonuclease activity, cap binding, and virion RNA promoter binding. *J. Virol.* **80**:7789–7798.

79. Hatada, E., and R. Fukuda. 1992. Binding of influenza A virus NS1 protein to dsRNA in vitro. *J. Gen. Virol.* **73**(Pt. 12):3325–3329.

80. Hatada, E., S. Saito, and R. Fukuda. 1999. Mutant influenza viruses with a defective NS1 protein cannot block the activation of PKR in infected cells. *J. Virol.* **73**: 2425–2433.

81. Hatada, E., T. Takizawa, and R. Fukuda. 1992. Specific binding of influenza A virus NS1 protein to the virus minus-sense RNA in vitro. *J. Gen. Virol.* **73**(Pt. 1):17–25.

82. Hayman, A., S. Comely, A. Lackenby, L. C. Hartgroves, S. Goodbourn, J. W. McCauley, and W. S. Barclay. 2007. NS1 proteins of avian influenza A viruses can act as antagonists of the human alpha/beta interferon response. *J. Virol.* **81**:2318–2327.

83. Hayman, A., S. Comely, A. Lackenby, S. Murphy, J. McCauley, S. Goodbourn, and W. Barclay. 2006. Variation in the ability of human influenza A viruses to induce and inhibit the IFN-beta pathway. *Virology* **347**:52–64.

84. Hemmi, H., O. Takeuchi, S. Sato, M. Yamamoto, T. Kaisho, H. Sanjo, T. Kawai, K. Hoshino, K. Takeda, and S. Akira. 2004. The roles of two IkappaB kinase-related kinases in lipopolysaccharide and double stranded RNA signaling and viral infection. *J. Exp. Med.* **199**:1641–1650.

85. Heuser, M., H. van der Kuip, B. Falini, C. Peschel, C. Huber, and T. Fischer. 1998. Induction of the pro-myelocytic leukaemia gene by type I and type II interferons. *Mediators Inflamm.* **7**:319–325.

86. Holzinger, D., C. Jorns, S. Stertz, S. Boisson-Dupuis, R. Thimme, M. Weidmann, J. L. Casanova, O. Haller, and G. Kochs. 2007. Induction of MxA gene expression by influenza A virus requires type I or type III interferon signaling. *J. Virol.* **81**:7776–7785.

87. Horimoto, T., N. Fukuda, K. Iwatsuki-Horimoto, Y. Guan, W. Lim, M. Peiris, S. Sugii, T. Odagiri, M. Tashiro, and Y. Kawaoka. 2004. Antigenic differences between H5N1 human influenza viruses isolated in 1997 and 2003. *J. Vet. Med. Sci.* **66**:303–305.

88. Hornung, V., J. Ellegast, S. Kim, K. Brzozka, A. Jung, H. Kato, H. Poeck, S. Akira, K. K. Conzelmann, M. Schlee, S. Endres, and G. Hartmann. 2006. 5′-Triphosphate RNA is the ligand for RIG-I. *Science* **314**:994–997.

89. Isaacs, A., and D. C. Burke. 1958. Mode of action of interferon. *Nature* **182**:1073–1074.

90. Ishihama, A. 1996. A multi-functional enzyme with RNA polymerase and RNase activities: molecular anatomy of influenza virus RNA polymerase. *Biochimie* **78**:1097–1102.

91. Jennings, S., L. Martinez-Sobrido, A. Garcia-Sastre, F. Weber, and G. Kochs. 2005. Thogoto virus ML protein suppresses IRF3 function. *Virology* **331**:63–72.

92. Jiao, P., G. Tian, Y. Li, G. Deng, Y. Jiang, C. Liu, W. Liu, Z. Bu, Y. Kawaoka, and H. Chen. 2008. A single-amino-acid substitution in the NS1 protein changes the pathogenicity of H5N1 avian influenza viruses in mice. *J. Virol.* **82**:1146–1154.

93. Johnson, N. P., and J. Mueller. 2002. Updating the accounts: global mortality of the 1918–1920 "Spanish" influenza pandemic. *Bull. Hist. Med.* **76**:105–115.

94. Joseph, T., J. McAuliffe, B. Lu, H. Jin, G. Kemble, and K. Subbarao. 2007. Evaluation of replication and pathogenicity of avian influenza a H7 subtype viruses in a mouse model. *J. Virol.* **81**:10558–10566.

95. Juang, Y. T., W. Lowther, M. Kellum, W. C. Au, R. Lin, J. Hiscott, and P. M. Pitha. 1998. Primary activation of interferon A and interferon B gene transcription by interferon regulatory factor 3. *Proc. Natl. Acad. Sci. USA* **95**:9837–9842.

96. Kanayama, A., R. B. Seth, L. Sun, C. K. Ea, M. Hong, A. Shaito, Y. H. Chiu, L. Deng, and Z. J. Chen. 2004. TAB2 and TAB3 activate the NF-kappaB pathway through binding to polyubiquitin chains. *Mol. Cell.* **15**:535–548.

97. Kash, J. C., T. M. Tumpey, S. C. Proll, V. Carter, O. Perwitasari, M. J. Thomas, C. F. Basler, P. Palese, J. K. Taubenberger, A. Garcia-Sastre, D. E. Swayne, and M. G. Katze. 2006. Genomic analysis of increased host immune and cell death responses induced by 1918 influenza virus. *Nature* **443**:578–581.

98. Kato, H., O. Takeuchi, S. Sato, M. Yoneyama, M. Yamamoto, K. Matsui, S. Uematsu, A. Jung, T. Kawai, K. J. Ishii, O. Yamaguchi, K. Otsu, T. Tsujimura, C. S. Koh, C. Reis e Sousa, Y. Matsuura, T. Fujita, and S. Akira. 2006. Differential roles of MDA5 and RIG-I helicases in the recognition of RNA viruses. *Nature* **441**:101–105.

99. Kawai, T., K. Takahashi, S. Sato, C. Coban, H. Kumar, H. Kato, K. J. Ishii, O. Takeuchi, and S. Akira. 2005. IPS-1, an adaptor triggering RIG-I- and Mda5-mediated type I interferon induction. *Nat. Immunol.* **6**:981–988.

100. Kim, M. J., S. Y. Hwang, T. Imaizumi, and J. Y. Yoo. 2008. Negative feedback regulation of RIG-I-mediated antiviral signaling by interferon-induced ISG15 conjugation. *J. Virol.* **82**:1474–1483.

101. Kim, T. W., K. Staschke, K. Bulek, J. Yao, K. Peters, K. H. Oh, Y. Vandenburg, H. Xiao, W. Qian, T. Hamilton, B. Min, G. Sen, R. Gilmour, and X. Li. 2007. A critical role for IRAK4 kinase activity in Toll-like receptor-mediated innate immunity. *J. Exp. Med.* **204**:1025–1036.

102. Kobasa, D., S. M. Jones, K. Shinya, J. C. Kash, J. Copps, H. Ebihara, Y. Hatta, J. H. Kim, P. Halfmann, M. Hatta, F. Feldmann, J. B. Alimonti, L. Fernando, Y. Li, M. G. Katze, H. Feldmann, and Y. Kawaoka. 2007. Aberrant innate immune response in lethal infection of macaques with the 1918 influenza virus. *Nature* **445**:319–323.

103. Kochs, G., A. Garcia-Sastre, and L. Martinez-Sobrido. 2007. Multiple anti-interferon actions of the influenza A virus NS1 protein. *J. Virol.* **81**:7011–7021.

104. Kochs, G., I. Koerner, L. Thiel, S. Kothlow, B. Kaspers, N. Ruggli, A. Summerfield, J. Pavlovic, J. Stech, and P. Staeheli. 2007. Properties of H7N7 influenza A virus strain SC35M lacking interferon antagonist NS1 in mice and chickens. *J. Gen. Virol.* **88**:1403–1409.

105. Kochs, G., F. Weber, S. Gruber, A. Delvendahl, C. Leitz, and O. Haller. 2000. Thogoto virus matrix protein is encoded by a spliced mRNA. *J. Virol.* **74**:10785–10789.

106. Koerner, I., G. Kochs, U. Kalinke, S. Weiss, and P. Staeheli. 2007. Protective role of beta interferon in host defense against influenza A virus. *J. Virol.* **81**:2025–2030.

107. Koyama, S., K. J. Ishii, H. Kumar, T. Tanimoto, C. Coban, S. Uematsu, T. Kawai, and S. Akira. 2007. Differential role of TLR- and RLR-signaling in the immune responses

to influenza A virus infection and vaccination. *J. Immunol.* 179:4711–4720.

108. Krug, R. M. 1981. Priming of influenza viral RNA transcription by capped heterologous RNAs. *Curr. Top. Microbiol. Immunol.* 93:125–149.

109. Leahy, M. B., J. T. Dessens, and P. A. Nuttall. 1997. In vitro polymerase activity of Thogoto virus: evidence for a unique cap-snatching mechanism in a tick-borne orthomyxovirus. *J. Virol.* 71:8347–8351.

110. Lee, T. G., N. Tang, S. Thompson, J. Miller, and M. G. Katze. 1994. The 58,000-dalton cellular inhibitor of the interferon-induced double-stranded RNA-activated protein kinase (PKR) is a member of the tetratricopeptide repeat family of proteins. *Mol. Cell. Biol.* 14:2331–2342.

111. Le Goffic, R., V. Balloy, M. Lagranderie, L. Alexopoulou, N. Escriou, R. Flavell, M. Chignard, and M. Si-Tahar. 2006. Detrimental contribution of the Toll-like receptor (TLR)3 to influenza A virus-induced acute pneumonia. *PLoS Pathog.* 2:e53.

112. Le Goffic, R., J. Pothlichet, D. Vitour, T. Fujita, E. Meurs, M. Chignard, and M. Si-Tahar. 2007. Cutting edge: influenza A virus activates TLR3-dependent inflammatory and RIG-I-dependent antiviral responses in human lung epithelial cells. *J. Immunol.* 178:3368–3372.

113. Lenschow, D. J., C. Lai, N. Frias-Staheli, N. V. Giannakopoulos, A. Lutz, T. Wolff, A. Osiak, B. Levine, R. E. Schmidt, A. Garcia-Sastre, D. A. Leib, A. Pekosz, K. P. Knobeloch, I. Horak, and H. W. Virgin IV. 2007. IFN-stimulated gene 15 functions as a critical antiviral molecule against influenza, herpes, and Sindbis viruses. *Proc. Natl. Acad. Sci. USA* 104:1371–1376.

114. Li, M., and B. Wang. 2007. Homology modeling and examination of the effect of the D92E mutation on the H5N1 nonstructural protein NS1 effector domain. *J. Mol. Model.* 13:1237–1244.

115. Li, M. L., P. Rao, and R. M. Krug. 2001. The active sites of the influenza cap-dependent endonuclease are on different polymerase subunits. *EMBO J.* 20:2078–2086.

116. Li, S., J. Y. Min, R. M. Krug, and G. C. Sen. 2006. Binding of the influenza A virus NS1 protein to PKR mediates the inhibition of its activation by either PACT or double-stranded RNA. *Virology* 349:13–21.

117. Li, Y., Z. Y. Chen, W. Wang, C. C. Baker, and R. M. Krug. 2001. The 3′-end-processing factor CPSF is required for the splicing of single-intron pre-mRNAs in vivo. *RNA* 7:920–931.

118. Li, Y., Y. Yamakita, and R. M. Krug. 1998. Regulation of a nuclear export signal by an adjacent inhibitory sequence: the effector domain of the influenza virus NS1 protein. *Proc. Natl. Acad. Sci. USA* 95:4864–4869.

119. Liu, J., P. A. Lynch, C. Y. Chien, G. T. Montelione, R. M. Krug, and H. M. Berman. 1997. Crystal structure of the unique RNA-binding domain of the influenza virus NS1 protein. *Nat. Struct. Biol.* 4:896–899.

120. Loeb, K. R., and A. L. Haas. 1992. The interferon-inducible 15-kDa ubiquitin homolog conjugates to intracellular proteins. *J. Biol. Chem.* 267:7806–7813.

121. Loo, Y. M., J. Fornek, N. Crochet, G. Bajwa, O. Perwitasari, L. Martinez-Sobrido, S. Akira, M. A. Gill, A. Garcia-Sastre, M. G. Katze, and M. Gale, Jr. 2008. Distinct RIG-I and MDA5 signaling by RNA viruses in innate immunity. *J. Virol.* 82:335–345.

122. Lu, G., J. T. Reinert, I. Pitha-Rowe, A. Okumura, M. Kellum, K. P. Knobeloch, B. Hassel, and P. M. Pitha. 2006. ISG15 enhances the innate antiviral response by inhibition of IRF-3 degradation. *Cell. Mol. Biol. (Noisy-le-grand)* 52:29–41.

123. Lu, J., E. B. O'Hara, B. A. Trieselmann, P. R. Romano, and T. E. Dever. 1999. The interferon-induced double-stranded RNA-activated protein kinase PKR will phosphorylate serine, threonine, or tyrosine at residue 51 in eukaryotic initiation factor 2alpha. *J. Biol. Chem.* 274:32198–32203.

124. Lu, Y., X. Y. Qian, and R. M. Krug. 1994. The influenza virus NS1 protein: a novel inhibitor of pre-mRNA splicing. *Genes Dev.* 8:1817–1828.

125. Lu, Y., M. Wambach, M. G. Katze, and R. M. Krug. 1995. Binding of the influenza virus NS1 protein to double-stranded RNA inhibits the activation of the protein kinase that phosphorylates the elF-2 translation initiation factor. *Virology* 214:222–228.

126. Ludwig, S., X. Wang, C. Ehrhardt, H. Zheng, N. Donelan, O. Planz, S. Pleschka, A. Garcia-Sastre, G. Heins, and T. Wolff. 2002. The influenza A virus NS1 protein inhibits activation of Jun N-terminal kinase and AP-1 transcription factors. *J. Virol.* 76:11166–11171.

127. Malakhov, M. P., O. A. Malakhova, K. I. Kim, K. J. Ritchie, and D. E. Zhang. 2002. UBP43 (USP18) specifically removes ISG15 from conjugated proteins. *J. Biol. Chem.* 277:9976–9981.

128. Malakhova, O. A., K. I. Kim, J. K. Luo, W. Zou, K. G. Kumar, S. Y. Fuchs, K. Shuai, and D. E. Zhang. 2006. UBP43 is a novel regulator of interferon signaling independent of its ISG15 isopeptidase activity. *EMBO J.* 25:2358–2367.

129. Malakhova, O. A., M. Yan, M. P. Malakhov, Y. Yuan, K. J. Ritchie, K. I. Kim, L. F. Peterson, K. Shuai, and D. E. Zhang. 2003. Protein ISGylation modulates the JAK-Stat signaling pathway. *Genes Dev.* 17:455–460.

130. Malathi, K., B. Dong, M. Gale, Jr., and R. H. Silverman. 2007. Small self-RNA generated by RNase L amplifies antiviral innate immunity. *Nature* 448:816–819.

131. Malathi, K., J. M. Paranjape, E. Bulanova, M. Shim, J. M. Guenther-Johnson, P. W. Faber, T. E. Eling, B. R. Williams, and R. H. Silverman. 2005. A transcriptional signaling pathway in the IFN system mediated by 2′-5′-oligoadenylate activation of RNase L. *Proc. Natl. Acad. Sci. USA* 102:14533–14538.

132. Marcus, P. I., J. M. Rojek, and M. J. Sekellick. 2005. Interferon induction and/or production and its suppression by influenza A viruses. *J. Virol.* 79:2880–2890.

133. Marion, R. M., T. Aragon, A. Beloso, A. Nieto, and J. Ortin. 1997. The N-terminal half of the influenza virus NS1 protein is sufficient for nuclear retention of mRNA and enhancement of viral mRNA translation. *Nucleic Acids Res.* 25:4271–4277.

134. McBeath, A. J., B. Collet, R. Paley, S. Duraffour, V. Aspehaug, E. Biering, C. J. Secombes, and M. Snow. 2006. Identification of an interferon antagonist protein encoded by segment 7 of infectious salmon anaemia virus. *Virus Res.* 115:176–184.

135. Melen, K., L. Kinnunen, R. Fagerlund, N. Ikonen, K. Y. Twu, R. M. Krug, and I. Julkunen. 2007. Nuclear and nucleolar targeting of influenza A virus NS1 protein: striking

differences between different virus subtypes. *J. Virol.* **81**:5995–6006.

136. Mellor, H., and C. G. Proud. 1991. A synthetic peptide substrate for initiation factor-2 kinases. *Biochem. Biophys. Res. Commun.* **178**:430–437.

137. Melville, M. W., W. J. Hansen, B. C. Freeman, W. J. Welch, and M. G. Katze. 1997. The molecular chaperone hsp40 regulates the activity of P58[IPK], the cellular inhibitor of PKR. *Proc. Natl. Acad. Sci. USA* **94**:97–102.

138. Meylan, E., J. Curran, K. Hofmann, D. Moradpour, M. Binder, R. Bartenschlager, and J. Tschopp. 2005. Cardif is an adaptor protein in the RIG-I antiviral pathway and is targeted by hepatitis C virus. *Nature* **437**:1167–1172.

139. Mibayashi, M., L. Martinez-Sobrido, Y. M. Loo, W. B. Cardenas, M. Gale, Jr., and A. Garcia-Sastre. 2007. Inhibition of retinoic acid-inducible gene I-mediated induction of beta interferon by the NS1 protein of influenza A virus. *J. Virol.* **81**:514–524.

140. Mikulasova, A., E. Vareckova, and E. Fodor. 2000. Transcription and replication of the influenza A virus genome. *Acta Virol.* **44**:273–282.

141. Min, J. Y., and R. M. Krug. 2006. The primary function of RNA binding by the influenza A virus NS1 protein in infected cells: inhibiting the 2′-5′ oligo (A) synthetase/RNase L pathway. *Proc. Natl. Acad. Sci. USA* **103**:7100–7105.

142. Muster, T., J. Rajtarova, M. Sachet, H. Unger, R. Fleischhacker, I. Romirer, A. Grassauer, A. Url, A. Garcia-Sastre, K. Wolff, H. Pehamberger, and M. Bergmann. 2004. Interferon resistance promotes oncolysis by influenza virus NS1-deletion mutants. *Int. J. Cancer* **110**:15–21.

143. Nemeroff, M. E., S. M. Barabino, Y. Li, W. Keller, and R. M. Krug. 1998. Influenza virus NS1 protein interacts with the cellular 30 kDa subunit of CPSF and inhibits 3′ end formation of cellular pre-mRNAs. *Mol. Cell* **1**:991–1000.

144. Nguyen, H. H., F. W. van Ginkel, H. L. Vu, M. J. Novak, J. R. McGhee, and J. Mestecky. 2000. Gamma interferon is not required for mucosal cytotoxic T-lymphocyte responses or heterosubtypic immunity to influenza A virus infection in mice. *J. Virol.* **74**:5495–5501.

145. Nguyen, L. H., L. Espert, N. Mechti, and D. M. Wilson, III. 2001. The human interferon- and estrogen-regulated *ISG20/HEM45* gene product degrades single-stranded RNA and DNA in vitro. *Biochemistry* **40**:7174–7179.

146. Noah, D. L., K. Y. Twu, and R. M. Krug. 2003. Cellular antiviral responses against influenza A virus are countered at the posttranscriptional level by the viral NS1A protein via its binding to a cellular protein required for the 3′ end processing of cellular pre-mRNAS. *Virology* **307**:386–395.

147. Norton, G. P., T. Tanaka, K. Tobita, S. Nakada, D. A. Buonagurio, D. Greenspan, M. Krystal, and P. Palese. 1987. Infectious influenza A and B virus variants with long carboxyl terminal deletions in the NS1 polypeptides. *Virology* **156**:204–213.

148. Obenauer, J. C., J. Denson, P. K. Mehta, X. Su, S. Mukatira, D. B. Finkelstein, X. Xu, J. Wang, J. Ma, Y. Fan, K. M. Rakestraw, R. G. Webster, E. Hoffmann, S. Krauss, J. Zheng, Z. Zhang, and C. W. Naeve. 2006. Large-scale sequence analysis of avian influenza isolates. *Science* **311**:1576–1580.

149. Oganesyan, G., S. K. Saha, B. Guo, J. Q. He, A. Shahangian, B. Zarnegar, A. Perry, and G. Cheng. 2006. Critical role of TRAF3 in the Toll-like receptor-dependent and -independent antiviral response. *Nature* **439**:208–211.

150. Okumura, A., G. Lu, I. Pitha-Rowe, and P. M. Pitha. 2006. Innate antiviral response targets HIV-1 release by the induction of ubiquitin-like protein ISG15. *Proc. Natl. Acad. Sci. USA* **103**:1440–1445.

151. Opitz, B., A. Rejaibi, B. Dauber, J. Eckhard, M. Vinzing, B. Schmeck, S. Hippenstiel, N. Suttorp, and T. Wolff. 2007. IFNbeta induction by influenza A virus is mediated by RIG-I which is regulated by the viral NS1 protein. *Cell. Microbiol.* **9**:930–938.

152. Osterhaus, A. D., G. F. Rimmelzwaan, B. E. Martina, T. M. Bestebroer, and R. A. Fouchier. 2000. Influenza B virus in seals. *Science* **288**:1051–1053.

153. Ozaki, H., and H. Kida. 2007. Extensive accumulation of influenza virus NS1 protein in the nuclei causes effective viral growth in vero cells. *Microbiol. Immunol.* **51**:577–580.

154. Palese, P., and M. L. Shaw. 2006. Orthomyxoviridae: *the Viruses and Their Replication*, 5th ed., vol. 1. Lippincott, Williams & Wilkins, Philadelphia, PA.

155. Panne, D., T. Maniatis, and S. C. Harrison. 2007. An atomic model of the interferon-beta enhanceosome. *Cell* **129**:1111–1123.

156. Peiris, J. S., M. D. de Jong, and Y. Guan. 2007. Avian influenza virus (H5N1): a threat to human health. *Clin. Microbiol. Rev.* **20**:243–267.

157. Perry, A. K., E. K. Chow, J. B. Goodnough, W. C. Yeh, and G. Cheng. 2004. Differential requirement for TANK-binding kinase-1 in type I interferon responses to Toll-like receptor activation and viral infection. *J. Exp. Med.* **199**:1651–1658.

158. Pichlmair, A., J. Buse, S. Jennings, O. Haller, G. Kochs, and P. Staeheli. 2004. Thogoto virus lacking interferon-antagonistic protein ML is strongly attenuated in newborn Mx1-positive but not Mx1-negative mice. *J. Virol.* **78**:11422–11424.

159. Pichlmair, A., O. Schulz, C. P. Tan, T. I. Naslund, P. Liljestrom, F. Weber, and C. Reis e Sousa. 2006. RIG-I-mediated antiviral responses to single-stranded RNA bearing 5′-phosphates. *Science* **314**:997–1001.

160. Plotch, S. J., M. Bouloy, I. Ulmanen, and R. M. Krug. 1981. A unique cap(m7GpppXm)-dependent influenza virion endonuclease cleaves capped RNAs to generate the primers that initiate viral RNA transcription. *Cell* **23**:847–858.

161. Price, G. E., A. Gaszewska-Mastarlarz, and D. Moskophidis. 2000. The role of alpha/beta and gamma interferons in development of immunity to influenza A virus in mice. *J. Virol.* **74**:3996–4003.

162. Qian, X. Y., C. Y. Chien, Y. Lu, G. T. Montelione, and R. M. Krug. 1995. An amino-terminal polypeptide fragment of the influenza virus NS1 protein possesses specific RNA-binding activity and largely helical backbone structure. *RNA* **1**:948–956.

163. Qiu, Y., and R. M. Krug. 1994. The influenza virus NS1 protein is a poly(A)-binding protein that inhibits nuclear export of mRNAs containing poly(A). *J. Virol.* **68**:2425–2432.

164. Qiu, Y., M. Nemeroff, and R. M. Krug. 1995. The influenza virus NS1 protein binds to a specific region in human U6 snRNA and inhibits U6-U2 and U6-U4 snRNA interactions during splicing. *RNA* 1:304–316.

165. Quinlivan, M., D. Zamarin, A. Garcia-Sastre, A. Cullinane, T. Chambers, and P. Palese. 2005. Attenuation of equine influenza viruses through truncations of the NS1 protein. *J. Virol.* 79:8431–8439.

166. Rebouillat, D., and A. G. Hovanessian. 1999. The human 2′,5′-oligoadenylate synthetase family: interferon-induced proteins with unique enzymatic properties. *J. Interferon Cytokine Res.* 19:295–308.

167. Saha, S. K., E. M. Pietras, J. Q. He, J. R. Kang, S. Y. Liu, G. Oganesyan, A. Shahangian, B. Zarnegar, T. L. Shiba, Y. Wang, and G. Cheng. 2006. Regulation of antiviral responses by a direct and specific interaction between TRAF3 and Cardif. *EMBO J.* 25:3257–3263.

168. Salomon, R., E. Hoffmann, and R. G. Webster. 2007. Inhibition of the cytokine response does not protect against lethal H5N1 influenza infection. *Proc. Natl. Acad. Sci. USA* 104:12479–12481.

169. Salomon, R., P. Staeheli, G. Kochs, H. L. Yen, J. Franks, J. E. Rehg, R. G. Webster, and E. Hoffmann. 2007. Mx1 gene protects mice against the highly lethal human H5N1 influenza virus. *Cell Cycle* 6:2417–2421.

170. Salvatore, M., C. F. Basler, J. P. Parisien, C. M. Horvath, S. Bourmakina, H. Zheng, T. Muster, P. Palese, and A. Garcia-Sastre. 2002. Effects of influenza A virus NS1 protein on protein expression: the NS1 protein enhances translation and is not required for shutoff of host protein synthesis. *J. Virol.* 76:1206–1212.

171. Sandrock, M., M. Frese, O. Haller, and G. Kochs. 2001. Interferon-induced rat Mx proteins confer resistance to Rift Valley fever virus and other arthropod-borne viruses. *J. Interferon Cytokine Res.* 21:663–668.

172. Sato, M., N. Tanaka, N. Hata, E. Oda, and T. Taniguchi. 1998. Involvement of the IRF family transcription factor IRF-3 in virus-induced activation of the IFN-beta gene. *FEBS Lett.* 425:112–116.

173. Sato, Y., K. Yoshioka, C. Suzuki, S. Awashima, Y. Hosaka, J. Yewdell, and K. Kuroda. 2003. Localization of influenza virus proteins to nuclear dot 10 structures in influenza virus-infected cells. *Virology* 310:29–40.

174. Satterly, N., P. L. Tsai, J. van Deursen, D. R. Nussenzveig, Y. Wang, P. A. Faria, A. Levay, D. E. Levy, and B. M. Fontoura. 2007. Influenza virus targets the mRNA export machinery and the nuclear pore complex. *Proc. Natl. Acad. Sci. USA* 104:1853–1858.

175. Seo, S. H., E. Hoffmann, and R. G. Webster. 2002. Lethal H5N1 influenza viruses escape host anti-viral cytokine responses. *Nat. Med.* 8:950–954.

176. Seth, R. B., L. Sun, C. K. Ea, and Z. J. Chen. 2005. Identification and characterization of MAVS, a mitochondrial antiviral signaling protein that activates NF-kappaB and IRF 3. *Cell* 122:669–682.

177. Sharma, S., B. R. tenOever, N. Grandvaux, G. P. Zhou, R. Lin, and J. Hiscott. 2003. Triggering the interferon antiviral response through an IKK-related pathway. *Science* 300:1148–1151.

178. Smith, E. J., I. Marie, A. Prakash, A. Garcia-Sastre, and D. E. Levy. 2001. IRF3 and IRF7 phosphorylation in virus-infected cells does not require double-stranded RNA-dependent protein kinase R or Ikappa B kinase but is blocked by vaccinia virus E3L protein. *J. Biol. Chem.* 276:8951–8957.

179. Solorzano, A., R. J. Webby, K. M. Lager, B. H. Janke, A. Garcia-Sastre, and J. A. Richt. 2005. Mutations in the NS1 protein of swine influenza virus impair anti-interferon activity and confer attenuation in pigs. *J. Virol.* 79:7535–7543.

180. Stadler, M., M. K. Chelbi-Alix, M. H. Koken, L. Venturini, C. Lee, A. Saib, F. Quignon, L. Pelicano, M. C. Guillemin, C. Schindler, and H. de The. 1995. Transcriptional induction of the PML growth suppressor gene by interferons is mediated through an ISRE and a GAS element. *Oncogene* 11:2565–2573.

181. Stertz, S., J. Dittmann, J. C. Blanco, L. M. Pletneva, O. Haller, and G. Kochs. 2007. The antiviral potential of interferon-induced cotton rat Mx proteins against orthomyxovirus (influenza), rhabdovirus, and bunyavirus. *J. Interferon Cytokine Res.* 27:847–855.

182. Stertz, S., M. Reichelt, J. Krijnse-Locker, J. Mackenzie, J. C. Simpson, O. Haller, and G. Kochs. 2006. Interferon-induced, antiviral human MxA protein localizes to a distinct subcompartment of the smooth endoplasmic reticulum. *J. Interferon Cytokine Res.* 26:650–660.

183. Szretter, K. J., S. Gangappa, X. Lu, C. Smith, W. J. Shieh, S. R. Zaki, S. Sambhara, T. M. Tumpey, and J. M. Katz. 2007. Role of host cytokine responses in the pathogenesis of avian H5N1 influenza viruses in mice. *J. Virol.* 81:2736–2744.

184. Takahasi, K., M. Yoneyama, T. Nishihori, R. Hirai, H. Kumeta, R. Narita, M. Gale, Jr., F. Inagaki, and T. Fujita. 2008. Nonself RNA-sensing mechanism of RIG-I helicase and activation of antiviral immune responses. *Mol. Cell* 29:428–440.

185. Talon, J., C. M. Horvath, R. Polley, C. F. Basler, T. Muster, P. Palese, and A. Garcia-Sastre. 2000. Activation of interferon regulatory factor 3 is inhibited by the influenza A virus NS1 protein. *J. Virol.* 74:7989–7996.

186. Tan, S. L., and M. G. Katze. 1998. Biochemical and genetic evidence for complex formation between the influenza A virus NS1 protein and the interferon-induced PKR protein kinase. *J. Interferon Cytokine Res.* 18: 757–766.

187. Thimme, R., M. Frese, G. Kochs, and O. Haller. 1995. Mx1 but not MxA confers resistance against tick-borne Dhori virus in mice. *Virology* 211:296–301.

188. Treanor, J. J., M. H. Snyder, W. T. London, and B. R. Murphy. 1989. The B allele of the NS gene of avian influenza viruses, but not the A allele, attenuates a human influenza A virus for squirrel monkeys. *Virology* 171:1–9.

189. Tumpey, T. M., K. J. Szretter, N. Van Hoeven, J. M. Katz, G. Kochs, O. Haller, A. Garcia-Sastre, and P. Staeheli. 2007. The Mx1 gene protects mice against the pandemic 1918 and highly lethal human H5N1 influenza viruses. *J. Virol.* 81:10818–10821.

190. Turan, K., M. Mibayashi, K. Sugiyama, S. Saito, A. Numajiri, and K. Nagata. 2004. Nuclear MxA proteins form a complex with influenza virus NP and inhibit the transcription of the engineered influenza virus genome. *Nucleic Acids Res.* 32:643–652.

191. Twu, K. Y., R. L. Kuo, J. Marklund, and R. M. Krug. 2007. The H5N1 influenza virus NS genes selected after

1998 enhance virus replication in mammalian cells. *J. Virol.* **81**:8112–8121.

192. Vincent, A. L., W. Ma, K. M. Lager, B. H. Janke, R. J. Webby, A. Garcia-Sastre, and J. A. Richt. 2007. Efficacy of intranasal administration of a truncated NS1 modified live influenza virus vaccine in swine. *Vaccine* **25**:7999–8009.

193. Wang, W., and R. M. Krug. 1996. The RNA-binding and effector domains of the viral NS1 protein are conserved to different extents among influenza A and B viruses. *Virology* **223**:41–50.

194. Wang, W., and R. M. Krug. 1998. U6atac snRNA, the highly divergent counterpart of U6 snRNA, is the specific target that mediates inhibition of AT-AC splicing by the influenza virus NS1 protein. *RNA* **4**:55–64.

195. Wang, W., K. Riedel, P. Lynch, C. Y. Chien, G. T. Montelione, and R. M. Krug. 1999. RNA binding by the novel helical domain of the influenza virus NS1 protein requires its dimer structure and a small number of specific basic amino acids. *RNA* **5**:195–205.

196. Wang, X., C. F. Basler, B. R. Williams, R. H. Silverman, P. Palese, and A. Garcia-Sastre. 2002. Functional replacement of the carboxy-terminal two-thirds of the influenza A virus NS1 protein with short heterologous dimerization domains. *J. Virol.* **76**:12951–12962.

197. Wang, X., E. R. Hinson, and P. Cresswell. 2007. The interferon-inducible protein viperin inhibits influenza virus release by perturbing lipid rafts. *Cell Host Microbe* **2**:96–105.

198. Wang, X., M. Li, H. Zheng, T. Muster, P. Palese, A. A. Beg, and A. Garcia-Sastre. 2000. Influenza A virus NS1 protein prevents activation of NF-kappaB and induction of alpha/beta interferon. *J. Virol.* **74**:11566–11573.

199. Wathelet, M. G., C. H. Lin, B. S. Parekh, L. V. Ronco, P. M. Howley, and T. Maniatis. 1998. Virus infection induces the assembly of coordinately activated transcription factors on the IFN-beta enhancer in vivo. *Mol. Cell* **1**:507–518.

200. Weaver, B. K., K. P. Kumar, and N. C. Reich. 1998. Interferon regulatory factor 3 and CREB-binding protein/p300 are subunits of double-stranded RNA-activated transcription factor DRAF1. *Mol. Cell. Biol.* **18**:1359–1368.

201. Weber, F., O. Haller, and G. Kochs. 1996. Nucleoprotein viral RNA and mRNA of Thogoto virus: a novel "cap-stealing" mechanism in tick-borne orthomyxoviruses? *J. Virol.* **70**:8361–8367.

202. Weber, F., V. Wagner, S. B. Rasmussen, R. Hartmann, and S. R. Paludan. 2006. Double-stranded RNA is produced by positive-strand RNA viruses and DNA viruses but not in detectable amounts by negative-strand RNA viruses. *J. Virol.* **80**:5059–5064.

203. Wolff, T., R. E. O'Neill, and P. Palese. 1998. NS1-binding protein (NS1-BP): a novel human protein that interacts with the influenza A virus nonstructural NS1 protein is relocalized in the nuclei of infected cells. *J. Virol.* **72**:7170–7180.

204. Xu, L. G., Y. Y. Wang, K. J. Han, L. Y. Li, Z. Zhai, and H. B. Shu. 2005. VISA is an adapter protein required for virus-triggered IFN-beta signaling. *Mol. Cell* **19**:727–740.

205. Yang, Y. L., L. F. Reis, J. Pavlovic, A. Aguzzi, R. Schafer, A. Kumar, B. R. Williams, M. Aguet, and C. Weissmann. 1995. Deficient signaling in mice devoid of double-stranded RNA-dependent protein kinase. *EMBO J.* **14**:6095–6106.

206. Yin, C., J. A. Khan, G. V. Swapna, A. Ertekin, R. M. Krug, L. Tong, and G. T. Montelione. 2007. Conserved surface features form the double-stranded RNA binding site of non-structural protein 1 (NS1) from influenza A and B viruses. *J. Biol. Chem.* **282**:20584–20592.

207. Yoneyama, M., W. Suhara, Y. Fukuhara, M. Fukuda, E. Nishida, and T. Fujita. 1998. Direct triggering of the type I interferon system by virus infection: activation of a transcription factor complex containing IRF-3 and CBP/p300. *EMBO J.* **17**:1087–1095.

208. Yuan, W., J. M. Aramini, G. T. Montelione, and R. M. Krug. 2002. Structural basis for ubiquitin-like ISG 15 protein binding to the NS1 protein of influenza B virus: a protein-protein interaction function that is not shared by the corresponding N-terminal domain of the NS1 protein of influenza A virus. *Virology* **304**:291–301.

209. Yuan, W., and R. M. Krug. 2001. Influenza B virus NS1 protein inhibits conjugation of the interferon (IFN)-induced ubiquitin-like ISG15 protein. *EMBO J.* **20**:362–371.

210. Zhao, C., C. Denison, J. M. Huibregtse, S. Gygi, and R. M. Krug. 2005. Human ISG15 conjugation targets both IFN-induced and constitutively expressed proteins functioning in diverse cellular pathways. *Proc. Natl. Acad. Sci. USA* **102**:10200–10205.

211. Zhao, T., L. Yang, Q. Sun, M. Arguello, D. W. Ballard, J. Hiscott, and R. Lin. 2007. The NEMO adaptor bridges the nuclear factor-kappaB and interferon regulatory factor signaling pathways. *Nat. Immunol.* **8**:592–600.

212. Zhou, A., J. Paranjape, T. L. Brown, H. Nie, S. Naik, B. Dong, A. Chang, B. Trapp, R. Fairchild, C. Colmenares, and R. H. Silverman. 1997. Interferon action and apoptosis are defective in mice devoid of 2′,5′-oligoadenylate-dependent RNase L. *EMBO J.* **16**:6355–6363.

213. Zurcher, T., R. M. Marion, and J. Ortin. 2000. Protein synthesis shut-off induced by influenza virus infection is independent of PKR activity. *J. Virol.* **74**:8781–8784.

Cellular Signaling and Innate Immune
Responses to RNA Virus Infections
Edited by A. R. Brasier et al.
© 2009 ASM Press, Washington, D.C.

Friedemann Weber
Richard M. Elliott

18

Bunyaviruses and Innate Immunity

The family *Bunyaviridae* is one of the most ubiquitous and widespread virus groups in the world (28, 29). Members of this family are able to infect arthropods, plants, and mammals, including humans. More than 350 named virus isolates are contained within the *Bunyaviridae* (29, 73), and several members cause significant disease in humans or domestic animals. Despite being recognized as an emerging threat, relatively little is known about their virulence mechanisms. Here we try to summarize the current state of knowledge about how the viruses of the *Bunyaviridae* succeed in establishing infection in the face of a powerful innate immune system.

THE FAMILY *BUNYAVIRIDAE*

Members of the *Bunyaviridae* are classified into five genera: *Orthobunyavirus, Phlebovirus, Hantavirus, Nairovirus,* and *Tospovirus.* In this chapter the term "bunyavirus" refers to a member of the *Bunyaviridae* family, while the terms "orthobunyavirus," "phlebovirus," etc., refer to viruses in the eponymous genus. With the exception of the plant-pathogenic tospoviruses, all

members of the family are able to infect mammalian hosts. The majority of these viruses are transmitted by arthropod vectors and are maintained in nature by a propagative cycle involving blood-feeding arthropods and susceptible vertebrate hosts. Many different arthropods, including mosquitoes, midges, sandflies, ticks, and bedbugs, can act as vectors, though each individual virus is usually restricted to a single or limited number of arthropod species. Exceptionally, the hantaviruses are maintained as persistent infections of rodents and are directly transmitted to humans via aerosolized rodent excreta (see references 29 and 92 for reviews).

The bunyavirus particle consists of a lipid envelope about 100 nm in diameter, containing two viral glycoproteins, that surrounds the tripartite RNA genome. The three single-stranded RNA segments are of negative polarity and are encapsidated by the nucleocapsid (N) protein in the form of ribonucleoprotein complexes. The genome encodes four structural proteins common to all members of the family: the viral polymerase (L) on the large (L) segment, the glycoproteins (Gn and Gc) on the medium (M) segment, and the N protein on the smallest (S) segment. Viruses within the

Friedemann **Weber**, Abteilung Virologie, Institut für Medizinische Mikrobiologie und Hygiene, Universität Freiburg, D-79008 Freiburg, Germany. **Richard M. Elliott,** Centre for Biomolecular Sciences, School of Biology, University of St. Andrews, St. Andrews, KY16 9ST, Scotland, United Kingdom.

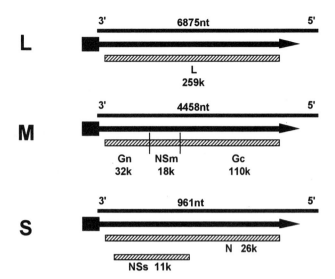

Figure 1 Bunyamwera orthobunyavirus coding strategy (not to scale). The three genomic RNA segments L, M, and S are shown as solid lines with their lengths in nucleotides (nt) shown above. mRNAs are indicated by arrows, with boxes depicting a nontemplated primer at the 5' end. Gene products are shown as hatched boxes, and protein designations and sizes (kDa) are indicated.

Orthobunyavirus, *Phlebovirus*, and *Tospovirus* genera also encode nonstructural proteins, either on the M segment (termed NSm) or on the S segment (NSs). The phlebovirus NSs protein and both the tospovirus NSm and NSs proteins are encoded in an ambisense strategy (see below). Figure 1 shows the genomic organization of Bunyamwera virus (BUNV), the prototype of both the family *Bunyaviridae* and the *Orthobunyavirus* genus, which is equipped with the maximal set of six viral proteins, namely L, Gn, Gc, NSm, N, and NSs. Viral replication takes place in the cytoplasm and assembled virus particles bud into Golgi-derived vesicles (29).

GENOME REPLICATION AND TRANSCRIPTION

In common with other negative-sense viruses, the first step in the infectious cycle after virion uncoating in the cytoplasm is primary transcription of the genomic segments by the virion-associated RNA polymerase to generate viral mRNAs (Fig. 2A). This is a primer-dependent process whereby the capped 5' ends of cytoplasmic cellular mRNAs are cleaved by an endonuclease function in the L protein to generate short heterogeneous oligonucleotides, about 12 to 18 nucleotides in length, that prime viral mRNA synthesis. Hence, bunyavirus mRNAs contain nontemplated nucleotides at their 5' ends. mRNA

transcription terminates at a signal 50 to 150 bases before the end of the genomic template RNA, but the viral mRNAs are not demonstrably polyadenylated. After translation of viral proteins, RNA synthesis switches to replication mode, which involves the primer-independent synthesis of a full-length, complementary positive-sense RNA, the antigenome, that acts as a template for the synthesis of progeny negative-sense genomes. Both the genome and antigenome RNAs are only found encapsidated by N protein as RNPs, whereas mRNAs are naked to allow ribosomal access. Encapsidation of genomes and antigenomes is presumably a way by which these viruses minimize double-stranded RNA formation in the infected cell.

In ambisense genome segments, open reading frames (ORFs) are found in both negative- and positive-sense orientations in the genomic RNA that are separated by an intergenic region that is usually predicted to form a stem-loop structure (Fig. 2B). Primary transcription produces a subgenomic mRNA for translation of the protein(s) encoded in the negative sense; transcription terminates in the intergenic region. Transcription of a second subgenomic mRNA, for the protein encoded in the positive sense, uses the antigenome as a template, thus providing the virus with some degree of temporal control for gene expression, as this transcriptional event occurs after the onset of genome replication.

Analyses of the terminal nucleotide sequences of bunyavirus RNAs revealed (i) that the sequences were invertedly complementary between the 5' and 3' ends; (ii) that within a virus the sequences of the L, M, and S segments were conserved; and (iii) that genus-specific consensus sequences could be defined (Table 1). Intriguingly, the 5' nucleotides of hantavirus and nairovirus genomic RNAs were found to be U residues, whereas dogma is that viral RNA polymerases only initiate transcription with a purine (A or G) residue (8). Furthermore, the U residue on hantavirus RNAs was shown to be monophosphorylated, rather than a triphosphorylated residue as is usually found on an exposed 5' end. This prompted Kolakofsky and colleagues to propose a "prime-and-realign" mechanism for genome replication (38) whereby RNA synthesis initiates with a G residue internally on the template, the nascent RNA then slides backwards along the reiterated terminal sequence (Table 1), and the resulting overhanging G is cleaved by the endonuclease activity of the L protein (Fig. 3). Although an attractive model to explain the presence of a 5' monophosphorylated U (5' pU), it has not proved possible so far to test the model experimentally. However, as will be seen later, 5' pU may help in preventing recognition of the genome RNA by cellular sensors.

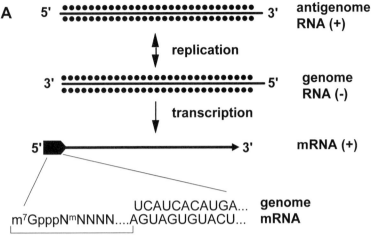

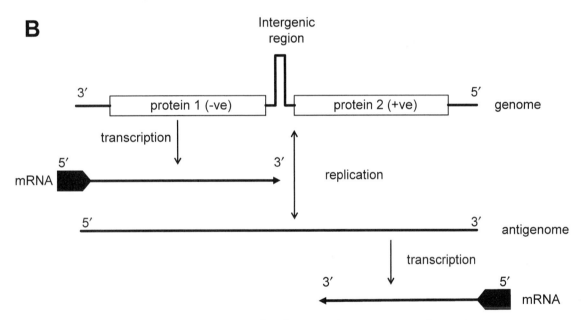

Figure 2 Bunyavirus transcription and replication. (A) Negative-sense bunyavirus genome segments. The genome RNA and the positive-sense complementary RNA known as the antigenome RNA are only found as ribonucleoprotein complexes and are encapsidated by N protein (•). The mRNA species contain host-derived primer sequences at their 5′ ends (•) and are truncated at the 3′ end relative to the viral RNA template; the mRNAs are not polyadenylated, nor are they encapsidated by N protein. The sequence at the 5′ end of an orthobunyavirus mRNA is shown. (B) Ambisense-sense bunyavirus genome segments. The genome RNA encodes proteins in both negative- and positive-sense orientations, separated by an intergenic region that can form a hairpin structure. The proteins are translated from subgenomic mRNAs, with the mRNA encoding protein 2 transcribed from the antigenome RNA after the onset of genome replication.

DISEASES CAUSED BY THE *BUNYAVIRIDAE*

Members of this diverse family cause disease in humans and their domestic animals and plant crops (Table 2). It is estimated that tospoviruses, which can infect more than 650 plant species, cause losses exceeding 1 billion U.S. dollars per year (87). There are four major types of human disease caused by members of the *Bunyaviridae*: fever, encephalitis, hemorrhagic fever, and severe respiratory illness. Many bunyaviruses that infect humans or animals can be economically important though they

Table 1 Consensus 3′ and 5′ terminal nucleotide sequences of bunyaviral genome RNAs

Virus	Sequence
Orthobunyavirus . . . . . 3′ UCAUCACAUGA . . . . UCGUGUGAUGA 5′	
Hantavirus . . . . 3′ AUCAUCAUCUG . . . . AUGAUGAU 5′	
Nairovirus . . . . 3′ AGAGUUUCU . . . . AGAAACUCU 5′	
Phlebovirus . . . . 3′ UGUGUUUC . . . . GAAACACA 5′	
Tospovirus . . . . 3′ UCUCGUUAG . . . . CUAACGAGA 5′	

cause a relatively mild febrile illness, self-limiting in course, and are rarely fatal. A prime example is Oropouche fever, which is the second most common arboviral disease after dengue fever in Brazil (84). Oropouche virus causes an acute febrile illness associated with headache, myalgia, arthralgia, and prostration. Recurrent epidemics have involved tens of thousands of patients and are economically significant due to man-hours lost from work. In the past 40 years Oropouche fever has emerged as a major public health problem in tropical areas of Central and South America (6).

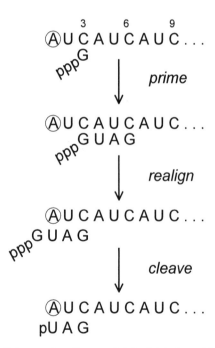

Figure 3 Prime-and-realign model for initiation of hantavirus genome synthesis. The template RNA is shown as the top strand with the 3′ nucleotide circled. Genome synthesis is initiated with GTP that aligns with the C residue at position 3 of the template. Following elongation of a few nucleotides, the nascent genomic strand realigns on the template by backwards slippage over the conserved sequence in the template. The initiating, now overhanging, GTP is then cleaved from the nascent strand, leaving a monophosphorylated U as the 5′ nucleotide in the genomic RNA. (Redrawn from reference 38.)

In the midwestern United States, La Crosse virus (LACV) causes severe encephalitis and aseptic meningitis in children and young adults (64). Around 75 to 100 cases of La Crosse encephalitis requiring hospitalization are reported annually (20), and up to 57% of these patients need to be admitted to the intensive care unit. More than 10% of the hospitalized patients will have long-lasting neurological deficits (64), with severe economic and social consequences (107). However, since less than 1.5% of LACV infections are clinically apparent, it is estimated that more than 300,000 infections occur annually in the midwestern United States alone (20). Related viruses including California encephalitis, Jamestown Canyon, and snowshoe hare have also been associated with human illness in North America, while Inkoo and Tahyna viruses cause human disease in Europe.

Naples and Sicilian sandfly fevers are rapid-onset febrile illnesses with associated nonspecific symptoms (headache, photophobia, myalgia, etc.) but no rash, and are nonfatal. The causative phleboviruses are transmitted by *Phlebotomus* (sandfly) species in the Mediterranean basin. Toscana virus (TOSV) causes aseptic meningitis in central Italy, and there is serological evidence for spread to neighboring countries. Punta Toro virus (PTV) has been repeatedly isolated from Panama and Colombia, where it causes acute febrile illness of short duration.

Viruses in all four genera that infect humans can cause hemorrhagic fevers. An orthobunyavirus, originally designated Garissa virus, was isolated from hemorrhagic fever cases during a large disease outbreak in East Africa in 1997 and 1998 (14). Genetic analysis revealed that the L and S genome segments were almost identical to those of BUNV (a relatively benign human pathogen), while the M segment was clearly distinct from that of BUNV. Subsequent nucleotide sequence analyses showed that Garissa virus was identical to a previously isolated virus called Ngari (NRIV), which, interestingly, had not been associated with hemorrhagic disease, and in turn that the M segment was very similar to that of Batai virus (18, 116). Thus, NRIV is likely to be a reassortant virus (the exact phlyogenetic relationships of the complete genomes of BUNV, Batai virus, and NRIV remain to be established),

Table 2 Selected important pathogens in the family *Bunyaviridae*

Genus	Virus	Disease	Vector	Distribution
Orthobunyavirus	Akabane	Cattle: abortion and congenital defects	Midge	Africa, Asia, Australia
	Cache Valley	Sheep, cattle: congenital defects	Mosquito	North America
	La Crosse	Humans: encephalitis	Mosquito	North America
	Ngari	Humans: hemorrhagic fever	Mosquito	Africa
	Oropouche	Humans: fever	Midge	South America
	Tahyna	Humans: fever	Mosquito	Europe
Hantavirus	Hantaan	Humans: severe HFRS; fatality rate 5 to 15%	Field mouse	Eastern Europe, Asia
	Seoul	Humans: moderate HFRS; fatality rate, 1%	Rat	Worldwide
	Puumala	Humans: mild HFRS; fatality rate, 0.1%	Bank vole	Western Europe
	Sin Nombre	Humans: hantavirus cardiopulmonary syndrome; fatality rate, 50%	Deer mouse	North America
Nairovirus	Crimean-Congo hemorrhagic fever	Humans: hemorrhagic fever; fatality rate, 20 to 80%	Tick, culicoid fly	Eastern Europe, Africa, Asia
	Nairobi sheep disease	Sheep, goats: fever, hemorrhagic gastroenteritis, abortion	Tick, culicoid fly, mosquito	Africa, Asia
Phlebovirus	Rift Valley fever	Humans: encephalitis, hemorrhagic fever, retinitis; fatality rate, 1 to 10% Domestic ruminants: necrotic hepatitis, hemorrhage, abortion	Mosquito	Africa
	Naples sandfly fever	Humans: fever	Sandfly	Europe, Africa
	Sicilian sandfly fever	Humans: fever	Sandfly	Europe, Africa
Tospovirus	Tomato spotted wilt	Plants: over 650 species, various symptoms	Thrips	Worldwide

illustrating the potential of genome reassortment to generate viruses with enhanced pathogenicity.

Rift Valley fever virus (RVFV) is a serious emerging pathogen affecting humans and livestock in sub-Saharan Africa, Egypt, Yemen, and Saudi Arabia. Since the first description of an outbreak in Kenya in 1931 (24), there have been recurrent epidemics, killing thousands of animals and hundreds of humans and causing significant economic losses (7). RVFV is primarily a mosquito-transmitted disease of cattle, sheep, and other ruminants in sub-Saharan Africa (35), causing necrotic hepatitis, hemorrhage, and abortion. Humans are often infected by close contact with infected animal material (72), causing a range of symptoms including temporarily incapacitating febrile illness, retinal disease, and in about 1% of cases, hemorrhagic fever; half of the patients with this last symptom will die (7, 35). The severity of RVFV zoonosis as well as the virus's capability to cause major epidemics in livestock and humans have prompted authorities to list RVFV as a notifiable disease and a potential biological weapon (12).

Crimean-Congo hemorrhagic fever virus (CCHFV) is a severe tick-borne nairovirus infection found in Africa, Asia, and Eastern Europe (30). CCHFV was originally described in the Crimea in the 1940s during an outbreak of hemorrhagic fever and later shown to be antigenically identical to a virus isolated from a febrile patient in the Congo. CCHFV can infect many mammalian species, as well as ostriches, though without causing disease. Human infections are acquired by tick bite, via contact with infected animal blood, or nosocomially; the overall case fatality rate is about 30%, with higher death rates associated with nosocomially acquired infections, presumably because of the higher viral inoculum.

Hantaviruses are associated with two major disease syndromes in man—hemorrhagic fever with renal syndrome (HFRS) and hantavirus pulmonary syndrome (HPS). HFRS has a wide distribution and ranges in severity from a mild illness, nephropathia epidemica, in Fenno-Scandinavia to the severe form in the Balkans and Asia. In China there are about 200,000 hospitalized cases per annum of HFRS caused by Hantaan (HTNV) and Seoul viruses, with about 5 to 15% mortality. HPS is caused by a number of newly discovered hantaviruses in the Americas. The first virus associated with HPS, now called Sin Nombre virus (SNV), made a dramatic appearance in 1993 in the Four Corners region of the United States. The identification and isolation of the virus, followed by investigation of environmental factors resulting in its entry into the human population, have become a classic paradigm for emerging viruses (54).

VIRAL PATHOGENESIS AND ADAPTIVE IMMUNE RESPONSES

Few members of the *Bunyaviridae* have been studied extensively in experimental animal hosts, largely because most of the important pathogens, such as RVFV, HTNV, SNV, and CCHFV, require special containment facilities or because there is no animal model that mimics human disease (40). Recent efforts from several groups, however, have led to a significant improvement of the situation.

A well-established animal model for pathogenicity of arthropod-transmitted bunyaviruses does exist for LACV virus in mice (41). In 16- to 18-day-old mice, LACV is both neuroinvasive (i.e., it can cross the blood-brain barrier) and neurovirulent (i.e., it infects cells of the central nervous system [CNS]). The neuroinvasiveness of LACV was mapped to the M segment (52). Subcutaneous inoculation mimics the route of transmission by the arthropod vector and the virus is distributed though the bloodstream to tissues such as striated muscle, in which it can replicate (41). Peripheral replication then leads to a plasma viremia that lasts approximately 1 week. During this time blood-sucking vector mosquitoes can acquire the virus and are able to spread the virus to uninfected animals. Due to the high levels of plasma viremia, the virus can cross the blood-brain barrier, invade the CNS, and infect neurons (10, 51). Infection of CNS neurons is so specific and ubiquitous (10) that LACV served as a model system to reveal the ability of neurons to produce type I interferons (IFNs) (25). High-level virus replication in the CNS leads to Bcl-2-mediated inhibition of apoptosis of the infected cells (79) and, eventually, fatal encephalitis.

For TOSV, a mouse model was recently established that mimics symptomatic CNS infection of humans (23). By passaging a virus isolate from a human clinical case in mouse brains, a highly virulent neuroadapted strain was selected that was able to efficiently infect neurons. Similar to LACV, the mouse-adapted TOSV induced apoptosis of infected cells, but unlike LACV it was not able to cross the blood-brain barrier by itself. After

intracerebral inoculation, a large number of neurons were infected. The extensive brain tissue injury caused by apoptosis of infected neurons leads to symptoms and eventually death.

Thus, for both LACV and TOSV, animal models suggest that the acute encephalitis and disease are directly caused by the virus's killing of nonrenewable neurons, and that inflammation or other immune responses are not involved in pathogenesis.

Bunyaviruses such as RVFV, HTNV, SNV, and CCHFV belong to the viral hemorrhagic fever group, producing systemic and often fatal disease (40). The clinical picture of the classic viral hemorrhagic fever syndrome usually presents as an initial fever, myalgia, and prostration, often with an abrupt onset. For RVFV and CCHFV, hepatic involvement is common, as are thrombocytopenia and lymphopenia. Among animals, sheep, goats, and cattle are most susceptible to RVFV. The acute form is most common in young animals, where it causes high mortality, mostly by liver destruction, whereas abortion is often the only sign of infection in adults. Experimentally infected rhesus monkeys develop viremia and liver damage, and 20% proceed to fatal disease with hemorrhagic fever (81). Interestingly, an attenuated recombinant RVFV lacking the NSm proteins was found to cause delayed but severe neurologic symptoms in rats, whereas animals infected with wild-type RVFV all died from fulminant hepatitis (9).

For CCHFV, the only host displaying obvious disease is man (30). The course of infection can be divided into four stages, namely incubation, prehemorrhagic (fever, myalgia, nausea, diarrhea), hemorrhagic, and convalescent. CCHFV is known to infect mononuclear phagocytes, endothelial cells, and hepatocytes (19). This could suggest that the dysregulation of the vascular system and the immune response as well as the typical hepatitis are directly caused by the virus. However, a recent study demonstrated that infection of epithelial cells by CCHFV did not cause any damage or leakage (22). Thus, at least the hemorrhages and coagulation disturbances are most probably caused indirectly, possibly by the high levels of the proinflammatory cytokines interleukin-6 (IL-6) and tumor necrosis factor-α (TNF-α) that can be detected in CCHFV patients and correlate with disease severity (31, 76). TNF-α is known to elevate vascular permeability (32, 90).

In contrast to the arthropod-transmitted bunyaviruses, hantaviruses persist lifelong in the natural rodent host. They inhabit the lungs, salivary glands, and kidneys and are maintained by cyclical transmission via aerosolized excreta (65). Experimental studies revealed that infection results first in a short acute stage with viremia, followed by a chronic stage with viral shedding in the rodent urine (58). Newborn animals are usually protected against vertical transmission by maternal antibodies (67, 119). Furthermore, nude rats lacking a T-cell response succumb rapidly to Seoul hantavirus infection (26). This suggests that downregulation of virus titers to the persistence level is, at least in part, mediated by the adaptive immune system and results in apathogenic infection of the rodent host (26). Interestingly, two recent studies revealed an important role of regulatory CD4[+] T cells for the maintenance of hantaviruses in their natural hosts. These cells were found to express substantial amounts of transforming growth factor-β in infected animals, thus suppressing the proinflammatory immune responses and fostering virus persistence (27, 93). By contrast, in incidentally infected humans, hantaviruses can cause acute infections that result in the often fatal HFRS or HPS, diseases that are not completely understood with respect to pathogenesis. However, and importantly, an animal model for HPS involving Syrian hamsters and Andes virus has now been established (47).

A recent temporal pathogenesis study based on this model revealed that lymphocyte apoptosis was the earliest marker of disease, followed by viremia, multiorgan infection, and pulmonary edema (109). In line with patient studies (118), lung tissue examinations showed viral antigen in endothelial cells, but no microanatomical gaps between adjacent cells, suggesting that vascular leakage, the pathological hallmark of HFRS and HPS, could be directly attributable to infection. Possibly the increased number of vacuoles observed inside infected endothelial cells is a sign of enforced transcellular transport of plasma (109). On the other hand, the massive infiltrates of mononuclear cells that are present in the lung tissue may also contribute to vascular leakage by producing proinflammatory cytokines. Just as it is described for CCHFV, high levels of IL-6 and TNF-α are present in hantavirus patients (61, 66). In vitro, IL-6 (39) and the chemokines RANTES and IP-10 (100) are induced in endothelial cells by HTNV and SNV. Hantavirus infection and a Th1 immune response may therefore cooperate to trigger immune-mediated increases in vascular permeability (100).

It could be generalized that bunyavirus infection usually commences with a transient, high-level plasma viremia. The host then clears the virus and survives, succumbs to infection (arthropod-transmitted bunyaviruses and hantaviruses in humans), or becomes chronically infected (hantaviruses in rodents). Effective humoral immunity is possible, since a single infection elicits high-titer neutralizing antibodies that can protect the organism against a rechallenge (47, 78). Cell-mediated immunity

may also play a role, since CD4[+] T cells contribute to protection against LACV (94), and for hantaviruses cross-reactive cell-mediated immune responses have been observed in laboratory animals (5) and in humans (108). Experimental vaccination using cDNA-containing plasmids, attenuated virus, or killed virus has been shown to be effective against RVFV (68, 71, 110), LACV (77, 94), and HTNV (46, 47), and such vaccines may be useful against other bunyaviruses as well. Protective effects have mostly been described for neutralizing antibodies against the envelope proteins (46, 94, 110), whereas immune reactions against the N protein have no or weaker effects (46, 74, 110).

INNATE IMMUNE RESPONSES

An age-dependent susceptibility to many bunyavirus diseases exists for humans as well as for livestock and laboratory animals (35, 41, 55). As has been demonstrated for LACV, newborn animals develop higher titers of viremia and are readily killed, whereas older animals often undergo mild or inapparent infection (51, 78). This may suggest that the initial host resistance is mediated by the type I IFN (IFN-α/β) system, which is not fully matured in young animals (82). Indeed, growth of several bunyaviruses can be inhibited by administering IFNs (4, 62, 63, 69, 70, 102, 103). Moreover, mice lacking a functional type I IFN receptor are highly susceptible to infection with BUNV, LACV, Dugbe nairovirus (DUGV), HTNV, and RVFV (10, 13, 15, 45, 111, 115). IFNs are an important part of the innate immune system. They are synthesized by the infected cell and establish a first line of defense by priming neighboring cells to express antiviral factors, thereby limiting the extent of virus spread (89). The best-characterized antiviral factors induced by type I IFNs are the protein kinase R (PKR) (37), the 2',5'-oligoadenylate synthetase/RNase L system (49), and the Mx proteins (43), which interfere with viral transcription, translation, or genome replication. Strong antibunyaviral activity has been demonstrated for MxA, whereas PKR has a modest effect against BUNV, but not DUGV, and RNase L appears not to have any effect (15, 99).

Mx proteins are large IFN-induced GTPases. Growth of several members of the family, including LACV, RVFV, CCHFV, DUGV, and HTNV, is reduced if the human MxA protein is expressed in the cell (3, 16, 36, 53). For LACV, inhibition by MxA was also shown in vivo, whereas in the DUGV animal model MxA was not sufficient for protection (15, 45), suggesting that other IFN-stimulated genes are also important for the antibunyaviral effect of IFN. The inhibition of LACV replication has been studied in depth, and it has been shown that the Mx protein sequesters the viral N protein, thereby depleting the viral RNA synthesis machinery of a factor that is essential for replicating the genome (56).

Several in vivo studies demonstrate that IFNs and proinflammatory cytokines determine the outcome of infection. A late onset of the IFN response was correlated with increased susceptibility of rhesus monkeys to RVFV (70), and a mouse model of RVFV infection revealed very little IFN production (13). For the related phlebovirus PTV, it was shown that virulence in Syrian hamsters depends on the ability of virus strains to delay IFN induction (2, 80). Similarly, human-pathogenic hantaviruses suppress IFN induction or signaling more efficiently than their nonpathogenic relatives (39, 57, 97). Interestingly, a recent study involving hantavirus-infected patients showed that type I IFN levels did not increase during the course of disease and that IFN-λ levels were even decreased (98). In mice, PTV induced large amounts of IL-6 (42). Curiously, it was found that IL-6 production was drastically lowered in animals lacking Toll-like receptor (TLR) 3, but that the TLR3$^{-/-}$ mice were much more resistant to PTV disease than wild-type animals, although virus titers were comparable. On the other hand, IL-6-deficient animals were more, not less, susceptible to PTV infection, suggesting that IL-6 plays a protective role but is detrimental if released in excess. Involvement of proinflammatory cytokines in pathogenesis has also been shown for hantaviruses and CCHFV, as mentioned above (66).

Thus, taken together, an efficient IFN response appears to be important for the host to control the critical first viremic phase of bunyavirus infection. Only then has the adaptive immune system enough time to develop neutralizing antibodies for clearing the virus and protecting the host against recurrent infection. However, some cytokines, such as IL-6 or TNF-α, and chemokines that are induced partly by the same signaling pathways as type I IFNs can have both a protective and a detrimental effect, depending on the timing, location, and amount being made.

How are the cytokines induced which on the one hand can mediate host protection (IFN-α/β) but on the other are responsible for severe symptoms (IL-6, TNF-α)? Very little information is available with regard to this important point, but based on current data it is clear that bunyaviruses have evolved various strategies to avoid or block cytokine induction or signaling (see below). Nevertheless, such viral countermeasures modulate rather than abrogate cytokine induction (43, 44, 112, 113) and may only affect certain branches of the signaling pathways. Induction may be triggered by several bunyavirus molecules. Similar to other negative-strand RNA

viruses, bunyaviruses produce very little double-stranded RNA, the classical inducer of IFN and other cytokines (114). However, TLR3 induction by PTV and PKR activation by BUNV indicate the presence of small but biologically active amounts of double-stranded RNA in infected cells (42, 99). A more important signaling pathway may be triggered by recognition of the genomic single-stranded RNA carrying a 5′ triphosphate group by retinoic acid-inducible gene I (RIG-I), as shown for influenza and rabies viruses (48, 83). However, this needs to be formally proven for bunyaviruses. Moreover, for hantaviruses that do not contain a triphosphate group at the 5′ ends of their genomic RNAs (38), results from one of us indicate that hantaviral RNAs indeed do not activate RIG-I (42a). At least in the case of hantaviruses it is therefore conceivable that other signaling pathways are more important for virus recognition. A recent study suggests that cytokine induction by SNV can occur independently of IFN regulatory factor (IRF) 3, IRF7, RIG-I, melanoma differentiation-associated gene 5 (MDA-5), or Toll–IL-1-resistance (TIR) domain-containing adaptor-inducing IFN-β (TRIF) or myeloid differentiation factor 88 (MyD88)-dependent TLR pathways (86). Cytokine induction is also independent of viral replication (85) and most probably occurs through an uncharacterized pattern recognition receptor (PRR) located at the cell surface (86). Indications of noncanonical virus recognition were also found for other enveloped viruses (21, 75) and may be mediated by glycoprotein binding. Interestingly, for alphaviruses it was found that mosquito cell-derived virus preparations were poor IFN inducers in myeloid dendrititc cells, whereas mammalian cell-derived viruses exhibited strong induction (95). This striking difference is most probably caused by differences in glycosylation, cholesterol content, or RNA modifications. A similar scenario is conceivable for those bunyaviruses that use arthropods as vectors. Another molecular pattern that could induce cytokine production is the viral nucleocapsid, as previously shown for two other negative-sense RNA viruses, vesicular stomatitis virus and measles virus (104, 105). Thus, bunyaviruses may activate cytokine production both by well-described and by noncanonical pathways, and future studies will have to address their relative contributions.

VIRAL COUNTERMEASURES

The cytokine induction profile of any given virus in a given cell type is defined (i) by the presence or absence of particular molecular patterns that could act as IFN elicitors and (ii) by the strength and specificity of its IFN-antagonistic factors. The first IFN antagonists described for bunyaviruses were the NSs proteins of the orthobunyavirus BUNV and the phlebovirus RVFV (13, 17, 111). The NSs of the orthobunyavirus LACV acts in a similar manner by abrogating the transcription of IFN genes (10), and reassortment and overexpression studies suggest that the phlebovirus PTV also expresses an IFN-antagonistic NSs protein (80). Although different in sequence, size, and mode of expression, both the orthobunyavirus and phlebovirus NSs proteins act by blocking host-cell RNA polymerase II (RNAP II). The NSs protein of RVFV interacts with the p44 subunit of the essential transcription factor TFIIH; it forms filamentous structures with p44 in the nucleus that also contain the XPD subunit of TFIIH. In this way NSs sequesters p44 and XPD, thus blocking assembly of TFIIH and resulting in a decrease in overall host transcription (59). BUNV NSs acts by interfering with phosphorylation of the carboxy-terminal domain of the large subunit of RNAP II, which also results in a decrease in host transcription (106). The BUNV NSs protein interacts with the MED8 component of Mediator, a protein complex necessary for mRNA production. The interacting domain on NSs was mapped to the C-terminal region, and a recombinant virus in which the interacting domain in NSs was deleted had strongly reduced ability to inhibit host protein expression. In addition, the virus expressing the truncated NSs protein was unable to inhibit the IFN response and behaved similarly to a virus lacking NSs entirely (60). For LACV the molecular target of NSs is unknown, but reverse transcription-PCR analyses and in vivo labeling indicate that, like BUNV NSs, it interferes with host cell transcription and causes a general shutoff of host cell synthesis (11).

Apparently, as bunyaviruses replicate in the cytoplasm and do not need ongoing cellular transcription for cap-snatching (88), blocking RNAP II function by NSs is the method of choice for orthobunyaviruses and phleboviruses to counteract innate immune responses. As the NSs of phleboviruses is about three times the size of the NSs of orthobunyaviruses, it is conceivable that additional functions are encoded on the former.

Hantaviruses were in general not considered to express an NSs equivalent protein, though a short ORF within the N ORF is conserved in the S segments of Puumala-like and SNV-like viruses, but not in the S segments of HTNV or Seoul virus (96). Some evidence for the expression of an NSs protein has been obtained for Puumala virus and Tula hantavirus, but the protein, however, is only weakly active (50). For the pathogenic New York-1 hantavirus, but not for the apathogenic Prospect Hill hantavirus, it was found that the cytoplasmic tail domain of the Gn (G1) glycoprotein downregulates IFN induction (1). Moreover, HTNV and New York-1 hantavirus virus were insensitive to IFN if applied >12 h after infection (1), most probably because the viral glycoproteins also downregulate IFN-induced Stat1/2 activation, as shown for Andes and Prospect Hill hantaviruses (97). For CCHFV, neither an NSs nor any other IFN antagonist has been identified so far, but the profound susceptibility to IFN and MxA (3, 4) implies virulence mechanisms similar to other bunyaviruses.

OPEN QUESTIONS AND OUTLOOK

Despite the significant economic and medical impact of this huge RNA virus family, the interactions of bunyaviruses with the innate immune system are only incompletely understood. Neither induction pathways nor viral IFN antagonists have been systematically characterized, let alone compared between the different family members. Such data are urgently needed, as virus-innate immune system interactions are critical determinants of pathogenesis. Moreover, viruses with targeted deletions in their IFN-antagonistic functions are excellent candidates for live virus vaccines. They can be grown to high titers in IFN-deficient cell cultures (117) but are attenuated in vivo since they elicit robust innate and adaptive immune responses. This concept has been proven for influenza viruses (33, 34, 91, 101) and several other viruses (reviewed in reference 112), and will most likely also apply to bunyaviruses. Also, given the susceptibility of bunyaviruses to the antiviral effect of IFN, interfering with viral IFN antagonism or enhancing PRR recognition by pharmacological means could be a therapeutic possibility in the future. Thus, a better understanding of the interplay between bunyaviruses and the innate immune response, as with most other viruses, can help in the design of new strategies for prevention and therapy.

References

1. **Alff, P. J., I. N. Gavrilovskaya, E. Gorbunova, K. Endriss, Y. Chong, E. Geimonen, N. Sen, N. C. Reich, and E. R. Mackow.** 2006. The pathogenic NY-1 hantavirus G1 cytoplasmic tail inhibits RIG-I- and TBK-1-directed interferon responses. *J. Virol.* 80:9676–9686.
2. **Anderson, G. W., Jr., M. V. Slayter, W. Hall, and C. J. Peters.** 1990. Pathogenesis of a phleboviral infection (Punta Toro virus) in golden Syrian hamsters. *Arch. Virol.* 114: 203–212.
3. **Andersson, I., L. Bladh, M. Mousavi-Jazi, K. E. Magnusson, A. Lundkvist, O. Haller, and A. Mirazimi.** 2004. Human MxA protein inhibits the replication of Crimean-Congo hemorrhagic fever virus. *J. Virol.* 78: 4323–4329.

4. Andersson, I., A. Lundkvist, O. Haller, and A. Mirazimi. 2006. Type I interferon inhibits Crimean-Congo hemorrhagic fever virus in human target cells. *J. Med. Virol.* 78:216–222.

5. Asada, H., K. Balachandra, M. Tamura, K. Kondo, and K. Yamanishi. 1989. Cross-reactive immunity among different serotypes of virus causing haemorrhagic fever with renal syndrome. *J. Gen. Virol.* 70:819–825.

6. Azevedo, R. S., M. R. Nunes, J. O. Chiang, G. Bensabath, H. B. Vasconcelos, A. Y. Pinto, L. C. Martins, H. A. Monteiro, S. G. Rodrigues, and P. F. Vasconcelos. 2007. Reemergence of Oropouche fever, northern Brazil. *Emerg. Infect. Dis.* 13:912–915.

7. Balkhy, H. H., and Z. A. Memish. 2003. Rift Valley fever: an uninvited zoonosis in the Arabian Peninsula. *Int. J. Antimicrob. Agents.* 21:153–157.

8. Banerjee, A. K. 1980. 5′ terminal cap structure in eukaryotic messenger ribonucleic acids. *Microbiol. Rev.* 44:175–205.

9. Bird, B. H., C. G. Albarino, and S. T. Nichol. 2007. Rift Valley fever virus lacking NSm proteins retains high virulence in vivo and may provide a model of human delayed onset neurologic disease. *Virology* 362:10–15.

10. Blakqori, G., S. Delhaye, M. Habjan, C. D. Blair, I. Sanchez-Vargas, K. E. Olson, G. Attarzadeh-Yazdi, R. Fragkoudis, A. Kohl, U. Kalinke, S. Weiss, T. Michiels, P. Staeheli, and F. Weber. 2007. La Crosse bunyavirus nonstructural protein NSs serves to suppress the type I interferon system of mammalian hosts. *J. Virol.* 81:4991–4999.

11. Blakqori, G., and F. Weber. 2005. Efficient cDNA-based rescue of La Crosse bunyaviruses expressing or lacking the nonstructural protein NSs. *J. Virol.* 79:10420–10428.

12. Borio, L. T. Inglesby, C. J. Peters, A. L. Schmaljohn, J. M. Hughes, P. B. Jahrling, T. Ksiazek, K. M. Johnson, A. Meyerhoff, T. O'Toole, M. S. Ascher, J. Bartlett, J. G. Breman, E. M. Eitzen Jr., M. Hamburg, J. Hauer, D. A. Henderson, R. T. Johnson, G. Kwik, M. Layton, S. Lillibridge, G. J. Nabel, M. T. Osterholm, T. M. Perl, P. Russell, K. Tonat; Working Group on Civilian Biodefense. 2002. Hemorrhagic fever viruses as biological weapons: medical and public health management. *JAMA* 287:2391–2405.

13. Bouloy, M., C. Janzen, P. Vialat, H. Khun, J. Pavlovic, M. Huerre, and O. Haller. 2001. Genetic evidence for an interferon-antagonistic function of Rift Valley fever virus nonstructural protein NSs. *J. Virol.* 75:1371–1377.

14. Bowen, M. D., S. G. Trappier, A. J. Sanchez, R. F. Meyer, C. S. Goldsmith, S. R. Zaki, L. M. Dunster, C. J. Peters, T. G. Ksiazek, and S. T. Nichol. 2001. A reassortant bunyavirus isolated from acute hemorrhagic fever cases in Kenya and Somalia. *Virology* 291:185–190.

15. Boyd, A., J. K. Fazakerley, and A. Bridgen. 2006. Pathogenesis of Dugbe virus infection in wild-type and interferon-deficient mice. *J. Gen. Virol.* 87:2005–2009.

16. Bridgen, A., D. A. Dalrymple, F. Weber, and R. M. Elliott. 2004. Inhibition of Dugbe nairovirus replication by human MxA protein. *Virus. Res.* 99:47–50.

17. Bridgen, A., F. Weber, J. K. Fazakerley, and R. M. Elliott. 2001. Bunyamwera bunyavirus nonstructural protein NSs is a nonessential gene product that contributes to viral pathogenesis. *Proc. Natl. Acad. Sci. USA* 98:664–669.

18. Briese, T., B. Bird, V. Kapoor, S. T. Nichol, and W. I. Lipkin. 2006. Batai and Ngari viruses: M segment reassortment and association with severe febrile disease outbreaks in East Africa. *J. Virol.* 80:5627–5630.

19. Burt, F. J., R. Swanepoel, W. J. Shieh, J. F. Smith, P. A. Leman, P. W. Greer, L. M. Coffield, P. E. Rollin, T. G. Ksiazek, C. J. Peters, and S. R. Zaki. 1997. Immunohistochemical and in situ localization of Crimean-Congo hemorrhagic fever (CCHF) virus in human tissues and implications for CCHF pathogenesis. *Arch. Pathol. Lab. Med.* 121:839–846.

20. Calisher, C. H. 1994. Medically important arboviruses of the United States and Canada. *Clin. Microbiol. Rev.* 7:89–116.

21. Collins, S. E., R. S. Noyce, and K. L. Mossman. 2004. Innate cellular response to virus particle entry requires IRF3 but not virus replication. *J. Virol.* 78:1706–1717.

22. Connolly-Andersen, A. M., K. E. Magnusson, and A. Mirazimi. 2007. Basolateral entry and release of Crimean-Congo hemorrhagic fever virus in polarized MDCK-1 cells. *J. Virol.* 81:2158–2164.

23. Cusi, M. G., G. Gori Savellini, C. Terrosi, G. Di Genova, M. Valassina, M. Valentini, S. Bartolommei, and C. Miracco. 2005. Development of a mouse model for the study of Toscana virus pathogenesis. *Virology* 333:66–73.

24. Daubney, R., J. R. Hudson, and P. C. Garnham. 1931. Enzootic hepatitis or Rift Valley fever: an undescribed virus disease of sheep, cattle and man from East Africa. *J. Pathol. Bacteriol.* 34:545–549.

25. Delhaye, S., S. Paul, G. Blakqori, M. Minet, F. Weber, P. Staeheli, and T. Michiels. 2006. Neurons produce type I interferon during viral encephalitis. *Proc. Natl. Acad. Sci. USA* 103:7835–7840.

26. Dohmae, K., M. Okabe, and Y. Nishimune. 1994. Experimental transmission of hantavirus infection in laboratory rats. *J. Infect. Dis.* 170:1589–1592.

27. Easterbrook, J., M. Zink, and S. Klein. 2007. Regulatory T cells enhance persistence of the zoonotic pathogen Seoul virus in its reservoir host. *Proc. Natl. Acad. Sci. USA* 104:15502–15507.

28. Elliott, R. M. 1997. Emerging viruses: the Bunyaviridae. *Mol. Med.* 3:572–577.

29. Elliott, R. M. (ed.). 1996. *The Bunyaviridae.* Plenum Press, New York, NY.

30. Ergonul, O. 2006. Crimean-Congo haemorrhagic fever. *Lancet Infect. Dis.* 6:203–214.

31. Ergonul, O., S. Tuncbilek, N. Baykam, A. Celikbas, and B. Dokuzoguz. 2006. Evaluation of serum levels of interleukin (IL)-6, IL-10, and tumor necrosis factor-alpha in patients with Crimean-Congo hemorrhagic fever. *J. Infect. Dis.* 193:941–944.

32. Feldmann, H., H. Bugany, F. Mahner, H. D. Klenk, D. Drenckhahn, and H. J. Schnittler. 1996. Filovirus-induced endothelial leakage triggered by infected monocytes/macrophages. *J. Virol.* 70:2208–2214.

33. Ferko, B., J. Stasakova, J. Romanova, C. Kittel, S. Sereinig, H. Katinger, and A. Egorov. 2004. Immunogenicity and protection efficacy of replication-deficient influenza A viruses with altered NS1 genes. *J. Virol.* 78:13037–13045.

34. Fernandez-Sesma, A., S. Marukian, B. J. Ebersole, D. Kaminski, M. S. Park, T. Yuen, S. C. Sealfon, A. Garcia-Sastre, and T. M. Moran. 2006. Influenza virus

evades innate and adaptive immunity via the NS1 protein. *J. Virol.* **80:**6295–6304.

35. **Flick, R., and M. Bouloy.** 2005. Rift Valley fever virus. *Curr. Mol. Med.* **5:**827–834.

36. **Frese, M., G. Kochs, H. Feldmann, C. Hertkorn, and O. Haller.** 1996. Inhibition of bunyaviruses, phleboviruses, and hantaviruses by human MxA protein. *J. Virol.* **70:** 915–923.

37. **Garcia, M. A., J. Gil, I. Ventoso, S. Guerra, E. Domingo, C. Rivas, and M. Esteban.** 2006. Impact of protein kinase PKR in cell biology: from antiviral to antiproliferative action. *Microbiol. Mol. Biol. Rev.* **70:**1032–1060.

38. **Garcin, D., M. Lezzi, M. Dobbs, R. M. Elliott, C. Schmaljohn, C. Y. Kang, and D. Kolakofsky.** 1995. The 5′ ends of Hantaan virus (Bunyaviridae) RNAs suggest a prime-and-realign mechanism for the initiation of RNA synthesis. *J. Virol.* **69:**5754–5762.

39. **Geimonen, E., S. Neff, T. Raymond, S. S. Kocer, I. N. Gavrilovskaya, and E. R. Mackow.** 2002. Pathogenic and nonpathogenic hantaviruses differentially regulate endothelial cell responses. *Proc. Natl. Acad. Sci. USA* **99:** 13837–13842.

40. **Geisbert, T. W., and P. B. Jahrling.** 2004. Exotic emerging viral diseases: progress and challenges. *Nat. Med.* **10:** S110–S121.

41. **Gonzales-Scarano, F., K. Bupp, and N. Nathanson.** 1996. Pathogenesis of diseases caused by viruses of the Bunyavirus genus, p. 227–251. *In* R. M. Elliott (ed.), *The Bunyaviridae.* Plenum Press, New York, NY.

42. **Gowen, B. B., J. D. Hoopes, M. H. Wong, K. H. Jung, K. C. Isakson, L. Alexopoulou, R. A. Flavell, and R. W. Sidwell.** 2006. TLR3 deletion limits mortality and disease severity due to phlebovirus infection. *J. Immunol.* **177:** 6301–6307.

42a. **Habjan, M., I. Andersson, J. Klingström, M. Schümann, A. Martin, P. Zimmermann, V. Wagner, A. Pichlmair, U. Schneider, E. Mühlberger, A. Mirazimi, and F. Weber.** 2008. Processing of genome 5′ termini as a strategy of negative-strand RNA viruses to avoid RIG-I-dependent interferon induction. *PLoS ONE* **3:**e2032.

43. **Haller, O., G. Kochs, and F. Weber.** 2007. Interferon, Mx, and viral countermeasures. *Cytokine Growth Factor Rev.* **18:**425–433.

44. **Haller, O., G. Kochs, and F. Weber.** 2006. The interferon response circuit: induction and suppression by pathogenic viruses. *Virology* **344:**119–130.

45. **Hefti, H. P., M. Frese, H. Landis, C. Di Paolo, A. Aguzzi, O. Haller, and J. Pavlovic.** 1999. Human MxA protein protects mice lacking a functional alpha/beta interferon system against La crosse virus and other lethal viral infections. *J. Virol.* **73:**6984–6991.

46. **Hooper, J. W., K. I. Kamrud, F. Elgh, D. Custer, and C. S. Schmaljohn.** 1999. DNA vaccination with hantavirus M segment elicits neutralizing antibodies and protects against Seoul virus infection. *Virology* **255:**269–278.

47. **Hooper, J. W., T. Larsen, D. M. Custer, and C. S. Schmaljohn.** 2001. A lethal disease model for hantavirus pulmonary syndrome. *Virology* **289:**6–14.

48. **Hornung, V., J. Ellegast, S. Kim, K. Brzozka, A. Jung, H. Kato, H. Poeck, S. Akira, K. K. Conzelmann, M. Schlee, S. Endres, and G. Hartmann.** 2006. 5′-Triphosphate RNA is the ligand for RIG-I. *Science* **314:**994–997.

49. **Hovanessian, A. G., and J. Justesen.** 2007. The human 2′-5′oligoadenylate synthetase family: unique interferon-inducible enzymes catalyzing 2′-5′ instead of 3′-5′ phosphodiester bond formation. *Biochimie* **89:**779–788.

50. **Jaaskelainen, K. M., P. Kaukinen, E. S. Minskaya, A. Plyusnina, O. Vapalahti, R. M. Elliott, F. Weber, A. Vaheri, and A. Plyusnin.** 2007. Tula and Puumala hantavirus NSs ORFs are functional and the products inhibit activation of the interferon-beta promoter. *J. Med. Virol.* **79:**1527–1536.

51. **Janssen, R., F. Gonzalez-Scarano, and N. Nathanson.** 1984. Mechanisms of bunyavirus virulence. Comparative pathogenesis of a virulent strain of La Crosse and an avirulent strain of Tahyna virus. *Lab. Invest.* **50:**447–455.

52. **Janssen, R. S., N. Nathanson, M. J. Endres, and F. Gonzalez-Scarano.** 1986. Virulence of La Crosse virus is under polygenic control. *J. Virol.* **59:**1–7.

53. **Kanerva, M., K. Melen, A. Vaheri, and I. Julkunen.** 1996. Inhibition of Puumala and Tula hantaviruses in Vero cells by MxA protein. *Virology* **224:**55–62.

54. **Khaiboullina, S., S. Morzunov, and S. St. Jeor.** 2005. Hantaviruses: molecular biology, evolution and pathogenesis. *Curr. Mol. Med.* **5:**773–790.

55. **Kim, G. R., and K. T. McKee, Jr.** 1985. Pathogenesis of Hantaan virus infection in suckling mice: clinical, virologic, and serologic observations. *Am. J. Trop. Med. Hyg.* **34:**388–395.

56. **Kochs, G., C. H. Janzen, H. Hohenberg, and O. Haller.** 2002. Antivirally active MxA protein sequesters La Crosse virus nucleocapsid protein into perinuclear complexes. *Proc. Natl. Acad. Sci. USA* **99:**3153–3158.

57. **Kraus, A. A., M. J. Raftery, T. Giese, R. Ulrich, R. Zawatzky, S. Hippenstiel, N. Suttorp, D. H. Kruger, and G. Schonrich.** 2004. Differential antiviral response of endothelial cells after infection with pathogenic and nonpathogenic hantaviruses. *J. Virol.* **78:**6143–6150.

58. **Lee, H. W., P. W. Lee, L. J. Baek, C. K. Song, and I. W. Seong.** 1981. Intraspecific transmission of Hantaan virus, etiologic agent of Korean hemorrhagic fever, in the rodent *Apodemus agrarius. Am. J. Trop. Med. Hyg.* **30:**1106–1112.

59. **Le May, N., S. Dubaele, L. P. De Santis, A. Billecocq, M. Bouloy, and J. M. Egly.** 2004. TFIIH transcription factor, a target for the Rift Valley hemorrhagic fever virus. *Cell* **116:**541–550.

60. **Leonard, V. H., A. Kohl, T. J. Hart, and R. M. Elliott.** 2006. Interaction of Bunyamwera orthobunyavirus NSs protein with mediator protein MED8: a mechanism for inhibiting the interferon response. *J. Virol.* **80:**9667–9675.

61. **Linderholm, M., C. Ahlm, B. Settergren, A. Waage, and A. Tarnvik.** 1996. Elevated plasma levels of tumor necrosis factor (TNF)-alpha, soluble TNF receptors, interleukin (IL)-6, and IL-10 in patients with hemorrhagic fever with renal syndrome. *J. Infect. Dis.* **173:**38–43.

62. **Livonesi, M. C., R. L. de Sousa, S. J. Badra, and L. T. Figueiredo.** 2007. In vitro and in vivo studies of the interferon-alpha action on distinct orthobunyaviruses. *Antiviral Res.* **75:**121–128.

63. **Luby, J. P.** 1975. Sensitivities of neurotropic arboviruses to human interferon. *J. Infect. Dis.* **132:**361–367.

64. **McJunkin, J. E., E. C. de los Reyes, J. E. Irazuzta, M. J. Caceres, R. R. Khan, L. L. Minnich, K. D. Fu, G. D. Lovett,**

T. Tsai, and A. Thompson. 2001. La Crosse encephalitis in children. *N. Engl. J. Med.* **344:**801–807.

65. Meyer, B. J., and C. S. Schmaljohn. 2000. Persistent hantavirus infections: characteristics and mechanisms. *Trends Microbiol.* **8:**61–67.

66. Mori, M., A. L. Rothman, I. Kurane, J. M. Montoya, K. B. Nolte, J. E. Norman, D. C. Waite, F. T. Koster, and F. A. Ennis. 1999. High levels of cytokine-producing cells in the lung tissues of patients with fatal hantavirus pulmonary syndrome. *J. Infect. Dis.* **179:**295–302.

67. Morita, C., S. Morikawa, K. Sugiyama, T. Komatsu, H. Ueno, and T. Kitamura. 1993. Inability of a strain of Seoul virus to transmit itself vertically in rats. *Jpn. J. Med. Sci. Biol.* **46:**215–219.

68. Morrill, J. C., L. Carpenter, D. Taylor, H. H. Ramsburg, J. Quance, and C. J. Peters. 1991. Further evaluation of a mutagen-attenuated Rift Valley fever vaccine in sheep. *Vaccine* **9:**35–41.

69. Morrill, J. C., G. B. Jennings, T. M. Cosgriff, P. H. Gibbs, and C. J. Peters. 1989. Prevention of Rift Valley fever in rhesus monkeys with interferon-alpha. *Rev. Infect. Dis.* **11**(Suppl. 4):S815–S825.

70. Morrill, J. C., G. B. Jennings, A. J. Johnson, T. M. Cosgriff, P. H. Gibbs, and C. J. Peters. 1990. Pathogenesis of Rift Valley fever in rhesus monkeys: role of interferon response. *Arch. Virol.* **110:**195–212.

71. Morrill, J. C., C. A. Mebus, and C. J. Peters. 1997. Safety and efficacy of a mutagen-attenuated Rift Valley fever virus vaccine in cattle. *Am. J. Vet. Res.* **58:**1104–1109.

72. Morrill, J. C., and C. J. Peters. 2003. Pathogenicity and neurovirulence of a mutagen-attenuated Rift Valley fever vaccine in rhesus monkeys. *Vaccine* **21:**2994–3002.

73. Nichol, S., B. J. Beaty, R. M. Elliott, R. Goldbach, A. Plyusnin, C. S. Schmaljohn, and R. B. Tesh. 2005. Bunyaviridae, p. 695–716. *In* C. M. Fauquet, M. A. Mayo, J. Maniloff, U. Desselberger, and L. A. Ball (ed.), *Virus Taxonomy. VIIIth Report of the International Committee on Taxonomy of Viruses.* Elsevier Academic Press, Amsterdam, The Netherlands.

74. Operschall, E., T. Schuh, L. Heinzerling, J. Pavlovic, and K. Moelling. 1999. Enhanced protection against viral infection by co-administration of plasmid DNA coding for viral antigen and cytokines in mice. *J. Clin. Virol.* **13:**17–27.

75. Paladino, P., D. T. Cummings, R. S. Noyce, and K. L. Mossman. 2006. The IFN-independent response to virus particle entry provides a first line of antiviral defense that is independent of TLRs and retinoic acid-inducible gene I. *J. Immunol.* **177:**8008–8016.

76. Papa, A., S. Bino, E. Velo, A. Harxhi, M. Kota, and A. Antoniadis. 2006. Cytokine levels in Crimean-Congo hemorrhagic fever. *J. Clin. Virol.* **36:**272–276.

77. Pavlovic, J., J. Schultz, H. P. Hefti, T. Schuh, and K. Molling. 2000. DNA vaccination against La Crosse virus. *Intervirology* **43:**312–321.

78. Pekosz, A., C. Griot, K. Stillmock, N. Nathanson, and F. Gonzalez-Scarano. 1995. Protection from La Crosse virus encephalitis with recombinant glycoproteins: role of neutralizing anti-G1 antibodies. *J. Virol.* **69:**3475–3481.

79. Pekosz, A., J. Phillips, D. Pleasure, D. Merry, and F. Gonzalez-Scarano. 1996. Induction of apoptosis by La Crosse virus infection and role of neuronal differentiation

and human *bcl-2* expression in its prevention. *J. Virol.* **70:**5329–5335.

80. Perrone, L. A., K. Narayanan, M. Worthy, and C. J. Peters. 2007. The S segment of Punta Toro virus (Bunyaviridae, Phlebovirus) is a major determinant of lethality in the Syrian hamster and codes for a type I interferon antagonist. *J. Virol.* **81:**884–892.

81. Peters, C. J., D. Jones, R. Trotter, J. Donaldson, J. White, E. Stephen, and T. W. Slone, Jr. 1988. Experimental Rift Valley fever in rhesus macaques. *Arch. Virol.* **99:**31–44.

82. Pfeifer, K., H. Ushijima, B. Lorenz, W. E. Muller, and H. C. Schroder. 1993. Evidence for age-dependent impairment of antiviral 2',5'-oligoadenylate synthetase/ribonuclease L-system in tissues of rat. *Mech. Ageing. Dev.* **67:**101–114.

83. Pichlmair, A., O. Schulz, C. P. Tan, T. I. Naslund, P. Liljestrom, F. Weber, and C. Reis e Sousa. 2006. RIG-I-mediated antiviral responses to single-stranded RNA bearing 5' phosphates. *Science* **314:**997–1001.

84. Pinheiro, F. P., A. P. Travassos da Rosa, J. F. Travassos da Rosa, R. Ishak, R. B. Freitas, M. L. Gomes, J. W. LeDuc, and O. F. Oliva. 1981. Oropouche virus. I. A review of clinical, epidemiological, and ecological findings. *Am. J. Trop. Med. Hyg.* **30:**149–160.

85. Prescott, J., C. Ye, G. Sen, and B. Hjelle. 2005. Induction of innate immune response genes by Sin Nombre hantavirus does not require viral replication. *J. Virol.* **79:**15007–15015.

86. Prescott, J. B., P. R. Hall, V. S. Bondu-Hawkins, C. Ye, and B. Hjelle. 2007. Early innate immune responses to Sin Nombre hantavirus occur independently of IFN regulatory factor 3, characterized pattern recognition receptors, and viral entry. *J. Immunol.* **179:**1796–1802.

87. Prins, M., and R. Goldbach. 1998. The emerging problem of tospovirus infection and nonconventional methods of control. *Trends Microbiol.* **6:**31–35.

88. Raju, R., and D. Kolakofsky. 1988. La Crosse virus infection of mammalian cells induces mRNA instability. *J. Virol.* **62:**27–32.

89. Randall, R., and S. Goodbourn. 2008. Interferons and viruses: an interplay between induction, signalling, antiviral responses and virus countermeasures. *J. Gen. Virol.* **89:**1–47.

90. Remick, D., and S. Kunkel. 1993. Pathophysiologic alterations induced by tumor necrosis factor. *Int. Rev. Exp. Pathol.* **34**(Pt. B):7–25.

91. Richt, J. A., P. Lekcharoensuk, K. M. Lager, A. L. Vincent, C. M. Loiacono, B. H. Janke, W. H. Wu, K. J. Yoon, R. J. Webby, A. Solorzano, and A. Garcia-Sastre. 2006. Vaccination of pigs against swine influenza viruses by using an NS1-truncated modified live-virus vaccine. *J. Virol.* **80:**11009–11018.

92. Schmaljohn, C. S., and J. W. Hooper. 2001. Bunyaviridae: the viruses and their replication, p.1581–1602. *In* D. M. Knipe, P. M. Howley, D. E. Griffin, R. A. Lamb, M. A. Martin, B. Roizman, and S. E. Straus (ed.), *Fields Virology,* 4th ed. Lippincott, Williams & Wilkins, Philadelphia, PA.

93. Schountz, T., J. Prescott, A. Cogswell, L. Oko, K. Mirowsky-Garcia, A. Galvez, and B. Hjelle. 2007. Regulatory T cell-like responses in deer mice persistently infected with Sin Nombre virus. *Proc. Natl. Acad. Sci. USA* **104:**15496–15501.

94. Schuh, T., J. Schultz, K. Moelling, and J. Pavlovic. 1999. DNA-based vaccine against La Crosse virus: protective

immune response mediated by neutralizing antibodies and CD4$^+$ T cells. *Hum. Gene. Ther.* **10:**1649–1658.

95. Shabman, R. S., T. E. Morrison, C. Moore, L. White, M. S. Suthar, L. Hueston, N. Rulli, B. Lidbury, J. P. Ting, S. Mahalingam, and M. T. Heise. 2007. Differential induction of type I interferon responses in myeloid dendritic cells by mosquito and mammalian-cell-derived alphaviruses. *J. Virol.* **81:**237–247.

96. Spiropoulou, C., S. Morzunov, H. Feldmann, A. Sanchez, C. Peters, and S. Nichol. 1994. Genome structure and variability of a virus causing hantavirus pulmonary syndrome. *Virology* **200:**715–723.

97. Spiropoulou, C. F., C. G. Albarino, T. G. Ksiazek, and P. E. Rollin. 2007. Andes and Prospect Hill hantaviruses differ in early induction of interferon although both can downregulate interferon signaling. *J. Virol.* **81:**2769–2776.

98. Stoltz, M., C. Ahlm, A. Lundkvist, and J. Klingstrom. 2007. Lambda interferon (IFN-λ) in serum is decreased in hantavirus-infected patients, and in vitro-established infection is insensitive to treatment with all IFNs and inhibits IFN-γ-induced nitric oxide production. *J. Virol.* **81:**8685–8691.

99. Streitenfeld, H., A. Boyd, J. K. Fazakerley, A. Bridgen, R. M. Elliott, and F. Weber. 2003. Activation of PKR by Bunyamwera virus is independent of the viral interferon antagonist NSs. *J. Virol.* **77:**5507–5511.

100. Sundstrom, J. B., L. K. McMullan, C. F. Spiropoulou, W. C. Hooper, A. A. Ansari, C. J. Peters, and P. E. Rollin. 2001. Hantavirus infection induces the expression of RANTES and IP-10 without causing increased permeability in human lung microvascular endothelial cells. *J. Virol.* **75:**6070–6085.

101. Talon, J., M. Salvatore, R. E. O'Neill, Y. Nakaya, H. Zheng, T. Muster, A. Garcia-Sastre, and P. Palese. 2000. Influenza A and B viruses expressing altered NS1 proteins: a vaccine approach. *Proc. Natl. Acad. Sci. USA* **97:**4309–4314.

102. Tamura, M., H. Asada, K. Kondo, M. Takahashi, and K. Yamanishi. 1987. Effects of human and murine interferons against hemorrhagic fever with renal syndrome (HFRS) virus (Hantaan virus). *Antiviral Res.* **8:**171–178.

103. Temonen, M., H. Lankinen, O. Vapalahti, T. Ronni, I. Julkunen, and A. Vaheri. 1995. Effect of interferon-alpha and cell differentiation on Puumala virus infection in human monocyte/macrophages. *Virology* **206:**8–15.

104. tenOever, B. R., M. J. Servant, N. Grandvaux, R. Lin, and J. Hiscott. 2002. Recognition of the measles virus nucleocapsid as a mechanism of IRF-3 activation. *J. Virol.* **76:**3659–3669.

105. tenOever, B. R., S. Sharma, W. Zou, Q. Sun, N. Grandvaux, I. Julkunen, H. Hemmi, M. Yamamoto, S. Akira, W. C. Yeh, R. Lin, and J. Hiscott. 2004. Activation of TBK1 and IKK epsilon kinases by vesicular stomatitis virus infection and the role of viral ribonucleoprotein in the development of interferon antiviral immunity. *J. Virol.* **78:**10636–10649.

106. Thomas, D., G. Blakqori, V. Wagner, M. Banholzer, N. Kessler, R. M. Elliott, O. Haller, and F. Weber. 2004. Inhibition of RNA polymerase II phosphorylation

by a viral interferon antagonist. *J. Biol. Chem.* **279:**31471–31477.

107. Utz, J. T., C. S. Apperson, J. N. MacCormack, M. Salyers, E. J. Dietz, and J. T. McPherson. 2003. Economic and social impacts of La Crosse encephalitis in western North Carolina. *Am. J. Trop. Med. Hyg.* **69:**509–518.

108. Van Epps, H. L., C. S. Schmaljohn, and F. A. Ennis. 1999. Human memory cytotoxic T-lymphocyte (CTL) responses to Hantaan virus infection: identification of virus-specific and cross-reactive CD8$^+$ CTL epitopes on nucleocapsid protein. *J. Virol.* **73:**5301–5308.

109. Wahl-Jensen, V., J. Chapman, L. Asher, R. Fisher, M. Zimmerman, T. Larsen, and J. W. Hooper. 2007. Temporal analysis of Andes virus and Sin Nombre virus infections of Syrian hamsters. *J. Virol.* **81:**7449–7462.

110. Wallace, D. B., C. E. Ellis, A. Espach, S. J. Smith, R. R. Greyling, and G. J. Viljoen. 2006. Protective immune responses induced by different recombinant vaccine regimes to Rift Valley fever. *Vaccine* **24:**7181–7189.

111. Weber, F., A. Bridgen, J. K. Fazakerley, H. Streitenfeld, R. E. Randall, and R. M. Elliott. 2002. Bunyamwera bunyavirus nonstructural protein NSs counteracts the induction of alpha/beta interferon. *J. Virol.* **76:**7949–7955.

112. Weber, F., and O. Haller. 2007. Viral suppression of the interferon system. *Biochimie* **89:**836–842.

113. Weber, F., G. Kochs, and O. Haller. 2004. Inverse interference: how viruses fight the interferon system. *Viral Immunol.* **17:**498–515.

114. Weber, F., V. Wagner, S. B. Rasmussen, R. Hartmann, and S. R. Paludan. 2006. Double-stranded RNA is produced by positive-strand RNA viruses and DNA viruses but not in detectable amounts by negative-strand RNA viruses. *J. Virol.* **80:**5059–5064.

115. Wichmann, D., H. J. Grone, M. Frese, J. Pavlovic, B. Anheier, O. Haller, H. D. Klenk, and H. Feldmann. 2002. Hantaan virus infection causes an acute neurological disease that is fatal in adult laboratory mice. *J. Virol.* **76:**8890–8899.

116. Yanase, T., T. Kato, M. Yamakawa, K. Takayoshi, T. Nakamura, T. Kokuba, and T. Tsuda. 2006. Genetic characterization of Batai virus indicates a genomic reassortment between orthobunyaviruses in nature. *Arch. Virol.* **151:**2253–2260.

117. Young, D., L. Andrejeva, A. Livingstone, S. Goodbourn, R. Lamb, P. Collins, R. Elliott, and R. Randall. 2003. Virus replication in engineered human cells that do not respond to interferons. *J. Virol.* **77:**2174–2181.

118. Zaki, S. R., P. W. Greer, L. M. Coffield, C. S. Goldsmith, K. B. Nolte, K. Foucar, R. M. Feddersen, R. E. Zumwalt, G. L. Miller, A. S. Khan, R. E. Rollin, K. G. Ksiazek, S. T. Nichol, B. W. J. Mahy, and C. J. Peters. 1995. Hantavirus pulmonary syndrome. Pathogenesis of an emerging infectious disease. *Am. J. Pathol.* **146:**552–579.

119. Zhang, X. K., I. Takashima, and N. Hashimoto. 1988. Role of maternal antibody in protection from hemorrhagic fever with renal syndrome virus infection in rats. *Arch. Virol.* **103:**253–265.

Cellular Signaling and Innate Immune
Responses to RNA Virus Infections
Edited by A. R. Brasier et al.
© 2009 ASM Press, Washington, D.C.

Juan C. de la Torre

Arenaviruses

19

Arenaviruses merit significant interest both as tractable model systems to study acute and persistent viral infections and as clinically important human pathogens including several causative agents of hemorrhagic fever (HF) disease. Evidence indicates that arenavirus persistence and pathogenesis are facilitated by the virus's ability to overcome the host innate immune response, which is elicited within minutes of infection and provides the host with critical protection during the window of time required to place in motion the more sophisticated adaptive response. This adaptive response, mediated by B and T cells, is very important in antiviral defense and generates memory, but it develops rather slowly and does not reach full efficacy for days or weeks. Moreover, the quality and magnitude of this adaptive response are significantly influenced by the early innate response. The elucidation of the mechanisms underlying the interactions between arenaviruses and the host innate immune response is essential for a better understanding of arenavirus pathogenesis and development of antiviral therapies to combat human arenaviral diseases.

LCMV INFECTION OF THE MOUSE: THE ROSETTA STONE OF VIRUS-HOST INTERACTION

Studies using the prototypic arenavirus lymphocytic choriomeningitis virus (LCMV) have led to major advances in virology and immunology that apply universally to other microbial infections and viral infections of humans (84, 128). These concepts include (i) virus-induced immunopathological disease, (ii) major histocompatibility complex restriction and T-cell-mediated killing, (iii) hallmarks and kinetics of antigen-specific T-cell immunity, (iv) effectiveness of adaptive immune T-cell therapy in clearing viral infection, and (v) disease initiated by virus-disrupted synthesis or activity of cellular factors in the absence of the classic hallmarks of inflammation and cytolysis.

The outcome of LCMV infection of its natural host, the mouse, varies dramatically depending on the species, age, immunocompetence, and genetic background of the host, as well as the route of infection and the strain and dose of infecting virus (15, 16, 84, 128). Peripheral inoculation of adult immunocompetent mice with many

Juan C. de la Torre, Molecular Integrative Neurosciences Department IMM-6, The Scripps Research Institute, 10550 North Torrey Pines Rd., La Jolla, CA 92037.

LCMV strains results in an acute infection that induces a protective immune response that mediates virus clearance in 10 to 14 days, a process predominantly mediated by virus-specific CD8$^+$ cytotoxic T lymphocytes (CTLs). However, the antiviral response against LCMV delivered by intracerebral inoculation has predominantly immunopathologic consequences resulting in lethal choriomeningitis, whereas intravenous inoculation of a high dose of specific LCMV variants, like clone 13 (Cl13), causes persistent infection associated with generalized immune suppression (1–3). Cl13-induced immunosuppression correlates with the virus's ability to very efficiently infect dendritic cells (DCs) in the T-cell area (white pulp) of the spleen and affect their function, including antigen presentation to naïve T cells and B cells, resulting in a generalized immunosuppression and persistent viral infection. In contrast, the viruses that do not cause immunosuppression do not infect DCs and are restricted to the red pulp of the spleen (112, 114). The precise mechanisms by which Cl13 and other immunosuppressive isolates of LCMV target DCs and block their function are currently unclear. Cl13 and its nonimmunosuppressive parental Armstrong strain differ at two amino acid positions within the virus GP1 (N176D and F260L) and one amino acid position in the virus L polymerase (K1079Q) (102, 104). There is some evidence linking the immunosuppressive potential of Cl13 to the single point mutation F260L in GP1, but a contribution of the additional N176D substitution in GP1 to virus-induced immunosuppression, as well as a role for the L polymerase, cannot be ruled out yet.

On the other hand, neonatal infection with LCMV leads to thymic deletion of virus-specific CTLs (57, 98), which results in a lifelong persistent infection that can be associated with a variety of pathologies, including growth hormone deficiency syndrome and behavioral abnormalities, depending on the genetic background of both the virus and the host (23, 25).

ARENAVIRUS-INDUCED DISEASE IN HUMANS

Arenaviruses cause chronic infections of rodents, with a worldwide distribution including the Americas (15). Asymptomatically infected animals move freely in their natural habitat and may invade human habitation. Humans are most likely infected through mucosal exposure to aerosols or by direct contact between infectious materials and abraded skin. These infections are common and in some cases severe. Some arenaviruses, chiefly Lassa virus (LASV), cause HF, which represents a serious public health problem in endemic areas (15, 71, 92). Moreover, in recent years, increased air travel between Africa and other areas has led to the importation of cases of Lassa fever (LF) into the United States, Europe, Japan, and Canada (38, 48, 50). Because of their severe morbidity and high mortality, the lack of immunization or effective treatment, their ease of introduction into a susceptible population, and the likelihood that aerosolized forms of the viruses would be highly infectious to humans, HF arenaviruses are included within category A biological agents and are considered to be viruses capable of posing a significant threat to human health.

Outside of category A biological agents, LCMV is a proven teratogen in humans, and evidence indicates that LCMV is a neglected human pathogen of clinical significance (14, 52, 76). Notably, transplant-associated infections by LCMV with a fatal outcome have been recently documented in the United States (34, 93) and Australia (86).

Individuals succumbing to LASV infection generate minimal or no anti-LASV immune responses (71), and histological examination of LF patients shows surprisingly minimal cellular damage and infiltration of inflammatory cells (71). These findings suggest that morbidity and mortality associated with LASV infection, and likely other HF arenavirus infections, are at least partly associated with the failure of the host's innate immune response to restrict virus replication and to facilitate the initiation of an effective adaptive immune response. Accordingly, the extent of viremia is a highly predictive factor for the outcome of LF patients (71). Notably, transplant-associated infections by LCMV showed remarkable similarities with cases of LF (34), reinforcing the critical role of the quality and magnitude of the host innate response.

There are no licensed arenavirus vaccines, and current antiarenavirus therapy is limited to the use of ribavirin (56, 72, 73). However, ribavirin is only partially effective and is often associated with significant secondary effects, including anemia and birth defects. The significance of arenaviruses in human health together with the very limited existing armamentarium to combat these infections underscore the importance of developing novel effective antiarenavirus therapies and vaccines, tasks that will be helped by a better understanding of the mechanisms by which arenaviruses counteract the early host innate immune response.

MOLECULAR AND CELL BIOLOGY OF ARENAVIRUSES

Virion and Genome Organization

Arenaviruses are enveloped viruses with a bisegmented negative-strand RNA genome and a life cycle restricted to the cell cytoplasm. Virions are pleomorphic but

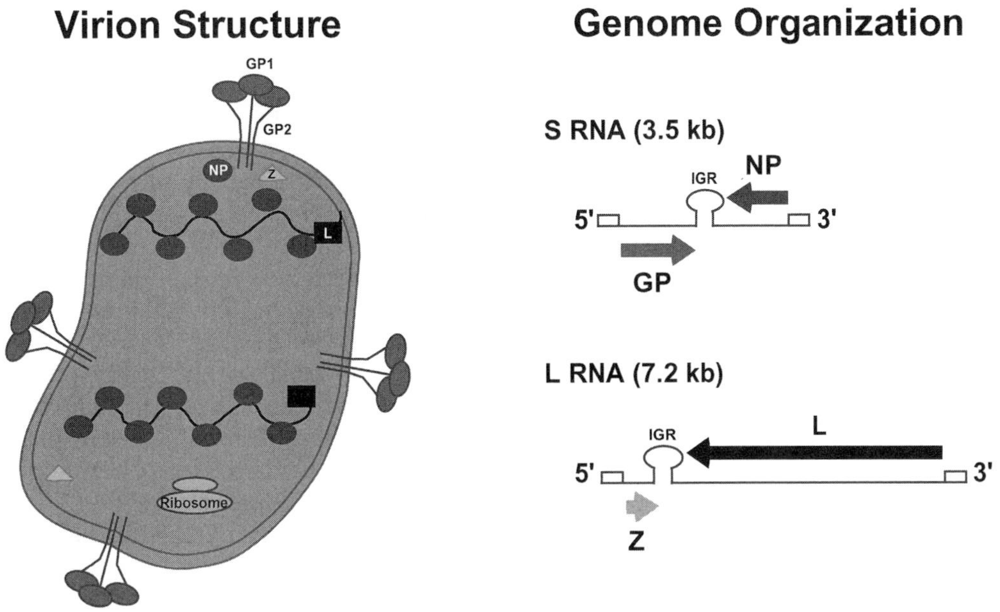

Virion Structure

Genome Organization

S RNA (3.5 kb)

L RNA (7.2 kb)

Figure 1 Scheme of the LCMV genome organization. LCMV has a bisegmented, negative-strand RNA genome. Each segment uses an ambisense strategy to direct the synthesis of two viral gene products. The S (small, ca. 3.5 kb) segment codes for the viral nucleoprotein (NP) and precursor surface viral glycoprotein (GPC), which is posttranslationally cleaved by the cellular protease S1P to generate GP1 and GP2. GP1/GP2 complexes form the spikes that decorate the surface of the virions. The L (large, ca. 7.5 kb) segment codes for the viral RdRp (L protein) and a small (ca. 12 kDa) RING finger protein called Z.

often spherical and covered with surface glycoprotein spikes. Virions contain the L and S genomic RNAs as helical nucleocapsid structures that are organized into circular configurations (127). Ribosome-like structures can be incorporated into virions, but their identity and roles in arenavirus biology remain to be determined (15).

Each genomic RNA segment, L (ca. 7.5 kb) and S (ca. 3.5 kb), uses an ambisense coding strategy to direct the synthesis of two polypeptides in opposite orientation, separated by a noncoding intergenic region (IGR) with a predicted folding of a stable hairpin structure (15, 77) (Fig. 1). The S RNA encodes the viral glycoprotein precursor, GPC (ca. 75 kDa), and the nucleoprotein, NP (ca. 63 kDa), whereas the L RNA encodes the viral RNA-dependent RNA polymerase (RdRp, or L polymerase) (ca. 200 kDa) and a small RING finger protein, Z (ca. 11 kDa).

Arenaviruses exhibit a high degree of sequence conservation at the genome 3′ termini (17 out of 19 nucleotides are identical), and as with other negative-strand RNA viruses, arenaviruses' genome termini exhibit terminal complementarity, with the 5′ and 3′ ends of both L and S genome segments predicted to form panhandle structures. This terminal complementarity may reflect the

presence at the 5′ ends of *cis*-acting signal sequences that provide a nucleation site for RNA encapsidation, required to generate the nucleocapsid (NC) templates recognized by the virus polymerase. Terminal complementarity may also be a consequence of strong similarities between the genome and antigenome promoters used by the virus polymerases. For several arenaviruses, an additional nontemplated G residue has been detected on the 5′ end of their genome RNAs (42, 100).

The IGRs are predicted to form a stable hairpin structure. Transcription termination of the S-derived NP and GP mRNAs occurs at multiple sites within the predicted stem of the IGR, suggesting that a structural motif promotes the release of the virus polymerase from the template RNA (77).

Arenavirus Proteins

The NP is the most abundant viral polypeptide in both infected cells and virions. NP is the main structural element of the viral RNP and plays an essential role in viral RNA synthesis. The viral glycoprotein precursor GPC is posttranslationally proteolytically processed by the S1P cellular protease to yield the two mature virion glycoproteins, GP1 (40 to 46 kDa) and GP2 (35 kDa) (8, 97). GPC contains a 58-amino-acid signal peptide

(SSP) that is expressed as a stable polypeptide in infected cells and that remains stably associated to the GP complex (GPcx). Besides its role in targeting the nascent polypeptide to the endoplasmic reticulum, this SSP likely serves additional roles in the trafficking and function of the viral envelope glycoproteins (31–33, 39, 126). GP1 mediates virus interaction with host cell surface receptors and is located at the top of the spike, away from the membrane, and is held in place by ionic interactions with the N terminus of the transmembrane GP2 (15, 83) (Fig. 1).

The *Arenavirus* L protein has the characteristic sequence motifs conserved among the RdRp (L proteins) of negative-strand RNA viruses (103, 122). Detailed sequence analysis and secondary structure predictions have been documented for the LASV L polymerase (122). These studies identified several regions of strong α-helical content and a putative coiled-coil domain at the N terminus, whose functional roles remain to be determined.

The *Arenavirus* RING finger protein Z has no homologue among other known negative-strand RNA viruses. Z is a structural component of the virion (106). In LCMV-infected cells Z has been shown to interact with several cellular proteins including the promyelocytic leukemia (PML) protein (11) and the eukaryotic translation initiation factor 4E (eIF4E) (18, 55), which have been proposed to contribute to the noncytolytic nature of LCMV infection and repression of cap-dependent translation, respectively. Biochemical studies suggested that Z might be the arenavirus counterpart of the matrix (M) protein found in other negative-strand RNA viruses (105, 106). The expression of Z during the progression from early to late phases of the LCMV life cycle appears to be highly regulated, and therefore Z might play different roles during the life cycle of LCMV.

Arenavirus Life Cycle

Cell Attachment and Entry

Consistent with a broad host range and cell-type tropism, a highly conserved and widely expressed cell surface protein, α-dystroglycan (aDG) has been identified as a main receptor for LCMV, Lassa fever, and several other arenaviruses (59). However, many arenaviruses appear to use an alternative receptor (61, 115). Upon receptor binding, arenavirus virions are internalized by uncoated vesicles and released into the cytoplasm by a pH-dependent membrane fusion step that is mediated by GP-2 (12, 26, 27). This fusion event is mediated by GP2, which is structurally similar to the fusion active membrane proximal portions of the GP of other enveloped viruses (40).

RNA Replication and Transcription

The fusion between viral and cellular membranes releases the viral RNP into the cytoplasm, which is followed by viral RNA synthesis. LCMV mRNAs have extra nontemplated nucleotides and a cap structure at their 5′ ends, but the origins of both the cap and 5′ nontemplated nucleotide extensions remain to be determined. Transcription termination of subgenomic nonpolyadenylated viral mRNAs was mapped to multiple sites within the distal side of the IGR (15, 77), suggesting that the IGR acts as a bona fide transcription termination signal for the virus polymerase. The basic steps of *Arenavirus* RNA replication and gene transcription are illustrated in Fig. 2, using the S segment as an example; the same scheme applies also to the L segment. The NP and L coding regions are transcribed into a genomic complementary mRNA, whereas the GPC and Z coding regions are not translated directly from genomic RNA, but rather from genomic sense mRNAs that are transcribed using as templates the corresponding antigenome RNA species, which also function as replicative intermediates.

Assembly and Budding

For most enveloped viruses, a matrix (M) protein is involved in organizing the virion components prior to assembly. Interestingly, arenaviruses do not have an obvious counterpart of M. However, cross-linking studies showed complex formation between NP and Z, suggesting a possible role of Z in virion morphogenesis (105, 106). Production of infectious virions requires the correct processing of GPC (8, 97), which necessitates the structural integrity of the GP2 cytoplasmic tail (60).

Arenavirus Reverse Genetics

The inability to genetically manipulate the *Arenavirus* genome has hampered studies aimed at understanding its molecular and cell biology, as well as the role played by each viral gene product in virus-host interactions during both acute and persistent infections and associated diseases. The development of reverse genetics systems for several arenaviruses, including the prototypic LCMV (46, 62, 65), have provided investigators with a novel and powerful approach for the investigation of the molecular and cell biology of arenaviruses, as well as arenavirus-host interactions and associated disease.

Reverse genetics approaches identified NP and L as the minimal viral *trans*-acting factors required for efficient RNA synthesis mediated by the polymerase of LCMV (62). Similar findings have been now documented for LASV (46) and the New World arenavirus Tacaribe virus (TCRV) (65). Biochemical and genetic

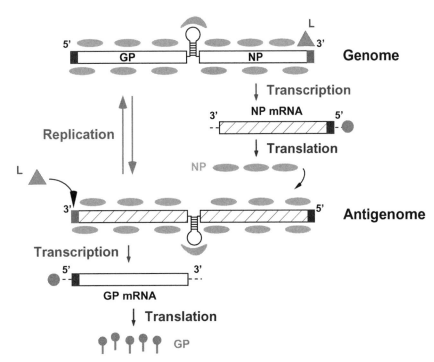

Figure 2 Basic aspects of *Arenavirus* RNA replication and gene transcription illustrated for the S segment. Once the virus RNP is delivered into the cytoplasm of the infected cell, the polymerase associated with the virus RNP initiates transcription from the genome promoter located at the genome 3′ end. Primary transcription results in synthesis of NP and L mRNA from the S and L segments, respectively. Subsequently, the virus polymerase can adopt a replicase mode and moves across the IGR to generate a copy of the full-length antigenome RNA (agRNA). This agRNA will serve as a template for the synthesis of the GP (agS) and Z (agL) mRNAs. The agRNA species also serve as templates for the amplification of the corresponding genome RNA species.

evidence indicated that oligomerization of L is required for the activity of the *Arenavirus* polymerase (107). The genomic promoter recognized by the virus polymerase requires both sequence specificity within the highly conserved 3′-terminal 19 nucleotides of *Arenavirus* genomes and the integrity of the predicted panhandle structure formed via sequence complementarity between the 5′ and 3′ termini of viral genome RNAs (90). Detailed structure-function studies using the LASV minigenome (MG) system (47) revealed that the LASV promoter regulates transcription and replication in a coordinated manner and is composed of two functional elements: a sequence-specific region from residue 1 to 12 and a variable complementary region from residue 13 to 19. The first region appears to interact with the replication complex mainly via base-specific interactions, while in the second region base pairing between the 3′ and 5′ termini of the promoter determines its activity. Likewise, in reverse genetics studies we have provided direct experimental confirmation that the IGR is a bona fide

transcription termination signal (95) and that intracellular levels of NP do not determine the balance between virus RNA replication and transcription (96).

Z was not required for RNA replication and transcription of an LCMV MG, but rather Z exhibited a dose-dependent inhibitory effect on both transcription and replication of LCMV MG (21–23, 63). This inhibitory effect of Z has been also reported for TCRV (65) and LASV (46). For most enveloped negative-strand RNA viruses, a matrix (M) protein is involved in organizing the virion components prior to assembly and often plays a central role in virus budding. Arenaviruses do not have an obvious counterpart of M, but consistent with its proposed role in virion morphogenesis (105, 106), evidence indicates that Z is an M-like protein and the main driving force of arenavirus budding (89, 117, 121). Z budding activity is mediated by its proline-rich late (L) domain motifs: PTAP and PPPY, similar to those known to control budding of several other viruses including human immunodeficiency virus and Ebola, via interaction with specific host-cell proteins (37). Consistent with this observation, Z exhibits features characteristic of bona fide budding proteins: the ability to bud from cells by itself and to efficiently substitute for other L domains. Targeting of Z to the plasma membrane, the location of *Arenavirus* budding, strictly required its myristoylation (91). This role of Z as the arenavirus counterpart of the M proteins found in many other negative-strand RNA viruses is consistent with recent ultrastructural data on arenavirus virions determined by cryoelectron microscopy (83).

The ability to rescue infectious LCMV entirely from cloned cDNAs (36, 108) has opened new avenues for the investigation of arenavirus-host interactions that influence a variable infection outcome, ranging from virus control and clearance by the host defenses to long-term chronic infection associated with subclinical disease to severe acute disease including HF.

THE TYPE I IFN RESPONSE DURING ARENAVIRUS INFECTION

The interferon (IFN) response, resulting in the induction of IFN-stimulated genes (ISGs), plays a central role in the host innate immune defense against viral infection (10, 41). Accordingly, viruses have developed a plethora of strategies to disrupt the IFN-mediated antiviral defense of the host, and viral gene products responsible for these disruptions are often major virulence factors. Recent work has started to elucidate the mechanisms by which viruses induce, and evade, the different pathways leading to the induction of IFN and ISGs. Viruses have developed a variety of mechanisms to subvert the host IFN response at different levels, including blocking IFN induction and its signaling or interfering with the action of one or more host antiviral gene products induced by type I IFNs (41, 64).

IFN regulatory factor 3 (IRF3) has emerged as one of the critical transcriptional regulators for the primary transcriptional induction of ISGs. In addition, IRF3 cooperates with nuclear factor-κB (NF-κB) in the transcriptional stimulation of the IFN-β promoter, which leads to IFN-dependent induction of ISGs and proinflammatory genes. IRF3 is a transcription factor present in a nonactive form in the cytoplasm of most cell types that becomes activated via pattern recognition receptors (PRRs) in response to pathogen-associated molecular patterns (PAMPs). IRF3 is one of the cellular factors most commonly targeted by viruses to inhibit the IFN response, illustrating the critical role of IRF3 in the IFN system (109). There is a growing list of viral IFN-antagonist proteins that target IRF3 (45) via different mechanisms,

including inhibition of IRF3 activation, inhibition of the transcriptional function of activated IRF3, or targeting proteins that interact with IRF3. Notably, these anti-IRF3 genes from different viral families usually do not share significant homology, suggesting independent acquisition during virus evolution.

Infection of the mouse with many viruses, including LCMV, is characterized by an early transient burst of type I IFN production that peaks shortly before the peak of production of infectious progeny and often helps to limit viral spread. Different viral components can serve as potent PAMPs that upon recognition by specific cellular PRRs can trigger the initiation of a powerful innate immune response that poses an extraordinary challenge to virus multiplication. To counteract this, many viruses target the IFN system, a key player in the orchestration of an effective innate immune response. Both nonstructural (e.g., influenza virus NS1) and structural (e.g., Ebola virus VP35) viral proteins have been involved in counteracting the IFN response. The outcome of this battle between IFN's antiviral effects and the virus's IFN-counteracting measures has an important influence on the outcome of infection. The biological relevance of the anti-IFN activity exerted by these viral products is illustrated by the highly attenuated virus phenotype in vivo associated with mutations affecting the anti-IFN activity of these viral proteins.

The contribution of different PRRs to LCMV recognition has not been elucidated. Likewise, the identity of the specific cell types, among the many capable, responsible for the early burst of type I IFN production in LCMV-infected mice has not been unequivocally established, but it appears to involve a complex network of cellular interactions including DCs (79, 80) and the specialized organization of the splenic marginal zone (24, 67). It is plausible that this burst of type I IFN production is driven by cells that are not infected with LCMV, but responding to signals produced by cells targeted by LCMV at the very early times of infection. The initially infected cell types could include fibroblasts and monocytes capable of producing IFN upon sensing viral RNA.

It should be noted that arenaviruses, including LCMV, are vertically transmitted in their natural hosts, which results in lifelong persistence due to T-cell tolerance. This chronic infection is associated with relatively high levels of virus RNA replication in many cell types capable of recognizing viral RNA via retinoic acid-inducible gene I (RIG-I), which should result in a robust induction of type I IFN production. However, only modest levels of IFN induction can be detected in these mice (17), suggesting that RIG-I and other cytoplasmic

PRRs that normally detect the presence of replicating RNA viruses are functionally impaired due to mechanisms that remain to be determined.

LCMV Infection Inhibits Induction of Type I IFN but Does Not Prevent the IFN-Mediated Antiviral State within Cells

Evidence indicates that arenavirus infection can interfere with the capacity of the host to initiate an effective innate immune response, which likely contributes to arenaviral persistence and pathogenesis. The mechanisms by which arenaviruses disrupt the host innate immune response are little understood. Recent studies using a well-established cell-based bioassay of type I IFNs (70, 87, 88) showed that infection with LCMV interfered with the production of type I IFNs by A549 cells characteristically observed in response to lipofectamine (LF)-based DNA transfection, as determined by inhibition of a recombinant Newcastle disease virus expressing green fluorescent protein (rNDV-GFP) in previously transfected human A549 cells (70, 87, 88). This inhibitory effect, however, can be prevented if the transfecting DNA expresses an IFN-counteracting protein such as the NS1 of influenza virus (78).

Treatment of LCMV-infected cultured cells with the nucleoside analog ribavirin results, in most cases, in efficient virus clearance (101). A549 cells cured from LCMV by ribavirin treatment recovered the ability to create an antiviral state in response to LF/DNA transfection (70), supporting the conclusion that active LCMV replication was required for the suppression of the IFN production.

Several viral proteins have been reported to be able to both prevent the induction of IFN production and interfere with the IFN-mediated establishment of an antiviral state within cells. LCMV-infected and mock-infected control cells exhibited similar levels of upregulation of ISG mRNAs in response to IFN-β treatment (70). These findings indicate that LCMV infection does not prevent Type I IFN signaling, but rather blocks production of endogenous IFN-β in response to viral infection.

The LCMV NP Suffices to Mediate Inhibition of Type I IFN Production

The inhibition of type I IFN production in LCMV-infected cells could reflect an effect associated with the activity of a specific LCMV gene product or caused by the global effect of the ongoing infection on cell physiology. Cell-based assays using cotransfection with plasmids expressing each individual LCMV gene product (L, NP, GP, and Z), together with reporter plasmids encoding GFP fused to the chloramphenicol acetyltransferase gene under the control of either an IFN-β or an IFN-stimulated response element (ISRE) promoter (70, 78), showed that cells expressing NP, but not GP, L, or Z, inhibited activation of both IFN-β and ISRE promoters induced following Sendai virus (SeV) infection (70). LCMV NP caused inhibition levels similar to those observed with influenza NS1, a validated IFN-antagonist protein (118). LCMV NP represents the first example of an IFN-counteracting viral protein encoded in the *Arenaviridae* family, as well as the first nucleoprotein from a negative-strand RNA virus with IFN-production-inhibitory properties. Recently, severe acute respiratory syndrome coronavirus NP has also been described to inhibit type I IFN activation (58).

LCMV NP Inhibits IRF3-Dependent Promoter Activation

Activation of the ISRE promoter following SeV infection is known to be mediated by IRF3. It would therefore be predicted that NP, but not GP, Z, or L, would be able to inhibit an IRF3-dependent promoter. Accordingly, results from cotransfection assays using expression plasmids encoding each one of the LCMV proteins together with an IRF3-dependent reporter plasmid encoding firefly luciferase showed a marked inhibition of SeV-mediated activation of the IRF3-dependent promoter in cells transfected with NP, but not with GP, L, or Z (70). Notably, LCMV NP inhibited the activity of the IRF3-dependent promoter to the same degree observed in cells expressing the influenza NS1 protein or a dominant negative form of IRF3 (IRF3DN). This observation is in contrast to the IRF3-activating properties associated with nucleoprotein complexes of other negative-strand RNA viruses, including measles virus and vesicular stomatitis virus (VSV) (111, 116).

IRF3 is a latent transcription factor present in an inactive state in the cytoplasm. Activation of IRF3 is mediated by hyperphosphorylation of its autoinhibitory carboxy-terminal domain (111). This hyperphosphorylation is thought to promote IRF3 dimerization, nuclear accumulation, and interaction with cyclic AMP-responsive element binding protein (CBP)/P300 (111, 123). Consistent with these data, SeV-induced nuclear localization of a GFP-tagged IRF3 (7) was similarly, and severely, inhibited in cells expressing either influenza NS1 protein or LCMV NP, but not in cells expressing LCMV L, Z, or GP protein (70). Likewise, SeV-induced nuclear translocation of IRF3 was similarly impaired in LCMV-infected cells (70). These findings indicate that inhibition of type I IFN production by LCMV is being blocked at an early step in the IRF3 activation pathway.

OLD WORLD AND NEW WORLD ARENAVIRUS NPs INHIBIT TYPE I IFN INDUCTION

The family *Arenaviridae* consists of one unique genus (*Arenavirus*) with more than 20 recognized virus species that are classified into two distinct groups: Old World and New World arenaviruses (15). This classification was originally established based on serological cross-reactivity, but it is well supported by recent sequence-based phylogenetic studies. Genetically, Old World arenaviruses constitute a single lineage, while New World arenaviruses segregate into clades A, B, and C.

The viral genetic determinants contributing to pathogenesis in humans associated with some arenaviral infections remain largely unknown, but evidence suggests that virus-induced impairment of the host innate immune response is a contributing factor. The fact that NP is the gene product with the highest degree of conservation among arenaviruses raised the question of whether the IFN-antagonistic activity of the LCMV NP was shared by other arenaviruses, including those involved in human disease.

With the exception of TCRV NP, SeV-mediated activation of IRF3 and IFN-β promoters was inhibited by NPs of a variety of both Old World and New World arenaviruses tested (69). Consistent with this finding, all NPs assayed, with exception of TCRV NP, were able to inhibit nuclear accumulation of IRF3 at levels comparable to LCMV NP.

Type I IFN Response Is Not Inhibited in TCRV-Infected Cells

The inability of TCRV NP to inhibit the transcriptional activity of IRF3 and production of IFN-β in transfection-based assays correlated with the situation found in virally infected cells. Thus, results based on the use of a validated IFN bioassay (70, 87, 88) revealed that in contrast to LCMV-infected cells, TCRV-infected A549 cells were, upon LF-based DNA transfection, highly refractory to multiplication of an rNDV-GFP (69). These findings suggested that TCRV infection did not prevent induction of type I IFNs in response to LF/DNA transfection. Consistent with these observations, treatment of Vero cells with tissue culture supernatants from TCRV-infected cells but not from LCMV-infected cells caused the establishment of an effective antiviral state that resulted in a dramatic inhibition of replication of a VSV expressing GFP (rVSV-GFP), a virus known to be highly susceptible to type I IFNs (116). Interestingly, nuclear translocation of IRF3 was observed in TCRV-infected cells in the absence of SeV infection (69), which further reinforced the view that in contrast to LCMV,

infection with TCRV permits activation of IRF3 and subsequent induction of type I IFN production.

MECHANISMS UNDERLYING THE IFN-ANTAGONISTIC PROPERTIES OF ARENAVIRAL NP

Most viral infections are thought to produce double-stranded RNA that provides a pathogen signature that is detected by both membrane-bound TLR3 and non-TLR cytosolic sensors to mediate activation of the main IFN regulatory transcription factors IRF3 and NF-κB, and subsequent IFN-β gene expression (125). Two cytosolic RNA helicases, RIG-I and melanoma differentiation-associated gene 5 (MDA-5), have been identified so far as sentinels for detection of viral RNA (54, 75, 125). These two related helicases are apparently nonredundant and seem to have a degree of virus specificity (54). RIG-I and MDA-5 are ubiquitously expressed in most tissues and are inducible by IFN, which facilitates amplification of the sensing system. Recent work has identified IFN-β promoter stimulator-1 (IPS-1) as a protein required to mediate RIG-I and MDA-5 activation signals to downstream factors leading to IRF3 activation (69, 96, 126). Downstream molecules interacting with IPS-1 have not been documented yet, but IPS-1 contributes to activation of two IκB kinase (IKK)-related kinases, IKK-ε and TRAF-associated NF-κB activator (TANK)-binding kinase-1 (TBK1), known to phosphorylate IRF3 (30, 35, 120). Phosphorylation of IRF3 results in its dimerization and nuclear translocation to initiate transcription of the IFN-β promoter and several other IFN-responsive elements (81). IRF3 can activate transcription of numerous ISGs, including ISG54 and ISG56 (43, 82, 94).

Findings with TCRV-infected cells indicated that arenaviruses have the potential to trigger IRF3 activation and IFN production via triggers and pathways that remain unknown. Influenza virus NS1 (78) and Ebola virus VP35 (19) inhibit activation of IRF3 by targeting cellular proteins involved in activation of this transcription factor. Similarly, arenavirus NP-mediated inhibition of nuclear translocation and transcriptional activation of IRF3 and subsequent inhibition of IFN production are likely to involve the interaction of NP with host-cell proteins involved in IRF3 activation.

THE TYPE I IFN-COUNTERACTING ACTIVITY OF LCMV NP

Viral infections are frequently associated with suppression of the host immune response, which varies in both severity and duration. These virally induced immune defects are likely to jeopardize an effective host response to a secondary infection by an opportunistic pathogen, which has significant implications for human health. This is well illustrated by the high morbidity associated with secondary infections by opportunistic pathogens following morbillivirus (measles virus [MV]) infection of non-vaccinated individuals in some developing countries (44).

Defects in both innate (49, 110) and adaptive (85) immunity contribute to viral-induced immunosuppression and increased frequency of opportunistic infections. The host IFN response to infection plays a critical role linking innate and adaptive immunity (41). During acute viral infections, type I IFN production is typically characterized by an early transient burst shortly before the peak in viral infection, which helps contain viral spread. The majority of cells are capable of producing type I IFNs in response to virus infection, but in whole organisms plasmacytoid DCs (pDCs), also known as IFN-producing cells, are considered to be the main IFN responders following viral infection (5, 74), which triggers the development of a systemic antiviral state aimed at controlling viral growth. Type I IFN production by cells, including pDCs, is initiated upon activation of a subset of pathogen recognition receptor pathways in the host, which include TLR3, 7, and 9 and the intracellular RIG-I pathways that have evolved to sense viral components (4, 51, 124, 125). Additionally, type I IFNs may also be upregulated by the autologous Jak/Stat pathway in response to increased levels of the cytokine to establish an antiviral state in uninfected cells (68).

Like many other viruses, LCMV induces systemic type I IFNs early after infection (9). The cellular source of this early wave of systemic type I IFNs during LCMV infection remains unclear, and it is still detected in the absence of pDCs (66). This early systemic type I IFN response is rapidly silenced during both acute and chronic LCMV infection, indicating that the production and presence of high viral load in blood and several tissues is not sufficient to maintain enhanced levels of type I IFNs in serum. The IFN response by pDCs usually peaks within the first 24 h following viral stimulation, and in vitro studies showed that after this first wave of type I IFN production, pDCs become less responsive in producing type I IFNs (20, 53). However, the regulation of the type I IFN response in vivo in the context of a productive chronic viral infection accompanied by high levels of viral components, as in the case of LCMV and other arenaviruses, is unclear. Due to currently unknown mechanisms, the different pathways that could induce type I IFNs in pDCs appeared to be only minimally or not at all activated during LCMV persistence in the mouse. Nevertheless, mice persistently infected with LCMV are capable of producing

type I IFNs in response to certain TLR ligands. Thus, LCMV-infected mice exhibited reduced levels (three- to fivefold), compared with control mice, of IFN production in response to the TLR3 ligand poly(I:C), which is known to induce production of type I IFNs via IRF3 signaling (29). Likewise, ex vivo-isolated macrophages from LCMV-infected mice exhibited a similar reduced ability to produce type I IFNs upon poly(I:C) stimulation. In contrast, and consistent with previous findings (28), LCMV-infected mice exhibited an overproduction of IFN in response to lipopolysaccharide that resulted in high mortality within 24 h after stimulation. Whether LCMV, and viruses in general, can differentially modulate the ability of different components of the innate immune system to produce type I IFNs upon in vivo TLR stimulation warrants further investigation.

LCMV infection affects pDCs at multiple levels, including decreased pDC numbers likely related to the virus's ability to block DC development from early undifferentiated progenitors (113) and to redirect the differentiation of immature bone marrow pDCs into CD11b+DCs (129). In addition, LCMV infection can alter pDC functions by preventing their production of type I IFNs, which may be related to the arenavirus NP's IFN-counteracting activity. This type I IFN-counteracting activity of NP might contribute to the modulation of the kinetics, magnitude, or duration of this early type I IFN response, which may influence both virus multiplication and propagation, as well as the ensuing host immune responses and thereby the outcome of infection. LCMV variants exhibiting an immune-suppressive phenotype (e.g., Cl13) could exert other effects including functional impairment, as well as possible tolerance and loss of DCs, which would likely prevent the induction of an effective CTL response, thus facilitating virus persistence. In addition, recent data have shown that programmed death-1 (PD-1) is a critical molecule for sustaining suppression of CD8+ T cells during Cl13 persistence (6), whereas early upregulation of interleukin-10 production by antigin-presenting cells was found to contribute to initiation of T-cell inactivation and viral persistence in Cl13-infected mice (13).

USE OF REVERSE GENETICS TO STUDY ARENAVIRUS-HOST INNATE IMMUNITY INTERACTIONS

Recently developed reverse genetics systems for LCMV allowed the rescue of recombinant LCMV entirely from cloned cDNAs (36, 97, 108), which should facilitate the engineering of rLCMVs bearing specific mutations in NP that disrupt the IFN-antagonistic activity of NP while retaining other NP functions required for virus multiplication. This, in turn, would allow assessment of the biological significance of NP's IFN-counteracting activity in the context of LCMV infection of the mouse, its natural host. LCMV infection of the mouse represents a unique and powerful model system to study the virus-host immune response interactions, for several reasons.

1. The mouse is the natural host of LCMV.
2. There is currently detailed knowledge of the anti-LCMV host immune response, including immunity from major histocompatibility complex restriction, kinetics of generation expansion and contraction of virus-specific CD8 and CD4 T cells, CD4 T-cell help for CD8 T cells, loss of T-cell function during persistent infections, parameters of memory cell numbers, and plasma cell migration and residence.
3. There is the possibility of a variety of infection outcomes depending on experimental settings, including both host and viral factors.
4. Notably, many of the viral immunology principles initially established with LCMV have been extended to other microbial infections and to viral infections of humans.

It would be feasible to determine whether the IFN-counteracting activity associated with the virus NP plays an important role in modulating the very early phases of infection and thereby influencing the kinetics of virus multiplication and the overall outcome of infection. It would also be possible to assess whether mutations affecting NP's IFN-counteracting activity result in an attenuated virus phenotype in cultured cells and mice competent in IFN production and signaling, which would be consistent with the observation that influenza viruses with mutant NS1 proteins unable to inhibit IFN production are attenuated in cultured cells and mouse models of infection (99, 119). Two issues of particular importance that could be examined using rLCMV that lack NP's IFN-counteracting activity would be:

1. The virus's ability to replicate and induce a specific antivirus immune response in the immunocompetent adult mouse. Inoculation of immunocompetent adult mice with most strains of LCMV induces a robust immune response that effectively eliminates the virus between days 8 and 10 postinfection. This response involves elements of both innate and adaptive immunity, but the virus-specific CD8[+] CTL response plays a central role in virus clearance. The kinetics of virus replication

and clearance, as well as the IFN response and the composition and magnitude of the anti-LCMV T-cell responses, are very well established. This would permit determination of whether the loss of the anti-IFN activity of NP influences any of these parameters.

2. The ability of viruses with a Cl13-like immuno-suppressive phenotype to establish a transient persistent infection in the adult mouse. Infection of wild-type mice with high doses (10^6 PFU) of an immunosuppressive variant of LCMV-ARM called clone 13, LCMV(Cl13), causes impaired DC and T-cell functions that results in a transient generalized immunosuppression and establishment of a persistent infection; hence the phenotype of Cl13 is termed CTL[-]/Pi[+]. Virus clearance takes place between days 60 and 100 postinfection and correlates with the recovery of normal host immune responses. This model provides an excellent opportunity to determine whether the IFN-antagonistic activity of NP is required to modulate very early steps of virus-host innate immune response to favor, in conjunction with other viral properties, the establishment of viral persistence.

References

1. **Ahmed, R., and M. B. Oldstone.** 1988. Organ-specific selection of viral variants during chronic infection. *J. Exp. Med.* **167:**1719–1724.
2. **Ahmed, R., A. Salmi, L. D. Butler, J. M. Chiller, and M. B. Oldstone.** 1984. Selection of genetic variants of lymphocytic choriomeningitis virus in spleens of persistently infected mice. Role in suppression of cytotoxic T lymphocyte response and viral persistence. *J. Exp. Med.* **160:**521–540.
3. **Ahmed, R., R. S. Simon, M. Matloubian, S. R. Kolhekar, P. J. Southern, and D. M. Freedman.** 1988. Genetic analysis of in vivo-selected viral variants causing chronic infection: importance of mutation in the L RNA segment of lymphocytic choriomeningitis virus. *J. Virol.* **62:**3301–3308.
4. **Akira, S., and H. Hemmi.** 2003. Recognition of pathogen-associated molecular patterns by TLR family. *Immunol. Lett.* **85:**85–95.
5. **Asselin-Paturel, C., and G. Trinchieri.** 2005. Production of type I interferons: plasmacytoid dendritic cells and beyond. *J. Exp. Med.* **202:**461–465.
6. **Barber, D. L., E. J. Wherry, D. Masopust, B. Zhu, J. P. Allison, A. H. Sharpe, G. J. Freeman, and R. Ahmed.** 2006. Restoring function in exhausted CD8 T cells during chronic viral infection. *Nature* **439:**682–687.
7. **Basler, C. F., A. Mikulasova, L. Martinez-Sobrido, J. Paragas, E. Muhlberger, M. Bray, H. D. Klenk, P. Palese, and A. Garcia-Sastre.** 2003. The Ebola virus VP35 protein inhibits activation of interferon regulatory factor 3. *J. Virol.* **77:**7945–7956.

8. Beyer, W. R., D. Popplau, W. Garten, D. von Laer, and O. Lenz. 2003. Endoproteolytic processing of the lymphocytic choriomeningitis virus glycoprotein by the subtilase SKI-1/S1P. *J. Virol.* 77:2866–2872.

9. Biron, C. A., L. P. Cousens, M. C. Ruzek, H. C. Su, and T. P. Salazar-Mather. 1998. Early cytokine responses to viral infections and their roles in shaping endogenous cellular immunity. *Adv. Exp. Med. Biol.* 452:143–149.

10. Bonjardim, C. A. 2005. Interferons (IFNs) are key cytokines in both innate and adaptive antiviral immune responses—and viruses counteract IFN action. *Microbes Infect.* 7:569–578.

11. Borden, K. L., E. J. Campbell Dwyer, and M. S. Salvato. 1998. An arenavirus RING (zinc-binding) protein binds the oncoprotein promyelocyte leukemia protein (PML) and relocates PML nuclear bodies to the cytoplasm. *J. Virol.* 72:758–766.

12. Borrow, P., and M. B. Oldstone. 1994. Mechanism of lymphocytic choriomeningitis virus entry into cells. *Virology* 198:1–9.

13. Brooks, D. G., M. J. Trifilo, K. H. Edelmann, L. Teyton, D. B. McGavern, and M. B. Oldstone. 2006. Interleukin-10 determines viral clearance or persistence in vivo. *Nat. Med.* 12:1301–1309.

14. Buchmeier, M. J., and A. J. Zajac. 1999. Lymphocytic choriomeningitis virus, p. 575–605. *In* R. Ahmed and I. Chen (ed.), *Persistent Viral Infections*. John Wiley & Sons, Inc., New York, NY.

15. Buchmeier, M. J., C. J. Peters, and J. C. de la Torre. 2007. Arenaviridae: the viruses and their replication, p. 1792–1827. *In* D. M. Knipe, and P. M. Howley (ed.), *Fields Virology*, 5th ed., vol. 2. Lippincott, Williams & Wilkins, Philadelphia, PA.

16. Buchmeier, M. J., R. M. Welsh, F. J. Dutko, and M. B. Oldstone. 1980. The virology and immunobiology of lymphocytic choriomeningitis virus infection. *Adv. Immunol.* 30:275–331.

17. Bukowski, J. F., C. A. Biron, and R. M. Welsh. 1983. Elevated natural killer cell-mediated cytotoxicity, plasma interferon, and tumor cell rejection in mice persistently infected with lymphocytic choriomeningitis virus. *J. Immunol.* 131:991–996.

18. Campbell Dwyer, E. J., H. Lai, R. C. MacDonald, M. S. Salvato, and K. L. Borden. 2000. The lymphocytic choriomeningitis virus RING protein Z associates with eukaryotic initiation factor 4E and selectively represses translation in a RING-dependent manner. *J. Virol.* 74:3293–3300.

19. Cardenas, W. B., Y. M. Loo, M. Gale, Jr., A. L. Hartman, C. R. Kimberlin, L. Martinez-Sobrido, E. O. Saphire, and C. F. Basler. 2006. Ebola virus VP35 protein binds double-stranded RNA and inhibits alpha/beta interferon production induced by RIG-I signaling. *J. Virol.* 80:5168–5178.

20. Cella, M., D. Jarrossay, F. Facchetti, O. Alebardi, H. Nakajima, A. Lanzavecchia, and M. Colonna. 1999. Plasmacytoid monocytes migrate to inflamed lymph nodes and produce large amounts of type I interferon. *Nat. Med.* 5:919–923.

21. Cornu, T. I., and J. C. de la Torre. 2002. Characterization of the arenavirus RING finger Z protein regions required for Z-mediated inhibition of viral RNA synthesis. *J. Virol.* 76:6678–6688.

22. Cornu, T. I., and J. C. de la Torre. 2001. RING finger Z protein of lymphocytic choriomeningitis virus (LCMV) inhibits transcription and RNA replication of an LCMV S-segment minigenome. *J. Virol.* 75:9415–9426.

23. Cornu, T. I., H. Feldmann, and J. C. de la Torre. 2004. Cells expressing the RING finger Z protein are resistant to arenavirus infection. *J. Virol.* 78:2979–2983.

24. Dalod, M., T. P. Salazar-Mather, L. Malmgaard, C. Lewis, C. Asselin-Paturel, F. Briere, G. Trinchieri, and C. A. Biron. 2002. Interferon alpha/beta and interleukin 12 responses to viral infections: pathways regulating dendritic cell cytokine expression in vivo. *J. Exp. Med.* 195:517–528.

25. de la Torre, J. C., and M. B. A. Oldstone. 1996. The anatomy of viral persistence: mechanisms of persistence and associated disease. *Adv. Virus Res.* 46:311–343.

26. Di Simone, C., and M. J. Buchmeier. 1995. Kinetics and pH dependence of acid-induced structural changes in the lymphocytic choriomeningitis virus glycoprotein complex. *Virology* 209:3–9.

27. Di Simone, C., M. A. Zandonatti, and M. J. Buchmeier. 1994. Acidic pH triggers LCMV membrane fusion activity and conformational change in the glycoprotein spike. *Virology* 198:455–465.

28. Doughty, L., K. Nguyen, J. Durbin, and C. Biron. 2001. A role for IFN-alpha beta in virus infection-induced sensitization to endotoxin. *J. Immunol.* 166:2658–2664.

29. Doyle, S., S. Vaidya, R. O'Connell, H. Dadgostar, P. Dempsey, T. Wu, G. Rao, R. Sun, M. Haberland, R. Modlin, and G. Cheng. 2002. IRF3 mediates a TLR3/TLR4-specific antiviral gene program. *Immunity* 17:251–263.

30. Ehrhardt, C., C. Kardinal, W. J. Wurzer, T. Wolff, C. von Eichel-Streiber, S. Pleschka, O. Planz, and S. Ludwig. 2004. Rac1 and PAK1 are upstream of IKK-epsilon and TBK-1 in the viral activation of interferon regulatory factor-3. *FEBS Lett.* 567:230–238.

31. Eichler, R., O. Lenz, T. Strecker, M. Eickmann, H. D. Klenk, and W. Garten. 2003. Identification of Lassa virus glycoprotein signal peptide as a *trans*-acting maturation factor. *EMBO Rep.* 4:1084–1088.

32. Eichler, R., O. Lenz, T. Strecker, M. Eickmann, H. D. Klenk, and W. Garten. 2004. Lassa virus glycoprotein signal peptide displays a novel topology with an extended endoplasmic reticulum luminal region. *J. Biol. Chem.* 279:12293–12299.

33. Eichler, R., O. Lenz, T. Strecker, and W. Garten. 2003. Signal peptide of Lassa virus glycoprotein GP-C exhibits an unusual length. *FEBS Lett.* 538:203–206.

34. Fischer, S. A., M. B. Graham, M. J. Kuehnert, C. N. Kotton, A. Srinivasan, F. M. Marty, J. A. Comer, J. Guarner, C. D. Paddock, D. L. DeMeo, W. J. Shieh, B. R. Erickson, U. Bandy, A. DeMaria, Jr., J. P. Davis, F. L. Delmonico, B. Pavlin, A. Likos, M. J. Vincent, T. K. Sealy, C. S. Goldsmith, D. B. Jernigan, P. E. Rollin, M. M. Packard, M. Patel, C. Rowland, R. F. Helfand, S. T. Nichol, J. A. Fishman, T. Ksiazek, and S. R. Zaki. 2006. Transmission of lymphocytic choriomeningitis virus by organ transplantation. *N. Engl. J. Med.* 354:2235–2249.

35. Fitzgerald, K. A., S. M. McWhirter, K. L. Faia, D. C. Rowe, E. Latz, D. T. Golenbock, A. J. Coyle, S. M. Liao, and T. Maniatis. 2003. IKKepsilon and TBK1 are essential

components of the IRF3 signaling pathway. *Nat. Immunol.* **4**:491–496.

36. Flatz, L., A. Bergthaler, J. C. de la Torre, and D. D. Pinschewer. 2006. Recovery of an arenavirus entirely from RNA polymerase I/II-driven cDNA. *Proc. Natl. Acad. Sci. USA* **103**:4663–4668.

37. Freed, E. O. 2002. Viral late domains. *J. Virol.* **76**:4679–4687.

38. Freedman, D. O., and J. Woodall. 1999. Emerging infectious diseases and risk to the traveler. *Med. Clin. North Am.* **83**:865–883, v.

39. Froeschke, M., M. Basler, M. Groettrup, and B. Dobberstein. 2003. Long-lived signal peptide of lymphocytic choriomeningitis virus glycoprotein pGP-C. *J. Biol. Chem.* **278**:41914–41920.

40. Gallaher, W. R., C. DiSimone, and M. J. Buchmeier. 2001. The viral transmembrane superfamily: possible divergence of Arenavirus and Filovirus glycoproteins from a common RNA virus ancestor. *BMC Microbiol.* **1**:1.

41. Garcia-Sastre, A., and C. A. Biron. 2006. Type 1 interferons and the virus-host relationship: a lesson in detente. *Science* **312**:879–882.

42. Garcin, D., and D. Kolakofsky. 1992. Tacaribe arenavirus RNA synthesis in vitro is primer dependent and suggests an unusual model for the initiation of genome replication. *J. Virol.* **66**:1370–1376.

43. Grandvaux, N., M. J. Servant, B. tenOever, G. C. Sen, S. Balachandran, G. N. Barber, R. Lin, and J. Hiscott. 2002. Transcriptional profiling of interferon regulatory factor 3 target genes: direct involvement in the regulation of interferon-stimulated genes. *J. Virol.* **76**:5532–5539.

44. Guilbert, J. J. 2003. The world health report 2002—reducing risks, promoting healthy life. *Educ. Health (Abingdon)* **16**:230.

45. Haller, O., G. Kochs, and F. Weber. 2006. The interferon response circuit: induction and suppression by pathogenic viruses. *Virology* **344**:119–130.

46. Hass, M., U. Golnitz, S. Muller, B. Becker-Ziaja, and S. Gunther. 2004. Replicon system for Lassa virus. *J. Virol.* **78**:13793–13803.

47. Hass, M., M. Westerkofsky, S. Muller, B. Becker-Ziaja, C. Busch, and S. Gunther. 2006. Mutational analysis of the Lassa virus promoter. *J. Virol.* **80**:12414–12419.

48. Holmes, G. P., J. B. McCormick, S. C. Trock, R. A. Chase, S. M. Lewis, C. A. Mason, P. A. Hall, L. S. Brammer, G. I. Perez-Oronoz, M. K. McDonnell, et al. 1990. Lassa fever in the United States. Investigation of a case and new guidelines for management. *N. Engl. J. Med.* **323**:1120–1123.

49. Hosmalin, A., and P. Lebon. 2006. Type I interferon production in HIV-infected patients. *J. Leukoc. Biol.* **80**:984–993.

50. Isaacson, M. 2001. Viral hemorrhagic fever hazards for travelers in Africa. *Clin. Infect. Dis.* **33**:1707–1712.

51. Ito, T., R. Amakawa, and S. Fukuhara. 2002. Roles of Toll-like receptors in natural interferon-producing cells as sensors in immune surveillance. *Hum. Immunol.* **63**:1120–1125.

52. Jahrling, P. B., and C. J. Peters. 1992. Lymphocytic choriomeningitis virus. A neglected pathogen of man. *Arch. Pathol. Lab. Med.* **116**:486–488.

53. Jarrossay, D., G. Napolitani, M. Colonna, F. Sallusto, and A. Lanzavecchia. 2001. Specialization and complementarity

in microbial molecule recognition by human myeloid and plasmacytoid dendritic cells. *Eur. J. Immunol.* **31**:3388–3393.

54. Kato, H., O. Takeuchi, S. Sato, M. Yoneyama, M. Yamamoto, K. Matsui, S. Uematsu, A. Jung, T. Kawai, K. J. Ishii, O. Yamaguchi, K. Otsu, T. Tsujimura, C. S. Koh, C. Reis e Sousa, Y. Matsuura, T. Fujita, and S. Akira. 2006. Differential roles of MDA5 and RIG-I helicases in the recognition of RNA viruses. *Nature* **441**:101–105.

55. Kentsis, A., E. C. Dwyer, J. M. Perez, M. Sharma, A. Chen, Z. Q. Pan, and K. L. Borden. 2001. The RING domains of the promyelocytic leukemia protein PML and the arenaviral protein Z repress translation by directly inhibiting translation initiation factor eIF4E. *J. Mol. Biol.* **312**:609–623.

56. Kilgore, P. E., T. G. Ksiazek, P. E. Rollin, J. N. Mills, M. R. Villagra, M. J. Montenegro, M. A. Costales, L. C. Paredes, and C. J. Peters. 1997. Treatment of Bolivian hemorrhagic fever with intravenous ribavirin. *Clin. Infect. Dis.* **24**:718–722.

57. King, C. C., B. D. Jamieson, K. Reddy, N. Bali, R. J. Concepcion, and R. Ahmed. 1992. Viral infection of the thymus. *J. Virol.* **66**:3155–3160.

58. Kopecky-Bromberg, S. A., L. Martinez-Sobrido, M. Frieman, R. A. Baric, and P. Palese. 2007. Severe acute respiratory syndrome coronavirus open reading frame (ORF) 3b, ORF 6, and nucleocapsid proteins function as interferon antagonists. *J. Virol.* **81**:548–557.

59. Kunz, S., P. Borrow, and M. B. Oldstone. 2002. Receptor structure, binding, and cell entry of arenaviruses. *Curr. Top. Microbiol. Immunol.* **262**:111–137.

60. Kunz, S., K. H. Edelmann, J.-C. de la Torre, R. Gorney, and M. B. Oldstone. 2003. Mechanisms for lymphocytic choriomeningitis virus glycoprotein cleavage, transport, and incorporation into virions. *Virology* **314**:168–178.

61. Kunz, S., N. Sevilla, J. M. Rojek, and M. B. Oldstone. 2004. Use of alternative receptors different than alpha-dystroglycan by selected isolates of lymphocytic choriomeningitis virus. *Virology* **325**:432–445.

62. Lee, K. J., I. S. Novella, M. N. Teng, M. B. Oldstone, and J. C. de La Torre. 2000. NP and L proteins of lymphocytic choriomeningitis virus (LCMV) are sufficient for efficient transcription and replication of LCMV genomic RNA analogs. *J. Virol.* **74**:3470–3477.

63. Lee, K. J., M. Perez, D. D. Pinschewer, and J. C. de la Torre. 2002. Identification of the lymphocytic choriomeningitis virus (LCMV) proteins required to rescue LCMV RNA analogs into LCMV-like particles. *J. Virol.* **76**:6393–6397.

64. Levy, D. E., and A. Garcia-Sastre. 2001. The virus battles: IFN induction of the antiviral State and mechanisms of viral evasion. *Cytokine Growth Factor Rev.* **12**:143–156.

65. Lopez, N., R. Jacamo, and M. T. Franze-Fernandez. 2001. Transcription and RNA replication of Tacaribe virus genome and antigenome analogs require N and L proteins: Z protein is an inhibitor of these processes. *J. Virol.* **75**:12241–12251.

66. Louten, J., N. van Rooijen, and C. A. Biron. 2006. Type 1 IFN deficiency in the absence of normal splenic architecture during lymphocytic choriomeningitis virus infection. *J. Immunol.* **177**:3266–3272.

67. Malmgaard, L., T. P. Salazar-Mather, C. A. Lewis, and C. A. Biron. 2002. Promotion of alpha/beta interferon induction during in vivo viral infection through alpha/beta interferon receptor/Stat1 system-dependent and -independent pathways. *J. Virol.* **76:**4520–4525.

68. Marie, I., J. E. Durbin, and D. E. Levy. 1998. Differential viral induction of distinct interferon-alpha genes by positive feedback through interferon regulatory factor-7. *EMBO J.* **17:**6660–6669.

69. Martinez-Sobrido, L., P. Giannakas, B. Cubitt, A. Garcia-Sastre, and J. C. de la Torre. 2007. Differential inhibition of type I interferon induction by arenavirus nucleoproteins. *J. Virol.* **81:**12696–12703.

70. Martinez-Sobrido, L., E. I. Zuniga, D. Rosario, A. Garcia-Sastre, and J. C. de la Torre. 2006. Inhibition of the type I interferon response by the nucleoprotein of the prototypic arenavirus lymphocytic choriomeningitis virus. *J. Virol.* **80:**9192–9199.

71. McCormick, J. B., and S. P. Fisher-Hoch. 2002. Lassa fever, p. 75–110. *In* M. B. Oldstone (ed.), *Arenaviruses I*, vol. 262. Springer-Verlag, Berlin, Germany.

72. McCormick, J. B., I. J. King, P. A. Webb, C. L. Scribner, R. B. Craven, K. M. Johnson, L. H. Elliott, and R. Belmont-Williams. 1986. Lassa fever. Effective therapy with ribavirin. *N. Engl. J. Med.* **314:**20–26.

73. McKee, K. T., Jr., J. W. Huggins, C. J. Trahan, and B. G. Mahlandt. 1988. Ribavirin prophylaxis and therapy for experimental Argentine hemorrhagic fever. *Antimicrob. Agents Chemother.* **32:**1304–1309.

74. McKenna, K., A. S. Beignon, and N. Bhardwaj. 2005. Plasmacytoid dendritic cells: linking innate and adaptive immunity. *J. Virol.* **79:**17–27.

75. Melchjorsen, J., S. B. Jensen, L. Malmgaard, S. B. Rasmussen, F. Weber, A. G. Bowie, S. Matikainen, and S. R. Paludan. 2005. Activation of innate defense against a paramyxovirus is mediated by RIG-I and TLR7 and TLR8 in a cell-type-specific manner. *J. Virol.* **79:**12944–12951.

76. Mets, M. B., L. L. Barton, A. S. Khan, and T. G. Ksiazek. 2000. Lymphocytic choriomeningitis virus: an underdiagnosed cause of congenital chorioretinitis. *Am. J. Ophthalmol.* **130:**209–215.

77. Meyer, B. J., J. C. de la Torre, and P. J. Southern. 2002. Arenaviruses: genomic RNAs, transcription, and replication. *Curr. Top. Microbiol. Immunol.* **262:**139–149.

78. Mibayashi, M., L. Martinez-Sobrido, Y. M. Loo, W. B. Cardenas, M. Gale, Jr., and A. Garcia-Sastre. 2007. Inhibition of retinoic acid-inducible gene I-mediated induction of beta interferon by the NS1 protein of influenza A virus. *J. Virol.* **81:**514–524.

79. Montoya, M., M. J. Edwards, D. M. Reid, and P. Borrow. 2005. Rapid activation of spleen dendritic cell subsets following lymphocytic choriomeningitis virus infection of mice: analysis of the involvement of type 1 IFN. *J. Immunol.* **174:**1851–1861.

80. Montoya, M., G. Schiavoni, F. Mattei, I. Gresser, F. Belardelli, P. Borrow, and D. F. Tough. 2002. Type I interferons produced by dendritic cells promote their phenotypic and functional activation. *Blood* **99:**3263–3271.

81. Mori, M., M. Yoneyama, T. Ito, K. Takahashi, F. Inagaki, and T. Fujita. 2004. Identification of Ser-386 of interferon regulatory factor 3 as critical target for inducible phosphorylation that determines activation. *J. Biol. Chem.* **279:**9698–9702.

82. Nakaya, T., M. Sato, N. Hata, M. Asagiri, H. Suemori, S. Noguchi, N. Tanaka, and T. Taniguchi. 2001. Gene induction pathways mediated by distinct IRFs during viral infection. *Biochem. Biophys. Res. Commun.* **283:**1150–1156.

83. Neuman, B. W., B. D. Adair, J. W. Burns, R. A. Milligan, M. J. Buchmeier, and M. Yeager. 2005. Complementarity in the supramolecular design of arenaviruses and retroviruses revealed by electron cryomicroscopy and image analysis. *J. Virol.* **79:**3822–3830.

84. Oldstone, M. B. 2002. Biology and pathogenesis of lymphocytic choriomeningitis virus infection, p. 83–118. *In* M. B. Oldstone (ed.), *Arenaviruses*, vol. 263. Springer-Verlag, Berlin, Germany.

85. Oldstone, M. B. 2006. Viral persistence: parameters, mechanisms and future predictions. *Virology* **344:**111–118.

86. Palacios, G., J. Druce, L. Du, T. Tran, C. Birch, T. Briese, S. Conlan, P. L. Quan, J. Hui, J. Marshall, J. F. Simons, M. Egholm, C. D. Paddock, W. J. Shieh, C. S. Goldsmith, S. R. Zaki, M. Catton, and W. I. Lipkin. 2008. A new arenavirus in a cluster of fatal transplant-associated diseases. *N. Engl. J. Med.* **358:**991–998.

87. Park, M. S., A. Garcia-Sastre, J. F. Cros, C. F. Basler, and P. Palese. 2003. Newcastle disease virus V protein is a determinant of host range restriction. *J. Virol.* **77:**9522–9532.

88. Park, M. S., M. L. Shaw, J. Munoz-Jordan, J. F. Cros, T. Nakaya, N. Bouvier, P. Palese, A. Garcia-Sastre, and C. F. Basler. 2003. Newcastle disease virus (NDV)-based assay demonstrates interferon-antagonist activity for the NDV V protein and the Nipah virus V, W, and C proteins. *J. Virol.* **77:**1501–1511.

89. Perez, M., R. C. Craven, and J. C. de la Torre. 2003. The small RING finger protein Z drives arenavirus budding: implications for antiviral strategies. *Proc. Natl. Acad. Sci. USA* **100:**12978–12983.

90. Perez, M., and J. C. de la Torre. 2003. Characterization of the genomic promoter of the prototypic arenavirus lymphocytic choriomeningitis virus. *J. Virol.*, **77:**1184–1194.

91. Perez, M., D. L. Greenwald, and J. C. de la Torre. 2004. Myristoylation of the RING finger Z protein is essential for arenavirus budding. *J. Virol.* **78:**11443–11448.

92. Peters, C. J. 2002. Human infection with arenaviruses in the Americas, p. 65–74. *In* M. B. Oldstone (ed.), *Arenaviruses I*, vol. 262. Springer-Verlag, Berlin, Germany.

93. Peters, C. J. 2006. Lymphocytic choriomeningitis virus—an old enemy up to new tricks. *N. Engl. J. Med.* **354:**2208–2211.

94. Peters, K. L., H. L. Smith, G. R. Stark, and G. C. Sen. 2002. IRF-3-dependent, NFkappa B- and JNK-independent activation of the 561 and IFN-beta genes in response to double-stranded RNA. *Proc. Natl. Acad. Sci. USA* **99:**6322–6327.

95. Pinschewer, D. D., M. Perez, and J. C. de la Torre. 2005. Dual role of the lymphocytic choriomeningitis virus intergenic region in transcription termination and virus propagation. *J. Virol.* **79:**4519–4526.

96. Pinschewer, D. D., M. Perez, and J. C. de la Torre. 2003. Role of the virus nucleoprotein in the regulation

of lymphocytic choriomeningitis virus transcription and RNA replication. *J. Virol.* **77**:3882–3887.

97. **Pinschewer, D. D., M. Perez, A. B. Sanchez, and J. C. de la Torre.** 2003. Recombinant lymphocytic choriomeningitis virus expressing vesicular stomatitis virus glycoprotein. *Proc. Natl. Acad. Sci. USA* **100**:7895–7900.

98. **Pircher, H., K. Burki, R. Lang, H. Hengartner, and R. M. Zinkernagel.** 1989. Tolerance induction in double specific T-cell receptor transgenic mice varies with antigen. *Nature* **342**:559–561.

99. **Quinlivan, M., D. Zamarin, A. Garcia-Sastre, A. Cullinane, T. Chambers, and P. Palese.** 2005. Attenuation of equine influenza viruses through truncations of the NS1 protein. *J. Virol.* **79**:8431–8439.

100. **Raju, R., L. Raju, D. Hacker, D. Garcin, R. Compans, and D. Kolakofsky.** 1990. Nontemplated bases at the 5′ ends of Tacaribe virus mRNAs. *Virology* **174**:53–59.

101. **Ruiz-Jarabo, C. M., C. Ly, E. Domingo, and J. C. de la Torre.** 2003. Lethal mutagenesis of the prototypic arenavirus lymphocytic choriomeningitis virus (LCMV). *Virology* **308**:37–47.

102. **Salvato, M., P. Borrow, E. Shimomaye, and M. B. Oldstone.** 1991. Molecular basis of viral persistence: a single amino acid change in the glycoprotein of lymphocytic choriomeningitis virus is associated with suppression of the antiviral cytotoxic T-lymphocyte response and establishment of persistence. *J. Virol.* **65**:1863–1869.

103. **Salvato, M., E. Shimomaye, and M. B. Oldstone.** 1989. The primary structure of the lymphocytic choriomeningitis virus L gene encodes a putative RNA polymerase. *Virology* **169**:377–384.

104. **Salvato, M., E. Shimomaye, P. Southern, and M. B. Oldstone.** 1988. Virus-lymphocyte interactions. IV. Molecular characterization of LCMV Armstrong (CTL+) small genomic segment and that of its variant, Clone 13 (CTL-). *Virology* **164**:517–522.

105. **Salvato, M. S.** 1993. Molecular biology of the prototype arenavirus, lymphocytic choriomeningitis virus, p. 133–156. *In* M. S. Salvato (ed.), *The Arenaviridae*, vol. 1. Plenum Press, New York, NY.

106. **Salvato, M. S., K. J. Schweighofer, J. Burns, and E. M. Shimomaye.** 1992. Biochemical and immunological evidence that the 11 kDa zinc-binding protein of lymphocytic choriomeningitis virus is a structural component of the virus. *Virus Res.* **22**:185–198.

107. **Sanchez, A. B., and J. C. de la Torre.** 2005. Genetic and biochemical evidence for an oligomeric structure of the functional L polymerase of the prototypic arenavirus lymphocytic choriomeningitis virus. *J. Virol.* **79**:7262–7268.

108. **Sanchez, A. B., and J. C. de la Torre.** 2006. Rescue of the prototypic arenavirus LCMV entirely from plasmid. *Virology* **350**:370–380.

109. **Sato, M., N. Tanaka, N. Hata, E. Oda, and T. Taniguchi.** 1998. Involvement of the IRF family transcription factor IRF-3 in virus-induced activation of the IFN-beta gene. *FEBS Lett.* **425**:112–116.

110. **Schlender, J., V. Hornung, S. Finke, M. Gunthner-Biller, S. Marozin, K. Brzozka, S. Moghim, S. Endres, G. Hartmann, and K. K. Conzelmann.** 2005. Inhibition of Toll-like receptor 7- and 9-mediated alpha/beta interferon production in human plasmacytoid dendritic cells by respiratory syncytial virus and measles virus. *J. Virol.* **79**:5507–5515.

111. **Servant, M. J., B. ten Oever, C. LePage, L. Conti, S. Gessani, I. Julkunen, R. Lin, and J. Hiscott.** 2001. Identification of distinct signaling pathways leading to the phosphorylation of interferon regulatory factor 3. *J. Biol. Chem.* **276**:355–363.

112. **Sevilla, N., S. Kunz, A. Holz, H. Lewicki, D. Homann, H. Yamada, K. P. Campbell, J. C. de La Torre, and M. B. Oldstone.** 2000. Immunosuppression and resultant viral persistence by specific viral targeting of dendritic cells. *J. Exp. Med.* **192**:1249–1260.

113. **Sevilla, N., D. B. McGavern, C. Teng, S. Kunz, and M. B. Oldstone.** 2004. Viral targeting of hematopoietic progenitors and inhibition of DC maturation as a dual strategy for immune subversion. *J. Clin. Invest.* **113**:737–745.

114. **Smelt, S. C., P. Borrow, S. Kunz, W. Cao, A. Tishon, H. Lewicki, K. P. Campbell, and M. B. Oldstone.** 2001. Differences in affinity of binding of lymphocytic choriomeningitis virus strains to the cellular receptor alpha-dystroglycan correlate with viral tropism and disease kinetics. *J. Virol.* **75**:448–457.

115. **Spiropoulou, C. F., S. Kunz, P. E. Rollin, K. P. Campbell, and M. B. Oldstone.** 2002. New World arenavirus clade C, but not clade A and B viruses, utilizes alpha-dystroglycan as its major receptor. *J. Virol.* **76**:5140–5146.

116. **Stojdl, D. F., B. D. Lichty, B. R. tenOever, J. M. Paterson, A. T. Power, S. Knowles, R. Marius, J. Reynard, L. Poliquin, H. Atkins, E. G. Brown, R. K. Durbin, J. E. Durbin, J. Hiscott, and J. C. Bell.** 2003. VSV strains with defects in their ability to shutdown innate immunity are potent systemic anti-cancer agents. *Cancer Cell* **4**:263–275.

117. **Strecker, T., R. Eichler, J. Meulen, W. Weissenhorn, H. Dieter Klenk, W. Garten, and O. Lenz.** 2003. Lassa virus Z protein is a matrix protein and sufficient for the release of virus-like particles [corrected]. *J. Virol.* **77**:10700–10705.

118. **Talon, J., C. M. Horvath, R. Polley, C. F. Basler, T. Muster, P. Palese, and A. Garcia-Sastre.** 2000. Activation of interferon regulatory factor 3 is inhibited by the influenza A virus NS1 protein. *J. Virol.* **74**:7989–7996.

119. **Talon, J., M. Salvatore, R. E. O'Neill, Y. Nakaya, H. Zheng, T. Muster, A. Garcia-Sastre, and P. Palese.** 2000. Influenza A and B viruses expressing altered NS1 proteins: a vaccine approach. *Proc. Natl. Acad. Sci. USA* **97**:4309–4314.

120. **tenOever, B. R., S. Sharma, W. Zou, Q. Sun, N. Grandvaux, I. Julkunen, H. Hemmi, M. Yamamoto, S. Akira, W. C. Yeh, R. Lin, and J. Hiscott.** 2004. Activation of TBK1 and IKK-ε kinases by vesicular stomatitis virus infection and the role of viral ribonucleoprotein in the development of interferon antiviral immunity. *J. Virol.* **78**:10636–10649.

121. **Urata, S., T. Noda, Y. Kawaoka, H. Yokosawa, and J. Yasuda.** 2006. Cellular factors required for Lassa virus budding. *J. Virol.* **80**:4191–4195.

122. **Vieth, S., A. E. Torda, M. Asper, H. Schmitz, and S. Gunther.** 2004. Sequence analysis of L RNA of Lassa virus. *Virology* **318:**153–168.

123. **Weaver, B. K., K. P. Kumar, and N. C. Reich.** 1998. Interferon regulatory factor 3 and CREB-binding protein/p300 are subunits of double-stranded RNA-activated transcription factor DRAF1. *Mol. Cell. Biol.* **18:**1359–1368.

124. **Yoneyama, M., and T. Fujita.** 2004. [RIG-I: critical regulator for virus-induced innate immunity]. *Tanpakushitsu Kakusan Koso* **49:**2571–2578.

125. **Yoneyama, M., M. Kikuchi, T. Natsukawa, N. Shinobu, T. Imaizumi, M. Miyagishi, K. Taira, S. Akira, and T. Fujita.** 2004. The RNA helicase RIG-I has an essential function in double-stranded RNA-induced innate antiviral responses. *Nat. Immunol.* **5:**730–737.

126. **York, J., V. Romanowski, M. Lu, and J. H. Nunberg.** 2004. The signal peptide of the Junin arenavirus envelope glycoprotein is myristoylated and forms an essential subunit of the mature G1-G2 complex. *J. Virol.* **78:**10783–10792.

127. **Young, P. R., and C. R. Howard.** 1983. Fine structure analysis of Pichinde virus nucleocapsids. *J. Gen. Virol.* **64**(Pt. 4):**833–842.

128. **Zinkernagel, R. M.** 2002. Lymphocytic choriomeningitis virus and immunology. *Curr. Top. Microbiol. Immunol.* **263:**1–5.

129. **Zuniga, E. I., D. B. McGavern, J. L. Pruneda-Paz, C. Teng, and M. B. Oldstone.** 2004. Bone marrow plasmacytoid dendritic cells can differentiate into myeloid dendritic cells upon virus infection. *Nat. Immunol.* **5:**1227–1234.

Cellular Signaling and Innate Immune
Responses to RNA Virus Infections
Edited by A. R. Brasier et al.
© 2009 ASM Press, Washington, D.C.

Brenda L. Fredericksen
Michael Gale, Jr.

Regulation of Innate Immunity by the *Flaviviridae*

20

THE *FLAVIVIRIDAE*

The *Flaviviridae* family comprises RNA viruses that have a major impact on public health and animal disease. Members of this family include arthropod-borne viruses, blood-borne agents, and viruses of zoonotic transmission and/or important agents of agricultural animal diseases. The *Flaviviridae* also represent important emergent viruses that will continue to impact public health and animal health as the world population increases and as global weather patterns change. Recent studies demonstrate that innate antiviral immune programs of mammalian cells are essential for control of virus replication and spread during infection by *Flaviviridae* members and that various family members direct strategies of innate immune evasion during infection. This chapter provides an overview of the *Flaviviridae* and their regulation of host-cell innate immune defenses.

The viruses that comprise the *Flaviviridae* family are small enveloped, single-stranded, positive-sense RNA viruses. The *Flaviviridae* are divided into three genera of more closely related viruses: *Flavivirus, Pestivirus,* and *Hepacivirus* (Fig. 1). The genomes of these viruses are between 9.6 and 12.5 kb in length and consist of a single open reading frame bounded by a 5' and 3' nontranslated region (NTR). The 5' and 3' NTRs contain conserved sequences and predicted secondary RNA structures that are likely required for negative-strand synthesis, genome amplification, translation, and packaging. Upon introduction into the host cell, the viral genome is translated by host cell machinery to generate a single viral polyprotein. Translation of the *Flavivirus* genome occurs in a cap-dependent manner, while both *Pestivirus* and *Hepacivirus* utilize an internal ribosome entry site (IRES) to initiate synthesis of the viral polyprotein. The viral polyprotein is co- and posttranslationally cleaved by a combination of host and virally encoded proteases to generate the individual viral proteins. The viral structural proteins, which are required for virus assembly, attachment, and entry, are encoded in the 5' portion of the genome, while the remainder of the genome encodes the nonstructural (NS) proteins. The NS proteins are multifunctional, playing critical roles in viral replication and interacting with host-cell factors to modify the cellular environment to promote efficient replication (15, 16, 28, 44, 111).

Flavivirus

Members of the the *Flavivirus* genus are among the most important arthropod-borne viruses globally and include

Brenda L. Fredericksen, Department of Cell Biology and Molecular Genetics, University of Maryland, College Park, MD 20742. Michael Gale, Jr., Department of Immunology, University of Washington School of Medicine, Seattle, WA 98195.

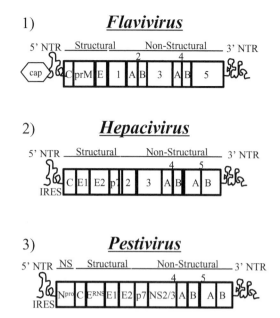

Figure 1 Genome and protein-coding organization of the *Flaviviridae* genera. The positions of the nontranslated region (NTR) and the polyprotein encoding the structural and non-structural (NS) proteins are shown. Cap denotes the 5′ cap present on *Flavivirus* genome RNA; IRES denotes the internal ribosome entry site found within the *Hepacivirus* and *Pestivirus* genome RNA. The positions of the mature viral proteins within each viral polyprotein are indicated. The structural proteins are abbreviated as follows: C, capsid; prM/M, membrane; E, envelope; E1, envelope protein 1; E2, envelope protein 2; E^{RNS}, envelope. The NS proteins are designated 1 to 5 with A and B subtypes, as shown.

pathogenic viruses such as yellow fever virus, dengue virus (DV), Japanese encephalitis virus (JEV), St. Louis encephalitis virus, tick-borne encephalitis virus (TBEV), and West Nile virus (WNV) (44). The spread of the yellow fever virus, the type member for which the family derived its name, from West Africa to the Americas is the earliest documented case of an emerging disease (16). Flaviviruses continue to represent some of the most globally significant examples of emerging and resurging viruses. These include the reemergence and spread of JEV throughout much of Asia and Oceania, the resurgence of DV in tropical and subtropical regions of the world, the increase and spread of TBEV throughout Europe and Asia, and the introduction of WNV into the Western hemisphere (84). Despite the fact that approximately 90% of flaviviral infections are subclinical, these viruses account for significant human disease burden worldwide.

JEV

Epidemics of JEV were first recognized in Japan in the 1870s. The virus was subsequently found throughout eastern and southern Asia, and more recently the virus has spread to parts of India, Indonesia, and Australia (84). Clinical manifestations of JEV infection range from a nonspecific febrile illness to severe neurological disease, including meningoencephalitis with Parkinsonian features, aseptic meningitis, or poliomyelitis-like flaccid paralysis. Though clinical disease may be found in all age groups, the majority of cases occur in infants and children. Approximately 50,000 cases of severe disease associated with JEV infection are reported annually, making JEV a leading cause of severe viral central nervous system (CNS) infections in Asia and Australia (123). Mortality associated with severe JEV infections ranges from 25 to 50%, and approximately 50% of cases result in permanent neurological sequelae, which include persistent motor deficits and severe cognitive and language impairment (16, 84, 123).

DV

Dengue virus causes more human disease than any other arbovirus, infecting approximately 50 million people annually. Infections can be associated with a full spectrum of clinical symptoms ranging from mild (dengue fever) to severe (dengue hemorrhagic fever [DHF] and dengue shock syndrome). DF, or breakbone fever, is typically a nonfatal but often debilitating febrile illness that commonly occurs in older children and adults. DHF and dengue shock syndrome, the severe forms of DV infection, are defined by vascular leakage and hemorrhagic manifestations (41, 67, 73). The incidence of dengue viral infection has been estimated to have increased 30-fold over the past 50 years. Prior to 1970 only five countries reported cases of DHF. In contrast, DV is now endemic in more than 100 countries and causes 500,000 cases of DHF annually, resulting in more than 25,000 deaths.

WNV

WNV has been endemic for many years in areas of Asia, the Middle East, and Africa. In these regions WNV has not been considered a public health concern since infections were generally asymptomatic or associated with a mild febrile illness in children. However, recent WNV epidemics in developed countries in Europe and North America have been associated with significantly higher rates of morbidity and mortality among mammals and birds than was previously documented (45, 66, 99). Since its introduction into North America in 1999, WNV has spread rapidly to every state within the continental United States, as well as parts of Canada, Mexico, and the Caribbean (46). As with other encephalitic flaviviruses, WNV symptoms range from a self-limiting mild febrile illness, known as West Nile fever, to severe neuroinvasive diseases, such as meningitis, encephalitis, and acute flaccid paralysis. Yearly outbreaks within the United States continue to be associated with severe disease, with 30% of reported cases resulting in neurological infections. In addition, there is increasing evidence of a high prevalence of neurological sequelae associated with WNV infections irrespective of the severity of the initial infection (16a, 65).

TBEV

TBEV and antigenically closely related viruses, such as Langat virus (8), are endemic in northern Asia as well as central and western Europe. These viruses cause approximately 10,000 hospitalizations per year (25). Severe neurological infections of TBEV typically present as a biphasic disease (39, 40). The first, or acute, phase of infection presents as a typical flu-like illness that lasts approximately 4 days. Resolution of the initial symptoms is followed by a symptom-free interval, after which 20 to 30% of patients develop a second phase of disease involving neurological symptoms. Severe disease outcomes associated with TBEV infection include meningoencephalitis, poliomyelitis, and polyradiculoneuritis. The severity of the disease is highly strain dependent, with mortality rates at high as 30% and neuropsychiatric sequelae in 10 to 20% of infections.

Hepacivirus Genus

Hepatitis C virus (HCV) was first identified in 1989 as the causative agent of non-A, non-B hepatitis. Based on sequence analysis of the viral genome, HCV was placed in a new genus, *Hepacivirus*, of the *Flaviviridae* (21, 28). HCV is a major public health concern, with an estimated 200 million people chronically infected worldwide and as many as 3 to 4 million new infections occurring every year (2). Approximately 60 to 75% of all HCV exposures are thought to result in chronic infection, which is typically asymptomatic for 10 to 30 years. HCV is a hepatotropic virus that replicates in hepatocytes (78). Long-term infection with HCV can result in severe liver disease, such as cirrhosis, hepatic fibrosis, and hepatocellular carcinoma (3, 9). HCV is now the most common cause of liver failure and liver transplantation in the United States and Europe (127). Current treatment for HCV consists of a combination therapy with pegylated interferon (IFN)-α and ribavirin. However, this treatment is relatively toxic and only effective for 50 to 60% of patients (88). The low response rate to therapy and the high frequency of viral persistence has been linked to HCV control of innate immune defenses (35).

Pestivirus Genus

The *Pestivirus* genus consists of a small group of animal viruses including classical swine fever virus (CSFV) and bovine viral diarrhea virus (BVDV), which cause significant economic losses worldwide. As with HCV, pestiviruses are able to establish persistent infections, though in the case of BVDV and CSFV, persistent infection results from transmission of the virus to a fetus during the first trimester of pregnancy. Infection of the fetus prior to the development of immune competence results in an immunotolerant animal that remains viremic for life. These persistently infected animals serve as a reservoir for transmission of the virus to uninfected animals, which can result in fatal infections.

HOST CELL DETECTION OF RNA VIRUSES

PRRs

The ability to sense an invading pathogen is critical to the cell's ability to respond in an appropriate manner to clear the infection. The cell utilizes a group of proteins known as pathogen recognition receptors (PRRs) to detect the presence of pathogen-associated molecular patterns (PAMPs) within products of viral replication (54, 61, 108). Upon sensing the invading virus, PRRs activate multiple distinct signaling pathways that result in the induction a number of latent transcription factors (108). The activation of these transcription factors in turn leads to a reprogramming of the cell's gene expression profile and induction of a wide variety of genes that establish an antiviral state. Interferon regulatory factor 3 (IRF3) and nuclear factor-κB (NF-κB) are two constitutively expressed latent transcription factors that are essential for establishing of an antiviral state and triggering the proinflammatory response, respectively.

Two classes of PRRs, Toll-like receptors (TLRs) and the RIG-I-like helicases retinoic acid-inducible gene I (RIG-I) and melanoma differentiation-associated gene 5 (MDA-5), have been shown to be involved in signaling the activation of IRF3 and NF-κB in response to the *Flaviviridae* (Fig. 2) (108). TLRs are expressed on the cell surface or within endocytic vesicles in a cell-type-dependent manner (61, 108). Leucine-rich repeats located within the ectodomain of TLRs are presumably responsible for detecting viral PAMPs, including viral double-stranded (ds) RNA by TLR3 (108). In particular, the endosomal localization of TLR3 indicates that it is most likely responsible for the detection of either extracellular dsRNA or dsRNA within vesicles, including exogenous dsRNA that might enter the cell through endocytosis. Upon detection of the viral pathogen, TLR3 recruits the

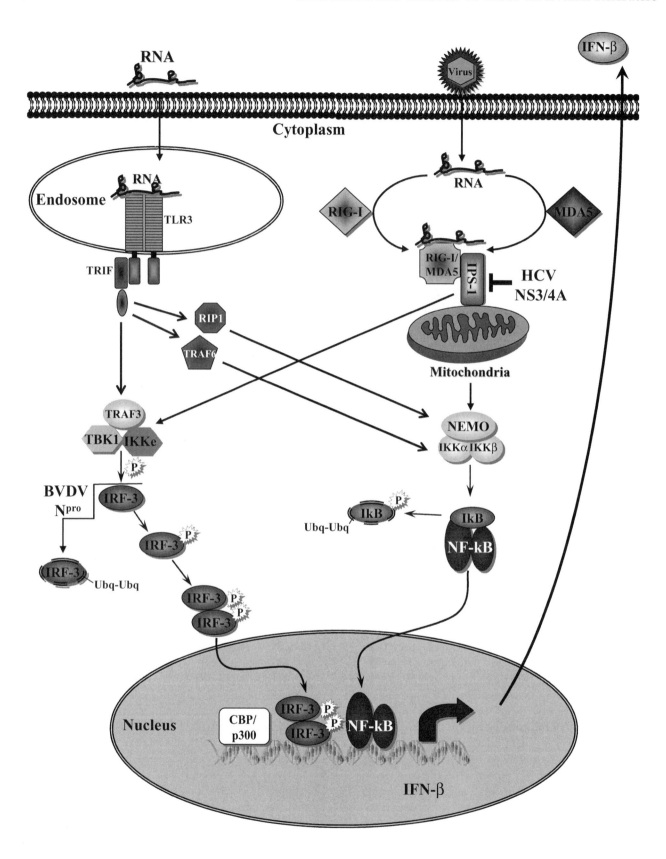

adaptor molecule TRIF (50, 96, 132, 133). TRIF in turn initiates the formation of a signaling complex that activates the noncanonical IκB kinases (IKKs) TBK1 and IKK-ε, which have been shown to phosphorylate IRF3 (4, 29). Phosphorylated IRF3 forms homodimers that are retained in the nucleus and interact with the CBP/p300 coactivator to induce the expression of IRF3 target genes, which initiates the immediate establishment of an initial antiviral state to combat or counteract viral replication (74, 75, 137, 138). TRIF also interacts with the signaling molecules TRAF6 and RIP-1, which mediate activation of NF-κB and the induction of proinflammatory cytokines (22). IRF3 and NF-κB are also critical to the induction of the cell's IFN response, as these transcription factors function cooperatively to mediate expression of IFN-β. Binding of secreted IFN-β to the IFN-α/β receptor amplifies and expands the antiviral response by triggering the activation of the Janus kinase and signal transducers and activators of transcription, or Jak-Stat, signal transduction pathway. Activation of the Jak/Stat pathway leads to the induced expression of a wide variety of IFN-stimulated genes (ISGs), which are responsible for conferring the antiproliferative, antiviral, and proapoptotic actions that serve to limit virus infection.

The recently defined cytoplasmic PRRs of virally derived PAMPs, RIG-I and MDA-5, have also been shown to be critical for the induction of IRF3 and NF-κB in response to RNA viruses in vitro and in vivo (58, 59, 108, 124, 136). RIG-I and MDA-5 contain two N-terminal caspase activation and recruitment domains (CARDs) followed by a DExD/H box RNA helicase domain. Binding of viral RNAs to the helicase domain is postulated to induce conformational changes that liberate the CARDs, allowing initiation of a signaling cascade (95, 135). The RIG-I and MDA-5 signaling pathways converge with the CARD-containing adaptor molecule IFN-β promoter stimulator-1 (IPS-1), also called MAVS, VISA, and Cardif (62, 89, 120, 131). IPS-1 functions to recruit a macromolecular signaling complex to the outer membrane of the mitochondria, which leads to the activation of both IRF3 and NF-κB and the subsequent induction of IFN-β. Recent studies demonstrate that IPS-1 is essential for innate immune signaling triggered by a

variety of viral genera (81) and that the processes of IPS-1 signaling are targeted for regulation by pathogenic RNA viruses (82, 90, 134).

Cellular Sensors of *Flaviviridae* Infection

RIG-I functions as a PRR for several members of the *Flaviviridae*, including HCV, JEV, DV, and WNV. Characterization of a subclone of Huh7 cells that were highly permissive for HCV replication identified a mutation in RIG-I that resulted in a defective innate immune response within this cell line (11, 124). Complementation with wild-type RIG-I restored the cells' ability to respond to and inhibit HCV replication, demonstrating that RIG-I is essential for detecting HCV infection. In addition, biochemical studies demonstrated that RIG-I binds specifically to the NTRs but not to a predicted linear region within the HCV genome (109). RIG-I has also been implicated in binding to RNA with free terminal 5′ triphosphates (ppp), thus defining 5′ ppp and dsRNA motifs as RIG-I substrates (51, 100). The 3′ NTR of HCV contains structural elements that are essential for viral translation, replication, and genome packaging. Mutations within this region are deleterious to the virus and this region is therefore highly conserved, thus providing a stable agonist for RIG-I detection. These observations support the model that RIG-I functions as the primary sensor for HCV RNA, which leads to activation of the innate antiviral response.

RIG-I has also been determined to function as a PRR for members of the *Flavivirus* genus. Exogenous expression of dominant negative forms of RIG-I reduced the virus-stimulated induction of the IFN-β promoter in cells infected with either JEV or DV (17). Other studies using murine embryo fibroblasts (MEFs) from RIG-I null mice confirmed that RIG-I is required to induce expression of IFN-β in response to JEV infection (59). Furthermore, while MDA-5 null mice displayed no phenotypic differences when infected with JEV, RIG-I-deficient mice exhibited a marked decrease in serum IFN-α levels and increased susceptibility to infection when compared with wild-type control mice. Thus, the RIG-I signaling pathway plays an essential role in initiating the antiviral response to JEV.

Figure 2 PRR signaling and viral control of the innate immune response to *Flaviviridae* infection. Viral RNA recognition by the TLR3 and RIG-I/MDA-5 pathways are shown. Recognition and binding of viral RNA products by TLRs, RIG-I, or MDA-5 induces downstream signaling of IRF3 and NF-κB activation, leading to the expression of a wide range of genes, including the production and secretion of IFN-β. The sites of regulation by BVDV N^{pro} and HCV NS3/4A proteins are indicated. Ubq-Ubq, protein modification by virus-induced ubiquitin chain conjugation; P, protein phosphorylation; NEMO, NF-κB essential modulator.

In contrast to JEV, RIG-I null MEFs retained the ability, albeit in reduced capacity, to induce IRF3 target genes in response to DV (81). DV infection of MDA-5 null MEFs also induced IRF3, suggesting that signaling in response to DV is more complex and involves both RIG-I and MDA-5. In support of this, a double deletion of both RIG-I and MDA-5 expression completely ablated IFN-β and ISG expression upon DV infection (81). Moreover, deletion of IPS-1 abrogated the induction of IFN-β and other IRF3-responsive genes, demonstrating that the antiviral response to DV is mediated through both RIG-I- and MDA-5-dependent signaling pathways. These observations indicate that the DV genome encodes agonists of both the RIG-I and MDA-5 pathways that independently stimulate induction of the innate immune response.

Recognition of WNV infection by the host cell has also been shown to involve multiple PRRs, including RIG-I, MDA-5, and in some cells TLR3, though the involvement of TLR3 results in increased pathogenesis rather than protection (130). Cells lacking RIG-I are still capable of sensing WNV infection; however, the onset of the innate antiviral response was delayed (31). This suggests that the RIG-I pathway mediates the initial activation of the antiviral response to WNV; nonetheless, distinct secondary pathways are also clearly involved. Despite the fact that the cells were still capable of mounting an innate immune response, ablation of RIG-I resulted in enhanced WNV replication (31). This indicates that RIG-I is critical for the establishment of an initial innate antiviral immune response necessary to constrain WNV replication. It is likely that RIG-I primarily detects motifs within the genomes of JEV, DV, and WNV, as it does with HCV, since *Flavivirus* genomes are capped (15) and therefore do not present naked 5′ ppp motifs as viral PAMPs.

MDA-5 represents a likely candidate for a secondary PRR of WNV infection. Indeed, WNV-induced activation of IRF3 and IRF3 target gene expression were abrogated upon disruption both the MDA-5 and RIG-I signaling pathways in cultured cells. Furthermore, IPS-1 null MEFs were refractory to WNV-mediated activation of IRF3, confirming the role of both RIG-I and MDA-5 in responding to WNV infections (32). This suggests a model for the response to WNV infection in which RIG-I functions as the primary PRR for sensing infection, thereby triggering an initial innate immune response to constrain virus replication. However, at later times during infection MDA-5 serves as a secondary PRR, working in concert with RIG-I to maintain and/or amplify the induction of the antiviral genes in an attempt to control viral replication (32). Therefore, in contrast to what

has been previously reported for other viruses, in which RIG-I or MDA-5 independently triggers innate immunity (38, 58, 81, 124), both PRRs are essential to mounting an innate immune response in response to WNV and DV infections.

While WNV infection triggers antiviral defenses in the host cell (33), in vitro studies have shown that TLR3 is not required for the processes of IRF3 activation and IFN-β production in cultured cells (31). However, Wang et al. (130) demonstrated that viral loads are higher in the blood of WNV-infected TLR3 null mice compared with wild-type mice. Nonetheless, despite increased viral loads, WNV virulence was attenuated in TLR3 null mice. The enhanced virulence in wild-type mice was attributed to an increased permeability of the blood-brain barrier caused by a TLR3-mediated inflammatory response. Therefore, stimulation of the TLR3 pathway by WNV in vivo leads to increased pathogenesis rather than protection. This suggests that highly virulent strains of WNV may have evolved to more efficiently stimulate the TLR3-mediated inflammatory response, leading to enhanced neuroinvasion and increased neurovirulence.

EVASION OF IRF3 BY MEMBERS OF THE FLAVIVIRUS FAMILY

IRF3 is a central component in the establishment of an antiviral state and is absolutely required for viral induction of IFN-β expression (113). Not surprisingly, a wide range of viral genera include members that have evolved to specifically block or evade the processes of IRF3 activation and/or function. The unique strategies utilized by the various member of the *Flaviviridae* to circumvent the triggering and action of IRF3 are discussed below.

Hepaciviruses

As stated above, motifs within the HCV genome are capable of stimulating RIG-I signaling, which leads to the activation of IRF3 transcriptional activity and the subsequent induction of type I IFNs. However, studies with the HCV replicon system and more recently with the JFH1 HCV infectious clone have demonstrated that HCV replication imposes a blockade on the RIG-I signaling pathway, which prevents activation of IRF3 in infected cells (reviewed in reference 35). A systematic analysis of the HCV proteins demonstrated that expression of the viral NS3/4A protein complex, an essential viral protease and RNA helicase, was sufficient to block IRF3 nuclear translocation and induction of IRF3-responsive promoters, including IFN-β (30). Mutational analysis of NS3/4A demonstrated that ablation of the viral protease activity completely abrogated the protein's ability to antagonize IRF3 activation, while mutants lacking the RNA helicase domain activity functioned normally (55). In addition, treatment of cells with peptidomimetic active-site inhibitors of NS3/4A restored the cells' ability to activate IRF3 in response to viral infection (30, 55). The determination that the proteolytic activity of NS3/4A is essential for antagonizing IRF3 activation strongly suggested that the HCV protease impedes the RIG-I pathway by selectively targeting an essential signaling component for cleavage. Indeed, it was subsequently demonstrated that HCV NS3/4A cleaves the adaptor molecule IPS-1, which is an essential component necessary for transmitting the activating signals from both RIG-I and MDA-5. Cleavage of IPS-1 results in its release from the mitochondrial membrane and a complete inhibition of its ability to relay the signals necessary to stimulate activation of both IRF3 and NF-κB (82, 89). In addition, HCV NS3/4A has also been shown to cleave TRIF, the essential adaptor molecule in the TLR3 pathway (72). TRIF cleavage by NS3/4A is proposed to ablate dsRNA signaling induced by TLR3 agonists that might accumulate during HCV infection. However, this observation is not universal (24), and TRIF cleavage has yet to be demonstrated to occur during HCV infection in vivo. These studies do define a common role of the viral NS3/4A protease in targeting and inactivating cellular factors involved in signaling IRF3 activation and innate antiviral immunity. These studies also suggest that inhibitors of NS3/4A proteolytic activity would have dual efficacy in treating HCV infection, in that they would both impede processing of the viral polyprotein and possibly restore the native induction of the innate immune response to HCV (82). Among the *Flaviviridae*, the viral NS3/4A protease of GB virus type B, another member of the *Hepacivirus* genus, has also been shown to disrupt RIG-I signaling by cleaving IPS-1 (31). However, this mechanism for disruption of the IFN response appears to be specific to the hepaciviruses, as serine proteases from a flavivirus and a pestivirus were unable to disrupt activation of the IFN-β promoter in side-by-side comparisons.

Pestiviruses

Nonpathogenic BVDV and attenuated strains of CSFV are capable of establishing lifelong persistent infections. However, the mechanism(s) by which these viruses evade the innate and adaptive immune responses are poorly understood. Pestiviruses have long been known to enhance the replication of other viruses by inhibiting induction of IFN-α/β, but only recently has the molecular basis for this ability to block the innate response be-

come clear (36, 52, 106). The N-terminal protease, N^{pro}, which is unique to pestiviruses, has recently been demonstrated to function as an agonist of IFN induction. The only previously established function for N^{pro} is an autoproteolytic activity that results in cleavage of the protein from the downstream nucleocapid protein C. Ectopic expression of either BVDV or CSFV N^{pro} leads to reduced levels of IRF3 expression (7, 20, 48, 70, 106, 117). However, unlike HCV, the proteolytic activity of N^{pro} was not essential to the protein's ability to impede the innate immune response (20, 36, 48, 106). The N^{pro} proteins of both cytopathic (cp) and noncytopathic (ncp) strains of BVDV have been shown to preferentially induce the degradation of cytoplasmic forms of IRF3 (20, 48). However, both cp and ncp BVDV strains induce the phosphorylation and nuclear translocation of IRF3 (5, 6). To counteract the activation of IRF3, ncpBVDV N^{pro} has also evolved to prevent binding of IRF3 to DNA, thus imposing a complete shutdown of IRF3 activity. Interestingly, the N^{pro} of cpBVDV does not prevent binding of activated IRF3 to DNA (20). Thus, cpBVDV may not be able to impose a complete blockage on IRF3 transcriptional activity in all cell types, which may account for the controversial in vivo and cell-line-specific induction of IFN by these virus strains (5, 6, 14, 18, 36, 116).

The mechanism by which CSFV N^{pro} reduces IRF3 levels in infected cells remains controversial. An initial report indicated that N^{pro} expression had no effect on IRF3 protein stability but rather inhibited the constitutive transcription of IRF3. In contrast, it has also been reported that CSFV N^{pro} induced the proteasomal degradation of IRF3 similarly to BVDV N^{pro} (7, 70).

An additional mechanism for the evasion of IRF3 activation has also been reported for BVDV. Expression of the secreted form of E^{rns}, a viral envelope protein of BVDV, has been reported to block the ability of the cell to respond to dsRNA treatment (53). The observation that E^{RNS} RNase activity was required for dsRNA signaling inhibition suggests that the protein functions by degrading potential extracellular agonists. Therefore, BVDV appears to direct multiple processes for preventing IRF3 activation. By degrading extracellular agonists released from dying cells, BVDV would prevent neighboring cells from sensing the infection and the establishment of an antiviral state to block viral spread.

Flaviviruses

In contrast to the hepaciviruses and pestiviruses, members of the *Flavivirus* genus cause acute infection and as such do not undergo selective pressure to evolve a mechanism for the long-term control of the innate immune response. In general, flaviviruses have not been shown to impose a nonspecific blockade on the IRF3 pathway. However, WNV has evolved a unique mechanism to avoid activation of IRF3 transcriptional activity. In human cells infected with WNV, activation of IRF3 does not occur until approximately 12 to 16 h postinfection, with maximal activation occurring much later (33). This is in sharp contrast to a variety of other viruses that have been shown to induce IRF3 activation within 3 to 6 h postinfection (26, 93, 103, 119, 122, 126, 138). By delaying the activation of IRF3, WNV is able to replicate virtually unchallenged by the host cell at early times postinfection, providing a substantial advantage to the virus. Though WNV has been shown to block TLR3-mediated activation of IRF3 in HeLa cells (115), most cell lines infected with WNV remain responsive to soluble, intracellular, and virally encoded forms of dsRNA ligand, including those ligands thought to function as TLR3 agonists (31). Therefore, the delayed activation of IRF3 is not due to a WNV-directed block against TLR3 signaling or a block imposed on either the activation or the transcriptional activity of IRF3. This suggests that unlike other members of the *Flaviviridea*, which have evolved strategies to block the IRF3 pathway, WNV simply evades detection by the host cell at early times postinfection. The mechanism by which WNV avoids detection by PRRs early in infection remains to be determined. One possible explanation is that high levels of the WNV agonist(s) are required for efficient activation of IRF3, such that activation does not occur until sufficient levels of the viral agonist(s) have accumulated. Alternatively, WNV may have evolved to specifically mask IRF3 agonist(s) produced early in infection, thus blocking the accessibility of the viral agonist(s) to PRRs until the virus has established a productive infection.

The significance of the delaying of IRF3 activation to WNV replication was demonstrated by the observation that exogenous stimulation of IRF3 prior to infection severely attenuated viral replication (31). Furthermore, deletion of IRF3 enhanced WNV replication and spread in vitro. This suggests that WNV is sensitive to the antiviral actions of the IRF3 pathway and that the ability to delay the activation of the host response is critical for WNV to achieve high levels early after infection. The importance of IRF3 to WNV pathogenesis was recently assessed in vivo (23). Mice lacking IRF3 exhibited an increased susceptibility to WNV infection, with increased viral levels in the blood, peripheral organs, and CNS. Furthermore, the absence of IRF3 also resulted in an expanded tissue tropism and earlier entry into the CNS. Interestingly, the increased level of WNV replication in IRF3$^{-/-}$ mice did not correlate with a significant alteration in systemic levels of type I IFNs, suggesting that the protective effect of IRF3 is likely due to antiviral actions of direct target genes. Indeed, examination of WNV replication in macrophages and cortical neuronal cells derived from IRF3-deficient mice demonstrated that IRF3 signaling triggers distinct cell-type-dependent mechanisms to inhibit WNV replication through IFN-dependent and -independent pathways. Taken together, these data demonstrated that IRF3 is required to control WNV replication and spread.

REGULATION OF IFN-α/β SIGNALING BY *FLAVIVIRIDAE*

Examination of IFN signaling in infected cells has revealed strategies by which *Flaviviridae* members attenuate IFN-α/β receptor signaling actions, thus limiting the downstream expression of ISGs that otherwise control virus replication and spread. IFN-α/β signaling is directed by the Jak-Stat pathway to induce ISG expression. Upon engagement of the IFN-α/β receptor, IFNs trigger the action of receptor-bound Tyk2 and Jak1 protein kinases, which catalyze the phosphorylation and dimerization of Stat1 and Stat2, resulting in their nuclear translocation and association with IRF9 to generate the IFN-stimulated gene factor-3 (ISGF3) complex. ISGF3 is a potent transactivator of transcription that binds to the IFN-stimulated response element (ISRE) on ISGs to direct their IFN-induced expression (Fig. 3) (118). As described below, studies have linked *Flaviviridae* control of IFN signaling with viral pathogenesis, thus defining the control of Jak-Stat pathway function as a virulence determinant.

Regulation of the Jak-Stat Pathway Is a Common Theme among the *Flaviviridae*

The NS proteins of various members of the flavivirus family have been identified as IFN antagonists whose actions disrupt signaling processes of the Jak-Stat pathway (reviewed in reference 111). The impact of Jak-Stat pathway control is to confer resistance to the antiviral actions of IFN through attenuation of ISG expression. As a result, the virus can replicate more productively, resulting in higher viral loads in blood and tissues, rapid viral spread, enhanced dissemination to target tissue of infection, and increased transmission between hosts (110).

Dengue Virus NS Proteins Disrupt Stat Activation

Cells infected with pathogenic strains of DV are relatively resistant to the antiviral actions of IFNs (56). A systematic analysis of the 10 DV proteins identified the viral NS2A, NS4A, and NS4B proteins as potential

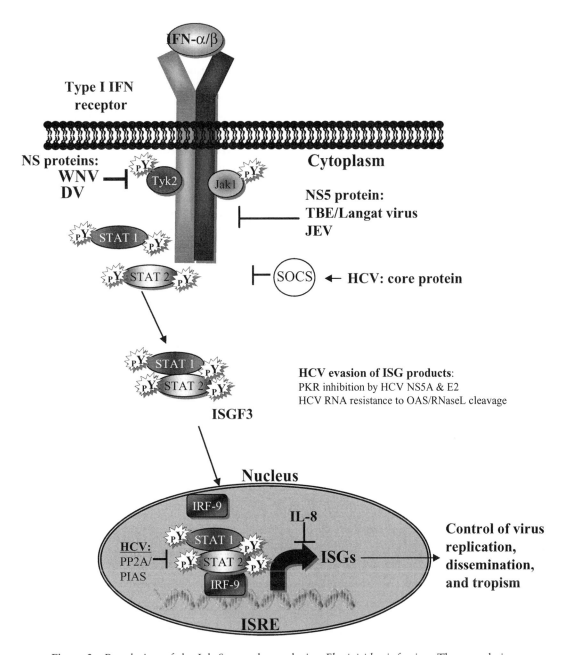

Figure 3 Regulation of the Jak-Stat pathway during *Flaviviridae* infection. The two chains of the IFN-α/β receptor are shown associated with Tyk2 or Jak1. IFN binding to the receptor triggers the phosphorylation of receptor-associated Stat1 and Stat2, which form heterodimers that translocate to the cell nucleus for interaction with IRF9. The trimeric complex of Stat1, Stat2, and IRF9 binds to the ISRE of target genes to induce ISG expression. ISG products function to control virus replication and cell-to-cell virus spread, and have a major impact on tissue tropism. Sites of Jak-Stat pathway regulation by HCV core, NS5A, and E2 proteins are indicated. Control points imposed by WNV, DV, TBE/Langat virus, and JEV NS proteins are also shown. pY, protein tyrosine phosphorylation.

IFN antagonists. These studies used an IFN-inducible reporter gene assay and analysis of a reporter virus system to identify DV protein suppressors of IFN actions, thus providing both biochemical and viral-genetic analyses of IFN signaling regulation. In these experiments *trans* expression of DV NS proteins blocked the IFN-induced expression of an ISRE-driven reporter gene and rescued the replication of an IFN-sensitive reporter virus in cultured cells treated with high-dose IFN (92). Further evaluation of the DV NS proteins identified the mature NS4B protein, in combination with the viral NS4A protein, as factors that block IFN-induced nuclear import of Stat1 through processes that likely disrupt Stat phosphorylation (92). These observations were validated in subsequent studies with dendritic cells, which constitute a major target for DV infection, capable of producing high levels of viral particles irrespective of the presence of IFNs (49). Analysis of dendritic cells demonstrated that DV imposes a blockade to IFN-induced activation of the Jak-Stat pathway and suppression of Stat1 activation, possibly by blocking the activation of the Tyk2 protein kinase. Similar studies have revealed that DV infection may reduce the overall level of expression of Stat2, though the mechanisms of this reduction remain to be determined (56). DV circulates as distinct serotypes that exhibit differential pathogenesis and sensitivity to the antiviral actions of IFN (92). The virulence determinants that impart varied pathogenic phenotypes of DV strains are unknown but could likely reflect a differential ability of viral proteins to suppress the Jak-Stat pathway.

The NS5 Protein of JEV and Langat Virus Blocks the Jak-Stat Pathway

JEV infection of cells also associates with a blockade of IFN signaling through the Jak-Stat pathway. Biochemical studies demonstrated that JEV imparts broad effects on IFN signaling by disrupting the activation of Stat1, Stat2, and Stat3 through a process linked with inhibition of Tyk2 function (77). Further analyses provided genetic evidence to link Stat deregulation to the actions of the viral NS5 protein. NS5, when expressed alone, was able to block IFN signaling of Tyk2 and Stat activation to attenuate ISG expression (Fig. 3) (76). Structure-function studies mapped this phenotype to within the NS5 amino terminus. NS5 is the flavivirus RNA-dependent RNA polymerase that copies viral RNA during the process of virus replication (15). The NS5 polymerase activity is localized to the protein's carboxy-terminal domain. That regulation of Jak-Stat signaling is mediated by the NS5 amino terminus suggests that control of IFN signaling by JEV may occur independently of NS5 polymerase function. Importantly, NS5

control of Jak-Stat signaling could be reversed by treatment of cells with a tyrosine phosphatase inhibitor (76), thereby implicating protein tyrosine phosphatase activity modulation as a possible mechanism by which NS5 regulates IFN actions.

The NS5 protein of Langat virus, a member of the TBEV group of viruses, has similarly been identified as an antagonist of IFN actions. Langat virus was found to be sensitive to pretreatment of cells with IFN, wherein virus replication was suppressed when infection was initiated after cells were exposed to IFN. However, like the aforementioned flaviviruses, Langat virus could resist IFN's antiviral effects when cells were treated after infection was already established. Examination of the infected cells revealed that the virus blocked the IFN-induced phosphorylation of both Tyk2 and Jak1 protein kinases. Systematic analysis of the viral NS proteins identified the NS5 protein as the IFN antagonist (8). Protein interaction studies showed that Langat virus NS5 could associate with IFN receptor complexes to block Jak-Stat signaling during infection of dendritic cells. A recent study has now assigned this regulation to the RNA polymerase domain within the carboxy-terminal region of NS5 (97). Thus, NS5 has been implicated as a JEV and TBEV/Langat virus IFN antagonist through its ability to modulate Jak-Stat signaling. The separation of function in each case, wherein distinct NS5 domains of each virus are thought to confer this regulation, is not understood. It is possible that these data represent differences in function of NS5 that are attributed to specific insect vector and host relationships. Alternatively, the data might implicate a "cross talk" of NS5 domains, in which one domain may govern the actions of the other, that is revealed through the analysis of NS5 deletion mutants. A direct comparison of NS5 of TBEV/Langat virus and JEV is needed to resolve and understand these distinctions.

Control of IFN Signaling by WNV

Pathogenic variants of WNV have been shown to impose a blockade of Jak-Stat signaling in infected cells (111). Examination of WNV-infected cells or cells harboring a WNV subgenomic replicon demonstrated that WNV infection or viral genome replication is associated with a blockade of IFN-induced Jak1 and Tyk2 phosphorylation, causing abrogation of Stat activation (42, 80). Various studies have since reported that the viral NS proteins, including NS2A, NS2B3, NS4A, and NS4B, may each block IFN signaling through disruption of Stat activation (79, 80, 91). In terms of NS4B, genetic analysis identified a cluster of charged residues in the protein's amino terminus that when mutated to alanine residues released the suppression of Jak-Stat signaling conferred by NS4B alone or in the context of a WNV subgenomic replicon (27). However, when these mutations were placed in the context of an infectious virus clone, they failed to relieve the viral suppression of Jak-Stat signaling. A caveat to these studies was that the NS4B expression construct and WNV subgenomic replicon were derived from a genetic lineage distinct from the WNV infectious clone, and second-site differences between construct/replicon and the clonal virus could possibly mitigate the effects of mutation in NS4B. Taken together, these studies collectively demonstrated that WNV can evade IFN actions in part through viral suppression of Jak-Stat signaling, perhaps by targeting the processes of Tyk2 and/or Jak1 activation triggered when IFN occupies the IFN receptor chains. However, the true contribution of the specific viral NS proteins and the structural proteins that mediate this regulation are not defined, nor are the mechanisms of action of signaling suppression understood.

Subsequent studies have compared a pathogenic WNV strain (WNV-TX02) to a nonpathogenic WNV strain (WNV-MAD78). These studies examined each virus infection in vitro in the context of wild-type MEFs or those lacking a functional IFN-α/β receptor (63). Growth of the nonpathogenic WNV-MAD78 strain was attenuated in wild-type cells and was associated with an inability of the virus to antagonize IFN signaling when directly compared with the pathogenic WNV-TX02 strain (63). Moreover, the WNV-MAD78 strain was avirulent in wild-type mice, but virulence was unmasked upon infection of mice lacking a functional IFN-α/β receptor. Thus, IFN-α/β are essential for immunity against WNV infection. This work serves to identify the control of JAK-Stat signaling through the IFN-α/β receptor as a factor influencing the outcome of WNV infection.

When wild-type mice were infected with pathogenic WNV, the virus was typically not detected in peripheral organs outside of the CNS. However, WNV infection of mice lacking a functional IFN-α/β receptor resulted in peripheral dissemination of the virus to heart, liver, kidney, and other sites not usually involved in infection (110). Peripheral dissemination of WNV in IFN-α/β receptor-deficient mice was associated with significantly higher viral load in serum and in the CNS. The increased CNS viral load was linked with a reduction in the survival of infected neurons (110). These results demonstrate the critical role IFN-α/β plays in controlling WNV replication and dissemination in addition to protecting the CNS. The ISG effector proteins of the IFN response that actually control WNV replication and dissemination have not been fully defined (see

below). Studies to examine the impact of WNV and IFN on global gene expression have revealed that ISGs are differentially modulated by different strains of WNV (129), suggesting that pathogenic and nonpathogenic strains may direct distinct host responses to infection that associate with protection or pathogenesis.

Recent studies have implicated the 2′,5′-oligoadenylate synthetase 1b (OAS1b) gene, RNase L, and dsRNA-dependent protein kinase (PKR) in resistance to WNV infection. Susceptibility to WNV infection in mice was recently mapped to a nonsense mutation in OAS1b, a member of the IFN-activated OAS/RNase L system, that results in the expression of a truncated and presumably dominant negative form of the protein (86, 98). The OAS family of proteins are IFN-induced dsRNA-dependent enzymes that typically catalyze the polymerization of ATP into 2′,5′-linked oligoadenylates, which in turn bind and activate the latent endoribonuclease RNase L (121). Once activated, RNase L blocks viral replication by cleaving RNA substrates in the cell, wherein these small RNAs serve to stimulate RIG-I and MDA-5 signaling (85, 114, 121). Overexpression studies confirmed that full-length OAS1b but not the truncated form protected susceptible cells from infection and limited the cell-to-cell spread of WNV (114). Moreover, constitutive expression of OAS1b limited viral replication by preventing the accumulation of positive-strand viral RNA within the cell (57); however, the inhibitory effects of OAS1b appeared to be independent of RNase L activity (57, 112, 114). Nonetheless, RNase L clearly plays an important role in controlling WNV, as suppression or ablation of RNase L expression enhanced viral replication and yields in vitro and in vivo (112, 114). Furthermore, mice lacking RNase L were more susceptible than wild-type control mice to WNV infection (112).

Another IFN-induced mediator of intracellular resistance, PKR, also appears to play a role in controlling WNV infections (37, 112). Once activated by dsRNA, PKR functions to both inhibit protein synthesis and activate the transcription factors NF-κB and IRF1, which may contribute to IFN-α/β production in some cell types (68, 69, 107). Inhibition of PKR activity in vitro resulted in decreased levels of IFN in response to WNV infection (37). However, serum levels of IFN in WNV-infected PKR/RNase L$^{-/-}$ mice were comparable to those detected in wild-type mice (112). Nonetheless, the PKR/RNase L$^{-/-}$ mice were more susceptible to WNV infection, with increased viral levels in the blood and peripheral organs, altered viral tropism, earlier entry into the brain, and higher viral loads in the CNS. PKR and RNase L appear to have synergistic antiviral activities, as PKR/RNase L$^{-/-}$ mice were more vulnerable to lethal

WNV infection than mice lacking RNase L alone. Ex vivo studies indicate that PKR and RNase L mediate the IFN-dependent inhibition of WNV in a cell-type-dependent manner, suggesting that other effector molecules are also involved in controlling WNV (112). Moreover, IFN-α/β receptor-deficient mice exhibited a more severe phenotype than PKR/RNase $L^{-/-}$ mice in similar studies (110, 112), further suggesting that additional IFN-induced antiviral effector molecules play a role in governing WNV infection.

HCV Can Resist IFN Actions

Local IFN production in hepatic tissue, the site of HCV infection, is likely to influence HCV replication and may impart antiviral effects that contribute to the resolution of acute infection (10). However, HCV successfully mediates a chronic infection in most cases. Chronic HCV infection is treated with IFN-α-based therapy. Therapeutic resolution of infection, known as a "sustained virologic response," occurs in about 50% of patients overall, albeit the response to IFN varies among patients and among viral genotypes (87). The overall low response rate of HCV (particularly genotype 1 HCV) to IFN therapy indicates that HCV can evade or resist IFN actions in vivo, both locally in the infected tissues and more globally in the context of IFN therapy. Assessment of IFN-α/β receptor signaling processes has revealed mechanisms by which HCV proteins can antagonize IFN signaling (Fig. 3). In vitro studies have associated HCV protein expression with inhibition of Stat1 function (12, 47) in part through increased levels of protein phosphatase 2A within HCV-infected liver tissue. The elevated protein phosphatase 2A levels have been suggested to induce the actions of the protein inhibitor of activated Stat to confer the hypomethylation and inactivation of Stat1 (105). In addition, the HCV core protein may induce the expression of suppressor of cytokine signaling (SOCS)-3, as has been demonstrated in cultured cells (13). The SOCS proteins function to mediate a classical negative-feedback loop on IFN-α/β receptor signaling events by inhibiting signal transduction actions of the Jak-Stat pathway (1). Thus, increased SOCS-3 levels would render an overall suppression of IFN signaling to attenuate ISG expression.

A recent study compared hepatic gene expression profiles in IFN-responsive and -nonresponsive HCV patients (19). Increased basal expression of the gene encoding the ubiquitin-specific protein (USP) 18 was found to correlate with nonresponse to therapy. USP18 is a cellular protease that cleaves protein conjugates generated by the ubiquitin-like protein ISG15, a known ISG effector protein (71). Therefore, elevated expression of USP18 would serve to counter ISG15's antiviral actions (104). Furthermore, USP18 expression was shown to have a negative impact on the anti-HCV actions of IFN in an in vitro system of HCV infection (104). These observations support the idea that USP18 may negatively influence the response to IFN therapy in HCV patients. It is important to note that hundreds of ISGs with pleiotropic effects are induced during IFN therapy (118). Therefore, HCV resistance to IFN is likely mediated through a combination of host factors and viral processes that cooperatively reduce the potency of the IFN response.

Various studies have shown that expression of the HCV NS5A protein alone can suppress IFN-α actions sufficiently to rescue the replication of an IFN-sensitive virus in cultured cells (83), suggesting that NS5A is an antagonist of IFN. Suppression of IFN action may be partially due to NS5A's induction of expression and secretion of interleukin (IL)-8, a proinflammatory chemokine that has been shown to antagonize IFN (64, 101). In support of this, serum IL-8 levels have been found to be elevated in IFN-nonresponder patients with chronic HCV (102). However, it is not clear how IL-8 suppresses IFN actions and what role, if any, other proinflammatory cytokines play in governing IFN actions.

The HCV NS5A and E2 proteins have both been identified as inhibitors of the antiviral ISG PKR (34, 94, 125). When activated, PKR phosphorylates the eukaryotic initiation factor 2-α (eIF2α) subunit to inhibit mRNA translation and stimulates the transcriptional transactivator activity of NF-κB and IRF1 (69). Inhibition of PKR would allow HCV to evade both translational control programs of IFN that otherwise limit virus replication, and the signaling actions of PKR that serve to amplify the innate antiviral response to infection (60). However, NS5A control of PKR is not a stable phenotype. The constant evolution of quasispecies during the course of chronic HCV infection sometimes leads to mutations within NS5A that result in the abrogation of its ability to modulate PKR (128). Similarly, HCV resistance to IFN may also involve a viral genetic disposition to evade the actions of RNase L. Genetic studies show that RNase L preferentially cleaves HCV RNA only at certain UU and UA dinucleotides (43), and that genotype 1 HCV sequences in general have fewer RNase L cleavage sites than other HCV genotypes (43). This distinction provides a genetic link between HCV genotypes 1a and 1b and resistance to OAS/RNase L actions and to IFN therapy in general. Thus, HCV may utilize multiple mechanisms to evade the antiviral effects of IFN.

CONCLUSION

The *Flaviviridae* represent major pathogens of public health, agriculture, and environmental importance. Molecular studies have revealed common and unique mechanisms by which members of the different *Flaviviridae* genera evade host innate immunity and IFN actions that would otherwise limit infection. The studies discussed above have elucidated key virus-host interactions that regulate the signaling processes that trigger IFN production, IFN signaling, or ISG effector functions, which ultimately result in the modulation of the host's ability to respond to infection. Further studies are required to fully define the spectrum of PRRs and their signaling pathways that induce antiviral immunity in target cells, and to define the ISGs that confer the antiviral actions of IFN against the spectrum of *Flaviviridae*.

Studies in the Fredericksen laboratory are supported by National Institutes of Health (NIH) grant K22AI063028; the Gale laboratory is supported in part by NIH grants AI060389, AI057568, AI40035 (Project 4), and DA024563 and a grant from the Burroughs Wellcome Fund.

References

1. Alexander, W. S. 2002. Suppressors of cytokine signalling (SOCS) in the immune system. *Nat. Rev. Immunol.* **2:**410–416.
2. Alter, M. J. 2007. Epidemiology of hepatitis C virus infection. *World J. Gastroenterol.* **13:**2436–2441.
3. Alter, M. J. H., H. S. Margolis, K. Krawczynski, F. Judson, A. Mares, W. J. Alexander, P.-Y. Hu, J. K. Miller, et al. 1992. The natural history of community-acquired hepatitis C in the United States. *N. Engl. J. Med.* **321:**1494–1500.
4. Au, W. C., and P. M. Pitha. 2001. Recruitment of multiple interferon regulatory factors and histone acetyltransferase to the transcriptionally active interferon A promoters. *J. Biol. Chem.* **276:**41629–41637.
5. Baigent, S. J., S. Goodbourn, and J. W. McCauley. 2004. Differential activation of interferon regulatory factors-3 and -7 by non-cytopathogenic and cytopathogenic bovine viral diarrhoea virus. *Vet. Immunol. Immunopathol.* **100:**135–144.
6. Baigent, S. J., G. Zhang, M. D. Fray, H. Flick-Smith, S. Goodbourn, and J. W. McCauley. 2002. Inhibition of beta interferon transcription by noncytopathogenic bovine viral diarrhea virus is through an interferon regulatory factor 3-dependent mechanism. *J. Virol.* **76:**8979–8988.
7. Bauhofer, O., A. Summerfield, Y. Sakoda, J. D. Tratschin, M. A. Hofmann, and N. Ruggli. 2007. Classical swine fever virus Npro interacts with interferon regulatory factor 3 and induces its proteasomal degradation. *J. Virol.* **81:**3087–3096.
8. Best, S. M., K. L. Morris, J. G. Shannon, S. J. Robertson, D. N. Mitzel, G. S. Park, E. Boer, J. B. Wolfinbarger, and M. E. Bloom. 2005. Inhibition of interferon-stimulated JAK-Stat signaling by a tick-borne flavivirus and identification of NS5 as an interferon antagonist. *J. Virol.* **79:**12828–12839.
9. Bialek, S. R., and N. A. Terrault. 2006. The changing epidemiology and natural history of hepatitis C virus infection. *Clin. Liver Dis.* **10:**697–715.
10. Bigger, C. B., K. M. Brasky, and R. E. Lanford. 2001. DNA microarray analysis of chimpanzee liver during acute resolving hepatitis C virus infection. *J. Virol.* **75:** 7059–7066.
11. Blight, K. J., J. A. McKeating, and C. M. Rice. 2002. Highly permissive cell lines for subgenomic and genomic hepatitis C virus RNA replication. *J. Virol.* **76:** 13001–13014.
12. Blindenbacher, A., F. H. Duong, L. Hunziker, S. T. Stutvoet, X. Wang, L. Terracciano, D. Moradpour, H. E. Blum, T. Alonzi, M. Tripodi, N. La Monica, and M. H. Heim. 2003. Expression of hepatitis C virus proteins inhibits interferon alpha signaling in the liver of transgenic mice. *Gastroenterology* **124:**1465–1475.
13. Bode, J. G., S. Ludwig, C. Ehrhardt, U. Albrecht, A. Erhardt, F. Schaper, P. C. Heinrich, and D. Haussinger. 2003. IFN-alpha antagonistic activity of HCV core protein involves induction of suppressor of cytokine signaling-3. *FASEB J.* **17:**488–490.
14. Brackenbury, L. S., B. V. Carr, Z. Stamataki, H. Prentice, E. A. Lefevre, C. J. Howard, and B. Charleston. 2005. Identification of a cell population that produces alpha/beta interferon in vitro and in vivo in response to noncytopathic bovine viral diarrhea virus. *J. Virol.* **79:**7738–7744.
15. Brinton, M. A. 2002. The molecular biology of West Nile virus: a new invader of the Western hemisphere. *Annu. Rev. Microbiol.* **56:**371–402.
16. Calisher, C. H., and E. A. Gould. 2003. Taxonomy of the virus family Flaviviridae. *Adv. Virus Res.* **59:**1–19.
16a. Carson, P. J., P. Konewko, K. S. Wold, P. G. Mariani, S. Goli, P. Bergloff, and R. D. Crosby. 2006. Long-term clinical and neuropsychological outcomes of West Nile virus infection. *Clin. Infect. Dis.* **43:**723–730.
17. Chang, T. H., C. L. Liao, and Y. L. Lin. 2006. Flavivirus induces interferon-beta gene expression through a pathway involving RIG-I-dependent IRF3 and PI3K-dependent NF-kappaB activation. *Microbes Infect.* **8:**157–171.
18. Charleston, B., L. S. Brackenbury, B. V. Carr, M. D. Fray, J. C. Hope, C. J. Howard, and W. I. Morrison. 2002. Alpha/beta and gamma interferons are induced by infection with noncytopathic bovine viral diarrhea virus in vivo. *J. Virol.* **76:**923–927.
19. Chen, L., I. Borozan, J. Feld, J. Sun, L. L. Tannis, C. Coltescu, J. Heathcote, A. M. Edwards, and I. D. McGilvray. 2005. Hepatic gene expression discriminates responders and nonresponders in treatment of chronic hepatitis C viral infection. *Gastroenterology* **128:**1437–1444.
20. Chen, Z., R. Rijnbrand, R. K. Jangra, S. G. Devaraj, L. Qu, Y. Ma, S. M. Lemon, and K. Li. 2007. Ubiquitination and proteasomal degradation of interferon regulatory factor-3 induced by Npro from a cytopathic bovine viral diarrhea virus. *Virology* **366:**277–292.
21. Choo, Q. L., G. Kuo, A. J. Weiner, L. R. Overby, D. W. Bradley, and M. Houghton. 1989. Isolation of a cDNA clone derived from a blood-borne non-A, non-B viral hepatitis genome. *Science* **244:**359–362.

22. Cusson-Hermance, N., S. Khurana, T. H. Lee, K. A. Fitzgerald, and M. A. Kelliher. 2005. Rip1 mediates the Trif-dependent toll-like receptor 3- and 4-induced NF-κB activation but does not contribute to interferon regulatory factor 3 activation. *J. Biol. Chem.* 280:36560–36566.

23. Daffis, S., M. A. Samuel, B. C. Keller, M. Gale, Jr., and M. S. Diamond. 2007. Cell-specific IRF3 responses protect against West Nile virus infection by interferon-dependent and -independent mechanisms. *PLoS Pathog.* 3:e106.

24. Dansako, H., M. Ikeda, and N. Kato. 2007. Limited suppression of the interferon-beta production by hepatitis C virus serine protease in cultured human hepatocytes. *FEBS J.* 274:4161–4176.

25. Dumpis, U., D. Crook, and J. Oksi. 1999. Tick-borne encephalitis. *Clin. Infect. Dis.* 28:882–890.

26. Elco, C. P., J. M. Guenther, B. R. Williams, and G. C. Sen. 2005. Analysis of genes induced by Sendai virus infection of mutant cell lines reveals essential roles of interferon regulatory factor 3, NF-kappaB, and interferon but not Toll-like receptor 3. *J. Virol.* 79:3920–3929.

27. Evans, J. D., and C. Seeger. 2007. Differential effects of mutations in NS4B on West Nile virus replication and inhibition of interferon signaling. *J. Virol.* 81:11809–11816.

28. Fields, B. N. 1990. *Virology.* Raven Press, New York, NY.

29. Fitzgerald, K. A., S. M. McWhirter, K. L. Faia, D. C. Rowe, E. Latz, D. T. Golenbock, A. J. Coyle, S. M. Liao, and T. Maniatis. 2003. IKKepsilon and TBK1 are essential components of the IRF3 signaling pathway. *Nat. Immunol.* 4:491–496.

30. Foy, E., K. Li, C. Wang, R. Sumpter, M. Ikeda, S. M. Lemon, and M. Gale, Jr. 2003. Regulation of interferon regulatory factor-3 by the hepatitis C virus serine protease. *Science* 300:1145–1148.

31. Fredericksen, B. L., and M. Gale, Jr. 2006. West Nile virus evades activation of interferon regulatory factor 3 through RIG-I-dependent and -independent pathways without antagonizing host defense signaling. *J. Virol.* 80:2913–2923.

32. Fredericksen, B. L., B. C. Keller, J. Fornek, M. G. Katze, and M. Gale, Jr. 2008. Establishment and maintenance of the innate antiviral response to West Nile virus involves both RIG-I and MDA5 signaling through IPS-1. *J. Virol.* 82:609–616.

33. Fredericksen, B. L., M. Smith, M. G. Katze, P. Y. Shi, and M. Gale, Jr. 2004. The host response to West Nile virus infection limits viral spread through the activation of the interferon regulatory factor 3 pathway. *J. Virol.* 78:7737–7747.

34. Gale, M., Jr., C. M. Blakely, B. Kwieciszewski, S.-L. Tan, M. Dossett, M. J. Korth, S. J. Polyak, D. R. Gretch, and M. G. Katze. 1998. Control of PKR protein kinase by hepatitis C virus nonstructural 5A protein: molecular mechanisms of kinase regulation. *Mol. Cell. Biol.* 18:5208–5218.

35. Gale, M., Jr., and E. M. Foy. 2005. Evasion of intracellular host defense by hepatitis C virus. *Nature* 436:939–945.

36. Gil, L. H., I. H. Ansari, V. Vassilev, D. Liang, V. C. Lai, W. Zhong, Z. Hong, E. J. Dubovi, and R. O. Donis. 2006. The amino-terminal domain of bovine viral diarrhea virus Npro protein is necessary for alpha/beta interferon antagonism. *J. Virol.* 80:900–911.

37. Gilfoy, F. D., and P. W. Mason. 2007. West Nile virus-induced interferon production is mediated by the double-stranded RNA-dependent protein kinase PKR. *J. Virol.* 81:11148–11158.

38. Gitlin, L., W. Barchet, S. Gilfillan, M. Cella, B. Beutler, R. A. Flavell, M. S. Diamond, and M. Colonna. 2006. Essential role of mda-5 in type I IFN responses to polyriboinosinic:polyribocytidylic acid and encephalomyocarditis picornavirus. *Proc. Natl. Acad. Sci. USA* 103:8459–8464.

39. Gritsun, T. S., V. A. Lashkevich, and E. A. Gould. 2003. Tick-borne encephalitis. *Antiviral Res.* 57:129–146.

40. Gritsun, T. S., P. A. Nuttall, and E. A. Gould. 2003. Tick-borne flaviviruses. *Adv. Virus Res.* 61:317–371.

41. Gubler, D. J. 2006. Dengue/dengue haemorrhagic fever: history and current Status. *Novartis Found. Symp.* 277:3–16.

42. Guo, J. T., J. Hayashi, and C. Seeger. 2005. West Nile virus inhibits the signal transduction pathway of alpha interferon. *J. Virol.* 79:1343–1350.

43. Han, J. Q., G. Wroblewski, Z. Xu, R. H. Silverman, and D. J. Barton. 2004. Sensitivity of hepatitis C virus RNA to the antiviral enzyme ribonuclease L is determined by a subset of efficient cleavage sites. *J. Interferon Cytokine Res.* 24:664–676.

44. Harris, E., K. L. Holden, D. Edgil, C. Polacek, and K. Clyde. 2006. Molecular biology of flaviviruses. *Novartis Found. Symp.* 277:23–39.

45. Hayes, C. G. 2001. West Nile virus: Uganda, 1937, to New York City, 1999. *Ann. N. Y. Acad. Sci.* 951:25–37.

46. Hayes, E. B., N. Komar, R. Nasci, S. Montgomery, D. O'Leary, and G. Campbell. 2005. Epidemiology and transmission dynamics of West Nile virus disease. *Emerg. Infect. Dis.* 11:1167–1173.

47. Heim, M. H., D. Moradpour, and H. E. Blum. 1999. Expression of hepatitis C virus proteins inhibits signal transduction through the Jak-Stat pathway. *J. Virol.* 73:8469–8475.

48. Hilton, L., K. Moganeradj, G. Zhang, Y. H. Chen, R. E. Randall, J. W. McCauley, and S. Goodbourn. 2006. The NPro product of bovine viral diarrhea virus inhibits DNA binding by interferon regulatory factor 3 and targets it for proteasomal degradation. *J. Virol.* 80:11723–11732.

49. Ho, L. J., L. F. Hung, C. Y. Weng, W. L. Wu, P. Chou, Y. L. Lin, D. M. Chang, T. Y. Tai, and J. H. Lai. 2005. Dengue virus type 2 antagonizes IFN-alpha but not IFN-gamma antiviral effect via down-regulating Tyk2-Stat signaling in the human dendritic cell. *J. Immunol.* 174:8163–8172.

50. Hoebe, K., X. Du, P. Georgel, E. Janssen, K. Tabeta, S. O. Kim, J. Goode, P. Lin, N. Mann, S. Mudd, K. Crozat, S. Sovath, J. Han, and B. Beutler. 2003. Identification of Lps2 as a key transducer of MyD88-independent TIR signalling. *Nature* 424:743–748.

51. Hornung, V., J. Ellegast, S. Kim, K. Brzózka, A. Jung, H. Kato, H. Poeck, S. Akira, K. K. Conzelmann, M. Schlee, S. Endres, and G. Hartmann. 2006. 5′-Triphosphate RNA is the ligand for RIG-I. *Science* 314:994–997.

52. Horscroft, N., D. Bellows, I. Ansari, V. C. Lai, S. Dempsey, D. Liang, R. Donis, W. Zhong, and Z. Hong. 2005. Establishment of a subgenomic replicon for bovine

viral diarrhea virus in Huh-7 cells and modulation of interferon-regulated factor 3-mediated antiviral response. *J. Virol.* 79:2788–2796.

53. Iqbal, M., E. Poole, S. Goodbourn, and J. W. McCauley. 2004. Role for bovine viral diarrhea virus E^rns glycoprotein in the control of activation of beta interferon by double-stranded RNA. *J. Virol.* 78:136–145.

54. Janeway, C. A., Jr., and R. Medzhitov. 2002. Innate immune recognition. *Annu. Rev. Immunol.* 20:197–216.

55. Johnson, C. L., D. M. Owen, and M. Gale, Jr. 2007. Functional and therapeutic analysis of hepatitis C virus NS3/4A protease control of antiviral immune defense. *J. Biol. Chem.* 14:10792–10803.

56. Jones, M., A. Davidson, L. Hibbert, P. Gruenwald, J. Schlaak, S. Ball, G. R. Foster, and M. Jacobs. 2005. Dengue virus inhibits alpha interferon signaling by reducing STAT2 expression. *J. Virol.* 79:5414–5420.

57. Kajaste-Rudnitski, A., T. Mashimo, M. P. Frenkiel, J. L. Guenet, M. Lucas, and P. Despres. 2006. The 2′,5′-oligoadenylate synthetase 1b is a potent inhibitor of West Nile virus replication inside infected cells. *J. Biol. Chem.* 281:4624–4637.

58. Kato, H., S. Sato, M. Yoneyama, M. Yamamoto, S. Uematsu, K. Matsui, T. Tsujimura, K. Takeda, T. Fujita, O. Takeuchi, and S. Akira. 2005. Cell type-specific involvement of RIG-I in antiviral response. *Immunity* 23:19–28.

59. Kato, H., O. Takeuchi, S. Sato, M. Yoneyama, M. Yamamoto, K. Matsui, S. Uematsu, A. Jung, T. Kawai, K. J. Ishii, O. Yamaguchi, K. Otsu, T. Tsujimura, C. S. Koh, C. Reis e Sousa, Y. Matsuura, T. Fujita, and S. Akira. 2006. Differential roles of MDA-5 and RIG-I helicases in the recognition of RNA viruses. *Nature* 441:101–105.

60. Katze, M. G., Y. He, and M. Gale, Jr. 2002. Viruses and interferon: a fight for supremacy. *Nature Reviews Immunology* 2:675–667.

61. Kawai, T., and S. Akira. 2006. Innate immune recognition of viral infection. *Nat. Immunol.* 7:131–137.

62. Kawai, T., K. Takahashi, S. Sato, C. Coban, H. Kumar, H. Kato, K. J. Ishii, O. Takeuchi, and S. Akira. 2005. IPS-1, an adaptor triggering RIG-I- and MDA-5-mediated type I interferon induction. *Nat. Immunol.* 6:981–988.

63. Keller, B. C., B. L. Fredericksen, M. A. Samuel, R. E. Mock, P. W. Mason, M. S. Diamond, and M. Gale, Jr. 2006. Resistance to alpha/beta interferon is a determinant of West Nile virus replication fitness and virulence. *J. Virol.* 80:9424–9434.

64. Khabar, K. S., F. Al Zoghaibi, M. N. Al Ahdal, T. Murayama, M. Dhalla, N. Mukaida, M. Taha, S. T. Al Sediairy, Y. Siddiqui, G. Kessie, and K. Matsushima. 1997. The alpha chemokine, interleukin 8, inhibits the antiviral action of interferon alpha. *J. Exp. Med.* 186:1077–1085.

65. Klee, A. L., B. Maidin, B. Edwin, I. Poshni, F. Mostashari, A. Fine, M. Layton, and D. Nash. 2004. Long-term prognosis for clinical West Nile virus infection. *Emerg. Infect. Dis.* 10:1405–1411.

66. Kramer, L. D., J. Li, and P. Y. Shi. 2007. West Nile virus. *Lancet Neurol.* 6:171–181.

67. Kroeger, A., M. Nathan, and J. Hombach. 2004. Dengue. *Nat. Rev. Microbiol.* 2:360–361.

68. Kumar, A., J. Haque, J. Lacoste, J. Hiscott, and B. R. G. Williams. 1994. Double-stranded RNA-dependent protein kinase activates transcription factor NF-κB by phosphorylating IκB. *Proc. Natl. Acad. Sci. USA* 91:6288–6292.

69. Kumar, A., Y.-L. Yang, V. Flati, S. Der, S. Kadereit, A. Deb, J. Haque, L. Reis, C. Weissmann, and B. R. G. Williams. 1997. Deficient cytokine signaling in mouse embryo fibroblasts with a targeted deletion in the PKR gene: role of IRF-1 and NF-kb. *EMBO J.* 16:406–416.

70. La Rocca, S. A., R. J. Herbert, H. Crooke, T. W. Drew, T. E. Wileman, and P. P. Powell. 2005. Loss of interferon regulatory factor 3 in cells infected with classical swine fever virus involves the N-terminal protease, N^pro. *J. Virol.* 79:7239–7247.

71. Lenschow, D. J., C. Lai, N. Frias-Staheli, N. V. Giannakopoulos, A. Lutz, T. Wolff, A. Osiak, B. Levine, R. E. Schmidt, A. Garcia-Sastre, D. A. Leib, A. Pekosz, K. P. Knobeloch, I. Horak, and H. W. Virgin IV. 2007. IFN-stimulated gene 15 functions as a critical antiviral molecule against influenza, herpes, and Sindbis viruses. *Proc. Natl. Acad. Sci. USA* 104:1371–1376.

72. Li, K., E. Foy, J. C. Ferreon, M. Nakamura, A. C. Ferreon, M. Ikeda, S. C. Ray, M. Gale, Jr., and S. M. Lemon. 2005. Immune evasion by hepatitis C virus NS3/4A protease-mediated cleavage of the Toll-like receptor 3 adaptor protein TRIF. *Proc. Natl. Acad. Sci. USA* 102:2992–2997.

73. Ligon, B. L. 2005. Dengue fever and dengue hemorrhagic fever: a review of the history, transmission, treatment, and prevention. *Semin. Pediatr. Infect. Dis.* 16:60–65.

74. Lin, R., C. Heylbroeck, P. M. Pitha, and J. Hiscott. 1998. Virus-dependent phosphorylation of the IRF3 transcription factor regulates nuclear translocation, transactivation potential, and proteasome-mediated degradation. *Mol. Cell. Biol.* 18:2986–2996.

75. Lin, R., Y. Mamane, and J. Hiscott. 1999. Structural and functional analysis of interferon regulatory factor 3: localization of the transactivation and autoinhibitory domains. *Mol. Cell. Biol.* 19:2465–2474.

76. Lin, R. J., B. L. Chang, H. P. Yu, C. L. Liao, and Y. L. Lin. 2006. Blocking of interferon-induced Jak-Stat signaling by Japanese encephalitis virus NS5 through a protein tyrosine phosphatase-mediated mechanism. *J. Virol.* 80:5908–5918.

77. Lin, R. J., C. L. Liao, E. Lin, and Y. L. Lin. 2004. Blocking of the alpha interferon-induced Jak-Stat signaling pathway by Japanese encephalitis virus infection. *J. Virol.* 78:9285–9294.

78. Lindenbach, B. D., and C. M. Rice. 2005. Unravelling hepatitis C virus replication from genome to function. *Nature* 436:933–938.

79. Liu, W. J., X. J. Wang, D. C. Clark, M. Lobigs, R. A. Hall, and A. A. Khromykh. 2006. A single amino acid substitution in the West Nile virus nonstructural protein NS2A disables its ability to inhibit alpha/beta interferon induction and attenuates virus virulence in mice. *J. Virol.* 80:2396–2404.

80. Liu, W. J., X. J. Wang, V. V. Mokhonov, P. Y. Shi, R. Randall, and A. A. Khromykh. 2005. Inhibition of interferon signaling by the New York 99 strain and Kunjin subtype of West Nile virus involves blockage of Stat1 and Stat2 activation by nonstructural proteins. *J. Virol.* 79:1934–1942.

81. Loo, Y. M., J. Fornek, N. Crochet, G. Bajwa, O. Perwistasari, L. Martinez-Sobrido, S. Akira, M. A. Gill,

A. García-Sastre, M. G. Katze, and M. J. Gale. 2008. Distinct RIG-I and MDA-5 signaling by RNA viruses in innate immunity. *J. Virol.* **82:**335–345.

82. Loo, Y. M., D. M. Owen, K. Li, A. K. Erickson, C. L. Johnson, P. M. Fish, D. S. Carney, T. Wang, H. Ishida, M. Yoneyama, T. Fujita, T. Saito, W. M. Lee, C. H. Hagedorn, D. T. Lau, S. A. Weinman, S. M. Lemon, and M. Gale, Jr. 2006. Viral and therapeutic control of IFN-beta promoter stimulator 1 during hepatitis C virus infection. *Proc. Natl. Acad. Sci. USA* **103:**6001–6006.

83. Macdonald, A., and M. Harris. 2004. Hepatitis C virus NS5A: tales of a promiscuous protein. *J. Gen. Virol.* **85:**2485–2502.

84. Mackenzie, J. S., D. J. Gubler, and L. R. Petersen. 2004. Emerging flaviviruses: the spread and resurgence of Japanese encephalitis, West Nile and dengue viruses. *Nat. Med.* **10:**S98–S109.

85. Malathi, K., B. Dong, M. Gale, Jr., and R. H. Silverman. 2007. Small self-RNA generated by RNase L amplifies antiviral innate immunity. *Nature* **448:**816–819.

86. Mashimo, T., M. Lucas, D. Simon-Chazottes, M. P. Frenkiel, X. Montagutelli, P. E. Ceccaldi, V. Deubel, J. L. Guenet, and P. Despres. 2002. A nonsense mutation in the gene encoding 2′-5′-oligoadenylate synthetase/L1 isoform is associated with West Nile virus susceptibility in laboratory mice. *Proc. Natl. Acad. Sci. USA* **99:**11311–11316.

87. McHutchison, J. G. 2004. Understanding hepatitis C. *Am. J. Manag. Care* **10:**S21-S29.

88. McHutchison, J. G., and K. Patel. 2002. Future therapy of hepatitis C. *Hepatology* **36:**S245-S252.

89. Meylan, E., J. Curran, K. Hofmann, D. Moradpour, M. Binder, R. Bartenschlager, and J. Tschopp. 2005. Cardif is an adaptor protein in the RIG-I antiviral pathway and is targeted by hepatitis C virus. *Nature* **437:**1167–1172.

90. Mibayashi, M., L. Martinez-Sobrido, Y. M. Loo, W. B. Cardenas, M. Gale, Jr., and A. Garcia-Sastre. 2007. Inhibition of retinoic acid-inducible gene I-mediated induction of beta interferon by the NS1 protein of influenza A virus. *J. Virol.* **81:**514–524.

91. Munoz-Jordan, J. L., M. Laurent-Rolle, J. Ashour, L. Martinez-Sobrido, M. Ashok, W. I. Lipkin, and A. Garcia-Sastre. 2005. Inhibition of alpha/beta interferon signaling by the NS4B protein of flaviviruses. *J. Virol.* **79:**8004–8013.

92. Munoz-Jordan, J. L., G. G. Sanchez-Burgos, M. Laurent-Rolle, and A. Garcia-Sastre. 2003. Inhibition of interferon signaling by dengue virus. *Proc. Natl. Acad. Sci. USA* **100:**14333–14338.

93. Navarro, L., K. Mowen, S. Rodems, B. Weaver, N. Reich, D. Spector, and M. David. 1998. Cytomegalovirus activates interferon immediate-early response gene expression and an interferon regulatory factor 3-containing interferon-stimulated response element-binding complex. *Mol. Cell. Biol.* **18:**3796–3802.

94. Noguchi, T., S. Satoh, T. Noshi, E. Hatada, R. Fukuda, A. Kawai, S. Ikeda, M. Hijikata, and K. Shimotohno. 2001. Effects of mutation in hepatitis C virus nonstructural protein 5A on interferon resistance mediated by inhibition of PKR kinase activity in mammalian cells. *Microbiol. Immunol.* **45:**829–840.

95. Onomoto, K., M. Yoneyama, and T. Fujita. 2007. Regulation of antiviral innate immune responses by RIG-I

96. Oshiumi, H., M. Matsumoto, K. Funami, T. Akazawa, and T. Seya. 2003. TICAM-1, an adaptor molecule that participates in Toll-like receptor 3-mediated interferon-beta induction. *Nat. Immunol.* **4:**161–167.

97. Park, G. S., K. L. Morris, R. G. Hallett, M. E. Bloom, and S. M. Best. 2007. Identification of residues critical for the interferon antagonist function of Langat virus NS5 reveals a role for the RNA-dependent RNA polymerase domain. *J. Virol.* **81:**6936–6946.

98. Perelygin, A. A., S. V. Scherbik, I. B. Zhulin, B. M. Stockman, Y. Li, and M. A. Brinton. 2002. Positional cloning of the murine flavivirus resistance gene. *Proc. Natl. Acad. Sci. USA* **99:**9322–9327.

99. Petersen, L. R., and J. T. Roehrig. 2001. West Nile virus: a reemerging global pathogen. *Emerg. Infect. Dis.* **7:**611–614.

100. Pichlmair, A., O. Schulz, C. P. Tan, T. I. Näslund, P. Liljeström, F. Weber, and C. Reis e Sousa. 2006. RIG-I-mediated antiviral responses to single-stranded RNA bearing 5′-phosphates. *Science* **314:**997–1001.

101. Polyak, S. J., K. S. Khabar, D. M. Paschal, H. J. Ezelle, G. Duverlie, G. N. Barber, D. E. Levy, N. Mukaida, and D. R. Gretch. 2001. Hepatitis C virus nonstructural 5A protein induces interleukin-8, leading to partial inhibition of the interferon-induced antiviral response. *J. Virol.* **75:**6095–6106.

102. Polyak, S. J., K. S. Khabar, M. Rezeiq, and D. R. Gretch. 2001. Elevated levels of interleukin-8 in serum are associated with hepatitis C virus infection and resistance to interferon therapy. *J. Virol.* **75:**6209–6211.

103. Preston, C. M., A. N. Harman, and M. J. Nicholl. 2001. Activation of interferon response factor-3 in human cells infected with herpes simplex virus type 1 or human cytomegalovirus. *J. Virol.* **75:**8909–8916.

104. Randall, G., L. Chen, M. Panis, A. K. Fischer, B. D. Lindenbach, J. Sun, J. Heathcote, C. M. Rice, A. M. Edwards, and I. D. McGilvray. 2006. Silencing of USP18 potentiates the antiviral activity of interferon against hepatitis C virus infection. *Gastroenterology* **131:**1584–1591.

105. Rehermann, B., and M. Nascimbeni. 2005. Immunology of hepatitis B virus and hepatitis C virus infection. *Nat. Rev. Immunol.* **5:**215–229.

106. Ruggli, N., J. D. Tratschin, M. Schweizer, K. C. McCullough, M. A. Hofmann, and A. Summerfield. 2003. Classical swine fever virus interferes with cellular antiviral defense: evidence for a novel function of N^pro. *J. Virol.* **77:**7645–7654.

107. Sadler, A. J., and B. R. Williams. 2007. Structure and function of the protein kinase R. *Curr. Top. Microbiol. Immunol.* **316:**253–292.

108. Saito, T., and M. Gale, Jr. 2006. Principles of intracellular viral recognition. *Curr. Opin. Immunol.* **19:**17–23.

109. Saito, T., R. Hirai, Y. M. Loo, D. Owen, C. L. Johnson, S. C. Sinha, S. Akira, T. Fujita, and M. Gale, Jr. 2006. Regulation of innate antiviral defenses through a shared repressor domain in RIG-I and LGP2. *Proc. Natl. Acad. Sci. USA.* **104:**582–587.

110. Samuel, M. A., and M. S. Diamond. 2005. Alpha/beta interferon protects against lethal West Nile virus infec-

tion by restricting cellular tropism and enhancing neuronal survival. *J. Virol.* 79:13350–13361.

111. Samuel, M. A., and M. S. Diamond. 2006. Pathogensis of West Nile virus infection: a balance between virulence, innate and adaptive immunity, and viral evasion. *J. Virol.* 80:9349–9360.

112. Samuel, M. A., K. Whitby, B. C. Keller, A. Marri, W. Barchet, B. R. Williams, R. H. Silverman, M. Gale, Jr., and M. S. Diamond. 2006. PKR and RNase L contribute to protection against lethal West Nile Virus infection by controlling early viral spread in the periphery and replication in neurons. *J. Virol.* 80:7009–7019.

113. Sato, M., H. Suemori, N. Hata, M. Asagiri, K. Ogasawara, K. Nakao, T. Nakaya, M. Katsuki, S. Noguchi, N. Tanaka, and T. Taniguchi. 2000. Distinct and essential roles of transcription factors IRF3 and IRF-7 in response to viruses for IFN-alpha/beta gene induction. *Immunity* 13:539–548.

114. Scherbik, S. V., J. M. Paranjape, B. M. Stockman, R. H. Silverman, and M. A. Brinton. 2006. RNase L plays a role in the antiviral response to West Nile virus. *J. Virol.* 80:2987–2999.

115. Scholle, F., and P. W. Mason. 2005. West Nile virus replication interferes with both poly(I:C)-induced interferon gene transcription and response to interferon treatment. *Virology* 342:77–87.

116. Schweizer, M., P. Matzener, G. Pfaffen, H. Stalder, and E. Peterhans. 2006. "Self" and "nonself" manipulation of interferon defense during persistent infection: bovine viral diarrhea virus resists alpha/beta interferon without blocking antiviral activity against unrelated viruses replicating in its host cells. *J. Virol.* 80:6926–6935.

117. Seago, J., L. Hilton, E. Reid, V. Doceul, J. Jeyatheesan, K. Moganeradj, J. McCauley, B. Charleston, and S. Goodbourn. 2007. The N^pro product of classical swine fever virus and bovine viral diarrhea virus uses a conserved mechanism to target interferon regulatory factor-3. *J. Gen. Virol.* 88:3002–3006.

118. Sen, G. C. 2001. Viruses and interferons. *Annu. Rev. Microbiol.* 55:255–281.

119. Servant, M. J., B. tenOever, C. LePage, L. Conti, S. Gessani, I. Juljunen, R. Lin, and J. Hiscott. 2001. Identification of distinct signaling pathways leading to the phosphorylation of interferon regulatory factor 3. *J. Biol. Chem.* 276:355–363.

120. Seth, R. B., L. Sun, C. K. Ea, and Z. J. Chen. 2005. Identification and characterization of MAVS, a mitochondrial antiviral signaling protein that activates NF-kappaB and IRF 3. *Cell* 122:669–682.

121. Silverman, R. H. 2007. Viral encounters with OAS and RNase L during the IFN antiviral response. *J. Virol.* 81:12720–12729.

122. Smith, E. J., I. Marie, A. Prakash, A. Garcia-Sastre, and D. E. Levy. 2001. IRF3 and IRF7 phosphorylation in virus-infected cells does not require double-stranded RNA-dependent protein kinase R or Ikappa B kinase but is blocked by vaccinia virus E3L protein. *J. Biol. Chem.* 276:8951–8957.

123. Solomon, T., and P. M. Winter. 2004. Neurovirulence and host factors in flavivirus encephalitis—evidence from clinical epidemiology. *Arch. Virol. Suppl.* 2004:161–170.

124. Sumpter, R., Y.-M. Loo, E. Foy, K. Li, M. Yoneyama, T. Fujita, S. M. Lemon, and M. J. Gale. 2005. Regulating intracellular anti-viral defense and permissiveness to hepatitis C virus RNA replication through a cellular RNA helicase, RIG-I. *J. Virol.* 79:2689–2699.

125. Taylor, D. R., S. T. Shi, P. R. Romano, G. N. Barber, and M. M. C. Lai. 1999. Inhibition of the interferon-inducible protein kinase PKR by HCV E2 protein. *Science* 285:107–110.

126. tenOever, B. R., M. J. Servant, N. Grandvaux, R. Lin, and J. Hiscott. 2002. Recognition of the measles virus nucleocapsid as a mechanism of IRF3 activation. *J. Virol.* 76:3659–3669.

127. Trinchet, J. C., N. Ganne-Carrie, P. Nahon, G. N'kontchou, and M. Beaugrand. 2007. Hepatocellular carcinoma in patients with hepatitis C virus-related chronic liver disease. *World J. Gastroenterol.* 13:2455–2460.

128. Tsuchihara, K., T. Tanaka, M. Hijikata, S. Kuge, H. Toyoda, A. Nomoto, N. Yamamoto, and K. Shimotohno. 1997. Specific interaction of polypyrimidine tract-binding protein with the extreme 3′-terminal structure of the hepatitis C virus genome, the 3′X. *J. Virol.* 71:6720–6726.

129. Venter, M., T. G. Myers, M. A. Wilson, T. J. Kindt, J. T. Paweska, F. J. Burt, P. A. Leman, and R. Swanepoel. 2005. Gene expression in mice infected with West Nile virus strains of different neurovirulence. *Virology* 342:119–140.

130. Wang, T., T. Town, L. Alexopoulou, J. F. Anderson, E. Fikrig, and R. A. Flavell. 2004. Toll-like receptor 3 mediates West Nile virus entry into the brain causing lethal encephalitis. *Nat. Med.* 10:1366–1373.

131. Xu, L. G., Y. Y. Wang, K. J. Han, L. Y. Li, Z. Zhai, and H. B. Shu. 2005. VISA is an adapter protein required for virus-triggered IFN-beta signaling. *Mol. Cell* 19:727–740.

132. Yamamoto, M., S. Sato, H. Hemmi, K. Hoshino, T. Kaisho, H. Sanjo, O. Takeuchi, M. Sugiyama, M. Okabe, K. Takeda, and S. Akira. 2003. Role of adaptor TRIF in the MyD88-independent Toll-like receptor signaling pathway. *Science* 301:640–643.

133. Yamamoto, M., S. Sato, K. Mori, K. Hoshino, O. Takeuchi, K. Takeda, and S. Akira. 2002. Cutting edge: a novel Toll/IL-1 receptor domain-containing adapter that preferentially activates the IFN-beta promoter in the Toll-like receptor signaling. *J. Immunol.* 169:6668–6672.

134. Yang, Y., Y. Liang, L. Qu, Z. Chen, M. Yi, K. Li, and S. M. Lemon. 2007. Disruption of innate immunity due to mitochondrial targeting of a picornaviral protease precursor. *Proc. Natl. Acad. Sci. USA* 104:7253–7258.

135. Yoneyama, M., and T. Fujita. 2007. RIG-I family RNA helicases: cytoplasmic sensor for antiviral innate immunity. *Cytokine Growth Factor Rev.* 18:545–551.

136. Yoneyama, M., M. Kikuchi, T. Natsukawa, N. Shinobu, T. Imaizumi, M. Miyagishi, K. Taira, S. Akira, and T. Fujita. 2004. The RNA helicase RIG-I has an essential function in double-stranded RNA-induced innate antiviral responses. *Nat. Immunol.* 5:730–737.

137. Yoneyama, M., W. Suhara, and T. Fujita. 2002. Control of IRF3 activation by phosphorylation. *J. Interferon Cytokine Res.* 22:73–76.

138. Yoneyama, M., W. Suhara, Y. Fukuhara, M. Fukuda, E. Nishida, and T. Fujita. 1998. Direct triggering of the type I interferon system by virus infection: activation of a transcription factor complex containing IRF3 and CBP/p300. *EMBO J.* 17:1087–1095.

Cellular Signaling and Innate Immune
Responses to RNA Virus Infections
Edited by A. R. Brasier et al.
© 2009 ASM Press, Washington, D.C.

Stanley M. Lemon

Evasion of Innate Host Antiviral Defenses by Picornaviruses

21

THE PICORNAVIRUSES

The family *Picornaviridae* comprises a large group of viruses with small, single-stranded, positive-sense RNA genomes that range from 7.2 to 8.4 kb in length. These viruses share a common overall genomic organization, particle structure, and replication scheme (82). The family includes a number of medically important viruses, such as poliovirus (PV) and hepatitis A virus (HAV); viruses of significance to veterinary medicine, such as foot-and-mouth disease virus (FMDV); and encephalomyocarditis virus (EMCV), a murine virus with broad capabilities to infect other animal species that has been used widely in studies of the interaction of viruses with the interferon (IFN) system.

The shared replication schemes of these viruses have features that likely shape their recognition by cellular pathogen-associated molecular pattern receptors. Picornaviruses enter cells through a process of receptor-mediated endocytosis and release their messenger-sense RNA genome into the cytoplasm, where it supports translation of a polyprotein encoded by a single large open reading frame (Fig. 1). Initiation of viral translation on the genomic RNA occurs by a cap-independent process involving the recruitment of 40S ribosome subunits to a highly structured internal ribosome entry site (IRES) in the lengthy 5′ nontranslated segment of the genome. The viral polyprotein is subsequently processed, primarily through a carefully coordinated process directed by virally encoded proteases, giving rise to three or four structural proteins and six to eight nonstructural proteins (82) (Fig. 2). Many of these have functions that are conserved across all genera in the family *Picornaviridae* and that are required for replication of the virus: e.g., 1A to 1D (previously known as VP1 to VP4; Fig. 2) comprise the four structural proteins that typically form the nonenveloped viral capsid, while 3B (also known as VPg) is a protein primer for RNA synthesis, $3C^{pro}$ is the major cysteine protease responsible for polyprotein processing, and $3D^{pol}$ is the conserved RNA-dependent RNA polymerase. Viral RNA synthesis is mediated by a multiprotein viral replicase complex within double-membrane vesicles derived largely from endoplasmic reticulum (ER) membranes (6, 15). This leads first to synthesis of a complementary negative-strand intermediate, and subsequently to nonconservative amplification of multiple positive-strand

Stanley M. Lemon, Center for Hepatitis Research, Institute for Human Infections and Immunity and the Department of Microbiology, University of Texas Medical Branch, Galveston, TX. 77555.

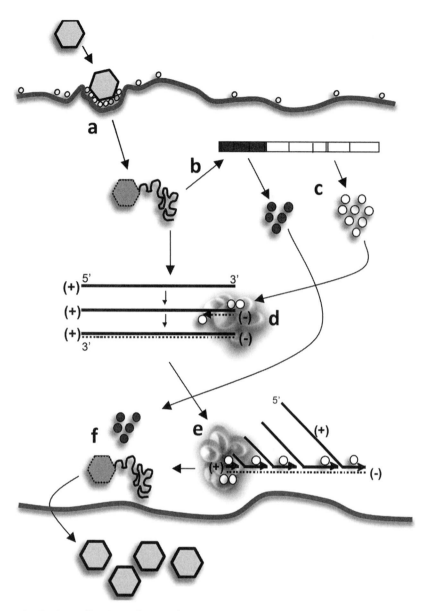

Figure 1 Basic replication scheme of picornaviruses: (a) virus uptake occurs through a process of receptor-mediated endocytosis leading to release of the positive-strand RNA genome into the cytoplasm; (b) cap-independent translation of the genomic RNA leads to expression of a large polyprotein; (c) the polyprotein is processed into structural (shaded) and nonstructural (not shaded) proteins by a well-coordinated series of viral protease-mediated cleavages; (d) nonstructural proteins assemble as replicase complexes at the 3′ end of the genomic RNA within double-membrane vesicles derived from a reordering of intracellular membranes, and initiate the synthesis of complementary negative-strand RNA, leading to production of dsRNA; (e) multiple copies of positive-strand genomic RNA are transcribed from the negative-strand intermediates; (f) newly synthesized positive-strand progeny assemble with structural proteins to form new viral particles, which are released from the cell through both lytic and nonlytic pathways. Replication occurs within the cytoplasm and involves a number of cellular proteins, some of which are normally localized in the nucleus.

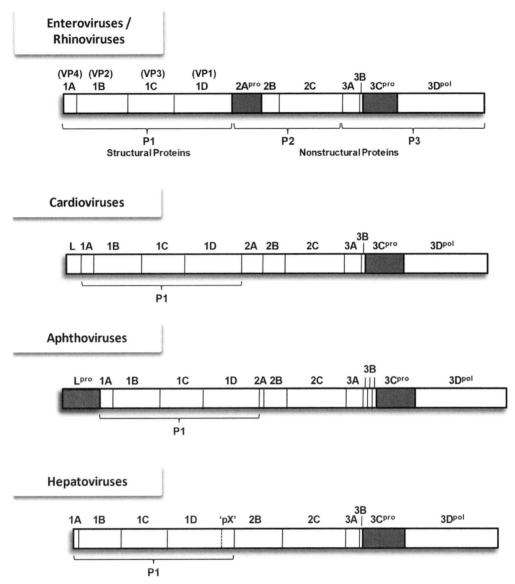

Figure 2 Polyprotein organization of viruses within the five major picornaviral genera, showing the general order of the individual mature proteins (L-VP4-VP2-VP3-VP1–2A-2B-2C-3A-3B-3C^{pro}-3D^{pol}) that are produced by proteolytic processing directed by virally encoded proteases (shaded proteins). All members of the family express a cysteine protease (3C^{pro}) that is responsible for most polyprotein processing events. Some genera express additional proteases (2A^{pro} in enteroviruses and rhinoviruses and L^{pro} in the aphthoviruses) that act in *cis* on the polyprotein and also target cellular proteins that are critical for cellular transcription and/or translation. Precursor proteins may also have specific proteolytic activities, such as 3ABC in the hepatoviruses, which targets MAVS/IPS-1 for cleavage. The amino acid residues flanking each cleavage site are shown for each of the polyproteins and are generally conserved within each picornaviral genus. 3D^{pol} is the conserved RNA-dependent RNA polymerase. (Modified from reference 58.)

progeny genomes (Fig. 1). Virus assembly occurs close to the location of RNA synthesis, and newly assembled particles exit the cell either through cell lysis or through poorly characterized nonlytic mechanisms.

Despite these and numerous other features shared in common by these viruses, there is substantial diversity between and even within the nine distinct genera that are currently considered to comprise this virus family (82)

Table 1 Abbreviated list of important members of the family *Picornaviridae*

Genus	Virus[a]	Host	Disease
Enterovirus	Poliovirus (PV)	Humans	Encephalitis, paralysis, myocarditis, pericarditis, meningitis
	Coxsackievirus (CVA, CVB)	Humans, other	
	Echoviruses (EV)	Humans, other	
Rhinovirus	Rhinoviruses (HRV)	Humans	Rhinitis
Cardiovirus	Encephalomyocarditis virus (EMCV)	Mice, many other	Encephalitis, myocarditis, demyelinating disease
	Theiler's virus (ThV)	Mice	
Aphthovirus	Foot-and-mouth disease virus (FMDV)	Cloven-hoofed animals	Ulcerative, vesicular disease
Hepatovirus	Hepatitis A virus (HAV)	Humans	Acute hepatitis
Parechovirus	Human parechovirus (HPeV)	Humans	Enteritis, meningoencephalitis
Erbovirus	Equine rhinitis virus B (ERBV)	Horses	Febrile respiratory disease
Kobuvirus	Aichi virus (AiV)	Humans	Enteritis
Teschovirus	Porcine teschovirus (PTV)	Pigs	Enteritis, encephalomyelitis

[a]Most genera include many serologically distinct viruses. Note that some picornaviral genera have further, subgenera groupings, and that there are numerous unassigned viruses within the family (82).

(Fig. 2). Five of these genera are of particular importance from a medical or veterinary perspective: *Enterovirus*, *Rhinovirus*, *Aphthovirus*, *Cardiovirus*, and *Hepatovirus* (Table 1). There is strong evidence that picornaviruses in each of these genera have evolved specific yet substantially different strategies to manage and survive their interactions with the host cell. The remarkable diversity that is evident in these adaptive strategies across the various picornaviral genera suggests that they are relatively recent additions in the evolution of the picornaviruses. A considerable amount of information has been amassed concerning how these viruses interact with various innate immune responses. However, it is clear that even among these relatively well-studied picornaviral genera, the surface has only been scratched and there is much more to be learned about how these pathogens both activate and confound innate cellular antiviral responses. Much less is known about the other four picornaviral genera—*Parechovirus*, *Teschovirus*, *Erbovirus*, and *Kobuvirus* (Table 1)—which will not be considered further in this chapter.

Cellular Sensing of Picornaviral Infections

All picornaviruses utilize a protein primer (3B or VPg; Fig. 2) for initiation of both positive- and negative-strand viral RNA synthesis (67). This leads to the presence of a small viral protein (VPg) covalently linked to the 5′ end of newly synthesized viral RNAs. While 3B is removed from the end of genomic RNAs prior to translation, the VPg-priming mechanism ensures that all picornaviral RNAs are devoid of free 5′ triphosphates. These nonself RNAs are thus likely to be poorly recognized by the intracellular pathogen recognition receptor (PRR) retinoic acid-inducible gene I (RIG-I), which

preferentially recognizes RNAs that terminate in a 5′ triphosphate (39, 70). Consistent with this, studies suggest that another cytoplasmic helicase, the melanoma differentiation-associated gene 5 protein (MDA-5), and not RIG-I, is the major intracellular receptor that senses EMCV infection and induces the activation of signaling pathways leading to antiviral responses (32, 43). MDA-5$^{-/-}$ knockout mice demonstrate a selective impairment in IFN responses to EMCV, in contrast to RIG-I$^{-/-}$ knockout mice, which have normal IFN responses to this virus. Other studies suggest the involvement of Toll-like receptor 3 (TLR3), TLR7, and TLR8 in recognition of picornavirus infection by different types of cells (2,81,89,90). However, there are relatively few studies of the cellular pathways responsible for activation of IFN-β synthesis in cells infected with picornaviruses other than the cardioviruses.

The results obtained with EMCV may well extend to other picornaviruses, since all picornaviral RNAs lack a 5′ triphosphate. However, it is wise to be cautious in making this assumption in the absence of experimental data. There are significant differences in the structures of cardioviral, enteroviral, and hepatoviral RNAs in terms of the presence and lengths of pyrimidine-rich tracts located within the 5′ nontranslated RNA segment of the genome. Some cardiovirus and aphthovirus genomes, for example, contain lengthy pure poly(C) tracts (up to 146 Cs in EMCV), while hepatovirus RNAs contain a shorter, pyrimidine-rich segment. No clear function for these elements has been demonstrated in viral replication. How the presence of such RNA segments might influence the sensing of viral RNAs by PRRs has not been studied but could be important. Deletions of the poly(C) tract significantly attenuate the virulence of EMCV in mice, but viruses with short poly(C) tracts may still be virulent (24, 35, 74).

A number of other features of picornaviruses may help mask their presence from the host. The 3′ end of the positive-strand, messenger-sense picornaviral RNA is polyadenylated, while the 3′ end of the negative-sense RNA (produced in substantially lesser abundance) terminates in at least two adenosine residues. Like the absence of free 5′ triphosphates, these features may help disguise picornaviral RNAs produced during infection as "self." In addition, viral RNA synthesis takes place within membranous vesicles (6, 15), and this sequestered location is likely to hinder the access of PRRs to these nonself, viral RNAs. Nonetheless, virus replication produces an abundance of double-stranded (ds) RNA (1, 95) that may serve as ligand for recognition by TLR3 or cytosolic RIG-I-like helicases, notably MDA-5 (86). Perhaps equally important for host recognition, there are multiple dsRNA segments within the positive-sense genomic RNA that might be recognized by these and other receptors. These include not only the highly structured IRES element that directs 5′ cap-independent translation of the picornaviral polyprotein, but also internal and 3′-terminal-complex stem-loop RNA structures required for genome replication (69, 88, 97). Whatever the source of ligand, there is longstanding evidence for frequent activation of the dsRNA-activated enzymes, protein kinase R (PKR) and 2′,5′-oligoadenylate synthetase (OAS), in cells infected with various picornaviruses (8, 80). Finally, most picornaviruses express less than a dozen different proteins (82); none of the four structural proteins of picornaviruses are glycosylated, and this might also lessen the probability of host sensing by TLRs.

Picornaviral Strategies To Confound Innate Host-Cell Defenses

Several features stand out in considering how the picornaviruses attempt to evade innate antiviral defenses. First, the regularity with which picornaviruses have been shown to attempt to block the activity of IFN regulatory factor 3 (IRF3) or nuclear factor-κB (NF-κB) (Table 2) attests to the importance of these transcription factors in the host-cell responses to viruses in general (38, 76, 85) and picornaviruses in particular. In addition, the frequency with which these viruses can be shown to antagonize the dsRNA-activated OAS/RNase L or PKR pathways (Table 2) also speaks to both the importance of these innate antiviral responses and the redundancy of innate immune responses. Striking among the mechanisms exploited by the picornaviruses to confound innate antiviral defenses is the action of virally encoded proteases.

These target a variety of critical cellular signaling molecules for destruction (Table 2). What is known about these evasion strategies is reviewed in the subsequent sections of this chapter for each of the major picornaviral genera. Not covered in significant detail, however, are the mechanisms by which picornaviruses elicit apoptosis, a late but variable event in some picornaviral infections (5, 11, 75). These cellular responses, and viral strategies to prevent or delay apoptosis, may also play important roles in the ultimate outcome of picornaviral infections.

APHTHOVIRUSES

FMDV L^pro-Mediated Suppression of IFN-α/β and NF-κB

FMDV is the causative agent of foot-and-mouth disease, a potentially devastating infection of cloven-hoofed animals including domesticated cattle, sheep, and swine. In cultured cells, the virus demonstrates impressive replicative capacity, completing its life cycle and producing abundant progeny virions within 4 to 5 h of cell entry. In part, it accomplishes this by rapidly inactivating the cap-dependent translation of cellular proteins via expression of its L^pro and 3C^pro proteases, which direct the sequential cleavage of eIF4GI (eukaryotic initiation factor 4GI), an essential cellular translation initiation factor (20, 21, 83). Viruses in other picornaviral genera either lack a leader (L) protein (*Enterovirus*, *Rhinovirus*, and *Hepatovirus spp.*) or express an L protein that lacks protease activity (*Cardiovirus spp.*) (Fig. 2). FMDV mutants in which L^pro has been deleted (so-called leaderless FMDV) remain viable, but they are attenuated, and unlike wild-type virus do not spread from the site of infection when inoculated by aerosol to susceptible cattle (10). The attenuated phenotype appears to be due to enhanced induction of type I IFNs by the mutant virus and subsequent inhibition of virus replication (20). In IFN-α/β-competent cells the leaderless virus displays markedly impaired replication capacity, while in BHK-21 cells (which fail to respond to type I IFNs) there is little difference in the replication capacity of wild-type and leaderless viruses. Furthermore, L^pro expression prevents IFN-α/β synthesis and also inhibits transcriptional activation of IFN-β and IFN-stimulated genes (ISGs) such as PKR, OAS, and Mx1 in cell culture (20).

It is likely that suppression of the IFN response by FMDV is due primarily to the general paralysis of cellular protein synthesis induced by L^pro cleavage of eIF4GI (20, 83). Importantly, viral protein synthesis is not blocked, and indeed is enhanced under these conditions, since the FMDV IRES does not require intact eIF4GI to direct translation initiation. FMDV's attack

Table 2 Strategies evolved by members of the *Picornaviridae* for evading innate host antiviral defenses

Genus	Virus[a]	Evasive mechanism(s)
Aphthovirus	Foot-and-mouth disease virus (FMDV)	IFN-α/β synthesis is inhibited at the transcriptional and translational level by L^{pro} (18).
		NF-κB activation is limited by L^{pro}-induced degradation of p65/RelA (17).
Cardiovirus	Encephalomyocarditis virus (EMCV), Theiler's virus (ThV)	L protein interacts with Ran-GTPase, disrupting nucleocytoplasmic trafficking, altering nuclear transport of IRF3, and inhibiting virus-activated IFN-α/β transcription (20, 71).
		EMCV infection results in progressive inactivation of RNase L by an unknown mechanism (80).
Enterovirus	Poliovirus (PV)	RNase L is inactivated by a complex stem-loop structure in the polyprotein-coding segment of group C enterovirus RNA, including PV (37).
		NF-κB activation is limited by 3C^{pro}-mediated cleavage of p65/RelA (64).
		Inhibition of ER-to-Golgi transport by the 3A protein reduces cytokine secretion and TNF receptor expression (22, 65).
		Nucleocytoplasmic trafficking is disrupted by 2A^{pro} cleavage of Nup98, potentially impairing transcriptional responses to infection (66).
		MDA-5 is cleaved late in the infection cycle due to virus activation of host-cell caspases (3).
Rhinovirus	Rhinoviruses (HRV)	As with PV, the HRV-14 3C^{pro} protease mediates cleavage of p65/RelA (64).
		As with PV, HRV-14 targets nuclear pore proteins to disrupt nucleocytoplasmic trafficking (34).
Hepatovirus	Hepatitis A virus (HAV)	3ABC^{pro} inhibits viral activation of IRF3 by cleaving MAVS/IPS-1 (96).

[a] Similar evasive mechanisms may occur with viruses that are closely related phylogenetically.

on cellular homeostasis is so profound that even antiviral responses that normally act, at least in part, independently of new protein synthesis, such as PKR, are unable to influence the course of virus infection. In porcine kidney (SK6) cells that are IFN-α/β competent, small interfering RNA-mediated knockdown of PKR does not cause an increase in yields of wild-type virus, suggesting that PKR normally has little impact on FMDV replication (20). However, PKR knockdown does enhance the yields of leaderless, L^{pro}-deficient FMDV, demonstrating that PKR can directly control FMDV replication in the absence of L^{pro}-mediated host-cell translational shutdown. The results of similar studies done in triply ISG-deficient PKR$^{-/-}$, RNase L$^{-/-}$, and Mx1$^{-/-}$ mouse embryonic fibroblasts suggest that PKR has greater capacity to control L^{pro}-deficient FMDV replication than the OAS/RNase L pathway (14).

As with the absence of IFN-α/β responses by FMDV-infected cells, the inability of PKR and other stress-activated kinases to limit wild-type FMDV replication has generally been attributed to L^{pro}-mediated translational shutdown (20). However, the underlying mechanism is likely to be more complex. Recent studies by de los Santos et al. (19) suggest that the p65/RelA subunit of NF-κB may be destabilized and degraded in an L^{pro}-dependent fashion during the course of FMDV infection. Although the specific mechanism by which this occurs is unknown, infection of porcine kidney cells with wild-type FMDV leads to a significant reduction in the cellular abundance of p65/RelA within 6 h of infection (19). This was not observed following infection with leaderless, L^{pro}-deficient FMDV, even at a 10-fold higher multiplicity of infection. Furthermore, while limited amounts of p65/RelA were observed within the nucleus of cells infected with the leaderless virus, this was never found in cells infected with wild-type virus. NF-κB-dependent promoter activity was correspondingly reduced in cells infected with the wild-type compared with L^{pro}-deficient, leaderless virus (19). Further studies will be needed to confirm these recent findings and to determine whether p65/RelA might be targeted directly for proteolysis by L^{pro} or perhaps degraded in a proteasome-dependent fashion in cells expressing L^{pro}. However, these results suggest a novel and potentially important additional function for L^{pro} in the pathogenesis of FMDV. The loss of p65/RelA expression could be an additional factor contributing to the lack of IFN-α/β responses in FMDV-infected cells. It could also limit PKR-mediated induction of proinflammatory cytokines.

The studies described above may seem to paint FMDV as a "take no prisoners" virus, one that survives innate antiviral host defenses by an L^{pro}-mediated blitzkrieg assault on cellular homeostasis. Yet, paradoxically, FMDV is not always so, at least in the intact animal host, as persistent infections are well documented in cattle and other species (57). Given the multiple, critical linkages that exist between innate immune responses and the subsequent development of robust adaptive B- and T-cell immunity (42), it seems likely that the mechanisms FMDV has evolved to evade innate antiviral defenses may play an important role in setting the stage for virus persistence. However, little is known about the specific mechanisms underlying persistent infections with FMDV, and it is not clear how important this is to maintenance of the virus in nature.

CARDIOVIRUSES

L Protein Disruption of Type I IFN Responses

The cardioviruses (e.g., EMCV and Theiler's virus [ThV]) cause a variety of disease manifestations in mice, and both have been used widely as models to study immune responses to viruses. ThV is of particular interest in that it can cause a persistent infection of the murine central nervous system that is associated with demyelination and is in some ways similar to multiple sclerosis. Like aphthoviruses, these viruses also express a leader (L) protein (Fig. 2). However, in contrast to the FMDV leader, the L protein of the cardioviruses lacks protease activity. Nonetheless, it plays an important role in virus evasion of IFN-mediated antiviral responses, and in the case of ThV, it is critical for establishment of persistent central nervous system infections (92). The L proteins of EMCV and ThV are both small proteins and share only 35% amino acid identity, yet they possess a common zinc-finger motif and a shared ability to inhibit the production of IFN-α/β by infected cells. The L protein of ThV specifically inhibits the synthesis of IFN-α/β mRNAs at the transcriptional level, and this is dependent upon the integrity of the zinc finger (92). Delhaye et al. (18) have shown that L protein suppression of virus-activated IFN-α/β transcription is due to disruption of the normal nucleocytoplasmic trafficking of cellular proteins, including IRF3. Despite substantial differences between their amino acid sequences, the L proteins of EMCV and ThV are functionally interchangeable, with

similar zinc-finger-dependent effects on nucleocytoplasmic trafficking (52, 68).

L protein-mediated disruption of nucleocytoplasmic trafficking of activated IRF3 appears to be particularly important in the pathogenesis of cardiovirus infections. IRF3, a transcription factor that plays a critical role in virus induction of the IFN-β promoter and subsequent host control of virus pathogenesis, is constitutively expressed within the cytoplasm in an inactive form (38, 53). When a virus infection is sensed through either the TLR3 or RIG-I/MDA-5 pathway (86), IRF3 is phosphorylated by the noncanonical IκB kinase homologs, IκB kinase-ε (IKK-ε) and TANK-binding kinase-1 (TBK1) (28). This leads to its dimerization and subsequent transport to the nucleus, where it activates the IFN-β promoter in coordination with the transcription factors NF-κB and ATF-2/c-Jun and several coactivator proteins (93). In addition to inducing transcription of the IFN-β gene, activated nuclear IRF3 also directly induces the expression of a small subset of ISGs with potential antiviral effects (33). Thus, in concert with the closely related transcription factor IRF7, which acts to amplify this process, IRF3 plays a pivotal role in the induction of cellular antiviral responses and control of virus infections by type I IFNs (38). The L proteins of cardioviruses target this response by disrupting the nucleocytoplasmic trafficking of IRF3 (and likely IRF7 as well) through a specific interaction with Ran-GTPase and disruption of the normal RanGDP/GTP gradient (18,71). While a proportion of cells infected with a ThV mutant in which the L protein zinc-finger motif has been ablated show clear evidence of nuclear IRF3 localization, this is not found in cells infected with wild-type ThV (18). These cells show instead an abnormal heterogenous nuclear staining for IRF3, which extends to the nucleoli. Thus, although cytoplasmic-to-nucleus transport of IRF3 is not ablated in the presence of the L protein, it appears to be disregulated, leading to aberrant intranuclear localization of IRF3 and a subsequent lack of IFN-β expression.

Despite these recent gains in understanding, the mechanisms underlying the suppression of IFN responses by the cardioviral L protein are incompletely understood. However, it is interesting that L proteins expressed by both aphthoviruses and cardioviruses, which do not share significant homology with each other, both act to disable innate antiviral responses, albeit through very different mechanisms (Table 2). Given the differences in these proteins, and their absence in enteroviruses, rhinoviruses, and hepatoviruses, it is reasonable to speculate that they have been late acquisitions in the evolution of the picornaviruses. The ways in which they function to disrupt innate antiviral responses are likely to represent relatively recent adaptations to hostile cellular environments. Yet it is not at all clear that these L proteins have evolved solely to confound innate antiviral responses. Although aphthovirus and cardiovirus mutants with disrupted L protein function retain the ability to replicate (18, 20), they are compromised since these proteins also serve to directly promote the replication of the virus. The FMDV L^{pro} protein promotes replication by knocking out cellular cap-dependent translation through its cleavage of eIF4GI (21), eliminating competition for 40S ribosome subunits and thus enhancing IRES-directed viral translation. Similarly, the disruption of nucleocytoplasmic trafficking by the cardioviral L protein is likely to promote viral replication by causing an efflux of nuclear proteins that facilitate viral protein translation and/or RNA synthesis (52). It is intriguing that many of the accessory functions ascribed to the L proteins of aphthoviruses and cardioviruses (e.g., host translational shutdown and disruption of normal nucleocytoplasmic transport) are common to the 2A^{pro} protein of PV (and probably rhinovirus as well, as described below) (4, 54, 66). This suggests that there has been a certain degree of parallel evolution among the different picornaviral genera as viruses in each have faced common challenges posed by host defenses.

RNase L, PKR, Stress-Activated Kinases, and Control of Cardiovirus Infections

EMCV has been used extensively as a model virus in studies of the IFN response and ways in which the expression of type I IFNs limits replication of viruses. A wealth of studies demonstrate that the RNase L-mediated degradation of viral RNA contributes substantially to the antiviral effects of IFN against EMCV (51, 79). The antiviral activity of RNase L is relatively specific for EMCV (and likely other picornaviruses), as vesicular stomatitis virus (VSV) RNA is resistant to RNase L cleavage. However, the activation of RNase L may also be associated with demonstrable degradation of cellular ribosomal RNAs (51). The role played by RNase L in control of EMCV infections is much less certain in the absence of IFN treatment. While early studies by Silverman et al. (80) revealed that EMCV infection of HeLa cells leads to activation of OAS and production of 2′,5′-oligoadenylate (2-5A), they also showed that EMCV continued to replicate efficiently in these cells in the absence of IFN treatment. This could be explained by the fact that EMCV infection led to a progressive inhibition of the ability of RNase L to be activated by 2-5A [ppp(A2′)nA]. The ability of 2-5A to activate RNase L was nearly completely inhibited by 6 h after infection with EMCV (12, 80). The inhibition of RNase L activation in EMCV-infected cells

could be abolished by IFN treatment (12, 80). As described above, the synthesis of endogenous type I IFNs is blocked by EMCV infection due to L protein-mediated disruption of nucleocytoplasmic trafficking of proteins (18, 71). However, the lack of RNase L activation cannot be attributed to an insufficient abundance of RNase L in the absence of IFN stimulation (51), as the basal expression of RNase L is substantial. No soluble inhibitor of RNase L activation could be identified in lysates of EMCV-infected cells (80), and these intriguing observations remain largely unexplained many years after they were reported. It is possible that the induction of a cellular ATP-binding cassette protein that is capable of suppressing the activity of RNase L, RNase L inhibitor (RLI, otherwise known as ABCE1), may contribute to this EMCV antagonism of RNase L activation (59, 60).

Early studies with PKR knockout mice (N-terminal truncation) provided little evidence that EMCV infections are naturally controlled by PKR. Infection of these PKR$^{-/-}$ mice with EMCV resulted in a systemic IFN response and median survival time similar to that observed in normal mice (98). Since the induction of IFN is inhibited in EMCV-infected cells, as described in the previous paragraph, the IFN response observed in infected mice is likely derived largely from noninfected immune cells, e.g., plasmacytoid dendritic cells. Treatment with exogenous IFN-α also resulted in an equivalent increase in survival time in both types of mice (98). However, a significantly greater increase in survival time was noted in wild-type versus PKR$^{-/-}$ mice treated with IFN-γ, suggesting that PKR does contribute to IFN-γ-mediated control of EMCV infection. At the cellular level, there was no difference in the yields of virus from PKR$^{-/-}$ and normal mouse fibroblasts (98). On the other hand, Yeung et al. (99) have demonstrated that repression of PKR expression shifted the course of EMCV infection in promonocytic U937 cells from an acute, lytic infection to a persistent infection. In these cells, reduced expression of PKR led to a delay in virus-induced apoptosis, setting the stage for viral persistence. Persistently infected cells demonstrated sustained expression of p53 and interleukin-1β (IL-1β)-converting enzyme and an absence of type I IFN expression (99). Overall, these results suggest that PKR may control resistance to EMCV in some, but not all, types of cells.

There is clear evidence that EMCV infection activates other stress-related kinases. Moran et al. (63) have shown that EMCV infection of RAW264.7 cells induces mitogen-activated protein kinase (MAPK) and leads to inducible nitric oxide synthetase (iNOS) expression and activation of NF-κB signaling pathways. These responses did not require replicating virus, but could be induced by exposure to capsid proteins and were independent of PKR (63). Iordanov et al. (40) have also demonstrated the PKR-independent activation of p38 MAPK leading to IL-6 expression in EMCV-infected mouse fibroblasts. However, they have proposed that MAPK is activated by the presence of dsRNA, rather than viral proteins, and that this occurs independently of RNase L as well as PKR (41). Further studies are needed to determine the extent to which cell survival or EMCV replication might be regulated by the activation of NF-κB through pathways independent of PKR, RNase L, or IRF3.

ENTEROVIRUSES

Poliovirus and Protein Kinase R

Human infections with PV, the prototype species within the genus *Enterovirus*, result in a wide range of pathology, from no apparent disease to severe destruction of motor neurons leading to paralysis. Growing evidence suggests that the success of PV as a pathogen may be due to its ability to evade or otherwise subvert cellular antiviral defenses. PKR is activated by newly synthesized viral dsRNA and becomes autophosphorylated within several hours of infection (8). Although Ransone and Dasgupta (73) reported that the phosphorylation of eIF2α by activated PKR was inhibited in PV-infected cells, different results were obtained by Black et al. (8), who observed the phosphorylation of eIFα within 1.5 to 3 h of infection with PV and demonstrated that PKR isolated from PV-infected cells was also capable of phosphorylating eIF2α in vitro. However, both immunoblot and pulse-chase analyses suggested that PKR was selectively degraded beginning about 3 h after PV infection (8). Further studies indicated that this was not due to either of the PV proteases, 2A^{pro} and 3C^{pro}, and thus might result from the activation of a cellular protease (7, 8). However, a more specific mechanism has not been established. The physiologic relevance of the regulation of PKR by PV also remains uncertain, as studies of a mouse-adapted PV strain in transgenic animals expressing a *trans*-dominant negative PKR mutant revealed a modest increase in survival time compared with infections in animals expressing wild-type PKR (77). Virus replication was comparable in both types of mice, but wild-type mice showed much greater histopathologic evidence of encephalomyelitis with significantly more severe inflammatory changes. These results argue against a significant role for PKR in control of PV infections, at least within the central nervous system, and suggest that virus-induced PKR signaling may result in inflammatory changes that are deleterious to the host (77).

Group C Enteroviruses and RNase L

Han et al. (37) recently identified a complex stem-loop structure within RNA encoding the 3C^{pro} protease of PV that appears to interact with and inactivate RNase L. Its presence within the genomic RNA of PV renders the RNA strongly resistant to degradation in the presence of recombinant RNase L and 2–5A, while the addition of stoichiometric quantities of synthetic RNA containing the stem-loop structure protects other viral RNAs in *trans* from degradation. The structure is phylogenetically conserved in other group C enteroviruses, including a number of type A coxsackieviruses (CVA) (37). Although its ability to functionally inhibit RNase L has not been demonstrated formally in other viruses, its absence in CVB3, a group B enterovirus, is associated with greater sensitivity of the genome to RNase L digestion in an in vitro assay (37). A detailed mutational analysis has demonstrated that the minimal requirements for inhibition of RNase L include two contiguous RNA sequences, each capable of folding into a stem-loop, with complementary bases within the loop segments allowing a "kissing" interaction between the stem-loops that appears to be necessary for efficient inhibition of RNase L (37). Further details of the putative interaction of this RNA structure with RNase L and the precise mechanism by which it inactivates the ribonuclease are not available. However, the presence of this RNase L-inhibitory sequence within the PV genome appears to have allowed PV to evolve a high content of UA and UU dinucleotides, which are potential sites of RNase L cleavage—112 dinucleotides per kilobase of open reading frame RNA—compared with hepatitis C virus (HCV) RNA, which is significantly more sensitive to RNase cleavage yet contains only 73 UA and UU dinucleotides per kilobase (36, 37).

Additional studies by Han et al. (37) demonstrated that RNase L becomes activated relatively late in the course of PV infection, 6 to 8 h after inoculation of HeLa cells with virus, when viral RNAs are abundantly present within cells. However, perhaps consistent with the intrinsic resistance of its RNA to digestion by activated RNase L, there was no enhancement in the ability of PV to replicate or produce new virus in HeLa cells expressing a dominant negative RNase L mutant in which there was no evidence of active RNase L. Surprisingly, however, such cells generated only very small PV plaques compared with wild-type HeLa cells or HeLa cells overexpressing an active form of RNase L (37). This contrasted with VSV, in which plaque size was reduced by overexpression of active RNase L and unchanged by expression of the dominant negative mutant. Although the mechanism underlying the decreased size of PV plaques in cells lacking active RNase L is uncertain, Han et al. speculate that the activation of RNase L late in the course of PV infection, when viral RNA is largely packaged in virions and protected from the ribonuclease, may lead to a cytopathic effect, perhaps by induction of apoptosis (79), and that this may facilitate release of virus to the extracellular environment and spread to adjacent cells (37). If so, this would represent an example of a virus not only evading a cellular antiviral response, but in fact coopting it in a perverse fashion so as to enhance its own survival.

The conservation of the RNase L-inhibitory RNA structure across a number of group C enteroviruses (37) suggests that this mechanism of evading the antiviral actions of RNase L is likely to be important to the enterovirus life cycle. Yet, surprisingly, when the RNA stem-loop structure is ablated by third-base changes that do not alter the encoded amino acid sequence, there is no reduction in the apparent fitness of the mutant virus in cell culture (37). It is possible that the advantage conferred by this innate immune escape mechanism might only be demonstrable in experimentally infected animals, but this has yet to be tested. Nonetheless, it is important to consider that the inhibition of RNase L activity by the PV stem-loop structure could have consequences beyond simply blocking this single innate antiviral defense mechanism. It could also reduce the intensity of MDA-5-induced responses by eliminating the amplification of MDA-5 signaling to IRF3 mediated by RNase L cleavage of viral and cellular RNAs (56).

PV 3A-Mediated Inhibition of Cytokine Secretion

The picornaviral 3A protein is an important component of the RNA replicase complex that associates tightly with cellular membranes. In the case of PV and most other picornaviruses, the 3A association appears to be primarily with membranes derived from the ER. In addition to, and distinct from, its role in RNA synthesis, Kirkegaard and colleagues have shown that the PV 3A protein also mediates specific inhibition of anterograde ER-to-Golgi trafficking of cellular proteins (23). Studies with a mutant PV that is deficient in its ability to inhibit ER-to-Golgi transport indicate that this accessory function of 3A results in an approximate threefold reduction in secretion of IFN-β as well as other virus-induced cytokines, IL-6 and IL-8 (22). The global inhibition of ER-to-Golgi transport also contributes to reduced plasma membrane expression of the tumor necrosis factor (TNF) receptor as well as class I major histocompatibility complex molecules, thereby potentially rendering the infected cell resistant to the proapoptotic effects of TNF

and reducing its susceptibility to attack by virus-specific cytotoxic T lymphocytes (17, 65). The ability of the 3A protein to inhibit ER-to-Golgi traffic is independent of its function in viral replication, as it can be eliminated by insertion of a single additional amino acid in the 3A sequence with minimal consequences to the replication capacity of PV (22, 23). This suggests that this particular 3A function has likely evolved as a specific adaptation to host immune responses. In further support of this hypothesis, the 3A proteins of several other picornaviruses, including closely related viruses, such as enterovirus 71 and rhinovirus type 14 (HRV-14), and more distantly related viruses from other genera, such as FMDV, ThV, and HAV, lack a similar capacity to disrupt ER-to-Golgi transport (16). However, the 3A protein of CVB3 inhibits ER-to-Golgi transport as efficiently as the 3A protein of PV (16).

PV Suppression of NF-κB Activity by 3C^pro Cleavage of p65/RelA

The transcription factor NF-κB can be activated by dsRNA produced during viral infections through either TLRs, RIG-I-like helicases, or PKR-dependent pathways. Once activated, it can suppress virus infections in multiple ways, including its accessory role in IFN-β induction, stimulation of proinflammatory cytokines, and under some conditions induction of apoptosis (76). Although its role in control of picornaviral infections is not well characterized, Neznanov et al. (64) found that infection of HeLa cells with PV leads to the activation and nuclear translocation of NF-κB within 3 to 4 h. This is accompanied by degradation of IκB-α, indicating that activation occurred through the canonical pathway. However, beginning about 3 h after infection, the cytoplasmic abundance of the p65/RelA component of NF-κB is sharply reduced, while a smaller, amino-terminal cleavage product can be detected in lysates of PV-infected cells (64). Strong but circumstantial evidence suggests that this occurs as a result of 3C^pro-mediated cleavage of p65/RelA. Cleavage occurs at a consensus 3C^pro cleavage site at residue 480, near the carboxy terminus and within the transactivation domain of p65/RelA (64). Cells infected with echovirus type 1, like PV a member of the genus *Enterovirus*, and HRV-14, classified within the genus *Rhinovirus* but quite closely related to PV phylogenetically, demonstrate a pattern of p65/RelA processing similar to that observed in PV-infected cells (64). This suggests that the 3C^pro proteases expressed by these closely related viruses share an ability to target p65/RelA for cleavage. The ablation of p65/RelA expression by PV is reminiscent of the degradation of p65/RelA induced by the L^pro protease of

FMDV (19). However, in the case of L^{pro}, there is thus far no evidence for direct proteolytic cleavage of p65/RelA by the viral protease. The fact that these picornaviruses should have evolved different mechanisms to target p65/RelA for degradation supports an important role for NF-κB in the control of these infections.

PV Targeting of IFN-Inducible Nuclear Pore Proteins

As described above, L protein-mediated disruption of nucleocytoplasmic trafficking has been associated with the lack of induction of IFN synthesis in cardiovirus infections (18, 71). Nucleocytoplasmic trafficking is also altered in PV-infected cells (4), but this appears to be mediated by a different mechanism and has not been as clearly linked to evasion of innate antiviral cellular defenses. In PV-infected cells, particular components of the multiprotein nuclear pore complex, Nup62, Nup98, and Nup153, are targeted for degradation (4, 66). This results in the inhibition of nuclear transport and cytoplasmic relocalization of some nuclear proteins, some of which may directly facilitate PV replication. Park et al. (66) have shown recently that one of the nuclear pore proteins, Nup98, is rapidly and specifically cleaved by the viral 2A^{pro} protease following infection with PV. Interestingly, Nup98 expression is induced by IFN-γ, and also targeted by the M protein of VSV, both of which suggest a role for Nup98 in host defense against virus infection (25). Overexpression of Nup98 is capable of reversing an M protein-mediated block in nuclear export of mRNA (25, 26), suggesting the possibility that 2A^{pro} proteolysis of Nup98 may somehow interfere with innate antiviral defenses at the level of host-cell transcription. However, further work will be required to determine the role played by Nup98 in cellular defenses against PV. Nuclear pore proteins are also targeted for degradation in HRV-14-infected cells (34).

PV Infection and Proteolytic Cleavage of MDA-5

As described above, studies with EMCV suggest that MDA-5, and not RIG-I, is likely to be the major intracellular PRR for the picornaviruses (32, 43). Unlike RIG-I, which is constitutively expressed in many cell types and upregulated in expression by type I IFNs, MDA-5 is generally not detectable in the absence of IFN stimulation. Although PV infection induces the synthesis of IFN-β by HeLa cells, MDA-5 does not reach a reproducibly detectable level of abundance in PV-infected HeLa cells (3). This may be due, at least in part, to proteolytic cleavage of MDA-5, which occurs in association with the activation of caspases during the induction of apoptosis relatively late in the course of PV infection (3).

MDA-5 cleavage can be demonstrated approximately 4 to 6 h after PV infection of HeLa cells pretreated with poly(I:C). The degradation of MDA-5 in PV-infected HeLa cells can be blocked by treatment with either Z-VAD, a caspase inhibitor, or MG-132 and epoxomicin, inhibitors of proteasome-associated proteases (3). It thus appears to be due directly or indirectly to the coordinated action of multiple cellular proteases. Consistent with this, efforts to demonstrate direct cleavage by either of the two PV-encoded proteases, 2A^{pro} or 3C^{pro}, were unsuccessful. However, MDA-5 was not cleaved in poly(I:C)-treated HeLa cells infected with a mutant PV containing specific amino acid substitutions within 2A^{pro} or 3C^{pro} that are known to prevent *trans*-cleavage of cellular proteins by these proteases (3). This may reflect the role played by 2A^{pro}- and 3C^{pro}-mediated cleavage of cellular proteins in the induction of apoptosis by PV (5), and is consistent with MDA-5 cleavage by apoptosis-associated proteases late in the course of PV infection. However, RNase L activation, which occurs within 6 h of infection of HeLa cells with PV (37), may also lead to apoptosis and thus may play a role in inducing the degradation of MDA-5. The PV-induced cleavage of MDA-5 is of uncertain importance to the virus life cycle, as its prevention by chemical inhibition of the proteasome does not increase the yield of PV from infected HeLa cells (3). Nonetheless, MDA-5 cleavage appears to be associated with only a limited number of picornaviruses (PV, HRV-1a, and EMCV) and was not observed in HeLa cells infected with HRV-16 or echovirus type 1 (3). This suggests that it may be a specific adaptive response of the virus.

RHINOVIRUSES

HRV Interactions with Innate Host Defenses

The rhinoviruses comprise a large group of serologically distinct viruses that cause limited upper respiratory tract infections, for the most part in humans. Although typically mild, these infections can cause acute exacerbations of asthma or chronic obstructive pulmonary disease in susceptible individuals. Like PV, human rhinoviruses (HRVs) have a limited host species range. However, the genus also includes bovine rhinoviruses that specifically infect cattle (82). As a genus, rhinoviruses are closely related to the enteroviruses phylogenetically, with which they share a common polyprotein processing scheme (Fig. 2). However, they differ in terms of their increased sensitivity to low pH, optimal growth at lower temperatures in cell culture, and tissue tropism in vivo.

Relatively little information exists concerning the interactions of HRVs with innate host defenses. A recently published study claiming the inhibition of IRF3 activation by HRV was unfortunately retracted. What is known is that infection of well-differentiated primary human airway epithelial cells results in the induction of IFN-β secretion and synthesis of a large number of ISGs, at least in part through paracrine signaling involving the Janus kinases (Jaks) and signal transducers and activators of transcription (Stats) (13). Interestingly, IFN-β induction could be blocked by prior treatment of the cells with 2-aminopurine, an inhibitor of PKR. However, this led to only minimal increases in viral yield (13). Other evidence supports the secretion of cytokines including TNF-α and IP-10 by HRV-infected monocytes and macrophages; TNF-α secretion was dependent upon activation of NF-κB (44, 45). The production of such cytokines during HRV infections in vivo could play an important role in the exacerbation of obstructive airway disease.

A limited number of studies suggest that HRV possesses some, but not all, of the mechanisms evolved by PV to counter innate host defenses. Thus, while HRV-14 infection appears to deregulate nucleocytoplasmic trafficking in a fashion similar to PV (34), and the HRV-14 3C^pro protease mediates cleavage of p65/RelA like the PV protease (64), the HRV-14 3A protein is not capable of disrupting ER-to-Golgi traffic as efficiently as that of PV (16). There are also differences between the various rhinoviruses, since HRV-1a infection appears to lead to MDA-5 cleavage while HRV-16 infection does not (3). While it is not surprising to find shared mechanisms of immune evasion among HRVs and PV given their close phylogenetic relationship, it is likely that there are significant distinctions and it is clear that there is much work to be done to specifically define the ways in which HRVs might counter innate antiviral defenses.

HEPATOVIRUSES

HAV 3ABC-Mediated Proteolysis of MAVS/IPS-1

HAV, the prototype species of the genus *Hepatovirus*, infects only humans and several higher primate species. It possesses a strong tropism for the hepatocyte and is a common cause of acute viral hepatitis. In contrast to HCV, a flavivirus that typically causes chronic viral hepatitis, HAV is incapable of establishing persistent infections and is always eliminated by host defenses. Thus, a comparison of the ways in which HAV and HCV, both positive-strand RNA viruses, interact with innate host defenses may be instructive in learning how innate immune responses might regulate virus persistence in the human liver.

HCV expresses a serine protease, NS3/4A, that disrupts the virus activation of IRF3 by proteolytically targeting mitochondrial antiviral signaling protein (MAVS; also known as IPS-1, Cardif, or VISA), preventing the activation of IRF3 following the sensing of HCV RNA by the cytoplasmic RIG-I receptor (29, 62). NS3/4A cleaves immediately upstream of the transmembrane domain of MAVS/IPS-1, disrupting its association with the outer mitochondrial membrane and ablating signaling to the downstream kinases responsible for activation of IRF3 (50, 62). This block at a proximal point in the signaling pathway also inhibits activation of NK-κB. Remarkably, NS3/4A also targets the TLR3 adaptor protein, TRIF (Toll-IL-1 receptor domain-containing adaptor including IFN-β), for proteolysis, disrupting dsRNA activation of IRF3 and NF-κB through this pathway as well (49). While the role of TLR3 signaling remains uncertain, the RIG-I/MAVS/IRF3 pathway appears to be the major signal transduction pathway for activation of IFN-α/β synthesis by HCV RNA, at least in cultured hepatoma cells (55, 84). The disruption of RIG-I signaling by the NS3/4A cleavage of MAVS/IPS-1 has thus been proposed to contribute to the unique capacity of HCV to establish chronic intrahepatic infections (30).

Although HAV is incapable of causing persistent infection in vivo, liver-derived cells infected with HAV or containing a self-replicating HAV RNA replicon demonstrate a block in activation of IRF3 and IFN-β expression through the RIG-I pathway when challenged with viruses such as Sendai virus (9, 27, 96). Work in our laboratory has shown that this is due to the ability of HAV to strongly downregulate the expression of MAVS, which also ablates signaling through the MDA-5 pathway, which is more likely to sense HAV infections (96). In a remarkable parallel to HCV, the 3ABC precursor of the only protease expressed by HAV, the 3C^pro cysteine protease (Fig. 2), targets MAVS/IPS-1 for proteolysis (Color Plate 4). The cleavage of MAVS/IPS-1 occurs 80 residues upstream of the HCV NS3/4A cleavage site and also results in the release of MAVS/IPS-1 from the mitochondrial membrane, thereby ablating its ability to function in signal transduction. 3ABC cleavage is dependent upon a transmembrane domain within 3A that directs the HAV protease specifically to mitochondria, a surprising feature of the 3ABC protease that is unique to HAV among the picornaviruses (96).

These results are of interest from the perspective of the picornaviruses, but they also are helpful in understanding HCV pathogenesis. They show that virus-mediated proteolysis of MAVS/IPS-1 is not, by itself, sufficient for the establishment of persistent intrahepatic infection with a positive-strand RNA virus. HCV infection also targets the TLR3 and PKR pathways (31, 49), and these evasive mechanisms may be equally important for the long-term persistence that typifies HCV infection. Whether HAV might also target these redundant dsRNA-activated antiviral pathways is not known. However, 3ABC-mediated proteolysis of MAVS/ IPS-1 seems likely to contribute to the lengthy (3 to 5 weeks) and clinically silent incubation period of hepatitis A, during which there is extensive virus replication within the liver and an increasing viremia (47, 78). 3ABC-mediated destruction of MAVS/IPS-1 may also facilitate the ability of HAV to establish persistent, noncytopathic infections in cultured cells, a feature that typifies HAV despite its sensitivity to IFN (72, 91).

SUMMARY AND FUTURE DIRECTIONS

The studies described above reveal that the picornaviruses have evolved multiple strategies for evading innate antiviral host defenses, but they are in no way complete for even one member of this large virus family. It is likely that this field of investigation will remain active for a long time, and additional picornaviral strategies for evasion of innate host defenses are certain to be uncovered in the future. It is also likely that strategies of evasion discovered for one member of the family will be found to have been adopted by viruses in other picornaviral genera, albeit perhaps executed by different viral proteins. This is already evident, for example, with disruption of nucleocytoplasmic trafficking of cellular proteins, which is caused by the L protein of the cardioviruses but 3C^{pro} of some enteroviruses and rhinoviruses (18, 34, 66, 71). On the other hand, there will be specialized mechanisms of evasion found in only certain groups of viruses, as, for example, the RNase L-confounding RNA stem-loop that appears to be restricted to the group C enteroviruses (37). This diversity of evasive maneuvers is likely to reflect, at least in part, differences in the tissue tropisms of various picornaviruses, as well as the redundant nature of innate antiviral defenses.

This latter feature of the host response will continue to complicate the assessment of the relevance of any single apparent mechanism of evasion. For example, the lack of a demonstrable effect of PKR on the yields of PV in infected mouse brains may raise questions about the physiologic relevance of PV-induced degradation of PKR noted in vitro (8, 77). Similarly, the fact that ablating the stem-loop that confers RNase L resistance to group C enteroviral RNAs has little effect on yield of virus in cell culture might raise questions about the importance of this particular evasive mechanism (37). However, both sets of observations may well reflect

only the redundancy of dsRNA-activated antiviral defenses that are thrown up by the cell.

Two specific areas of rapidly advancing research suggest the possibility of novel mechanisms of viral evasion and/or subversion that may be exploited by picornaviruses and are worthy of particular note. First, recent studies suggest that the cytoplasmic double-membrane vesicles on which picornaviruses such as PV replicate their RNA contain autophagy-related proteins, including microtubule-associated protein light chain 3 (LC3) and lysosome-associated membrane protein 1 (LAMP-1), and that the recruitment of LC3 and the formation of the membrane-bound viral replicase complex by viral proteins may recapitulate early steps in autophagy (87). Autophagy, the process by which a cell forms vesicles that engulf parts of its own cytoplasm and targets their contents for lysosomal degradation, plays critical roles in development, maintenance of cellular homeostasis, and the response to infection. Autophagy is also important to both innate and adaptive immune responses and has recently been shown to contribute to the sensing of viral RNAs by TLRs within plasmacytoid dendritic cells (46, 48). While much more needs to be learned about both of these processes, it is reasonable to speculate that the subversion of early steps in autophagy to support the replication of some picornaviruses might also impact the ability of cells to sense virus infection through TLRs and to activate innate antiviral responses. A second area of expanding knowledge relates to the inhibition of cytoplasmic stress granule formation by picornaviruses (94). These cytoplasmic bodies serve as repositories for translationally inactive cellular mRNAs and mRNA-protein complexes, and they contribute to the regulation of mRNA translation when cells encounter a variety of different types of stress. Stress granules assemble early in the course of PV infection (61, 94), but their formation is blocked in later stages by 3C^{pro}-mediated cleavage of G3PB, a cellular protein that has a pivotal role in stress granule formation (94). Expression of a cleavage-resistant G3PB mutant preserved the ability of the cell to assemble stress granules late in the infection cycle, and also reduced the yields of PV. This finding raises the possibility that stress granule formation may in some way contribute to antiviral defense. Although roles for stress granule formation and autophagy in innate cellular defenses against picornaviruses remain speculative, these are likely to be productive lines of future investigation.

This work was supported in part by a grant from the National Institute of Allergy and Infectious Diseases, U19-AI40035. The author thanks Kui Li and Peter Mason for helpful comments on the manuscript.

References

1. Agol, V. I., A. V. Paul, and E. Wimmer. 1999. Paradoxes of the replication of picornaviral genomes. *Virus Res.* 62:129–147.

2. Al-Salleeh, F., and T. M. Petro. 2007. TLR3 and TLR7 are involved in expression of IL-23 subunits while TLR3 but not TLR7 is involved in expression of IFN-beta by Theiler's virus-infected RAW264.7 cells. *Microbes Infect.* 9:1384–1392.

3. Barral, P. M., J. M. Morrison, J. Drahos, P. Gupta, D. Sarkar, P. B. Fisher, and V. R. Racaniello. 2007. MDA-5 is cleaved in poliovirus-infected cells. *J. Virol.* 81:3677–3684.

4. Belov, G. A., P. V. Lidsky, O. V. Mikitas, D. Egger, K. A. Lukyanov, K. Bienz, and V. I. Agol. 2004. Bidirectional increase in permeability of nuclear envelope upon poliovirus infection and accompanying alterations of nuclear pores. *J. Virol.* 78:10166–10177.

5. Belov, G. A., L. I. Romanova, E. A. Tolskaya, M. S. Kolesnikova, Y. A. Lazebnik, and V. I. Agol. 2003. The major apoptotic pathway activated and suppressed by poliovirus. *J. Virol.* 77:45–56.

6. Bienz, K., D. Egger, T. Pfister, and M. Troxler. 1992. Structural and functional characterization of the poliovirus replication complex. *J. Virol.* 66:2740–2747.

7. Black, T. L., G. N. Barber, and M. G. Katze. 1993. Degradation of the interferon-induced 68,000-M_r protein kinase by poliovirus requires RNA. *J. Virol.* 67:791–800.

8. Black, T. L., B. Safer, A. Hovanessian, and M. G. Katze. 1989. The cellular 68,000-M_r protein kinase is highly autophosphorylated and activated yet significantly degraded during poliovirus infection: implications for translational regulation. *J. Virol.* 63:2244–2251.

9. Brack, K., I. Berk, T. Magulski, J. Lederer, A. Dotzauer, and A. Vallbracht. 2002. Hepatitis A virus inhibits cellular antiviral defense mechanisms induced by double-stranded RNA. *J. Virol.* 76:11920–11930.

10. Brown, C. C., M. E. Piccone, P. W. Mason, T. S. C. McKenna, and M. J. Grubman. 1996. Pathogenesis of wild-type and leaderless foot-and-mouth disease virus in cattle. *J. Virol.* 70:5638–5641.

11. Buenz, E. J., and C. L. Howe. 2006. Picornaviruses and cell death. *Trends Microbiol.* 14:28–36.

12. Cayley, P. J., M. Knight, and I. M. Kerr. 1982. Virus-mediated inhibition of the ppp(A2′p)nA system and its prevention by interferon. *Biochem. Biophys. Res. Commun.* 104:376–382.

13. Chen, Y., E. Hamati, P. K. Lee, W. M. Lee, S. Wachi, D. Schnurr, S. Yagi, G. Dolganov, H. Boushey, P. Avila, and R. Wu. 2006. Rhinovirus induces airway epithelial gene expression through double-stranded RNA and IFN-dependent pathways. *Am. J. Respir. Cell Mol. Biol.* 34:192–203.

14. Chinsangaram, J., M. Koster, and M. J. Grubman. 2001. Inhibition of L-deleted foot-and-mouth disease virus replication by alpha/beta interferon involves double-stranded RNA-dependent protein kinase. *J. Virol.* 75:5498–5503.

15. Cho, M. W., N. Teterina, D. Egger, K. Bienz, and E. Ehrenfeld. 1994. Membrane rearrangement and vesicle induction by recombinant poliovirus 2C and 2BC in human cells. *Virology* 202:129–145.

16. Choe, S. S., D. A. Dodd, and K. Kirkegaard. 2005. Inhibition of cellular protein secretion by picornaviral 3A proteins. *Virology* 337:18–29.

17. Deitz, S. B., D. A. Dodd, S. Cooper, P. Parham, and K. Kirkegaard. 2000. MHC I-dependent antigen presentation is inhibited by poliovirus 3A. *Proc. Natl. Acad. Sci. USA* 97:13790–13795.

18. Delhaye, S., V. van Pesch, and T. Michiels. 2004. The leader protein of Theiler's virus interferes with nucleocytoplasmic trafficking of cellular proteins. *J. Virol.* 78:4357–4362.

19. de Los Santos, T., F. Diaz-San Segundo, and M. J. Grubman. 2007. Degradation of nuclear factor kappa B during foot-and-mouth disease virus infection. *J. Virol.* 81:12803–12815.

20. de Los Santos, T., S. de Avila Botton, R. Weiblen, and M. J. Grubman. 2006. The leader proteinase of foot-and-mouth disease virus inhibits the induction of beta interferon mRNA and blocks the host innate immune response. *J. Virol.* 80:1906–1914.

21. Devaney, M. A., V. N. Vakharia, R. E. Lloyd, E. Ehrenfeld, and M. J. Grubman. 1988. Leader protein of foot-and-mouth disease virus is required for cleavage of the p200 component of the cap-binding protein complex. *J. Virol.* 62:4407–4409.

22. Dodd, D. A., T. H. Giddings, Jr., and K. Kirkegaard. 2001. Poliovirus 3A protein limits interleukin-6 (IL-6), IL-8, and beta interferon secretion during viral infection. *J. Virol.* 75:8158–8165.

23. Doerdens, J. R., T. H. Giddings, Jr., and K. Kirkegaard. 1997. Inhibition of ER-to-Golgi traffic by poliovirus 3A: genetic and ultrastructural analysis. *J. Virol.* 71:9054–9064.

24. Duke, G. M., J. E. Osorio, and A. C. Palmenberg. 1990. Attenuation of Mengo virus through genetic engineering of the 5′ noncoding poly(C) tract. *Nature* 343:474–476.

25. Enninga, J., D. E. Levy, G. Blobel, and B. M. Fontoura. 2002. Role of nucleoporin induction in releasing an mRNA nuclear export block. *Science* 295:1523–1525.

26. Faria, P. A., P. Chakraborty, A. Levay, G. N. Barber, H. J. Ezelle, J. Enninga, C. Arana, J. van Deursen, and B. M. Fontoura. 2005. VSV disrupts the Rae1/mrnp41 mRNA nuclear export pathway. *Mol. Cell* 17:93–102.

27. Fensterl, V., D. Grotheer, I. Berk, S. Schlemminger, A. Vallbracht, and A. Dotzauer. 2005. Hepatitis A virus suppresses RIG-I-mediated IRF3 activation to block induction of beta interferon. *J. Virol.* 79:10968–10977.

28. Fitzgerald, K. A., S. M. McWhirter, K. L. Faia, D. C. Rowe, E. Latz, D. T. Golenbock, A. J. Coyle, S. M. Liao, and T. Maniatis. 2003. IKKepsilon and TBK1 are essential components of the IRF3 signaling pathway. *Nat. Immunol.* 4:491–496.

29. Foy, E., K. Li, C. Wang, R. Sumter, M. Ikeda, S. M. Lemon, and M. Gale, Jr. 2003. Regulation of interferon regulatory factor-3 by the hepatitis C virus serine protease. *Science* 300:1145–1148.

30. Gale, M., Jr., and E. M. Foy. 2005. Evasion of intracellular host defence by hepatitis C virus. *Nature* 436:939–945.

31. Gale, M., Jr., and M. G. Katze. 1998. Molecular mechanisms of interferon resistance mediated by viral-directed inhibition of PKR, the interferon-induced protein kinase. *Pharmacol. Ther.* 78:29–46.

32. Gitlin, L., W. Barchet, S. Gilfillan, M. Cella, B. Beutler, R. A. Flavell, M. S. Diamond, and M. Colonna. 2006. Essential role of mda-5 in type I IFN responses to polyriboinosinic:polyribocytidylic acid and encephalomyocarditis picornavirus. *Proc. Natl. Acad. Sci. USA* 103:8459–8464.

33. Grandvaux, N., M. J. Servant, B. tenOever, G. C. Sen, S. Balachandran, G. N. Barber, R. Lin, and J. Hiscott. 2002. Transcriptional profiling of interferon regulatory factor 3 target genes: direct involvement in the regulation of interferon-stimulated genes. *J. Virol.* 76:5532–5539.

34. Gustin, K. E., and P. Sarnow. 2002. Inhibition of nuclear import and alteration of nuclear pore complex composition by rhinovirus. *J. Virol.* 76:8787–8796.

35. Hahn, H., and A. C. Palmenberg. 1995. Encephalomyocarditis viruses with short poly(C) tracts are more virulent than their mengovirus counterparts. *J. Virol.* 69:2697–2699.

36. Han, J. Q., and D. J. Barton. 2002. Activation and evasion of the antiviral 2′-5′ oligoadenylate synthetase/ribonuclease L pathway by hepatitis C virus mRNA. *RNA* 8:512–525.

37. Han, J. Q., H. L. Townsend, B. K. Jha, J. M. Paranjape, R. H. Silverman, and D. J. Barton. 2007. A phylogenetically conserved RNA structure in the poliovirus open reading frame inhibits the antiviral endoribonuclease RNase L. *J. Virol.* 81:5561–5572.

38. Hiscott, J. 2007. Triggering the innate antiviral response through IRF3 activation. *J. Biol. Chem.* 282:15325–15329.

39. Hornung, V., J. Ellegast, S. Kim, K. Brzózka, A. Jung, H. Kato, H. Poeck, S. Akira, K. K. Conzelmann, M. Schlee, S. Endres, and G. Hartmann. 2006. 5′-Triphosphate RNA is the ligand for RIG-I. *Science* 314:994–997.

40. Iordanov, M. S., J. M. Paranjape, A. Zhou, J. Wong, B. R. Williams, E. F. Meurs, R. H. Silverman, and B. E. Magun. 2000. Activation of p38 mitogen-activated protein kinase and c-Jun NH$_2$-terminal kinase by double-stranded RNA and encephalomyocarditis virus: involvement of RNase L, protein kinase R, and alternative pathways. *Mol. Cell. Biol.* 20:617–627.

41. Iordanov, M. S., J. Wong, J. C. Bell, and B. E. Magun. 2001. Activation of NF-kappaB by double-stranded RNA (dsRNA) in the absence of protein kinase R and RNase L demonstrates the existence of two separate dsRNA-triggered antiviral programs. *Mol. Cell. Biol.* 21:61–72.

42. Kabelitz, D., and R. Medzhitov. 2007. Innate immunity—cross-talk with adaptive immunity through pattern recognition receptors and cytokines. *Curr. Opin. Immunol.* 19:1–3.

43. Kato, H., O. Takeuchi, S. Sato, M. Yoneyama, M. Yamamoto, K. Matsui, S. Uematsu, A. Jung, T. Kawai, K. J. Ishii, O. Yamaguchi, K. Otsu, T. Tsujimura, C. S. Koh, C. Reis e Sousa, Y. Matsuura, T. Fujita, and S. Akira. 2006. Differential roles of MDA-5 and RIG-I helicases in the recognition of RNA viruses. *Nature* 441:101–105.

44. Korpi-Steiner, N. L., M. E. Bates, W. M. Lee, D. J. Hall, and P. J. Bertics. 2006. Human rhinovirus induces robust IP-10 release by monocytic cells, which is independent of viral replication but linked to type I interferon receptor ligation and Stat1 activation. *J. Leukoc. Biol.* 80:1364–1374.

45. Laza-Stanca, V., L. A. Stanciu, S. D. Message, M. R. Edwards, J. E. Gern, and S. L. Johnston. 2006. Rhinovirus

replication in human macrophages induces NF-kappaB-dependent tumor necrosis factor alpha production. *J. Virol.* 80:8248–8258.

46. Lee, H. K., J. M. Lund, B. Ramanathan, N. Mizushima, and A. Iwasaki. 2007. Autophagy-dependent viral recognition by plasmacytoid dendritic cells. *Science* 315:1398–1401.

47. Lemon, S. M., L. N. Binn, R. Marchwicki, P. C. Murphy, L.-H. Ping, R. W. Jansen, L. V. S. Asher, J. T. Stapleton, D. G. Taylor, and J. W. LeDuc. 1990. In vivo replication and reversion to wild-type of a neutralization-resistant variant of hepatitis A virus. *J. Infect. Dis.* 161:7–13.

48. Levine, B., and V. Deretic. 2007. Unveiling the roles of autophagy in innate and adaptive immunity. *Nat. Rev. Immunol.* 7:767–777.

49. Li, K., E. Foy, J. C. Ferreon, M. Nakamura, A. C. M. Ferreon, M. Ikeda, S. C. Ray, M. Gale, Jr., and S. M. Lemon. 2005. Immune evasion by hepatitis C virus NS3/4A protease-mediated cleavage of the TLR3 adaptor protein TRIF. *Proc. Natl. Acad. Sci. USA* 102:2992–2997.

50. Li, X. D., L. Sun, R. B. Seth, G. Pineda, and Z. J. Chen. 2005. Hepatitis C virus protease NS3/4A cleaves mitochondrial antiviral signaling protein off the mitochondria to evade innate immunity. *Proc. Natl. Acad. Sci. USA* 102:17717–17722.

51. Li, X. L., J. A. Blackford, and B. A. Hassel. 1998. RNase L mediates the antiviral effect of interferon through a selective reduction in viral RNA during encephalomyocarditis virus infection. *J. Virol.* 72:2752–2759.

52. Lidsky, P. V., S. Hato, M. V. Bardina, A. G. Aminev, A. C. Palmenberg, E. V. Sheval, V. Y. Polyakov, F. J. van Kuppeveld, and V. I. Agol. 2006. Nucleocytoplasmic traffic disorder induced by cardioviruses. *J. Virol.* 80:2705–2717.

53. Lin, R., C. Heylbroeck, P. M. Pitha, and J. Hiscott. 1998. Virus-dependent phosphorylation of the IRF3 transcription factor regulates nuclear translocation, transactivation potential, and proteasome-mediated degradation. *Mol. Cell. Biol.* 18:2986–2996.

54. Lloyd, R. E. 2006. Translational control by viral proteinases. *Virus Res.* 119:76–88.

55. Loo, Y. M., D. M. Owen, K. Li, A. L. Erickson, C. L. Johnson, P. Fish, D. S. Carney, T. Wang, H. Ishida, M. Yoneyama, T. Fujita, T. Saito, W. M. Lee, C. H. Hagedorn, D. T. Lau, S. A. Weinman, S. M. Lemon, and M. Gale, Jr. 2006. Viral and therpeutic control of interferon beta promoter stimulator 1 during hepatitis C virus infection. *Proc. Natl. Acad. Sci. USA* 103:6001–6006.

56. Malathi, K., B. Dong, M. Gale, Jr., and R. H. Silverman. 2007. Small self-RNA generated by RNase L amplifies antiviral innate immunity. *Nature* 448:816–819.

57. Malirat, V., P. A. De Mello, B. Tiraboschi, E. Beck, I. Gomes, and I. E. Bergmann. 1994. Genetic variation of foot-and-mouth disease virus during persistent infection in cattle. *Virus Res.* 34:31–48.

58. Martin, A., and S. M. Lemon. 2002. The molecular biology of hepatitis A virus, p. 23–50. *In* J.-H. Ou (ed.), *Hepatitis Viruses.* Kluwer Academic Publishers, Norwell, MA.

59. Martinand, C., T. Salehzada, M. Silhol, B. Lebleu, and C. Bisbal. 1998. RNase L inhibitor (RLI) antisense constructions block partially the down regulation of the

2–5A/RNase L pathway in encephalomyocarditis-virus-(EMCV)-infected cells. *Eur. J. Biochem.* **254:**248–255.

60. Martinand, C., T. Salehzada, M. Silhol, B. Lebleu, and C. Bisbal. 1998. The RNase L inhibitor (RLI) is induced by double-stranded RNA. *J. Interferon Cytokine Res.* **18:**1031–1038.

61. Mazroui, R., R. Sukarieh, M. E. Bordeleau, R. J. Kaufman, P. Northcote, J. Tanaka, I. Gallouzi, and J. Pelletier. 2006. Inhibition of ribosome recruitment induces stress granule formation independently of eukaryotic initiation factor 2alpha phosphorylation. *Mol. Biol. Cell* **17:**4212–4219.

62. Meylan, E., J. Curran, K. Hofmann, D. Moradpour, M. Binder, R. Bartenschlager, and J. Tschopp. 2005. Cardif is an adaptor protein in the RIG-I antiviral pathway and is targeted by hepatitis C virus. *Nature* **437:**1167–1172.

63. Moran, J. M., M. A. Moxley, R. M. Buller, and J. A. Corbett. 2005. Encephalomyocarditis virus induces PKR-independent mitogen-activated protein kinase activation in macrophages. *J. Virol.* **79:**10226–10236.

64. Neznanov, N., K. M. Chumakov, L. Neznanova, A. Almasan, A. K. Banerjee, and A. V. Gudkov. 2005. Proteolytic cleavage of the p65-RelA subunit of NF-kappaB during poliovirus infection. *J. Biol. Chem.* **280:**24153–24158.

65. Neznanov, N., A. Kondratova, K. M. Chumakov, B. Angres, B. Zhumabayeva, V. I. Agol, and A. V. Gudkov. 2001. Poliovirus protein 3A inhibits tumor necrosis factor (TNF)-induced apoptosis by eliminating the TNF receptor from the cell surface. *J. Virol.* **75:**10409–10420.

66. Park, N., P. Katikaneni, T. Skern, and K. E. Gustin. 2008. Differential targeting of nuclear pore complex proteins in poliovirus-infected cells. *J. Virol.* **82:**1647–1655.

67. Paul, A. V., J. H. Van Boom, D. Filippov, and E. Wimmer. 1998. Protein-primed RNA synthesis by purified poliovirus RNA polymerase. *Nature* **393:**280–284.

68. Paul, S., and T. Michiels. 2006. Cardiovirus leader proteins are functionally interchangeable and have evolved to adapt to virus replication fitness. *J. Gen. Virol.* **87:**1237–1246.

69. Pelletier, J., and N. Sonenberg. 1988. Internal initiation of translation of eukaryotic mRNA directed by a sequence derived from poliovirus RNA. *Nature* **334:**320–325.

70. Pichlmair, A., O. Schulz, C. P. Tan, T. I. Naslund, P. Liljestrom, F. Weber, M. Reiser, and C. Reis e Sousa. 2006. RIG-I-mediated antiviral responses to single-stranded RNA bearing 5′ phosphates. *Science* **314:**999–1001.

71. Porter, F. W., Y. A. Bochkov, A. J. Albee, C. Wiese, and A. C. Palmenberg. 2006. A picornavirus protein interacts with Ran-GTPase and disrupts nucleocytoplasmic transport. *Proc. Natl. Acad. Sci. USA* **103:**12417–12422.

72. Provost, P. J., P. A. Giesa, W. J. McAleer, and M. R. Hilleman. 1981. Isolation of hepatitis A virus *in vitro* in cell culture directly from human specimens. *Proc. Soc. Exper. Biol. Med.* **167:**201–206.

73. Ransone, L. J., and A. Dasgupta. 1988. A heat-sensitive inhibitor in poliovirus-infected cells which selectively blocks phosphorylation of the alpha subunit of eucaryotic initiation factor 2 by the double-stranded RNA-activated protein kinase. *J. Virol.* **62:**3551–3557.

74. Rieder, E., T. Bunch, F. Brown, and P. W. Mason. 1993. Genetically engineered foot-and-mouth disease viruses

75. Romanova, L. I., G. A. Belov, P. V. Lidsky, E. A. Tolskaya, M. S. Kolesnikova, A. G. Evstafieva, A. B. Vartapetian, D. Egger, K. Bienz, and V. I. Agol. 2005. Variability in apoptotic response to poliovirus infection. *Virology* **331:**292–306.

76. Santoro, M. G., A. Rossi, and C. Amici. 2003. NF-kappaB and virus infection: who controls whom. *EMBO J.* **22:**2552–2560.

77. Scheuner, D., M. Gromeier, M. V. Davies, A. J. Dorner, B. Song, R. V. Patel, E. J. Wimmer, R. E. McLendon, and R. J. Kaufman. 2003. The double-stranded RNA-activated protein kinase mediates viral-induced encephalitis. *Virology* **317:**263–274.

78. Schulman, A. N., J. L. Dienstag, D. R. Jackson, J. H. Hoofnagle, R. J. Gerety, R. H. Purcell, and L. F. Barker. 1976. Hepatitis A antigen particles in liver, bile, and stool of chimpanzees. *J. Infect. Dis.* **134:**80–84.

79. Silverman, R. H. 2007. Viral encounters with 2′,5′-oligoadenylate synthetase and RNase L during the interferon antiviral response. *J. Virol.* **81:**12720–12729.

80. Silverman, R. H., P. J. Cayley, M. Knight, C. S. Gilbert, and I. M. Kerr. 1982. Control of the ppp(a2′p)nA system in HeLa cells. Effects of interferon and virus infection. *Eur. J. Biochem.* **124:**131–138.

81. So, E. Y., M. H. Kang, and B. S. Kim. 2006. Induction of chemokine and cytokine genes in astrocytes following infection with Theiler's murine encephalomyelitis virus is mediated by the Toll-like receptor 3. *Glia* **53:**858–867.

82. Stanway, G., F. Brown, P. Christian, T. Hovi, T. Hyypiä, A. M. Q. King, N. J. Knowles, S. M. Lemon, P. D. Minor, M. A. Pallansch, A. Palmenberg, and T. Skern. 2005. Family *Picornaviridae*, p. 757–778. *In* C. M. Fauquet, M. A. Mayo, J. Maniloff, U. Desselberger, and L. A. Ball (ed.), *Virus Taxonomy. Eighth Report of the International Committee on Taxonomy of Viruses.* Elsevier/Academic Press, London, U.K.

83. Strong, R., and G. J. Belsham. 2004. Sequential modification of translation initiation factor eIF4GI by two different foot-and-mouth disease virus proteases within infected baby hamster kidney cells: identification of the 3C^pro cleavage site. *J. Gen. Virol.* **85:**2953–2962.

84. Sumpter, R., Jr., M. Y. Loo, E. Foy, K. Li, M. Yoneyama, T. Fujita, S. M. Lemon, and M. J. Gale, Jr. 2005. Regulating intracellular antiviral defense and permissiveness to hepatitis C virus RNA replication through a cellular RNA helicase, RIG-I. *J. Virol.* **79:**2689–2699.

85. Takeuchi, O., and S. Akira. 2007. Recognition of viruses by innate immunity. *Immunol. Rev.* **220:**214–224.

86. Takeuchi, O., and S. Akira. 2007. Signaling pathways activated by microorganisms. *Curr. Opin. Cell Biol.* **19:**185–191.

87. Taylor, M. P., and K. Kirkegaard. 2007. Modification of cellular autophagy protein LC3 by poliovirus. *J. Virol.* **81:**12543–12553.

88. Todd, S., and B. L. Semler. 1996. Structure-infectivity analysis of the human rhinovirus genomic RNA 3′ noncoding region. *Nucleic Acids Res.* **24:**2133–2142.

89. Triantafilou, K., G. Orthopoulos, E. Vakakis, M. A. Ahmed, D. T. Golenbock, P. M. Lepper, and M. Triantafilou. 2005. Human cardiac inflammatory responses

triggered by Coxsackie B viruses are mainly Toll-like receptor (TLR) 8-dependent. *Cell. Microbiol.* **7:**1117–1126.

90. **Triantafilou, K., E. Vakakis, G. Orthopoulos, M. A. Ahmed, C. Schumann, P. M. Lepper, and M. Triantafilou.** 2005. TLR8 and TLR7 are involved in the host's immune response to human parechovirus 1. *Eur. J. Immunol.* **35:** 2416–2423.

91. **Vallbracht, A., L. Hofmann, K. G. Wurster, and B. Flehmig.** 1984. Persistent infection of human fibroblasts by hepatitis A virus. *J. Gen. Virol.* **65:**609–615.

92. **van Pesch, V., O. van Eyll, and T. Michiels.** 2001. The leader protein of Theiler's virus inhibits immediate-early alpha/beta interferon production. *J. Virol.* **75:**7811–7817.

93. **Wathelet, M. G., C. H. Lin, B. S. Parekh, L. V. Ronco, P. M. Howley, and T. Maniatis.** 1998. Virus infection induces the assembly of coordinately activated transcription factors on the IFN-beta enhancer in vivo. *Mol. Cell* **1:**507–518.

94. **White, J. P., A. M. Cardenas, W. E. Marissen, and R. E. Lloyd.** 2007. Inhibition of cytoplasmic mRNA stress granule formation by a viral proteinase. *Cell Host Microbe* **2:**295–305.

95. **Wimmer, E., C. U. T. Hellen, and X. Cao.** 1993. Genetics of poliovirus. *Annu. Rev. Genet.* **27:**353–436.

96. **Yang, Y., Y. Liang, L. Qu, Z. Chen, M. Yi, K. Li, and S. M. Lemon.** 2007. Disruption of innate immunity due to mitochondrial targeting of a picornaviral protease precursor. *Proc. Natl. Acad. Sci. USA* **104:**7253–7258.

97. **Yang, Y., R. Rijnbrand, K. L. McKnight, E. Wimmer, A. Paul, A. Martin, and S. M. Lemon.** 2002. Sequence requirements for viral RNA replication and VPg uridylylation directed by the internal *cis*-acting replication element (*cre*) of human rhinovirus type 14. *J. Virol.* **76:** 7485–7494.

98. **Yang, Y. L., L. F. Reis, J. Pavlovic, A. Aguzzi, R. Schafer, A. Kumar, B. R. Williams, M. Aguet, and C. Weissmann.** 1995. Deficient signaling in mice devoid of double-stranded RNA-dependent protein kinase. *EMBO J.* **14:** 6095–6106.

99. **Yeung, M. C., D. L. Chang, R. E. Camantigue, and A. S. Lau.** 1999. Inhibitory role of the host apoptogenic gene PKR in the establishment of persistent infection by encephalomyocarditis virus in U937 cells. *Proc. Natl. Acad. Sci. USA* **96:**11860–11865.

Cellular Signaling and Innate Immune
Responses to RNA Virus Infections
Edited by A. R. Brasier et al.
© 2009 ASM Press, Washington, D.C.

William B. Klimstra
Kate D. Ryman

22

Togaviruses

INTRODUCTION TO THE *TOGAVIRIDAE* FAMILY

The family *Togaviridae* contains two genera, the *Alphavirus* genus and the *Rubivirus* genus (Table 1). Alphaviruses include more than 25 types and are maintained in nature by alternating growth in arthropod (typically mosquito) and vertebrate hosts, in a classical arbovirus transmission cycle (reviewed in reference 42). In contrast, the *Rubivirus* genus contains only one member, rubella virus, which is not an arbovirus and for which humans are the only vertebrate host (reviewed in reference 16). For the most part, this chapter describes the interaction between the alphaviruses and the alpha/beta interferon (IFN-α/β) system, as relatively little research has been performed with rubella virus in this area, thwarted by the lack of a small-animal disease model. However, similarities in replication strategies employed by the viruses in these genera suggest that much of our knowledge of alphavirus/IFN-α/β interactions may also pertain to rubella virus.

Alphaviruses of the Old World, including Sindbis (SINV), Semliki Forest (SFV), Chikungunya (CHIKV), O'nyong-nyong, Ockelbo, Ross River (RRV), and Barmah Forest viruses, are responsible for millions of cases of serious, though primarily not life-threatening, illness in humans. In recent years, an epidemic of Chikungunya fever of unprecedented magnitude has afflicted inhabitants of the Indian Ocean territories, with an estimated 1.5 million people infected in a large geographic area encompassing the East African equatorial coasts, the Indian Ocean islands, and most of the Indian peninsula (21). The acute phase of the these diseases is characterized by fever, chills, headache, eye pain, generalized myalgia, arthralgia, diarrhea, vomiting, and rash (9, 26, 97, 102, 127). Symptoms of arthralgia/arthritis sometimes persist for years.

Closely related to these viruses are important human pathogens of the New World, including Venezuelan (VEEV), eastern (EEEV), and western equine encephalitis viruses. Similar symptoms to those described above are manifest in the acute phase of infection with most of these viruses (133), but as their names suggest, the New World alphaviruses can cause acute encephalitis in humans and equines. Due to their potential for use as agents of bioterrorism/biowarfare, EEEV and VEEV have been declared Select Agents by U.S. Department of Agriculture and Centers for Disease Control and Prevention.

In contrast to the mosquito-borne alphaviruses described above, rubella virus is spread by respiratory,

William B. Klimstra and Kate D. Ryman, Department of Microbiology & Immunology, Center for Molecular & Tumor Virology, Louisiana State University Health Sciences Center, Shreveport, LA 71130.

Table 1 Disease-causing representatives of the family *Togaviridae*[a]

Virus (abbreviation)	Vertebrate reservoir	Human (animal) disease	Geographic distribution
Rubivirus genus			
Rubella	Humans	Rubella fever, congenital defects	Worldwide
Alphavirus genus			
Barmah Forest (BF)	Birds	Fever, arthritis, rash	Australia
Chikungunya (CHIK)	Primates	Fever, arthritis, rash, hemorrhagic fever	Africa, Southeast Asia, Philippines
Eastern equine encephalitis (EEE)	Birds	Encephalitis (horse, pheasant, pigeon, emu, turkey)	North America, South America, Caribbean
Everglades (EVE)	Mammals	Encephalitis	Florida
Getah (GET)	Mammals	(Horse)	
Highlands J (HJ)	Birds	(Horse, turkey, emu, pheasant)	North America
Mayaro (MAY)	?	Fever, arthritis	South America
O'nyong-nyong (ONN)	?	Fever, arthritis, rash	East Africa
Ross River (RR)	Mammals	Fever, arthritis, rash	Australia, South Pacific
Semliki Forest (SF)	?	Fever, encephalitis (horse)	Africa
Sindbis (SIN)	Birds	Fever, arthritis, rash	Australia, Africa, Europe, Middle East
UNA (UNA)	?	(Horse)	South America
Venezuelan equine encephalitis (VEE)	Mammals	Fever, encephalitis (horse)	North America, South America
Western equine encephalitis (WEE)	Birds, mammals	Fever, encephalitis (horse, emu)	North America, South America

[a] Adapted from references 16 and 42.

congenital, or perinatal routes and causes a self-limiting disease in children and adults characterized by mild fever, lymphadenopathy, and morbiliform rash. If acquired early during pregnancy, the virus can also be a significant cause of congenital defects. However, incidence of the disease has been greatly reduced in developed countries through use of an effective vaccine (16).

RELEVANT ASPECTS OF TOGAVIRUS REPLICATION

The alphaviruses are relatively uncomplicated, positive-sense RNA viruses, encoding only seven proteins. Their small genetic budget does not allow for the encoding of proteins with dedicated luxury function, but necessitates the use of one protein for a multitude of purposes. Their replication is very tightly regulated and small changes alter susceptibility to host immune responses. Here we review aspects of alphavirus infection and replication pertinent to virus virulence/attenuation and induction or evasion of the innate immune response.

Virion Structure and Genome Organization
Togavirus virions are small and enveloped (reviewed in reference 125), with an icosahedral nucleocapsid

composed of capsid (C) protein monomers, cloaked in a lipid envelope studded with membrane-anchored glycoprotein components, E1 and E2. The alphavirus E2 glycoprotein interacts with cell surface attachment receptors such as heparan sulfate (66) and the C-type lectin DC-SIGN (64), while E1 mediates pH-dependent fusion with endosomal membranes (63, 98). The encapsidated alphavirus genome resembles a cellular mRNA, consisting of a single-stranded (ss), positive-sense RNA molecule of approximately 12 kb, with a 5′-terminal methyl-guanylate cap structure and 3′ polyadenylation (124) (Fig. 1). The genome is divided into two open reading frames encoding the four nonstructural (nsPs) and three structural proteins as two polyproteins (124). The 9.5-kb rubella virus genome is capped, polyadenylated, and similarly organized, but the 5′-proximal open reading frame encodes only two nsPs (P150 and P90) (reviewed in reference 16).

Genome Replication and Synthesis of Subgenomic mRNA
Although relatively simplistic, togavirus genome replication and RNA transcription are tightly and temporally regulated. Replication processes have been most extensively studied for the alphaviruses SINV and SFV

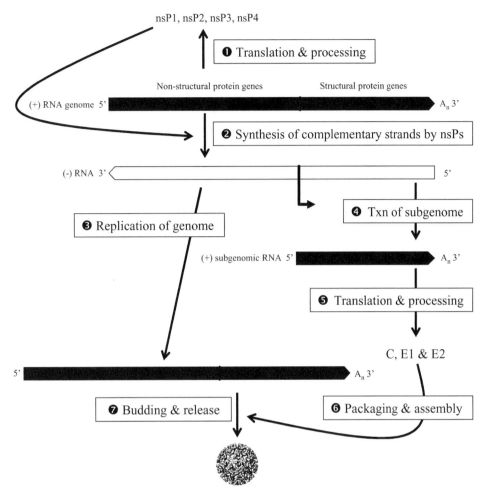

nsP1, nsP2, nsP3, nsP4

❶ Translation & processing

Non-structural protein genes   Structural protein genes

(+) RNA genome 5' A_n 3'

❷ Synthesis of complementary strands by nsPs

(-) RNA 3'   5'

❸ Replication of genome   **❹ Txn of subgenome**

(+) subgenomic RNA 5' A_n 3'

❺ Translation & processing

C, E1 & E2

5' A_n 3'

❼ Budding & release   **❻ Packaging & assembly**

Figure 1 Diagrammatic representation of alphavirus replication in permissive cells. The replication cycle is depicted as a series of temporally regulated steps: (1) translation and processing of nonstructural polyprotein; (2) synthesis of complementary, negative-sense RNA; (3) synthesis of progeny genomes; (4) synthesis of subgenomic RNAs; (5) translation and processing of structural polyprotein C/PE2/6KE1; (6) packaging of progeny genomes into nucleocapsids; and (7) envelopment of progeny virions at the plasma membrane.

and extrapolated to other members of the *Togaviridae*. Infection is initiated when the four nsPs are translated as a polyprotein from the incoming mRNA-like genome and sequentially cleaved to form a series of replication complexes. Early in infection, nsP complexes transcribe genome-length, negative-sense RNA copies, creating partially double-stranded (ds) RNA replicative intermediates (69, 70). After approximately 3 h, there is a switch to preferential synthesis of positive-sense RNAs, which continues to generate progeny genomes throughout the infection cycle (71, 121). Two positive-sense RNA species are synthesized from promoters encoded in the negative-sense sequence: 42S RNA progeny genomes and 26S subgenomic RNA. The subgenomic

RNA, synthesized from the high-level internal 26S promoter at a 5- to 20-fold molar excess over genomic, is identical in sequence to the 3' one-third of the genome and encodes the viral structural proteins. The 26S mRNA is also translated as a polyprotein; the capsid protein is cleaved in the cytoplasm and the remaining polyprotein is processed and cleaved in the secretory pathway to yield the E1 and E2 glycoproteins in preparation for packaging and egress.

Importantly, alphavirus RNA replication occurs not in the open cytoplasm but associated with cytoplasmic surfaces of endosomal vesicles known as cytopathic vacuoles, modified from the intracellular membranes of infected host cells (31, 43, 68, 93). Membrane association of the replicase is mediated by nsP1 via membrane phosphatidylserine and other anionic phospholipids (3). The surfaces of cytopathic vacuoles have small invaginations or "spherules," sequestering the nsPs and nascent RNA and creating a microenvironment for RNA replication, synthesis of structural proteins, and assembly of genome-C complexes. It is tempting to speculate that this compartmentalization of the viral replicative intermediates and proteins provides concealment from the host's pathogen surveillance mechanisms (discussed further below), as recently proposed for another positive-sense RNA virus, West Nile (50).

The specificity of alphavirus RNA replication is achieved via recognition of *cis*-acting conserved sequence elements (CSEs) present in the termini of the viral genome and the negative-sense genome template. The secondary structures of these *cis*-acting elements are conserved for all members of the *Alphavirus* genus and are mostly conserved in rubella virus. A 19-nucleotide CSE in the 3' nontranslated region (NTR) immediately upstream of the poly(A) tail is the core promoter for synthesis of genome-length negative-sense RNA replicative intermediates (91), and appears to interact with the 5' NTR via translation initiation factors to initiate replication and/or translation (29). The complement of the 5' NTR in the minus-strand RNA acts as a promoter for the synthesis of positive-sense daughter genomes (89, 91). A third *cis*-acting element that is essential for replication, the 51-nucleotide CSE, is also found at the 5' end of the genomic RNA in the nsP1-encoding gene. The primary sequence and two-stem-loop secondary structure of this CSE are highly conserved among the alphaviruses (90) and serve as a replication enhancer (29). A variety of alphavirus mutants have been generated with defects in *cis*-acting CSEs or nsP genes that significantly modulate virus virulence. In some cases, virulence changes have been linked to an altered innate immune system interaction (see, e.g., reference 135).

THE ROLE OF IFN-α/β IN ALPHAVIRUS INFECTION IN VIVO

Very little is known regarding rubella virus pathogenesis due to the lack of a small-animal model. Therefore, the following section will summarize our current understanding of alphavirus pathogenesis, developed primarily using mice. Many alphaviruses can infect and cause disease of varying severity in mice, and this model has proven particularly amenable to experimental investigation (42). A typical pathogenic sequence for an encephalitic New World alphavirus has been delineated through subcutaneous infection of adult mice with VEEV, mimicking the bite of an infected mosquito vector. After deposition of virus in the subcutaneous tissues, the virus infects dendritic cells (DCs) and macrophages in the skin at the site of inoculation and "hijacks" these cells to access the regional draining lymph node (DLN), where the virus amplifies and seeds a serum viremia (5, 18, 20, 75). The virus then disseminates systemically; infects other lymphoid tissues, skeletal muscle, and connective tissues distant from the site of inoculation; and crosses the blood-brain barrier by infection of the olfactory neuroepithelium (18). Infected neurons in the central nervous system (CNS) undergo apoptotic degeneration directly following infection and in areas of astrogliosis (40, 56).

Wild-type, Old World alphaviruses tend to be attenuated/avirulent in immunocompetent adult mice, although they cause severe, often fatal disease in suckling animals (41, 86, 108, 110). With SINV, for example, subcutaneous inoculation of adult mice is subclinical, with self-limiting virus replication and little tissue damage. Virus clearance from serum and tissues begins several days postinfection with the appearance of serum immunoglobulin M (41, 120). Similar to VEEV, DCs are primary targets for SINV infection in adult mice at the site of inoculation and migrate to the DLN, but the number of detectably infected cells is greatly reduced within 12 h postinfection and dissemination beyond the DLN is limited (110), correlating with the avirulent phenotype. In model systems in which substantial virus spread occurs, osteoblasts, skeletal muscle, and connective tissue are infected in the periphery, followed by neuroinvasion and replication within neurons (48, 60, 65, 74, 118, 120). Unlike human infections, Old World alphaviruses can be manipulated to cause encephalitis in mice by varying the age of the mouse, the strain of the virus, or the inoculation route such that CNS replication is prominent, but limited replication of the virus in peripheral tissues occurs (23, 61, 65, 120, 128). However, virus control during the acute phase of primary infection does not require the presence of T or B lymphocytes (41, 134), and early studies indicated a role for the IFN-α/β response (101, 103).

Genetically modified mice in which IFN-α/β signaling has been abrogated by targeted gene disruption of the IFN-α/β receptor subunit, IFNAR1 (IFNAR1$^{-/-}$), are unable to establish an antiviral state in response to IFN-α/β induction and are highly susceptible to infection by SINV (110, 111), CHIKV (19a), and SFV (87), succumbing to the infection within a few days despite the presence of an otherwise intact immune system (15, 110). In the absence of IFN-α/β signaling, SINV replicates extensively in DCs and macrophages in the

DLN; disseminates efficiently by high-titer serum viremia to similar cells in spleen, liver, and other tissues; and induces a dysregulated, proinflammatory cytokine cascade (110, 111). Thus, in the absence of IFN-α/β signaling, a normally benign SINV infection becomes rapidly fatal, reflecting disseminated replication in myeloid-lineage cells (110). These studies highlight the extreme sensitivity of SINV to IFN-α/β-mediated antiviral activity in vivo, particularly within DCs and macrophages, and demonstrate the pivotal role of IFN-α/β in protection against alphavirus disease.

In IFNAR1⁻/⁻ mice, the VEEV disease course becomes extremely rapid, with all animals dying within 2 days, compared with 7 to 9 days in normal counterparts (40, 135). In normal mice, IFN-α/β apparently mediates partial clearance of virus from the periphery, but fails to protect the CNS from pathology or the host from death (18). A single nucleotide substitution in the 5′ NTR of the wild-type VEEV genome produces an avirulent virus, exhibiting lower peak titers and earlier clearance (135). In contrast, wild-type and mutant viruses are equally virulent for and replicate similarly in IFNAR1⁻/⁻ mice. IFN-α/β induction in vivo and in vitro is similar for the wild-type and mutant viruses, but the mutant virus is significantly more sensitive to the antiviral actions of IFN-α/β priming (135). These results suggest that VEEV remains virulent in the face of a functional IFN-α/β response by evading and/or disabling the inducion or activity of IFN-α/β-induced antiviral effectors. Thus, whereas the permissivity of DCs and macrophages to SINV and VEEV 5′-NTR mutant virus infection is strictly controlled by the IFN-α/β-mediated antiviral response (110, 135), wild-type VEEV is able to infect and replicate in these cells throughout the body during the lymphoid phase (75), despite rapid and substantial induction of IFN-α/β (17).

To counter the host's potent innate immune system, many pathogenic viruses have developed diverse mechanisms to evade or antagonize the IFN-α/β response (reviewed in reference 8), either to directly combat antiviral activities and/or to subvert the transition from innate to adaptive immunity (24). An ever-increasing number of viruses are being found to encode viral gene products that (i) repress transcriptional activation of IFN genes; (ii) interfere with IFN-α/β signaling through the IFNAR receptor; and/or (iii) directly inhibit IFN-α/β-controlled antiviral gene products, such as the dsRNA-dependent protein kinase, PKR. Several key observations suggest that the relative sensitivities of alphaviruses to IFN-α/β-mediated antiviral activity are primary determinants of virulence and attenuation. Alphaviruses with little or no ability to evade or antagonize mammalian IFN-α/β, such as SINV, are extremely attenuated in mice. On the other hand, a mouse-virulent alphavirus, such as VEEV, is relatively much more resistant to the antiviral activity of IFN-α/β (see below). Mutants of VEEV with increased sensitivity to IFN-α/β are attenuated in mice. It is likely that the high degree of virulence of New World alphaviruses for adult mice (and perhaps humans) is determined, at least in part, by antagonism and/or avoidance of the IFN-α/β response. In order to understand the mechanisms involved, we must consider the interactions between the alphaviruses and the IFN-α/β system at the cellular and molecular level. The current status of our knowledge in these areas is described in the following sections, focusing on IFN-α/β induction and effector mechanisms.

IFN-α/β INDUCTION BY ALPHAVIRUSES

As described in other chapters, DCs and other cells bear sensory machinery that can recognize the presence of pathogen-associated molecular patterns (PAMPs) of invading microbes through pattern recognition receptors (PRRs). PRRs include extracellular and endosomal receptors such as Toll-like receptors (TLRs) (55) and C-type lectin scavenger receptors (80), as well as intracellular receptors that appear to recognize intermediates of RNA virus replication such as PKR (58) and the RNA helicases retinoic acid-inducible gene I (RIG-I) and melanoma differentiation-associated gene 5 (MDA-5) (62, 136). Viral PAMPS include viral glycoproteins (C-type lectins, TLR2) and viral ssRNA (TLR7) or dsRNA (TLR3, PKR, RIG-I, and MDA-5) or ssRNA with an uncapped 5′ triphosphate (96) (reviewed in references 37, 62, 83, and 99). Via PAMP detection, cell signaling pathways trigger PRR-bearing cells such as DCs to secrete cytokines (including type I IFNs). DCs also differentiate, mature, and migrate to the peripheral lymphoid tissues, where they stimulate and shape adaptive immune responses (55; reviewed in references 59 and 79). A distinction is generally made between the cell-surface/endosomal PRRs (C-type lectins, TLR3, TLR7, and others) and the cytoplasmic PRRs (PKR, RIG-I, and MDA-5) in that the functional cell surface PRRs are expressed only by subsets of DCs (e.g., TLR7, primarily found on plasmacytoid DCs [pDCs]) or macrophages (e.g., some C-type lectin receptors), whereas cytoplasmic PRRs are expressed by most cells (reviewed in references 62, 83, and 99). Therefore, viral induction of IFN-α/β may exhibit cell-type specificity, depending upon the cells infected and the pathway(s) triggered.

Several signaling pathways resulting in IFN-α/β induction have been elucidated for the various PRRs. The canonical induction pathway was originally defined as dsRNA-replicative intermediates in the cytoplasm activating the IFN regulatory factor (IRF) 3 transcription factor, which then translocates to the nucleus, where it activates IFN-β gene transcription (116). This has recently been shown to involve interaction of viral RNA with the RIG-I or MDA-5 RNA helicase PRRs, followed by recruitment of the IPS-1 adaptor molecule and activation of TANK binding kinase-1 (TBK1) or IKK-ε in lymphoid cells, which may directly phosphorylate IRF3 (92). Interestingly, recent studies indicate that, while MDA-5 may interact with dsRNA, RIG-I is activated by interaction with uncapped RNA molecules containing 5′ triphosphates (37, 52, 96). IRF3 is constitutively expressed; however, IRF7, which shares some activation characteristics with IRF3, exhibits low expression in nonlymphoid cells and is upregulated by IFN signaling (92). PKR, a dsRNA-binding protein, appears to augment the induction of IFN-α/β by signaling through the nuclear factor-κB (NF-κB) pathway and may be a downstream modulator of TLR3 signaling (58). With the exception of TLR3, TLRs signal through the canonical TLR pathway, including MyD88–IRAK-1 adaptor proteins, to activate NF-κB and induce the IFN-β promoter. In contrast, TLR3 signals through the TRIF/TRAF6 pathway to activate both NF-κB and TBK1-IRF3 (reviewed in reference 51).

IFN-α/β Induction by Alphaviruses in Primary Fibroblasts

For most alphaviruses, it is known that type I IFN is induced in vitro by infection of a number of cell types (6, 25, 42, 78, 84; K. D. Ryman and W. B. Klimstra, unpublished data), and in animal models, high levels of IFN-α/β are associated with systemic virus replication (7, 12, 65, 110, 113, 120, 135). One group reported that induction of IFN-α/β by SFV in cultured myeloid DCs (mDCs) occurred in the absence of MyD88, implying independence of TLRs other than TLR3, but was dependent upon IRF3 (49). Furthermore, IFN-α/β induction was only slightly impaired in MDA-5−/− macrophages infected with SINV, suggesting that this molecule was not vital to responses against alphaviruses (37). We previously reported that heterogeneous cultures of bone marrow-derived mDCs lacking PKR exhibited delayed IFN-α/β induction in response to SINV infection, although maximum levels were relatively unaffected, suggestive that PKR had a role in the timing of IFN-α/β induction (113). Aside from these studies, the alphavirus-expressed PAMPs and the particular PRRs and pathways associated with IFN-α/β induction remain relatively uncharacterized. More recent evidence

from our laboratory suggests that PKR may be involved in IFN-α/β induction in commonly used fibroblast cultures following alphavirus infection (C. W. Burke, C. L. Gardner, K. D. Ryman, and W. B. Klimstra, submitted for publication), similar to findings with the flavivirus West Nile (36).

We have examined the particular signaling cascades involved in IFN-α/β induction after SINV infection of normal cells compared with cells deficient in pathways associated with the major intracellular PRRs, using TBK1−/− and wild-type murine embryonic fibroblasts (MEFs) since the RIG-I/MDA-5 pathways both signal through this kinase (reviewed in reference 51). Importantly, MEFs do not appear to support signaling through TLR3 or TLR7 (Burke et al., submitted), providing a more defined experimental system. Moreover, null mutation for TBK1 is embryonic lethal; therefore, only MEFs are available (11). Consistent with reported studies (81), IFN-α/β induction by the negative-sense paramyxovirus, Sendai virus (SeV) control was robust after MEF infection and almost completely dependent upon TBK1. In contrast, IFN-α/β induction after infection with SINV, VEEV, or EEEV measured by bioassay was minimal for wild-type, cytopathic effect (CPE)-inducing strains (Fig. 2). When induction of IFN-α/β by SINV and SeV infection was compared using enzyme-linked immunosorbent assay for IFN-β or IFN-α, or a reporter assay for IFN-β promoter activity, SINV did not induce detectable IFN measurable by any assay. In contrast, SeV induced high levels of IFN-β (but not IFN-α) as measured by all three methods (Burke et al., submitted).

The hypothesis that the wild-type alphaviruses fail to induce detectable IFN-α/β in MEFs due to rapid host shutoff and development of CPE was tested using a SINV nsP2 mutant that was defective in transcription and translation shutoff and resulted in IFN-α/β production in infected 3T3 fibroblasts (28, 30). This virus induced IFN-α/β in the primary MEFs later after infection in TBK1-dependent fashion (Fig. 2) (Burke et al., submitted). To more completely define the induction pathway, we examined the activity of the cytoplasmic RNA helicases RIG-I and MDA-5 in IFN-α/β induction by the noncytopathic SINV by small interfering RNA (siRNA)-mediated knockdown of their mRNAs. With the SeV-infected control, IFN-α/β induction was completely abrogated by RIG-I knockdown, but MDA-5 knockdown had no effect, which is consistent with reported studies (36, 123). With the noncytopathic SINV, RIG-I knockdown did not decrease IFN-α/β induction and MDA-5 knockdown inhibited induction approximately 50%. Based upon our previous data suggesting

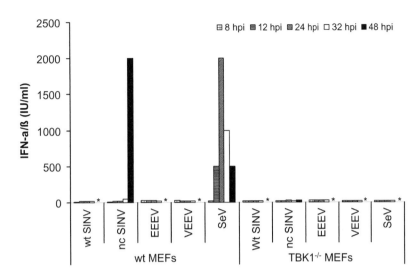

Figure 2 Induction of IFN-α/β as measured by a biological assay of supernatants collected from virus-infected (multiplicity of infection = 3) MEF cultures at different times postinfection. nc SINV is the Sindbis virus strain TR339 with the noncytopathic nsP2 726 G mutation (30). wt SINV is the TR339 strain of Sindbis virus. SeV is a Sendai virus stock acquired from the American Type Culture Collection. Asterisks indicate times at which samples were not taken due to complete cytopathic effect. hpi, hours postinfection.

a role for PKR in IFN-α/β induction by wild-type virus in DCs (113), we examined the activity of PKR in induction by the noncytopathic SINV in the MEF cultures. IFN-α/β induction was significantly reduced in a dose-dependent manner by treatment of MEFs with a PKR inhibitor, or after infection of PKR$^{-/-}$ MEFs. These data suggest that in MEFs, PLR and MDA-5 act as PRRs for SINV and signal through TBK1.

Overall, SINV, VEEV, and EEEV fail to induce significant type I IFNs in the MEF system, and this differs dramatically from induction by SeV, which is dependent upon RIG-I recognition. Moreover, use of a SINV strain defective in host transcription/translation shutoff allowed cells to survive longer and produce IFN-α/β in an MDA-5/PKR/TBK1-dependent manner. The failure of RIG-I or MDA-5 to recognize the replicative intermediates of wild-type SINV could be due to the avoidance of detection by SINV-associated dsRNA/5′ triphosphates, as has been suggested for mouse hepatitis virus (138), or due to antagonism of these pathways by the viruses (described below). However, results with MEF cultures are not consistent with IFN-α/β induction profiles in vivo or in cultured DCs (see below).

IFN-α/β Induction by Alphaviruses in Mice and Primary DCs

In a direct comparison of IFN-α/β induction in vivo by SINV, VEEV, or EEEV propagation-competent viruses, the highest levels of biologically active IFN-α/β are produced by VEEV, followed by SINV, while very low levels are produced by EEEV (C. W. Burke, C. L. Gardner, K. D. Ryman, and W. B. Klimstra, unpublished data) (Fig. 3), consistent with previous studies examining each virus individually (1, 18, 65, 120). Furthermore, IFN-α/β induction was detectable by 8 to 12 h postinfection. Interestingly, similar induction profiles were observed when mice were infected with nonpropagative "replicon" particles from whose genomes the structural protein genes had been deleted (Fig. 3). These results illustrate two important points: (i) the induction mechanisms triggered by wild-type SINV and VEEV in MEFs (in which infection fails to elicit IFN-α/β production) and intact hosts are very different; and (ii) EEEV does not produce large amounts of IFN-α/β in vivo regardless of the presence or absence of the capsid protein in its genome (comparing virus versus replicon particle inoculation), which has been associated with IFN-α/β antagonism (see below).

As described above, important infected cell types in mice following alphavirus inoculation include DCs and macrophages during the early stages of infection, osteoblasts during the serum viremic phase, and neurons during terminal disease stages; however, only DCs have been analyzed in detail for mechanisms of IFN-α/β induction. Several groups have demonstrated IFN-α/β induction after alphavirus infection of bone marrow-derived

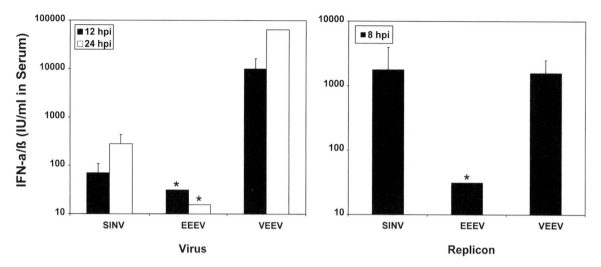

Figure 3 Production of IFN-α/β measured by biological assay of serum after infection of CD-1 mice with similar doses (pfu or IU) of SINV (TR339 strain), EEEV, or VEEV virus or replicon particles. In EEEV-infected mice, IFN levels were below the limit of detection (*). hpi, hours postinfection.

mDCs (49, 113, 119). SINV, VEEV, RRV, and SFV each produce biological IFN-α/β after infection of these cells, which for SFV was MyD88 independent but required IRF3 (49) and for RRV was independent of TLR3 (119). In these latter studies, some IFN-α/β was also induced by infection with partially UV-inactivated viruses (49, 119), suggesting that completion of the replication cycle was not required. Notably, EEEV failed to induce IFN-α/β in myeloid cells; however, rates of infection were extremely low (C. L. Gardner, C. W. Burke, P. Glass, M. Parker, W. B. Klimstra, and K. D. Ryman, submitted for publication). We have more recently completed similar studies with granulocyte-macrophage colony-stimulating factor/Flt3-ligand-differentiated bone marrow cells, which contain a significant proportion of pDCs (13). In these studies, IFN-α/β induction was detected rapidly after infection with SINV (C. W. Burke, C. L. Gardner, K. D. Ryman, and W. B. Klimstra, unpublished), similar to results with intact mice, suggesting that pDCs may be important for early antitogavirus responses in vivo.

Antagonism of IFN-α/β Induction by Alphaviruses

Several studies have indicated that alphaviruses are capable of antagonizing induction of stress and innate immune responses, including IFN-α/β, in infected cells (2, 14, 30). With the Old World viruses such as SINV and SFV, activities of the nsP2 protein were implicated in this ability (14, 30). In contrast, the capsid proteins of New World viruses VEEV and EEEV were found to be

responsible for these effects (2, 34). The alphavirus nsP2 protein is the primary protease responsible for cleavage of the nsP1–4 polyprotein, into individual proteins, driving formation of replication complexes with different activities involved in viral RNA synthesis. The multifunctional nsP2 protein contains nucleoside triphosphatase (106), RNA triphosphatase (130), RNA helicase (38), and papain-like protease (46) domains, and a large portion of total nsP2 is translocated to the nucleus during infection (33, 94). The capsid protein also contains a (chymotrypsin-like) protease domain (19), as well as RNA-binding and self-assembly domains (95). Interestingly, these proteins were also found to be mediators of host translation shutoff and cytopathic effect (CPE) in the cell culture systems used for analysis of IFN-α/β induction (2, 33, 34). In these studies, Old World SINV and SFV mutants selected for nsP2 noncytopathogenicity (30) or defective nsP2 nuclear localization signal (14), respectively, or New World alphavirus vectors that did not express the capsid protein (2, 34) were used. The nsP2 noncytopathic mutant did not shut off host-cell transcription or translation activities and induced IFN-α/β, while parental viruses rapidly shut off macromolecular synthesis and avoided IFN-α/β induction. Similarly, the capsid proteins of EEEV and VEEV were shown to promote host-cell transcription shutoff (2, 34), both in the context of replicating virus vectors and overexpression of individual proteins including, with the EEEV capsid protein, a conventional antagonism assay in which the growth of the IFN-sensitive

Newcastle disease virus was increased by transient protein expression (2). Interestingly, the nsP2 proteins of New World alphaviruses and the capsid proteins of Old World alphaviruses did not exhibit the rapid transcription shutoff activity (34), suggesting that the different virus groups evolved these activities independently (33). However, the capsid proteins of both New World and Old World viruses may contribute to host translation shutoff by stimulating the activation of PKR and phosphorylation of eukaryotic translation initiation factor (eIF) 2α (2, 22).

For both the capsid and nsP2 proteins, the primary effect appears to be an interruption of host transcription resulting in a general failure of cells to express stress-responsive genes including the IFNs (2, 33, 34). Effects of the nsP2 protein on host-cell transcription have been mapped to a carboxy-terminal fragment (33), while effects of the capsid protein have been mapped to a positively charged amino-terminal fragment (2, 2a, 32). Surprisingly, these segments of the proteins are outside of protease or other functionally characterized domains (2, 32, 33). At least with the SFV nsP2 mutant, antagonism involves nuclear translocation of the protein (14). The VEEV capsid protein is also partially localized to the nucleus (34) and to nuclear pore structures, an activity mediated by the region implicated in transcriptional shutoff (32). As indicated above, the authors of most of these studies inferred that IFN antagonism and a general disruption of host-cell transcription are overlapping activities. However, the SFV mutant that lacks efficient nuclear translocation does not appear to differ from the wild-type virus in effects upon host transcription early after infection, although the mutant exhibits a less pronounced inhibition of cellular DNA synthesis (105) and a reduced rate of viral RNA and polypeptide accumulation (14). Notably, wild-type SFV also blocked accumulation of tumor necrosis factor-α in cultured cell supernatants while the mutant did not, implying an effect upon host responses outside of the IFN-α/β pathways, but did not interfere with nuclear translocation of IRF3 or NF-κB transcription factors (14). It remains to be determined if the capacity of the alphavirus nsP2 proteins to inhibit IFN-α/β induction in cultured cells is an indirect effect of a general block to transcription or if more specific effects upon particular stress response pathways are involved. Further characterization of the mechanism through which nsP2 and capsid proteins affect gene expression will be extremely informative in this regard.

An important feature of any discussion of IFN-α/β antagonism is a consideration of cell types involved. All of the studies implicating particular alphavirus proteins in antagonism of IFN-α/β induction were performed using highly permissive cultured cell lines such as 3T3 MEFs (14, 30), in which alphavirus infection causes substantial shutoff of host transcription and translation as early as 4 h postinfection. In contrast with these results, when cultures of heterogeneous primary bone marrow-derived mDCs were infected with wild-type SINV and global transcription analyses were performed 12 h postinfection, more than 400 genes were upregulated >2-fold over mock, some by >50- to 100-fold, including IFN-β and multiple IFN-α subtypes. Moreover, high levels of biologically active IFN-α/β were present in cell supernatants (112). Interestingly, CPE, an indirect effect of the arrest of host macromolecular synthesis, was not prominent in these cells (112, 113). Similarly, analyses of transcription in CNS tissue from mice infected with VEEV (117) indicated robust induction of IFN-α/β and other stress-responsive genes. These data suggest that the capacity of some alphaviruses to "antagonize" IFN responses is cell type dependent and may be related to the rapid shutoff of host processes characteristic of infection of highly permissive immortalized fibroblastic and similar cell lines. Additionally, IFN-α/β is highly induced within 12 h in the serum of mice infected with wild-type SINV or VEEV (18, 65, 110; Burke et al., unpublished). As indicated above, some EEEV strains induce only low levels of IFN-α/β in vivo (1; Gardner et al., submitted); however, it is unclear whether or not this reflects antagonism of induction in infected cells (2), the failure of some EEEV strains to productively infect myeloid cells (132; Gardner et al., submitted), or the cumulative effect of both evasion mechanisms. Further research will be required to ascertain whether or not the inhibition of IFN-α/β induction by nsP2 or capsid proteins of alphaviruses is a function of a generalized effect upon highly permissive host-cell processes or a more specific interaction with components of the IFN-α/β-inductive pathways.

Another putative mechanism of induction antagonism may arise from the sylvatic transmission cycle of the alphaviruses related to the attachment receptor utilized by viruses that have replicated in a mosquito host. Several recent studies have established that a number of arthropod-vectored viruses can utilize the C-type lectins DC-SIGN or L-SIGN as attachment receptors during infection of DCs derived from mammalian hosts (64, 88, 126). With the alphaviruses, carbohydrate structures on the surface proteins of particles produced by replication in insect cells are efficient ligands for these receptors, greatly increasing infection efficiency for cultured DCs (64), whereas alphavirus carbohydrate structures derived from mammalian cells generally do not promote efficient attachment to C-type lectins (64; W. B. Klimstra, D. L. Browning, and K. D. Ryman, submitted for publication). This is assumed to arise from the limited carbohydrate processing achieved by insect cell enzymes, resulting in high-mannose carbohydrate structures as final modifications of insect cell carbohydrates, which has been documented for such structures present upon SINV E1 and E2 proteins derived from these cells (53, 54). A recent study indicated that infection of murine bone marrow-derived mDCs by insect cell-derived RRV, VEEV, or Barmah Forest virus was associated with a reduction of the magnitude of IFN-α/β induction versus analogous mammalian cell-derived viruses (119), associated with higher rates of infection by the insect cell-derived viruses. The authors suggested that viruses delivered by an arthropod vector may interact with DCs via the incompletely processed carbohydrate modifications on the E1 or E2 protein and, as a consequence, stimulate less IFN-α/β induction in the first round of infection, thereby increasing initial replication within the infected host. However, effects of carbohydrate modification upon IFN-α/β induction in vivo were not evaluated. We have been unable to demonstrate any significant differences in serum IFN-α/β levels after infection of mice with equal particle doses of mammalian versus mosquito cell-derived SINV viruses (W. B. Klimstra, D. L. Browning, and K. D. Ryman, unpublished data). These results suggest that the effect of differential carbohydrate processing upon IFN-α/β induction may be cell type specific and that the response of cells initially producing IFN-α/β in vivo (such as pDCs) may not be as strongly affected by carbohydrate structures of the infecting alphavirus as bone marrow-derived mDCs in vitro. Furthermore, in our studies (64) and those of Shabman et al. (119), the infectivity for myeloid cells was greatly enhanced for mosquito cell-derived virus. Considering the IFN-α/β- suppressive effect of host-cell shutoff documented in highly permissive cultured cell lines, the effect of infection efficiency upon mDC shutoff is potentially a factor in the outcome of these studies.

Another, possibly similar, mechanism of antagonism was suggested for RRV particles that gained entry of macrophages by an antibody-dependent enhancement of infection (ADE) (73, 77). During ADE infection, which has been observed with other alphaviruses (100) and human immunodeficiency virus (129) and has been implicated as one mechanism underlying the development of dengue hemorrhagic fever/dengue shock syndrome (see, e.g., reference 85), low levels of incompletely neutralizing virus-specific antibodies attach to virus particles and facilitate increased rates of cell infection due to

efficient antibody attachment to Fc receptors present upon myeloid cells, particularly macrophages. ADE of RRV infectivity for cultured RAW 264.7 macrophages reduced the accumulation of stress-responsive gene mRNAs, including tumor necrosis factor-α and IFN-β, and also appeared to reduce the abundance of several stress-responsive transcription factor complexes involving Stat1 and NF-κB (77). The effects upon cell stress responses required virus replication and were ablated by UV inactivation of the virus preparations.

Importantly, none of the antagonism mechanisms outlined above have been shown to occur in the intact, infected host. Therefore, a priority for validation of these studies is detailed characterization of virus-induced IFN-α/β production in mice and primary tissues derived from them that are important for initial virus replication, dissemination within the host, and manifestation of disease. It is also noteworthy that each of the mechanisms described has a relationship to the infection efficiency of cultured cell models used, either through the use of highly permissive cultured cells or the enhancement of infectivity of poorly infected cells by manipulation of carbohydrate modifications or reaction with infection-promoting antibody. Further research is clearly required to determine whether or not antagonism of IFN-α/β gene induction by alphaviruses is associated with general shutoff of host transcription/translation that is increased as a function of infection efficiency or, perhaps, more subtle and diverse effects upon particular pathways within infected cells.

EFFECTORS OF THE ANTIALPHAVIRUS RESPONSE AND THEIR ANTAGONISM

It has long been known that pretreatment, or "priming," of responsive cells with type I IFNs induces an antiviral state against togaviruses. Indeed, the potency of IFN-α/β-mediated antiviral activity against SINV and SFV replication in cell culture led to the common use of these viruses as indicators in biological IFN-α/β assays. Whereas virus rapidly arrests host macromolecular synthesis with cytopathic consequences in unstimulated, cultured cells, cellular transcription and protein synthesis are restored in IFN-α/β-primed cells. Coordinately, production of progeny virions is dramatically reduced. Early studies suggested that virus replication is inhibited almost immediately after entry, with lower levels of nonstructural proteins translated in IFN-treated cells (27, 122). Moreover, late in infection IFN-α/β treatment appears to interfere with virion morphogenesis and egress (104). However, these antiviral effects can be alleviated

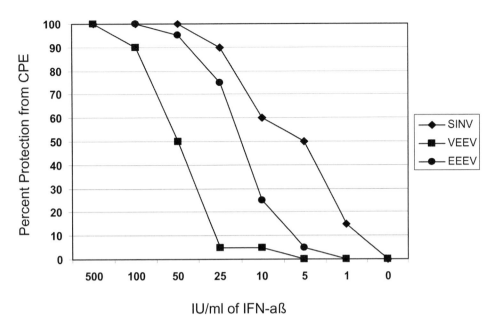

Figure 4 Relative sensitivity of the replication of SINV (TR339 strain), VEEV, or EEEV to IFN-α/β priming of Swiss 3T3 murine fibroblasts. Cells were treated with the indicated IFN concentration for 12 h prior to infection (multiplicity of infection = 3), and CPE was measured at 24 h postinfection.

by treatment with an RNA Pol II inhibitor such as actinomycin D or alpha-amanitin, implying that establishment of the refractory state is dependent upon de novo transcription of IFN-α/β-stimulated genes (ISGs) whose protein products possess antialphaviral activity (27, 82).

The intrinsic sensitivity of alphaviruses to the antiviral effects of IFN-α/β may be of primary importance in controlling the apparent tropism and virulence of each alphavirus. Direct comparison of the sensitivity of SINV, EEEV, and VEEV to IFN-α/β-mediated priming has revealed that SINV replication in MEFs is inhibited at 10- to 20-fold lower concentrations of IFN-α/β than VEEV (Fig. 4). Interestingly, EEEV exhibits an intermediate IFN-α/β sensitivity phenotype. When considered alongside pathogenesis studies involving mice deficient in IFN-α/β responses, it is clear that the IFN-mediated innate immune responses can control the extent of alphavirus replication, the apparent tissue tropism, and the severity of disease. Indeed, the resistance of a given virus to this innate immune mechanism may be a critical virulence determinant.

Effectors of the IFN-α/β-Inducible Antiviral Response
Although considerable evidence indicates that the IFN-α/β response constitutes a vital component of the host's antiviral defense against alphaviruses, mechanistic details

of host antiviral effectors and viral antagonism are only now coming to light. It is known that IFN-α/β exerts its multiple properties by inducing a number of cellular ISGs, which appear to be poorly induced by virus infection in the absence of IFN-α/β signaling through its receptor (112). However, the contribution of individual IFN-induced proteins to the generation of the antiviral environment is difficult to assess, because various effector proteins appear to have overlapping and cumulative antiviral activities and many of these proteins are also constitutively expressed by unprimed cells. Genes whose transcription is upregulated after IFN-α/β treatment of cultured cells and whose products potentially contribute to these effects number in the hundreds (112); however, few have been assessed for potential antiviral activity. In this section, we will consider first the best-known antiviral proteins—PKR, RNase L, and Mx—before describing the activities of other, newly discovered antiviral proteins.

The PKR Pathway
Arguably the best-characterized antiviral protein is PKR, which is constitutively expressed in unstimulated cells and modestly induced by IFN-α/β signaling (4). Upon interaction with dsRNA, PKR autophosphorylates, dimerizes, and catalyzes the phosphorylation of several substrates. The antiviral effects of PKR have been

primarily attributed to phosphorylation of the α subunit of eIF2α, which results in the inhibition of GTP-eIF2-tRNAiMet ternary complex formation and global suppression of translation initiation for both viral and cellular proteins. Infection with SINV or SFV leads to extensive activation of endogenous PKR and the phosphorylation of a large proportion of the available eIF2α molecules (39, 113, 131). Although it is generally assumed that PKR is activated by dsRNA viral replicative intermediates in infected cells, accumulating SFV and EEEV capsid proteins also trigger the activation of PKR and eIF2α phosphorylation in the absence of viral RNA replication (2, 22). In order to evade this cellular defense mechanism, the alphaviruses have evolved tolerance to translational shutoff, apparently allowing the cell's inoperative translational apparatus to be redirected for the synthesis of viral proteins. Indeed, once shutoff has occurred, virtually the only proteins synthesized by the cell are virus-encoded structural proteins, expressed from the subgenomic mRNA. Translation enhancer elements have been identified in the 5′ termini of SINV and SFV 26S mRNAs, which facilitate the continued expression of alphavirus structural proteins during the translation inhibition imposed by eIF2α phosphorylation, by briefly stalling the scanning ribosome (131). Furthermore, SFV infection induces transient formation of stress granules containing cellular TIA-1/R proteins, which sequester cellular mRNAs but disassemble near the viral replicase.

Despite the ability of the alphaviruses to evade the antiviral effects of PKR in highly permissive cultured cells, early control of SINV replication was found to be defective in primary bone marrow-derived mDCs generated from mice with targeted disruption of the PKR gene, suggesting that the PKR pathway suppresses virus replication in these relatively refractory cells independently of IFN-α/β induction (113). In vivo, PKR plays a minor antiviral role against SINV by suppressing replication early in infection, primarily in DCs of the DLN (113). However, the presence or absence of PKR has little impact on IFN-α/β-induced antiviral activity against SINV in vitro or in vivo (112, 113), or likely any alphavirus.

2′,5′-OAS/RNase L Pathway

IFN-inducible 2′,5′-oligoadenylate synthetase (2′,5′-OAS) binds dsRNA and is activated to synthesize 2′,5′-linked oligoadenylates that activate constitutive RNase L to cleave host and viral ssRNA. In the absence of RNase L, the synthesis of minus-strand templates and progeny genomes continues unabated in SINV-infected MEFs, facilitating the establishment of a persistent, noncytopathic infection (115). The authors infer that RNase L plays a role, perhaps indirectly, in the formation of

stable replication complexes during alphavirus infection and that this may represent a host antiviral response to suppress progeny genome production and structural protein synthesis. Although the specific effects on minus-strand synthesis were not evaluated, in primary mDCs or in vivo the absence of 2′,5′-OAS/RNase L activity, alone or combined with the absence of PKR, was not observed to affect SINV infection (113). Priming of IFN-α/β-mediated antiviral activity against SINV and autocrine effects of induced IFN-α/β occurred independently of RNase L (112, 113, 115). Consequently, IFN-α/β-competent cells and animals are protected from SINV infection by an IFN-α/β-induced, PKR/RNase L-independent, "alternative" pathway, and any diminution of antiviral activity resulting from the absence of PKR and/or RNase L is largely obscured by the potency of the residual effector(s) (112, 113).

Mx GTPase Pathway

The Mx family of GTPases are synthesized under the stringent control of IFN-α/β and exhibit potent antiviral activity, primarily demonstrated against negative-strand RNA viruses. Most commercially available mouse strains are Mx deficient. However, overexpression of the human MxA protein in the cytoplasm of cells that do not normally express this protein conferred significant protection against SFV infection (67). Progeny viral yields were greatly reduced from MxA-transfected cells and the accumulation of both genomic and subgenomic RNA species was prevented, suggesting that SFV was inhibited early in its replication cycle prior to packaging and egress. Through generation of MxA-transgenic IFNAR1−/− mice the antiviral activity of human MxA was assessed in vivo in the absence of generalized ISG expression. In this context, MxA conferred partial resistance to SFV, reducing viral titers and mortality rates and extending average survival times (47). Although these results suggest that MxA contributes to the human antiviral response to potentially fatal infections by alphaviral pathogens, activity against other alphaviruses and their ability to antagonize/evade this activity remain to be tested.

PKR/RNase L/Mx-Independent Alternative Pathways

Several lines of evidence established above indicate the existence of IFN-α/β signaling-dependent antiviral pathway(s), able to function independently of PKR, OAS/RNase L, and/or Mx, with the ability to suppress SINV infection (112, 113) (Fig. 5). First, SINV infection of mice triply deficient (TD) in RNase L, PKR, and Mx was subclinical, whereas infection of IFNAR1−/− mice

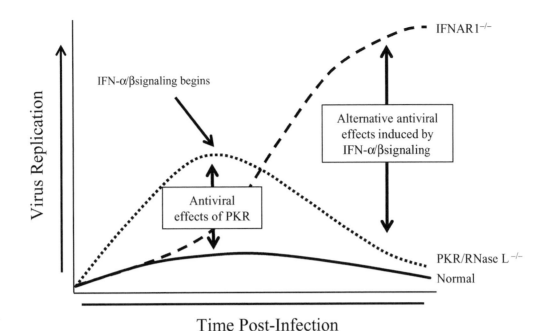

Figure 5 Diagrammatic representation of the PKR-mediated and PKR/RNase L/Mx-independent antiviral effects in SINV-infected primary DCs and mice in vivo. Constitutively expressed PKR significantly suppresses early virus replication; however, secreted IFN-α/β acts via an alternative pathway in the presence or absence of PKR, RNase L, and Mx to protect cells and clear virus infection. In the absence of the IFN-α/β-induced alternative pathway, virus replication continues unabated, resulting in cell death.

was rapidly fatal, indicating that systemic dissemination of virus is restricted by an alternative IFNAR-dependent mechanism. Second, antiviral activity was observed in vitro in mDCs from PKR$^{-/-}$ or TD mice at approximately 12 h postinfection. The delay in this effect presumably reflects the requirement for synthesis of IFN-α/β and expression of the ISG effector(s). Third, this effect was sensitive to neutralization of IFN-α/β and was not observed in IFNAR1$^{-/-}$ mDCs. Finally, priming of TD-derived mDCs with IFN-α/β induced residual antiviral activity against SINV that appeared nearly as potent as that observed in cells from wild-type mice. Thus, IFN-α/β can limit SINV replication via a PKR/RNase L/Mx-independent alternative pathway that requires IFN-α/β-mediated induction or activation of effector(s).

The "alternative" antiviral effect(s) that inhibits SINV replication in the absence of PKR, RNase L, and Mx was shown to require an IFN-α/β-mediated signaling cascade and new gene transcription, since the inhibitory effect was almost completely abrogated by inhibition of RNA Pol II (112). Global gene regulation in virus-infected or IFN-α/β-treated DCs was assessed and data mining was used to produce a set of 44 IFN-α/β-inducible genes proposed as candidate effectors of PKR/RNase

L-independent, IFN-α/β-induced anti-SINV activity (112). These genes were induced by at least fourfold in virus-infected and IFN-α/β-primed DCs derived from either normal or TD mice, but were significantly less induced by virus infection of IFNAR1$^{-/-}$ cells, correlating with the conditions under which antiviral activity was observed.

Candidate genes initially tested for antiviral activity were those that could conceivably have a role in translation inhibition, including known translation inhibitors, RNases, GTPases, and RNA-sequestering proteins, as well as genes that have been more recently identified as inhibiting SINV replication either in mice (ISG15) (72, 107) or when overexpressed in cultured cells (zinc finger activated protein [ZAP]) (10). By way of expression of candidate antiviral genes from a second subgenomic promoter in wild-type SINV, inserted downstream of the structural protein genes (109), the potential for these gene products to inhibit SINV replication and attenuate virus infection was tested in vivo (137). Neonatal mice were inoculated subcutaneously with in vitro-transcribed virus genomes, thereby avoiding the possibility that expression of genes with strong antiviral activities would be disproportionately attenuated by mutation or deletion during in vitro packaging reactions. Significant attenuation of disease compared with vectors expressing similarly sized control proteins was observed with ZAP, viperin, and ISG20, with ZAP expression exhibiting the highest degree of protection (137). Expression of ISG15, ISG56, ISG49, or ISG54 genes did not significantly alter the outcome of infection (137). However, a similar experiment expressing ISG15 from a less virulent SINV vector demonstrated significant antiviral activity for this protein in IFNAR1$^{-/-}$ mice (72), and increased virulence was observed in ISG15$^{-/-}$ animals (35). This discrepancy may be attributable to the fact that ISG15, a ubiquitin homolog, acts relatively late in infection and therefore has little or no effect if mortality occurs rapidly.

When overexpressed in vitro in a stable tetracycline-inducible MEF culture system, both ISG20 and ZAP potently inhibited SINV replication and progeny virion production, whereas viperin, ISG56, and ISG15 gene expression exhibited modest replication inhibition and ISG54 and ISG49 expression had little to no effect (137). siRNA-mediated knockdown of gene mRNAs in fibroblast cultures followed by SINV replication assessment confirmed this hierarchy of antiviral activity in IFN-α/β-primed cells. It was further demonstrated that (i) constitutive, endogenous expression of ZAP and ISG20 in unstimulated MEFs was partially protective; and (ii) combined suppression of ZAP and ISG20 produced a cumulative effect on cell permissivity (137), establishing that multiple IFN-α/β-inducible effectors act in concert to suppress alphavirus replication. Recent studies suggest that maximal antialphaviral activity of ZAP requires synergism with another, as yet unidentified, IFN-α/β-induced factor (76). Exploration of the antiviral mechanism employed by ZAP has revealed that steady-state levels of alphaviral RNA and, consequently, early translation of the nsPs are greatly reduced in ZAP-overexpressing cells (10). ZAP appears to bind specific viral RNA sequences via CCCH-type zinc-finger motifs (44) and recruit the exosome to mediate mRNA degradation (45).

The relative sensitivities of the more virulent alphaviruses to these newly identified IFN-α/β-inducible antiviral effectors have yet to be rigorously evaluated, although VEEV replicon genome replication has been shown to be sensitive to ZAP (10). Since alpha viruses exhibit dramatically increased resistance to the antiviral state established in IFN-α/β-primed cells, it seems likely that differences in their effects on maintenance of the antiviral state, as well as differences in their sensitivity to (and possibly specific antagonism of) individual effectors, will be observed.

Mechanism of IFN-α/β Antiviral Activity

A conundrum in the study of the antiviral state is the fact that, in general, the functions of host cells are preserved during times at which virus replication is severely inhibited. This implies that the antiviral state comprises activities that are capable of subtly distinguishing self from nonself and profoundly blocking nonself. These represent points in virus replication cycles at which nonself can be identified, as well as identification of virus vulnerabilities to specific inhibitory mechanisms. Several studies have focused upon determining the point(s) in the alphavirus replication cycle that is inhibited by IFN-α/β priming. One potent IFN-α/β-inducible antiviral activity acts at the level of translation initiation of the incoming alphavirus genome, largely independently of PKR (112, 126a). IFN-α/β treatment of PKR$^{-/-}$ cells results in transcriptional upregulation of a factor(s) that allows mRNAs generated in the nucleus to be preferentially translated over mRNAs introduced across the cytoplasmic membrane, and this effect is not specific to SINV sequences. Translation of VEEV and EEEV RNA genomes was also significantly inhibited; however, encephalomyocarditis virus internal ribosome entry site-driven translation initiation is resistant to this activity (126a). Virus delivery of RNA into the cytoplasm or its transcription within that compartment represents a point of distinction between the replication cycle of most RNA (or cytoplasmic) viruses and normal host processes. Therefore, this may represent an important and very early point of virus vulnerability to antiviral strategies that may be exploited for therapeutic benefit. An understanding of the mechanism by which IFN-α/β-induced effectors specifically inhibit the translation of infecting viruses while sparing nucleus-produced mRNA may promote the development of a new generation of antiviral drugs.

PERSPECTIVES

Upon virus infection of a host cell, a cascade of events occurs, resulting in the induction of innate immune cytokines including IFN-α/β. This IFN then acts to reduce virus replication in infected cells and increase the resistance of surrounding uninfected cells to initiation of infection. A great deal of research is currently devoted to investigation of viral mechanisms of antagonism of IFN-α/β induction and/or signaling pathways as a strategy through which viruses increase their replicative fitness within vertebrate hosts (reviewed in reference 8). Antiviral activity of IFN-α/β is a critical determinant of the outcome of togavirus infection both in vitro and in vivo. Furthermore, the characteristics of a given togavirus's interaction with this system are highly correlated with tropism and virulence in mice. For example, SINV, which induces substantial IFN-α/β in vivo and is avirulent in adult mice, causes uniform and rapid mortality in the absence of this response. VEEV induces high levels of IFN-α/β in vivo, yet causes mortality in normal adult mice in spite of this host response. In contrast, EEEV avoids the induction of IFN-α/β in adult normal mice and causes mortality. In vitro, SINV is the most sensitive to IFN-α/β priming, while VEEV is most resistant and EEEV gives an intermediate phenotype. A simple model derived from these results would be that SINV exhibits minimal antagonism of inductive or effector phases of the response, while VEEV effectively antagonizes the effector response and EEEV primarily antagonizes the inductive response. As described above, putative mechanisms of antagonism of IFN-α/β-inductive phases have been described for each of these viruses using in vitro systems. Important research goals include determining (i) whether or not other togaviruses antagonize inductive or effector responses; (ii) whether or not the mechanisms of antagonism identified in in vitro experimental systems actually function in relevant tissues in infected animals; (iii) the mechanism(s) through which proteins such as nsP2 or capsid result in blockage of cellular gene expression; (iv) whether or not individual mechanisms of antagonism or resistance are specific to particular hosts in the alternating vertebrate reservoir-arthropod vector-human cycle of the vector-borne togaviruses; and (v) which IFN-α/β-induced antiviral effectors are most important to inhibition of different togaviruses. A greater understanding of the interactions of togaviruses with the IFN-α/β system should lead to identification of pathogen vulnerabilities that can be exploited for antiviral drug and live-attenuated vaccine design.

References

1. Aguilar, P. V., S. Paessler, A. S. Carrara, S. Baron, J. Poast, E. Wang, A. C. Moncayo, M. Anishchenko, D. Watts, R. B. Tesh, and S. C. Weaver. 2005. Variation in interferon sensitivity and induction among strains of eastern equine encephalitis virus. *J. Virol.* **79:**11300–11310.

2. Aguilar, P. V., S. C. Weaver, and C. F. Basler. 2007. Capsid protein of eastern equine encephalitis virus inhibits host cell gene expression. *J. Virol.* **81:**3866–3876.

2a. Aguilar, P. V., L. W. Leung, E. Wang, S. C. Weaver, and C. F. Basler. 2008. A five amino acid deletion of the Eastern equine encephalitis virus capsid protein attenuates replication in mammalian but not mosquito cells. *J. Virol.* Epub ahead of print.

3. Ahola, T., A. Lampio, P. Auvinen, and L. Kaariainen. 1999. Semliki Forest virus mRNA capping enzyme requires association with anionic membrane phospholipids for activity. *EMBO J.* **18:**3164–3172.

4. Al Khatib, K., B. R. Williams, R. H. Silverman, W. Halford, and D. J. Carr. 2003. The murine double-stranded RNA-dependent protein kinase PKR and the murine 2′,5′-oligoadenylate synthetase-dependent RNase L are required for IFN-beta-mediated resistance against herpes simplex virus type 1 in primary trigeminal ganglion culture. *Virology* **313:**126–135.

5. Aronson, J. F., F. B. Grieder, N. L. Davis, P. C. Charles, T. Knott, K. Brown, and R. E. Johnston. 2000. A single-site mutant and revertants arising in vivo define early steps in the pathogenesis of Venezuelan equine encephalitis virus. *Virology* **270:**111–123.

6. Atkins, G. J., and C. L. Lancashire. 1976. The induction of interferon by temperature-sensitive mutants of Sindbis virus: its relationship to double-stranded RNA synthesis and cytopathic effect. *J. Gen. Virol.* **30:**157–165.

7. Baron, S., C. E. Buckler, R. V. McCloskey, and R. L. Kirschstein. 1966. Role of interferon during viremia. I. Production of circulating interferon. *J. Immunol.* **96:**12–16.

8. Basler, C. F., and A. Garcia-Sastre. 2002. Viruses and the type I interferon antiviral system: induction and evasion. *Int. Rev. Immunol.* **21:**305–337.

9. Beard, J. R., M. Trent, G. A. Sam, and V. C. Delpech. 1997. Self-reported morbidity of Barmah Forest virus infection on the north coast of New South Wales. *Med. J. Aust.* **167:**525–528.

10. Bick, M. J., J. W. Carroll, G. Gao, S. P. Goff, C. M. Rice, and M. R. Macdonald. 2003. Expression of the zinc-finger antiviral protein inhibits alphavirus replication. *J. Virol.* **77:**11555–11562.

11. Bonnard, M., C. Mirtsos, S. Suzuki, K. Graham, J. Huang, M. Ng, A. Itie, A. Wakeham, A. Shahinian, W. J. Henzel, A. J. Elia, W. Shillinglaw, T. W. Mak, Z. Cao, and W. C. Yeh. 2000. Deficiency of T2K leads to apoptotic liver degeneration and impaired NF-kappaB-dependent gene transcription. *EMBO J.* **19:**4976–4985.

12. Bradish, C. J., K. Allner, and H. B. Maber. 1972. Infection, interaction and the expression of virulence by defined strains of Semliki forest virus. *J. Gen. Virol.* **16:**359–372.

13. Brawand, P., D. R. Fitzpatrick, B. W. Greenfield, K. Brasel, C. R. Maliszewski, and T. De Smedt. 2002. Murine plasmacytoid pre-dendritic cells generated from Flt3 ligand-supplemented bone marrow cultures are immature APCs. *J. Immunol.* **169:**6711–6719.

14. Breakwell, L., P. Dosenovic, G. B. Karlsson Hedestam, M. D'Amato, P. Liljestrom, J. Fazakerley, and G. M. McInerney. 2007. Semliki Forest virus nonstructural protein 2 is involved in suppression of the type I interferon response. *J. Virol.* **81:**8677–8684.

15. Byrnes, A. P., J. E. Durbin, and D. E. Griffin. 2000. Control of Sindbis virus infection by antibody in interferon-deficient mice. *J. Virol.* **74:**3905–3908.

16. Chantler, J., J. S. Wolinsky, and A. Tingle. 2001. Rubella virus, p. 963–990. *In* D. M. Knipe and P. M. Howley (ed.), *Fields Virology*, 4th ed. Lippincott, Williams & Wilkins, Philadelphia, PA.

17. Charles, P. C., J. Trgovcich, N. L. Davis, and R. E. Johnston. 2001. Immunopathogenesis and immune modulation of Venezuelan equine encephalitis virus-induced disease in the mouse. *Virology* **284:**190–202.

18. Charles, P. C., E. Walters, F. Margolis, and R. E. Johnston. 1995. Mechanism of neuroinvasion of Venezuelan equine encephalitis virus in the mouse. *Virology* 208:662–671.

19. Choi, H. K., L. Tong, W. Minor, P. Dumas, U. Boege, M. G. Rossmann, and G. Wengler. 1991. Structure of Sindbis virus core protein reveals a chymotrypsin-like serine proteinase and the organization of the virion. *Nature* 354:37–43.

19a. Couderc, T., F. Chretien, C. Shilte, O. Disson, M. Brigitte, F. Guivel-Benhassine, Y. Touret, G. Barau, N. Cayet, I. Schuffenecker, P. Despres, F. Arenzana-Seisedos, A. Michault, M. L. Albert, and M. Lecuit. 2008. A mouse model for Chikungunya: young age and inefficient type-I interferon signaling are risk factors for severe disease. *PLoS. Pathog.* 4:e29.

20. Davis, N. L., F. B. Grieder, J. F. Smith, G. F. Greenwald, M. L. Valenski, D. C. Sellon, P. C. Charles, and R. E. Johnston. 1994. A molecular genetic approach to the study of Venezuelan equine encephalitis virus pathogenesis. *Arch. Virol. Suppl.* 9:99–109.

21. Enserink, M. 2007. Epidemiology. Tropical disease follows mosquitoes to Europe. *Science* 317:1485.

22. Favre, D., E. Studer, and M. R. Michel. 1996. Semliki Forest virus capsid protein inhibits the initiation of translation by upregulating the double-stranded RNA-activated protein kinase (PKR). *Biosci. Rep.* 16:485–511.

23. Fazakerley, J. K. 2004. Semliki forest virus infection of laboratory mice: a model to study the pathogenesis of viral encephalitis. *Arch. Virol. Suppl.* 179–190.

24. Finlay, B. B., and G. McFadden. 2006. Anti-immunology: evasion of the host immune system by bacterial and viral pathogens. *Cell* 124:767–782.

25. Finter, N. B. 1964. Interferon production and sensitivity of Semliki Forest virus variants. *J. Hyg. (Lond.)* 62:337–349.

26. Flexman, J. P., D. W. Smith, J. S. Mackenzie, J. R. Fraser, S. P. Bass, L. Hueston, M. D. Lindsay, and A. L. Cunningham. 1998. A comparison of the diseases caused by Ross River virus and Barmah Forest virus. *Med. J. Aust.* 169:159–163.

27. Friedman, R. M. 1968. Inhibition of arbovirus protein synthesis by interferon. *J. Virol.* 2:1081–1085.

28. Frolov, I., E. Agapov, T. A. Hoffman, Jr., B. M. Pragai, M. Lippa, S. Schlesinger, and C. M. Rice. 1999. Selection of RNA replicons capable of persistent noncytopathic replication in mammalian cells. *J. Virol.* 73:3854–3865.

29. Frolov, I., R. Hardy, and C. M. Rice. 2001. Cis-acting RNA elements at the 5' end of Sindbis virus genome RNA regulate minus- and plus-strand RNA synthesis. *RNA* 7:1638–1651.

30. Frolova, E. I., R. Z. Fayzulin, S. H. Cook, D. E. Griffin, C. M. Rice, and I. Frolov. 2002. Roles of nonstructural protein nsP2 and alpha/beta interferons in determining the outcome of Sindbis virus infection. *J. Virol.* 76:11254–11264.

31. Froshauer, S., J. Kartenbeck, and A. Helenius. 1988. Alphavirus RNA replicase is located on the cytoplasmic surface of endosomes and lysosomes. *J. Cell Biol.* 107:2075–2086.

32. Garmashova, N., S. Atasheva, W. Kang, S. C. Weaver, E. Frolova, and I. Frolov. 2007. Analysis of Venezuelan equine encephalitis virus capsid protein function in the inhibition of cellular transcription. *J. Virol.* 81:13552–13565.

33. Garmashova, N., R. Gorchakov, E. Frolova, and I. Frolov. 2006. Sindbis virus nonstructural protein nsP2 is cytotoxic and inhibits cellular transcription. *J. Virol.* 80:5686–5696.

34. Garmashova, N., R. Gorchakov, E. Volkova, S. Paessler, E. Frolova, and I. Frolov. 2007. The Old World and New World alphaviruses use different virus-specific proteins for induction of transcriptional shutoff. *J. Virol.* 81:2472–2484.

35. Giannakopoulos, N. V., J. K. Luo, V. Papov, W. Zou, D. J. Lenschow, B. S. Jacobs, E. C. Borden, J. Li, H. W. Virgin, and D. E. Zhang. 2005. Proteomic identification of proteins conjugated to ISG15 in mouse and human cells. *Biochem. Biophys. Res. Commun.* 336:496–506.

36. Gilfoy, F. D., and P. W. Mason. 2007. West Nile virus-induced interferon production is mediated by the double-stranded RNA-dependent protein kinase PKR. *J. Virol.* 81:11148–11158.

37. Gitlin, L., W. Barchet, S. Gilfillan, M. Cella, B. Beutler, R. A. Flavell, M. S. Diamond, and M. Colonna. 2006. Essential role of mda-5 in type I IFN responses to polyriboinosinic:polyribocytidylic acid and encephalomyocarditis picornavirus. *Proc. Natl. Acad. Sci. USA* 103:8459–8464.

38. Gomez de Cedron, M., N. Ehsani, M. L. Mikkola, J. A. Garcia, and L. Kaariainen. 1999. RNA helicase activity of Semliki Forest virus replicase protein NSP2. *FEBS Lett.* 448:19–22.

39. Gorchakov, R., E. Frolova, B. R. Williams, C. M. Rice, and I. Frolov. 2004. PKR-dependent and -independent mechanisms are involved in translational shutoff during Sindbis virus infection. *J. Virol.* 78:8455–8467.

40. Grieder, F. B., and S. N. Vogel. 1999. Role of interferon and interferon regulatory factors in early protection against Venezuelan equine encephalitis virus infection. *Virology* 257:106–118.

41. Griffin, D. E. 1976. Role of the immune response in age-dependent resistance of mice to encephalitis due to Sindbis virus. *J. Infect. Dis.* 133:456–464.

42. Griffin, D. E. 2001. Alphaviruses, p. 917–962. *In* D. M. Knipe and P. M. Howley (ed.), *Fields Virology*, 4th ed. Lippincott, Williams & Wilkins, Philadelphia, PA.

43. Grimley, P. M., J. G. Levin, I. K. Berezesky, and R. M. Friedman. 1972. Specific membranous structures associated with the replication of group A arboviruses. *J. Virol.* 10:492–503.

44. Guo, X., J. W. Carroll, M. R. Macdonald, S. P. Goff, and G. Gao. 2004. The zinc finger antiviral protein directly binds to specific viral mRNAs through the CCCH zinc finger motifs. *J. Virol.* 78:12781–12787.

45. Guo, X., J. Ma, J. Sun, and G. Gao. 2007. The zinc-finger antiviral protein recruits the RNA processing exosome to degrade the target mRNA. *Proc. Natl. Acad. Sci. USA* 104:151–156.

46. Hardy, W. R., and J. H. Strauss. 1989. Processing the nonstructural polyproteins of Sindbis virus: nonstructural proteinase is in the C-terminal half of nsP2 and functions both in *cis* and in *trans*. *J. Virol.* 63:4653–4664.

47. Hefti, H. P., M. Frese, H. Landis, C. Di Paolo, A. Aguzzi, O. Haller, and J. Pavlovic. 1999. Human MxA protein protects mice lacking a functional alpha/beta interferon system against La Crosse virus and other lethal viral infections. *J. Virol.* 73:6984–6991.

48. Heise, M. T., D. A. Simpson, and R. E. Johnston. 2000. Sindbis-group alphavirus replication in periosteum and endosteum of long bones in adult mice. *J. Virol.* 74:9294–9299.

49. Hidmark, A. S., G. M. McInerney, E. K. Nordstrom, I. Douagi, K. M. Werner, P. Liljestrom, and G. B. Karlsson Hedestam. 2005. Early alpha/beta interferon production by myeloid dendritic cells in response to UV-inactivated virus requires viral entry and interferon regulatory factor 3 but not MyD88. *J. Virol.* 79:10376–10385.

50. Hoenen, A., W. Liu, G. Kochs, A. A. Khromykh, and J. M. Mackenzie. 2007. West Nile virus-induced cytoplasmic membrane structures provide partial protection against the interferon-induced antiviral MxA protein. *J. Gen. Virol.* 88:3013–3017.

51. Honda, K., A. Takaoka, and T. Taniguchi. 2006. Type I interferon [corrected] gene induction by the interferon regulatory factor family of transcription factors. *Immunity* 25:349–360.

52. Hornung, V., J. Ellegast, S. Kim, K. Brzózka, A. Jung, H. Kato, H. Poeck, S. Akira, K. K. Conzelmann, M. Schlee, S. Endres, and G. Hartmann. 2006. 5′-Triphosphate RNA is the ligand for RIG-I. *Science* 314:994–997.

53. Hsieh, P., and P. W. Robbins. 1984. Regulation of asparagine-linked oligosaccharide processing. Oligosaccharide processing in *Aedes albopictus* mosquito cells. *J. Biol. Chem.* 259:2375–2382.

54. Hsieh, P., M. R. Rosner, and P. W. Robbins. 1983. Host-dependent variation of asparagine-linked oligosaccharides at individual glycosylation sites of Sindbis virus glycoproteins. *J. Biol. Chem.* 258:2548–2554.

55. Iwasaki, A., and R. Medzhitov. 2004. Toll-like receptor control of the adaptive immune responses. *Nat. Immunol.* 5:987–995.

56. Jackson, A. C., and J. P. Rossiter. 1997. Apoptotic cell death is an important cause of neuronal injury in experimental Venezuelan equine encephalitis virus infection of mice. *Acta Neuropathol.* 93:349–353.

57. Jiang, Z., T. W. Mak, G. Sen, and X. Li. 2004. Toll-like receptor 3-mediated activation of NF-kappaB and IRF3 diverges at Toll-IL-1 receptor domain-containing adapter inducing IFN-beta. *Proc. Natl. Acad. Sci. USA* 101:3533–3538.

58. Jiang, Z., M. Zamanian-Daryoush, H. Nie, A. M. Silva, B. R. Williams, and X. Li. 2003. Poly(I-C)-induced Toll-like receptor 3 (TLR3)-mediated activation of NFkappa B and MAP kinase is through an interleukin-1 receptor-associated kinase (IRAK)-independent pathway employing the signaling components TLR3-TRAF6-TAK1-TAB2-PKR. *J. Biol. Chem.* 278:16713–16719.

59. Jin, Y., L. Fuller, G. Ciancio, G. W. Burke III, A. G. Tzakis, C. Ricordi, J. Miller, and V. Esquenazi. 2004. Antigen presentation and immune regulatory capacity of immature and mature-enriched antigen presenting (dendritic) cells derived from human bone marrow. *Hum. Immunol.* 65:93–103.

60. Johnson, R. T. 1965. Virus invasion of the central nervous system: a study of Sindbis virus infection in the mouse using fluorescent antibody. *Am. J. Pathol.* 70:929–943.

61. Johnson, R. T., H. F. McFarland, and S. E. Levy. 1972. Age-dependent resistance to viral encephalitis: studies of infections due to Sindbis virus in mice. *J. Infect. Dis.* 125:257–262.

62. Kang, D. C., R. V. Gopalkrishnan, Q. Wu, E. Jankowsky, A. M. Pyle, and P. B. Fisher. 2002. *mda-5*: an interferon-inducible putative RNA helicase with double-stranded RNA-dependent ATPase activity and melanoma growth-suppressive properties. *Proc. Natl. Acad. Sci. USA* 99:637–642.

63. Kielian, M. 2002. Structural surprises from the flaviviruses and alphaviruses. *Mol. Cell* 9:454–456.

64. Klimstra, W. B., E. M. Nangle, M. S. Smith, A. D. Yurochko, and K. D. Ryman. 2003. DC-SIGN and L-SIGN can act as attachment receptors for alphaviruses and distinguish between mosquito cell- and mammalian cell-derived viruses. *J. Virol.* 77:12022–12032.

65. Klimstra, W. B., K. D. Ryman, K. A. Bernard, K. B. Nguyen, C. A. Biron, and R. E. Johnston. 1999. Infection of neonatal mice with Sindbis virus results in a systemic inflammatory response syndrome. *J. Virol.* 73:10387–10398.

66. Klimstra, W. B., K. D. Ryman, and R. E. Johnston. 1998. Adaptation of Sindbis virus to BHK cells selects for use of heparan sulfate as an attachment receptor. *J. Virol.* 72:7357–7366.

67. Landis, H., A. Simon-Jodicke, A. Kloti, C. Di Paolo, J. J. Schnorr, S. Schneider-Schaulies, H. P. Hefti, and J. Pavlovic. 1998. Human MxA protein confers resistance to Semliki Forest virus and inhibits the amplification of a Semliki Forest virus-based replicon in the absence of viral structural proteins. *J. Virol.* 72:1516–1522.

68. Lee, J. Y., J. A. Marshall, and D. S. Bowden. 1994. Characterization of rubella virus replication complexes using antibodies to double-stranded RNA. *Virology* 200:307–312.

69. Lemm, J. A., and C. M. Rice. 1993. Assembly of functional Sindbis virus RNA replication complexes: requirement for coexpression of P123 and P34. *J. Virol.* 67:1905–1915.

70. Lemm, J. A., and C. M. Rice. 1993. Roles of nonstructural polyproteins and cleavage products in regulating Sindbis virus RNA replication and transcription. *J. Virol.* 67:1916–1926.

71. Lemm, J. A., T. Rumenapf, E. G. Strauss, J. H. Strauss, and C. M. Rice. 1994. Polypeptide requirements for assembly of functional Sindbis virus replication complexes: a model for the temporal regulation of minus- and plus-strand RNA synthesis. *EMBO J.* 13:2925–2934.

72. Lenschow, D. J., N. V. Giannakopoulos, L. J. Gunn, C. Johnston, A. K. O'Guin, R. E. Schmidt, B. Levine, and H. W. Virgin IV. 2005. Identification of interferon-stimulated gene 15 as an antiviral molecule during Sindbis virus infection in vivo. *J. Virol.* 79:13974–13983.

73. Lidbury, B. A., and S. Mahalingam. 2000. Specific ablation of antiviral gene expression in macrophages by antibody-dependent enhancement of Ross River virus infection. *J. Virol.* 74:8376–8381.

74. Lidbury, B. A., C. Simeonovic, G. E. Maxwell, I. D. Marshall, and A. J. Hapel. 2000. Macrophage-induced muscle pathology results in morbidity and mortality for Ross River virus-infected mice. *J. Infect. Dis.* 181:27–34.

75. MacDonald, G. H., and R. E. Johnston. 2000. Role of dendritic cell targeting in Venezuelan equine encephalitis virus pathogenesis. *J. Virol.* 74:914–922.

76. Macdonald, M. R., E. S. Machlin, O. R. Albin, and D. E. Levy. 2007. The zinc-finger antiviral protein acts synergistically with an interferon-induced factor for maximal activity against alphaviruses. *J. Virol.* **81**:13509–13518.

77. Mahalingam, S., and B. A. Lidbury. 2002. Suppression of lipopolysaccharide-induced antiviral transcription factor (Stat-1 and NF-kappa B) complexes by antibody-dependent enhancement of macrophage infection by Ross River virus. *Proc. Natl. Acad. Sci. USA* **99**:13819–13824.

78. Marcus, P. I., and F. J. Fuller. 1979. Interferon induction by viruses. II. Sindbis virus: interferon induction requires one-quarter of the genome—genes G and A. *J. Gen. Virol.* **44**:169–177.

79. Marland, G., A. B. Bakker, G. J. Adema, and C. G. Figdor. 1996. Dendritic cells in immune response induction. *Stem Cells* **14**:501–507.

80. McGreal, E. P., L. Martinez-Pomares, and S. Gordon. 2004. Divergent roles for C-type lectins expressed by cells of the innate immune system. *Mol. Immunol.* **41**:1109–1121.

81. McWhirter, S. M., K. A. Fitzgerald, J. Rosains, D. C. Rowe, D. T. Golenbock, and T. Maniatis. 2004. IFN-regulatory factor 3-dependent gene expression is defective in *Tbk1*-deficient mouse embryonic fibroblasts. *Proc. Natl. Acad. Sci. USA* **101**:233–238.

82. Mecs, E., J. A. Sonnabend, E. M. Martin, and K. H. Fantes. 1967. The effect of interferon on the synthesis of RNA in chick cells infected with Semliki forest virus. *J. Gen. Virol.* **1**:25–40.

83. Meylan, E., J. Tschopp, and M. Karin. 2006. Intracellular pattern recognition receptors in the host response. *Nature* **442**:39–44.

84. Monlux, W. S., A. J. Luedke, and J. Bowne. 1972. Central nervous system response of horses to Venezuelan equine encephalomyelitis vaccine (TC-83). *J. Am. Vet. Med. Assoc.* **161**:265–269.

85. Morens, D. M., and S. B. Halstead. 1990. Measurement of antibody-dependent infection enhancement of four dengue virus serotypes by monoclonal and polyclonal antibodies. *J. Gen. Virol.* **71**(Pt. 12):2909–2914.

86. Morrison, T. E., A. C. Whitmore, R. S. Shabman, B. A. Lidbury, S. Mahalingam, and M. T. Heise. 2006. Characterization of Ross River virus tropism and virus-induced inflammation in a mouse model of viral arthritis and myositis. *J. Virol.* **80**:737–749.

87. Muller, U., U. Steinhoff, L. F. Reis, S. Hemmi, J. Pavlovic, R. M. Zinkernagel, and M. Aguet. 1994. Functional role of type I and type II interferons in antiviral defense. *Science* **264**:1918–1921.

88. Navarro-Sanchez, E., R. Altmeyer, A. Amara, O. Schwartz, F. Fieschi, J. L. Virelizier, F. Arenzana-Seisdedos, and P. Despres. 2003. Dendritic-cell-specific ICAM3-grabbing non-integrin is essential for the productive infection of human dendritic cells by mosquito-cell-derived dengue viruses. *EMBO Rep.* **4**:723–728.

89. Niesters, H. G., and J. H. Strauss. 1990. Defined mutations in the 5′ nontranslated sequence of Sindbis virus RNA. *J. Virol.* **64**:4162–4168.

90. Niesters, H. G., and J. H. Strauss. 1990. Mutagenesis of the conserved 51-nucleotide region of Sindbis virus. *J. Virol.* **64**:1639–1647.

91. Ou, J. H., E. G. Strauss, and J. H. Strauss. 1981. Comparative studies of the 3′-terminal sequences of several alpha virus RNAs. *Virology* **109**:281–289.

92. Paz, S., Q. Sun, P. Nakhaei, R. Romieu-Mourez, D. Goubau, I. Julkunen, R. Lin, and J. Hiscott. 2006. Induction of IRF3 and IRF7 phosphorylation following activation of the RIG-I pathway. *Cell. Mol. Biol. (Noisy-le-grand)* **52**:17–28.

93. Peranen, J., and L. Kaariainen. 1991. Biogenesis of type I cytopathic vacuoles in Semliki Forest virus-infected BHK cells. *J. Virol.* **65**:1623–1627.

94. Peranen, J., M. Rikkonen, P. Liljestrom, and L. Kaariainen. 1990. Nuclear localization of Semliki Forest virus-specific nonstructural protein nsP2. *J. Virol.* **64**:1888–1896.

95. Perera, R., K. E. Owen, T. L. Tellinghuisen, A. E. Gorbalenya, and R. J. Kuhn. 2001. Alphavirus nucleocapsid protein contains a putative coiled coil alpha-helix important for core assembly. *J. Virol.* **75**:1–10.

96. Pichlmair, A., O. Schulz, C. P. Tan, T. I. Naslund, P. Liljestrom, F. Weber, and C. Reis e Sousa. 2006. RIG-I-mediated antiviral responses to single-stranded RNA bearing 5′-phosphates. *Science* **314**:997–1001.

97. Pinheiro, F. P., R. B. Freitas, J. F. Travassos da Rosa, Y. B. Gabbay, W. A. Mello, and J. W. LeDuc. 1981. An outbreak of Mayaro virus disease in Belterra, Brazil. I. Clinical and virological findings. *Am. J. Trop. Med. Hyg.* **30**:674–681.

98. Pletnev, S. V., W. Zhang, S. Mukhopadhyay, B. R. Fisher, R. Hernandez, D. T. Brown, T. S. Baker, M. G. Rossmann, and R. J. Kuhn. 2001. Locations of carbohydrate sites on alphavirus glycoproteins show that E1 forms an icosahedral scaffold. *Cell* **105**:127–136.

99. Pollara, G., D. R. Katz, and B. M. Chain. 2005. LIGHTing up dendritic cell activation: immune regulation and viral exploitation. *J. Cell. Physiol.* **205**:161–162.

100. Porterfield, J. S. 1986. Antibody-dependent enhancement of viral infectivity. *Adv. Virus Res.* **31**:335–355.

101. Postic, B., C. J. Schleupner, J. A. Armstrong, and M. Ho. 1969. Two variants of Sindbis virus which differ in interferon induction and serum clearance. I. The phenomenon. *J. Infect. Dis.* **120**:339–347.

102. Proll, S., G. Dobler, M. Pfeffer, T. Jelinek, H. D. Nothdurft, and T. Loscher. 1999. [Persistent arthralgias in Ross-River-Virus disease after travel to the South Pacific]. *Dtsch. Med. Wochenschr.* **124**:759–762.

103. Rabinovich, S., and C. Liu. 1968. Interferon content of organs and serum after intravenous inoculation of mice with Sindbis virus. *Exp. Med. Surg.* **26**:117–123.

104. Rebello, M. C., M. E. Fonseca, J. O. Marinho, and M. A. Rebello. 1993. Interferon action on Mayaro virus replication. *Acta Virol.* **37**:223–231.

105. Rikkonen, M. 1996. Functional significance of the nuclear-targeting and NTP-binding motifs of Semliki Forest virus nonstructural protein nsP2. *Virology* **218**:352–361.

106. Rikkonen, M., J. Peranen, and L. Kaariainen. 1994. ATPase and GTPase activities associated with Semliki Forest virus nonstructural protein nsP2. *J. Virol.* **68**:5804–5810.

107. Ritchie, K. J., C. S. Hahn, K. I. Kim, M. Yan, D. Rosario, L. Li, J. C. de la Torre, and D. E. Zhang. 2004.

Role of ISG15 protease UBP43 (USP18) in innate immunity to viral infection. *Nat. Med.* **10:**1374–1378.

108. Ryman, K. D., C. L. Gardner, C. W. Burke, K. C. Meier, J. M. Thompson, and W. B. Klimstra. 2007. Heparan sulfate binding can contribute to the neurovirulence of neuroadapted and nonneuroadapted Sindbis viruses. *J. Virol.* **81:**3563–3573.

109. Ryman, K. D., W. B. Klimstra, and R. E. Johnston. 2004. Attenuation of Sindbis virus variants incorporating uncleaved PE2 glycoprotein is correlated with attachment to cell-surface heparan sulfate. *Virology* **322:**1–12.

110. Ryman, K. D., W. B. Klimstra, K. B. Nguyen, C. A. Biron, and R. E. Johnston. 2000. Alpha/beta interferon protects adult mice from fatal Sindbis virus infection and is an important determinant of cell and tissue tropism. *J. Virol.* **74:**3366–3378.

111. Ryman, K. D., K. C. Meier, C. L. Gardner, and W. B. Klimstra. 2007. Non-pathogenic Sindbis virus causes viral hemorrhagic fever in the absence of alpha/beta and gamma interferons. *Virology* **368:**273–285.

112. Ryman, K. D., K. C. Meier, E. M. Nangle, S. L. Ragsdale, N. L. Korneeva, R. E. Rhoads, M. R. Macdonald, and W. B. Klimstra. 2005. Sindbis virus translation is inhibited by a PKR/RNase L-independent effector induced by alpha/beta interferon priming of dendritic cells. *J. Virol.* **79:**1487–1499.

113. Ryman, K. D., L. J. White, R. E. Johnston, and W. B. Klimstra. 2002. Effects of PKR/RNase L-dependent and alternative antiviral pathways on alphavirus replication and pathogenesis. *Viral Immunol.* **15:**53–76.

114. Reference deleted.

115. Sawicki, D. L., R. H. Silverman, B. R. Williams, and S. G. Sawicki. 2003. Alphavirus minus-strand synthesis and persistence in mouse embryo fibroblasts derived from mice lacking RNase L and protein kinase R. *J. Virol.* **77:**1801–1811.

116. Schafer, S. L., R. Lin, P. A. Moore, J. Hiscott, and P. M. Pitha. 1998. Regulation of type I interferon gene expression by interferon regulatory factor-3. *J. Biol. Chem.* **273:**2714–2720.

117. Schoneboom, B. A., J. S. Lee, and F. B. Grieder. 2000. Early expression of IFN-alpha/beta and iNOS in the brains of Venezuelan equine encephalitis virus-infected mice. *J. Interferon Cytokine Res.* **20:**205–215.

118. Schoub, B. D., C. J. Dommann, S. Johnson, C. Downie, and P. L. Patel. 1990. Encephalitis in a 13-year-old boy following 17D yellow fever vaccine. *J. Infect.* **21:**105–106.

119. Shabman, R. S., T. E. Morrison, C. Moore, L. White, M. S. Suthar, L. Hueston, N. Rulli, B. Lidbury, J. P. Ting, S. Mahalingam, and M. T. Heise. 2007. Differential induction of type I interferon responses in myeloid dendritic cells by mosquito and mammalian-cell-derived alphaviruses. *J. Virol.* **81:**237–247.

120. Sherman, L. A., and D. E. Griffin. 1990. Pathogenesis of encephalitis induced in newborn mice by virulent and avirulent strains of Sindbis virus. *J. Virol.* **64:**2041–2046.

121. Shirako, Y., and J. H. Strauss. 1994. Regulation of Sindbis virus RNA replication: uncleaved P123 and nsP4

function in minus-strand RNA synthesis, whereas cleaved products from P123 are required for efficient plus-strand RNA synthesis. *J. Virol.* **68:**1874–1885.

122. Sonnabend, J. A., E. M. Martin, E. Mecs, and K. H. Fantes. 1967. The effect of interferon on the synthesis and activity of an RNA polymerase isolated from chick cells infected with Semliki forest virus. *J. Gen. Virol.* **1:**41–48.

123. Strahle, L., J. B. Marq, A. Brini, S. Hausmann, D. Kolakofsky, and D. Garcin. 2007. Activation of the beta interferon promoter by unnatural Sendai virus infection requires RIG-I and is inhibited by viral C proteins. *J. Virol.* **81:**12227–12237.

124. Strauss, E. G., C. M. Rice, and J. H. Strauss. 1984. Complete nucleotide sequence of the genomic RNA of Sindbis virus. *Virology* **133:**92–110.

125. Strauss, J. H., and E. G. Strauss. 1994. The alphaviruses: gene expression, replication, and evolution. *Microbiol. Rev.* **58:**491–562.

126. Tassaneetrithep, B., T. H. Burgess, A. Granelli-Piperno, C. Trumpfheller, J. Finke, W. Sun, M. A. Eller, K. Pattanapanyasat, S. Sarasombath, D. L. Birx, R. M. Steinman, S. Schlesinger, and M. A. Marovich. 2003. DC-SIGN (CD209) mediates dengue virus infection of human dendritic cells. *J. Exp. Med.* **197:**823–829.

126a.Tesfay, M., J. Yin, C. L. Gardner, M. Khoretonenko, N. Korneeva, R. E. Rhoads, K. D. Ryman, and W. B. Klimstra. 2008. Interferon alpha/beta inhibits cap-dependent translation of viral but not cellular mRNA by a PKR-independent mechanism. *J. Virol.* **82:**2620–2630.

127. Tesh, R. B., D. M. Watts, K. L. Russell, C. Damodaran, C. Calampa, C. Cabezas, G. Ramirez, B. Vasquez, C. G. Hayes, C. A. Rossi, A. M. Powers, C. L. Hice, L. J. Chandler, B. C. Cropp, N. Karabatsos, J. T. Roehrig, and D. J. Gubler. 1999. Mayaro virus disease: an emerging mosquito-borne zoonosis in tropical South America. *Clin. Infect. Dis.* **28:**67–73.

128. Trgovcich, J., J. F. Aronson, and R. E. Johnston. 1996. Fatal Sindbis virus infection of neonatal mice in the absence of encephalitis. *Virology* **224:**73–83.

129. Trischmann, H., D. Davis, and P. J. Lachmann. 1995. Lymphocytotropic strains of HIV type 1 when complexed with enhancing antibodies can infect macrophages via Fc gamma RIII, independently of CD4. *AIDS Res. Hum. Retroviruses* **11:**343–352.

130. Vasiljeva, L., A. Merits, P. Auvinen, and L. Kaariainen. 2000. Identification of a novel function of the alphavirus capping apparatus. RNA 5′-triphosphatase activity of Nsp2. *J. Biol. Chem.* **275:**17281–17287.

131. Ventoso, I., M. A. Sanz, S. Molina, J. J. Berlanga, L. Carrasco, and M. Esteban. 2006. Translational resistance of late alphavirus mRNA to eIF2alpha phosphorylation: a strategy to overcome the antiviral effect of protein kinase PKR. *Genes Dev.* **20:**87–100.

132. Vogel, P., W. M. Kell, D. L. Fritz, M. D. Parker, and R. J. Schoepp. 2005. Early events in the pathogenesis of eastern equine encephalitis virus in mice. *Am. J. Pathol.* **166:**159–171.

133. Watts, D. M., J. Callahan, C. Rossi, M. S. Oberste, J. T. Roehrig, M. T. Wooster, J. F. Smith, C. B. Cropp, E. M. Gentrau, N. Karabatsos, D. Gubler, and C. G. Hayes.

1998. Venezuelan equine encephalitis febrile cases among humans in the Peruvian Amazon River region. *Am. J. Trop. Med. Hyg.* **58:**35–40.

134. **Wesselingh, S. L., B. Levine, R. J. Fox, S. Choi, and D. E. Griffin.** 1994. Intracerebral cytokine mRNA expression during fatal and nonfatal alphavirus encephalitis suggests a predominant type 2 T cell response. *J. Immunol.* **152:**1289–1297.

135. **White, L. J., J. G. Wang, N. L. Davis, and R. E. Johnston.** 2001. Role of alpha/beta interferon in Venezuelan equine encephalitis virus pathogenesis: effect of an attenuating mutation in the 5′ untranslated region. *J. Virol.* **75:**3706–3718.

136. **Yoneyama, M., M. Kikuchi, T. Natsukawa, N. Shinobu, T. Imaizumi, M. Miyagishi, K. Taira, S. Akira, and T. Fujita.** 2004. The RNA helicase RIG-I has an essential function in double-stranded RNA-induced innate antiviral responses. *Nat. Immunol.* **5:**730–737.

137. **Zhang, Y., C. W. Burke, K. D. Ryman, and W. B. Klimstra.** 2007. Identification and characterization of interferon-induced proteins that inhibit alphavirus replication. *J. Virol.* **81:**11246–11255.

138. **Zhou, H., and S. Perlman.** 2007. Mouse hepatitis virus does not induce beta interferon synthesis and does not inhibit its induction by double-stranded RNA. *J. Virol.* **81:**568–574.

Cellular Signaling and Innate Immune
Responses to RNA Virus Infections
Edited by A. R. Brasier et al.
© 2009 ASM Press, Washington, D.C.

Krishna Narayanan
Shinji Makino

Coronaviruses and Arteriviruses

23

The ability to defend against invading pathogens depends upon the detection of the pathogen and mounting of a rapid and effective response by the infected host. The antiviral innate immune response is triggered after the host senses a danger signal like viral infection in the cellular microenvironment. Over the past decade, several families of sensors of viral infection have been identified. These sensors are called pattern recognition receptors (PRRs) because they recognize conserved structural moieties present in viruses, known as pathogen-associated molecular patterns (PAMPs) (87). The two major, well-characterized PRRs are the Toll-like receptors (TLRs) and the retinoic acid-inducible gene I (RIG-I)-like RNA helicases (RLHs). These receptors form a key sensing arm of the innate antiviral immune response that upon stimulation by ligands such as PAMPs initiates an intracellular signaling cascade leading to the production of innate immune mediators like type I interferons (IFNs) and proinflammatory cytokines. The innate immunity, through the action of mediators like type I IFNs, serves as the immediate responder to viral infection by controlling virus spread and buying time for the host to develop an effective antigen-specific adaptive immune response.

In order to replicate efficiently and establish a productive infection in the host, it seems highly reasonable that many viruses have developed strategies to inhibit the production or action of IFNs to disarm the early innate immune response. A critical factor that determines the host range, virulence, and pathogenicity of a virus is the kinetics and efficiency by which the virus antagonizes the IFN response in different susceptible hosts. A clear understanding of the molecular mechanisms of immune evasion is essential to identify the determinants of virulence and pathogenesis of specific viruses and develop effective antiviral approaches.

A great deal of information is available regarding the interactions of coronaviruses with the adaptive immune system; however, only a limited number of studies have investigated the role of innate immune responses in modulating coronavirus and arterivirus infections. The emergence of a severe acute respiratory syndrome coronavirus (SARS-CoV) epidemic in 2003 has ignited a renewed interest in studying how this family of viruses circumvents the innate immune responses and its role in viral pathogenesis. The purpose of this chapter is to review the literature to summarize what is known about (i) the interactions of coronaviruses and arteriviruses

Krishna Narayanan and Shinji Makino, Department of Microbiology and Immunology, The University of Texas Medical Branch at Galveston, Galveston, TX 77555-1019.

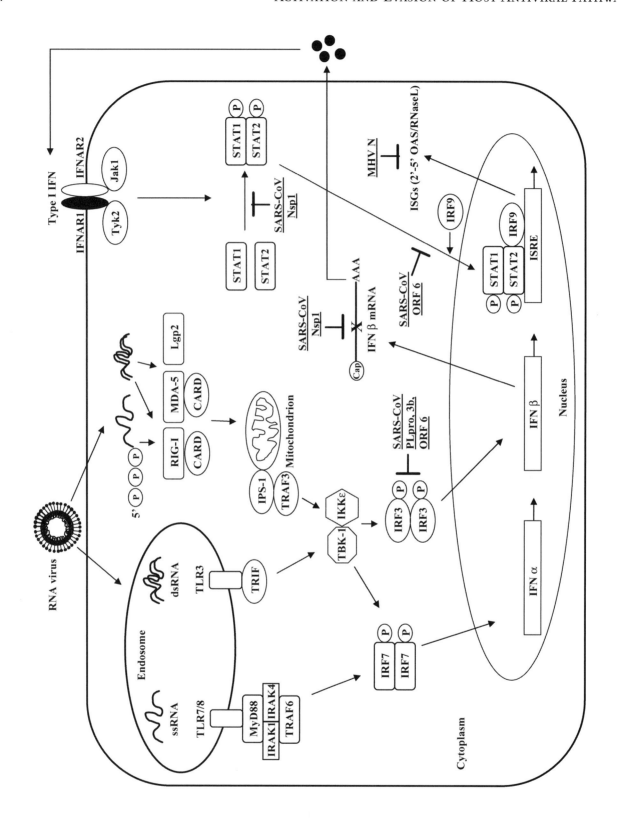

with the innate immune response, including the type I IFN system; and (ii) the strategies employed by these viruses to evade or inhibit the antiviral IFN response.

THE INNATE ANTIVIRAL IMMUNE RESPONSE

The key components of the antiviral innate immune response include cellular effectors like natural killer (NK) cells, dendritic cells (DCs), and macrophages, and soluble mediators like type I IFNs, chemokines, and cytokines. NK cells can suppress viral replication by direct perforin-mediated killing of infected cells or indirectly via the production of IFN-γ, a type II IFN. DCs, including myeloid (mDC) and plasmacytoid (pDC) cells, and macrophages are critical players in sensing PAMPs and inducing an immediate response to viral infection by producing high levels of type I IFNs and other proinflammatory cytokines and chemokines (3, 15, 23, 26, 42, 44, 76). DCs and macrophages also function as antigen-presenting cells and stimulate the acquired arm of the immune response by producing immunoregulatory cytokines that regulate the activity of NK cells, B cells, and T cells (50).

The IFNs are secreted proteins that have multifunctional roles in antiviral defense, immune system activation, cell growth, and apoptosis (34). Initially identified as mediators of the antiviral state (41), IFNs also play an important role in integrating the innate and adaptive immune responses through the induction of a large number of inducible genes called IFN-stimulated genes (ISGs) (84). The IFNs are now classified into three distinct groups based on amino acid sequences and receptor recognition. Type I IFNs (IFN-α/β) consist of IFN-β and a large number of IFN-α isotypes and are produced in direct response to viral infection. After IFN-α/β is secreted from the infected cell, it is responsible for inducing an antiviral state in the neighboring cell to counter

viral spread (84). IFN-α/β triggers a signaling cascade through a common type I IFN receptor, IFNAR, via activation and transphosphorylation of the receptor-associated kinases Janus kinase 1 (Jak1) and tyrosine kinase 2 (Tyk2), resulting in the phosphorylation and activation of the signal transducers and activators of transcription (Stat) molecules Stat1 and Stat2. The phosphorylated Stat1-Stat2 heterodimer interacts with IFN regulatory factor (IRF) 9 to form a heterotrimeric complex called IFN-stimulated gene factor-3 (ISGF3), which binds to specific promoter sequences called IFN-stimulated response elements (ISREs) to induce ISGs that have multiple functions, including antiviral effects and regulation of the adaptive immune response (Fig. 1) (1, 34, 36, 75). IFN-β also signals through the type I IFN receptor to activate IRF7, a transcription factor important for the induction of IFN-α, resulting in the amplification of the IFN-α/β response. While IFN-α is produced predominantly by leukocytes, with pDCs being the major source of IFN-α in humans, IFN-β is produced by most cell types, including fibroblasts (23, 26, 34). Type II IFNs consist of the product of the IFN-γ gene and are produced by activated T cells and NK cells. IFN-γ activates a signaling pathway through a distinct receptor, IFNGR, to induce a different set of genes regulated by promoters containing gamma-activated sequences (GASs) (34). Type III IFN, a newly discovered IFN family, consists of IFN-λ1, IFN-λ2, and IFN-λ3 (94). Several lines of evidence suggest that type III IFNs also have antiviral activity in vitro (5, 6). The type III IFNs signal through a unique receptor (interleukin [IL]28Rα/IL-10Rβ) to trigger IFN-inducible genes that are also involved in the establishment of the antiviral state (6, 94). Current knowledge suggests that IFN-λ's are produced mainly by antigen-presenting cells like pDCs and mDCs and have similar functional characteristics as IFN-α/β. IFNs act in both autocrine and paracrine fashion to establish the antiviral state.

Figure 1 Simplified scheme of the type I IFN activation and signaling pathways and their inhibition by coronaviruses. Internalization of RNA viruses exposes the genomic RNA to cytoplasmic RNA sensors like RIG-I, MDA-5, and LGP2 that recognize dsRNA and 5′-triphosphate ssRNA produced during the life cycle of RNA viruses. In the endosomes, RNA sensors like TLR3 and TLR7/8 recognize dsRNA and ssRNA, respectively. TLR3 and TLR7/8 signal through the adaptor proteins TRIF and MyD88, respectively, to activate IFN-α/β production. RIG-I and MDA-5 trigger signaling cascades via IPS-1, an adaptor protein on mitochondrial membrane, to activate IRF3, leading to IFN-β gene expression and production of IFN-β protein. The released IFN-α/β proteins bind to the cognate type I IFN receptors in the same cell or neighboring cells to activate a Stat-dependent pathway to trigger the induction of ISGs with ISRE promoter sequences. The different steps in the IFN activation and signaling pathways that are inhibited or targeted by coronaviral proteins, as discussed in the text, are indicated. IKK, Ikappa B kinase; IRAK, IL-1 receptor-associated; TBK, TANK (TRAF family member-associated NF-κB activator)-binding kinase; TRAF, tumor necrosis factor (TNF) receptor-associated factor.

Type I IFNs exert their antiviral action by inducing/ activating a large number of effector proteins with well-defined functions that limit viral replication. These proteins include double-stranded (ds) RNA-activated protein kinase R (PKR), the 2′,5′-oligoadenylate synthetase (OAS)/RNase L system, adenosine deaminase (ADAR), and Mx GTPases (34). These effector proteins, through their interference with cellular or viral processes, establish an antiviral state in target cells by inhibiting or impairing the replication of viruses.

As the effectiveness of the IFN response is dependent upon the rapid production of IFNs and also regulatory factors that control the extent of the response, viruses have developed several strategies to escape, delay, and/or antagonize the different stages of the IFN pathway in order to establish an infection. The mechanisms of IFN antagonism employed by viruses fall into four broad categories: (i) escaping detection, where the virus hides or sequesters the PAMPs like dsRNAs from the host-cell sensors that activate IFN production; (ii) actively blocking IFN production by specifically interfering with the signaling molecules involved in type I IFN gene induction or by general downregulation of host-cell transcription/mRNA levels and/or protein synthesis; (iii) inhibiting the IFN signaling pathway at the level of IFN receptor engagement or downstream signaling events, thereby preventing the activation of the antiviral IFN-induced genes; and (iv) encoding factors that downregulate the activity of the IFN-induced antiviral proteins like PKR and OAS.

PAMPS AND PRRS INVOLVED IN TYPE I IFN INDUCTION

Viruses consist of structural components like nucleic acids, glycoproteins, lipid envelope, and dsRNA, generated as a by-product of the replication cycle, that constitute the PAMPs that are recognized by host PRRs to initiate the IFN pathway. The nature of PAMP-PRR interaction depends upon the type of the infected cell, the expression levels of PRRs in different cell types, the cellular compartment where the host senses the invading virus, and the stage of the virus life cycle when the host senses the infection. TLRs like TLR3, TLR7, TLR8, and TLR9 are the PRRs that are responsible for recognizing viruses on the cell surface as well as in the endosomal compartment after virus entry by detecting viral nucleic acids, including dsRNA and single-stranded (ss) RNA (Fig. 1) (87, 88). In contrast, PRRs of the RLH family like RIG-I, melanoma differentiation-associated gene 5 (MDA-5), and LGP2 detect viruses in the cytoplasm by sensing the presence of dsRNA, a viral replicative intermediate, and viral ssRNAs containing an unprotected 5′

triphosphate group (87, 88). Upon stimulation, both TLRs and the cytoplasmic RLHs signal through adaptor proteins that converge on common signaling molecules and transcription factors to trigger type I IFN induction (Fig. 1). The adaptor proteins utilized by the TLR pathway are TIR-domain-containing adaptor inducing IFN-β (TRIF) and/or MyD88 (87). The RLH proteins interact with the mitochondrial adaptor protein IFN-β promoter stimulator 1 (IPS-1; also known as MAVS, VISA, and Cardif) through their caspase activation and recruitment domains (CARDs) (Fig. 1) (87). Both pathways ultimately lead to the activation of the transcription factors IRF3, nuclear factor-κB (NF-κB), and ATF-2/c-Jun, which along with the coactivator CBP/p300 induce type I IFN gene transcription (87, 88).

GENOME FEATURES AND REPLICATION STRATEGIES OF CORONAVIRUSES AND ARTERIVIRUSES

Coronaviruses and arteriviruses are a group of enveloped animal RNA viruses that are united in the order *Nidovirales* (14). Coronaviruses, belonging to the family *Coronaviridae*, possess the largest known RNA genomes, ranging in size from 26 to 32 kb. Coronaviruses are classified into three groups, 1 to 3, and they include human coronaviruses (HCoV) (HCoV-NL63, HCoV-OC43, HCoV-229E, SARS-CoV) and animal coronaviruses, including infectious bronchitis virus (IBV), mouse hepatitis virus (MHV), bovine coronavirus (BCoV), porcine transmissible gastroenteritis virus (TGEV), feline infectious peritonitis virus, and turkey coronavirus. Also, many coronaviruses have been isolated from bats (51, 58, 105). Coronaviruses are associated with upper respiratory, gastrointestinal, and neurological infections in humans and animals (27, 48, 61, 101, 103). Arteriviruses, in the family *Arteriviridae*, have a much smaller RNA genome (13 to 16 kb). Viruses in this family include equine arteritis virus, porcine reproductive respiratory syndrome virus (PRRSV), simian hemorrhagic fever virus, and lactate dehydrogenase-elevating virus of mice (14, 78). Arteriviruses cause several clinically relevant and economically important diseases in horses, donkeys, and swine (7, 70, 78).

Coronaviruses and arteriviruses possess a positive-sense ssRNA genome. The coronavirus genome is encapsidated by the nucleocapsid (N) protein to form a helical ribonucleoprotein (RNP) complex in virus particles, while the arterivirus genome is enclosed within an icosahedral capsid structure. In both virus families, the incoming genome has a 5′ cap structure and a 3′ poly(A) tail and functions as an mRNA. The genome also contains untranslated regions (UTRs) at the 5′ and 3′

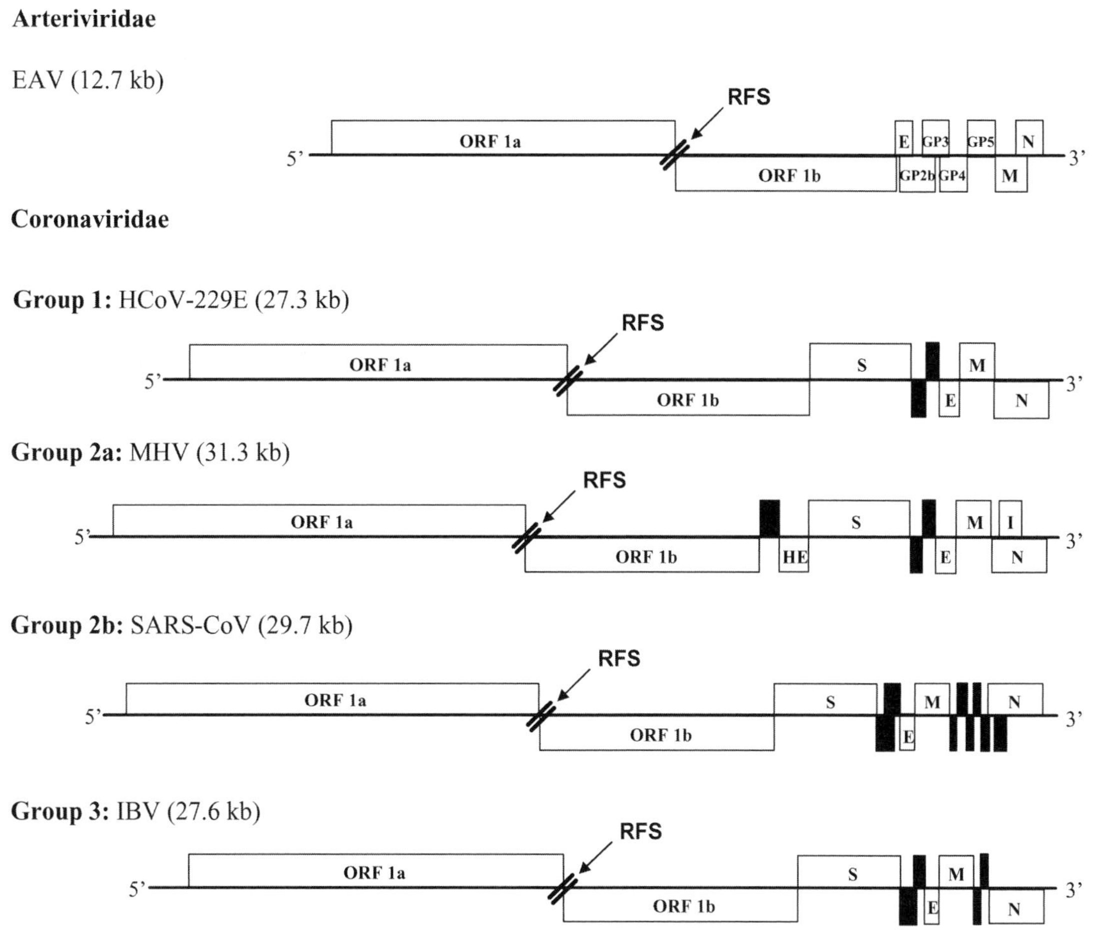

Figure 2 Genome organization of selected members in the *Arteriviridae* and *Coronaviridae* families. The replicase ORFs, ORF1a and ORF1b, and the names of the major virion structural genes are indicated. The accessory genes are shown as black boxes. The figure is not drawn to scale. EAV, equine arteritis virus; RFS, ribosomal frameshift.

termini that flank multiple open reading frames (ORFs), whose numbers vary among the different viruses in the *Nidovirales* order (Fig. 2). The two 5'-most ORFs, ORF1a and ORF1b, are called gene 1 and occupy almost two-thirds to three-quarters of the genome. The ORFs downstream of ORF1b encode the virus structural and accessory proteins (Fig. 2) (49).

Virus entry constitutes the first step in the life cycle; it involves binding of the virus to the receptor on the cell surface and is a critical determinant of the tropism of coronavirus and arterivirus infections. In coronaviruses, the viral protein responsible for receptor recognition is the S protein, and the cellular receptors have been identified for several members in the family (49, 57, 103). After virus attachment, the fusion of the virus envelope at the target cell membrane or in endosomes releases the

genome into the cytosol, where it serves as an mRNA for the translation of viral replicase proteins from gene 1 (Fig. 3). Gene 1 of nidoviruses is made up two large ORFs, ORF1a and ORF1b. During gene 1 protein translation, a ribosomal frameshift event occurs at the junction of ORF1a and ORF1b (Fig. 2), resulting in the production of two large polyprotein precursors, pp1a and pp1ab, both of which are proteolytically processed by viral proteinases to generate the mature gene 1 proteins, many of which have functions in RNA replication and transcription. In both coronavirus- and arterivirus-infected cells, many of these gene 1 proteins are localized on double-membrane vesicles (DMVs), derived from cellular membranes, which are induced during viral replication (11, 33, 35, 68, 79) (Fig. 3). The viral replication complexes are formed on the DMVs that serve as the sites of

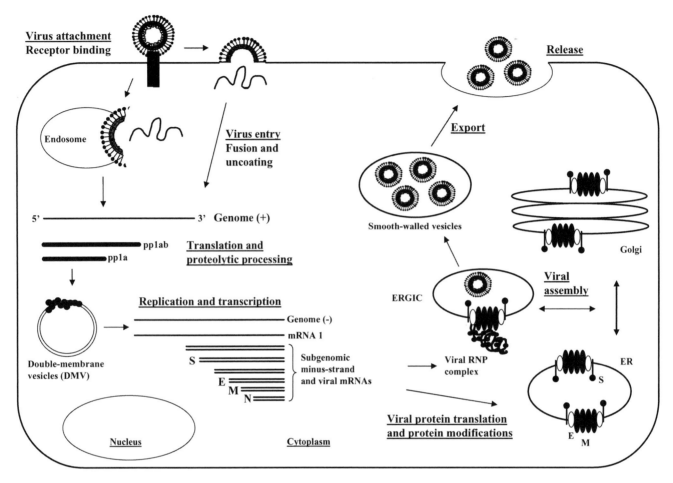

Figure 3 Schematic representation of the nidovirus life cycle. Virus entry is triggered by attachment to the cellular receptor and fusion of the viral envelope to the plasma membrane or the endosomes, which releases the viral genome into the cytoplasm. The genome is translated to produce precursor polyproteins, pp1a and pp1ab, which are proteolytically processed to produce mature viral proteins necessary for replication and transcription. The replication machinery, localized on DMVs, produces the viral genomic RNA, minus-strand RNAs, and subgenomic mRNAs. Viral mRNAs are translated to produce proteins that undergo modifications in cellular compartments like ER/Golgi. Viral assembly and budding occurs on an ER/Golgi intermediate compartment (ERGIC), where the viral envelope proteins and viral RNPs are assembled into virus particles. The mature virus particles are exported in vesicles to the cell surface and released into the extracellular environment by fusion with the cell membrane. (Adapted from Fig. 6 in reference 29.)

viral RNA synthesis. Nidovirus RNA replication machinery generates mRNA 1, which is the intracellular form of the viral genomic RNA, subgenomic mRNAs, and minus-strand RNAs, each of which corresponds to the positive-sense mRNAs in infected cells (10, 49, 78). All the nidovirus mRNAs have a 3′-coterminal nested set structure (Fig. 3). Usually, with a few exceptions, only the 5′-most ORF on each subgenomic mRNA is translated (14). The genome-length mRNA, mRNA 1, is used to translate gene 1 proteins, while the subgenomic mRNAs are used

to produce the structural proteins necessary for virus particle assembly and the accessory proteins whose functions in the viral life cycle have not been clearly demonstrated. Among viruses in the order *Nidovirales*, mechanisms of coronavirus assembly and release are well characterized. The intracellular site of coronavirus assembly and budding is an intermediate compartment (ERGIC) located between the endoplasmic reticulum (ER) and Golgi (89). Detailed studies using MHV have shown that the viral envelope proteins M, E, and S are posttranslationally modified and specific interactions between the envelope proteins and viral RNPs drive the packaging of viral RNA and budding of virus particle (45, 47, 62, 65–67, 95). The mature coronavirus particles are transported to the cell surface in large vesicular structures using the host secretory apparatus and released into the extracellular space via fusion with the cell membrane (Fig. 3).

INDUCTION OF THE TYPE I IFN RESPONSE

The Type I IFN Response in Arterivirus Infections

The type I IFN response to arterivirus infections has been investigated predominantly in studies using PRRSV. The consensus opinion from these studies, performed using PRRSV infections in pigs and in vitro in alveolar macrophages and MARC-145 cells, is that PRRSV is a poor inducer of IFN-β and IFN-α and possibly actively suppresses the induction of type I IFNs (21, 55, 60). The mechanism by which PRRSV inhibits the induction of IFN is unclear, while it has been suggested that the inhibition of IFN production may be at a posttranscriptional step. It is possible that virus-mediated downregulation of general host protein expression results in the inhibition of IFN production (55).

The Type I IFN Response in Coronavirus Infections

Understanding of coronavirus interactions with the type I IFN response requires analysis both in coronavirus-infected animals as well as in coronavirus-infected cultured cells. At the animal level, a clear picture has not emerged from studies investigating the regulation of type I IFN responses in coronavirus infections, which could be explained primarily by inherent differences in the virulence, level of replication, and tissue tropism of the different groups of coronaviruses as well as the different animal models used to generate the information. Likewise, a consensus opinion regarding the induction of type I IFN responses to coronavirus infections in cultured cells has not been generated.

Infection with group 1 coronaviruses like TGEV and porcine respiratory coronavirus in newborn piglets causes early and high IFN-α production in lung and intestinal secretions and in several other organs (18, 73). pDCs seem to be the major IFN-α-producing cells in TGEV-infected pigs (18). TGEV M protein plays a major role in the induction of IFN-α in vitro in lymphoid cells; virus-like pseudoparticles expressing only TGEV M and E proteins and lacking viral genome were able to induce IFN-α production in porcine leukocytes (8, 9, 17), and single amino acid mutations in M protein abrogated IFN-α induction by TGEV (53). HCoV-229E, another

group 1 coronavirus, induces IFN-β transcription in human monocyte-derived macrophages (19).

Analysis of the type I IFN response in group 3 coronaviruses, which include avian coronaviruses like IBV (69), has been limited. It was shown that type I IFNs and IFN-induced antiviral genes like Mx are induced in response to IBV infection in chickens (98).

Type I IFN responses in group 2a and 2b coronaviruses, like MHV and SARS-CoV, respectively, have received considerable attention, and there is accumulating evidence obtained from both in vivo and in vitro studies that highlight the modulation of the IFN-α/β response by group 2 coronaviruses. Infection with BCoV, a group 2a coronavirus, has been shown to induce IFN-α/β production in infected cattle, both locally in the upper respiratory tract and systemically (90). An in vitro study showed that BCoV-like pseudoparticles expressing only M and E proteins induced IFN-α production in porcine peripheral blood mononuclear cells (PBMCs) and further demonstrated that the interferogenicity of coronavirus pseudoparticles in leukocytes was determined by M protein (9). These observations, combined with data obtained using TGEV (9), provide evidence that inert viral structures in coronaviruses can trigger IFN-α production, suggesting that viral replication and/or gene expression is not required for IFN induction in lymphoid cells. In contrast, BCoV infection in the intestinal loops of newborn calves failed to induce type I IFN genes at 18 h postinfection (2). Various strains of MHV cause hepatic and central nervous system (CNS) diseases in mice and have been used as a model for hepatitis, viral encephalitis, and demyelination (103). Infection with a neurotropic strain of MHV (MHV-JHM) induces IFN-β mRNA and protein production in primary neuronal cultures as well as in the CNS of mice (72, 74). Cervantes-Barragan et al. showed that infection of murine bone marrow-derived and splenic pDCs with a CNS-tropic strain of MHV (MHV-A59) induces a rapid and high level of IFN-α production (16). pDCs recognize RNA viruses via the TLR7/MyD88 pathway to induce type I IFN production (44). Consistent with this notion, pDCs derived from TLR-deficient (TLR7$^{-/-}$) and MyD88-deficient (MyD88$^{-/-}$) mice fail to produce significant IFN-α in response to MHV infection, demonstrating that the recognition of MHV by pDCs is determined solely by the TLR7/MyD88 pathway (16). In contrast to these observations, several studies have shown that MHV infection is a poor inducer of type I IFN expression in fibroblast cells (31, 97, 109) and mDCs (16). Roth-Cross et al. reported that MHV-infected murine L2 cells show delayed activation of IFN-β transcription, yet they fail to release IFN-β protein (74).

Several studies have reported the lack of type I IFN induction upon infection of SARS-CoV in cultured cells (29, 80, 82, 83, 96). For example, SARS-CoV infection does not trigger significant type I IFN induction in PBMCs, monocyte-derived macrophages, conventional DCs, and lung epithelial cells, although some proinflammatory cytokines and chemokines, like IL-8, IFN-inducible protein-10 (IP-10), monocyte chemoattractant protein-1 (MCP-1), and macrophage inflammatory protein-1α (MIP-1α), are produced from these infected cells (19, 54, 92, 93, 107, 110). In contrast, Cervantes-Barragan et al. reported that SARS-CoV infection in pDCs rapidly induced a significant amount of IFN-α production and also caused the upregulation of ISGs such as ISG56 and MxA (16). Other studies have also reported a delayed induction of IFN-α mRNA in immature DCs and PBMCs (13, 82). It should be noted that PBMCs, DCs, and macrophages have been reported to be only abortively infected by SARS-CoV, suggesting that these cells mount type I IFN responses in the absence of productive viral replication (19, 92, 107). Also, it has been reported that UV-inactivated SARS-CoV could induce IFN-α in immature DCs (82), and cocultivation of SARS-CoV-infected Vero cells with normal PBMCs also triggers efficient IFN expression (13). Devaraj et al. reported that SARS-CoV infection in the African green monkey kidney cell line MA104 induced a weak and delayed upregulation of IFN-β mRNA (25). Based on a majority of data obtained from studies in cell culture, the overall theme that emerges is a lack of a robust type I IFN response or a delay in type I IFN expression in cells productively infected with SARS-CoV, which allows the virus to escape the action of IFNs.

Analysis of type I IFN responses in SARS-CoV-infected individuals and experimental infections in animal models presents a more complex picture. Several studies have reported an unusual lack of type I IFN response in the lungs and peripheral blood of SARS patients (52, 64, 71, 104), while some have reported that SARS patients have robust type I and type II IFN responses and show high levels of IFN-stimulated chemokines and antiviral ISGs in the plasma (12). Several groups reported a lack of detection of IFN-α or IFN-β in the lungs of SARS-CoV-infected mice (32, 59, 91), while de Lang et al. reported the induction of high levels of type I IFNs and ISGs in the lungs of SARS-CoV-infected macaques and also showed the activation of Stat1 in the nuclei of cells in the lungs (24). The authors of the latter study speculate that pDCs could be the source of IFN-β production in SARS-infected macaques (24).

The importance of the type I IFN system in controlling coronavirus infections has been suggested in some

in vitro and animal studies. Cervantes-Barragan et al. showed that infection of type I IFN receptor-deficient (IFNAR$^{-/-}$) mice with a low dose of MHV-A59 is lethal, while wild-type mice survived the infection with no signs of clinical illness (16). Additionally, pDCs lacking IFNAR supported a higher level of MHV-A59 replication as compared with wild-type pDCs (16). Similarly, type I IFN signaling was shown to be important for the control of the neurotropic MHV-JHM infection in the CNS of mice (40). Furthermore, SARS-CoV infection in Stat1 knockout mice, which are resistant to the antiviral effects of the IFN signaling pathway, showed a higher level of virus replication in the lungs compared with normal mice and exhibited severe morbidity and mortality (39). Several lines of evidence also suggest that SARS-CoV is susceptible to prophylactic treatment with type I IFNs in vitro (22, 38, 81, 85) as well as in vivo (37), while MHV has been shown to be only moderately sensitive to type I IFNs in vitro (74, 86).

EVASION OR INHIBITION OF THE TYPE I IFN RESPONSE

Type I IFN Evasion Mechanisms in Coronaviruses

Our understanding of the molecular mechanisms of type I IFN evasion strategies utilized by nidoviruses is still preliminary. Very little is known about the inhibition of the type I IFN pathway by arteriviruses, while significant advances have been made, primarily using MHV and SARS-CoV, toward understanding how coronaviruses evade type I IFN induction and the type I IFN signaling pathway (Fig. 3).

There are several steps in the coronavirus life cycle that can be the target for the host antiviral-sensing machinery to initiate the type I IFN cascade. Virus entry into the endosomes exposes the viral genomic RNA to endosome-associated TLRs like TLR3 and TLR7. Indeed, Cervantes-Barragan et al. reported the role of TLR7 in sensing MHV and SARS-CoV infection in pDCs (16). After the viral RNA is released into the cytoplasm, the RLH family of cytoplasmic viral sensors, such as RIG-I and MDA-5, might recognize PAMPs like dsRNAs and unprotected ssRNAs containing 5′ triphosphates, yet there is no report that demonstrates the role of the RIG-I/MDA-5 pathway in sensing coronavirus infections. Based on the current information, largely obtained from studies in cell culture, about the potential type I IFN evasion mechanisms employed by coronaviruses, it is evident that these viruses possibly target the different arms of the type I IFN pathway, by (i) avoiding detection by masking viral RNAs through compartmentalization of viral replication in DMVs or modification of the 5′ end of viral RNA; (ii) producing proteins that actively block pathways that lead to type I IFN production; (iii) expressing proteins that block the mounting of the antiviral response through inhibition of the IFN signaling pathway and/or block the action of antiviral genes; and (iv) employing posttranscriptional mechanisms to downregulate host mRNA levels and/or host protein synthesis to block or delay induction of the type I IFN response. There are published reports that provide evidence supporting each of these mechanisms in coronavirus infections; however, it is important to note that most of the data, obtained primarily using expression systems, are preliminary and vary according to the cell types, experimental design, and assay systems used by different investigators.

Stealth Mechanisms in Coronaviruses

A few studies have suggested that coronaviruses are able to avoid detection by host PRRs through the sequestration of viral PAMPs (96, 109). Zhou and Perlman demonstrated that NF-κB and IRF3 are not activated during MHV infection (109). They also showed that MHV infection does not block the activation of these transcription factors after superinfection with Sendai virus or treatment with a synthetic dsRNA analog, poly (I:C) (109). Furthermore, MHV infection does not inhibit the production of IFN-β triggered by the RIG-I, MDA-5, and TLR pathways (109). In a similar study, Versteeg et al. showed that both MHV and SARS-CoV infection do not induce IRF3 nuclear translocation and IFN-α gene transcription but cannot block the induction of IFN upon treatment with poly(I:C) or superinfection with Sendai virus (96). Versteeg et al. and others have detected the presence of significant amounts of dsRNA in coronavirus-infected cells using a dsRNA-specific antibody (96, 100). Versteeg et al. speculate that viral PAMPs like dsRNA might be hidden from host PRRs due to their compartmentalization in DMVs that are induced during coronavirus replication. Versteeg et al. further proposed that capping of the 5′ end of viral RNAs, possibly by the virally encoded 2′-O-methyltransferase enzyme, prevents RIG-I from recognizing the capped viral RNAs, as RIG-I has been shown to detect free 5′ triphosphates (96). However, other studies and data from our laboratory show the induction of type I IFNs in MHV- and SARS-CoV-infected cells (25, 63, 74, 111), suggesting that stealth mechanisms alone cannot fully explain the inhibition of IFN gene induction by coronaviruses. It is possible that the expression and accessibility levels of the PRRs to coronaviral RNAs or virus-specific protein(s) could differ among cell types and cell lines, which could account for these varied observations.

Inhibition of IFN Induction by Coronaviruses

Spiegel et al. reported that SARS-CoV infection actively blocks the activation of the IFN-β promoter for up to 16 h in human embryonic kidney (HEK) 293 fibroblast cells (80). The authors demonstrated that IRF3, the essential transcription factor for IFN-β promoter activation, translocates to the nucleus early in infection, but IRF3 nuclear translocation does not occur late in infection. At either times postinfection, IRF3 fails to undergo hyperphosphorylation and dimerization and could not interact with the nuclear coactivator CBP (80). In a separate study, Spiegel et al. showed a delayed upregulation of IFN-α expression in SARS-CoV-infected HEK 293 cells and speculated that the virus might have developed a mechanism to inhibit or delay the activation of the IFN promoter (82).

Devaraj et al. showed that the papain-like protease (PLpro) domain of the SARS-CoV replicase protein, nsp3, one of the gene 1 proteins, is an IFN antagonist (25). In this report, using a plasmid-based expression system, the authors showed that PLpro interacts with IRF3 and inhibits its phosphorylation and nuclear translocation, thereby affecting the downstream activation of the type I IFN gene (25). Furthermore, this study also showed the association of nsp3 with IRF3 in SARS-CoV-infected MA104 cells (25). These data suggest that SARS-CoV nsp3 could potentially trap IRF3 in the cytoplasm and prevent its nuclear translocation in infected cells (Fig. 1).

Kopecky-Bromberg et al. identified the SARS-CoV accessory proteins 3b and ORF6 and the structural protein N as antagonists of IFN induction and signaling pathways (46). Expression of each of these three proteins in HEK 293 cells inhibits the induction of IFN-β-, ISRE-, and IRF3-promoter-regulated reporter plasmids upon stimulation with a known potent IFN inducer, Sendai virus (46). The authors showed that these proteins operate by inhibiting the phosphorylation and nuclear translocation of IRF3 in response to Sendai virus infection (46) (Fig. 1). Furthermore, N protein expression also inhibits the Sendai virus-induced NF-κB promoter activation, suggesting the inhibition of the function of another key transcription factor that is involved in IFN-β gene induction. However, recombinant SARS-CoV, in which each of the accessory gene ORFs, 3a, 3b, 6, 7a, 7b, 8b, and 9b, is deleted, still failed to induce IFN-β or NF-κB promoters (29), suggesting that 3b and ORF6 gene products alone might not be responsible for the inhibition of IFN induction in infected cells.

Inhibition of IFN Signaling and IFN-Induced Effector Protein Functions by Coronaviruses

A few studies have suggested that coronaviruses encode proteins that inhibit the IFN-dependent signaling pathways as well as interfere with the function of antiviral ISGs. Ye et al. reported that MHV-A59 is resistant to type I IFN treatment and that N protein is an IFN antagonist (106). Using the heterologous recombinant vaccinia virus (VV) lacking the IFN antagonist E3L, which is sensitive to IFN, they showed that insertion of the MHV-A59 N gene into VVΔE3L virus results in generation of IFN-resistant VV (106). They also showed that N protein inhibits the activity of the IFN-stimulated protein 2′,5′-OAS/RNase L (106) (Fig. 1).

Zust et al. showed that MHV nonstructural protein (nsp) 1, the most N-terminal gene 1 protein, interferes with the type I IFN system by inhibiting or counteracting IFN signaling and/or the antiviral functions of IFN-induced proteins (111). In this highly significant study, the authors showed that an MHV nsp1 mutant carrying a 99-nucleotide deletion at the C-terminal region of nsp1 (MHV-nsp1Δ99) is highly attenuated in wild-type mice, while the replication and the spread of this virus was similar to wild-type MHV in IFNAR$^{-/-}$ mice, demonstrating for the first time the IFN-regulatory role of a coronavirus protein during virus infection in vivo (111). This study also found that MHV-nsp1Δ99 virus was more susceptible to IFN-α pretreatment in macrophages, which provides further evidence for the IFN signaling-inhibitory activity of MHV nsp1 protein (111). Wathelet et al. showed that SARS-CoV nsp1 protein expression inhibits the IFN-induced activation of ISRE promoters and also decreases the IFN-induced phosphorylation levels of Stat1 (99) (Fig. 1). Using a recombinant SARS-CoV nsp1 mutant virus, the authors showed that SARS-CoV nsp1 protein also plays a role in the inhibition of IFN-dependent signaling and induction of specific ISGs (99).

Expression studies have assigned an IFN signaling-inhibitory function to the SARS-CoV accessory proteins 3b and ORF6, in addition to their role in blocking IFN induction (46). Both 3b and ORF6 proteins inhibited the induction of ISRE promoters in response to IFN-β treatment (46). While the mechanism of action of 3b protein is not clear, Frieman et al. demonstrated that ORF6 protein blocks the nuclear translocation of Stat1 after IFN treatment by disrupting the nuclear import complex formation via its interaction with an import factor, karyopherin α2 (30). The authors also showed that Stat1 localizes to the nucleus in cells infected with SARS-CoV lacking ORF6, while wild-type SARS-CoV

blocks Stat1 nuclear localization (30) (Fig. 1). The role of 3b and ORF6 proteins in the inhibition of Stat1-mediated IFN response genes in SARS-CoV-infected cells remains to be demonstrated.

Posttranscriptional Mechanisms of Evasion in Coronaviruses

There are several examples of viruses that suppress general host gene expression to block host innate immune responses (20, 28, 56, 77, 102). Some studies have shown that coronaviruses employ posttranscriptional strategies to suppression general host gene expression in infected cells (43, 74). Roth-Cross et al. showed that MHV infection in L2 cells causes a significant decrease in the production of IFN-β that is induced by replication of heterologous viruses like Sendai virus and Newcastle disease virus. Because coinfection of MHV with Sendai virus or Newcastle disease virus does not affect IFN-β mRNA production or stability, the authors speculated that the mechanism of MHV-mediated inhibition of IFN-β production is at the posttranscriptional level (74).

Our laboratory has reported that SARS-CoV nsp1 protein expression induces host mRNA degradation and host translational suppression, resulting in general host gene expression, including IFN-β mRNA accumulation (43) (Fig. 1). Consistent with this notion, SARS-CoV replication in Vero E6 cells and HEK 293 cells stably expressing the SARS-CoV receptor protein angiotensin-converting enzyme 2 (ACE2) (293/ACE2 cells) promoted the degradation of host mRNAs (43). Further, we have shown that a recombinant SARS-CoV nsp1 mutant virus carrying mutations in the C-terminal region of the nsp1 gene induced high levels of IFN-β mRNA accumulation and high titers of type I IFN protein production in 293/ACE2 cells (63). These data clearly demonstrate the role of SARS-CoV nsp1 in suppressing type I IFN expression in infected cells and also provide further proof of the importance of the SARS-CoV nsp1 protein in the evasion of host IFN responses.

Expression of HCoV-229E and MHV nsp1 proteins also suppresses host gene expression, and the C-terminal region of MHV nsp1 is responsible for this activity (111), suggesting a functional similarity between nsp1 of groups 1 (HCoV-229E), 2a (MHV), and 2b (SARS-CoV) coronaviruses. Coronavirus nsp1 gene is located at the 5′ end of the viral genome and is produced as the first mature viral protein in the cytoplasm upon viral entry into the host cells, suggesting that immediate expression of nsp1 after infection could provide a growth advantage to the virus in IFN-competent cells.

CONCLUDING REMARKS

Innate immunity plays a critical role in regulating coronavirus infections in experimental as well as natural systems. The discovery of novel host proteins and pathways regulating host-pathogen interactions has opened a productive research area for coronavirus investigators that has already led to the identification of several viral proteins and unique strategies of immune evasion. As expected from the large coding capacity of the coronavirus genome, several genes with redundant inhibitory functions targeting the crucial steps in the type I IFN pathway have been identified. It has also raised several important questions about the contribution of these mechanisms in controlling the outcome of coronavirus infections in vivo. As most of the immune evasion functions of coronavirus proteins have been identified in vitro, it is very important to validate these results using mutant coronaviruses, which are generated by reverse genetics systems (4, 108), in suitable animal models. The availability of transgenic and other valuable natural mouse models for MHV, SARS-CoV, and HCoV-229E enables researchers to investigate the role of nonstructural and accessory proteins in virulence and pathogenesis. Using such systems, the potential identification of some common, evolutionarily conserved molecular mechanisms of immune evasion among different groups of coronaviruses is an exciting possibility in the future.

References

1. Aaronson, D. S., and C. M. Horvath. 2002. A road map for those who don't know JAK-Stat. *Science* **296:** 1653–1655.
2. Aich, P., H. L. Wilson, R. S. Kaushik, A. A. Potter, L. A. Babiuk, and P. Griebel. 2007. Comparative analysis of innate immune responses following infection of newborn calves with bovine rotavirus and bovine coronavirus. *J. Gen. Virol.* **88:**2749–2761.
3. Akira, S., and H. Hemmi. 2003. Recognition of pathogen-associated molecular patterns by TLR family. *Immunol. Lett.* **85:**85–95.
4. Almazan, F., M. L. Dediego, C. Galan, D. Escors, E. Alvarez, J. Ortego, I. Sola, S. Zuniga, S. Alonso, J. L. Moreno, A. Nogales, C. Capiscol, and L. Enjuanes. 2006. Construction of a severe acute respiratory syndrome coronavirus infectious cDNA clone and a replicon to study coronavirus RNA synthesis. *J. Virol.* **80:**10900–10906.
5. Ank, N., H. West, C. Bartholdy, K. Eriksson, A. R. Thomsen, and S. R. Paludan. 2006. Lambda interferon (IFN-lambda), a type III IFN, is induced by viruses and IFNs and displays potent antiviral activity against select virus infections in vivo. *J. Virol.* **80:**4501–4509.
6. Ank, N., H. West, and S. R. Paludan. 2006. IFN-lambda: novel antiviral cytokines. *J. Interferon Cytokine Res.* **26:**373–379.
7. Balasuriya, U. B., and N. J. MacLachlan. 2004. The immune response to equine arteritis virus: potential lessons for other arteriviruses. *Vet. Immunol. Immunopathol.* **102:**107–129.
8. Baudoux, P., L. Besnardeau, C. Carrat, P. Rottier, B. Charley, and H. Laude. 1998. Interferon alpha inducing property of coronavirus particles and pseudoparticles. *Adv. Exp. Med. Biol.* **440:**377–386.
9. Baudoux, P., C. Carrat, L. Besnardeau, B. Charley, and H. Laude. 1998. Coronavirus pseudoparticles formed with recombinant M and E proteins induce alpha interferon synthesis by leukocytes. *J. Virol.* **72:**8636–8643.
10. Brian, D. A., and R. S. Baric. 2005. Coronavirus genome structure and replication. *Curr. Top. Microbiol. Immunol.* **287:**1–30.
11. Brockway, S. M., C. T. Clay, X. T. Lu, and M. R. Denison. 2003. Characterization of the expression, intracellular localization, and replication complex association of the putative mouse hepatitis virus RNA-dependent RNA polymerase. *J. Virol.* **77:**10515–10527.
12. Cameron, M. J., L. Ran, L. Xu, A. Danesh, J. F. Bermejo-Martin, C. M. Cameron, M. P. Muller, W. L. Gold, S. E. Richardson, S. M. Poutanen, B. M. Willey, M. E. DeVries, Y. Fang, C. Seneviratne, S. E. Bosinger, D. Persad, P. Wilkinson, L. D. Greller, R. Somogyi, A. Humar, S. Keshavjee, M. Louie, M. B. Loeb, J. Brunton, A. J. McGeer, and D. J. Kelvin. 2007. Interferon-mediated immunopathological events are associated with atypical innate and adaptive immune responses in patients with severe acute respiratory syndrome. *J. Virol.* **81:**8692–8706.
13. Castilletti, C., L. Bordi, E. Lalle, G. Rozera, F. Poccia, C. Agrati, I. Abbate, and M. R. Capobianchi. 2005. Coordinate induction of IFN-alpha and -gamma by SARS-CoV also in the absence of virus replication. *Virology* **341:**163–169.
14. Cavanagh, D. 1997. Nidovirales: a new order comprising Coronaviridae and Arteriviridae. *Arch. Virol.* **142:** 629–633.
15. Cella, M., D. Jarrossay, F. Facchetti, O. Alebardi, H. Nakajima, A. Lanzavecchia, and M. Colonna. 1999. Plasmacytoid monocytes migrate to inflamed lymph nodes and produce large amounts of type I interferon. *Nat. Med.* **5:**919–923.
16. Cervantes-Barragan, L., R. Zust, F. Weber, M. Spiegel, K. S. Lang, S. Akira, V. Thiel, and B. Ludewig. 2007. Control of coronavirus infection through plasmacytoid dendritic-cell-derived type I interferon. *Blood* **109:**1131–1137.
17. Charley, B., and H. Laude. 1988. Induction of alpha interferon by transmissible gastroenteritis coronavirus: role of transmembrane glycoprotein E1. *J. Virol.* **62:**8–11.
18. Charley, B., S. Riffault, and K. Van Reeth. 2006. Porcine innate and adaptive immune responses to influenza and coronavirus infections. *Ann. N. Y. Acad. Sci.* **1081:** 130–136.
19. Cheung, C. Y., L. L. Poon, I. H. Ng, W. Luk, S. F. Sia, M. H. Wu, K. H. Chan, K. Y. Yuen, S. Gordon, Y. Guan, and J. S. Peiris. 2005. Cytokine responses in severe acute respiratory syndrome coronavirus-infected macrophages in vitro: possible relevance to pathogenesis. *J. Virol.* **79:** 7819–7826.

20. Chinsangaram, J., M. E. Piccone, and M. J. Grubman. 1999. Ability of foot-and-mouth disease virus to form plaques in cell culture is associated with suppression of alpha/beta interferon. *J. Virol.* **73:**9891–9898.

21. Chung, H. K., J. H. Lee, S. H. Kim, and C. Chae. 2004. Expression of interferon-alpha and Mx1 protein in pigs acutely infected with porcine reproductive and respiratory syndrome virus (PRRSV). *J. Comp. Pathol.* **130:**299–305.

22. Cinatl, J., B. Morgenstern, G. Bauer, P. Chandra, H. Rabenau, and H. W. Doerr. 2003. Treatment of SARS with human interferons. *Lancet* **362:**293–294.

23. Colonna, M., G. Trinchieri, and Y. J. Liu. 2004. Plasmacytoid dendritic cells in immunity. *Nat. Immunol.* **5:**1219–1226.

24. de Lang, A., T. Baas, T. Teal, L. M. Leijten, B. Rain, A. D. Osterhaus, B. L. Haagmans, and M. G. Katze. 2007. Functional genomics highlights differential induction of antiviral pathways in the lungs of SARS-CoV-infected macaques. *PLoS Pathog.* **3:**e112.

25. Devaraj, S. G., N. Wang, Z. Chen, Z. Chen, M. Tseng, N. Barretto, R. Lin, C. J. Peters, C. T. Tseng, S. C. Baker, and K. Li. 2007. Regulation of IRF3-dependent innate immunity by the papain-like protease domain of the severe acute respiratory syndrome coronavirus. *J. Biol. Chem.* **282:**32208–32221.

26. Diebold, S. S., M. Montoya, H. Unger, L. Alexopoulou, P. Roy, L. E. Haswell, A. Al-Shamkhani, R. Flavell, P. Borrow, and C. Reis e Sousa. 2003. Viral infection switches non-plasmacytoid dendritic cells into high interferon producers. *Nature* **424:**324–328.

27. Drosten, C., S. Gunther, W. Preiser, S. van der Werf, H. R. Brodt, S. Becker, H. Rabenau, M. Panning, L. Kolesnikova, R. A. Fouchier, A. Berger, A. M. Burguiere, J. Cinatl, M. Eickmann, N. Escriou, K. Grywna, S. Kramme, J. C. Manuguerra, S. Muller, V. Rickerts, M. Sturmer, S. Vieth, H. D. Klenk, A. D. Osterhaus, H. Schmitz, and H. W. Doerr. 2003. Identification of a novel coronavirus in patients with severe acute respiratory syndrome. *N. Engl. J. Med.* **348:**1967–1976.

28. Ferran, M. C., and J. M. Lucas-Lenard. 1997. The vesicular stomatitis virus matrix protein inhibits transcription from the human beta interferon promoter. *J. Virol.* **71:**371–377.

29. Frieman, M., M. Heise, and R. Baric. 2008. SARS coronavirus and innate immunity. *Virus Res.* **133:**101–112.

30. Frieman, M., B. Yount, M. Heise, S. A. Kopecky-Bromberg, P. Palese, and R. S. Baric. 2007. Severe acute respiratory syndrome coronavirus ORF6 antagonizes Stat1 function by sequestering nuclear import factors on the rough endoplasmic reticulum/Golgi membrane. *J. Virol.* **81:**9812–9824.

31. Garlinghouse, L. E., Jr., A. L. Smith, and T. Holford. 1984. The biological relationship of mouse hepatitis virus (MHV) strains and interferon: in vitro induction and sensitivities. *Arch. Virol.* **82:**19–29.

32. Glass, W. G., K. Subbarao, B. Murphy, and P. M. Murphy. 2004. Mechanisms of host defense following severe acute respiratory syndrome-coronavirus (SARS-CoV) pulmonary infection of mice. *J. Immunol.* **173:**4030–4039.

33. Goldsmith, C. S., K. M. Tatti, T. G. Ksiazek, P. E. Rollin, J. A. Comer, W. W. Lee, P. A. Rota, B. Bankamp, W. J. Bellini, and S. R. Zaki. 2004. Ultrastructural characterization of SARS coronavirus. *Emerg. Infect. Dis.* **10:**320–326.

34. Goodbourn, S., L. Didcock, and R. E. Randall. 2000. Interferons: cell signalling, immune modulation, antiviral response and virus countermeasures. *J. Gen. Virol.* **81:**2341–2364.

35. Gosert, R., A. Kanjanahaluethai, D. Egger, K. Bienz, and S. C. Baker. 2002. RNA replication of mouse hepatitis virus takes place at double-membrane vesicles. *J. Virol.* **76:**3697–3708.

36. Grandvaux, N., B. R. tenOever, M. J. Servant, and J. Hiscott. 2002. The interferon antiviral response: from viral invasion to evasion. *Curr. Opin. Infect. Dis.* **15:**259–267.

37. Haagmans, B. L., T. Kuiken, B. E. Martina, R. A. Fouchier, G. F. Rimmelzwaan, G. van Amerongen, D. van Riel, T. de Jong, S. Itamura, K. H. Chan, M. Tashiro, and A. D. Osterhaus. 2004. Pegylated interferon-alpha protects type 1 pneumocytes against SARS coronavirus infection in macaques. *Nat. Med.* **10:**290–293.

38. Hensley, L. E., L. E. Fritz, P. B. Jahrling, C. L. Karp, J. W. Huggins, and T. W. Geisbert. 2004. Interferon-beta 1a and SARS coronavirus replication. *Emerg. Infect. Dis.* **10:**317–319.

39. Hogan, R. J., G. Gao, T. Rowe, P. Bell, D. Flieder, J. Paragas, G. P. Kobinger, N. A. Wivel, R. G. Crystal, J. Boyer, H. Feldmann, T. G. Voss, and J. M. Wilson. 2004. Resolution of primary severe acute respiratory syndrome-associated coronavirus infection requires Stat1. *J. Virol.* **78:**11416–11421.

40. Ireland, D. D., S. A. Stohlman, D. R. Hinton, R. Atkinson, and C. C. Bergmann. 2008. Type I interferons are essential in controlling neurotropic coronavirus infection irrespective of functional CD8 T cells. *J. Virol.* **82:**300–310.

41. Isaacs, A., and J. Lindenmann. 1957. Virus interference. I. The interferon. *Proc. R. Soc. Lond. B Biol. Sci.* **147:**258–267.

42. Ito, T., Y. H. Wang, and Y. J. Liu. 2005. Plasmacytoid dendritic cell precursors/type I interferon-producing cells sense viral infection by Toll-like receptor (TLR) 7 and TLR9. *Springer Semin. Immunopathol.* **26:**221–229.

43. Kamitani, W., K. Narayanan, C. Huang, K. Lokugamage, T. Ikegami, N. Ito, H. Kubo, and S. Makino. 2006. Severe acute respiratory syndrome coronavirus nsp1 protein suppresses host gene expression by promoting host mRNA degradation. *Proc. Natl. Acad. Sci. USA* **103:**12885–12890.

44. Kato, H., S. Sato, M. Yoneyama, M. Yamamoto, S. Uematsu, K. Matsui, T. Tsujimura, K. Takeda, T. Fujita, O. Takeuchi, and S. Akira. 2005. Cell type-specific involvement of RIG-I in antiviral response. *Immunity* **23:**19–28.

45. Klumperman, J., J. K. Locker, A. Meijer, M. C. Horzinek, H. J. Geuze, and P. J. Rottier. 1994. Coronavirus M proteins accumulate in the Golgi complex beyond the site of virion budding. *J. Virol.* **68:**6523–6534.

46. Kopecky-Bromberg, S. A., L. Martinez-Sobrido, M. Frieman, R. A. Baric, and P. Palese. 2007. Severe acute respiratory syndrome coronavirus open reading frame (ORF) 3b, ORF 6, and nucleocapsid proteins function as interferon antagonists. *J. Virol.* **81:**548–557.

47. Krijnse-Locker, J., M. Ericsson, P. J. Rottier, and G. Griffiths. 1994. Characterization of the budding compartment of mouse hepatitis virus: evidence that transport from the RER to the Golgi complex requires only one vesicular transport step. *J. Cell Biol.* **124:**55–70.

48. Ksiazek, T. G., D. Erdman, C. S. Goldsmith, S. R. Zaki, T. Peret, S. Emery, S. Tong, C. Urbani, J. A. Comer, W. Lim, P. E. Rollin, S. F. Dowell, A. E. Ling, C. D. Humphrey, W. J. Shieh, J. Guarner, C. D. Paddock, P. Rota, B. Fields, J. DeRisi, J. Y. Yang, N. Cox, J. M. Hughes, J. W. LeDuc, W. J. Bellini, and L. J. Anderson. 2003. A novel coronavirus associated with severe acute respiratory syndrome. *N. Engl. J. Med.* **348:**1953–1966.

49. Lai, M. M., and D. Cavanagh. 1997. The molecular biology of coronaviruses. *Adv. Virus Res.* **48:**1–100.

50. Laiosa, C. V., M. Stadtfeld, and T. Graf. 2006. Determinants of lymphoid-myeloid lineage diversification. *Annu. Rev. Immunol.* **24:**705–738.

51. Lau, S. K., P. C. Woo, K. S. Li, Y. Huang, H. W. Tsoi, B. H. Wong, S. S. Wong, S. Y. Leung, K. H. Chan, and K. Y. Yuen. 2005. Severe acute respiratory syndrome coronavirus-like virus in Chinese horseshoe bats. *Proc. Natl. Acad. Sci. USA* **102:**14040–14045.

52. Lau, Y. L., and J. S. Peiris. 2005. Pathogenesis of severe acute respiratory syndrome. *Curr. Opin. Immunol.* **17:**404–410.

53. Laude, H., J. Gelfi, L. Lavenant, and B. Charley. 1992. Single amino acid changes in the viral glycoprotein M affect induction of alpha interferon by the coronavirus transmissible gastroenteritis virus. *J. Virol.* **66:**743–749.

54. Law, H. K., C. Y. Cheung, H. Y. Ng, S. F. Sia, Y. O. Chan, W. Luk, J. M. Nicholls, J. S. Peiris, and Y. L. Lau. 2005. Chemokine up-regulation in SARS-coronavirus-infected, monocyte-derived human dendritic cells. *Blood* **106:**2366–2374.

55. Lee, S. M., S. K. Schommer, and S. B. Kleiboeker. 2004. Porcine reproductive and respiratory syndrome virus field isolates differ in in vitro interferon phenotypes. *Vet. Immunol. Immunopathol.* **102:**217–231.

56. Le May, N., S. Dubaele, L. Proietti De Santis, A. Billecocq, M. Bouloy, and J. M. Egly. 2004. TFIIH transcription factor, a target for the Rift Valley hemorrhagic fever virus. *Cell* **116:**541–550.

57. Li, W., M. J. Moore, N. Vasilieva, J. Sui, S. K. Wong, M. A. Berne, M. Somasundaran, J. L. Sullivan, K. Luzuriaga, T. C. Greenough, H. Choe, and M. Farzan. 2003. Angiotensin-converting enzyme 2 is a functional receptor for the SARS coronavirus. *Nature* **426:**450–454.

58. Li, W., Z. Shi, M. Yu, W. Ren, C. Smith, J. H. Epstein, H. Wang, G. Crameri, Z. Hu, H. Zhang, J. Zhang, J. McEachern, H. Field, P. Daszak, B. T. Eaton, S. Zhang, and L. F. Wang. 2005. Bats are natural reservoirs of SARS-like coronaviruses. *Science* **310:**676–679.

59. McCray, P. B., Jr., L. Pewe, C. Wohlford-Lenane, M. Hickey, L. Manzel, L. Shi, J. Netland, H. P. Jia, C. Halabi, C. D. Sigmund, D. K. Meyerholz, P. Kirby, D. C. Look, and S. Perlman. 2007. Lethal infection of K18-hACE2 mice infected with severe acute respiratory syndrome coronavirus. *J. Virol.* **81:**813–821.

60. Miller, L. C., W. W. Laegreid, J. L. Bono, C. G. Chitko-McKown, and J. M. Fox. 2004. Interferon type I response in porcine reproductive and respiratory syndrome virus-infected MARC-145 cells. *Arch. Virol.* **149:**2453–2463.

61. Murray, R. S., B. Brown, D. Brian, and G. F. Cabirac. 1992. Detection of coronavirus RNA and antigen in multiple sclerosis brain. *Ann. Neurol.* **31:**525–533.

62. Narayanan, K., A. Maeda, J. Maeda, and S. Makino. 2000. Characterization of the coronavirus M protein and nucleocapsid interaction in infected cells. *J. Virol.* **74:**8127–8134.

63. Narayanan, K., C. Huang, K. Lokugamage, W. Kamitani, T. Ikegami, C. T. Tseng, and S. Makino. 2008 Severe acute respiratory syndrome coronavirus nsp1 suppresses host gene expression, including that of type I interferon, in infected cells. *J. Virol.* **82:**4471–4479.

64. Nicholls, J. M., L. L. Poon, K. C. Lee, W. F. Ng, S. T. Lai, C. Y. Leung, C. M. Chu, P. K. Hui, K. L. Mak, W. Lim, K. W. Yan, K. H. Chan, N. C. Tsang, Y. Guan, K. Y. Yuen, and J. S. Peiris. 2003. Lung pathology of fatal severe acute respiratory syndrome. *Lancet* **361:**1773–1778.

65. Oostra, M., C. A. de Haan, R. J. de Groot, and P. J. Rottier. 2006. Glycosylation of the severe acute respiratory syndrome coronavirus triple-spanning membrane proteins 3a and M. *J. Virol.* **80:**2326–2336.

66. Opstelten, D. J., M. C. Horzinek, and P. J. Rottier. 1993. Complex formation between the spike protein and the membrane protein during mouse hepatitis virus assembly. *Adv. Exp. Med. Biol.* **342:**189–195.

67. Opstelten, D. J., M. J. Raamsman, K. Wolfs, M. C. Horzinek, and P. J. Rottier. 1995. Envelope glycoprotein interactions in coronavirus assembly. *J. Cell Biol.* **131:**339–349.

68. Pedersen, K. W., Y. van der Meer, N. Roos, and E. J. Snijder. 1999. Open reading frame 1a-encoded subunits of the arterivirus replicase induce endoplasmic reticulum-derived double-membrane vesicles which carry the viral replication complex. *J. Virol.* **73:**2016–2026.

69. Pei, J., M. J. Sekellick, P. I. Marcus, I. S. Choi, and E. W. Collisson. 2001. Chicken interferon type I inhibits infectious bronchitis virus replication and associated respiratory illness. *J. Interferon Cytokine Res.* **21:**1071–1077.

70. Plagemann, P. G., and V. Moennig. 1992. Lactate dehydrogenase-elevating virus, equine arteritis virus, and simian hemorrhagic fever virus: a new group of positive-strand RNA viruses. *Adv. Virus Res.* **41:**99–192.

71. Reghunathan, R., M. Jayapal, L. Y. Hsu, H. H. Chng, D. Tai, B. P. Leung, and A. J. Melendez. 2005. Expression profile of immune response genes in patients with severe acute respiratory syndrome. *BMC Immunol.* **6:**2.

72. Rempel, J. D., S. J. Murray, J. Meisner, and M. J. Buchmeier. 2004. Differential regulation of innate and adaptive immune responses in viral encephalitis. *Virology* **318:**381–392.

73. Riffault, S., C. Carrat, L. Besnardeau, C. La Bonnardiere, and B. Charley. 1997. In vivo induction of interferon-alpha in pig by non-infectious coronavirus: tissue localization and in situ phenotypic characterization of interferon-alpha-producing cells. *J. Gen. Virol.* **78(Pt. 10):**2483–2487.

74. Roth-Cross, J. K., L. Martinez-Sobrido, E. P. Scott, A. Garcia-Sastre, and S. R. Weiss. 2007. Inhibition of the alpha/beta interferon response by mouse hepatitis virus at multiple levels. *J. Virol.* **81:**7189–7199.

75. Samuel, C. E. 2001. Antiviral actions of interferons. *Clin. Microbiol. Rev.* **14**:778–809, table of contents.

76. Siegal, F. P., N. Kadowaki, M. Shodell, P. A. Fitzgerald-Bocarsly, K. Shah, S. Ho, S. Antonenko, and Y. J. Liu. 1999. The nature of the principal type 1 interferon-producing cells in human blood. *Science* **284**:1835–1837.

77. Smiley, J. R. 2004. Herpes simplex virus virion host shut-off protein: immune evasion mediated by a viral RNase? *J. Virol.* **78**:1063–1068.

78. Snijder, E. J., and J. J. Meulenberg. 1998. The molecular biology of arteriviruses. *J. Gen. Virol.* **79**(Pt. 5):961–979.

79. Snijder, E. J., Y. van der Meer, J. Zevenhoven-Dobbe, J. J. Onderwater, J. van der Meulen, H. K. Koerten, and A. M. Mommaas. 2006. Ultrastructure and origin of membrane vesicles associated with the severe acute respiratory syndrome coronavirus replication complex. *J. Virol.* **80**:5927–5940.

80. Spiegel, M., A. Pichlmair, L. Martinez-Sobrido, J. Cros, A. Garcia-Sastre, O. Haller, and F. Weber. 2005. Inhibition of beta interferon induction by severe acute respiratory syndrome coronavirus suggests a two-step model for activation of interferon regulatory factor 3. *J. Virol.* **79**:2079–2086.

81. Spiegel, M., A. Pichlmair, E. Muhlberger, O. Haller, and F. Weber. 2004. The antiviral effect of interferon-beta against SARS-coronavirus is not mediated by MxA protein. *J. Clin. Virol.* **30**:211–213.

82. Spiegel, M., K. Schneider, F. Weber, M. Weidmann, and F. T. Hufert. 2006. Interaction of severe acute respiratory syndrome-associated coronavirus with dendritic cells. *J. Gen. Virol.* **87**:1953–1960.

83. Spiegel, M., and F. Weber. 2006. Inhibition of cytokine gene expression and induction of chemokine genes in non-lymphatic cells infected with SARS coronavirus. *Virol. J.* **3**:17.

84. Stark, G. R., I. M. Kerr, B. R. Williams, R. H. Silverman, and R. D. Schreiber. 1998. How cells respond to interferons. *Annu. Rev. Biochem.* **67**:227–264.

85. Stroher, U., A. DiCaro, Y. Li, J. E. Strong, F. Aoki, F. Plummer, S. M. Jones, and H. Feldmann. 2004. Severe acute respiratory syndrome-related coronavirus is inhibited by interferon-alpha. *J. Infect. Dis.* **189**:1164–1167.

86. Taguchi, F., and S. G. Siddell. 1985. Difference in sensitivity to interferon among mouse hepatitis viruses with high and low virulence for mice. *Virology* **147**:41–48.

87. Takeuchi, O., and S. Akira. 2007. Signaling pathways activated by microorganisms. *Curr. Opin. Cell Biol.* **19**:185–191.

88. Thompson, A. J., and S. A. Locarnini. 2007. Toll-like receptors, RIG-I-like RNA helicases and the antiviral innate immune response. *Immunol. Cell Biol.* **85**:435–445.

89. Tooze, J., S. Tooze, and G. Warren. 1984. Replication of coronavirus MHV-A59 in sac- cells: determination of the first site of budding of progeny virions. *Eur. J. Cell Biol.* **33**:281–293.

90. Traven, M., K. Naslund, N. Linde, B. Linde, A. Silvan, C. Fossum, K. O. Hedlund, and B. Larsson. 2001. Experimental reproduction of winter dysentery in lactating cows using BCV—comparison with BCV infection in milk-fed calves. *Vet. Microbiol.* **81**:127–151.

91. Tseng, C. T., C. Huang, P. Newman, N. Wang, K. Narayanan, D. M. Watts, S. Makino, M. M. Packard,

S. R. Zaki, T. S. Chan, and C. J. Peters. 2007. Severe acute respiratory syndrome coronavirus infection of mice transgenic for the human angiotensin-converting enzyme 2 virus receptor. *J. Virol.* **81**:1162–1173.

92. Tseng, C. T., L. A. Perrone, H. Zhu, S. Makino, and C. J. Peters. 2005. Severe acute respiratory syndrome and the innate immune responses: modulation of effector cell function without productive infection. *J. Immunol.* **174**:7977–7985.

93. Tseng, C. T., J. Tseng, L. Perrone, M. Worthy, V. Popov, and C. J. Peters. 2005. Apical entry and release of severe acute respiratory syndrome-associated coronavirus in polarized Calu-3 lung epithelial cells. *J. Virol.* **79**:9470–9479.

94. Uze, G., and D. Monneron. 2007. IL-28 and IL-29: newcomers to the interferon family. *Biochimie* **89**:729–734.

95. Vennema, H., G. J. Godeke, J. W. Rossen, W. F. Voorhout, M. C. Horzinek, D. J. Opstelten, and P. J. Rottier. 1996. Nucleocapsid-independent assembly of coronavirus-like particles by co-expression of viral envelope protein genes. *EMBO J.* **15**:2020–2028.

96. Versteeg, G. A., P. J. Bredenbeek, S. H. van den Worm, and W. J. Spaan. 2007. Group 2 coronaviruses prevent immediate early interferon induction by protection of viral RNA from host cell recognition. *Virology* **361**:18–26.

97. Versteeg, G. A., O. Slobodskaya, and W. J. Spaan. 2006. Transcriptional profiling of acute cytopathic murine hepatitis virus infection in fibroblast-like cells. *J. Gen. Virol.* **87**:1961–1975.

98. Wang, X., A. J. Rosa, H. N. Oliverira, G. J. Rosa, X. Guo, M. Travnicek, and T. Girshick. 2006. Transcriptome of local innate and adaptive immunity during early phase of infectious bronchitis viral infection. *Viral Immunol.* **19**:768–774.

99. Wathelet, M. G., M. Orr, M. B. Frieman, and R. S. Baric. 2007. Severe acute respiratory syndrome coronavirus evades antiviral signaling: role of nsp1 and rational design of an attenuated strain. *J. Virol.* **81**:11620–11633.

100. Weber, F., V. Wagner, S. B. Rasmussen, R. Hartmann, and S. R. Paludan. 2006. Double-stranded RNA is produced by positive-strand RNA viruses and DNA viruses but not in detectable amounts by negative-strand RNA viruses. *J. Virol.* **80**:5059–5064.

101. Wege, H., S. Siddell, and V. ter Meulen. 1982. The biology and pathogenesis of coronaviruses. *Curr. Top. Microbiol. Immunol.* **99**:165–200.

102. Weidman, M. K., R. Sharma, S. Raychaudhuri, P. Kundu, W. Tsai, and A. Dasgupta. 2003. The interaction of cytoplasmic RNA viruses with the nucleus. *Virus Res.* **95**:75–85.

103. Weiss, S. R., and S. Navas-Martin. 2005. Coronavirus pathogenesis and the emerging pathogen severe acute respiratory syndrome coronavirus. *Microbiol. Mol. Biol. Rev.* **69**:635–664.

104. Wong, C. K., C. W. Lam, A. K. Wu, W. K. Ip, N. L. Lee, I. H. Chan, L. C. Lit, D. S. Hui, M. H. Chan, S. S. Chung, and J. J. Sung. 2004. Plasma inflammatory cytokines and chemokines in severe acute respiratory syndrome. *Clin. Exp. Immunol.* **136**:95–103.

105. Wong, S., S. Lau, P. Woo, and K. Y. Yuen. 2007. Bats as a continuing source of emerging infections in humans. *Rev. Med. Virol.* **17**:67–91.

106. **Ye, Y., K. Hauns, J. O. Langland, B. L. Jacobs, and B. G. Hogue.** 2007. Mouse hepatitis coronavirus A59 nucleocapsid protein is a type I interferon antagonist. *J. Virol.* **81:**2554–2563.

107. **Yilla, M., B. H. Harcourt, C. J. Hickman, M. McGrew, A. Tamin, C. S. Goldsmith, W. J. Bellini, and L. J. Anderson.** 2005. SARS-coronavirus replication in human peripheral monocytes/macrophages. *Virus Res.* **107:**93–101.

108. **Yount, B., K. M. Curtis, E. A. Fritz, L. E. Hensley, P. B. Jahrling, E. Prentice, M. R. Denison, T. W. Geisbert, and R. S. Baric.** 2003. Reverse genetics with a full-length infectious cDNA of severe acute respiratory syndrome coronavirus. *Proc. Natl. Acad. Sci. USA* **100:** 12995–13000.

109. **Zhou, H., and S. Perlman.** 2007. Mouse hepatitis virus does not induce beta interferon synthesis and does not inhibit its induction by double-stranded RNA. *J. Virol.* **81:**568–574.

110. **Ziegler, T., S. Matikainen, E. Ronkko, P. Osterlund, M. Sillanpaa, J. Siren, R. Fagerlund, M. Immonen, K. Melen, and I. Julkunen.** 2005. Severe acute respiratory syndrome coronavirus fails to activate cytokine-mediated innate immune responses in cultured human monocyte-derived dendritic cells. *J. Virol.* **79:**13800–13805.

111. **Zust, R., L. Cervantes-Barragan, T. Kuri, G. Blakqori, F. Weber, B. Ludewig, and V. Thiel.** 2007. Coronavirus non-structural protein 1 is a major pathogenicity factor: implications for the rational design of coronavirus vaccines. *PLoS Pathog.* **3:**e109.

Cellular Signaling and Innate Immune
Responses to RNA Virus Infections
Edited by A. R. Brasier et al.
© 2009 ASM Press, Washington, D.C.

Susana Guix
Mary K. Estes

Caliciviridae and *Astroviridae*

24

Viral infections of the gastrointestinal tract cause approximately 2 billion cases of diarrhea in children per year worldwide, resulting in approximately 18 million hospitalizations, and they are the third-leading cause of mortality worldwide (77). As many as 3 million pediatric deaths result every year from viral gastroenteritis and dehydration (17), with most of the cases occurring in developing countries.

Caliciviruses and astroviruses, along with rotavirus, are regarded as important causes of viral gastroenteritis. Although rotavirus remains the leading cause of severe diarrhea in children under 5 years of age (33), the human caliciviruses are responsible for the vast majority of gastroenteritis outbreaks in industrialized countries (5). Surveillance studies using recently developed molecular diagnostic methods have also highlighted the importance of human astroviruses and other underreported pathogens as important causes of pediatric acute gastroenteritis. Epidemiological information on human caliciviruses and astroviruses is providing data required to assess the economic burden of the disease and justify the development of prevention strategies, including vaccination.

From a molecular point of view, both caliciviruses and astroviruses share many features in their genomic organization and replication cycle. Both viral families are small, nonenveloped viruses with an icosahedral capsid of 27 to 40 nm in diameter, containing a plus-sense, single-stranded, polyadenylated RNA genome that is approximately 7,000 nucleotides in length. The calicivirus genome is covalently linked to a VPg protein on its 5' end, and although it has not been biologically demonstrated for any astrovirus, a putative VPg coding region has been identified in the astrovirus genome (2). Both astrovirus and calicivirus genomes contain the open reading frames coding for nonstructural proteins at the 5' end of their genomes and structural proteins at the 3' end. Their replication cycle takes place in the cytoplasm of infected cells in association with intracellular membranes, and in all astroviruses and caliciviruses studied to date, a subgenomic RNA is produced during infection and used for the synthesis of large amounts of capsid proteins (32, 59). A diagram of the general viral replication cycle for caliciviruses and astroviruses is depicted in Fig. 1.

During the last decade, there has been a tremendous advance in our knowledge of cell signaling pathways that lead to type I interferon (IFN) gene expression, and it is now clear that many viruses encode specific gene products that antagonize this response (27). This review

Susana Guix, Enteric Virus Laboratory, Department of Microbiology, University of Barcelona, Barcelona 08028, Spain. **Mary K. Estes**, Department of Molecular Virology and Microbiology, Baylor College of Medicine, Houston, TX 77030.

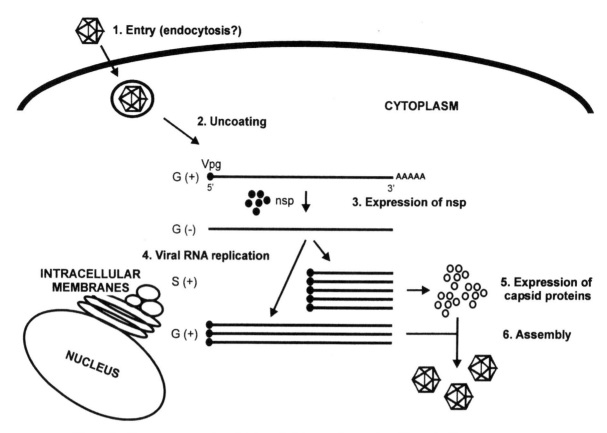

Figure 1 Current proposed model for calicivirus and astrovirus life cycle. The entry process in both calicivirus and astrovirus infection following receptor binding on the cell surface is poorly understood. Adsorptive endocytosis was described as the most probable mechanism by which astroviruses enter susceptible cells (19), and recently it has been shown that FCV enters the cell by clathrin-mediated endocytosis and that acidification is required for uncoating of the genome and access to the cytoplasm (75). Once uncoated in the cytoplasm, genomic RNA is translated as a polyprotein. Cleavage of the polyprotein results in the individual non-structural proteins (nsp) necessary for replication. Viral replication complexes assemble in close association with intracellular membranes, resulting in the formation of both genomic and subgenomic positive-sense RNAs. Large quantities of subgenomic RNAs are used to synthesize high levels of capsid proteins. While it is known that both genomic and subgenomic RNAs are encapsidated into virions of animal caliciviruses, it is still unclear whether this occurs for human caliciviruses and astroviruses.

summarizes what is known of the interactions between calicivirus and astrovirus replication and the host innate immune system, and discusses how these interactions may play a role in regulating virulence, pathogenesis, and the outcome of infection. This review will conclude with a few remarks meant to stimulate discussion and future studies to better understand the interplay between calicivirus and astrovirus infections and the cellular antiviral responses.

CALICIVIRUSES

The first member and prototype strain of the *Caliciviridae* family, the Norwalk virus (NV), was identified in 1972 from an outbreak of gastroenteritis in Norwalk, Ohio

(45). Progress in the understanding of the Norwalk-like virus biology, pathogenesis, classification, diagnosis, and vaccine development has been slow due to the inability to culture these viruses. Currently, four genera are established within the *Caliciviridae* family: *Norovirus, Sapovirus, Lagovirus,* and *Vesivirus.* The *Norovirus* genus is divided in five genogroups, with genogroups GI, GII, and GIV containing human strains, GIII containing bovine strains, and GV containing murine norovirus (MNV) strains. In addition to human strains, GI also contains porcine strains, and a novel enteric calicivirus, genetically related to human noroviruses of genogroup IV, was identified in a lion cub that died of severe hemorrhagic enteritis (57). The *Sapovirus* genus includes human as well as animal pathogens such as the porcine enteric calicivirus (PEC). *Lagovirus* and *Vesivirus* genera include caliciviruses of veterinary importance such as rabbit hemorrhagic disease virus and feline calicivirus (FCV), respectively. The *Vesivirus* genus also includes the San Miguel sea lion virus, which infects marine mammals. Recently, a novel calicivirus has been isolated from rhesus macaques and named Tulane virus. The complete genomic sequence suggests that Tulane virus may represent a new calicivirus genus with the proposed name *Recovirus* (23).

In this section we describe what is now known about the activation of innate responses after both human and nonhuman calicivirus infections, and specifically emphasize the effect that innate antiviral responses might have on restricting human norovirus propagation in vitro. To date, the role of innate immunity in controlling infections caused by caliciviruses has been demonstrated for MNV (46, 65) and PEC (12), but only limited data are available to confirm this for human noroviruses (10, 13).

Human Caliciviruses

Human caliciviruses, including noroviruses and sapoviruses, are a major cause of acute gastroenteritis (reviewed in references 21 and 38), and due to their high infectivity, their high stability in the environment, and their association with debilitating illness, they are listed as category B bioterrorism agents by the National Institute of Allergy and Infectious Diseases at the U.S. National Institutes of Health. Human norovirus infections cause vomiting, diarrhea, and nausea, beginning approximately 24 h after exposure and resolving 24 to 48 h later. Human noroviruses can infect all age groups, although the majority of outbreaks involve the elderly, school-aged children, and adults. Human sapoviruses cause mainly milder endemic pediatric gastroenteritis, although cases of human noroviruses in infants and young children are also common. Noroviruses also can cause persistent infections in immunosuppressed individuals, including transplant patients.

Since human norovirus infections have a short incubation period and a short disease time course, some researchers have hypothesized that host innate antiviral responses might play important roles in controlling virus-induced disease and pathogenesis. In addition, since cellular innate responses have been shown to limit viral replication in some animal caliciviruses (see below), it has been suggested that host innate immune responses may also play a role in determining susceptibility or resistance to infection. Unfortunately, the lack of a susceptible cell culture system and a small-animal

model to study human norovirus infection and disease hampers research to address these questions.

Early human challenge studies with the prototype NV suggested the existence of short-term immunity to homologous but not heterologous virus challenge (87). Long-term protection was not conferred by a single NV challenge (67), and long-term volunteer rechallenge showed that some individuals are either repeatedly susceptible or resistant to NV infection, leading to the identification of a genetically determined factor for susceptibility to NV infection (42, 55, 67). The *FUT2* gene, which encodes an α1,2-fucosyltransferase responsible for the secretor status according to the expression of certain types of histo-blood group antigens, has been determined to be a susceptibility allele for NV infection (55, 56), although not for all human norovirus strains. A recent challenge study using Snow Mountain virus, a GII norovirus, showed that unlike NV, which belongs to GI, Snow Mountain virus infection is not dependent upon blood group secretor status, and confirmed previous observations made with NV (30, 44) showing that prechallenge serum-specific immunoglobulin G titer is not predictive of the risk of infection (54). Although other factors such as mucosal immunoglobulin A response have been suggested to play a role in limiting infection and disease, current data are insufficient to draw definitive conclusions, and further studies are still required to better understand host factors related to protection against norovirus infection.

Recently, a human norovirus GII strain has been found to infect and cause mild diarrhea in gnotobiotic pigs, with virus shedding and immunofluorescent detection of both structural and nonstructural proteins in enterocytes (16). Systemic and intestinal cytokine and antibody responses have been analyzed in this experimental animal model (73). In the intestinal contents (IC) of inoculated pigs, a significant peak of IFN-α coincided with the period of viral shedding, viremia, and diarrhea, and a second peak was detected 21 days after inoculation. In serum, a single peak was detected after clearance of acute viral infection. The authors speculated that the early increase of IFN-α secretion in IC was the result of a local, early response of the innate immune system (epithelial or dendritic cells or macrophages) to viral infection. To explain the delayed peak of IFN-α in the IC and in the serum after clearance of acute viral infection, the authors hypothesized the possibility that the virus could replicate in macrophages and dendritic cells, as has been described for murine noroviruses (see below), resulting in a low level of viral expression and stimulation of the secretion of innate cytokines even after virus shedding was no longer detectable (73). Although this possibility is interesting, human noroviruses are thought to have an enteric tropism and there is still no evidence supporting their replication in macrophages or dendritic cells. Intestinal biopsies from challenged volunteers show villous broadening and blunting in the jejunum as well as the infiltration of mononuclear cells, but studies to confirm and determine the tissue and cell tropism of human norovirus infections both in vivo and in vitro are still pending. Using recombinant virus-like particles and human gastrointestinal biopsies, a preferential binding to epithelial cells of the pyloric region of the stomach and to enterocytes on duodenal villi has been observed (56), and although the virus does not replicate in any conventional cell line, differentiated human intestinal CaCo-2 cells bind significantly more virus-like particles than other cell lines and internalize a small proportion of particles (20, 82). Finally, different strains of human noroviruses have been reported to be able to be passaged several times with limited replication in a differentiated three-dimensional cell culture system derived from a human small intestinal cell line (74), further suggesting that human noroviruses show a tropism for epithelial cells of intestinal origin.

Although results using the three-dimensional cell culture system are promising, these results need to be confirmed by other laboratories, so it is still too soon to know whether this system is robust enough to be widely used to efficiently propagate human noroviruses in vitro and perform studies to characterize the cellular innate immune responses to viral infection. In the absence of a fully permissive cell culture system for these viruses, Chang et al. (13) generated a human hepatoma cell line (Huh7) stably transfected with a self-replicating NV RNA bearing an engineered neomycin resistance gene. Virus replicon systems have proved to be important tools in the screening of antivirals and the investigation of virus-host interactions. Similar to research performed with hepatitis C virus replicons, which allowed the identification of different viral mechanisms to evade cellular innate responses, the stable NV replicon system has been used to investigate interactions between NV and the activation of cellular antiviral responses. Expression of NV replicon RNA is significantly reduced in the presence of exogenous IFN-α in a dose-dependent manner. A nearly complete clearance of the replicon proteins and RNA is seen when 20 U/ml of IFN-α is added to the culture supernatant for 72 h. Since the NV replicon model does not allow study of the viral entry and assembly steps of the virus life cycle, the block of IFN on viral replication in this system occurs at the stages of viral genome stability, RNA transcription/replication, and/or translation. In addition, the expression of signal transducer and activator of transcription 1 (Stat1) detected by Western blot

analysis increased proportionately to the concentrations of IFN-α, suggesting a role for the Stat1 pathway in the inhibitory effects of exogenous IFN-α on RNA replication (13). Recently, IFN-γ and ribavirin have also been shown to act as antinoroviral agents, suggesting that they may be possible therapeutic options for noroviral gastroenteritis (10). Like IFN-α, both agents inhibited the replication of NV in the replicon-bearing cells, showing a reduction of the NV genome and proteins.

To examine whether NV replication exploits strategies to subvert the innate immune response, the effect of the NV replicon on the ability of Sendai virus to induce an IFN response was examined (13). The presence of NV replication does not inhibit the ability of the cell to mount an innate immune response upon infection with Sendai virus, suggesting that NV does not encode strong antiviral innate immunity mechanisms. However, this study did not analyze whether the presence and replication of the NV genome in the replicon-bearing cells result in the activation of a cellular IFN response, or whether NV encodes mechanisms to antagonize some sensing arms of the IFN pathway different from the ones activated after Sendai virus infection. Thus, IFN could potentially be an excellent antiviral drug to clear norovirus infections in the clinical setting, but it remains unclear if NV-infected cells both in vitro and in vivo are able to sense the presence of a foreign noroviral RNA and activate IFN responses. Microarray analysis using the NV replicon cell line and studies to determine whether certain early steps of the cellular IFN response such as activation of IFN regulatory factor 3 (IRF3) take place in these cells would help answer these questions. In addition, it would be interesting to analyze whether a type I IFN response is generated in human subjects after NV infection. Using the gnotobiotic pig model for human norovirus infection, a peak of IFN-α in the serum was only detected at late times after clearance of acute viral infection, although at early times after inoculation a significant peak of levels of IFN-α was observed in IC (73). Although it is clear that NV replication can be inhibited by exogenous IFN, it may not induce a strong IFN response, so activation of cellular antiviral mechanisms may or may not be the reason to explain the severe growth restrictions of NV in cell culture.

Efforts to develop a cell culture system for NV and to understand the block(s) of NV growth in vitro demonstrate that NV RNA isolated from human stools is fully infectious in cultured cells (34). Transfection of wild-type NV RNA into human hepatoma Huh7 cells leads to a single cycle of viral replication, with expression of viral antigens, viral replication, and release of viral particles into the medium. Overexpressing the human

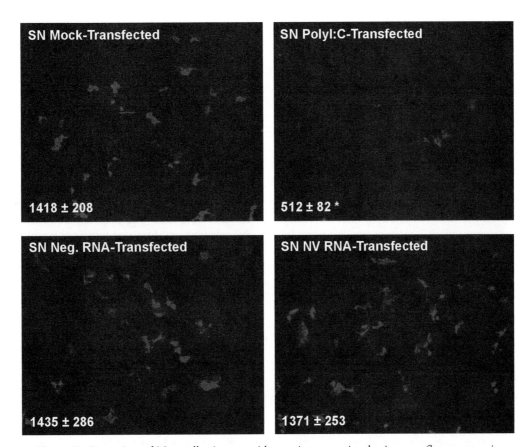

Figure 2 Detection of Norwalk virus capsid protein expression by immunofluorescence in Huh7 cells transfected with wild-type NV RNA isolated from human volunteer stool samples. Low-magnification images are shown. Prior to NV RNA transfection, cell monolayers were incubated for 3 h at 37°C with supernatant (SN) from cultures that had been mock-transfected, or transfected with poly(I:C), a carrier RNA used as a negative control, or wild-type NV RNA for 24 h. After NV RNA transfection, SNs used for the pretreatment of cells were further incubated with cells until immunostaining was performed at 48 h posttransfection. The number in each panel indicates the average number plus or minus standard deviation of NV-positive cells per transfected well after transfecting 3 wells of a 48-well plate for each condition. The asterisk denotes statistically significant differences ($P < 0.05$) after the average numbers of NV-positive cells in each condition were compared by Student's t test.

FUT2 enzyme in Huh7 cells enhances virus binding, and the cells remain permissive to NV replication after RNA transfection, but a second round of viral infection is not observed. Transfection of viral NV RNA infectivity in Huh7.5.1 cells, which contain an inactivating mutation in the retinoic acid-inducible gene I (RIG-I) (7, 76), does not result in an increase in virus replication efficiency or spreading, indicating that innate immune pathways activated by RIG-I do not play a role in limiting NV replication. RIG-I mediates antiviral responses by recognition of single-stranded RNA bearing 5′ phosphates (39, 69). These results are not surprising since, similar to other animal caliciviruses, human NV RNA 5′ ends are thought to be covalently linked to a VPg protein.

In summary, NV RNA transfection experiments suggest that cells are fully permissive to a complete infection after initial NV RNA transfection, but the infection does not spread throughout a culture of cells either due to the lack of a factor required during viral entry and/or uncoating or to the induction of cellular antiviral responses independent of RIG-I that would inhibit replication at the early stages of the replication cycle (34). To test the latter possibility, cells were treated with the supernatant from Huh7 cells transfected with poly(I:C) to induce IFN expression, and NV replication was inhibited after RNA transfection in naïve cells, while the supernatant of NV-transfected cultures did not have the same effect (Fig. 2) (35). These results indicate that although NV RNA replication may be sensitive to IFN, as shown by Chang et al. (13) using the NV replicon cell line, the lack of cell-to-cell spread after an initial round of NV replication is unlikely to be due to the activation of cellular antiviral responses that can block viral propagation. Although further experiments are required to confirm these observations, these results may also suggest that NV does not induce a strong type I IFN response in productively infected Huh7 cells. The ability to observe viral replication after transfecting the wild-type NV RNA into cells offers an excellent model to dissect virus-host cell interactions.

Animal Caliciviruses

Until a robust cell culture system or a small-animal model for human noroviruses becomes available, most researchers are relying on the study of animal caliciviruses that replicate in the laboratory to gain insight into different aspects of norovirus biology and pathogenesis and the mechanisms of immune control of norovirus infections. Recent studies suggest that cellular innate immune responses play an important role in the control of replication of MNV, which belongs to genogroup V within the genus *Norovirus*, and PEC, within the genus *Sapovirus*.

MNV

A norovirus that infects mice was first identified in 2003 in immunocompromised animals lacking recombination-activating gene 2 (RAG2) and Stat1 (46). Virus infection in Stat1$^{-/-}$ mice results in high levels of viral RNA in the proximal small intestine, lung, liver, spleen, brain, blood, and feces, accompanied by severe pneumonia and destruction of splenic and liver tissues. Today, MNVs are considered to be one of the most common naturally occurring infections of laboratory mice in North America. The seroprevalence of MNV in mouse research colonies in the United States and Canada is estimated to be as high as 22.1%, indicating that MNVs are endemic in mouse populations (40). MNV shares many biological and molecular properties with other noroviruses and with caliciviruses in general, and it represents the only member of the *Norovirus* genus that replicates in cell culture as well as in a small, genetically manipulable animal model. This has provided the first opportunity to study norovirus biology and virus-host interactions in detail (reviewed in reference 85). Although different strains may differ in their ability to grow in culture and to infect and/or persist in wild-type mice (78), the study of the MNV model system has revealed a tropism of MNV for cells of the hematopoietic lineage both in vivo and in cell culture, as well as an important role for innate immune responses in the control of infection and disease in mice.

After oral inoculation in Stat1$^{-/-}$ mice, MNV was detected in cells of the macrophage lineage by immunohistochemical staining, and virus was shown to replicate in murine primary bone marrow-derived macrophages and dendritic cells of wild-type, Stat1$^{-/-}$, and IFN-α/βR$^{-/-}$ (IFN-α/β receptor knockout) mice; several murine macrophage cell lines such as RAW 264.7; and Stat1$^{-/-}$ but not wild-type mouse embryonic fibroblasts. The virus yield was consistently higher in macrophages lacking Stat1, IFN-α/β/γR, or IFN-α/βR than in the wild-type counterparts, indicating a role of the antiviral molecules Stat1 and type I IFN in limiting MNV growth in vitro (46, 84). The importance of Stat1 and IFN in the control of MNV growth is mirrored in vivo, since Stat1$^{-/-}$ and IFN-α/β/γR$^{-/-}$ mice are highly susceptible to lethal MNV infection (46). Significant mortality is not observed in PKR$^{-/-}$ or iNOS$^{-/-}$ mice, and lack of PKR and iNOS in cultured bone marrow-derived macrophages does not enhance virus yields in vitro either, indicating that these two well-known mediators of IFN-induced antiviral pathways are not relevant in controlling MNV replication and disease (46, 84). Thus, other mediators of IFN and Stat1 pathways are likely required to control MNV growth in vitro and clinical disease induced by MNV in vivo.

Although MNV was originally isolated by serial passage from the brains of severely immunocompromised mice, and it is clear that Stat1-dependent antiviral activity is required to control disease, initial studies also showed that MNV can replicate in wild-type mice and cause asymptomatic infections (40, 46). Due to its ability to infect wild-type mice after peroral inoculation and spread naturally between mice in the same cage, MNV is considered to be an enteric virus. In wild-type mice, MNV seeds the proximal small intestine, replicates in cells of the lamina propria, and disseminates extraintestinally, but does not cause clinical disease (65). Even though MNV infection in wild-type mice is associated with histopathological alterations in the intestine (mild inflammation) and the spleen, caution must be taken when using MNV-infected mice as a small-animal model of human norovirus infection, since unlike MNV, human noroviruses cause clinical disease in immunocompetent hosts. In mice, Stat1-dependent pathways are critical in controlling viral replication at the primary site of infection, preventing high levels of viral dissemination to peripheral tissues, protecting lymphoid tissue from pathological cell death, and limiting the severity of MNV pathogenesis (65). Although it would be interesting to examine whether the severity and persistence of human norovirus infections are influenced by genetically determined variations in components of the innate immune system, this hypothesis has not been tested, and other factors such as the initial infectious dose, virus strain, or factors related to adaptive immunity could also be important. In the MNV model, adaptive immunity has been shown to be important to clear infection (46). Similar to wild-type mice, RAG$^{-/-}$ mice lacking T- and B-cell-dependent adaptive immunity are resistant to MNV-induced lethality, but contain higher levels of viral RNA in all organs analyzed and for longer periods of time.

In summary, the study of MNV has revealed an important role of Stat1 and type I IFN innate immune responses in the control of virus growth both in vitro and in vivo, as well as in the induction of disease and lethality in mice, but the molecular mechanisms underlying these observations are still unclear. Although molecular mediators of these antiviral responses such as PKR and iNOS are not critical, future research will focus on the identification of the factors responsible for the protective effect of Stat1 and IFN pathways on MNV-induced pathogenesis at the molecular level. Incubation of the human NV replicon cell line (13) with increasing concentrations of exogenous IFN results in upregulation of Stat1 expression, indicating that Stat1 is also involved in the antiviral effects of IFN against human NV replication. Both human and murine noroviruses may lack a strategy to counteract host cellular defense mechanisms involving Stat1.

PEC

Downregulation of Stat1 also plays a role in allowing propagation in cell culture of other animal caliciviruses such as PEC (12). Wild-type PEC causes diarrhea in pigs and is morphologically and genetically similar to the human sapoviruses (37). The Cowden strain of PEC was adapted to grow in cell culture by serial passage in the porcine kidney LLC-PK cell line, but the presence of IC, obtained from uninfected gnotobiotic pigs, in the culture medium is essential for viral replication even after more than 25 serial passages (11, 25, 68). Initial screening studies to identify the active factor(s) in IC required for PEC replication indicated that the positive effect was dependent on a cyclic AMP signaling pathway with involvement of the protein kinase A (PKA) (11). In addition, bile acids have been identified as essential active factors in the IC required for PEC growth, in association with activation of the PKA/cyclic AMP pathway and downregulation of Stat1. When added to the culture medium of LLC-PK, bile acids can inhibit IFN-mediated activation of Stat1, as measured by levels of phosphorylated Stat1 by immunoblot analysis (13). This observation provides a possible mechanism by which IC may

negatively regulate innate immunity and therefore allow replication of this enteric calicivirus. It is also consistent with recent findings on the role of Stat1 in MNV growth and pathogenesis, but it does not rule out the possible involvement of other components present in IC and their mechanisms of action in promoting viral growth.

The effect of certain human bile acids on replication efficiency of human noroviruses was investigated using the NV replicon developed in Huh7 hepatoma cells (13). No differences were observed in NV replication (9), suggesting that NV differs from PEC in its requirements to grow in cell culture. In addition, whereas bile acids are essential to support recovery of PEC from cells even after transfection of viral RNA (11, 13), human NV RNA is fully infectious after transfection in the absence of bile acids (34). These results suggest that while the effect(s) of bile acids on PEC replication are essential for events in the virus replication cycle subsequent to receptor binding, internalization, and uncoating, human NV RNA expression, replication, and virus assembly can occur unhindered in the absence of bile acids. Addition of certain bile acids in the culture medium after transfection of wild-type NV RNA does not result in an improvement of viral replication efficiency either (S. Guix and M. K. Estes, unpublished data).

Interestingly, although bile acids do not affect the replication of NV in replicon-harboring cells, Chang and George (9) observed that bile acids can inhibit the anti-NV effect induced by exogenous IFN in a dose-dependent manner, but only when cells are pretreated with bile acids for 24 h prior to addition of IFN. When cells were simultaneously incubated with bile acids and IFN, NV RNA levels were similar among the groups with or without bile acids. These results suggest that the effect of bile acids may require the upregulation of certain genes required to inhibit IFN-mediated responses. Future studies to characterize differences in gene expression pattern in cells expressing NV RNA may uncover a molecular basis for the design of new therapeutic options for norovirus infections.

FCV

FCV is an important respiratory pathogen that causes a variety of mild to severe syndromes in cats, with a high prevalence in the feline population (reviewed in reference 70). Specific current problems related to FCV include the high degree of strain variability associated with varied levels of virulence, the presence of persistently infected cats, the fact that available vaccines prevent disease without protecting against infection so that vaccinated animals may become carriers following subclinical infection, and the emergence of hypervirulent strains. One of these emerging strains is the virulent systemic feline calicivirus (VS-FCV), a novel pathogen that causes up to 67% mortality even in healthy adult cats (41). The molecular basis of this enhanced mortality remains unknown.

Even though FCV can replicate easily in cell culture, the ability of FCV to block antiviral responses at a cellular level has not been examined yet. Early experiments found FCV to be sensitive to both feline and human IFN-α (26). Different recombinant feline IFN (rFeIFN) subtypes have been expressed as heterologous proteins and have shown efficacy against FCV in vitro (6). The degree of antiviral activity of rFeIFN observed in vitro may be dependent on viral dose, virus strain, and host-cell system (61, 66). Several rFeIFN-α subtypes are effective in limiting FCV growth in a feline embryonic fibroblast cell line, but none of them exhibit antiviral effect against the virus in Crandell-Reese feline kidney cells, routinely used to propagate FCV in the laboratory (6). The reasons why these differences exist are unclear, and the molecular mechanisms of antiviral action of IFN on these cells have not been explored yet. These results highlight the fact that sensitivity to the antiviral activities of IFN may vary depending on the cell types used in vitro.

ASTROVIRUSES

Astroviruses were first identified by Appleton and Higgins in 1975, in association with an outbreak of gastroenteritis affecting infants in a maternity ward in England (3). Subsequently, they were found in association with gastroenteritis in a wide variety of young mammals and birds (59). Human astroviruses (HAstV) are one of the leading causes of acute viral gastroenteritis in infants and young children (80). Besides enteric disease in humans and animals, astrovirus infections have been associated with fatal hepatitis in ducks (29) and interstitial nephritis causing growth retardation in young chickens (43). Often they have been associated with asymptomatic infections, and coinfections with other enteric viruses are frequent. Today, astroviruses are divided in two genera: *Mamastrovirus,* which includes human, bovine, feline, porcine, ovine, and mink astroviruses; and *Avastrovirus,* which includes chicken, turkey, and duck astroviruses. Unlike human noroviruses, HAstV can be efficiently propagated in cell culture, provided trypsin is added to the culture media, as it is required to maintain virus infectivity. Among animal astroviruses, so far only bovine, porcine, duck, and chicken astroviruses have been serially propagated in established cell lines.

Despite the impact of astroviruses on human and animal health, very little is known about the mechanisms of pathogenesis (reviewed in reference 64) or the immune response to astrovirus infection (reviewed in reference 48). Although observational data from volunteer and epidemiological studies suggest that adaptive immunity is the key mediator for protection in humans, some animal models have recently highlighted the importance of innate immune responses in controlling astrovirus infections in vivo, especially primary infections in young animals (49). In addition, the HAstV capsid protein has recently been described to inhibit complement activation, and this is suggested as a mechanism of virus-induced pathogenesis by disruption of the innate immune system (8). Although none of these interactions between astroviruses and the innate immune signaling pathways has been fully characterized at a cellular level, this review will focus on describing the most important observations extrapolated from both human and animal studies.

Human Astroviruses

HAstV are common causes of diarrhea in children worldwide, and serologic studies indicate that most children are infected with HAstV and develop antibodies to the virus early in life (51, 52). The elderly and immunocompromised individuals represent high-risk groups as well. HAstV infection induces a mild, watery diarrhea that typically lasts 2 to 3 days, associated with vomiting, fever, anorexia, and abdominal pain. Compared with norovirus-induced gastroenteritis, the reported incubation period for HAstV disease is longer. According to adult volunteer studies, symptoms manifest 3 to 4 days postinfection, with viral shedding taking place before that (53). In general, HAstV diarrhea is milder than that caused by rotavirus or norovirus, and it resolves spontaneously, although in some cases HAstV infections require hospitalization (reviewed in reference 80).

The observation that the HAstV detection rate decreases with age, data from seroprevalence studies, and results from normal adult volunteer studies suggest that antibody responses to HAstV are important in limiting infection and in providing protection from reinfection (53, 60). Analysis of HAstV-specific T cells from normal intestinal biopsies also suggests that CD4$^+$ T cells are important in mucosal defense against recurrent astroviral infections (62). Overall, histological analysis of biopsies from infected individuals indicates that intestinal lesions and the inflammatory response in the lamina propria are relatively minor, suggesting that destruction of the intestinal epithelium and inflammation are not the main mechanisms causing diarrhea (64, 71). Interestingly, two recently described features of the HAstV capsid protein provide some insights into possible mechanisms to explain HAstV-induced pathogenesis. First, the HAstV capsid protein has been proposed to affect intestinal epithelial barrier permeability and contribute to diarrhea in vivo (63). Second, the ability of the viral capsid protein to inhibit complement activation has been proposed as a novel viral mechanism of innate immune evasion (8). Although diverse mechanisms of inhibition of complement activation have been described for many enveloped animal viruses, this is the first report describing such a phenomenon for a nonenveloped icosahedral virus. Interestingly, both of these pathogenic features seem to occur independently of viral replication.

Animal Astroviruses

Although viruses in the *Astroviridae* family infect many different species of young mammals and birds, only cattle, lambs, chickens, and turkeys have been experimentally infected to study different aspects of viral replication and to examine the role of immune responses in the control of infection. Some of these studies indicate that macrophages and the innate immune system may be involved in the response to astrovirus infection.

Histological studies in lambs, turkeys, and also humans infected with astrovirus indicate a tropism for enterocytes of the small intestine. Astrovirus also appears to target M cells overlying the Peyer's patches in experimentally infected calves (86). Inclusions of viral particles have been observed in lysosomes in macrophages in the lamina propria of gnotobiotic infected lambs (31), and viremia and spread to extraintestinal organs occurs in the turkey astrovirus (TAstV) animal model (50). Whether interaction with cells from the innate immune system, viremia, and extraintestinal spread are general features of all astrovirus infections, including HASTV, remains to be explored.

A major role for macrophages and the innate immune system in controlling primary infections has been demonstrated in astrovirus-infected turkeys, without the involvement of adaptive immune responses (49, 50). Although productive TAstV replication is not evident in macrophages, activation of nitric oxide (NO) production by viral particles is detected in vitro using the well-characterized chicken macrophage cell line HD11, and this response correlates with upregulation of the iNOS gene. Finally, increases of NO production have been confirmed in vivo, and inhibition of iNOS in experimentally infected embryos results in a 1,000-fold increase in total viral RNA titer isolated from the intestines. Results of studies designed to analyze the underlying mechanisms

that result in induction of NO activity in turkey macrophages in the absence of productive replication would be extremely interesting.

Despite differences between the TAstV model and infections in humans, especially the lack of adaptive immune responses observed after TAstV infection, the turkey animal model offers an excellent tool to examine astrovirus pathogenesis and immunity and to lay the foundation to better understand the role of innate immune responses in limiting replication in other animals and in humans. The acute nature of the gastroenteritis, the short duration of symptoms, and the occurrence of most clinical infections in infants and young children who may lack acquired immunity strongly suggest that innate immune responses may play a key role in controlling virus replication and limiting disease. While human adult volunteer studies suggest that preexisting antibodies are key mediators of protection from disease, innate immune responses may be more relevant to control of primary infections.

CLOSING REMARKS

Although no data are available regarding astrovirus sensitivity to IFN, it is clear that most caliciviruses are sensitive to exogenous IFN, and the Stat1 signaling pathway is important in controlling host susceptibility to some animal calicivirus infections and disease (12, 85). The most relevant findings related to in vivo and in vitro innate immune responses to both calicivirus and astrovirus infections are summarized in Table 1. However, many questions remain unanswered concerning whether calicivirus and astrovirus replication triggers a type I IFN response in productively infected cells and whether these viruses encode specific gene products that antagonize the sensing and signaling arms of these antiviral cellular pathways. As positive-sense RNA viruses, there are several steps in the replication cycle of caliciviruses and astroviruses (Fig. 1) where the cellular antiviral machinery could detect foreign viral components. During entry, endosome-associated Toll-like receptors (TLRs) may detect genomic RNA, and the cytoplasmic RNA sensors RIG-I and melanoma differentiation-associated gene 5 (MDA-5) might also detect the RNA as it enters into the cytoplasm. Since it is believed that the 5′ end of viral RNAs is covalently linked to a VPg protein, and it has recently been shown that RIG-I specifically binds to free phosphates at the 5′ end of the RNA (39, 69), viral RNAs may be protected from recognition by RIG-I. Consistent with this, Huh7.5.1, a hepatoma cell line deficient in RIG-I, does not show increased permissivity to NV replication efficiency after RNA transfection when

Table 1 Highlights of innate responses induced by caliciviruses and astroviruses

Virus	Model	Observations	Reference(s)
Human norovirus	Gnotobiotic pig model	Infection results in an early and late peak of IFN-α in intestinal contents and only a late peak of IFN-α in serum.	73
	Huh7 cells harboring an NV replicon	NV replication is inhibited by exogenous IFN-α and IFN-γ. Incubation with exogenous IFN-α inhibits viral RNA replication and induces upregulation of Stat1 expression. NV replication does not interfere with the induction of an IFN response by Sendai virus. Bile acids can inhibit the antiviral effects of exogenous IFN-α and IFN-γ.	9, 10, 13
	Huh7 cells transfected with wild-type NV RNA	RIG-I pathway does not play a role in limiting NV replication. Cellular responses induced by poly(I:C) can inhibit NV RNA replication.	34, 35
Murine norovirus	In vivo infection in mice	Stat1 and type I IFN responses limit initial MNV replication in the proximal small intestine, virus dissemination, clinical disease, and lethality. PKR and iNOS are not key mediators of the Stat1- and IFN-induced protection from disease.	46, 65
	In vitro infection using primary BMMΦ and BMDCs[a]	Stat1 and type I IFN responses are important in limiting MNV growth in vitro. PKR and iNOS are not involved in the cellular response that limits norovirus growth.	84
Porcine enteric calicivirus	In vitro infection in cultured cells	Bile acids-mediated downregulation of Stat1 response is essential for virus replication.	12
Feline calicivirus	In vitro infection in cultured cells	Virus replication is sensitive to feline and human IFN-α. Viral dose, virus strain, and host-cell system may affect the degree of antiviral activity of IFN.	6, 26
Human astrovirus	In vitro hemolytic complement activity assays	HAstV-1 virions and purified capsid protein both suppress the complement system.	8
Turkey astrovirus	In vivo infection in turkeys	Increased NO production after virus infection plays a role in limiting viral replication.	49
	In vitro infection in a chicken macrophage cell line	Virus can stimulate macrophages to produce NO through a replication-independent mechanism, with upregulation of iNOS expression.	49

[a]BMDC, bone-marrow-derived dendritic cells.

compared with Huh7 cells (34). Thus, the presence of VPg at the 5′ end of the viral RNA may be regarded as a mechanism to block its sensing by RIG-1. A specific role of TLRs or MDA-5 in detecting caliciviruses and astroviruses has not been reported yet.

Replication of the genomes of all well-characterized positive-strand RNA viruses, including plant, animal, and insect viruses, occurs in large replication complexes associated with intracellular membranes (1). Ultrastructural analysis of HAstV-infected intestinal CaCo-2 cells shows the presence of viral aggregates surrounded by a large number of double-membrane vesicles induced by virus infection (36, 58), and the astrovirus capsid proteins appear to associate with intracellular membranes where RNA replication and virus morphogenesis occur (58). In MNV-infected macrophage cells, particles have also been observed within or next to single- or double-membrane vesicles in the cytoplasm, accompanied by a complete rearrangement of intracellular membranes (84). Viruses could use double-membrane vesicles to anchor the replication complexes and sequester the produced RNAs from the cellular RNA-sensing machinery. This possibility has been considered and is being actively investigated for many coronaviruses, for which the mechanisms of evasion of early host innate immune responses remain incompletely understood (79, 88).

Some alternate mechanisms that inhibit cellular IFN responses previously identified in other viral families such as picornaviruses could also be applicable to caliciviruses and astroviruses (see chapter 21, this volume). In cell culture, FCV infection induces shutoff of host protein synthesis associated with cleavage of the host translation initiation factors (83), and a unique role for VPg in initiation and regulation of protein synthesis within the infected cell has been highlighted for many caliciviruses (14, 18). In vitro experiments with human NV showed that VPg can inhibit cap-dependent and internal ribosome entry site-driven translation of reporter mRNAs in a dose-dependent manner (18). This strategy appears to be nonspecific, but modification of the cellular protein expression pattern may result in the inhibition of IFN synthesis. The induction of host protein synthesis shutoff in astrovirus-infected cells has not been analyzed yet.

Inhibition of the intracellular trafficking of secreted proteins, such as IFN, may also constitute a viral anti-IFN mechanism. Characterization of the N-terminal nonstructural protein of human NV by transient expression led to the conclusion that it can disassemble the Golgi apparatus, causing disruption of intracellular protein trafficking (22, 24). In addition, NV nonstructural protein p22, which shows locational homology in the NV genome when compared with the poliovirus genome, may also function to mediate Golgi disruption by binding coat protein complex II (COPII) vesicles (72). Finally, other less explored mechanisms such as a block in mRNA expression by micro-RNAs may also be used by some viruses to downregulate the cellular antiviral IFN response.

New tools to study norovirus biology, such as recently developed infectious clones for human and murine noroviruses (4, 15, 47, 81), together with the establishment of a robust cell culture system for human noroviruses should soon provide the ability to investigate in detail the critical steps in the virus replication cycle that may positively or negatively interact with the cellular antiviral responses. Although knowledge of caliciviruses may not be extrapolated to viruses in the *Astroviridae* family, the high degree of similarity between the two virus families, along with the availability of astrovirus cell culture systems, a HAstV reverse genetics system (28), and a small-animal model to study astrovirus biology, can provide the foundation to start similar studies with important astrovirus pathogens. Understanding the interplay between caliciviruses and astroviruses and the IFN response, and how these molecular interactions play a role in determining virulence and the outcome of infection, could help us to design new strategies for prevention and therapy. For instance, in the case of noroviruses, therapies based on IFN might be applicable to prolonged persistent infections, and if the role of innate immune responses required for control of murine norovirus disease is confirmed for human noroviruses, live-virus vaccines that do not counteract the IFN response may be designed to elicit a robust innate immune response and induce protection against norovirus-induced disease. Thus, further exploration of the interplay between caliciviruses and astroviruses and the IFN response not only will help us to better understand viral pathogenesis but also can result in novel vaccination strategies and therapies.

We are grateful to Robert L. Atmar, Albert Bosch, and Rosa M. Pintó for critical reading of the manuscript and excellent suggestions.

References

1. **Ahlquist, P., A. O. Noueiry, W. M. Lee, D. B. Kushner, and B. T. Dye.** 2003. Host factors in positive-strand RNA virus genome replication. *J. Virol.* **77:**8181–8186.
2. **Al-Mutairy, B., J. E. Walter, A. Pothen, and D. K. Mitchell.** 2005. Genome prediction of putative genome-linked viral protein (VPg) of astroviruses. *Virus Genes* **31:**21–30.
3. **Appleton, H., and P. G. Higgins.** 1975. Letter: viruses and gastroenteritis in infants. *Lancet* **1:**1297.
4. **Asanaka, M., R. L. Atmar, V. Ruvolo, S. E. Crawford, F. H. Neill, and M. K. Estes.** 2005. Replication and packaging of Norwalk virus RNA in cultured mammalian cells. *Proc. Natl. Acad. Sci. USA* **102:**10327–10332.
5. **Atmar, R. L., and M. K. Estes.** 2006. The epidemiologic and clinical importance of norovirus infection. *Gastroenterol. Clin. North Am.* **35:**275–90.
6. **Baldwin, S. L., T. D. Powell, K. S. Sellins, S. V. Radecki, J. J. Cohen, and M. J. Milhausen.** 2004. The biological effects of five feline IFN-alpha subtypes. *Vet. Immunol. Immunopathol.* **99:**153–167.
7. **Blight, K. J., J. A. McKeating, and C. M. Rice.** 2002. Highly permissive cell lines for subgenomic and genomic hepatitis C virus RNA replication. *J. Virol.* **76:**13001–13014.
8. **Bonaparte, R. S., P. S. Hair, D. Banthia, D. M. Marshall, K. M. Cunnion, and N. K. Krishna.** 2008. Human astrovirus coat protein inhibits serum complement activation via C1, the first component of the classical pathway. *J. Virol.* **82:**817–827.

9. Chang, K. O., and D. W. George. 2007. Bile acids promote the expression of hepatitis C virus in replicon-harboring cells. *J. Virol.* **81:**9633–9640.

10. Chang, K. O., and D. W. George. 2007. Interferons and ribavirin effectively inhibit norwalk virus replication in replicon-bearing cells. *J. Virol.* **81:**12111–12118.

11. Chang, K. O., K. Y. Green, and L. J. Saif. 2002. Cell-culture propagation of porcine enteric calicivirus mediated by intestinal contents is dependent on the cyclic AMP signaling pathway. *Virology* **304:**302–310.

12. Chang, K. O., S. V. Sosnovtsev, G. Belliot, Y. Kim, L. J. Saif, and K. Y. Green. 2004. Bile acids are essential for porcine enteric calicivirus replication in association with down-regulation of signal transducer and activator of transcription 1. *Proc. Natl. Acad. Sci. USA* **101:**8733–8738.

13. Chang, K.O., S. V. Sosnovtsev, G. Belliot, A. D. King, and K. Y. Green. 2006. Stable expression of a Norwalk virus RNA replicon in a human hepatoma cell line. *Virology* **353:**463–473.

14. Chaudhry, Y., A. Nayak, M. E. Bordeleau, J. Tanaka, J. Pelletier, G. J. Belsham, L. O. Roberts, and I. G. Goodfellow. 2006. Caliciviruses differ in their functional requirements for eIF4F components. *J. Biol. Chem.* **281:**25315–25325.

15. Chaudhry, Y., M. A. Skinner, and I. G. Goodfellow. 2007. Recovery of genetically defined murine norovirus in tissue culture by using a fowlpox virus expressing T7 RNA polymerase. *J. Gen. Virol.* **88:**2091–2100.

16. Cheetham, S., M. Souza, T. Meulia, S. Grimes, M. G. Han, and L. J. Saif. 2006. Pathogenesis of a genogroup II human norovirus in gnotobiotic pigs. *J. Virol.* **80:**10372–10381.

17. Clark, B., and M. McKendrick. 2004. A review of viral gastroenteritis. *Curr. Opin. Infect. Dis.* **17:**461–469.

18. Daughenbaugh, K. F., C. S. Fraser, J. W. Hershey, and M. E. Hardy. 2003. The genome-linked protein VPg of the Norwalk virus binds eIF3, suggesting its role in translation initiation complex recruitment. *EMBO J.* **22:**2852–2859.

19. Donelli, G., F. Superti, A. Tinari, and M. L. Marziano. 1992. Mechanism of astrovirus entry into Graham 293 cells. *J. Med. Virol.* **38:**271–277.

20. Duizer, E., K. J. Schwab, F. H. Neill, R. L. Atmar, M. P. Koopmans, and M. K. Estes. 2004. Laboratory efforts to cultivate noroviruses. *J. Gen. Virol.* **85:**79–87.

21. Estes, M. K., B. V. Prasad, and R. L. Atmar. 2006. Noroviruses everywhere: has something changed? *Curr. Opin. Infect. Dis.* **19:**467–474.

22. Ettayebi, K., and M. E. Hardy. 2003. Norwalk virus nonstructural protein p48 forms a complex with the SNARE regulator VAP-A and prevents cell surface expression of vesicular stomatitis virus G protein. *J. Virol.* **77:**11790–11797.

23. Farkas, T., K. Sestak, C. Wei, and X. Jiang. 2008. Characterization of a rhesus monkey calicivirus representing a new genus of the *Caliciviridae. J. Virol.* **82:**5408–5416.

24. Fernandez-Vega, V., S. V. Sosnovtsev, G. Belliot, A. D. King, T. Mitra, A. Gorbalenya, and K. Y. Green. 2004. Norwalk virus N-terminal nonstructural protein is associated with disassembly of the Golgi complex in transfected cells. *J. Virol.* **78:**4827–4837.

25. Flynn, W.T., and L. J. Saif. 1988. Serial propagation of porcine enteric calicivirus-like virus in primary porcine kidney cell cultures. *J. Clin. Microbiol.* **26:**206–212.

26. Fulton, R. W., and L. J. Burge. 1985. Susceptibility of feline herpesvirus 1 and a feline calicivirus to feline interferon and recombinant human leukocyte interferons. *Antimicrob. Agents Chemother.* **28:**698–699.

27. García-Sastre, A., and C. A. Biron. 2006. Type 1 interferons and the virus-host relationship: a lesson in détente. *Science* **312:**879–882.

28. Geigenmuller, U., N. H. Ginzton, and S. M. Matsui. 1997. Construction of a genome-length cDNA clone for human astrovirus serotype 1 and synthesis of infectious RNA transcripts. *J. Virol.* **71:**1713–1717.

29. Gough, R. E., M. S. Collins, E. Borland, and L. F. Keymer. 1984. Astrovirus-like particles associated with hepatitis in ducklings. *Vet. Rec.* **114:**279.

30. Graham, D. Y., X. Jiang, T. Tanaka, A. R. Opekun, H. P. Madore, and M. K. Estes. 1994. Norwalk virus infection of volunteers: new insights based on improved assays. *J. Infect. Dis.* **170:**34–43.

31. Gray, E. W., K. W. Angus, and D. R. Snodgrass. 1980. Ultrastructure of the small intestine in astrovirus-infected lambs. *J. Gen. Virol.* **49:**71–82.

32. Green, K. Y. 2007. *Caliciviridae*, p. 949–979. *In* D. M. Knipe, P. M. Howley, D. E. Griffin, R. A. Lamb, M. A. Martin, B. Roizman, and S. E. Straus (ed.), *Fields Virology*, 5th ed. Lippincott, Williams & Wilkins, Philadelphia, PA.

33. Grimwood, K., and J. P. Buttery. 2007. Clinical update: rotavirus gastroenteritis and its prevention. *Lancet* **370:**302–304.

34. Guix, S., M. Asanaka, K. Katayama, S. E. Crawford, F. H. Neill, R. L. Atmar, and M. K. Estes. 2007. Norwalk virus RNA is infectious in mammalian cells. *J. Virol.* **81:**12238–12248.

35. Guix, S., M. Asanaka, K. Katayama, S. E. Crawford, F. H. Neill, R. L. Atmar, and M. K. Estes. 2007. Norwalk virus RNA is infectious in mammalian cells. Abstract. Third International Calicivirus Conference, Cancun, Mexico, 10–13 November 2007.

36. Guix, S., S. Caballero, A. Bosch, and R. M. Pintó. 2004. C-terminal nsP1a protein of human astrovirus colocalizes with the endoplasmic reticulum and viral RNA. *J. Virol.* **78:**13627–13636.

37. Guo, M., J. Hayes, K. O. Cho, A. V. Parwani, L. M. Lucas, and L. J. Saif. 2001. Comparative pathogenesis of tissue culture-adapted and wild-type Cowden porcine enteric calicivirus (PEC) in gnotobiotic pigs and induction of diarrhea by intravenous inoculation of wild-type PEC. *J. Virol.* **75:**9239–9251.

38. Hansman, G. S., T. Oka, K. Katayama, and N. Takeda. 2007. Human sapoviruses: genetic diversity, recombination, and classification. *Rev. Med. Virol.* **17:**133–141.

39. Hornung, V., J. Ellegast, S. Kim, K. Brzózka, A. Jung, H. Kato, H. Poeck, S. Akira, K. K. Conzelmann, M. Schlee, S. Endres, and G. Hartmann. 2006. 5′-Triphosphate RNA is the ligand for RIG-I. *Science* **314:**994–997.

40. Hsu, C. C., C. E. Wobus, E. K. Steffen, L. K. Riley, and R. S. Livingston. 2005. Development of a microsphere-based serologic multiplexed fluorescent immunoassay and a reverse transcriptase PCR assay to detect murine norovirus 1 infection in mice. *Clin. Diagn. Lab. Immunol.* **12:**1145–1151.

41. Hurley, K., P. Pesavento, N. C. Pedersen, A. M. Poland, E. Wilson, and J. Foley. 2004. An outbreak of virulent

systemic feline calicivirus disease. *J. Am. Vet. Med. Assoc.* **224:**241–249.

42. Hutson, A. M., F. Airaud, J. LePendu, M. K. Estes, and R. L. Atmar. 2005. Norwalk virus infection associates with secretor status genotyped from sera. *J. Med. Virol.* **77:**116–120.

43. Imada, T., S. Yamaguchi, M. Mase, K. Tsukamoto, M. Kubo, and A. Morooka. 2000. Avian nephritis virus (ANV) as a new member of the family Astroviridae and construction of infectious ANV cDNA. *J. Virol.* **74:**8487–8493.

44. Johnson, P. C., J. J. Mathewson, H. L. DuPont, and H. B. Greenberg. 1990. Multiple-challenge study of host susceptibility to Norwalk gastroenteritis in US adults. *J. Infect. Dis.* **161:**18–21.

45. Kapikian, A. Z., R. G. Wyatt, R. Dolin, T. S. Thornhill, A. R. Kalica, and R. M. Chanock. 1972. Visualization by immune electron microscopy of a 27-nm particle associated with acute infectious nonbacterial gastroenteritis. *J. Virol.* **10:**1075–1081.

46. Karst, S. M., C. E. Wobus, M. Lay, J. Davidson, and H. W. Virgin IV. 2003. Stat1-dependent innate immunity to a Norwalk-like virus. *Science* **299:**1575–1578.

47. Katayama, K., G. S. Hansman, T. Oka, S. Ogawa, and N. Takeda. 2006. Investigation of norovirus replication in a human cell line. *Arch. Virol.* **151:**1291–1308.

48. Koci, M. D. 2005. Immunity and resistance to astrovirus infection. *Viral Immunol.* **18:**11–16.

49. Koci, M. D., L. A. Kelley, D. Larsen, and S. Schultz-Cherry. 2004. Astrovirus-induced synthesis of nitric oxide contributes to virus control during infection. *J. Virol.* **78:**1564–1574.

50. Koci, M. D., L. A. Moser, L. A. Kelley, D. Larsen, C. C. Brown, and S. Schultz-Cherry. 2003. Astrovirus induces diarrhea in the absence of inflammation and cell death. *J. Virol.* **77:**11798–11808.

51. Koopmans, M. P., M. H. Bijen, S. S. Monroe, and J. Vinje. 1998. Age-stratified seroprevalence of neutralizing antibodies to astrovirus types 1 to 7 in humans in The Netherlands. *Clin. Diagn. Lab. Immunol.* **5:**33–37.

52. Kriston, S., M. M. Willcocks, M. J. Carter, and W. D. Cubitt. 1996. Seroprevalence of astrovirus types 1 and 6 in London, determined using recombinant virus antigen. *Epidemiol. Infect.* **117:**159–64.

53. Kurtz, J. B., T. W. Lee, J. W. Craig, and S. E. Reed. 1979. Astrovirus infection in volunteers. *J. Med. Virol.* **3:**221–230.

54. Lindesmith, L., C. Moe, J. LePendu, J. A. Frelinger, J. Treanor, and R. S. Baric. 2005. Cellular and humoral immunity following Snow Mountain virus challenge. *J. Virol.* **79:**2900–2909.

55. Lindesmith, L., C. Moe, S. Marionneau, N. Ruvoen, X. Jiang, L. Lindblad, P. Stewart, J. LePendu, and R. Baric. 2003. Human susceptibility and resistance to Norwalk virus infection. *Nat. Med.* **9:**548–553.

56. Marionneau, S., N. Ruvoen, B. Le Moullac-Vaidye, M. Clement, A. Cailleau-Thomas, G. Ruiz-Palacios, P. Huang, X. Jiang, and J. Le Pendu. 2002. Norwalk virus binds to histo-blood group antigens present on gastroduodenal epithelial cells of secretor individuals. *Gastroenterology* **122:**1967–1977.

57. Martella, V., M. Campolo, E. Lorusso, P. Cavicchio, M. Camero, A. L. Bellacicco, N. Decaro, G. Elia, G. Greco, M. Corrente, C. Desario, S. Arista, K. Banyai, M. Koopmans, and C. Buonavoglia. 2007. Norovirus in captive lion cub (*Panthera leo*). *Emerg. Infect. Dis.* **13:**1071–1073.

58. Méndez, E., G. Aguirre-Crespo, G. Zavala, and C. F. Arias. 2007. Association of the astrovirus structural protein VP90 with membranes plays a role in virus morphogenesis. *J. Virol.* **81:**10649–10658.

59. Méndez, E., and C. F. Arias. 2007. *Astroviridae*, p. 981–1000. *In* D. M. Knipe, P. M. Howley, D. E. Griffin, R. A. Lamb, M. A. Martin, B. Roizman, and S. E. Straus (ed.), *Fields Virology*, 5th ed. Lippincott, Williams & Wilkins, Philadelphia, PA.

60. Midthun, K., H. B. Greenberg, J. B. Kurtz, G. W. Gary, F. Y. Lin, and A. Z. Kapikian. 1993. Characterization and seroepidemiology of a type 5 astrovirus associated with an outbreak of gastroenteritis in Marin County, California. *J. Clin. Microbiol.* **31:**955–962.

61. Mochizuki, M., H. Nakatani, and M. Yoshida. 1994. Inhibitory effects of recombinant feline interferon on the replication of feline enteropathogenic viruses *in vitro*. *Vet. Microbiol.* **39:**145–152.

62. Molberg, O., E. M. Nilsen, L. M. Sollid, H. Scott, P. Brandtzaeg, E. Thorsby, and K. E. Lundin. 1998. CD4+ T cells with specific reactivity against astrovirus isolated from normal human small intestine. *Gastroenterology* **114:**115–122.

63. Moser, L. A., M. Carter, and S. Schultz-Cherry. 2007. Astrovirus increases epithelial barrier permeability independently of viral replication. *J. Virol.* **81:**11937–11945.

64. Moser, L. A., and S. Schultz-Cherry. 2005. Pathogenesis of astrovirus infection. *Viral Immunol.* **18:**4–10.

65. Mumphrey, S. M., H. Changotra, T. N. Moore, E. R. Heimann-Nichols, C. E. Wobus, M. J. Reilly, M. Moghadamfalahi, D. Shukla, and S. M. Karst. 2007. Murine norovirus 1 infection is associated with histopathological changes in immunocompetent hosts, but clinical disease is prevented by Stat1-dependent interferon responses. *J. Virol.* **81:**3251–3263.

66. Ohe, K., T. Takahashi, D. Hara, and M. Hara. 2008. Sensitivity of FCV to recombinant feline interferon (rFeIFN). *Vet. Res. Commun.* **32:**167–174.

67. Parrino, T. A., D. S. Schreiber, J. S. Trier, A. Z. Kapikian, and N. R. Blacklow. 1977. Clinical immunity in acute gastroenteritis caused by Norwalk agent. *N. Engl. J. Med.* **297:**86–89.

68. Parwani, A. W., W. T. Flynn, K. L. Gadfield, and L. J. Saif. 1991. Serial propagation of porcine enteric calicivirus in a continuous cell line. Effect of medium supplementation with intestinal contents or enzymes. *Arch. Virol.* **120:**115–122.

69. Pichlmair, A., O. Schulz, C. P. Tan, T. I. Naslund, P. Liljestrom, F. Weber, C. Reis e Sousa. 2006. RIG-I-mediated antiviral responses to single-stranded RNA bearing 5′-phosphates. *Science* **314:**997–1001.

70. Radford, A. D., K. P. Coyne, S. Dawson, C. J. Porter, and R. M. Gaskell. 2007. Feline calicivirus. *Vet. Res.* **38:**319–335.

71. Sebire, N. J., M. Malone, N. Shah, G. Anderson, H. B. Gaspar, and W. D. Cubitt. 2004. Pathology of astrovirus associated diarrhoea in a paediatric bone marrow transplant recipient. *J. Clin. Pathol.* **57:**1001–1003.

72. **Sharp, T. M., K. Katayama, S. Crawford, and M. K. Estes.** 2007. Norwalk virus nonstructural protein p22 mediates Golgi disruption by utilizing a novel ER exit signal to sequester COPII vesicles. Abstract. Third International Calicivirus Conference, Cancun, Mexico, 10–13 November 2007.

73. **Souza, M., S. M. Cheetham, M. S. Azevedo, V. Costantini, and L. J. Saif.** 2007. Cytokine and antibody responses in gnotobiotic pigs after infection with human norovirus genogroup II.4 (HS66 strain). *J. Virol.* **81:**9183–9192.

74. **Straub, T. M., K. Höner. zu Bentrup, P. Orosz-Coghlan, A. Dohnalkova, B. K. Mayer, R. A. Bartholomew, C. O. Valdez, C. J. Bruckner-Lea, C. P. Gerba, M. Abbaszadegan, and C. A. Nickerson.** 2007. *In vitro* cell culture infectivity assay for human noroviruses. *Emerg. Infect. Dis.* **13:**396–403.

75. **Stuart, A. D., and T. D. Brown.** 2006. Entry of feline calicivirus is dependent on clathrin-mediated endocytosis and acidification in endosomes. *J. Virol.* **80:**7500–7509.

76. **Sumpter, R., Jr., Y. M. Loo, E. Foy, K. Li, M. Yoneyama, T. Fujita, S. M. Lemon, and M. Gale, Jr.** 2005. Regulating intracellular antiviral defense and permissiveness to hepatitis C virus RNA replication through a cellular RNA helicase, RIG-I. *J. Virol.* **79:**2689–2699.

77. **Taterka, J. A., C. F. Cuff, and D. H. Rubin.** 1992. Viral gastrointestinal infections. *Gastroenterol. Clin. North Am.* **21:**303–330.

78. **Thackray, L. B., C. E. Wobus, K. A. Chachu, B. Liu, E. R. Alegre, K. S. Henderson, S. T. Kelley, and H. W. Virgin.** 2007. Murine noroviruses comprising a single genogroup exhibit biological diversity despite limited sequence divergence. *J. Virol.* **81:**10460–10473.

79. **Versteeg, G. A., P. J. Bredenbeek, S. H. van den Worm, and W. J. Spaan.** 2007. Group 2 coronaviruses prevent immediate early interferon induction by protection of viral RNA from host cell recognition. *Virology* **361:**18–26.

80. **Walter, J. E., and D. K. Mitchell.** 2003. Astrovirus infection in children. *Curr. Opin. Infect. Dis.* **16:**247–253.

81. **Ward, V. K., C. J. McCormick, I. N. Clarke, O. Salim, C. E. Wobus, L. B. Thackray, H. W. Virgin IV, and P. R. Lambden.** 2007. Recovery of infectious murine norovirus using pol II-driven expression of full-length cDNA. *Proc. Natl. Acad. Sci. USA* **104:**11050–11055.

82. **White, L. J., J. M. Ball, M. E. Hardy, T. N. Tanaka, N. Kitamoto, and M. K. Estes.** 1996. Attachment and entry of recombinant Norwalk virus capsids to cultured human and animal cell lines. *J. Virol.* **70:**6589–6597.

83. **Willcocks, M. M., M. J. Carter, and L. O. Roberts.** 2004. Cleavage of eukaryotic initiation factor eIF4G and inhibition of host-cell protein synthesis during feline calicivirus infection. *J. Gen. Virol.* **85:**1125–1130.

84. **Wobus, C. E., S. M. Karst, L. B. Thackray, K. O. Chang, S. V. Sosnovtsev, G. Belliot, A. Krug, J. M. Mackenzie, K. Y. Green, and H. W. Virgin IV.** 2004. Replication of *Norovirus* in cell culture reveals a tropism for dendritic cells and macrophages. *PLoS Biol.* **2(12):**e432.

85. **Wobus, C. E., L. B. Thackray, and H. W. Virgin IV.** 2006. Murine norovirus: a model system to study norovirus biology and pathogenesis. *J. Virol.* **80:**5104–5112.

86. **Woode, G. N., J. F. Pohlenz, N. E. Gourley, and J. A. Fagerland.** 1984. Astrovirus and Breda virus infections of dome cell epithelium of bovine ileum. *J. Clin. Microbiol.* **19:**623–630.

87. **Wyatt, R. G., R. Dolin, N. R. Blacklow, H. L. DuPont, R. F. Buscho, T. S. Thornhill, A. Z. Kapikian, and R. M. Chanock.** 1974. Comparison of three agents of acute infectious nonbacterial gastroenteritis by cross-challenge in volunteers. *J. Infect. Dis.* **129:**709–714.

88. **Zhou, H., and S. Perlman.** 2007. Mouse hepatitis virus does not induce beta interferon synthesis and does not inhibit its induction by double-stranded RNA. *J. Virol.* **81:**568–574.

Cellular Signaling and Innate Immune
Responses to RNA Virus Infections
Edited by A. R. Brasier et al.
© 2009 ASM Press, Washington, D.C.

Barbara Sherry
John T. Patton
Terence S. Dermody

Innate Immune Responses Elicited by Reovirus and Rotavirus

25

REOVIRIDAE VIRUSES

The *Reoviridae* family includes 10 genera of double-stranded (ds) RNA viruses that infect a wide variety of plants and animals (111). The mammalian orthoreoviruses (called simply reoviruses) are the type species of the *Orthoreovirus* genus, which also contains viruses that infect birds and reptiles. Three reovirus serotypes have been isolated from a broad range of mammalian hosts. Other genera of *Reoviridae* that infect animals are *Rotavirus*, *Orbivirus*, *Coltivirus*, and *Aquareovirus*. Rotaviruses are responsible for gastroenteritis in many animals, including humans, in which they are the primary cause of life-threatening infantile diarrhea. Orbiviruses are transmitted by arthropod vectors, and the prototype orbivirus, bluetongue virus, is an economically significant pathogen of sheep named for a symptom present in sick animals. Coltiviruses are also transmitted by arthropod vectors, and the prototype member, Colorado tick fever virus, can cause lethal neurologic disease in humans. Aquareoviruses infect fish and mollusks. The reoviruses and rotaviruses are the subject of this review.

General Features of Animal dsRNA Virus Replication

Viruses of the *Reoviridae* family are formed from one to three concentric protein shells that contain genomes consisting of 9 to 12 segments of dsRNA (5, 111). Attachment and internalization of these viruses lead to delivery of subvirion core particles into the cytoplasm. Transcription of the viral genome occurs within the cores, generating single-stranded RNAs that serve as templates for translation and replication. Gene-specific packaging (assortment) and replication as well as the formation of progeny cores occur within viral inclusion bodies (129). These non-membrane-bound structures are formed by the activity of viral nonstructural proteins, possibly acting in concert with cellular proteins. Progeny cores may migrate to secondary sites within the infected cell, where they undergo additional stages of morphogenesis to form mature virions. Mechanisms by which these viruses are released from cells are not well understood.

Barbara Sherry, Department of Molecular Biomedical Sciences, North Carolina State University, Raleigh, NC 27606. **John T. Patton**, Laboratory of Infectious Diseases, National Institute of Allergy and Infectious Diseases, National Institutes of Health, Bethesda, MD 20892. **Terence S. Dermody**, Departments of Pediatrics and Microbiology and Immunology and Elizabeth B. Lamb Center for Pediatric Research, Vanderbilt University School of Medicine, Nashville, TN 37232.

Pathogenesis of Reovirus Infections

Most mammalian species including humans serve as hosts for reovirus infection, but reovirus-induced disease is restricted to the very young (178). Reovirus infections of newborn mice have been used as a tractable experimental system for studies of viral pathogenesis. One of the best-characterized models of reovirus pathogenesis is infection of the murine central nervous system (CNS), in which serotype 1 (T1) and serotype 3 (T3) reoviruses display markedly different patterns of disease. Following oral inoculation of newborn mice, reovirus invades the CNS, yet the T1 and T3 strains use different routes of dissemination and manifest distinct pathologic consequences. T1 reovirus spreads to the CNS hematogenously and infects ependymal cells (171, 181), causing hydrocephalus (180). In contrast, T3 reovirus spreads to the CNS neurally and infects neurons (118, 171, 181), causing lethal encephalitis (167, 180). Studies using reassortant viruses generated from a cross of reovirus prototype strains type 1 Lang (T1L) and type 3 Dearing (T3D) show that the pathways of viral spread (171) and tropism for neural tissues (45, 181) segregate with the S1 gene, which encodes the viral attachment protein, $\sigma 1$ (99, 179). These findings suggest that the $\sigma 1$ protein determines the CNS cell types that serve as targets for reovirus infection, presumably through specific receptor binding.

Reovirus pathogenesis is not restricted to the CNS, and $\sigma 1$ is not the sole determinant of reovirus virulence. Reovirus infection causes pathology and physiologic dysfunction in a wide range of organs and tissues including the hepatobiliary system, myocardium, lungs, and endocrine tissues (178). Of these, reovirus myocarditis has become a particularly well-established experimental model of reovirus-induced disease. Myocarditis caused by reovirus infection is unusual in comparison with other viral etiologies of myocarditis in that the pathology is not mediated by T cells or components of the inflammatory response. Instead, reovirus cytopathicity is a direct cause of myocyte injury, which results from a complex interplay of the interferon (IFN) (6, 121, 158) and apoptotic pathways (40, 124). Efficiency of viral RNA synthesis correlates with the extent of myocardial damage (153). Accordingly, viral gene segments encoding proteins involved in viral transcription and genome replication play important roles in determining strain-specific differences in the capacity of reovirus to induce myocarditis (154, 155, 158).

Pathogenesis of Rotavirus Infections

Rotaviruses display a similarly broad host range, causing severe gastroenteritis in many animal species (134). In humans, rotaviruses are responsible for most cases of acute dehydrating diarrhea in infants and children under the age of 5 years (10). Globally, such infections result in ~500,000 deaths each year, the vast majority occurring in developing countries (182). Although any single rotavirus strain can usually infect a broad range of species, such strains rarely cause disease in species other than the natural host, suggesting species-specific evolution of virus strains. The attenuated behavior of rotavirus strains in heterologous hosts has formed the basis for using animal strains as live-virus vaccines in humans (79).

Rotaviruses infect mature enterocytes at the tips of the villi of the small intestine (10, 134). Infection leads to blunting and denuding of the villi and crypt hyperplasia, which likely account for the severe fluid loss in infected children (133). Animal studies have established that at least some virus strains can escape the gastrointestinal tract, yielding a systemic infection affecting various organs including the liver, spleen, and lung (36, 57). Anecdotal evidence and the detection of viral antigen, RNA, and virus in the blood of infected children indicate that rotavirus infections also can disseminate beyond the gut in humans (15). Some rotavirus strains appear to infect plasmacytoid dendritic cells (120), raising the possibility that these trafficking cells provide a means for the virus to escape the gastrointestinal tract to cause systemic infection.

The contribution of individual viral genes to rotavirus pathogenesis has been examined by analyzing disease phenotypes of reassortant rotaviruses using various animal model systems. In one extensive study using gnotobiotic piglets, analysis of reassortants generated from virulent porcine and avirulent human rotavirus strains demonstrated that four viral genes are linked to diarrheal disease (80). Two of these genes encode virion outer-shell proteins, notably attachment protein VP4 (gene 4) and glycoprotein VP7 (gene 9). These gene products are also determinants of rotavirus host range (spread) and disease in other studies (19, 117, 126). The third gene identified as a disease determinant in the gnotobiotic piglet study is that encoding NSP4 (gene 10). The NSP4 gene product is suggested to function as a viral endotoxin, capable of inducing diarrhea through Ca^{2+} mobilization in rotavirus-infected animals (55). The gnotobiotic piglet study also provided evidence that the gene 3 product, VP3, is a disease determinant (80), a surprising finding given that this protein is a minor inner-core component that functions as the viral RNA-capping enzyme (27).

Studies of newborn mice infected with simian rotavirus reassortants indicate that the gene 7 product, NSP3, influences virus growth in the intestine and extraintestinal spread (119). The function of NSP3 is poorly understood; it is dispensable for viral replication

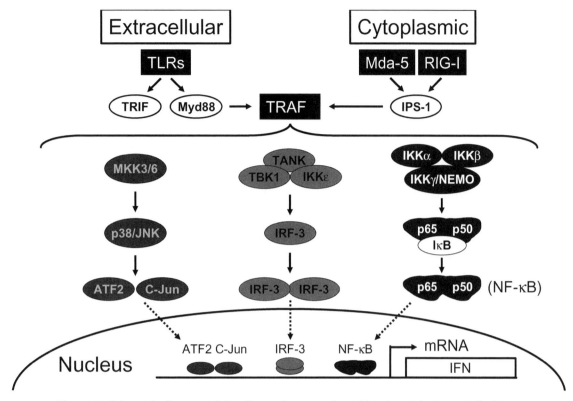

Figure 1 Schematic diagram of signaling pathways activated by virus infection. Viral PAMPs such as single-stranded or double-stranded RNA activate numerous signaling cascades through TLR-dependent (e.g., TLR3, TLR7, and TLR9) and TLR-independent (MDA-5 and RIG-I) pathways, leading to kinase activation through TRAF family members. TRIF and MyD88 link TLRs to the TRAF proteins, whereas IPS-1 links MDA-5 and RIG-I to TRAF3. TRAF-dependent induction of the kinases JNK, IKK-α, IKK-β, IKK-ϵ, TBK1, and IRAK-1 induces the binding of ATF-2/c-Jun, IRF3, and NF-κB (p50 and p65) to the IFN-β promoter. (Modified from reference 74.)

but may be involved in stimulating viral protein synthesis while suppressing host protein synthesis (116). Finally, a study of mouse pups infected with murine-human rotavirus reassortants suggests that the gene 5 product, NSP1, influences viral growth, spread, and virulence (20). The effect of NSP1 can be linked to its role as an antagonist of IFN expression. The collective results from these studies suggest that rotavirus pathogenesis reflects the combinatorial action of multiple viral gene products.

REOVIRUS INDUCTION OF INNATE IMMUNITY

Virus Activation of Signaling Pathways for Induction of Innate Immunity
Cellular mechanisms that detect viral infection and signal to activate innate immune response transcriptional programs have been the focus of much recent interest (74)

(Fig. 1). Toll-like receptors (TLRs) and other pattern recognition receptors (PRRs), including the nucleotide-binding oligomerization domain (NOD) proteins and RNA helicases such as melanoma differentiation-associated gene 5 (MDA-5) and retinoic acid-inducible gene I (RIG-I), recognize viral pathogen-associated molecular patterns (PAMPs). Recognition of viral PAMPs, such as CpG DNA or dsRNA, initiates signaling cascades by PRRs that activate cytoplasmic transcription factors, including activating transcription factor-2 (ATF-2)/c-Jun, IFN regulatory factor 3 (IRF3), and nuclear factor-κB (NF-κB) (115). Nuclear translocation of these transcription factors induces the expression of type I IFNs (IFN-α and -β) and other proinflammatory cytokines. In turn, IFNs induce transcription of IFN-stimulated genes (ISGs). The ISG IRF7, when activated by viral infection, further induces IFN transcription for a positive-feedback loop of IFN induction (108, 142). Other ISGs mediate antiviral, immunomodulatory, and cell-cycle-inhibitory effects. ISGs also regulate both the extrinsic and intrinsic pathways of apoptosis (26, 141). These pathways can limit viral dissemination by priming uninfected cells to produce antiviral proteins, such as protein kinase R (PKR) and 2′,5′-oligoadenylate synthetase (OAS), and induce apoptosis of infected cells to prevent further viral replication (26).

Reovirus Activation of NF-κB

A critical component of the intracellular signal transduction apparatus activated following reovirus infection is NF-κB, a family of structurally related transcription factors that play important roles in cell growth and survival (62). Reovirus activates NF-κB in cell culture beginning at 2 to 4 h postinfection and achieves maximal levels of activation at 8 to 10 h postinfection (34, 70). Electrophoretic mobility shift assays using antisera specific for p50 and p65 identified both of these subunits in the NF-κB complexes activated during reovirus infection. Concordantly, cells devoid of either p50 or p65 do not activate NF-κB following reovirus infection (34), supporting the involvement of these NF-κB subunits in the complexes activated by reovirus. A second phase of NF-κB regulation occurs in some cell types following viral RNA synthesis and involves downregulation of NF-κB signaling through a mechanism dependent on the inhibition of IκB-α degradation (29).

Mechanisms leading to NF-κB activation following reovirus infection are not completely understood. Viral disassembly is required for NF-κB activation, but subsequent events in viral replication are dispensable (33). This finding suggests that replication steps following formation of the disassembly intermediates but before commencement of RNA synthesis are responsible for activating NF-κB. Intriguingly, NF-κB activation following reovirus infection occurs over a much longer time course than that elicited by other NF-κB agonists such as tumor necrosis factor-α (TNF-α) (14), suggesting that the viral agonist is constitutively active, similar to Epstein-Barr virus latent membrane protein 1 (22) or human T-cell leukemia virus Tax (109).

The signal-induced IκB kinases (IKKs) IKK-α and IKK-β phosphorylate IκB proteins at regulatory serines, targeting these inhibitors for degradation (107, 177) (Fig. 1). IKK-α and IKK-β are components of a large (~700 to 900 kDa), multisubunit complex that contains a regulatory subunit called IKK-γ/NEMO (46, 110, 136, 186, 191). The IKK catalytic subunits are activated following stimulation with several NF-κB-inducing agonists, including TNF-α and interleukin-1 (46, 110, 191). Experiments using IKK subunit-specific small interfering RNAs (siRNAs) and cells deficient in individual IKK

subunits were used to demonstrate a key role for IKK-α but not IKK-β in reovirus-induced NF-κB activation (70). Despite the preferential usage of IKK-α, NF-κB activation is attenuated in cells lacking IKK-γ/NEMO (70), an essential regulatory subunit of IKK-β (137, 147, 186). Moreover, deletion of the gene encoding NF-κB-inducing kinase (NIK), which is known to modulate IKK-α function (101, 150, 184), has no inhibitory effect on NF-κB activation in reovirus-infected cells (70). These findings indicate a novel pathway of NF-κB activation during reovirus infection involving IKK-α and IKK-γ/NEMO.

NF-κB activation can either potentiate (1, 68, 91) or inhibit apoptosis (13, 102, 173), depending on the nature of the NF-κB agonist. Transient expression of a dominant negative form of IκB that constitutively blocks NF-κB nuclear translocation significantly reduces levels of apoptosis induced by reovirus (34), suggesting a key signal-transducing function for NF-κB in the reovirus-induced death response. In keeping with this hypothesis, mouse embryonic fibroblasts (MEFs) lacking the IKK (70) or NF-κB (34) subunits required for NF-κB signaling by reovirus undergo apoptosis substantially less efficiently than wild-type MEFs following reovirus infection. Together, these data indicate that NF-κB plays an essential role in the mechanism by which reovirus induces apoptosis of host cells.

Reovirus Activation of IRF3

Reovirus infection of primary cardiac myocytes (121) and epithelial and fibroblast cell lines (75) leads to activation of IRF3. Reovirus activates IRF3 in cell culture beginning at 1 h postinfection and achieves maximal levels of activation at 4 to 6 h postinfection (75). Like the activation of NF-κB by reovirus, activation of IRF3 requires viral disassembly but not viral RNA synthesis (75). However, in contrast to NF-κB, activation of IRF3 requires reovirus nucleic acid. Whereas reovirus particles devoid of genomic dsRNA are capable of activating NF-κB, empty particles fail to activate IRF3 (75). Nonetheless, these findings place the IRF3-activating events in reovirus replication within the same temporal window as those required for activation of NF-κB.

TLR3 and cytoplasmic RNA helicases MDA-5 and RIG-I function independently to engage dsRNA and activate IRF3 (18, 59, 104, 188) (Fig. 1). MDA-5 (92) and RIG-I (164, 188) are expressed at basal levels in most tissues and are induced by type I IFNs. Binding of dsRNA by the helicase domain of these proteins exposes an N-terminal caspase-associated recruitment domain (CARD) (164, 188), which binds to the CARD of the downstream effector IPS-1 (MAVS/VISA/Cardif) (93, 113, 151, 185).

IFN-β promoter stimulator-1 (IPS-1) recruits TNF receptor-associated factor-3 (TRAF3) and IKK-ε to activate NF-κB and IRF3, respectively. Reovirus dsRNA can activate TLR3 (3). However, reovirus pathogenesis is not altered in TLR3-deficient mice (51, 90), suggesting that TLR3 is not essential for the host response to reovirus.

Dominant negative inhibitor RIG-IC, which lacks the CARD domain of RIG-I required for signaling to IPS-1 (188), inhibits IRF3 activation by both reovirus infection and poly(I:C) transfection in 293T cells (75). In contrast, RIG-IKA, which contains a mutation in the Walker's ATP-binding motif in the RIG-I helicase domain (188), inhibits IRF3 activation by reovirus but not by poly(I:C) (75), suggesting involvement of RIG-I in the IRF3 response evoked by reovirus. Concordantly, transfection of siRNA specific for RIG-I strongly inhibits IRF3 activation by reovirus infection (Fig. 2A). However, siRNA specific for MDA-5 has no effect on reovirus-induced IRF3 activity (75). Transfection of siRNA specific for IPS-1 inhibits IRF3 activation following reovirus infection by approximately the same magnitude as RIG-I siRNA (Fig. 2A). Thus, reovirus activates IRF3 in 293T cells via a pathway dependent on RIG-I and IPS-1. However, experiments using MEFs lacking RIG-I or MDA-5 suggest that reovirus can use either pathway to induce ISGs (104), suggesting either that the choice of RNA helicase for reovirus induction of IFN is cell type specific or that reovirus can use MDA-5 for direct induction of a subset of ISGs.

RIG-I and IPS-1 activate both IRF3 and NF-κB in response to cytoplasmic viral RNA (93, 188). Transfection of full-length RIG-I modestly increases NF-κB activation following reovirus infection (75). However, in contrast to IRF3 activation, neither RIG-IC nor RIG-IKA inhibits reovirus-induced NF-κB activation (75). In agreement with these findings, siRNA specific for either RIG-I or IPS-1 has no effect on NF-κB activation by reovirus (Fig. 2B), but siRNA specific for IPS-1 inhibits NF-κB activation by poly(I:C) (75). These results indicate that reovirus activates NF-κB via a mechanism that is independent of RIG-I and IPS-1.

At a multiplicity of infection (MOI) of >1 PFU/cell, reovirus produces equivalent yields in the presence and absence of IRF3 (75), indicating that this transcription factor does not influence reovirus replication following a single-cycle synchronous infection. However, at an MOI of <1 PFU/cell under conditions that allow the virus to spread from cell to cell, yields are substantially less in IRF3$^{+/+}$ MEFs than in IRF3$^{-/-}$ MEFs (75), indicating a role for IRF3 in induction of an antiviral state in surrounding cells. At both high and low MOI, reovirus induces significantly greater levels of apoptosis in wild-type MEFs than in MEFs lacking IRF3 (75). These

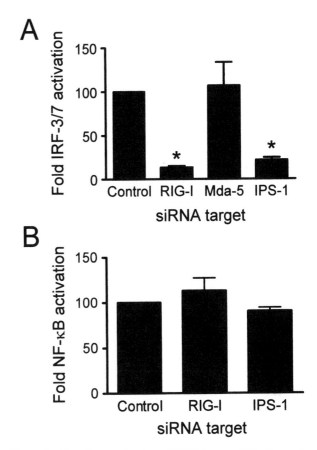

Figure 2 Reovirus activation of IRF3 but not NF-κB requires RIG-I and IPS-1. (A) 293T cells were cotransfected with a control siRNA or an siRNA specific for RIG-I, MDA-5, or IPS-1, and with IRF3 reporter plasmid p-55C1BLuc and control plasmid p-Renilla-Luc. Following 24 h incubation, cells were infected with T3D at an MOI of 100 PFU/cell. Luciferase activity in cell culture lysates was determined at 24 h postinfection. (B) 293T cells were cotransfected with a control siRNA or an siRNA specific for either RIG-I or IPS-1, and with NF-κB reporter plasmid pNF-κB-Luc and p-Renilla-Luc. Following 24 h incubation, cells were infected with T3D at an MOI of 100 PFU/cell. Luciferase activity in cell culture lysates was determined at 24 h postinfection. Results are presented as the ratio of normalized luciferase activity from infected cell lysates to that from mock-treated or mock-infected lysates for triplicate samples. For all panels, error bars indicate standard deviation. *, $P > 0.05$ as determined by Student's *t* test in comparison to mock treatment. (Modified from reference 75.)

findings indicate that IRF3 enhances the apoptotic response to reovirus and supports a requirement for IRF3-dependent gene expression in reovirus-induced apoptosis of cultured cells.

Reovirus-Induced Apoptosis and Innate Immunity

Transcription factors NF-κB and IRF3 are essential for both reovirus-induced apoptosis and innate immunity, suggesting an important functional overlap between these two antiviral processes. However, these cellular responses are also distinguishable. The cell-death and IFN pathways activated by reovirus are associated with different viral gene products. Additionally, the mitogen-activated protein kinase (MAPK) pathway has been implicated only in reovirus-induced apoptosis.

Strain-specific differences in the capacity of reovirus to induce apoptosis segregate with the reovirus S1 and M2 genes, encoding the σ1 and μ1 proteins (32, 135, 172). These viral gene products play important roles in attachment (99, 179) and membrane penetration (24, 25, 105, 123, 169), respectively. Experiments using pharmacologic inhibitors have identified steps in viral replication following disassembly but prior to RNA synthesis as the proximal mediators of death signaling during reovirus infection (33). Importantly, the viral receptors can be bypassed in the proapoptotic signaling process, provided that virions are delivered into the endocytic pathway for subsequent acid-dependent disassembly (38). These findings suggest that viral membrane-penetration protein μ1 serves as a proapoptotic effector, perhaps via its capacity to mediate membrane disruption and mitochondrial injury or by activating cellular sensors of viral infection (39). These data suggest that signaling molecules stimulated by μ1-mediated disassembly steps converge on NF-κB and IRF3, inducing apoptosis.

Genetic analyses using reassortant reoviruses identified the M1, L2, and S2 genes as determinants of strain-specific differences in induction of IFN-β in cardiac myocytes (158). These genes encode the viral core proteins μ2, λ2, and σ2, respectively, but the mechanism by which they modulate induction of IFN is unclear. These proteins are involved in virion core stability, viral inclusion body formation, and viral RNA synthesis (144), each of which could influence the induction of IFN. Thus, while genetic and biochemical analyses implicate the M2-encoded μ1 protein in reovirus-induced apoptosis, strain-specific differences in reovirus-induced IFN gene expression are associated with a different set of genes involved in many stages of the viral replication cycle. Possible explanations for this divergence include differences in required local concentrations or timing of NF-κB and IRF3 activation in apoptosis and IFN gene expression and possible cell-type-specific differences, all of which are currently under investigation.

Activation of MAPKs in Reovirus-Induced Apoptosis

MAPKs are important signal transducers that respond to a wide variety of stimuli (170). Several MAPKs transduce signals initiated by reovirus infection and are associated with reovirus-induced apoptosis, but their role in reovirus-induced IFN remains unexplored. Reovirus activates c-Jun NH$_2$-terminal kinase (JNK) by 10 to 12 h postinfection, and this activation is sustained for at least 20 to 30 h (31). Strain-specific differences in the capacity of reovirus to activate JNK and its downstream effector c-Jun correlate with the capacity of these strains to induce apoptosis, suggesting that JNK activation is required for apoptotic signaling. Additionally, cells lacking MAP kinase or ERK kinase (MEK) kinase 1 (MEKK1), an upstream activator of JNK, or engineered to express a kinase-inactive form of MEKK1 do not phosphorylate JNK or undergo apoptosis in response to reovirus infection (30). Pharmacologic inhibitors of JNK inhibit reovirus-induced apoptosis (30) but do not block reovirus growth (30, 122), indicating that JNK activity is not required for reovirus replication. Interestingly, although JNK phosphorylates and activates c-Jun in response to reovirus infection, c-Jun activation is not required for apoptosis. These data indicate that JNK contributes to apoptosis induction via a mechanism independent of its activation of c-Jun (30), possibly through its effect on mitochondrial signaling pathways.

The p38 MAPK is activated by reovirus between 4 and 8 h postinfection (112). However, p38 becomes downregulated at late times (24 to 48 h) postinfection (31). Pharmacologic inhibitors of p38 MAPK reduce reovirus growth in cells that express an activated Ras pathway, indicating that this pathway is important for replication of the virus (122). Inhibition of p38 also blocks reovirus-induced secretion of the proinflammatory cytokines interleukin-1β and TNF-α (112). However, pharmacologic inhibitors of p38 have no effect on reovirus-induced apoptosis (31), indicating that this pathway is distinct from NF-κB-mediated death signaling during reovirus infection (34). Finally, although reovirus infection activates the MAPK extracellular signal-regulated kinase (ERK) at early (10 to 30 min) and late (24 h) times postinfection, pharmacologic inhibitors of ERK do not inhibit reovirus-induced apoptosis (31). The role of ERK activation in reovirus replication is unknown.

IFN EFFECTOR RESPONSES ELICITED BY REOVIRUS

Jak-Stat Signaling

Binding of type I IFNs to the IFN-α/β receptor activates Janus kinase (Jak) and tyrosine kinase 2 (Tyk2) kinases for phosphorylation of signal transducer and activator of transcription 1 (Stat1) and Stat2 (145). Once activated, these transcription factors engage IRF9, forming a heterotrimeric complex. The complex translocates to

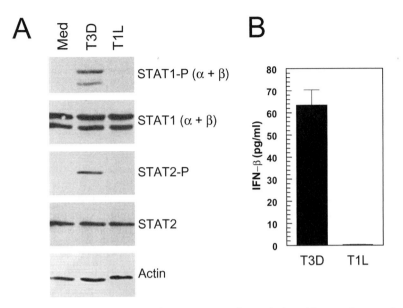

Figure 3 The IFN-β response in cardiac myocyte cultures induced by reovirus strains T1L and T3D. Murine primary cardiac myocyte cultures were prepared from fetal mouse hearts. At 2 days post-plating, the cultures were infected at an MOI of 10 PFU/cell and harvested at 4 h postinfection. (A) Immunoblot of total cell lysates using the indicated antibodies (Stat1-P denotes phosphorylation on Tyr-701; Stat2-P denotes phosphorylation on Tyr-689). (B) Enzyme-linked immunosorbent assay quantitation of IFN-β in culture supernatants. (Figure courtesy of Jennifer Zurney, North Carolina State University.)

the nucleus and binds to IFN-stimulated response elements, resulting in transcriptional activation of a large family of ISGs, including many with antiviral effects. Reovirus induces the synthesis and secretion of type I IFNs for subsequent activation of the Jak-Stat pathway and induction of ISGs (64, 192). Reovirus activation of Jak-Stat signaling has been explored in studies using several tissue types and primary cell cultures.

Reovirus activates the Jak-Stat pathway in the CNS of infected mice and in cell culture models (64). Strains type 3 Abney (T3A) and T3D, and to a lesser extent T1L, stimulate phosphorylation of Stat1 Tyr-701 in murine primary neuronal cultures. This activation results in Stat1 nuclear translocation and increases expression of total Stat1. T3A also stimulates phosphorylation of Stat1 Tyr-701 in the murine brain following intracranial inoculation (64). Interestingly, most Stat1 phosphorylation in the brain occurs in uninfected cells surrounding infected cells, suggesting that reovirus infection limits autocrine IFN signaling in neurons. Inhibition of signaling by IFN-α/β, but not IFN-γ, ablates T3A-induced Stat1 phosphorylation in primary neuronal cultures (64), indicating that signaling is exclusively via type I IFNs. Both T1L and T3A cause increased lethality following intracranial inoculation of newborn mice lacking Stat1 (64), demon-

strating that Stat1 signaling provides a critical protective role in reovirus-infected newborn mice.

Type I IFNs are an essential determinant of protection against reovirus-induced myocarditis in newborn mice and reovirus replication and spread in murine primary cardiac cell cultures (158). T3D, which fails to induce murine myocarditis despite gaining access to the heart, stimulates phosphorylation and nuclear translocation of Stat1 and Stat2 in murine primary cardiac myocyte and cardiac fibroblast cultures (192). Reovirus-induced phosphorylation of Stat1 and Stat2 is virus strain specific in murine primary cardiac myocyte cultures and, as expected, correlates with the level of IFN-β induction (Fig. 3). Reovirus-induced Stat phosphorylation is also cell type specific (192). Reovirus-induced Jak-Stat signaling upregulates IRF7 gene expression (163, 192), which likely contributes to further IFN gene expression by positive feedback, as reovirus induces much lower levels of IFN in cells generated from mice lacking the IFN-α/β receptor, IFN-αR1 (B. Sherry, unpublished data).

Key ISGs

Reovirus strains vary in sensitivity to the antiviral effects of IFN (88, 158). However, the roles of only a handful of ISGs in the reovirus-stimulated antiviral response have been directly addressed (139). Notably, the ISG PKR can be activated by reovirus, the IFN-stimulated antiviral state in L929 cells parallels PKR accumulation (139), and reovirus induces more cardiac damage in PKR null mice than in wild-type controls (162). Together, these data implicate PKR as an important component of the IFN-stimulated antiviral state evoked by reovirus. However, reovirus replication in cardiac tissue of PKR null mice is not significantly increased relative to that in wild-type mice (162). And, remarkably, phosphorylation of eukaryotic initiation factor 2α (eIF2α) can actually benefit reovirus replication, in part by promoting synthesis of the transcription factor ATF-4 as a stress response (160). Thus, the role of the ISG PKR in reovirus replication is unclear. The ISGs OAS and RNase L do not appear to be essential for the IFN-induced antireovirus state (139). Like PKR, these ISGs may enhance reovirus replication (161).

The ISGs required to limit reovirus replication remain elusive. Reovirus induces a large number of ISGs, presumably through both the classic IFN-mediated pathway and the activation of IRFs that directly induce ISG expression, such as IRF3. Microarray analyses have been used to assess global gene expression patterns in response to infection by reovirus strains T3A, T3D, and T3 clone 8 in L929 cells (160); T3D in HeLa cells with and without an intact NF-κB signaling axis (125); and T1L and T3A in human embryonic kidney (HEK) 293 cells (41, 132). Not surprisingly, while many ISGs are induced, those observed are dependent on virus strain, cell type, and time postinfection. Notably, Smith et al. (160) found that reovirus infection results in increased expression of p58^{IPK}, the cellular inhibitor of PKR, and a second eIF2α kinase, PERK (PKR-like endoplasmic reticulum kinase). Upregulation of these genes was found to be beneficial, rather than inhibitory, to reovirus replication (160). Further application of these global approaches for identification of reovirus-induced ISGs should provide additional ISG candidates responsible for IFN's effects on viral replication and possibly apoptosis induction.

TISSUE-SPECIFIC EFFECTS OF IFN INDUCTION AND SENSITIVITY

Type I IFN Regulation of Reovirus Pathogenesis
Prototype reovirus strains T1L and T3D are capable of infecting the murine heart. T1L replicates to relatively higher titers and causes tissue damage, while T3D replication is severely limited and damage is undetectable (157). A potently myocarditic reassortant virus, 8B, isolated from a mouse coinfected with T1L and T3D (157) has been used to further characterize reovirus-induced

murine myocarditis as a model for this important human disease (16, 54). Following peroral, intraperitoneal, or intramuscular inoculation, reovirus 8B induces extensive cardiac tissue damage with minimal inflammatory infiltrate (157; Sherry, unpublished). This pathology is directly attributable to viral infection of cardiac cells and not immune-mediated damage, as evidenced by the capacity of reovirus to induce myocarditis in nude mice (157) and SCID mice (156). Cardiac cell death is due to apoptosis and can be abrogated by inhibitors of calpain or caspase proteases (40, 42).

Reovirus strains differ in the extent of cardiac damage, and genetic approaches have revealed important roles for the M1, L1, and L2 genes in this process (154). The degree to which reovirus strains induce myocarditis in mice correlates with induction of cytopathic effect in murine primary cardiac myocyte cultures but not primary cardiac fibroblast cultures (11). Interestingly, this effect is most apparent using conditions that require viral spread through the culture, suggesting the involvement of cytokine-mediated defenses. Indeed, nonmyocarditic reovirus strains induce substantially more IFN-β and are more sensitive to its antiviral effects than myocarditic reovirus strains in primary cardiac myocyte cultures (158). Moreover, a nonmyocarditic reovirus strain induces cardiac lesions in mice depleted of IFN-α/β. The reovirus M1, S2, and L2 genes are associated with these IFN-related phenotypes (158).

Type I IFN regulation of reovirus-induced disease is tissue specific (124). Reovirus strain T3SA+ causes encephalitis but not myocarditis in wild-type mice. However, in mice lacking the NF-κB p50 subunit, encephalitis caused by T3SA+ is attenuated, while profound myocarditis is newly evident. The decreased encephalitis most likely reflects a requirement for NF-κB in reovirus-induced apoptosis (34, 70, 124). In contrast, the appearance of myocarditis suggests a protective role for NF-κB in the heart. In support of this contention, treatment of p50 null mice with IFN-β abrogates myocarditis induced by T3SA+ (124), consistent with a function for NF-κB in regulating IFN-β gene transcription (28, 58, 190). Thus, while NF-κB is required for apoptosis in the CNS, it is dispensable for apoptosis in the heart. In addition, either IFN-β does not influence spread of virus in the CNS or NF-κB is not required for reovirus induction of IFN-β at that site.

Type I IFNs also can influence reovirus pathogenesis in the enteric tract (90). T1L induces a self-limited infection following peroral inoculation of newborn mice and replicates minimally in the enteric tract when inoculated into adult mice. In contrast, following peroral inoculation of adult mice lacking IFN-αR1, T1L replicates to high titers in the enteric tract, resulting in a fatal outcome by 8 days postinoculation. Interestingly, while Peyer's patches and intestinal mucosa exhibit marked damage, nonlymphoid organs appear relatively normal. IFN-αR1 null mice reconstituted with wild-type bone marrow cells survive T1L infection, whereas wild-type mice reconstituted with IFN-αR1 bone marrow cells are more susceptible (90). These results indicate that hematopoietic cells must be capable of responding to type I IFNs to control T1L infection. Moreover, type I IFN production and responsiveness of epithelial cells (or other stromal cells) are not sufficient to limit local and systemic spread of reovirus T1L. Accordingly, conventional dendritic cells in Peyer's patches are a critical source of protective type I IFNs (90). Thus, Peyer's patch-derived type I IFNs are an important determinant of reovirus-induced pathology in the enteric tract.

IFN's Effects in Cardiac Myocytes and Fibroblasts

While recent data suggest that stem cells can repopulate the heart after ischemic or other types of damage, it is not apparent that stem cell differentiation is capable of restoring cardiac function after pathological insults (17). Viral myocarditis, in particular, is devastating because cardiac myocytes are nondividing cells critical for cardiac function, and their loss profoundly impairs cardiac function (56). Given the variety of viruses associated with myocarditis and their ubiquity in the human population (16), the heart would benefit from the equivalent of a blood-brain barrier. Type I IFNs provide at least part of that protection.

Cardiac myocytes are nondividing, highly differentiated cells. However, cardiac fibroblasts are readily replenishable, providing the scar tissue following cardiac myocyte loss. These very different cardiac functions are protected by distinct variations in basal and virus-induced expression of components of the IFN pathway (163, 192). Cardiac myocytes express higher basal levels of IFN-β, nuclear phosphorylated Stat1 and Stat2, and ISGs than do cardiac fibroblasts, consistent with a requirement for nondividing cardiac myocytes to respond rapidly to viral infection. Continuous autocrine IFN signaling maintains high basal expression of these components in cardiac myocytes, as Stat1 and Stat2 activation and ISG expression are dramatically reduced in mice lacking IFN-αR1. In contrast, cardiac fibroblasts express higher basal levels of IFN-αR1 and higher basal cytoplasmic levels of Jak1, Tyk2, Stat1, and Stat2 than do cardiac myocytes. High basal levels of these IFN signaling intermediaries position cardiac fibroblasts to respond efficiently to IFN, thus preventing these cells from serving as reservoirs for viral replication. In fact, cardiac fibroblasts are more responsive than cardiac

myocytes to the addition of IFN-β. Importantly, while high basal IFN-β expression in cardiac myocytes is sufficient for autocrine stimulation, it is insufficient for paracrine stimulation of adjacent fibroblasts, which respond only when virus upregulates IFN expression in infected cardiac myocytes. Together, these two cell types form an integrated network of cell-specific innate immune components for organ protection.

ANTAGONISM OF IFN ACTIVATION BY ROTAVIRUS

Replication of rotavirus, like reovirus, is restricted by pretreatment of permissive cells with either type I or type II IFNs (2, 9, 37). Similarly, preadministration of IFN to newborn calves and piglets limits rotavirus replication and reduces the duration of diarrhea (98, 148). In early studies, Vanden Broecke et al. (174) and Schwers et al. (149) observed that IFN production is delayed in newborn calves infected with low doses of rotavirus, even though this period is characterized by severe, transient diarrhea. In contrast, calves infected with high doses of virus produce IFN without delay and remain free of severe diarrhea. These studies were the first to suggest that rotaviruses have evolved mechanisms to suppress the IFN antiviral response. Indeed, a recent analysis of gene expression in infected calf-intestinal loops by quantitative real-time PCR confirms that rotaviruses suppress the activation of type I and type II IFNs and ISGs (2).

Features of NSP1

Rotavirus encodes several proteins with the potential to suppress IFN expression. These include proteins with dsRNA-binding (e.g., VP2 and NSP5) (97, 176) and RTPase (NSP2) (175) activities that could sequester and modify viral RNA targets sensed by RIG-I, MDA-5, TLR3, or PKR. However, recent reports point to NSP1 as the sole rotavirus gene product responsible for countering IFN-dependent innate immune responses.

NSP1, the 55-kDa product of the gene 5 RNA, is a cytoplasmic RNA-binding protein that is expressed in relatively low quantities in infected cells (21, 83). Of all rotavirus proteins, NSP1 shows the greatest sequence variation, with identities less than 40% between some virus strains (49, 82, 165). However, for strains isolated from the same host species, the NSP1 proteins are usually highly conserved (53). These observations, combined with phylogenetic analyses, suggest that NSP1 has coevolved with the natural host of the virus, leading to NSP1 proteins that are species specific in function (49, 95). Such cospeciation might explain the nonrandom assortment of gene 5 RNAs into progeny viruses produced

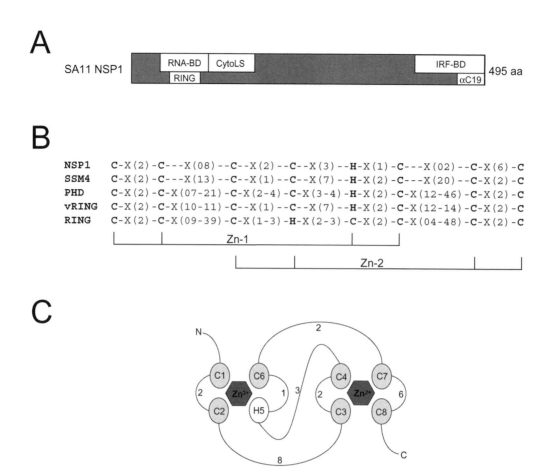

Figure 4   RING domain of rotavirus NSP1. (A) Schematic representation of structural and functional domains in the NSP1 protein of SA11 rotavirus. RING, RING finger; CytoLS, cytoplasm localization domain; RNA-BD, RNA-binding domain; αC19, recognition epitope of the C19 antibody used in detecting wild-type NSP1. (B) Organization of the consensus cysteine (C)- and histidine (H)-rich region of rotavirus NSP1 in comparison with the related domain of the SSM4 protein (4) and the consensus plant homeodomain (PHD), viral C4HC3-type RING domain, and classic RING domain (47). Residues coordinating binding of the two zinc atoms (Zn-1, Zn-2) of the PHD, vRING, and RING domains are indicated. X represents any amino acid. (C) Hypothetical organization of the RING domain of NSP1, including possible cross-bracing between the two zinc fingers. Note that the distances proposed between C6 and C7 and between C7 and C8 for NSP1 markedly differ from those of the PHD, vRING, and RING domains, suggesting that alternative organizations are possible.

in animals or cell lines coinfected with distinct rotavirus strains (20, 67).

Despite their notable sequence diversity, all NSP1 proteins possess an N-terminal domain comprising an identical array of eight cysteine and histidine residues (Fig. 4). These residues are proposed to form two cross-bridged zinc-binding fingers reminiscent of the RING (Really Interesting New Gene) superfamily of proteins (23, 35, 71, 96). Instead of the more classic C3HC4 RING signature, NSP1 proteins have a predicted C4HC3 arrangement typical of viral SSM4-related (viral RING) proteins

(4, 47). The reported zinc-binding activity of NSP1 is consistent with the presence of a RING domain (21). The RING domain is situated within a region of NSP1 associated with RNA-binding activity, and thus, this sequence may mediate binding to RNA (81) (Fig. 4). Immediately downstream of the RING domain is the cytoplasmic localization signal of NSP1, without which the protein accumulates in the nucleus (81). All NSP1 proteins share a near identical distribution of proline and cysteine residues throughout their sequence, suggesting that the overall structure of the protein is conserved (82).

NSP1 as a Factor in Virus Spread

Serial passage of rotaviruses at high MOI often leads to selection of variant viruses with one or more genome segments of atypical size (44, 85). The atypical size results from intragenic sequence rearrangements, most commonly a sequence duplication and less frequently a sequence deletion (166, 168). Sequence rearrangements within the gene 5 RNA almost always interrupt the ~500-amino-acid open reading frame for NSP1. As a result, rotaviruses with gene 5 rearrangements (RVg5re) usually produce C-terminal-truncated forms of the protein (NSP1ΔC) (94, 130). The most extreme gene 5 rearrangement described to date yields an NSP1ΔC product of only 40 amino acids (166). The capacity of RVg5re variants to grow in cell culture and their isolation from animals and children, albeit infrequently, indicates that NSP1 is a nonessential viral protein (52, 127). The nonessential nature of NSP1 has been confirmed in vitro using siRNAs that target the gene 5 mRNA (159).

Although NSP1 is not required for replication, rotaviruses producing NSP1ΔC have altered growth characteristics. In particular, RVg5re variants have small-plaque phenotypes, and in several cell lines the variants produce titers 10-fold less than those produced by wild-type rotavirus (8, 83, 130). Comparative analyses of the growth characteristics of mixtures of wild-type rotavirus and RVg5re strains reveal that under conditions in which the virus is required to spread from cell to cell to complete infection (e.g., low MOI infection), viruses expressing wild-type NSP1 are selected (85). In contrast, when the virus is not required to spread from cell to cell (e.g., high MOI infection), selective pressures favoring viruses expressing wild-type NSP1 are lost. The importance of NSP1 for viral spread is also apparent from studies of murine-rhesus reassortant rotaviruses indicating linkage between the origins of the gene 5 RNA and the capacity of the virus to grow and spread (20). The behavior of NSP1 is characteristic of viral proteins that subvert IFN-mediated innate immune responses that limit virus spread.

Degradation of IRF Proteins

Direct demonstration of the importance of NSP1 in suppressing IFN signaling emerged from assays performed with VSV-GFP, an IFN-sensitive recombinant vesicular stomatitis virus that expresses green fluorescent protein (8). This assay system showed that medium recovered from cell cultures infected with wild-type rotavirus does not inhibit the replication of VSV-GFP and notably lacks IFN. In contrast, medium from RVg5re-infected cells suppresses VSV-GFP replication, an effect overcome by the addition of neutralizing IFN-β antisera. These results

indicate that although rotaviruses are inherently capable of inducing IFN, this process is subverted by the activity of NSP1. Assays performed using siRNAs targeting host dsRNA-binding proteins suggest that IFN production in response to RVg5re infection of colon cancer cell lines is triggered by RIG-I and not TLR-3 or PKR (72, 187). Whether RIG-I serves as the primary sensor in all rotavirus infections, regardless of cell type, remains to be addressed. Although PKR may not initiate IFN expression in rotavirus-infected cells, it is nonetheless activated, thereby stimulating phosphorylation of eIF2α (61, 116). However, this phosphorylation event has no known effect on the replication of the virus, suggesting an as yet undiscovered mechanism by which rotaviruses avoid the antiviral responses of the host.

The mechanism by which NSP1 mediates IFN antagonism was initially suggested by studies showing that NSP1 expression in rotavirus-infected cells is associated with loss of IRF3 (7), the constitutively expressed IRF responsible for inducing IFN-β expression (78) (Fig. 5). Unlike wild-type NSP1, NSP1ΔC expression in RVg5re-infected cells has little effect on IRF3 levels. Transient expression assays confirm that NSP1 causes IRF3 degradation and that this effect is induced by wild-type, but not C-truncated, forms of NSP1. In the presence of the proteasome inhibitor MG132, IRF3 turnover is averted, suggesting that NSP1 induces IRF3 degradation through a proteasome-dependent process (7).

The presence of shared structural and sequence elements within the IRF protein family (76) led to studies addressing whether NSP1 directs the degradation of family members other than IRF3. The most notable of these showed that IRF7, the "master" regulator of type I IFN synthesis (77), accumulates minimally in rotavirus-infected cells (8). Subsequent transient expression assays demonstrated that NSP1 causes the proteasomal degradation of IRF7 and does so with efficiency similar to that seen for IRF3. In comparison, IRF7 accumulates to high levels in RVg5re-infected cells, pointing to a failure of truncated NSP1 to induce IRF7 degradation. Transient expression assays confirm that NSP1ΔC does not induce IRF7 degradation (8). In addition to IRF3 and IRF7, transient expression assays also show that NSP1 induces the proteasomal degradation of a third IRF family member, IRF5. This IFN-regulatory factor functions not only in the expression of type I IFNs but also in cell cycle regulation and apoptosis induction (76, 131). Predictably, NSP1ΔC does not induce IRF5 degradation (8).

Collectively, a compelling body of evidence indicates that NSP1 directs the degradation of multiple members of the IRF protein family through a common

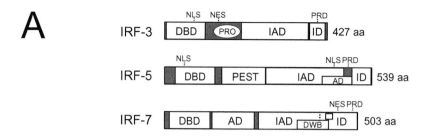

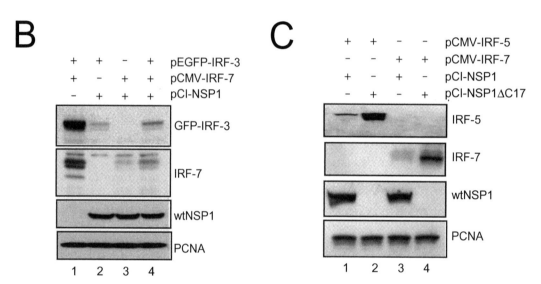

Figure 5 Members of the IRF protein family targeted for degradation by rotavirus NSP1. (A) Structural and functional domains of the IRF3, IRF5, and IRF7 targets. DBD, tryptophan-pentad repeat DNA-binding domain; PRO, proline-rich region; IAD, IRF-interactive (dimerization) domain; ID, autoinhibitory domain; PEST, proline-glutamate-serine-threonine-rich domain; AD, constitutive activation domain; DWB, region similar to domain B in dwarfin family proteins; NLS, nuclear localization domain; NES, nuclear export signal; PRD, phosphorylation-mediated response domain (73, 128, 131). (B) Immunoblot analysis of proteins transiently expressed in 293T cells showing that wild-type (wt) NSP1 induces the degradation of chimeric GFP-IRF3 protein and IRF7 (8). (C) Immunoblot analysis showing that wild-type wtNSP1, but not the C-truncated form (NSP1ΔC17), induces degradation of IRF5 and IRF7 (8).

proteasome-linked process. These findings establish NSP1 as a broad-spectrum antagonist of IFN-mediated innate immune responses regulated by the IRF protein family. In most fibroblasts, IRF5 and IRF7 are inducible products whose expression is upregulated by IFN-β, the gene product stimulated by IRF3. The activity of NSP1 has a two-pronged suppressive effect on type I IFN synthesis, attacking both the IRF3/IFN-β-mediated pathway that activates IRF7 expression and the IRF7 product itself. The capacity of NSP1 to degrade IRF3, IRF5, and IRF7 may be critical for rotavirus replication in trafficking cells (plasmacytoid dendritic cells) in

which IRF5 and IRF7 are expressed in a constitutive manner. Rotavirus replication in such cells may enable the virus to escape the gastrointestinal tract, providing an avenue to cause a more generalized infection of the host (134).

NSP1-IRF Interactions

Coimmunoprecipitation assays show that NSP1 forms complexes with IRF proteins in rotavirus-infected cells (7). Such interactions also have been demonstrated using glutathione S-transferase-tagged IRF3 as bait in pulldown assays of NSP1 from infected cells and using vectors expressing NSP1 and IRF3 in yeast two-hybrid screens (65, 66). Analysis of NSP1 mutants using a yeast two-hybrid approach indicates that the C terminus contains the IRF-binding domain (IRF-BD) of the protein (66). Interestingly, the IRF-BD is located within a region of NSP1 that is highly variable in sequence, a characteristic consistent with the idea that NSP1 has co-evolved with its host such that the IRF-BD is species specific in its recognition of IRF targets. The location of the IRF-BD at the C terminus suggests that failure of the NSP1ΔC proteins of the RVg5re variants to induce IRF degradation results from their inability to bind IRF targets. Mutagenesis studies indicate that the RING domain of NSP1 is important, but not sufficient, for interaction with IRF proteins (65, 66). Moreover, alteration of the conserved cysteine and histidine residues of the RING domain shows that, without structurally intact zinc fingers, NSP1 cannot induce degradation of the IRF proteins (65). Thus, IRF degradation requires (i) the physical interaction of NSP1 with the IRF target, (ii) an NSP1 protein with an intact C terminus and RING fingers, and (iii) proteasomal proteolysis.

NSP1 as an E3 Ubiquitin Ligase

Ubiquitination produces posttranslational modifications that lead to protein recognition and degradation by proteasomes (60). The 76-amino-acid ubiquitin (Ub) polypeptide is ligated onto the target protein through the action of three enzymes: E1, E2, and E3 (86, 87). E1, the Ub-activating enzyme, initiates the process by forming a covalent linkage with Ub in an ATP-dependent manner. E2, the Ub-conjugating enzyme, then accepts the Ub moiety from E1. By bridging the charged E2 enzyme and the target protein, the E3 Ub ligase facilitates transfer of Ub onto the target. In some cases, the Ub moiety becomes transiently linked to the E3 ligase during the transfer event, creating an E3-Ub complex. Repeated cycles of ubiquitination on a target protein create the poly-Ub chains that act as the signals for proteasomal proteolysis.

A common attribute of many E3 Ub ligases is a RING domain (23, 60). Since all NSP1 proteins contain a predicted RING domain, NSP1 could induce IRF degradation via Ub-dependent proteasomal proteolysis by serving as an E3 ligase. Proof of such an activity awaits evidence that NSP1 interacts not only with its IRF targets but also with the Ub-charged E2 enzyme to facilitate transfer of the Ub moiety from E2 onto an IRF protein. However, the observation that the turnover of NSP1 itself is mediated by proteasomal proteolysis suggests that NSP1 does indeed interact with one or more enzymes (E2 or E3) of the ubiquitination process.

Rotavirus NSP1 is an antagonist of the IFN signaling pathway, most likely functioning through an E3 Ub ligase activity that induces the ubiquitination and proteasomal proteolysis of several members of the IRF family. A few other viral E3 ligases have been identified that similarly function to subvert innate immune responses, including the Kaposi sarcoma herpesvirus K3 and K5 proteins (47, 48) and the poxvirus p28 protein (84). Although many viruses utilize E3 protein ligases to manipulate the host machinery in favor of viral replication, in most cases the E3 enzymes are of host origin.

ANTAGONISM OF THE IFN PATHWAY BY REOVIRUS

The σ3 Protein Sequesters dsRNA and Inhibits PKR

PKR is induced by IFN and activated by dsRNA-stimulated autophosphorylation. Activated PKR phosphorylates eIF2α to inhibit translation as a general antiviral response (139). Reovirus dsRNA activates PKR in vitro, and reovirus can activate PKR in infected cells (69, 138, 140). While it is not apparent that reovirus dsRNA is ever free in the cytoplasm, it is possible that some dsRNA exposure occurs as a consequence of aberrant viral disassembly (75). The reovirus S4 gene encodes a dsRNA-binding protein, σ3, which can inhibit dsRNA-mediated activation of PKR (103, 189). Indeed, the reovirus σ3 protein can restore vaccinia virus replication in cells infected with a vaccinia virus mutant lacking its cognate dsRNA-binding protein, E3L (12). The dsRNA-binding domain of σ3 is localized to a stretch of basic residues in an 85-amino-acid carboxy-terminal fragment of σ3 (43, 106, 114, 143).

The reovirus σ3 protein is complexed with membrane-penetration protein μ1 on the intact virus particle (100). Proteolysis of σ3 during viral disassembly is required for penetration of the virus into the cytosol (50, 63, 89, 183). In infected cells, newly synthesized σ3 and μ1 also colocalize, and interactions with μ1 may prevent σ3 sequestration of dsRNA and concomitant failure to inhibit activation of PKR (146, 189). Reovirus inhibition of host-cell protein synthesis varies between virus strains, and genetic analyses using reassortant viruses generated from these strains implicate the S4 gene (152), consistent with the known role of σ3 in modulation of PKR activity. However, other as yet unidentified cell factors have been implicated in reovirus inhibition of host-cell protein synthesis, and the role of σ3 and other reovirus proteins in modulation of these activities remains unexplored (161).

Possibility of Other Reovirus Antagonists of the IFN Pathway

There is no direct reovirus homolog of the rotavirus NSP1 protein, as is true for many reovirus-rotavirus protein comparisons. There also is no published evidence that reovirus disrupts IFN-β expression. Genetic analyses using reassortant reoviruses identified the M1, L2, and S2 genes as determinants of strain-specific differences in induction of IFN-β in cardiac myocytes (158). These genes encode the viral core proteins μ2, λ2, and σ2, respectively, but the mechanism by which they modulate induction of IFN is unclear. Importantly, the S4 gene, encoding reovirus dsRNA-binding protein σ3, was not identified in these studies, suggesting that either σ3 plays no role in inhibiting induction of IFN or it does so equivalently in the virus strains tested.

Many viruses encode proteins that interfere with Jak-Stat signaling. However, there is no published evidence for similar disruption by reoviruses. Like analysis of strain-specific differences in IFN-β induction, genetic analyses using reassortant reoviruses identified the M1, L2, and S2 genes as determinants of strain-specific differences in sensitivity to the antiviral effects of IFN-β in cardiac myocytes (158). Again, possible mechanisms for differential sensitivity remain speculative and await further exploration.

This work was supported by Public Health Service awards R01 AI62657 (B. S.) and R01 AI50080 (T. S. D.) and the Elizabeth B. Lamb Center for Pediatric Research. J. T. P. was supported by the Intramural Research Program of the National Institute of Allergy and Infectious Diseases, National Institutes of Health.

References

1. Abbadie, C., N. Kabrun, F. Bouali, B. Vandenbunder, and P. Enrietto. 1993. High levels of *c-rel* expression are associated with programmed cell death in the developing avian embryo and in bone marrow cells *in vitro*. *Cell* 75:899–912.

2. Aich, P., H. L. Wilson, R. S. Kaushik, A. A. Potter, L. A. Babiuk, and P. Griebel. 2007. Comparative analysis of innate immune responses following infection of newborn calves with bovine rotavirus and bovine coronavirus. *J. Gen. Virol.* 88:2749–2761.

3. Alexopoulou, L., A. C. Holt, R. Medzhitov, and R. A. Flavell. 2001. Recognition of double-stranded RNA and activation of NF-κB by Toll-like receptor 3. *Nature* 413:732–738.

4. Aravind, L., L. M. Iyer, and E. V. Koonin. 2003. Scores of RINGS but no PHDs in ubiquitin signaling. *Cell Cycle* 2:123–126.

5. Attoui, H., F. Mohd Jaafar, M. Belhouchet, P. Biagini, J. F. Cantaloube, P. de Micco, and X. de Lamballerie. 2005. Expansion of family Reoviridae to include nine-segmented dsRNA viruses: isolation and characterization of a new virus designated *Aedes pseudoscutellaris* reovirus assigned to a proposed genus (Dinovernavirus). *Virology* 343:212–223.

6. Azzam-Smoak, K., D. L. Noah, M. J. Stewart, M. A. Blum, and B. Sherry. 2002. Interferon regulatory factor-1, interferon-beta, and reovirus-induced myocarditis. *Virology* 298:20–29.

7. Barro, M., and J. T. Patton. 2005. Rotavirus nonstructural protein 1 subverts innate immune response by inducing degradation of IFN regulatory factor 3. *Proc. Natl. Acad. Sci. USA* 102:4114–4119.

8. Barro, M., and J. T. Patton. 2007. Rotavirus NSP1 inhibits expression of type I interferon by antagonizing the function of interferon regulatory factors IRF3, IRF5, and IRF7. *J. Virol.* 81:4473–4481.

9. Bass, D. M. 1997. Interferon γ and interleukin 1, but not interferon α, inhibit rotavirus entry into human intestinal cell lines. *Gastroenterology* 113:81–89.

10. Bass, E. S., D. A. Pappano, and S. G. Humiston. 2007. Rotavirus. *Pediatr. Rev.* 28:183–191.

11. Baty, C. J., and B. Sherry. 1993. Cytopathogenic effect in cardiac myocytes but not in cardiac fibroblasts is correlated with reovirus-induced acute myocarditis. *J. Virol.* 67:6295–6298.

12. Beattie, E., K. L. Denzler, J. Tartaglia, M. Perkus, E. Paoletti, and B. L. Jacobs. 1995. Reversal of the interferon-sensitive phenotype of a vaccinia virus lacking E3L by expression of the reovirus S4 gene. *J. Virol.* 69:499–505.

13. Beg, A., and D. Baltimore. 1996. An essential role for NF-κB in preventing TNF-α-induced cell death. *Science* 274:782–784.

14. Beg, A., T. Finco, P. Nantermet, and A. S. Baldwin. 1993. Tumor necrosis factor and interleukin-1 lead to phosphorylation and loss of IκBα: a mechanism for NF-κB activation. *Mol. Cell. Biol.* 13:3301–3310.

15. Blutt, S. E., D. O. Matson, S. E. Crawford, M. A. Staat, P. Azimi, B. L. Bennett, P. A. Piedra, and M. E. Conner. 2007. Rotavirus antigenemia in children is associated with viremia. *PLoS Med.* 4:e121.

16. Bowles, N. E., J. Ni, D. L. Kearney, M. Pauschinger, H. P. Schultheiss, R. McCarthy, J. Hare, J. T. Bricker, K. R. Bowles, and J. A. Towbin. 2003. Detection of viruses in myocardial tissues by polymerase chain reaction. Evidence of adenovirus as a common cause of myocarditis in children and adults. *J. Am. Coll. Cardiol.* 42:466–472.

17. Braun, T., and A. Martire. 2007. Cardiac stem cells: paradigm shift or broken promise? A view from developmental biology. *Trends Biotechnol.* 25:441–447.

18. Breiman, A., N. Grandvaux, R. Lin, C. Ottone, S. Akira, M. Yoneyama, T. Fujita, J. Hiscott, and E. F. Meurs. 2005. Inhibition of RIG-I-dependent signaling to the interferon pathway during hepatitis C virus expression and restoration of signaling by IKKepsilon. *J. Virol.* 79:3969–3978.

19. Bridger, J. C., G. I. Tauscher, and U. Desselberger. 1998. Viral determinants of rotavirus pathogenicity in pigs: evidence that the fourth gene of a porcine rotavirus confers diarrhea in the homologous host. *J. Virol.* 72:6929–6931.

20. Broome, R. L., P. T. Vo, R. L. Ward, H. F. Clark, and H. B. Greenberg. 1993. Murine rotavirus genes encoding outer capsid proteins VP4 and VP7 are not major

determinants of host range restriction and virulence. *J. Virol.* **67**:2448–2455.

21. **Brottier, P., P. Nandi, M. Bremont, and J. Cohen.** 1992. Bovine rotavirus segment 5 protein expressed in the baculovirus system interacts with zinc and RNA. *J. Gen. Virol.* **73**:1931–1938.

22. **Cahir McFarland, E. D., K. M. Izumi, and G. Mosialos.** 1999. Epstein-Barr virus transformation: involvement of latent membrane protein 1-mediated activation of NF-kappaB. *Oncogene* **18**:6959–6964.

23. **Capili, A. D., E. L. Edghill, K. Wu, and K. L. Borden.** 2004. Structure of the C-terminal RING finger from a RING-IBR-RING/TRIAD motif reveals a novel zinc-binding domain distinct from a RING. *J. Mol. Biol.* **340**:1117–1129.

24. **Chandran, K., D. L. Farsetta, and M. L. Nibert.** 2002. Strategy for nonenveloped virus entry: a hydrophobic conformer of the reovirus membrane penetration protein μ1 mediates membrane disruption. *J. Virol.* **76**:9920–9933.

25. **Chandran, K., J. S. Parker, M. Ehrlich, T. Kirchhausen, and M. L. Nibert.** 2003. The delta region of outer-capsid protein μ1 undergoes conformational change and release from reovirus particles during cell entry. *J. Virol.* **77**:13361–13375.

26. **Chawla-Sarkar, M., D. J. Lindner, Y. F. Liu, B. R. Williams, G. C. Sen, R. H. Silverman, and E. C. Borden.** 2003. Apoptosis and interferons: role of interferon-stimulated genes as mediators of apoptosis. *Apoptosis* **8**:237–249.

27. **Chen, D., C. L. Luongo, M. L. Nibert, and J. T. Patton.** 1999. Rotavirus open cores catalyze 5′-capping and methylation of exogenous RNA: evidence that VP3 is a methyltransferase. *Virology* **264**:120–130.

28. **Chu, W. M., D. Ostertag, Z. W. Li, L. Chang, Y. Chen, Y. Hu, B. Williams, J. Perrault, and M. Karin.** 1999. JNK2 and IKKβ are required for activating the innate response to viral infection. *Immunity* **11**:721–731.

29. **Clarke, P., S. M. Meintzer, L. A. Moffitt, and K. L. Tyler.** 2003. Two distinct phases of virus-induced nuclear factor kappa B regulation enhance tumor necrosis factor-related apoptosis-inducing ligand-mediated apoptosis in virus-infected cells. *J. Biol. Chem.* **278**:18092–18100.

30. **Clarke, P., S. M. Meintzer, Y. Wang, L. A. Moffitt, S. M. Richardson-Burns, G. L. Johnson, and K. L. Tyler.** 2004. JNK regulates the release of proapoptotic mitochondrial factors in reovirus-infected cells. *J. Virol.* **78**:13132–13138.

31. **Clarke, P., S. M. Meintzer, C. Widmann, G. L. Johnson, and K. L. Tyler.** 2001. Reovirus infection activates JNK and the JNK-dependent transcription factor c-Jun. *J. Virol.* **75**:11275–11283.

32. **Connolly, J. L., E. S. Barton, and T. S. Dermody.** 2001. Reovirus binding to cell surface sialic acid potentiates virus-induced apoptosis. *J. Virol.* **75**:4029–4039.

33. **Connolly, J. L., and T. S. Dermody.** 2002. Virion disassembly is required for apoptosis induced by reovirus. *J. Virol.* **76**:1632–1641.

34. **Connolly, J. L., S. E. Rodgers, P. Clarke, D. W. Ballard, L. D. Kerr, K. L. Tyler, and T. S. Dermody.** 2000. Reovirus-induced apoptosis requires activation of transcription factor NF-κB. *J. Virol.* **74**:2981–2989.

35. **Coscoy, L., and D. Ganem.** 2003. PHD domains and E3 ubiquitin ligases: viruses make the connection. *Trends Cell Biol.* **13**:7–12.

36. **Crawford, S. E., D. G. Patel, E. Cheng, Z. Berkova, J. M. Hyser, M. Ciarlet, M. J. Finegold, M. E. Conner, and M. K. Estes.** 2006. Rotavirus viremia and extraintestinal viral infection in the neonatal rat model. *J. Virol.* **80**:4820–4832.

37. **Dagenais, L., P. P. Pastoret, C. Van den Broecke, and J. Werenne.** 1981. Susceptibility of bovine rotavirus to interferon. *Arch. Virol.* **70**:377–379.

38. **Danthi, P., M. W. Hansberger, J. A. Campbell, J. C. Forrest, and T. S. Dermody.** 2006. JAM-A-independent, antibody-mediated uptake of reovirus into cells leads to apoptosis. *J. Virol.* **80**:1261–1270.

39. **Danthi, P., T. Kobayashi, G. H. Holm, M. W. Hansberger, T. W. Abel, and T. S. Dermody.** 2008. Reovirus apoptosis and virulence are regulated by host cell membrane-penetration efficiency. *J. Virol.* **82**:161–172.

40. **DeBiasi, R., C. Edelstein, B. Sherry, and K. Tyler.** 2001. Calpain inhibition protects against virus-induced apoptotic myocardial injury. *J. Virol.* **75**:351–361.

41. **DeBiasi, R. L., P. Clarke, S. M. Meintzer, R. M. Jotte, B. K. Kleinschmidt-Demasters, G. L. Johnson, and K. L. Tyler.** 2003. Reovirus-induced alteration in expression of apoptosis and DNA repair genes with potential roles in viral pathogenesis. *J. Virol.* **77**:8934–8947.

42. **DeBiasi, R. L., B. A. Robinson, B. Sherry, R. Bouchard, R. D. Brown, M. Rizeq, C. Long, and K. L. Tyler.** 2004. Caspase inhibition protects against reovirus-induced myocardial injury in vitro and in vivo. *J. Virol.* **78**:11040–11050.

43. **Denzler, K. L., and B. L. Jacobs.** 1994. Site-directed mutagenic analysis of reovirus σ3 protein binding to dsRNA. *Virology* **204**:190–199.

44. **Desselberger, U.** 1996. Genome rearrangements of rotaviruses. *Adv. Virus Res.* **46**:69–95.

45. **Dichter, M. A., and H. L. Weiner.** 1984. Infection of neuronal cell cultures with reovirus mimics in vitro patterns of neurotropism. *Ann. Neurol.* **16**:603–610.

46. **DiDonato, J. A., M. Hayakawa, D. M. Rothwarf, E. Zandi, and M. Karin.** 1997. A cytokine-responsive IκB kinase that activates the transcription factor NF-κB. *Nature* **388**:548–554.

47. **Dodd, R. B., M. D. Allen, S. E. Brown, C. M. Sanderson, L. M. Duncan, P. J. Lehner, M. Bycroft, and R. J. Read.** 2004. Solution structure of the Kaposi's sarcoma-associated herpesvirus K3 N-terminal domain reveals a novel E2-binding C4HC3-type RING domain. *J. Biol. Chem.* **279**:53840–53847.

48. **Duncan, L. M., S. Piper, R. B. Dodd, M. K. Saville, C. M. Sanderson, J. P. Luzio, and P. J. Lehner.** 2006. Lysine-63-linked ubiquitination is required for endolysosomal degradation of class I molecules. *EMBO J.* **25**:1635–1645.

49. **Dunn, S. J., T. L. Cross, and H. B. Greenberg.** 1994. Comparison of the rotavirus nonstructural protein NSP1 (NS53) from different species by sequence analysis and Northern blot hybridization. *Virology* **203**:178–183.

50. **Ebert, D. H., J. Deussing, C. Peters, and T. S. Dermody.** 2002. Cathepsin L and cathepsin B mediate reovirus disassembly in murine fibroblast cells. *J. Biol. Chem.* **277**:24609–24617.

51. **Edelmann, K. H., S. Richardson-Burns, L. Alexopoulou, K. L. Tyler, R. A. Flavell, and M. B. Oldstone.** 2004.

Does Toll-like receptor 3 play a biological role in virus infections? *Virology* **322:**231–238.

52. **Eiden, J., G. A. Lasonsky, J. Johnson, and R. H. Yolken.** 1985. Rotavirus RNA variation during chronic infection of immunocompromised children. *Pediatr. Infect. Dis.* **4:**632–637.

53. **El-Attar, L., W. Dhaliwal, C. R. Howard, and J. C. Bridger.** 2001. Rotavirus cross-species pathogenicity: molecular characterization of a bovine rotavirus pathogenic for pigs. *Virology* **291:**172–182.

54. **Ellis, C. R., and T. Di Salvo.** 2007. Myocarditis: basic and clinical aspects. *Cardiol. Rev.* **15:**170–177.

55. **Estes, M. K., G. Kang, C. Q. Zeng, S. E. Crawford, and M. Ciarlet.** 2001. Pathogenesis of rotavirus gastroenteritis. *Novartis Found. Symp.* **238:**82–96; discussion 96–100.

56. **Felker, G. M., R. E. Thompson, J. M. Hare, R. H. Hruban, D. E. Clemetson, D. L. Howard, K. L. Baughman, and E. K. Kasper.** 2000. Underlying causes and long-term survival in patients with initially unexplained cardiomyopathy. *N. Engl. J. Med.* **342:**1077–1084.

57. **Fenaux, M., M. A. Cuadras, N. Feng, M. Jaimes, and H. B. Greenberg.** 2006. Extraintestinal spread and replication of a homologous EC rotavirus strain and a heterologous rhesus rotavirus in BALB/c mice. *J. Virol.* **80:**5219–5232.

58. **Fitzgerald, K. A., S. M. McWhirter, K. L. Faia, D. C. Rowe, E. Latz, D. T. Golenbock, A. J. Coyle, S. M. Liao, and T. Maniatis.** 2003. IKKε and TBK1 are essential components of the IRF3 signaling pathway. *Nat. Immunol.* **4:**491–496.

59. **Foy, E., K. Li, R. J. Sumpter, Y. M. Loo, C. L. Johnson, C. Wang, P. M. Fish, M. Yoneyama, T. Fujita, S. M. Lemon, and M. Gale, Jr.** 2005. Control of antiviral defenses through hepatitis C virus disruption of retinoic acid-inducible gene-I signaling. *Proc. Natl. Acad. Sci. USA* **102:**2986–2991.

60. **Gao, G., and H. Luo.** 2006. The ubiquitin-proteasome pathway in viral infections. *Can. J. Physiol. Pharmacol.* **84:**5–14.

61. **Garcia, M. A., E. F. Meurs, and M. Esteban.** 2007. The dsRNA protein kinase PKR: virus and cell control. *Biochimie* **89:**799–811.

62. **Gilmore, T. D.** 2006. Introduction to NF-κB: players, pathways, perspectives. *Oncogene* **25:**6680–6684.

63. **Golden, J. W., J. Linke, S. Schmechel, K. Thoemke, and L. A. Schiff.** 2002. Addition of exogenous protease facilitates reovirus infection in many restrictive cells. *J. Virol.* **76:**7430–7443.

64. **Goody, R. J., J. D. Beckham, K. Rubtsova, and K. L. Tyler.** 2007. JAK-Stat signaling pathways are activated in the brain following reovirus infection. *J. Neurovirol.* **13:**373–383.

65. **Graff, J. W., J. Ewen, K. Ettayebi, and M. E. Hardy.** 2007. Zinc-binding domain of rotavirus NSP1 is required for proteasome-dependent degradation of IRF3 and autoregulatory NSP1 stability. *J. Gen. Virol.* **88:**613–620.

66. **Graff, J. W., D. N. Mitzel, C. M. Weisend, M. L. Flenniken, and M. E. Hardy.** 2002. Interferon regulatory factor 3 is a cellular partner of rotavirus NSP1. *J. Virol.* **76:**9545–9550.

67. **Graham, A., G. Kudesia, A. M. Allen, and U. Desselberger.** 1987. Reassortment of human rotavirus possessing genome rearrangements with bovine rotavirus: evidence for host cell selection. *J. Gen. Virol.* **68:**115–122.

68. **Grimm, S., M. K. Bauer, P. A. Baeuerle, and K. Schulze-Osthoff.** 1996. Bcl-2 down-regulates the activity of transcription factor NF-κB induced upon apoptosis. *J. Cell Biol.* **134:**13–23.

69. **Gupta, S. L., S. L. Holmes, and L. L. Mehra.** 1982. Interferon action against reovirus: activation of interferon induced protein kinase in mouse L 929 cells upon reovirus infection. *Virology* **12:**495–499.

70. **Hansberger, M. W., J. A. Campbell, P. Danthi, P. Arrate, K. N. Pennington, K. B. Marcu, D. W. Ballard, and T. S. Dermody.** 2007. IκB kinase subunits α and γ are required for activation of NF-κB and induction of apoptosis by mammalian reovirus. *J. Virol.* **81:**1360–1371.

71. **He, F., T. Umehara, K. Tsuda, M. Inoue, T. Kigawa, T. Matsuda, T. Yabuki, M. Aoki, E. Seki, T. Terada, M. Shirouzu, A. Tanaka, S. Sugano, Y. Muto, and S. Yokoyama.** 2007. Solution structure of the zinc finger HIT domain in protein FON. *Protein Sci.* **16:**1577–1587.

72. **Hirata, Y., A. H. Broquet, L. Menchen, and M. F. Kagnoff.** 2007. Activation of innate immune defense mechanisms by signaling through RIG-I/IPS-1 in intestinal epithelial cells. *J. Immunol.* **179:**5425–5432.

73. **Hiscott, J.** 2007. Triggering the innate antiviral response through IRF3 activation. *J. Biol. Chem.* **282:**15325–15329.

74. **Hiscott, J., T. L. Nguyen, M. Arguello, P. Nakhaei, and S. Paz.** 2006. Manipulation of the nuclear factor-κB pathway and the innate immune response by viruses. *Oncogene* **25:**6844–6867.

75. **Holm, G. H., J. Zurney, V. Tumilasci, P. Danthi, J. Hiscott, B. Sherry, and T. S. Dermody.** 2007. Retinoic acid-inducible gene-I and interferon-β promoter stimulator-1 augment proapoptotic responses following mammalian reovirus infection via interferon regulatory factor-3. *J. Biol. Chem.* **282:**21953–21961.

76. **Honda, K., and T. Taniguchi.** 2006. IRFs: master regulators of signalling by Toll-like receptors and cytosolic pattern-recognition receptors. *Nat. Rev. Immunol.* **6:**644–658.

77. **Honda, K., H. Yanai, H. Negishi, M. Asagiri, M. Sato, T. Mizutani, N. Shimada, Y. Ohba, A. Takaoka, N. Yoshida, and T. Taniguchi.** 2005. IRF7 is the master regulator of type-I interferon-dependent immune responses. *Nature* **434:**772–777.

78. **Honda, K., H. Yanai, A. Takaoka, and T. Taniguchi.** 2005. Regulation of the type I IFN induction: a current view. *Int. Immunol.* **17:**1367–1378.

79. **Hoshino, Y., R. W. Jones, J. Ross, and A. Z. Kapikian.** 2003. Construction and characterization of rhesus monkey rotavirus (MMU18006)- or bovine rotavirus (UK)-based serotype G5, G8, G9 or G10 single VP7 gene substitution reassortant candidate vaccines. *Vaccine* **21:**3003–3010.

80. **Hoshino, Y., L. J. Saif, S. Y. Kang, M. M. Sereno, W. K. Chen, and A. Z. Kapikian.** 1995. Identification of group A rotavirus genes associated with virulence of a porcine rotavirus and host range restriction of a human rotavirus in the gnotobiotic piglet model. *Virology* **209:**274–280.

81. **Hua, J., X. Chen, and J. T. Patton.** 1994. Deletion mapping of the rotavirus metalloprotein NS53 (NSP1): the conserved cysteine-rich region is essential for virus-specific RNA binding. *J. Virol.* **68:**3990–4000.

82. Hua, J., E. A. Mansell, and J. T. Patton. 1993. Comparative analysis of the rotavirus NS53 gene: conservation of basic and cysteine-rich regions in the protein and possible stem-loop structures in the RNA. *Virology* **196:**372–378.

83. Hua, J., and J. T. Patton. 1994. The carboxyl-half of the rotavirus nonstructural protein NS53 (NSP1) is not required for virus replication. *Virology* **198:**567–576.

84. Huang, J., Q. Huang, X. Zhou, M. M. Shen, A. Yen, S. X. Yu, G. Dong, K. Qu, P. Huang, E. M. Anderson, S. Daniel-Issakani, R. M. Buller, D. G. Payan, and H. H. Lu. 2004. The poxvirus p28 virulence factor is an E3 ubiquitin ligase. *J. Biol. Chem.* **279:**54110–54116.

85. Hundley, F., B. Biryahwaho, M. Gow, and U. Desselberger. 1985. Genome rearrangements of bovine rotavirus after serial passage at high multiplicity of infection. *Virology* **143:**88–103.

86. Hurley, J. H., S. Lee, and G. Prag. 2006. Ubiquitin-binding domains. *Biochem. J.* **399:**361–372.

87. Jackson, P. K., A. G. Eldridge, E. Freed, L. Furstenthal, J. Y. Hsu, B. K. Kaiser, and J. D. Reimann. 2000. The lore of the RINGs: substrate recognition and catalysis by ubiquitin ligases. *Trends Cell Biol.* **10:**429–439.

88. Jacobs, B. L., and R. E. Ferguson. 1991. The Lang strain of reovirus serotype 1 and the Dearing strain of reovirus serotype 3 differ in their sensitivities to β interferon. *J. Virol.* **65:**5102–5104.

89. Jané-Valbuena, J., L. A. Breun, L. A. Schiff, and M. L. Nibert. 2002. Sites and determinants of early cleavages in the proteolytic processing pathway of reovirus surface protein σ3. *J. Virol.* **76:**5184–5197.

90. Johansson, C., J. D. Wetzel, C. Mikacenic, J. P. He, T. S. Dermody, and B. Kelsall. 2007. Type I interferons produced by hematopoietic cells protect mice against lethal infection by mammalian reovirus. *J. Exp. Med.* **204:**1349–1358.

91. Jung, M., Y. Zhang, S. Lee, and A. Dritschilo. 1995. Correction of radiation sensitivity in ataxia telangiectasia cells by a truncated IκB-α. *Science* **268:**1619–1621.

92. Kato, H., O. Takeuchi, S. Sato, M. Yoneyama, M. Yamamoto, K. Matsui, S. Uematsu, A. Jung, T. Kawai, K. J. Ishii, O. Yamaguchi, K. Otsu, T. Tsujimura, C. S. Koh, C. Reis e Sousa, Y. Matsuura, T. Fujita, and S. Akira. 2006. Differential roles of MDA5 and RIG-I helicases in the recognition of RNA viruses. *Nature* **441:**101–105.

93. Kawai, T., K. Takahashi, S. Sato, C. Coban, H. Kumar, H. Kato, K. J. Ishii, O. Takeuchi, and S. Akira. 2005. IPS-1, an adaptor triggering RIG-I- and Mda5-mediated type I interferon induction. *Nat. Immunol.* **6:**981–988.

94. Kojima, K., K. Taniguchi, M. Kawagishi-Kobayashi, S. Matsuno, and S. Urasawa. 2000. Rearrangement generated in double genes, NSP1 and NSP3, of viable progenies from a human rotavirus strain. *Virus Res.* **67:**163–171.

95. Kojima, K., K. Taniguchi, and N. Kobayashi. 1996. Species-specific and interspecies relatedness of NSP1 sequences in human, porcine, bovine, feline, and equine rotavirus strains. *Arch. Virol.* **141:**1–12.

96. Kosarev, P., K. F. Mayer, and C. S. Hardtke. 2002. Evaluation and classification of RING-finger domains encoded by the *Arabidopsis* genome. *Genome Biol.* **3:**research0016.1-research0016.12.

97. Labbe, M., P. Baudoux, A. Charpilienne, D. Poncet, and J. Cohen. 1994. Identification of the nucleic acid binding domain of the rotavirus VP2 protein. *J. Gen. Virol.* **75:**3423–3430.

98. Lecce, J. G., J. M. Cummins, and A. B. Richards. 1990. Treatment of rotavirus infection in neonate and weanling pigs using natural human interferon alpha. *Mol. Biother.* **2:**211–216.

99. Lee, P. W., E. C. Hayes, and W. K. Joklik. 1981. Protein σ1 is the reovirus cell attachment protein. *Virology* **108:**156–163.

100. Liemann, S., K. Chandran, T. S. Baker, M. L. Nibert, and S. C. Harrison. 2002. Structure of the reovirus membrane-penetration protein, μ1, in a complex with its protector protein, σ3. *Cell* **108:**283–295.

101. Ling, L., Z. Cao, and D. V. Goeddel. 1998. NF-κB-inducing kinase activates IKK-α by phosphorylation of Ser-176. *Proc. Natl. Acad. Sci. USA* **95:**3792–3797.

102. Liu, Z.-G., H. Hsu, D. Goeddel, and M. Karin. 1996. Dissection of TNF receptor 1 effector functions: JNK activation is not linked to apoptosis while NF-κB activation prevents cell death. *Cell* **87:**565–576.

103. Lloyd, R. M., and A. J. Shatkin. 1992. Translational stimulation by reovirus polypeptide σ3: substitution for VA1 RNA and inhibition of phosphorylation of the α subunit of eukaryotic initiation factor 2. *J. Virol.* **66:**6878–6884.

104. Loo, Y. M., J. Fornek, N. Crochet, G. Bajwa, O. Perwitasari, L. Martinez-Sobrido, S. Akira, M. A. Gill, A. Garcia-Sastre, M. G. Katze, and M. Gale, Jr. 2008. Distinct RIG-I and MDA5 signaling by RNA viruses in innate immunity. *J. Virol.* **82:**335–345.

105. Lucia-Jandris, P., J. W. Hooper, and B. N. Fields. 1993. Reovirus M2 gene is associated with chromium release from mouse L cells. *J. Virol.* **67:**5339–5345.

106. Mabrouk, T., C. Danis, and G. Lemay. 1995. Two basic motifs of reovirus σ3 protein are involved in double-stranded RNA binding. *Biochem. Cell Biol.* **73:**137–145.

107. Maniatis, T. 1997. Catalysis by a multiprotein IκB kinase complex. *Science* **278:**818–819.

108. Marie, I., J. E. Durbin, and D. E. Levy. 1998. Differential viral induction of distinct interferon-alpha genes by positive feedback through interferon regulatory factor-7. *EMBO J.* **17:**6660–6669.

109. McKinsey, T. A., J. A. Brockman, D. C. Scherer, S. W. Al-Murrani, P. L. Green, and D. W. Ballard. 1996. Inactivation of IκBβ by the Tax protein of human T-cell leukemia virus type 1: a potential mechanism for constitutive induction of NF-κB. *Mol. Cell. Biol.* **16:**2083–2090.

110. Mercurio, F., H. Zhu, B. W. Murray, A. Shevchenko, B. L. Bennett, J. Li, D. B. Young, M. Barbosa, M. Mann, A. Manning, and A. Rao. 1997. IKK-1 and IKK-2: cytokine-activated IκB kinases essential for NF-κB activation. *Science* **278:**860–866.

111. Mertens, P. P. C., R. Duncan, H. Attoui, and T. S. Dermody. 2005. Reoviridae, p. 447–454. *In* C. M. Fauquet, M. A. Mayo, J. Maniloff, U. Desselberger, and L. A. Ball (ed.), *Virus Taxonomy: the Classification and Nomenclature of Viruses. The Eighth Report of the International Committee on Taxonomy of Viruses.* Elsevier/Academic Press, London, United Kingdom.

112. Meusel, T. R., and F. Imani. 2003. Viral induction of inflammatory cytokines in human epithelial cells follows a p38 mitogen-activated protein kinase-dependent but NF-κB-independent pathway. *J. Immunol.* **171:**3768–3774.

113. Meylan, E., J. Curran, K. Hofmann, D. Moradpour, M. Binder, R. Bartenschlager, and J. Tschopp. 2005. Cardif is an adaptor protein in the RIG-I antiviral pathway and is targeted by hepatitis C virus. *Nature* **437:**1167–1172.

114. Miller, J. E., and C. E. Samuel. 1992. Proteolytic cleavage of the reovirus σ3 protein results in enhanced double-stranded RNA-binding activity: identification of a repeated basic amino acid motif within the C-terminal binding region. *J. Virol.* **66:**5347–5356.

115. Mogensen, T. H., and S. R. Paludan. 2005. Reading the viral signature by Toll-like receptors and other pattern recognition receptors. *J. Mol. Med.* **83:**180–192.

116. Montero, H., C. F. Arias, and S. Lopez. 2006. Rotavirus nonstructural protein NSP3 is not required for viral protein synthesis. *J. Virol.* **80:**9031–9038.

117. Mori, Y., M. A. Borgan, M. Takayama, N. Ito, M. Sugiyama, and N. Minamoto. 2003. Roles of outer capsid proteins as determinants of pathogenicity and host range restriction of avian rotaviruses in a suckling mouse model. *Virology* **316:**126–134.

118. Morrison, L. A., R. L. Sidman, and B. N. Fields. 1991. Direct spread of reovirus from the intestinal lumen to the central nervous system through vagal autonomic nerve fibers. *Proc. Natl. Acad. Sci. USA* **88:**3852–3856.

119. Mossel, E. C., and R. F. Ramig. 2002. Rotavirus genome segment 7 (NSP3) is a determinant of extraintestinal spread in the neonatal mouse. *J. Virol.* **76:**6502–6509.

120. Narvaez, C. F., J. Angel, and M. A. Franco. 2005. Interaction of rotavirus with human myeloid dendritic cells. *J. Virol.* **79:**14526–14535.

121. Noah, D. L., M. A. Blum, and B. Sherry. 1999. Interferon regulatory factor 3 is required for viral induction of beta interferon in primary cardiac myocyte cultures. *J. Virol.* **73:**10208–10213.

122. Norman, K. L., K. Hirasawa, A. D. Yang, M. A. Shields, and P. W. Lee. 2004. Reovirus oncolysis: the Ras/RalGEF/p38 pathway dictates host cell permissiveness to reovirus infection. *Proc. Natl. Acad. Sci. USA* **101:**11099–11104.

123. Odegard, A. L., K. Chandran, X. Zhang, J. S. Parker, T. S. Baker, and M. L. Nibert. 2004. Putative autocleavage of outer capsid protein μ1, allowing release of myristoylated peptide μ1N during particle uncoating, is critical for cell entry by reovirus. *J. Virol.* **78:**8732–8745.

124. O'Donnell, S. M., M. W. Hansberger, J. L. Connolly, J. D. Chappell, M. J. Watson, J. M. Pierce, J. D. Wetzel, W. Han, E. S. Barton, J. C. Forrest, T. Valyi-Nagy, F. E. Yull, T. S. Blackwell, J. N. Rottman, B. Sherry, and T. S. Dermody. 2005. Organ-specific roles for transcription factor NF-κB in reovirus-induced apoptosis and disease. *J. Clin. Invest.* **115:**2341–2350.

125. O'Donnell, S. M., G. H. Holm, J. M. Pierce, B. Tian, M. J. Watson, R. S. Chari, D. W. Ballard, A. R. Brasier, and T. S. Dermody. 2006. Identification of an NF-κB-dependent gene network in cells infected by mammalian reovirus. *J. Virol.* **80:**1077–1086.

126. Offit, P. A., G. Blavat, H. B. Greenberg, and H. F. Clark. 1986. Molecular basis of rotavirus virulence: role of gene segment 4. *J. Virol.* **57:**46–49.

127. Oishi, I., T. Kimura, T. Murakami, K. Haruki, K. Yamazaki, Y. Seto, Y. Minekawa, and H. Funamoto. 1991. Serial observations of chronic rotavirus infection in an immunodeficient child. *Microbiol. Immunol.* **35:**953–961.

128. Ozato, K., P. Tailor, and T. Kubota. 2007. The interferon regulatory factor family in host defense: mechanism of action. *J. Biol. Chem.* **282:**20065–20069.

129. Patton, J. T., and E. Spencer. 2000. Genome replication and packaging of segmented double-stranded RNA viruses. *Virology* **277:**217–225.

130. Patton, J. T., Z. Taraporewala, D. Chen, V. Chizhikov, M. Jones, A. Elhelu, M. Collins, K. Kearney, M. Wagner, Y. Hoshino, and V. Gouvea. 2001. Effect of intragenic rearrangement and changes in the 3′ consensus sequence on NSP1 expression and rotavirus replication. *J. Virol.* **75:**2076–2086.

131. Paun, A., and P. M. Pitha. 2007. The IRF family, revisited. *Biochimie* **89:**744–753.

132. Poggioli, G. J., R. L. DeBiasi, R. Bickel, R. Jotte, A. Spalding, G. L. Johnson, and K. L. Tyler. 2002. Reovirus-induced alterations in gene expression related to cell cycle regulation. *J. Virol.* **76:**2585–2594.

133. Ramig, R. F. 2004. Pathogenesis of intestinal and systemic rotavirus infection. *J. Virol.* **78:**10213–10220.

134. Ramig, R. F. 2007. Systemic rotavirus infection. *Expert Rev. Anti Infect. Ther.* **5:**591–612.

135. Rodgers, S. E., E. S. Barton, S. M. Oberhaus, B. Pike, C. A. Gibson, K. L. Tyler, and T. S. Dermody. 1997. Reovirus-induced apoptosis of MDCK cells is not linked to viral yield and is blocked by Bcl-2. *J. Virol.* **71:**2540–2546.

136. Rothwarf, D. M., E. Zandi, G. Natoli, and M. Karin. 1998. IKK-γ is an essential regulatory subunit of the IκB kinase complex. *Nature* **395:**297–300.

137. Rudolph, D., W. C. Yeh, A. Wakeham, B. Rudolph, D. Nallainathan, J. Potter, A. J. Elia, and T. W. Mak. 2000. Severe liver degeneration and lack of NF-κB activation in NEMO/IKKγ-deficient mice. *Genes Dev.* **14:**854–862.

138. Samuel, C. E. 1979. Mechanism of interferon action: phosphorylation of protein synthesis initiation factor eIF-2 in interferon-treated human cells by a ribosome-associated kinase processing site specificity similar to hemin-regulated rabbit reticulocyte kinase. *Proc. Natl. Acad. Sci. USA* **76:**600–604.

139. Samuel, C. E. 1998. Reoviruses and the interferon system. *Curr. Top. Microbiol. Immunol.* **233:**125–145.

140. Samuel, C. E., R. Duncan, G. S. Knutson, and J. W. Hershey. 1984. Mechanism of interferon action: increased phosphorylation of protein synthesis initiation factor eIF-2α in interferon-treated, reovirus-infected mouse L_{929} fibroblasts *in vitro* and *in vivo*. *J. Biol. Chem.* **259:**13451–13457.

141. Sarkar, S. N., and G. C. Sen. 2004. Novel functions of proteins encoded by viral stress-inducible genes. *Pharmacol. Ther.* **103:**245–259.

142. Sato, M., N. Tanaka, N. Hata, E. Oda, and T. Taniguchi. 1998. Involvement of the IRF family transcription factor IRF3 in virus-induced activation of the IFN-β gene. *FEBS Lett.* **425:**112–116.

143. Schiff, L. A., M. L. Nibert, M. S. Co, E. G. Brown, and B. N. Fields. 1988. Distinct binding sites for zinc and

double-stranded RNA in the reovirus outer capsid protein σ3. *Mol. Cell. Biol.* 8:273–283.

144. **Schiff, L. A., M. L. Nibert, and K. L. Tyler.** 2007. Orthoreoviruses and their replication, p. 1853–1915. *In* D. M. Knipe and P. M. Howley (ed.), *Fields Virology,* 5th ed., vol. 2. Lippincott, Williams & Wilkins, Philadelphia, PA.

145. **Schindler, C., D. E. Levy, and T. Decker.** 2007. JAK-Stat signaling: from interferons to cytokines. *J. Biol. Chem.* 282:20059–20063.

146. **Schmechel, S., M. Chute, P. Skinner, R. Anderson, and L. Schiff.** 1997. Preferential translation of reovirus mRNA by a σ3-dependent mechanism. *Virology* 232:62–73.

147. **Schomer-Miller, B., T. Higashimoto, Y. K. Lee, and E. Zandi.** 2006. Regulation of IκB kinase (IKK) complex by IKKγ-dependent phosphorylation of the T-loop and C terminus of IKKβ. *J. Biol. Chem.* 281:15268–15276.

148. **Schwers, A., C. Vanden Broecke, M. Maenhoudt, J. M. Beduin, J. Werenne, and P. P. Pastoret.** 1985. Experimental rotavirus diarrhoea in colostrum-deprived newborn calves: assay of treatment by administration of bacterially produced human interferon (Hu-IFN α2). *Ann. Vet. Res.* 16:213–218.

149. **Schwers, A., C. Vanden Broecke, P. P. Pastoret, J. Werenne, L. Dagenais, and M. Maenhoudt.** 1983. Dose effect on experimental reproduction of rotavirus diarrhoea in colostrum-deprived newborn calves. *Vet. Rec.* 112:250.

150. **Senftleben, U., Y. Cao, G. Xiao, F. R. Greten, G. Krahn, G. Bonizzi, Y. Chen, Y. Hu, A. Fong, S. C. Sun, and M. Karin.** 2001. Activation by IKKα of a second, evolutionary conserved, NF-κB signaling pathway. *Science* 293:1495–1499.

151. **Seth, R. B., L. Sun, C. K. Ea, and Z. J. Chen.** 2005. Identification and characterization of MAVS, a mitochondrial antiviral signaling protein that activates NF-κB and IRF3. *Cell* 122:669–682.

152. **Sharpe, A. H., and B. N. Fields.** 1982. Reovirus inhibition of cellular RNA and protein synthesis: role of the S4 gene. *Virology* 122:381–391.

153. **Sherry, B., C. J. Baty, and M. A. Blum.** 1996. Reovirus-induced acute myocarditis in mice correlates with viral RNA synthesis rather than generation of infectious virus in cardiac myocytes. *J. Virol.* 70:6709–6715.

154. **Sherry, B., and M. A. Blum.** 1994. Multiple viral core proteins are determinants of reovirus-induced acute myocarditis. *J. Virol.* 68:8461–8465.

155. **Sherry, B., and B. N. Fields.** 1989. The reovirus M1 gene, encoding a viral core protein, is associated with the myocarditic phenotype of a reovirus variant. *J. Virol.* 63:4850–4856.

156. **Sherry, B., X. Y. Li, K. L. Tyler, J. M. Cullen, and H. W. Virgin IV.** 1993. Lymphocytes protect against and are not required for reovirus-induced myocarditis. *J. Virol.* 67:6119–6124.

157. **Sherry, B., F. J. Schoen, E. Wenske, and B. N. Fields.** 1989. Derivation and characterization of an efficiently myocarditic reovirus variant. *J. Virol.* 63:4840–4849.

158. **Sherry, B., J. Torres, and M. A. Blum.** 1998. Reovirus induction of and sensitivity to beta interferon in cardiac myocyte cultures correlate with induction of myocarditis and are determined by viral core proteins. *J. Virol.* 72:1314–1323.

159. **Silvestri, L. S., Z. F. Taraporewala, and J. T. Patton.** 2004. Rotavirus replication: plus-sense templates for double-stranded RNA synthesis are made in viroplasms. *J. Virol.* 78:7763–7774.

160. **Smith, J. A., S. C. Schmechel, A. Raghavan, M. Abelson, C. Reilly, M. G. Katze, R. J. Kaufman, P. R. Bohjanen, and L. A. Schiff.** 2006. Reovirus induces and benefits from an integrated cellular stress response. *J. Virol.* 80:2019–2033.

161. **Smith, J. A., S. C. Schmechel, B. R. Williams, R. H. Silverman, and L. A. Schiff.** 2005. Involvement of the interferon-regulated antiviral proteins PKR and RNase L in reovirus-induced shutoff of cellular translation. *J. Virol.* 79:2240–2250.

162. **Stewart, M. J., M. A. Blum, and B. Sherry.** 2003. PKR's protective role in viral myocarditis. *Virology* 314:92–100.

163. **Stewart, M. J., K. Smoak, M. A. Blum, and B. Sherry.** 2005. Basal and reovirus-induced beta interferon (IFN-β) and IFN-β-stimulated gene expression are cell type specific in the cardiac protective response. *J. Virol.* 79:2979–2987.

164. **Sumpter, R. J., Y. M. Loo, E. Foy, K. Li, M. Yoneyama, T. Fujita, S. M. Lemon, and M. J. Gale.** 2005. Regulating intracellular antiviral defense and permissiveness to hepatitis C virus RNA replication through a cellular RNA helicase, RIG-I. *J. Virol.* 79:2689–2699.

165. **Taniguchi, K., K. Kojima, N. Kobayashi, T. Urasawa, and S. Urasawa.** 1996. Structure and function of rotavirus NSP1. *Arch. Virol. Suppl.* 12:53–58.

166. **Taniguchi, K., K. Kojima, and S. Urasawa.** 1996. Nondefective rotavirus mutants with an NSP1 gene which has a deletion of 500 nucleotides, including a cysteine-rich zinc finger motif-encoding region (nucleotides 156 to 248), or which has a nonsense codon at nucleotides 153 to 155. *J. Virol.* 70:4125–4130.

167. **Tardieu, M., M. L. Powers, and H. L. Weiner.** 1983. Age-dependent susceptibility to reovirus type 3 encephalitis: role of viral and host factors. *Ann. Neurol.* 13:602–607.

168. **Tian, Y., O. Tarlow, A. Ballard, U. Desselberger, and M. A. McCrae.** 1993. Genomic concatemerization/deletion in rotaviruses: a new mechanism for generating rapid genetic change of potential epidemiological importance. *J. Virol.* 67:6625–6632.

169. **Tosteson, M. T., M. L. Nibert, and B. N. Fields.** 1993. Ion channels induced in lipid bilayers by subvirion particles of the nonenveloped mammalian reoviruses. *Proc. Natl. Acad. Sci. USA* 90:10549–10552.

170. **Turjanski, A. G., J. P. Vaque, and J. S. Gutkind.** 2007. MAP kinases and the control of nuclear events. *Oncogene* 26:3240–3253.

171. **Tyler, K. L., D. A. McPhee, and B. N. Fields.** 1986. Distinct pathways of viral spread in the host determined by reovirus S1 gene segment. *Science* 233:770–774.

172. **Tyler, K. L., M. K. Squier, S. E. Rodgers, S. E. Schneider, S. M. Oberhaus, T. A. Grdina, J. J. Cohen, and T. S. Dermody.** 1995. Differences in the capacity of reovirus strains to induce apoptosis are determined by the viral attachment protein σ1. *J. Virol.* 69:6972–6979.

173. **Van Antwerp, D., S. Martin, T. Kafri, D. Green, and I. Verma.** 1996. Suppression of TNF-α-induced apoptosis by NF-κB. *Science* 274:787–789.

174. **Vanden Broecke, C., A. Schwers, L. Dagenais, A. Goossens, M. Maenhoudt, P. P. Pastoret, and J. Werenne.**

1984. Interferon response in colostrum-deprived newborn calves infected with bovine rotavirus: its possible role in the control of the pathogenicity. *Ann. Vet. Res.* **15:**29–34.

175. Vasquez-Del Carpio, R., F. D. Gonzalez-Nilo, G. Riadi, Z. F. Taraporewala, and J. T. Patton. 2006. Histidine triad-like motif of the rotavirus NSP2 octamer mediates both RTPase and NTPase activities. *J. Mol. Biol.* **362:**539–554.

176. Vende, P., Z. F. Taraporewala, and J. T. Patton. 2002. RNA-binding activity of the rotavirus phosphoprotein NSP5 includes affinity for double-stranded RNA. *J. Virol.* **76:**5291–5299.

177. Verma, I. M., and J. Stevenson. 1997. IκB kinase: beginning, not the end. *Proc. Natl. Acad. Sci. USA* **94:**11758–11760.

178. Virgin, H. W., K. L. Tyler, and T. S. Dermody. 1997. Reovirus, p. 669–699. *In* N. Nathanson (ed.), *Viral Pathogenesis.* Lippincott-Raven, New York, NY.

179. Weiner, H. L., K. A. Ault, and B. N. Fields. 1980. Interaction of reovirus with cell surface receptors. I. Murine and human lymphocytes have a receptor for the hemagglutinin of reovirus type 3. *J. Immunol.* **124:**2143–2148.

180. Weiner, H. L., D. Drayna, D. R. Averill, Jr., and B. N. Fields. 1977. Molecular basis of reovirus virulence: role of the S1 gene. *Proc. Natl. Acad. Sci. USA* **74:**5744–5748.

181. Weiner, H. L., M. L. Powers, and B. N. Fields. 1980. Absolute linkage of virulence and central nervous system tropism of reoviruses to viral hemagglutinin. *J. Infect. Dis.* **141:**609–616.

182. Widdowson, M. A., M. I. Meltzer, X. Zhang, J. S. Bresee, U. D. Parashar, and R. I. Glass. 2007. Cost-effectiveness and potential impact of rotavirus vaccination in the United States. *Pediatrics* **119:**684–697.

183. Wilson, G. J., E. L. Nason, C. S. Hardy, D. H. Ebert, J. D. Wetzel, B. V. Venkataram Prasad, and T. S. Dermody. 2002. A single mutation in the carboxy terminus of reovirus outer-capsid protein σ3 confers enhanced kinetics of σ3 proteolysis, resistance to inhibitors of viral disassembly, and alterations in σ3 structure. *J. Virol.* **76:**9832–9843.

184. Xiao, G., E. W. Harhaj, and S. C. Sun. 2001. NF-κB-inducing kinase regulates the processing of NF-κB2 p100. *Mol. Cell. Biol.* **7:**401–409.

185. Xu, L. G., Y. Y. Wang, K. J. Han, L. Y. Li, Z. Zhai, and H. B. Shu. 2005. VISA is an adapter protein required for virus-triggered IFN-β signaling. *Mol. Cell* **19:**727–740.

186. Yamaoka, S., G. Courtois, C. Bessia, S. T. Whiteside, R. Weil, F. Agou, H. E. Kirk, R. J. Kay, and A. Israel. 1998. Complementation cloning of NEMO, a component of the IκB kinase complex essential for NF-κB activation. *Cell* **93:**1231–1240.

187. Yoneyama, M., and T. Fujita. 2007. Function of RIG-I-like receptors in antiviral innate immunity. *J. Biol. Chem.* **282:**15315–15318.

188. Yoneyama, M., M. Kikuchi, T. Natsukawa, N. Shinobu, T. Imaizumi, M. Miyagishi, K. Taira, S. Akira, and T. Fujita. 2004. The RNA helicase RIG-I has an essential function in double-stranded RNA-induced innate antiviral responses. *Nat. Immunol.* **5:**730–737.

189. Yue, Z., and A. J. Shatkin. 1997. Double-stranded RNA-dependent protein kinase (PKR) is regulated by reovirus structural proteins. *Virology* **234:**364–371.

190. Zamanian-Daryoush, M., T. H. Mogensen, J. A. DiDonato, and B. R. Williams. 2000. NF-κB activation by double-stranded-RNA-activated protein kinase (PKR) is mediated through NF-κB-inducing kinase and IκB kinase. *Mol. Cell. Biol.* **20:**1278–1290.

191. Zandi, E., D. M. Rothwarf, M. Delhase, M. Hayakawa, and M. Karin. 1997. The IκB kinase complex (IKK) contains two kinase subunits, IKKα and IKKβ, necessary for IκB phosphorylation and NF-κB activation. *Cell* **91:**243–252.

192. Zurney, J., K. E. Howard, and B. Sherry. 2007. Basal expression levels of IFNAR and Jak-Stat components are determinants of cell-type-specific differences in cardiac antiviral responses. *J. Virol.* **81:**13668–13680.

Summary and Perspectives

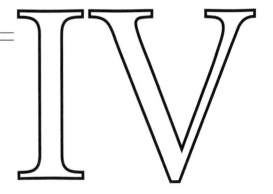

*Cellular Signaling and Innate Immune
Responses to RNA Virus Infections*
Edited by A. R. Brasier et al.
© 2009 ASM Press, Washington, D.C.

Allan R. Brasier
Adolfo García-Sastre
Stanley M. Lemon

Summary and Perspectives

26

Rapid activation of the innate immune response has a significant impact on the outcome or resolution of RNA virus infections. This occurs not only through the restrictions imposed on viral replication by the induction of interferon (IFN) α/β synthesis, but also through the controlling influence of cytokines and chemokines on the acquisition of antigen-specific adaptive immunity and inflammation at the site of infection. For this reason, pattern recognition receptors (PRRs) represent a critical first line of defense in the host response to RNA virus infections. They are responsible for continuously sensing various cellular compartments for the presence of molecular signals indicating invasion by an RNA virus. Significant strides have been made in identifying the key PRRs that respond to RNA virus infections, including Toll-like receptors (TLRs) and retinoic acid-inducible gene I (RIG-I)-like receptors (RLRs), and the downstream signaling pathways that they control. However, even though atomic-level-resolution structural models are beginning to emerge, exactly how molecular patterns activate PRRs is still not well understood on a molecular or structural level. A central

concept discussed herein includes ligand-induced conformational changes, but the nature of conformational change and its effect on inter- and intramolecular interactions are not yet understood. Moreover, it is increasingly clear that PRRs, specifically the TLRs, are expressed in a cell-type-specific pattern and that a fraction of them traffic between the endosomal compartment and cell surface. The underlying biological reasons and consequences of these phenomena are not well understood.

From examination of the PRR signaling pathways, a common mechanism for signal transduction involves the assembly of protein complexes via adaptor proteins that have kinase and, in some cases, ubiquitin ligase activity. It is increasingly clear that PRR signaling pathways share common adaptor proteins that facilitate assembly of signaling complexes that converge on two major groups of transcription factors, IFN regulatory factors (IRFs) and nuclear factor-κB (NF-κB). The molecular surfaces controlling complex assembly, adaptor utilization, and pathway specificity are not yet fully understood. Moreover, the composition of these multiprotein

---

Allan R. Brasier, Department of Medicine and the Sealy Center for Molecular Medicine, University of Texas Medical Branch, Galveston, TX 77555-1060. **Adolfo García-Sastre,** Department of Microbiology, Mount Sinai School of Medicine, New York, NY 10029. **Stanley M. Lemon,** Institute for Human Infections and Immunity, University of Texas Medical Branch, Galveston, TX 77555-0428.

signaling complexes and their temporal changes and posttranslational modifications will require new technologies for their study. Also to be understood are the extent and biological consequences of cross talk between these signaling pathways and others that control the expression of cellular genes involved in metabolism and other aspects of cellular physiology. These analyses will teach us important general principles about intracellular signal transduction, not only for the innate immune response.

Although IRF3 is a major regulator of type I IFN production, we now know that this protein also has other actions, such as controlling nuclear receptor function in *trans* and influencing monocyte and hepatocyte lipid metabolism, for which the role in disease pathogenesis is just beginning to be elucidated. The role of the NF-κB pathway, previously considered to be somewhat peripheral to the innate immune response, is now known to be an important mediator of immunopathogenesis in paramyxovirus and reovirus infections. Its role in pathogenesis of other viral infections appears certain but will require further elucidation. An emerging concept is that the NF-κB and IRF pathways are interconnected networks. For example, shared signaling adaptors, such as tumor necrosis factor (TNF) receptor-associated factor (TRAF), TIR domain-containing adapter including IFN-β (TRIF), myeloid differentiation primary response gene 88 (MyD88), and IKB kinase (IKK)-γ, play roles in controlling both IRF3 and NF-κB activation. NF-κB also participates in activating expression of the inducible IRF1 and IRF7 during the amplification of the innate immune response. These pathways are also being connected to other PRRs that were first recognized many years ago: the double-stranded RNA-activated enzymes protein kinase R (PKR) and 2',5'-oliogadenylate synthetase (OAS). The activation of PKR, itself an IFN-stimulated gene (ISG), leads in turn to activation of NF-κB, while the generation of small fragments of cellular RNAs mediated by RNase L following the activation of OAS plays an important role in amplifying signals induced by RIG-I and melanoma differentiation-associated gene 5 (MDA-5). These and other recent studies indicating that TLR upregulation is downstream of RIG-I, and vice versa, suggest that these networks communicate, are interdependent, and are strongly involved in positive-feedback control.

Much less well understood, but no less important, these pathways appear subject to intense negative regulation at almost every step. Recently recognized negative regulators include ubiquitin ligases, such as RNF125, and deubiquitinases such as debiquitinizing enzyme A (DUBA), as well as the mitochondrial membrane nucleotide binding oligomerization domain (NOD) protein NLRX1. Much more is to be learned about these negative regulators in the future, as they are likely to play critical roles in maintaining homeostasis and preventing autoimmunity.

The concept that the relative timing of PRR-induced signaling responses and rate of viral spread are critical determinants of disease pathogenesis and outcome is well entrenched. However, many of the data supporting this statement are derived from simplified in vitro models and generalizations thereof. Much more work needs to be done to determine the biological roles of PRR signaling pathways in modifying disease pathogenesis. A critical gap is in our understanding of the role played by innate immune responses in shaping the quality and durability of adaptive T-cell responses, for example, determining whether acute hepatitis C virus infections are resolved or go on to long-term persistence. These deficits in our knowledge base are largely due to limitations of human disease-relevant animal models. In addition, the roles of TLRs or their genetic polymorphisms in modifying human diseases are not well understood.

We have learned much from study of the mechanisms that viruses use to antagonize type I IFN signaling and its multiple antiviral actions. For the most part, viral-encoded proteins have evolved to specifically interact with key cellular signaling proteins. As a consequence, viral antagonists affect signaling by three major mechanisms: (i) inducing proteolysis of key signaling molecules, (ii) affecting subcellular compartmentalization or cytoplasmic-nuclear transfer of cellular proteins, or (iii) disrupting their target's ability to form productive signaling complexes. Further investigation of these interactions will undoubtedly provide important insight into the normal cellular regulation of these proteins and their mechanism of action. Many other viral evasion tactics target the activity of specific ISG effector molecules, such as RNase L or PKR.

Modulation by viruses of both induction and inhibition of the innate immune response has been shaped by coevolution of viruses with their hosts, to the point that it is likely that many viruses have achieved a delicate balance between induction and repression of the innate immune response that favors virus and host survival. In the extreme case, viruses with potent mechanisms for repressing the innate immune response are likely to kill their hosts, compromising their own transmission and survival, and therefore a potent inhibition of these responses might not provide a selective advantage for viruses. In contrast, certain aspects of the innate response may be used by viruses for their own advantage. For example, inhibition of viral replication by the innate immune response might favor viral persistence by hiding viral antigens from the adaptive immune system.

Finally, identifying virally encoded antagonists of the innate immune response and defining their mechanisms of action appear certain to stimulate new developments in vaccine and possibly antiviral drug development.

Recombinant viruses that lack IFN antagonism may produce highly effective and safe attenuated vaccines. Also, understanding the role of the PRRs in acquisition of adaptive immunity may stimulate applications of PRR agonists/antagonists and the development of better adjuvants capable of enhancing responses to those viral vaccines that elicit poor or inconsistent protective immunity.

Index